Lithium-Ion Batteries and Applications

A Practical and Comprehensive Guide to Lithium-Ion Batteries and Arrays, from Toys to Towns

Volume 1

Batteries

For the *Artech House Power Engineering Library*
go to the back of this book.

Lithium-Ion Batteries and Applications

A Practical and Comprehensive Guide to Lithium-Ion Batteries and Arrays, from Toys to Towns

Volume 1

Batteries

Davide Andrea

ARTECH HOUSE
BOSTON | LONDON
artechhouse.com

Library of Congress Cataloging-in-Publication Data

A catalog record for this book is available from the U.S. Library of Congress.

British Library Cataloguing in Publication Data

A catalog record for this book is available from the British Library.

ISBN-13: 978-1-63081-767-1

Cover design by John Gomes

685 Canton Street
Norwood, MA 02062

10 9 8 7 6 5 4 3 2 1

To Ann

CONTENTS

CHAPTER 4

CHAPTER 6

CHAPTER 7

PREFACE

WHAT THIS BOOK IS

This is a practical book to help in the selection, design, configuration, and operation of a *lithium-ion* (Li-ion) battery (whether ready-made or custom-designed) for a particular application. It goes down deep to the heart of the components inside a battery, yet it also explores the broad view of the application in which the battery is used. It steers the reader away from the common mistakes that would make a Li-ion battery dangerous.

WHAT THIS BOOK IS NOT

This is not a textbook; there are no numbered equations, no exercises, and little scientific rigor. It does not discuss cell chemistry and technology. It simply describes the functionality of the cell. It is not a complete how-to for each application. It only discusses aspects that relate to its battery. It is not a guide for battery management system design. It discusses off-the-shelf BMSs.

INTENDED AUDIENCE

This book is for the product designer, the project manager, the repair and maintenance technician, the purchasing agent, the member of a student racing team, and anyone dealing with Li-ion batteries and applications that use them.

The reader is expected to have a basic understanding of science and technology to be able to interpret graphs and block diagrams and to be able to read simple electrical schematic diagrams.

MOTIVATION

I wrote this book for two reasons:

- To accelerate the development of batteries as one component to address climate change and avoid the fate of Earth's thermal runaway;

- To enable the design of safe Li-ion batteries, which can otherwise be deadly.[1]

BEYOND THIS BOOK

This book's website, LiIonBook.com, offers additional materials and a contact form for additions and corrections. It also includes the unabridged Table of Contents.

The slides that I use when I teach classes based on this book[2] are freely available online.[3]

THE MOST IMPORTANT TAKEAWAY

If you only remember one thing from this book, let it be this:

THE BMS *MUST* BE ABLE TO SHUT OFF THE BATTERY CURRENT!

A BMS that is not allowed to interrupt the entirety of the battery current (directly or indirectly) does not protect its battery and is therefore pointless.

ORIENTATION

This is the first volume of a two-volume set and is incomplete without volume two, first edition.[4]

Volume 1 discusses batteries (Figure P.1(a)):

- Basic concepts, common misunderstandings, and introduction of new concepts and terms (Chapter 1);
- Lithium-ion cells (Chapter 2) and their arrangement (Chapter 3);
- Battery management systems (BMSs) (Chapter 4);
- Battery design (Chapter 5), and modular batteries and arrays (Chapter 6);
- Production and deployment (Chapter 7), and problems and solutions (Chapter 8).

Volume 2 discusses applications:

- The four classes of applications that constitute the majority of Li-ion battery usage (Figure P.1(b)):
  - Small batteries (Chapter 1);
  - Low-voltage batteries and battery arrays (telecom, residential, house power, etc.) (Chapter 2);
  - Traction battery packs for vehicles (UAVs, passenger EVs, marine, industrial, etc.) (Chapter 3);
  - High-voltage, stationary batteries, on-grid and off-grid (Chapter 4);

1. As I write this, Tallmadge D'Elia was killed by an incompetent Li-ion design (the Li-ion cell in his e-cigarette exploded and fragments penetrated his skull): "Vape pen explosion pierces Florida man's cranium killing him", BBC, May 17, 2018, www.bbc.com/news/world-us-canada-44149281 [1].
2. I taught classes in Mumbai and Santiago Chile; I am about to do so in Detroit.
3. www.slideshare.net/DavideAndrea.
4. This book refers to particular sections in Volume 2, first edition; the section numbering of other editions differs.

FIGURE P.1
Book flow: (a) Volume 1—batteries, and (b) Volume 2—applications.

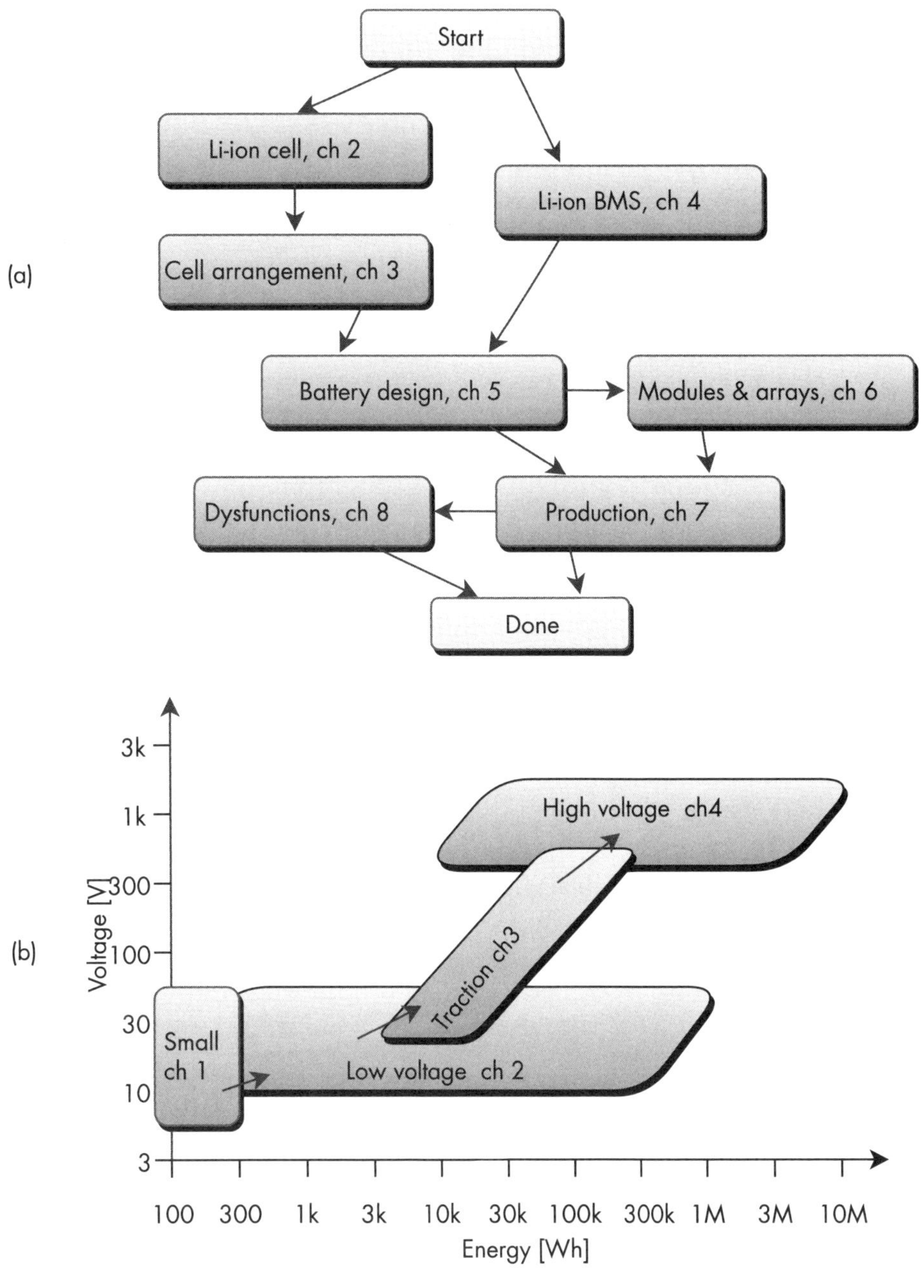

- Case studies of accidents (Chapter 5).

ABOUT THIS BOOK

This book is based on 16 years of experience building batteries and supporting Elithion's customers through the design of batteries for a wide variety of applications and requirements. This book compiles the challenges we encountered and the solutions we implemented.

In 2010 I wrote a successful book focused just on the Li-ion BMSs.[5] That book opened the door for other authors to cover that subject in greater depth, from the point of view of the BMS designer (see the "Resources" section in Volume 2). Therefore, rather than writing a second edition, I chose to write this book on the wider subject of Li-ion batteries and to do so from the point of view of the battery designer and user.

You may note that I am fond of rules of thumb, bullet points, and footnotes.[6] I hope that you won't let the casual tone detract from the breadth of the materials presented here and from the depth of the research that went into this book.

ABOUT ME

I have designed BMSs for large Li-ion battery packs since 2004, first for my electric vehicle, then in 2006 for Hybrids Plus, converting hybrid cars to plug-in hybrids. There I became aware of the industry's great need for an off-the-shelf BMS for large Li-ion batteries. In 2008 I started Elithion to meet that need.

My LiIonBMS.com website has directories of Li-ion products and educational materials. I publish articles in trade publications.

I was born in Italy in 1957 and have lived in Colorado (USA) since 1976. I received an EECS degree in 1982 from the University of Colorado. Today I live and work in Longmont, Colorado.

ABOUT MY COMPANY

Elithion is a small company offering off-the-shelf BMSs and offering consulting services to the Li-ion industry for a great variety of applications. Elithion has sold BMSs longer than any other company. At this point in my life, I would love it if someone would take over my company so that I could just devote myself to engineering and writing books.

ABOUT THE CONTRIBUTORS

Some 45 people were kind enough to review portions of this book for correctness, clarity, and completeness. I am grateful to all of you who have helped hone this book into this published version.[7]

Thank you for reading.

Reference

[1] "Vape pen explosion pierces Florida man's cranium killing him," BBC, May 17, 2018, www.bbc.com/news/world-us-canada-44149281.

5. *Battery Management Systems for Large Lithium-Ion Battery Packs,* Artech House, 2010.
6. Though not to the extent David Foster Wallace did in *Infinite Jest.*
7. Especially P. G. for his insights and Byron Azarm for volunteering innumerable hours.

CHAPTER 1

FUNDAMENTAL CONCEPTS

1.1 INTRODUCTION

Before we dive into the subject, it would be helpful to use the same set of battery-related terms and concepts. Even if you are already familiar with batteries, please go over this chapter, as there may be some discrepancy between your present understanding and the premises in this book.

At various times over my years working with batteries, I found myself needing particular standards, concepts, or terms that the battery industry did not have. In this book I present a few that other authors and I have recently coined, hoping that you too will find them useful and that you will adopt them as well. In particular, I would like to highlight the concept of maximum power time (MPT) (see Section 1.5.2) and the term inverger (see Section 1.8.4.2).

1.1.1 Tidbits

Some interesting items in this chapter include:

- An unprotected 18650 is a cell, not a battery (1.2.2.2);
- If you open a healthy lithium-ion cell, you won't see any lithium (1.2.2.4);
- There's no such thing as LiPo (1.2.2.5);
- C-rating is a bogus concept invented for marketing purposes (1.2.2.8);
- In most cases, Ohm's law doesn't apply (1.3.5);
- Inverter/charger is a mouthful; let's call it inverger (1.8.4.2);
- For high-power applications, let's compare cells using their maximum power time (1.5.2);
- "State of (something or other)" means different things to different people (1.6.1).

1.1.2 Orientation

This chapter starts by defining graphic symbols and terms.

- It then discusses misused terms and misunderstandings.
- Then it goes through all the electrical measures that are of interest in this field and proposes a new one: maximum power time.
- It lists the alphabet soup of state of (something or other).
- It describes some charts that we use when discussing batteries.
- Finally, it lists power converters commonly connected to batteries.

1.2 SYMBOLS AND TERMS

Let's start by defining some terms and symbols.

1.2.1 Symbols

This book uses standard symbols for electrical *schematic diagrams* and *block diagrams.* A difference between these two is that a schematic diagram shows the full path for the current (e.g., from a battery, through a switch, through a load and back to the negative terminal of the battery) while a block diagram shows connected by lines representing signal or power.

1.2.1.1 Schematic Diagram Symbols

Figure 1.1 shows some of the electrical and electronic symbols you'll see in the schematic diagrams in this book. They include wiring; note the use of ground symbols,

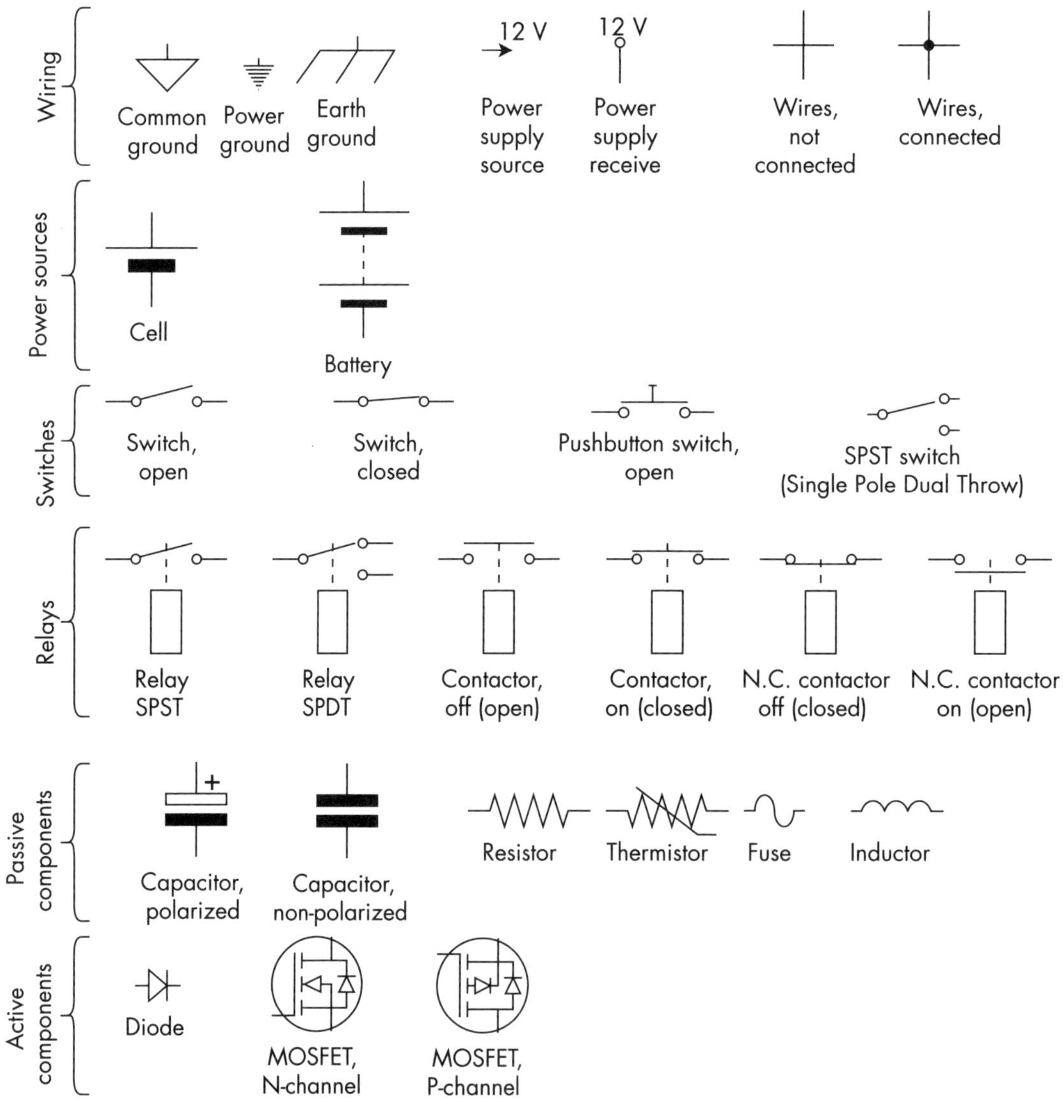

FIGURE 1.1 Electrical schematic diagram symbols.

and understand that all the ground symbols in a schematic diagram are connected. Doing so clarifies a diagram by removing a ground line routed all over the place. Similarly, all power supply symbols with a given legend (e.g., *12V*) are connected.

Relays are electromechanical switches that are operated by powering their coil. *Contactors* is the name we give to high-power relays.

Among the passive components, *fuses* are crucial in batteries. When they carry too much current, they blow and open the circuit.

Active components are solid-state semiconductors that are used to control an electrical current; (MOSFETs) are a type of transistor (see Section 5.12.4.1)

By convention, in a schematic diagram, higher voltages are at the top and lower voltages at the bottom, so that current flows down (except in power sources, where it flows up). If possible, inputs are on the left and outputs on the right, so that processing flows left to right (Figure 1.2).

Following this convention helps others read your diagrams just like writing sheet music with the treble on top and the bass at the bottom, proceeding from left to right, allows musicians to read a musical score.

1.2.1.2 Block Diagram Symbols

The symbols in this book are based on the IEC 60617 standard (Figure 1.3). These include power sources, storage, loads, and switches.

The converters may be unidirectional or bidirectional and convert between or within AC or DC.

In block diagrams, I try to place components so that power flows left to right; grayed-out boxes represent optional devices.

1.2.2 Terms

Different fields may use different terms for the same item. For example, an electrician may call *amperage* what we call *current*; a car technician may call *condenser* what we call a *capacitor*.

Similarly, people from different parts of the world may use different terms for the same item: what I, as an American, call a *battery*, a German may call an *accumulator*. To that end, this book offers an international dictionary specific to battery terms (see Volume 2, International Dictionary).

While a given term may not inherently be more correct than another one, technical discussions are more effective when all parties in that discussion use the same, clearly defined terms. For the sake of clear communications, when we are made aware that a term is ambiguous or has been misinterpreted, we strive to use a better term and encourage others to do so as well.

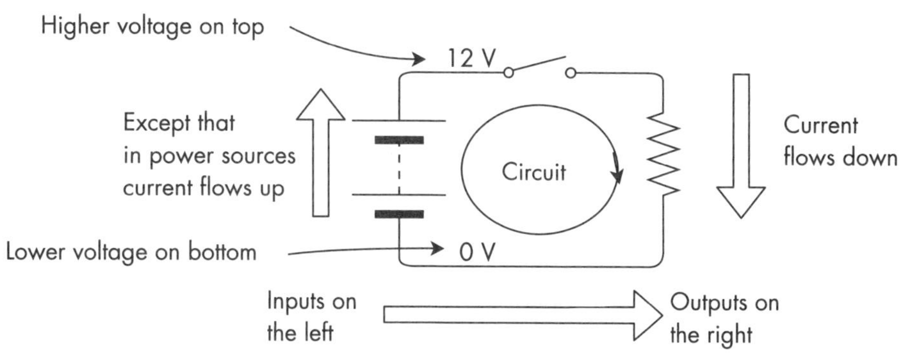

FIGURE 1.2
Electrical schematic diagram flow conventions.

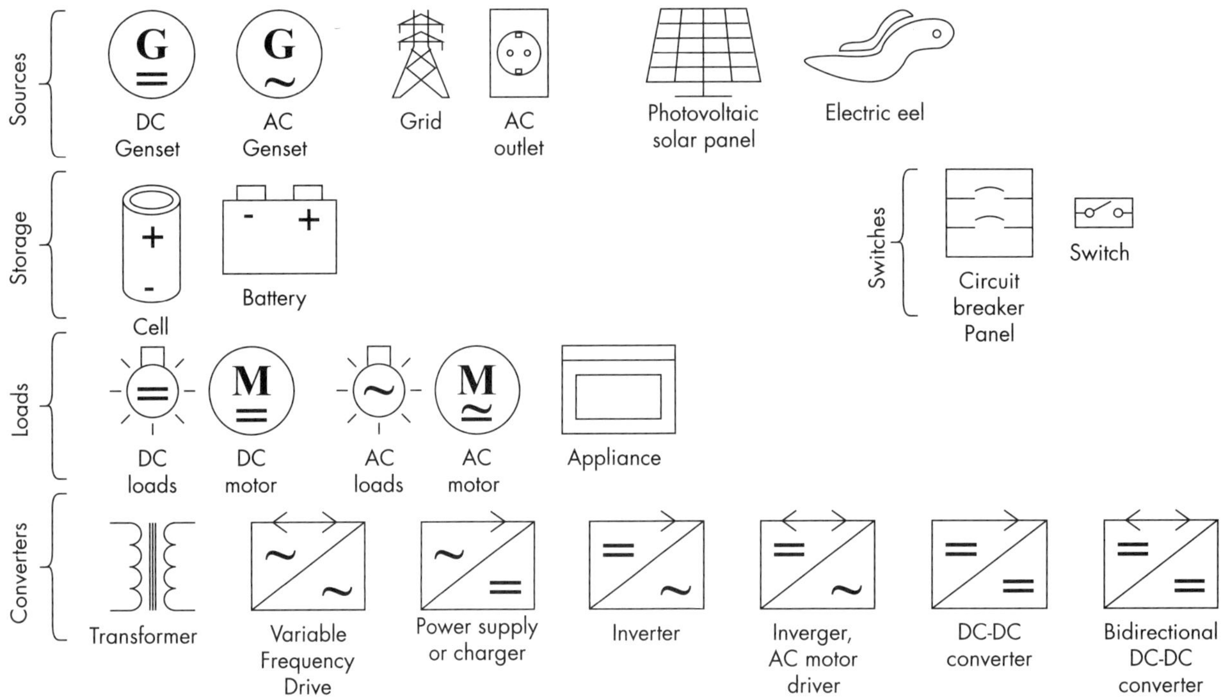

FIGURE 1.3 Block diagram symbols.

1.2.2.1 Neologisms

Please allow me to introduce these new terms, which I invite you to use as well.

- *Inverger*[1]: Combination inverter and charger that converts DC to AC or vice versa (see Section 1.8.4.2);
- *Total specific charge transferred (TSCT):* The total charge transferred by a cell or battery over its useful lifetime;
- *Emery:* Dimensionless unit for total specific charge transferred (see Section 1.4.1.2).
- *Maximum power time (MPT):* A constant that characterizes a particular cell or battery technology (see Section 1.5.2);
- *Discharge availability (DA):* How much charge a battery array can deliver; similar to state of charge, but measured in Ah and considering that some batteries may be disconnected at a given time.
- *Charge acceptance* (*CA*): How much charge a battery array can accept; similar to *depth of discharge.*

1. Inverger is a portmanteau of inverter and charger that my partner Carl Lawrence coined in 2007. The trademark changed hands a few times, and in 2017 it was allowed to expire. Now I'd like to reclaim the term and to put it in the public domain. I like it because it is fully descriptive (unlike combi) and is a single word (unlike charger/inverter).

1.2.2.2 Cell Versus Battery

A *cell*[2] is the most fundamental electrochemical device that stores DC charge, while a *battery*[3] is a device that includes two or more components—cells, a *battery management system (BMS),* connectors, thermistors, and an enclosure.

Examples of cells include a coin cell, a Li-ion cell, and a 1.5V alkaline cell (Figure 1.4(a)). Examples of batteries include a 9V alkaline battery (four 1.5V cells in series), a 12V car battery (six 2V cells in series), a cluster of 18650 cells, and a battery module (Figure 1.4(b)).

As an analogy, consider a train composed of a locomotive and 10 cars. Each car is not a train, it's just a car; you are not seeing 11 trains, you are seeing a single train that includes 10 cars and a locomotive. Similarly, you are not seeing 10 batteries; you are seeing a single battery that includes 10 cells and a BMS.

1.2.2.3 Anode and Cathode

The *cathode* is the electrode from which electric current flows out of a device; the *anode* is the electrode from which electric current flows into a device. In the case of a *primary* (nonrechargeable) cell or battery, the discharge current flows from the positive terminal, and therefore that terminal is the cathode (Figure 1.5(a)). But in a *secondary* (rechargeable) cell or battery, the direction of the current changes when switching between charging and discharging

- When discharging, the + terminal is the cathode (Figure 1.5(b);
- When charging, the + terminal is the anode (Figure 1.5(c));
- When idle, the + terminal is neither (Figure 1.5(d)).

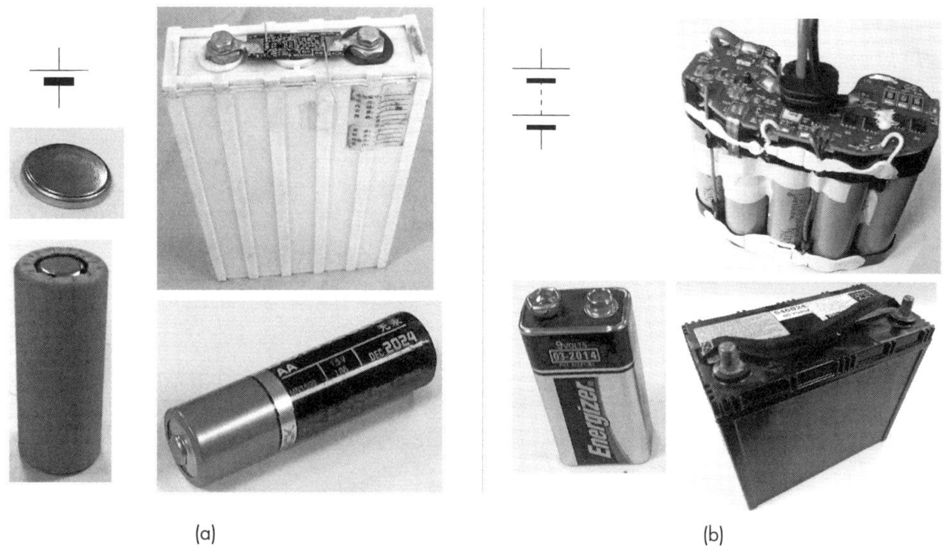

FIGURE 1.4
Cells versus batteries: (a) cells, and (b) batteries.

2. Technically, the correct term is galvanic cell.
3. Battery, as in "assault and battery," and similar expressions comes from the Latin bauttere, to beat, from which battery is used for a grouping of artillery pieces. A sense of "coordinated collection of similar items" is implied in the French word batterie. This was extended to apply to kitchenware, drum kits (in European languages), electrical generators, and piles of cells (Alessandro Volta invented the Voltaic pile, the first battery). Hence, we use the term battery to mean a collection of cells.

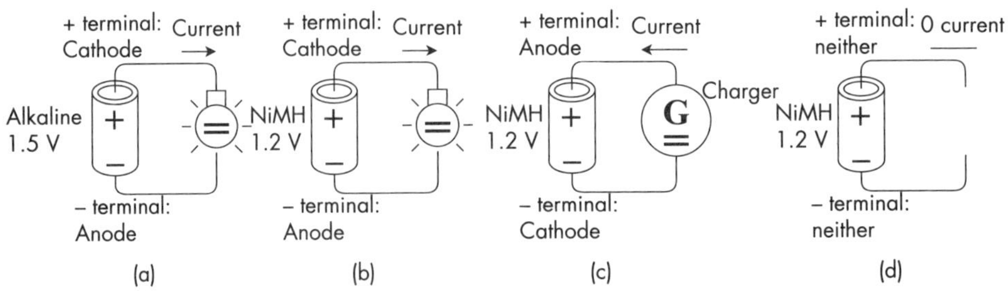

FIGURE 1.5
Cell terminal names and functions: (a) primary, (b) rechargeable, discharging, (c) rechargeable, charging, and (d) rechargeable, idle.

It is impractical to rename a terminal whenever the current direction changes. Therefore, technically, the terminals of a rechargeable cell or battery should be called *positive terminal* and *negative terminal*.

As this book concerns rechargeable cells and batteries, you won't find the terms anode and cathode anywhere else (unless referring to a diode).

1.2.2.4 Lithium Versus Li-ion

Do not conflate the terms *lithium* and Li-ion. By my count, there are five different battery technologies that use lithium, only one of which is Li-ion (Table 1.1).Therefore, to avoid confusion, say Li-ion because the termlithium is assumed to mean a lithium metal cell, a nonrechargeable coin cell that uses bulk metallic lithium. Rechargeable lithium metal cells are presently under development (see Volume 2, Section A.3.3).

1.2.2.5 "LiPo"

Technically, *lithium polymer* (*LiPo*) is a solid electrolyte polymer separator that includes a lithium compound. Such a cell uses a solid plastic ionic conductor instead of a liquid or gelled electrolyte.

No Li-ion cell using lithium polymer has ever left the lab [1]; none has ever gone into production [2]. Therefore, "LiPo" is just a meaningless buzzword.[4]

TABLE 1.1
Five Cell Technologies that Use Lithium

Chemistry	Rechargeable	Voltage	Contains Bulk Lithium Metal	Uses	Notes
Lithium metal	No	3, 3.6	Yes	Watches, hearing aids, desktop computers	9V batteries too
Li-ion	Yes	2.3~3.7	No	Phones, laptops, power tools, electric vehicles (EVs), solar	The subject of this book
Lithium metal polymer	Yes	3.7	Yes	EVs	In development
All-solid-state lithium	Yes	3.7	Yes	EVs	In development
Lithium AA	No	1.5	?	Cameras, flashlights	Alkaline replacement

4. LMP cells do use a lithium metal polymer, but they are lithium metal cells, not Li-ion cells, so, again, not LiPo.

Even though proper LiPo cells don't exist, cells by that name are seen everywhere because that term was appropriated by hobbyists and marketers who use "LiPo" as a shorthand for "Li-ion cell in a soft plastic package with a nominal voltage of 3.6V." Yet, "LiPo" is neither a cell chemistry nor a cell format, nor a voltage. Professionals and manufacturers call a pouch cell a pouch. They call a 3.6V cell NMC, LCO, or whatever chemistry is it. If they ever use the term LiPo it is with qualms, to acquiesce to the market.

You won't see the term LiPo anywhere else in this book.

1.2.2.6 "LiHV"

To the best of my understanding, *LiHV* is not a particular Li-ion cell chemistry; rather, it appears that a vendor of rebranded cells and batteries for the hobby market invented this marketing term. The cells appear to be standard Li-ion LCO pouch cells that are labeled by the vendor to indicate that they can be charged up to 4.35V instead of 4.2V. Of course, charging any Li-ion cell to a slightly higher voltage does indeed increase the capacity; it also reduces the lifetime. It seems that the only difference between a standard cell and a LiHV cell is the label—both appear to be the same Li-ion LCO pouch cell.

1.2.2.7 Li-ion Versus LiFePO4 or Li-ion Versus LiPo

The standard *LiCo* (*lithium-cobalt,* LCO) Li-ion cell in an 18650 format was developed first, so some may assume that it is the one and only Li-ion cell. Yet, the more recently developed cell chemistries (NMC, LFP, LTO) and formats (pouch, prismatic, and others) are no less of a Li-ion cell than that original 18650 LiCo cell because they all are based on Li-ion intercalation (see Section 2.2). Saying "LiFePO4 is safer than Li-ion" is as illogical as saying that Toyota, Corollas are nicer than cars, because the former is a subset of the latter.

1.2.2.8 C-Rating

C-rating is a marketing term, not an engineering parameter.[5] It was invented by salespeople selling Li-ion cells, not by the chemists and engineers that design them. As time goes by, the values for C-rating increase, not because cells are getting better (and they are), but because of a ratings race among vendors.

We see numbers such as "120C" in certain vendor websites. 120C means discharging a battery in 30 seconds! (see Section 1.4.3.4) Assuming that such fast discharge is even possible, the real question is: at what cost? Discharging in minutes damages the cells and gives little usable power to the load, because most power is just heating the cells. After all, you could get the highest current out of a battery if you discharge it across a short circuit! The current would be high, yes, but so what: the terminal voltage is 0V, and, therefore, the power into the load is 0W, which is useless.

I know of a particular Li-ion cell that is characterized at 0.5C by its manufacturer and rated for 5C by its vendors; that is a whole order of magnitude! When a cell that has been used at 5C dies prematurely, the cell manufacturer can justifiably reply "We never told you to use it at 5C!"

A given cell could be rated for

- A conservative value that ensures maximum cycle life (e.g., 0.5C);

---

5. Do not confuse the marketing term C-rating with the technical term C-rate.

- A reasonable compromise between cycle life and power generation (e.g., 1C);
- An aggressive value that is impressive compared to other cells (e.g., 5C);
- The current at the maximum power point (see Section 1.5) (e.g., 20C);
- The impractical value of short-circuit current (e.g., 40C).

There are a few ways for a cell manufacturer to get fantastic C-ratings:

- Either state a lower capacity or actually reduce the capacity. Since the specific current is relative to the capacity, halving the capacity doubles the C-rating.
- Report the short-circuit current as if it were the safe discharge current.
- Use a fully charged, overheated cells for just 1 second (the internal resistance decreases at high temperatures).

For example, the GEB 6619140 cell is rated for a 120C discharge (!). It achieves this rating by having a low energy density (only 56 Wh/kg), which increases the specific current. The fact is that using the same volume of EiG F-007 cells (with a C-rating of "only" 30C) and the same number of cells in series, one would get the same total pack resistance, yet 60% more capacity. Both packs would have the same weight and mass and would have the same efficiency and voltage sag at a given current. However, at that current, the discharge from a pack built with those EiG cells (rated at 30C) would last 60% longer than a pack built with those GEB cells (rated for 120C).

1.2.2.9 Hard Short Circuit Versus Soft Short Circuit

A *short circuit* is an unintended conductive path, which shunts across two networks and results in an unintended flow of current. A short circuit may be

- A *hard short*: Practically 0 Ω, able to carry unlimited current;
- A *soft short*: More than 0 Ω (practically, in the range of 100 Ω to 100 kΩ), which limits the current.

For a given current a hard short produces practically no heat. However, a soft short may limit the current while a hard short does not.

1.2.2.10 List of Terms

Here is a simple (and simplistic) explanation of the key terms used in this book. Unlike a glossary, it is not alphabetical; it is organized by subject and designed for a quick scan. As you scan it, stop when you see an unfamiliar term and look it up in the full glossary at the end of this book (see Volume 2, Glossary).

Measures

- *Charge*—number of electrons, measured in Ah; *capacity*—how much charge a cell or battery can store, also measured in Ah; *capacitance*—the same, but for a capacitor, measured in farad; *emery* (neologism)—a charge equivalent to the capacity of a battery.
- *Current*—a flow of charge, measured in A; specific current—same, but relative to the capacity of the cell or battery, typically measured in C (C-rate).
- *Resistance*—limits the current out of a cell and causes the voltage to sag under load; impedance—the same concept, but for AC.

- *Terminal voltage*—actual cell voltage, such as when under load; open-circuit voltage (OCV)—the voltage at rest, no load; voltage sag—the terminal voltage under load; IR drop—the change in voltage, the difference between the OCV and the terminal voltage; breakdown voltage—maximum voltage across an item, an arc occurs if exceeded.
- *Power density*—how much power a battery of a given volume can produce; specific power—the same, but for a given mass; energy density—how much energy a battery of a given volume can store; specific energy—the same, but for a given mass.
- *MPT*—how long it takes to discharge a cell or a battery at maximum power, as a measure of how well cells or batteries perform at high power (lower is better).
- *Power efficiency*—how much of the power generated internally by a battery is seen by a load; *energy efficiency*—the same, but over time; *Coulombic efficiency*—the portion of the charge placed into a cell during charge that can then be retrieved during discharge, nearly 100 % for Li-ion.

Li-ion cells

- *Li-ion cell*—stores charge by moving lithium ions between electrodes, forms the basis of a Li-ion battery.
- LCO, LFP, LTO—three-letter codes for various Li-ion chemistries, short for cobalt or LiCo, iron phosphate or LiFePO4 or Titanate, respectively; other codes include NMC, LMO, LNO, NCA.
- Pouch, small cylindrical, large cylindrical, small prismatic, large prismatic—various cell formats; that is, the shape of the cell.
- *Energy cells*—optimized for maximum capacity; power cells—optimized for minimum resistance, maximum ability to deliver power.
- *Cycle life*—each time a cell is charged and discharged, it degrades; calendar life—even if the cell is not used, it degrades.
- *Relaxation*—at rest, the cell voltage slowly approaches the OCV; hysteresis—how the cell voltage never quite relaxes completely to the OCV.
- *Self-discharge*—over time, a cell's SoC slowly drops.
- *Weak cell*—low-capacity or high-resistance cell.
- *Thermal runaway*—unstoppable process that destroys a cell, releasing smoke and fire.
- *Specification sheets*—describe the characteristics of a given cell; charging curves—describe the characteristics of a cell as it's being charged; discharging curves—the same, but for discharging.
- *Safe operating area (SOA)*—the range inside which it's OK to operate; maximum power point (MPP)—using the battery as hard as possible, beyond which it is diminishing returns.

Cell or battery state

- *State of charge (SoC)*—how much charge is left; discharge availability—the same, but for a battery array; depth of discharge (DoD)—, how much charge has been delivered; charge acceptance—the same, but for a battery array.

- *State of health (SoH)*—how well a battery is doing.
- *Balanced*—in a series string, all cells are at the same SoC; *imbalance*—how much SoC differs among cells; *balancing*—the act of restoring that balance; *bypass balancing*—balancing by wasting extra charge in heat; *charge transfer balancing*—balancing by moving charge between cells instead; *top balancing*—when the battery is at 100% SoC, so are all the cells; *mid balancing*—the same, but at 50% instead; *maintenance balancing*—slow balancing by the BMS; *gross balancing*—fast balancing performed manually when the series string is badly imbalanced; *redistribution*—high-power technology to keep all cells at the same SoC at all times.

Components

- *Wire*—single conductor; *cable*—multiple conductors in a single tube; *harness*—the same, though loose; *bus*—either a communication link or a high-power conductor to interconnect batteries and other devices.
- *Switch*—lets current through when closed, doesn't when open; Safety disconnect—manual switch to open the battery circuit; breaker, circuit breaker—switch that opens automatically in case of overcurrent; fuse—the same, but not resettable; transfer switch—switch that selects one of two power sources; bypass switch—the same, but one of the sources is a battery; combiner—selects one of two batteries or both.
- *Relay*—electrically controlled switch; contactor—the same, but larger; MOSFET—the same, but solid-state transistors; protector switch—any of the above, to disconnects the battery if required to protect the cells.
- *Connector*—lets you connect or disconnect a wire, cable, or printed circuit board (PCB).
- *PCB*—circuit board on which electronic components are mounted.
- *Resistor*—converts electrical power to heat; capacitor—stores a charge, smaller than a battery; supercapacitor, ultracapacitor—the same, but larger, still smaller than a battery.
- *Thermistor*—senses temperature; *current shunt*—senses current; *Hall effect sensor*—also senses current.

Battery

- *Series string, parallel block, series-first, parallel-first*—four basic ways of interconnecting cells to achieve the desired battery voltage and capacity.
- *Battery*—a collection of cells, BMSs, and other components, with one current, one SoC, one protector switch.
- *Single-switch battery*—the protector switch uses one contactor; dual-switch, single-port battery—the same, but two contactors or MOSFETs, one for charging, one for discharging; dual-port battery—the same, but separate input for charging and output for discharging.
- *Battery array*—two or more complete connected batteries; modular battery—a single battery subdivided into smaller modules; split battery—two batteries connected in series with a center tap; ganged batteries—a battery array in which the protector switches are all on or all off simultaneously.

- *Inrush current*—a pulse of excessive current when connecting a battery directly to a device; precharge—a technique to avoid the above.
- *Isolation*—no connection between the battery and earth ground; galvanic isolation—the same, but capable of withstanding a high voltage across it; isolation loss, ground fault—unintentional connection to earth ground; isolation test, ground fault test—detection of the above.

BMSs

- *BMS*—protects cells and manages the battery; *fault protector*—redundant BMS.
- *Protector BMS, protector circuit module (PCM)*—small BMS that includes a protector switch; battery management unit/battery monitor unit (BMU)—larger BMS without a protector switch, relies on the system to obey it.
- *Wired BMS*—connected to the cell through tap wires; distributed BMS—a cell board mounted on each cell to sense its state; master/slave BMS—BMS subdivided into modules.
- *Current limits*—a BMU tells the system to limit the current to these levels; charge current limit (CCL)—the same, but just for the charging current; discharge current limit (DCL)—the same, but for discharging.
- *CAN, RS-232, RS-485, USB, SMB*—various digital data links and buses.
- *Voltage translation, coulomb counting, Kalman filter*—techniques to evaluate the SoC indirectly; *fuel gauge*—the informal name for SoC evaluation.

Devices

- *Transformer*—converts an AC voltage to another AC voltage, isolates the two; *DC-DC converter*—the same, but for DC; *power supply, AC adapter*—converts AC to DC; *charger*—the same, but current limited; *EVSE*, *charging station, pedestal, shore power*—the same, but stationary and for a vehicle; *solar charge controller*—the same, but powered by solar panels; inverter—converts DC to AC, either line frequency or to drive a motor; *inverger* (neologism)—the same, but in either direction; *variable frequency drive* (*VFD*)—converts AC to AC to drive stationary motors.
- *Genset, AC generator*—converts fuel to AC power; grid—electric power from the power company; microgrid—the same, but much smaller and local; generator—converts mechanical power to electrical power; alternator—the same, in a vehicle; photovoltaic panel (PV), solar panel, solar array—generates power from the sun; wind generator—the same, but from wind.
- *Load*—something that uses power; critical load—must remain powered in case of grid failure; auxiliary load—may go off in case of grid failure.
- *Base transceiver station (BTS)*—telecommunication equipment at a base station.
- *Engine*—converts fuel to mechanical power; *motor*—converts electrical power to mechanical power; *traction motor*—the same, but for a vehicle; *motor driver*—converts DC power to the voltage required to drive a motor; *motor controller*—for a DC motor; *inverter*—for an AC motor; *quadrants*—four permutations of forward and reverse, drive and brake.

Application

- *Low voltage*—up to about 40 or 48V; *High voltage*—above 48V or above about 500, depending on the context.
- *Small battery*—low voltage, used in consumer products; traction battery—propels a vehicle; large stationary low-voltage battery—12 to 24V at a fixed location, such as for solar storage; high-voltage stationary battery—the same, but much higher voltage, such as connected to the power grid.
- *Energy battery*—Fully charged, used slowly; power battery—the same, but used fast; buffer battery—the same but kept at about 50% SoC; charge deplete—how an energy or power battery is used; charge sustain—the same, for a buffer battery.
- *Energy storage system (ESS)*—any form of energy storage; battery energy storage system (BESS)—the same, but using batteries; uninterruptible power supply (UPS)—the same, but with an AC input and an AC output; power bank—the same, but DC output.
- *Auxiliary power unit (APU)*—fuel-powered generator in a vehicle independent of the engine; electronic generator/EAPU—the same, but uses a battery; starter lighting ignition (SLI)—standard car battery.
- *Residential*—for a house; marine—for a boat or other vessel; telecom—for cell phones and other radios; agricultural—for farming; business—for large buildings where people work; industrial—the same, but for heavy-duty manufacturing.
- *EV*—propelled by electricity; battery EV (BEV)—the same, with a battery, no engine; plug-in hybrid EV (PHEV)—the same, with an engine as well; hybrid EV (HEV)—the same, but no power cord; micro-hybrid EV (MHEV)—the same, but only a starter motor; recreational vehicle (RV)—vehicle with living space; unmanned aerial vehicle (UAV)—drone; vehicle control unit (VCU)—computer that controls a vehicle.
- *Off-grid*—disconnected from the power grid; *grid-tied/on-grid*—connected to the power grid; *back-feed*—sending power back to the grid; *grid-interactive inverter*—able to back-feed or to power the local loads in case of grid failure.

Miscellaneous

- *Cycle*—charging and discharging the battery once.
- *Radar chart*—compares multiple parameters of cell technologies at a glance; Ragone plot—compares the power and energy of cell technologies at a glance.
- *Short circuit*—an unintentional connection that draws a high current; hard short—zero-resistance short; soft short—higher-resistance short, but still bad.
- *Pulse-width modulation (PWM)*—rapid turn on and off of a voltage, to reduce its average to the desired level.

1.3 COMMON MISUNDERSTANDINGS

Questions about the following items are common.[6]

6. Such questions come up all the time on Reddit, in the /r/Batteries sub, which I moderate.

1.3.1 Charging While Discharging

When people ask, "Can a battery charge and discharge at the same time?" the implicit question is, "Can I connect this battery to its charger while I am using it?" The answer is, of course, yes: nothing inherently prevents a battery from being simultaneously connected to a charger and a load. However:

- The battery design may physically prevent you from making both connections at the same time;
- A power bank may get too hot if its internal charger and its internal DC-DC converter (that powers the output) are powered at the same time.

The direct answer to the question, as asked, is that at a given moment, the battery itself cannot be both charging and discharging.

Specifically:

- While just the charger is connected the battery charges until the battery is full (Figure 1.6(a));
- While just the load is connected the battery discharges until the battery is empty (Figure 1.6(b));
- While both the charger and the load are connected, the battery current is the difference between the charger's current and the load current:
  - If the load draws less current than the charger can provide, the battery charges until the battery is full, at which point the current drops down to what the load needs (Figure 1.6(c));
  - If the load draws exactly as much current as the charger can provide, the battery current is zero; the battery is neither charging nor discharging (Figure 1.6(d));
  - If the load draws more current than the charger can provide, the battery provides the extra current (Figure 1.6(e)).

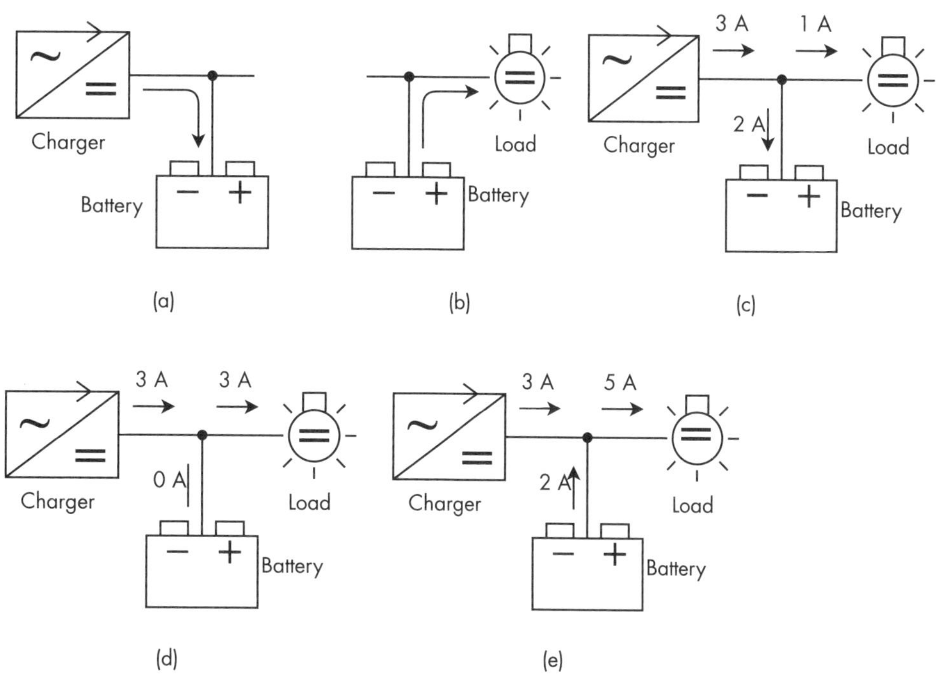

FIGURE 1.6
Charging while discharging: (a) charger only, (b) load only, (c) both, light load, (d) both, medium load, and (e) both, high load.

1.3.2 The Load Sets the Current

Someone may ask: "My cappuccino maker needs 12V, 0.5A, but the AC adapter pushes 12V, 2A; I am afraid that it will burn my machine. How do I reduce the current down to 0.5A?" The questioner does not understand that the AC adapter specifies the *maximum* current it can supply. A 1-amp AC adapter does not push 1A regardless. Instead, it allows *the load* to pull anywhere from 0 to 1 A. Therefore, yes, it's OK to plug in a 2-amp AC adapter into your toy that only needs 0.5A; just make sure that the voltages are the same.

As long as the source can provide enough current, the *load* determines the current (at the given voltage), *not* the source. Or, to put it in an imprecise yet poetic way:

> Voltage is pushed, current is pulled

Every electrician understands this fundamental concept innately, yet, frustratingly, not enough fresh graduates in electrical engineering understand it.

Some examples:

- A 120Vac outlet in your U.S. home is rated for 15A; however, that doesn't mean its current is 15A, it just means that it's rated for *up to* 15A. The outlet does not set the current; whatever you plug into the outlet sets the current—0.1A for a phone AC adapter, 1A for a work lamp (but only if it's turned on), 10A for a vacuum cleaner (only if it's running).
- The AC power adapter for your laptop computer is rated 18V and 5A; however, that doesn't mean its current is 5A, it just means that it's rated for *up to* 5A. The adapter does not set the current, your laptop computer does; if your computer specifies 18V, 3A, it means that it needs anywhere between 0 A and 3A, depending on what it's doing at the time. You can power that computer with any 18V AC power adapter that is rated 3 A or higher (including 18V, 10A).
- The power supply for an LED lamp is rated for a constant current (say, 2A); that means that it adjusts its output voltage to try to maintain a current of 2A through any LED lamp that happens to be connected to it. Yes, the current is a constant 2 A, but that doesn't mean that the power supply is pushing 2A; instead, the LED is pulling 2A *at that voltage*. The supply is still pushing a voltage, which just so happens to be the voltage that results in the LED pulling 2A. Thus, voltage is pushed, current is pulled even with constant current supplies..

1.3.3 No Voltage Across a Switch

All too often, degreed engineers place a voltmeter across a switch that is disconnected from everything and wonder why they see no voltage across it. They fail to grasp that to see a voltage across an open switch, *something* must supply that voltage. That switch must be part of a circuit that is powered by some power supply; no supply, no voltage to be seen. The switch cannot somehow magically generate a voltage out of thin air (see Section 8.12.6). As an analogy, if you don't hear any music from your empty CD player, it's not because the player is broken, it's because you did not insert a CD in the player.

1.3.4 Power Supply Versus Charger

A *power supply* generates a constant voltage. If connected directly to a battery, it fights the voltage of the battery, at best resulting in the supply shutting down, or at worst damaging one or both devices.

A *charger*, instead, generates a constant current to charge a battery; later, it generates the correct voltage to finish charging the battery as the current decreases to zero. The "charger" for a phone or a laptop computer is not a charger; it is a power supply (an *AC adapter*). The actual charger is inside the phone or laptop (see Volume 2, Section 1.2.1).

1.3.5 Ohm's Law

Ohm's law says that at any instant, the current through a purely resistive load is directly proportional to the voltage across it: if you double the voltage, the current through a resistive load doubles. The proportionality constant is the load resistance: $V = I \times R$.

This law is misapplied by people who miss the "purely resistive" limitation. If you have a hammer, everything looks like a nail.[7] If you know Ohm's law, everything looks like a resistor. The vast majority of real-world loads are *not* resistive, and Ohm's law does not apply to them! Attempting to apply Ohm's law to these loads gives bogus results.[8]

Look around your room and list every electrical product you see; go down the list, and mark each as purely resistive:

- Extension and power cords;
- Light switches, outlets.

Then mark the ones that are resistive only after they reach a constant temperature:[9]

- Lamps with incandescent light bulbs without a dimmer;
- Electric stoves and heaters.

Then mark the ones that are completely not resistive:

- Lamps with a dimmer;
- Anything with a motor: washing machine, hair drier, electric car, power tools;
- Anything with a power supply: computers, chargers, compact fluorescent lamp (CFL) and light-emitting diode (LED) lamps, inverter;
- Anything battery powered: phones, laptops, calculator, camera; even an old-fashioned flashlight;[10]
- Anything variable: lamps on dimmers, heaters with gradual control.

If you see a resistor, yes, Ohm's law applies to it, but a resistor is an electronic component, not a product. I asked you to list products.

---

7. The source of the "law of the instruments," or "Maslow's law" is uncertain.
8. There is a similar-looking, unnamed law that applies to reactive loads (but it's not Ohm's law) that can be used on ideal passive components: $\underline{V} = \underline{I} \times \underline{Z}$ (where $\underline{V}$, $\underline{I}$, and $\underline{Z}$ are complex phasors). Yet, real-word AC loads (transformers and AC motors) are not purely reactive, so even that law does not directly apply to them.
9. The resistance of an incandescent lamp or heater increases as the filament gets hotter. The resistance is constant only at a given temperature, such as when the current varies rapidly during which the filament remains at the same temperature. If the current varies slowly, then the temperature changes, and the lamp no longer has a constant resistance—Ohms law does not apply.
10. An old flashlight (torch) has an incandescent bulb. See the previous note.

Ohm's law applies to the first set (Figure 1.7(a)); note how the current versus voltage curve is a straight line. The slope changes with the resistance of the load, but it's always a straight line, indicating that the current is directly proportional to the voltage. Ohm's law *does not* apply to the second set; note how the current versus voltage line is wonky for DC loads (Figure 1.7(b)) or AC loads (Figure 1.7(c)).

When I bring this up, I am met with a lot of resistance[11] because of how deeply this incomplete understanding of Ohm's law has been so ingrained into students, to the point that even some seasoned electrical engineers vehemently disagree [3].

1.4 MEASURES

This section discusses units of measurements that are specifically intended for use with batteries[12] as well as units that take a particular meaning in this industry. The next chapter will discuss these measures specifically for Li-ion cells and batteries.

The following measures are often confused (see Section 1.4).

- Capacity versus charge, capacity versus capacitance;
- DC resistance versus AC impedance;
- A versus Ah;
- kWh versus kW.

Quite often, two different measures use the same units, and people get confused and incorrectly assume they are the same thing. For example, both charge and capacity are measured in Ah. Or two different measures use the same term:

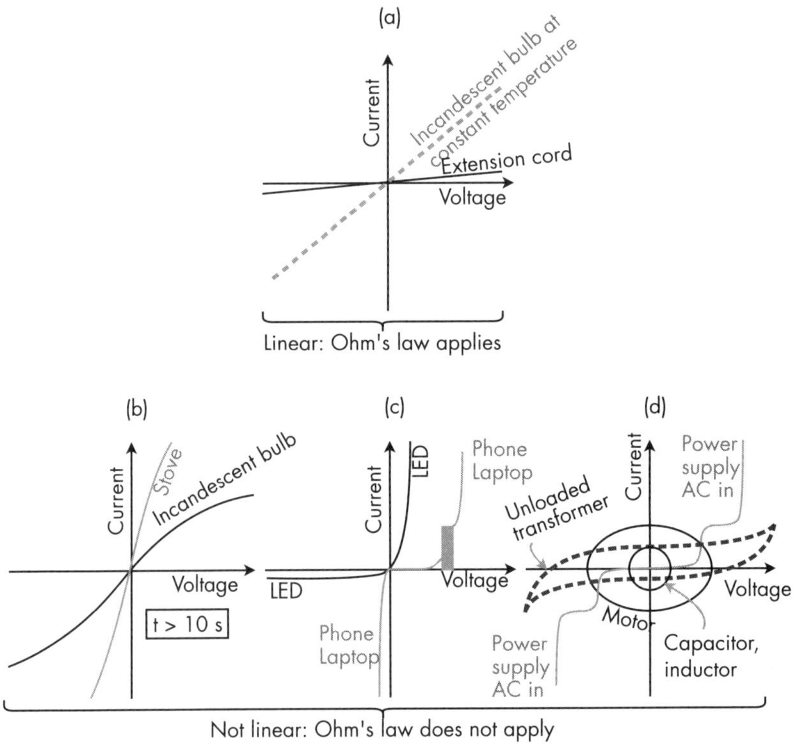

FIGURE 1.7
Current versus voltage: (a) resistive loads, AC or DC, (b) temperature-dependent loads, slow test, (c) nonresistive DC loads, and (d) nonresistive AC loads.

11. Har har har.
12. These measures apply to both cells and batteries. Still, for simplicity's sake, in this section, I say either cells or batteries when meaning both.

- In some English-speaking countries, pounds are used as units for both weight and force even though they are different phenomena;
- In German, Kapazität means both capacity and capacitance; Russian, French, and Italian have the same limitation.

In the following sections, I try to clarify some of the confusion that may arise between different measures that use the same units or terms.

1.4.1 Charge and Coulombic Efficiency

Both *capacity* and *charge* are measured in Ah, but they are different measures:

- The charge in a cell varies as it is charged or discharged;
- The capacity of a cell is effectively a fixed value, independent of its state of charge.[13]

When people say "How much capacity is left in a battery?" what they truly mean is "How much charge is left?"

Often, plots in scientific papers incorrectly show capacity on the X-axis (Figure 1.8(a)) rather than the correct measure: charge (Figure 1.8(b)). The capacity is actually at the bottom-right corner of the curve for the lowest current.

As an analogy, think of coffee in a coffee cup: the cup is capacity, the coffee is charge. As you drink, the quantity of coffee goes down, but the cup stays the same size. Similarly, as the battery is used the quantity of the charge goes down, but the battery's capacity remains the same.

1.4.1.1 Charge [Ah]

Charge is a quantity of electrical charges (e.g., electrons).

In physics, charge is measured in coulombs. In the battery field, we prefer to use amp-hours.[14] For a constant current:

$$\text{Charge [Ah]} = \text{Current [A]} \times \text{Time [h]} \tag{1.1}$$

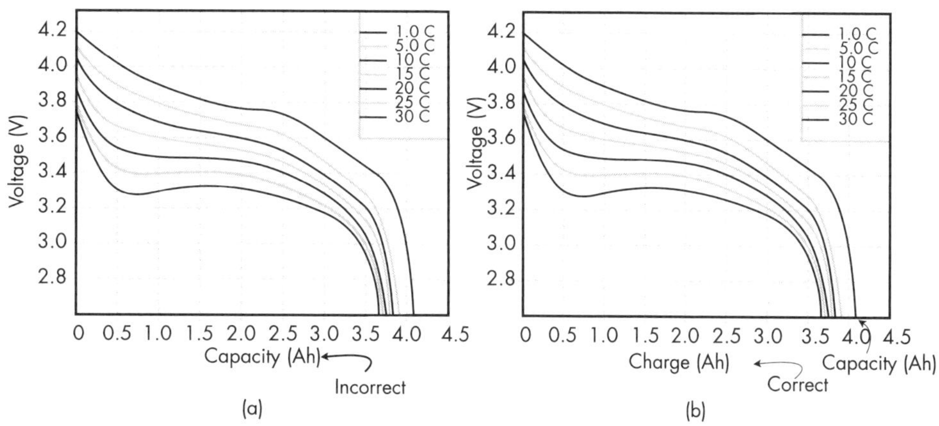

FIGURE 1.8
X-axis label: (a) technically incorrect, and (b) correct.

13. Capacity does decrease slowly as cells degrade. Effective capacity decreases at high current and depends on what we define as full and empty.
14. There is a charge of 3,600 coulomb in 1 Ah because there are 3,600 seconds in an hour.

A charge of 10 Ah can be transferred by a 1A current over 10 hours, a 2A current over 5 hours, or a 10A current over 1 hour. If the current is variable, we integrate it over time to get the charge.[15]

1.4.1.2 *Total Specific Charge Transferred [emery]*

Knowing the total charge that is transferred through a battery throughout its life helps us evaluate its usefulness. Consider these cases:

- A 200Ah battery that lasts only 100 full cycles;
- A 200Ah battery that lasts 250 cycles when only 50% of its range is used;
- A 100Ah battery that lasts 500 full cycles.

The last case is most valuable because it transferred 50 kAh during its life, while the first one only transferred 20 kAh and the second one only 25 kAh.

In more general terms that are independent of battery capacity, it's helpful to measure the *total transferred charge* relative to the battery capacity. Allow me to introduce a new measure and a new unit:

- *Total specific charge transferred* (*TSCT*): The ratio of the total transferred charge over the battery capacity;
- The *[emery]*[16]: A dimensionless unit for total specific charge transferred.

Total specific charge transferred is the product of the SoC range used in a cell times the number of cycles in the life of that cell. It is also the ratio of the total charge moved in the lifetime of a cell relative to its nominal cell capacity:

$$\begin{aligned}&\left[\text{specific\_charge}\right]=\text{number\_of\_cycles}\left[\text{-}\right]\times\text{SoC\_range}\left[\%\right]/100\left[\%\right]=\\&=\text{total\_transferred\_charge}\left[\text{Ah}\right]/\text{Capacity}\left[\text{Ah}\right]\end{aligned} \quad (1.2)$$

For example, regardless of cell capacity:

- If a cell is cycled 200 times from 0% to 100% SoC, the TSCT is 200 emery;
- If a cell is cycled 400 times from 0% to 100% SoC, the TSCT is 400 emery;
- If a cell is cycled 200 times from 30% to 80% SoC, the TSCT is 100 emery.

Also, regardless of SoC range or number of cycles:

- If a 10Ah cell transfers a total of 2,000 Ah, the TSCT is 200 emery;
- If a 10Ah cell transfers a total of 4,000 Ah, the TSCT is 400 emery;
- If a 20Ah cell transfers a total of 2,000 Ah, the TSCT is 100 emery.

By analogy, think of bucketfuls of water: without knowing the size of a zinc bucket, you can say that it is suitable for 10,000 bucketfuls during its life before it falls apart. Similarly, you can say that a cell is suitable for 10,000 emerys without knowing its capacity. You can also say that if you only fill a zinc bucket halfway, it will last

15. Charge [Ah] = ∫ instantaneous_current [A] di/dt, where the units of time are [hours].
16. This unit of measure was mistakenly coined by Isidor Buchmann, the author of the Battery University website; I asked Isidor how he came up with that unit, and he replied that he had meant to write "Energy units" and it was a typo. Mistake or not, I am keeping it because it's useful.

longer—30,000 uses. Similarly, you can say that if you only use 50% of a cell's SoC range, it will last longer—30,000 emerys.

Note that TSCT doesn't necessarily imply "over the lifetime of a cell," because it can mean over any specific period (3 years before the planned replacement) or for any number of cycles (3,650 cycles = 10 years assuming one shallow cycle per day).

TSCT will be useful later when calculating the advantage of oversizing cells to get the best value out of them (see Section 2.5.3.1).

1.4.1.3 Coulombic Efficiency [%]

Coulombic efficiency (also called *Faraday efficiency*, *Faradaic efficiency*, *current efficiency*, or *charge efficiency*[17]) is the ratio of the charge extracted from a battery starting from a given state of charge and the charge that is required to restore the battery to that same initial state of charge.

$$\text{Coulombic\_efficiency}[\%]=100[\%]\times\text{Charge out}[\text{Ah}]/\text{Charge in}[\text{Ah}] \tag{1.3}$$

For Li-ion, charge efficiency is practically 100%. Note that charge efficiency is not *energy* efficiency (see Section 1.4.4.7), which is certainly less than 100%.

1.4.1.4 Coulombic Inefficiency [%]

Coulombic inefficiency (*CIE*) is a measurement of loss of charge that may be correlated to capacity loss.

$$\text{CIE }[\%] = 100\ [\%] - \text{CE }[\%] \tag{1.4}$$

1.4.2 Capacity [Ah]

Capacity is the total charge that a cell or battery can store. It is measured in Ah. For example, if a cell can deliver 1A for 10 hours, the total charge is 10 Ah. Therefore, its capacity is 10 Ah.

Capacity may be viewed from three perspectives: nominal, actual, and effective:

- I buy a 100Ah cell: that's the nominal capacity;
- If I measure its capacity, I see it's 110 Ah: that's the actual capacity;
- When used in a go-cart, its voltage sags too much at 30% SoC: 77 Ah is the effective capacity.

The capacity of a battery is a slightly different concept than the capacity of a cell. Therefore, in total, there are six permutations of nominal, actual, and effective capacity for either a cell (see Section 2.5.1) or a battery (see Section 3.1.6). Let's look at all six.

Do not confuse capacity (for batteries) and *capacitance* (for capacitors) (see Section 1.4.7); unfortunately, many languages use the same term for both (German, Italian, etc.) (see Volume 2, "International Dictionary").

17. Some call it charge acceptance, but this is misleading because charge acceptance is how much charge a battery or an array can still accept (see Section 1.6.7). After much thought, I decided to go for coulombic efficiency because the more technically correct charge efficiency is used by the general public to mean other, ill-defined measures, and because charge is measured in coulomb, not farad. Also, my favorite authors, including Jeff Dahn, prefer coulombic efficiency.

1.4.2.1 *Nominal Cell Capacity*

The cell manufacturer specifies the *nominal capacity* of a cell. The nominal capacity is the charge released by the cell when discharged at a low current from the maximum to the minimum cell voltages. Reputable manufacturers specify a nominal capacity that can be met by all of the cells it ships.

1.4.2.2 *Actual Cell Capacity*

The *actual capacity* is a condition of a specific cell at a given moment in its life. It is measured by someone testing an actual cell, new or used, using the test limits specified by the manufacturer. A new cell should have an actual capacity that meets or exceeds the nominal capacity specified by its manufacturer. The actual capacity of an old cell is likely to be lower.

1.4.2.3 *Effective Cell Capacity*

The *effective capacity* of a cell is the charge it can store in actual use, in a given application, and at a given point in its life. A BMS typically measures it. It is measured the same way as nominal capacity, except

- Instead of low current, the current is the full current used by the application;
- Instead of the voltage limits specified by the manufacturer, the application sets the limits (this makes the capacity dependent on the defined SoC range).

The effective capacity of a Li-ion cell is reduced at high current levels or low temperature because it reaches the low-voltage cut-out sooner. Since not all the charge is available, the effective capacity is reduced to less than the actual capacity (Figure 1.9(a)).

If we could use a *Constant Current/Constant Voltage* (CCCV) load, we would be able to access all the charge in the cell: the effective capacity would be the same as the operational capacity (Figure 1.9(b)). However, a CCCV load is not practical.

To continue the coffee cup analogy:

- The portion of the coffee you can gulp is the effective capacity: if you drink fast, you leave a bit of coffee in the cup, so the "effective capacity" of the cup seems lower than its actual capacity;
- Yet, if later you take the time to sip, you can finish every drop in the cup.

For batteries other than Li-ion, the effective capacity is also lower because the charge efficiency is less than 100%, and, at high current, not all of the stored chemical energy is converted to electricity[18].

1.4.2.4 *Nominal Battery Capacity*

The *nominal battery capacity* is the charge released by a new, balanced battery (see Section 3.2.5) when discharged at a low current, from the maximum to the minimum cell voltage (defined by the cell manufacturer) (Figure 1.10(a), top-left corner). The battery manufacturer specifies the nominal capacity of a battery.

18. As quantified by the Peukert law.

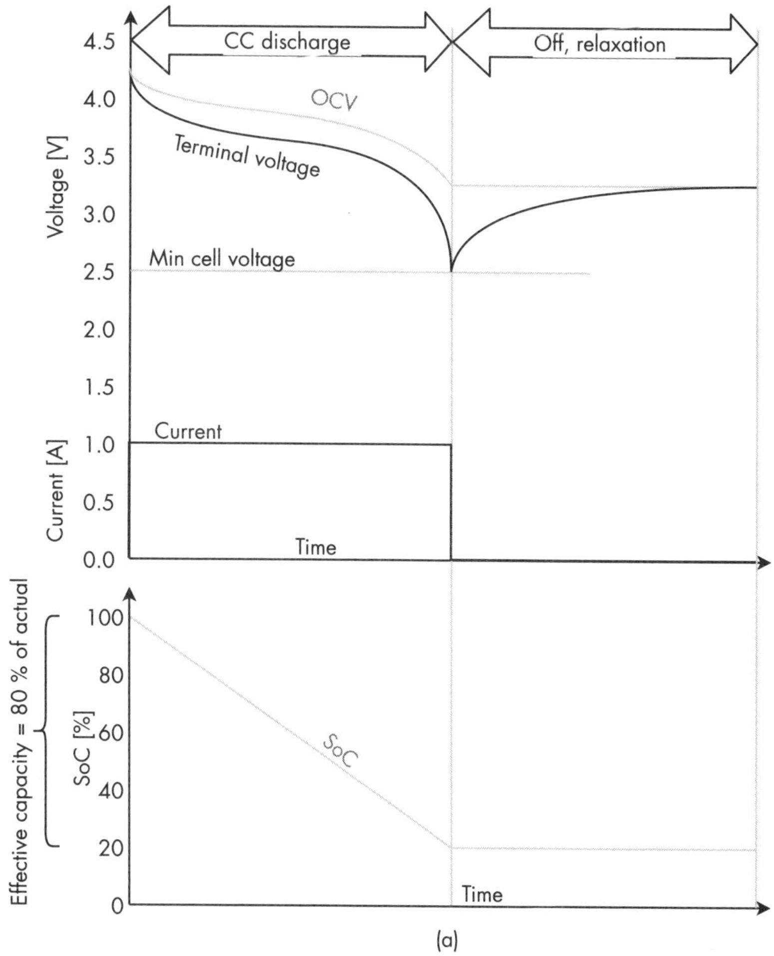

Figure 1.9
Draining a cell:
(a) partially, using a constant current load, and
(b) completely, using a CCCV load (impractical).

1.4.2.5 *Actual Battery Capacity*

The *actual battery capacity* is its capacity at a given point in its life. It is a measure of the condition of a particular battery at a given time (Figure 1.10(a)). It is measured at low current by someone testing the battery. The battery must be balanced first.

A new battery should have an actual capacity that meets or exceeds the nominal capacity specified by its manufacturer. An old, degraded, or unbalanced battery has a lower actual capacity.

1.4.2.6 *Effective Battery Capacity*

The *effective battery capacity* is the charge it can store in actual use, in a given application, at its present state of balance, and at a given point in its life (Figure 1.10(b)). A user may measure it manually or a BMS may measure it automatically. It is measured the same way as actual capacity, except that the current is the full current used by the application, using the application's SoC limits, and any battery imbalance is not corrected.

1.4.2.7 *Why Not Use Wh Instead of Ah?*

Often people ask: "Wouldn't rating batteries by their energy rather than their capacity make more sense?"

Yes, in a way.

FIGURE 1.9 (continued)

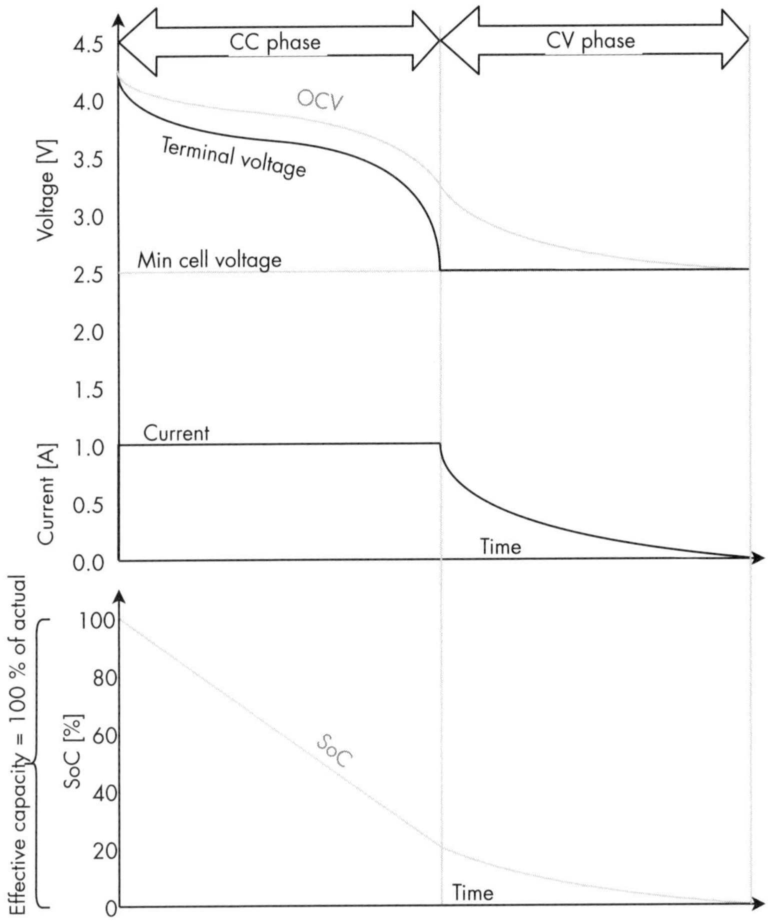

(b)

However, for a Li-ion battery, energy is strongly dependent on discharge rate and actual capacity isn't, so we prefer rating by capacity instead of energy.

The energy that can be extracted from a battery depends significantly on how fast the battery is discharged (Figure 1.10(b)):

- At a slow discharge (right edge of the graph), practically all of the charge and energy can be extracted from the battery;
- As the battery is discharged faster (in less time, toward the left in the graph), the energy that can be extracted decreases;
- When discharged at the maximum power point (see Section 1.5.1) the discharge time is equal to the maximum power time of the battery and the energy extracted is down to 50%;
- When discharged through a short circuit (the discharge time is 0.5 times the battery's MPT), the energy extracted is down to 0%.

Note that the battery is discharged completely (until its SoC is 0%), well past the point where the terminal voltage is too low. Therefore, the charge extracted is 100%, regardless of the discharge rate (the gray line at the top of the graph).

We said that the actual capacity of a Li-ion cell is constant regardless of the discharge rate. This is not to say that the *effective* capacity is 100% regardless of the discharge rate. On the contrary, the effective capacity decreases as the battery is

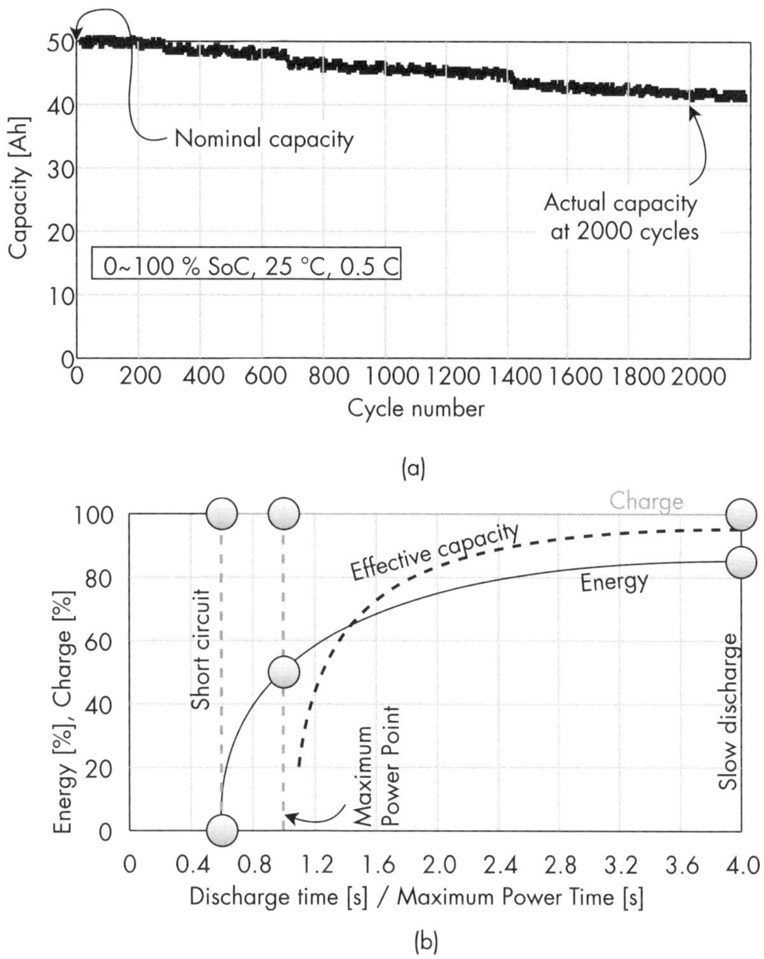

FIGURE 1.10
(a) Nominal and actual capacity, and (b) reduction in energy and effective capacity at shorter discharge times.

discharged faster. This is because the effective capacity is a measure of the charge extracted until the cell voltages have reached a minimum, which occurs while the SoC is still more than 0%. Therefore, the effective capacity does decrease (dashed curve in the graph). Note that at longer discharge times, which is the region where we normally operate batteries (the right portion of the graph), the effective capacity doesn't drop as much as the energy. Yes, the effective capacity does drop sharply if the battery is used close to its maximum power time, but few applications discharge a battery that quickly.

Therefore, rating batteries by their capacity is somewhat more meaningful rather than rating them by their energy.

1.4.3 Current and Specific Current

Current is the flow of electrical charges.

1.4.3.1 Current [A]

In a load, current flows from the more positive voltage to the more negative one (see Volume 2, Section A.2.2). Inside a source (e.g., a battery as it is discharging), it flows the opposite way. A rechargeable battery is a load while charging: current flows from its positive terminal to its negative terminal.

1.4.3.2 Sign of Current and Power

When a battery powers a load, power flows from the battery to the load; current flows from the battery's positive terminal through the load and back into the battery's negative terminal (Figure 1.11(a)). Conversely, when a charger charges a secondary battery, power flows from the charger to the battery; current flows from the charger's positive terminal through the battery and back into the charger's negative terminal (Figure 1.11(b)). In either case, there are two equal and opposite currents, one into one of the battery terminals and one out the other battery terminal. The net current between the battery and the load is neither positive nor negative. As seen in a schematic diagram, the current through the battery flows either up or down.

However, when discussing charging and discharging, we would like to describe the battery power and current as either positive or negative, even though doing so is technically incorrect because the current is truly a loop, not a one-way flow.

The question then becomes: Should we consider a discharging current positive (Figure 1.11(c)) or negative (Figure 1.11(d))? The difficulty in answering this question arises from our attempt to put a sign to a loop, which is a physical phenomenon that doesn't have a sign. It would be like attempting to ask: "Do the hands in an old-fashioned dial clock turn positive or negative?" There is no physical basis for this selection.

The beauty of a positive charging current is that it feels right, considering that a positive charging current increases the SoC.[19] The beauty of a positive discharging current is that it is consistent for both primary and secondary cells. Different industries and researchers use either option (Table 1.2). This book defines discharging current as positive,[20] as most applications do.

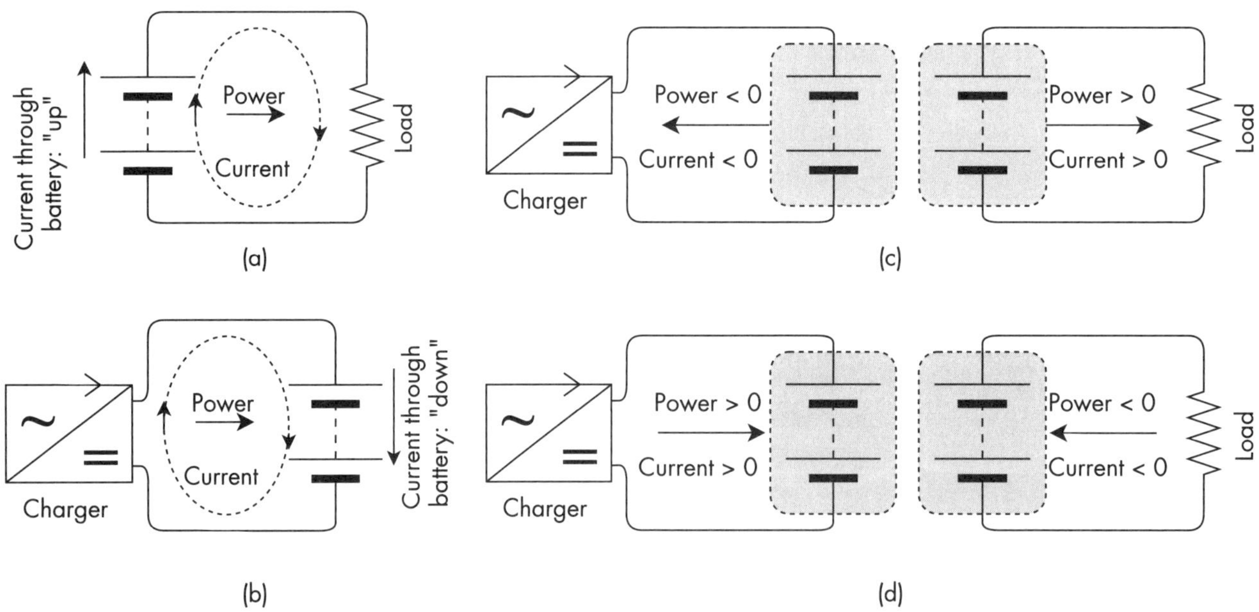

FIGURE 1.11 Sign for current: (a) loop current, discharging, (b) loop current, charging, (c) discharging power and current are reported as positive—used in this book, and (d) discharging power and current are reported as negative.

19. Yet one could just as easily say that a positive discharging current increases the DoD.
20. Apologies to Jeff Dahn, a renowned expert in Li-ion cells, who defines charging as positive.

TABLE 1.2
Which Industries and Researchers Use Which Convention for the Sign Battery Current.

		Charging +	Discharging +
Primary cell industry	As the cell can only discharge, discharging is positive		✔
Primary cell electrochemists	Same as above		✔
Secondary cell industry	For consistency with primary cell applications, discharging is positive		✔
Secondary cell electrochemists	Many, though not all, define charging as positive	✔	✔
Electrical engineers	Sourced current is positive, drained current is negative; from its point of view, the battery is the source of the discharging current, and therefore discharging current is positive, as is power		✔
Chargers	Use the same argument as above, except that the charger is seen as the source; therefore, charging is positive	✔	
Loads	Despite the argument above, loads such as motor drivers specify discharging as positive; if the load generates power (e.g., regen), it is seen as negative		✔
BMSs	Off-the-shelf BMS manufacturers use both conventions; the split is about 50/50	✔	✔
This book	Positive charging never occurred to me		✔

1.4.3.3 Continuous and Peak Current

When discussing current limits for components such as cells, batteries, fuses, or wires, we refer to two different levels:

- *Continuous current;*
- *Peak current.*

The continuous current limit is such that the component doesn't overheat and cells do not degrade too quickly. The peak current limit is higher than that, and it's for a short duration (typically 10 seconds).

Current peaks vary in level and duration. Higher peaks should be limited to a shorter duration than lower peaks. For example, if the continuous limit is 1A, and the peak limit is 5A (Figure 1.12):

- A 7A current should be limited to 5s;
- A 5A current to 10s;
- A 2A current to 25s;
- A 1A current can last indefinitely.

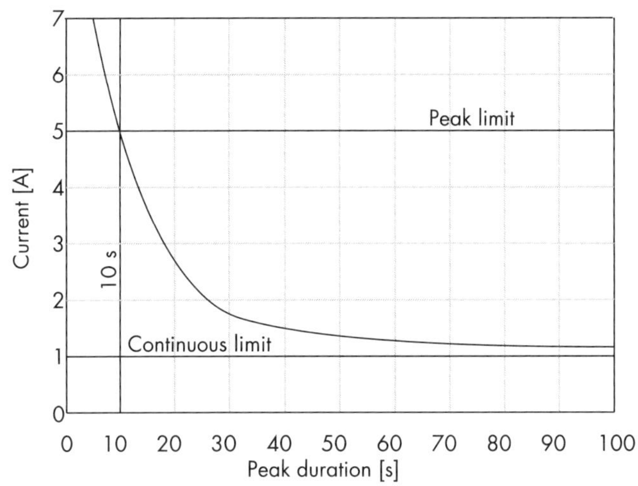

FIGURE 1.12 Maximum duration of peak current versus current.

1.4.3.4 *Specific Current, C-Rate [1/h], [h^{-1}]*

Specific current, also known as *C-rate*,[21] enables us to talk in general terms about current regardless of the capacity of a cell. It is measured in inverse hours—h^{-1}—though, informally, it is measured in C.

For example, a specific current of 2C is 200 A in a 100Ah cell, or 20A in a 10Ah cell. In the second case, the actual current is one-tenth, but its loading on the cells is the same: in both cases, the cells discharge just as fast.

Note that this measure works for any energy storage system, not just for batteries. For example, it can be used for *pumped hydro* (see Volume 2, Section 4.1) Bath County pumped hydro can transfer 45,000,000 m^3 of water between its two lakes in 15 hours at a rate of 820 m^3/s. That is a specific flow of 0.065 [1/h] (i.e., 0.065C). In this case, the term *specific flow* may be more appropriate than specific current.

When discussing self-discharge current, which is quite low, this book uses the units mC and μC.

1.4.4 Energy, Energy Density, Specific Energy, Efficiency

Energy is a measure of total heat or work.

1.4.4.1 *Energy [Wh], [J]*

In physics, energy is measured in joules. In power electronics, we prefer to use watt-hour for ease of calculations.[22]

The energy stored in a cell or battery is directly related to its capacity and voltage. Since the voltage is relatively constant, the energy of a cell is approximately

$$\text{Energy [Wh]} = \text{Voltage [V]} \times \text{Capacity [Ah]} \quad (1.5)$$

For voltage, we use the nominal OCV (at 50% SoC). For capacity, we use the nominal capacity.

21. Not to be confused with the marketing term "C-rating" or with Coulomb, a measure of charge.

22. There are 3,600 joules in 1 Wh because there are 3,600 seconds in an hour.

The actual energy is somewhat different from this value because the open circuit voltage changes with the state of charge and the terminal voltage depends on discharge rate.[23]

1.4.4.2 *Nominal Energy*

The nominal energy of a cell is the energy it may release at a low rate, from full to empty, as defined by the cell manufacturer. In effect, it is the area under its OCV vs. SoC curve.24 (Figure 1.13a)

1.4.4.3 *Actual Energy*

The *actual energy* is a condition of a specific cell at a given moment in its life, measured in the same way as nominal energy. For a battery, it must be balanced first (see Section 3.2.5).

1.4.4.4 *Effective Energy*

The *energy efficiency* of any real-world process is always less than 100% because some energy is wasted in heat.[25]

When discharging a cell, its energy is split into two parts:

- Useful work performed outside the cell;
- Internal heating of the cell.

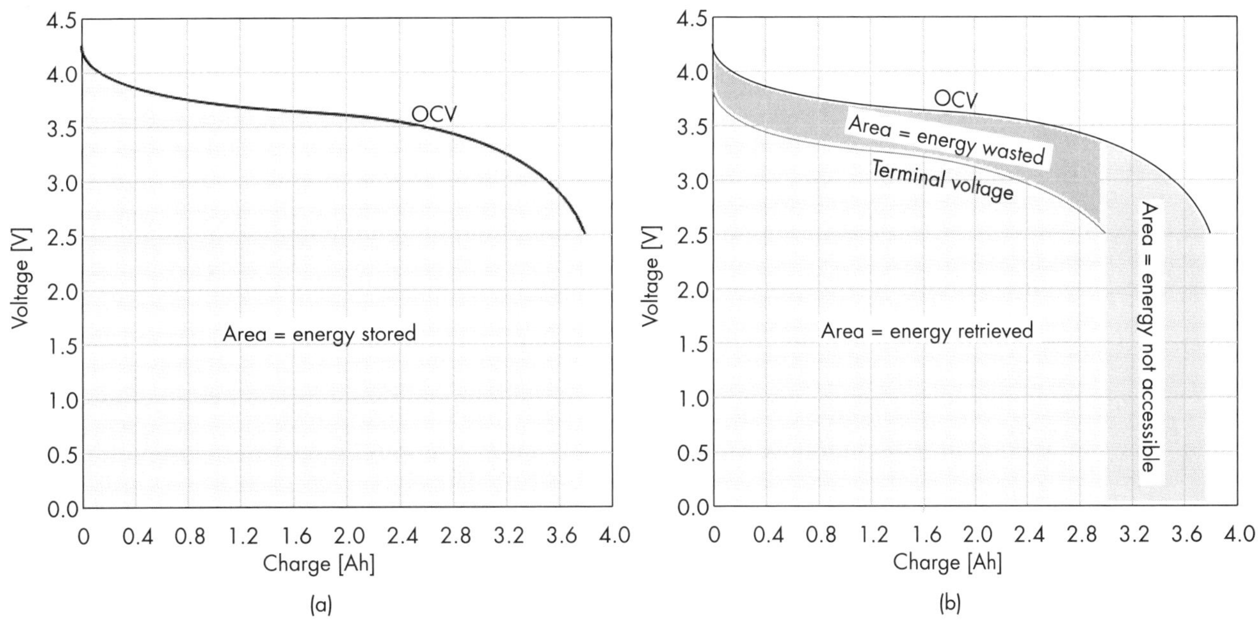

FIGURE 1.13 Graphic derivation of energy: (a) nominal energy, and (b) effective energy

23. The accurate equation is energy [Wh] = capacity [Ah] × ∫ instantaneous_voltage [V] dv/dt = ∫ instantaneous_power [W] dp/dt, where the units of time are [hours], measured over a full discharge.
24. Mathematically, the integral of the power over a full discharge cycle at a minimum current.
25. Due to Entropy.

At low current (long discharge time), most of the energy in the cell goes into external work. But at high current (fast discharge time), a higher proportion of the energy goes into heating the cell (lower efficiency).

The energy retrieved from a cell during discharge is the area under the "terminal voltage versus SoC" curve for the application (Figure 1.13(b)). The area between the curves is the energy wasted in heat. The terminal voltage is below the OCV due to the internal resistance, which is where the heat is wasted. Discharging stops when a cell terminal voltage reaches the minimum, at which point there's still charge left in the cell, whose energy is the area at the right end of the plot. This energy is not accessible at high current.

Just as effective capacity is affected by the limits of the application (SoC range, low voltage cutoff-point at high current), so is effective energy: defining a narrower range of operation reduces both the effective capacity and the effective energy. Additionally, effective energy is affected by the energy efficiency (which decreases at high current) and because, at high current, discharge is stopped earlier, before the cell is empty.

Appendix A describes how to accurately measure the energy stored in a cell (see Volume 2, Section A.3.8).

1.4.4.5 Specific Energy [Wh/kg], [J/kg]

Specific energy[26] allows us to compare energy storage systems (including cells and batteries), by specifying how much energy they can store in a unit of mass (gravimetric).

$$\text{Specific energy [Wh/kg]} = \text{Energy [Wh]/Mass [kg]} \quad (1.6)$$

For example, in many applications, a Li-ion battery is more desirable than a lead-acid battery of the same energy because the Li-ion battery is lighter and smaller. How much more desirable can be quantified by comparing their specific energy.

Knowing the specific energy of an energy storage system, you can calculate how much energy it can store for a given mass. Or, knowing the specific energies of two storage technologies, you can calculate how much more energy one technology can store compared to the other one, for a given mass.

1.4.4.6 Energy Density [Wh/l], [J/l]

Energy density is similar to specific energy, except that it is relative to volume instead of mass (volumetric instead of gravimetric). The unit for volume is liter[27] or equivalently, dm^3:

$$\text{Energy density [Wh/l]} = \text{Energy [Wh] / Volume [l]} \quad (1.7)$$

Energy density lets you calculate how much energy is contained in a given volume (in a storage system such as a battery or a fuel such as a tank of gasoline) or compare two technologies.

Energy density and specific energy are generally constant for any cell from the same family regardless of size, and, to a great extent, of a battery built from those cells.

26. Sometimes called gravimetric energy density, though this is not accurate: energy density is energy per unit volume.

27. The symbol is L. Unfortunately, the lowercase L looks like a 1 (the number one) or an I (uppercase letter i), which can be confusing, so some people use uppercase L.

1.4.4.7 *Energy Efficiency [%]*

Energy efficiency is the ratio of extracted energy over stored energy (which includes energy wasted into heat). For any battery (including Li-ion), it is less than 100%. It is worse at higher levels of current. Energy efficiency is not the *coulombic* efficiency (see Section 1.4.1.3), which for Li-ion is practically 100%. The one-way, *discharging energy efficiency* is the ratio of the energy that is extracted from a cell over the total energy produced by the cell during that same time. The latter includes both the heat inside the cell and the energy that is extracted from the cell:

$$\text{Discharging\_energy\_efficiency}[\%] = 100[\%] \times \text{Energy\_out}[\text{Wh}] / \text{Energy\_used}[\text{Wh}] \tag{1.8}$$

The one-way *charging energy efficiency* is the ratio of the energy that is stored inside a cell over the energy applied to the cell during that same time. The latter includes both the heat generated inside the battery and the energy that is stored in the cell:

$$\text{Charging\_energy\_efficiency}[\%] = 100[\%] \times \text{Energy\_stored}[\text{Wh}] / \text{Energy\_in}[\text{Wh}] \tag{1.9}$$

Round-trip energy efficiency[28] is the ratio of the energy that is extracted from a battery starting from a given state of charge and the energy that is required to restore the battery to that same state of charge:

$$\begin{aligned}&\text{Round\_trip\_energy\_efficiency}[\%] = \\ &100[\%] \times \text{Energy\_out}[\text{Wh}] / \text{Energy\_in}[\text{Wh}] = \\ &\text{Discharging\_energy\_efficiency}[\%] \times \text{Charging\_energy\_efficiency}[\%] / 100[\%]\end{aligned} \tag{1.10}$$

1.4.4.8 *Voltaic Efficiency, Inefficiency [%]*

Rather than round-trip energy efficiency, some prefer using *voltaic efficiency* of *voltaic inefficiency*, which is variously defined as

- The difference in the cell voltage during charging and discharging [4];
- The ratio of round-trip efficiency over coulombic efficiency [4];
- The ratio of the average discharge voltage to the average charge voltage [5].

1.4.5 Power, Power Density, Specific Power, Efficiency

Power is a measure of work done in a given time.

1.4.5.1 *Power [W]*

Power is measured in watts.[29]

28. Also known as *battery efficiency*.
29. For mechanical power, it may be measured in horsepower; 1 hp = 746W.

$$\text{Power}[\text{W}]=\text{Voltage}[\text{V}]\times\text{Current}[\text{A}]=\text{Energy}[\text{Wh}]/\text{Time}[\text{h}] \quad (1.11)$$

For a resistive load (see Section 1.3.5).

$$\text{Power}[\text{W}]=(\text{Voltage}[\text{V}])^2/\text{Resistance}[\text{W}]= (\text{Current}[\text{A}])^2\times\text{Resistance}[\text{W}] \quad (1.12)$$ [30]

For example, the power dissipated in a cell as heat is proportional to its internal resistance:

$$\text{Heat\_power [W]} = (\text{Current [A]})^2 \times \text{DC\_resistance } [\Omega] \quad (1.13)$$

A power source produces maximum power when operated at its maximum power point (see Section 1.5s).

There is no convention for the sign of discharging power. This book defines discharging power as positive, same as discharging current (see Section 1.4.3.2).

1.4.5.2 Specific Power [W/kg]

Specific power[31] measures the maximum power that an energy storage system or fuel of a given mass can deliver (gravimetric):

$$\text{Specific power [W/kg]} = \text{Power [W]} / \text{Mass [kg]} \quad (1.14)$$

Knowing the specific power of a given technology, one may calculate how much power it can deliver for a given mass. Or, knowing the specific power of two technologies, one may calculate how much more power one technology can deliver compared to the other one for a given mass.

1.4.5.3 Power Density [W/l]

Power density is similar to specific power, except that it is relative to volume instead of mass (volumetric instead of gravimetric):

$$\text{Power density [W/l]} = \text{Power [W]} / \text{Volume [l]} \quad (1.15)$$ [32]

Power density lets one calculate how much power a given technology can deliver for a given volume or compare two technologies.

1.4.5.4 Power Efficiency [%]

Power efficiency is the ratio of the power that is extracted from an energy storage system (such as a battery) at a given moment, and the total power released by it (which includes the power to heat itself):

30. Voltage and current can be measured directly, while power is calculated. Therefore, a calculated power suffers from the inaccuracies of both the voltage reading and the current reading.
31. Sometimes called gravimetric power density, though this is not accurate: power density is power per unit volume.
32. That's a lowercase L for liter.

$$\begin{aligned}&\text{Power\_efficiency}[\%]=100[\%]\\&\times\text{Power\_out}[\text{W}]/\text{Total\_power\_generated}[\text{W}]\end{aligned}\tag{1.16}$$

For any storage system (including Li-ion batteries), it is less than 100%. It is worse (lower) at higher current levels.

Power efficiency is not the *coulombic* efficiency (see Section 1.4.1.3, "Coulombic efficiency [%]"), which for Li-ion is practically 100%.

1.4.6 Resistance and Impedance [Ω]

Although both *resistance* and *impedance* are measured in ohms [Ω],[33] they measure different parameters:

- DC resistance of a cell determines the sag in the cell voltage when discharged continuously;
- AC impedance determines how stable a cell's terminal voltage is in the presence of an AC current; the lower the impedance, the more stable the voltage. As the impedance varies with frequency, whenever a value for impedance is specified, the frequency must also be specified (e.g., 1 kHz).

AC cell impedance is extremely valuable to cell manufacturers, researchers, and the more sophisticated battery designer (see Volume 2, Section A.3.5). However, it is of limited use to the average battery designer, to whom DC resistance matters the most.

Unfortunately, when most cell manufacturers specify cell resistance, they are talking about its *AC impedance at 1 kHz*. This is because

- Test equipment to measure it is readily available;
- An unusually high value is a good indicator of a manufacturing problem;
- Most Li-ion cell manufacturers do not think in terms of DC resistance.

The impedance at 1 kHz is unrelated to the DC resistance and is of little use to the battery designer. As a battery designer, what you need is the DC resistance because DC is what flows through the cells:

- It lets you compare cells to be used in a power application;
- It lets you estimate voltage sag under load;
- It lets you calculate the battery efficiency under load;
- It lets you calculate how much heat must be removed from the cells through cooling.

In truth, we do concern ourselves with AC impedance because that's what is behind the slow relaxation in the terminal voltage after the current changes. But this AC impedance is at a frequency of about 0.01 Hz, quite far from the 1,000 Hz where the cell manufacturers test the cells.

33. For batteries, milliohm [mΩ] is more practical than Ω.

1.4.7 Capacitance [F]

Two different measurements of capacitance (measured in farads [F]) are of concern to batteries.[34] While both are measured in farads [F], they are different measurements.

1. Battery capacitance to ground;
2. Internal cell capacitance.

The capacitance to ground only becomes evident when cells are mounted in a metal enclosure, especially if they have heat sinks between each cell. This capacitance has a few secondary effects: it hinders isolation measurements and affects electromagnetic emissions.

The internal capacitance of a cell determines the relaxation time in the cell voltage when a load is first applied or first removed.

We also use capacitance when talking about supercapacitors (see Volume 2, Section B.2).

1.5 MAXIMUM POWER POINT AND MAXIMUM POWER TIME

In high-power applications, a battery may be operated at its *maximum power point*. A related concept is *maximum power time*, a tool to compare cells and batteries for power applications.

1.5.1 Maximum Power Point

For a power source that can be modeled as a voltage source with a series resistance,[35] the *maximum ower point* (MPP) is the operating point of where it outputs as much power as it can. This is the good news. The bad news is that the efficiency is just 50%—the source heats itself with as much power as the load uses.

1.5.1.1 Battery MPP

Let's discharge a battery with a variable load and plot the power versus the resistance of that load (Figure 1.14(a)):

- At the left end of the graph, the load is a short circuit; the terminal voltage is 0V, so the output power is also 0W; the short-circuit current heats the battery with an immense power; the total power generated by the battery is just the power to heat itself because the load sees no power; and the efficiency is 0%.
- At the other end (far past the right end of the graph), at no load, the battery delivers no current, so the output power is also 0W; there is no power heating the cell; the total power is also zero, as is the efficiency; and the terminal voltage is at the OCV.
- Everywhere in between, the output power is greater than 0W and the efficiency is less than 100%.
- The output power peaks at the maximum power point (the circle in the plot) when the load resistance is equal to the internal resistance; at this point the internal heat power and the output power are the same: the efficiency is 50%. The terminal voltage is half of the OCV (half the voltage appears across the cell's

34. Do not confuse battery capacity [Ah] and capacitor capacitance [F].
35. Including batteries.

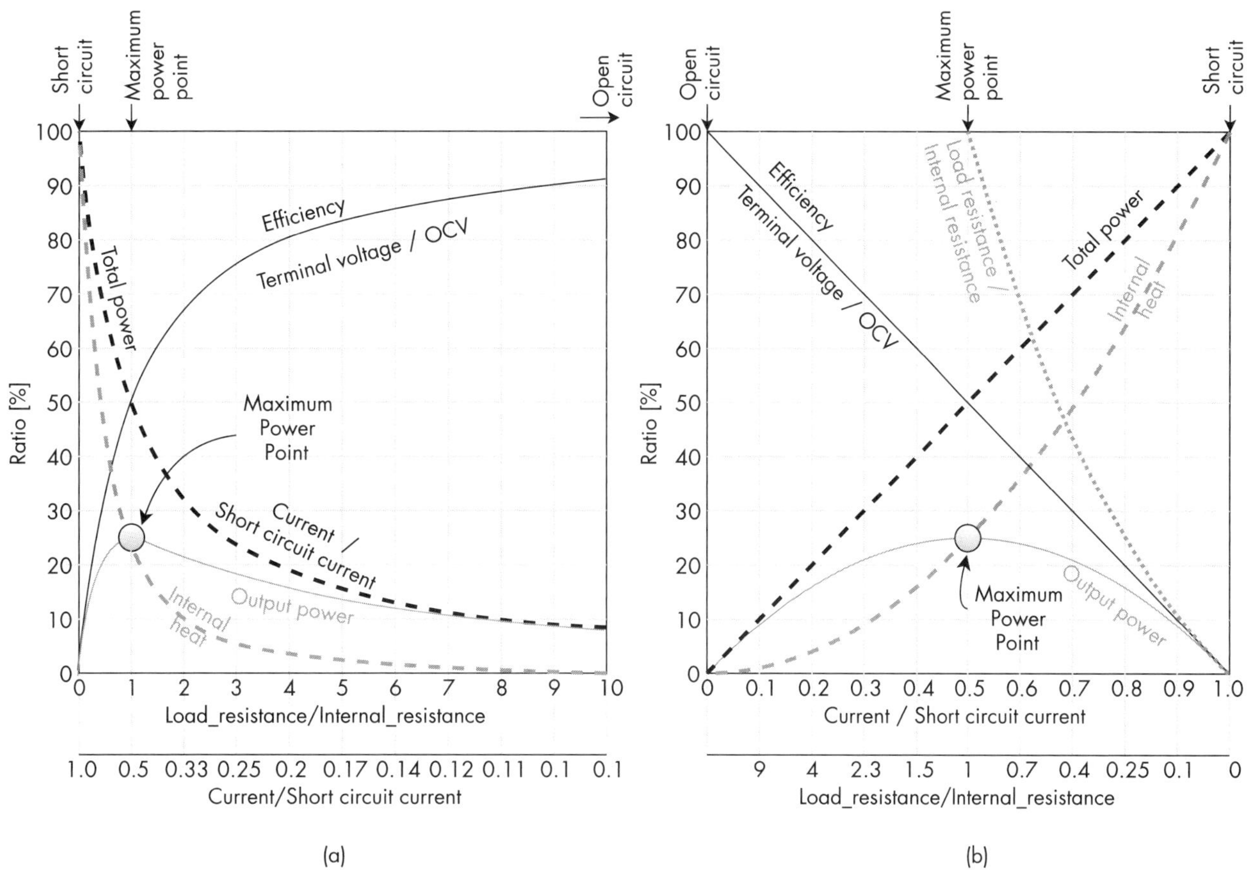

FIGURE 1.14 Maximum power point: (a) versus load resistance, and (b) versus load current.

internal resistance, half across the load), meaning that it is likely to be below the minimum cell voltage rating for that cell.

The second curve (Figure 1.14(b)) presents the same information but with the current on the X-axis. Note how, at the maximum power point, the load resistance is the same as the internal resistance.

The voltage, power, and current at the maximum power point are[36]

$$\text{Voltage}_{(\text{Maximum\_Power\_Point})}[\text{V}] = \text{Vocv}[\text{V}]/2 \tag{1.17}$$

$$\text{Power}_{(\text{Maximum\_Power\_Point})}[\text{W}] = (\text{OCV}/2)^2/\text{Internal\_resistance}[\Omega] \tag{1.18}$$

$$\begin{aligned}&\text{Current}_{(\text{Maximum\_Power\_Point})}[\text{A}]\\&= \text{Vocv}[\text{V}]/(2 \times \text{Internal\_resistance}[\Omega])\end{aligned} \tag{1.19}$$

36. The factor of 2 is because half the voltage is across the cell's internal resistance, and half is across the load.

Operation at the maximum power point is impractical in most applications because of the terrible efficiency, the amount of heat in the battery, and the damage to the cells. Yet, some applications operate close to this point, such as race vehicles[37] and UAVs (see Volume 2, Section 3.10).[38]

This discussion applies not just to batteries, but also to any source that can be closely modeled as an ideal voltage source with a fixed series resistance.

1.5.1.2 Photovoltaic MPP

Photovoltaic (PV) cells cannot be modeled as an ideal voltage source with a fixed series resistance, so the above conclusions do not apply to them.[39] For a PV cell, the efficiency at the maximum power point is on the order of 70%, which is much higher than the 50% for a battery. Note that this is the *electrical* power efficiency, not the photovoltaic conversion efficiency, which is lower.

1.5.1.3 Maximum Power Current

The *maximum power current* occurs when the cell is operated at the maximum power point. It is debatable whether operating at such levels of current is even possible.

1.5.2 Maximum Power Time

When selecting cells for a power battery, it is useful to have the ability to quickly compare various cell technologies and calculate the resulting pack resistance and efficiency, independently of capacity and voltage. We need a constant that characterizes a particular cell technology, and batteries that use that cell, derived from measurable parameters.

Allow me to propose *maximum power time* as such a constant (for the antecedent to MPT, see Volume 2, Section A.2.1).

1.5.3 MPT Definition

MPT is the time that it takes to fully discharge a storage system (including a cell or a battery, but also any other storage system) while operating at the maximum power point. This time is constant for given cells, for other cells using the same technology, and for any battery that is built using those cells, regardless of cell arrangement. It is somewhat related to power density, in the sense that for a given cell size and voltage, the MTP is approximately inversely proportional to the power density. Though, again, the beauty of MPT is that it is independent of size and voltage. For simplicity, in the following discussion, "cell," means either a cell or a battery.

Given a cell's DC resistance, capacity, and voltage, its maximum power time is.[40]

$$\begin{aligned}&\text{Max\_Power\_Time}[\text{h}] = \text{capacity}[\text{Ah}] \times 2 \times \\ &\text{DC\_resistance}[\Omega]/\text{nominal\_voltage}[\text{V}]\end{aligned} \quad (1.20)$$

Since the maximum power time of Li-ion cells ranges from 0.008 to 0.12 hours (equal to 30 to 440 s), seconds is a more practical unit than hours:

37. Race cars in ¼ mile sprints: the damage to the cells is secondary to the primary goal to break an acceleration record.
38. As mentioned earlier, unmanned aerial vehicles are also known as drones.
39. They are closer to current sources, and, more importantly, the internal series resistance increases with load current.
40. The factor of 2 is because the total resistance (internal plus external) is twice the cell's internal resistance.

$$\begin{aligned}&\text{Max\_Power\_Time}[\text{s}]=7200[\text{s/2h}]\times\text{capacity}[\text{Ah}]\\&\times\text{DC\_resistance}[\Omega]/\text{nominal\_voltage}[\text{V}]\end{aligned} \tag{1.21}$$

The MPT can be calculated from data in the specification sheets for a cell, or graphically from the discharge curves for a cell, or it can be measured empirically.

1.5.4 MPT Empirical Characterization

Having access to an actual cell, you can characterize its MPT.

1.5.4.1 Timing a Discharge Cycle

You can measure the actual MPT by seeing how long it takes to discharge a cell (from 100% to 0% SoC) into a maximum power point tracking (MPPT)[41] converter. As the cell discharges, this converter constantly changes the load to the cell to maximize the power from the cell. In other words, as the cell's internal resistance varies with temperature and SoC, the converter applies a load to the cell whose equivalent resistance is identical to the cell's internal resistance at the time. The cells should be cooled down to 25°C. This test is likely to damage the cell, so it may only be performed once.

1.5.4.2 From Voltage Sag

Assuming you know the cell capacity, you can derive the maximum power time from the voltage sag under load:

1. Bring the cell to 50% SoC and have it rest with zero current and at 25°C for at least 1 hour
2. Measure the OCV
3. Measure the voltage at high current:
    a. Apply a load to the cell to draw approximately 1C of specific current
    b. Measure the load current
    c. Wait 10 s for the cell voltage to settle and measure the loaded cell voltage
4. Calculate the cell DC resistance:

$$\begin{aligned}&\text{DC\_resistance}[\Omega]=\\&(\text{Open\_circuit\_voltage}[\text{V}]-\text{Loaded\_voltage}[\text{V}])/\text{Test\_current}[\text{A}]\end{aligned} \tag{1.22}$$

5. The load resistance at the maximum power point is the same as the internal DC resistance; therefore, the total resistance seen by the internal voltage source is twice the internal DC resistance:

$$\text{total\_resistance }[\Omega] = 2 \times \text{DC\_resistance }[\Omega] \tag{1.23}$$

41. Maximum power point tracking is typically used to maximize the power from solar panels.

6. At this point, the cell voltage, cell capacity, and cell resistance are known, and therefore the cell's maximum power time can be calculated.[42]

$$\begin{aligned}&\mathrm{MPT}[s] = 7200[s/2h] \times (1 - \mathrm{Loaded\_voltage}[V]) / \\ &\mathrm{Open\_Circuit\_Voltage}[V]/\mathrm{Specific\_current}[1/h]\end{aligned} \tag{1.24}$$

7. If discharged at the maximum power point, the current would be

$$\text{Specific current } [h^{-1}] = 1/\text{MPT } [h] = 3600/\text{MPT } [s] \tag{1.25}$$

For example, we have a 2.5 Ah 18650 cell, whose voltage at 50% SoC is 3.65V. When loaded with a resistor the current is 3.2A (which is 1.28C), and the voltage drops to 3.45V. Then

$$\begin{aligned}&\mathrm{Max\_Power\_Time}[s] = 7200\,s/2h \times \\ &(1 - 3.45V/3.65\,V)/1.28h^{-1} = 308[s]\end{aligned} \tag{1.26}$$

If operated at its maximum power point, that cell discharges from 100% to 0% in about 5 minutes.

Now, just to be clear, that's a specific current of

$$\text{Specific current } [h^{-1}] = 3600/\text{MPT } [s] = 3600/308 \text{ s} = 11.7\text{C} \tag{1.27}$$

That high current damages the average 18650 cell, so don't design for operation at the maximum power point if you want to get any decent cycle life out of a cell.

1.5.5 MPT Derivation from Specs

The MPT can be derived from a complete and truthful cell specification sheet.

1.5.5.1 From Specification Data

Again, the MPT is calculated as

$$\begin{aligned}&\mathrm{Max\_Power\_Time}[s] = 7200[s/2h] \times \mathrm{capacity}[Ah] \times \\ &\mathrm{DC\_resistance}[\Omega]/\mathrm{nominal\_voltage}[V]\end{aligned} \tag{1.28}$$

For example, the specifications of an A123 M1 cell (26650 size, LFP) are 3.3V, 2.3 Ah, 10 mΩ. Therefore, the maximum power time of those cells (and of batteries built from those cells, regardless of the arrangement) is

42. Max_Power_Time [s] = 3600 [s/h] × capacity [Ah] × total_resistance [Ω]/voltage [V] = 3600 [s/h] × capacity [Ah] × 2 × (Open_circuit_voltage [V] – Loaded_voltage [V])/Test_current [A]/Open_circuit_voltage [V] = 7200 [s/2h] × capacity [Ah] × (1 – Loaded voltage [V]/Open CircuitVoltage [V])/test_current [A] = 7200 [s/2h] × (1 – Loaded voltage [V]/Open CircuitVoltage [V])/Specific_current [1/h].

$$\text{Max\_Power\_Time}[\text{s}] = 7200\,\text{s}/2\text{h} \times 2.3Ah \times 10\ \text{m}\Omega/3.3\text{V} = 50\text{s} \tag{1.29}$$

and

$$\text{Specific current } [\text{h}^{-1}] = 3600/\text{MPT } [\text{s}] = 3600/50 \text{ s} = 72 \text{ C} \tag{1.30}$$

This tells us that if an A123 M1 cell (or a battery made of such cells) is discharged at the maximum power point, the discharge would last 50s, and the current would be 72C, which, of course, would be excessive.

1.5.5.2 From Discharge Curves

Unfortunately, few manufacturers specify true DC resistance. In the absence of this datum, the maximum power time may be derived from discharge curves. These curves plot the cell terminal voltage versus SoC at various specific currents (such as 0.5C, 1C, 2C, 5C…). From such a set of curves, pick two points at 50% SoC (for example, at 0.5C and 2C). Note the difference in the cell voltage at those two points. This is the delta-voltage [V]. Take the difference of the two specific currents. This is the delta specific current, in [1/ h]. Then, use those values to calculate the maximum power time:

$$\begin{aligned}\text{MPT}[\text{s}] = 7200[\text{s}/2\text{h}] \times \text{delta\_cell\_voltage}[\text{V}]/\\ \text{delta\_specific\_current}[1/\text{h}]/\text{nominal\_voltage}[\text{V}]\end{aligned} \tag{1.31}$$

For example, we can pick two points in the discharge curve for an LFP cell (Figure 1.15).

The delta voltage is 3.28 – 2.98 = 0.3 [V]. The delta specific current is 5C – 1C = 4 [1/h]. The nominal cell voltage is 3.28V. Then, the maximum power time of this cell (and of cells of any size using the same technology, and of batteries using these cells) is

$$\text{Max\_Power\_Time}[\text{s}] = 7200\,\text{s}/2\text{h} \times 0.3\text{V}/4\text{h}^{-1}/3.28\text{V} = 160\text{s} \tag{1.32}$$

1.5.6 Typical Values of MPT

It is useful to get a general idea of the maximum power time values for various storage technologies (Figure 1.16). Note that Li-ion cells span a wide range, going from the best (shortest maximum power time, at the left end) to the worse (at the right end) (see Section 2.6.1 for a table of MPT values for common cells). Only capacitors (on the far left) have a better MPT.

Advanced technology lead-acid batteries[43] have an admirable maximum power time.[44] For applications where weight is not an issue, they should be given serious consideration.

We'll go into more detail on the maximum power time for specific cells when we discuss cell selection (see Section 5.3.2).

43. Such as from the now-defunct Xtreme Power.
44. For discharge only; lead-acid is slow charging.

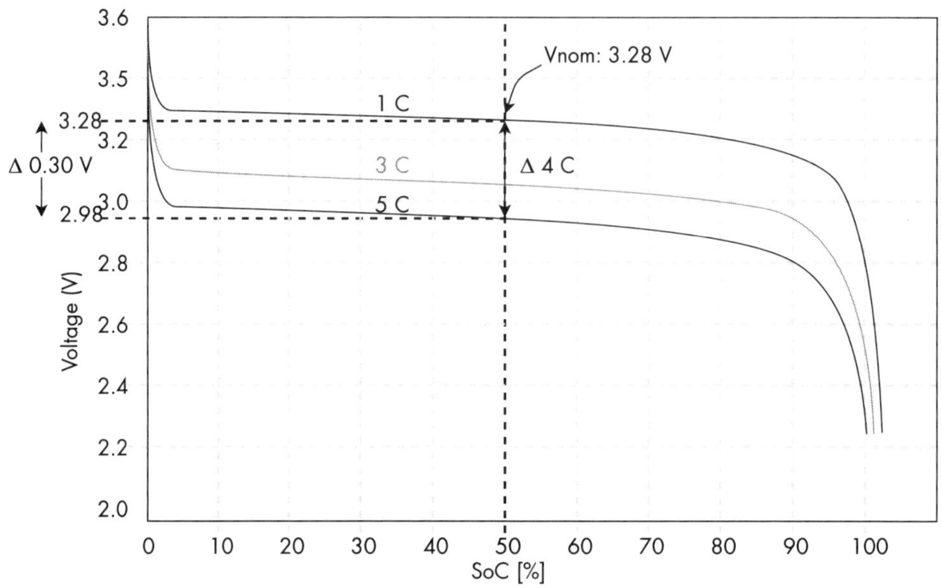

FIGURE 1.15 Graphical derivation of MPT from discharge curves.

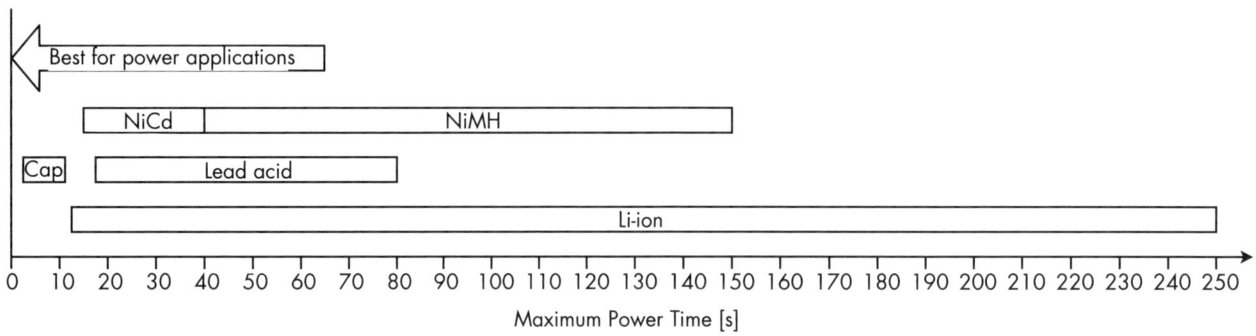

FIGURE 1.16 Maximum power time ranges for various electrical storage technologies.

1.5.7 MPT Conversions

Given the maximum power time of a given electrical storage technology, we can quickly calculate

- Internal series resistance;
- Energy efficiency;
- Voltage sag under load.

1.5.7.1 Internal Series Resistance Calculation

Knowing the maximum power time, we can calculate the nominal internal resistance.

Given the battery voltage and capacity:

$$\text{resistance}[\Omega]=\text{Max\_Power\_Time}[\text{s}]\times\text{nominal\_voltage}[\text{V}]/\text{capacity}[\text{Ah}]/7200[\text{s/2h}] \tag{1.33}$$

Given the voltage and energy:

$$\text{resistance}[\Omega]=\text{Max\_Power\_Time}[\text{s}]\times(\text{nominal\_voltage}[\text{V}])^2/\text{energy}[\text{Wh}]/7200[\text{s/2h}] \quad (1.34)$$

Given the energy and capacity:

$$\text{resistance}[\Omega]=\text{Max\_Power\_Time}[\text{s}]\times\text{energy}[\text{Wh}]/(\text{capacity}[\text{Ah}])^2/7200[\text{s/2h}] \quad (1.35)$$

For example, given the voltage and capacity of a battery that uses cells with a maximum power time of 36 seconds, one may calculate its resistance (Table 1.3). The values for resistance make sense: as the voltage increases, having more cells in series results in a higher resistance. Conversely, as the capacity increases, having more cells in parallel results a lower resistance.

1.5.7.2 *Energy Efficiency Calculation*

Given the maximum power time, the energy efficiency is easily derived.[45]

$$\text{energy\_efficiency}[\%]=100[\%]\times(1-\text{Max\_Power\_Time}[\text{s}]/2\times\text{actual\_discharge\_time}[\text{s}]) \quad (1.36)$$

The energy efficiency is a function of the actual discharge time (Table 1.4, Figure 1.17).

For example, if a battery is discharged 10 times slower than its maximum power time, the energy efficiency is 95%. If 100 times, 99%. Note that the energy efficiency is not directly proportional to the specific current: a 1C discharge of a cell with a low MPT is more efficient than a 1C discharge of a cell with a high MPT.

TABLE 1.3
Examples or Resistance Calculation for an MPT of 36 s

	10 Ah	**100 Ah**	**1,000 Ah**
10V	36 x 10/10/7200 = 5 mΩ	36 x 10/100/7200 = 0.5 mΩ	36 x 10/1000/7200 = 0.05 mΩ
20V	36 x 20/10/7200 = 10 mΩ	36 x 20/100/7200 = 1 mΩ	36 x 20/1000/7200 = 0.1 mΩ
50V	36 x 50/10/7200 = 25 mΩ	36 x 50/100/7200 = 2.5 mΩ	36 x 50/1000/7200 = 0.25 mΩ
100 V	36 x 100/10/7200 = 50 mΩ	36 x 100/100/ 7200 = 5 mΩ	36 x 100/1000/7200 = 5 mΩ
200V	36 x 200/10/7200 = 100 mΩ	36 x 200/100/7200 = 10 mΩ	36 x 200/1000/7200 = 1 mΩ
500V	36 x 500/10/7200 = 250 mΩ	36 x 500/100/7200 = 25 mΩ	36 x 500/1000/7200 = 2.5 mΩ

45. The factor of 2 is because the total resistance is twice the internal resistance when used at maximum power point.

Table 1.4
Energy Efficiency for a Given Discharge Time

Actual Discharge Time [s]/Maximum Power Time [s]	**Load Resistance [Ω]/Internal DC Resistance [Ω]**	**Energy Efficiency [%]**	**Notes**
0.5	0	0%	Short circuit
0.75	0.5	33.3%	
1	1	50.0%	Maximum power point
2.5	4	80.0%	
5	9	90.0%	90% efficiency point
10	19	95.0%	
25	49	98.0%	
50	99	99.0%	99 %efficiency point
100	199	99.5%	

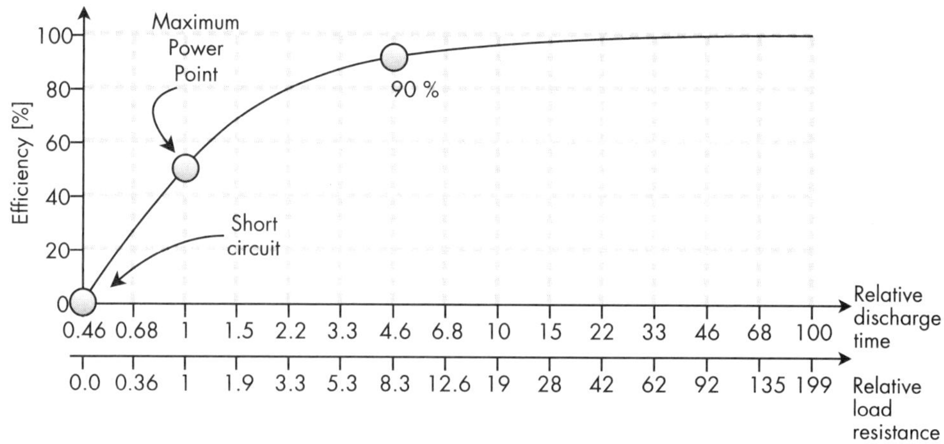

Figure 1.17
Energy efficiency versus relative discharge time.

1.5.7.3 Voltage Sag Calculation

Given the maximum power time of an electrical storage technology, the *voltage sag* is easily calculated for a given specific current:

$$\text{Voltage\_sag}[\%] = 100[\%] \times \text{Specific\_current}[1/\text{h}] \times \text{Maximum Power Time}[\text{s}]/7200[\text{s}/2\text{h}] \quad (1.37)$$

For example, let's take a battery built out of A123 M1 cells (with a maximum power time of 50s); regardless of the cell arrangement, with a 2-C load current, the voltage sags:

$$\text{Voltage\_sag}[\%] = 100\% \times 2\text{h}^{-1} \times 50\text{s}/7200\text{s} / 2\text{h} = 1.39\% \quad (1.38)$$

Table 1.5 lists sag for a given MPT and specific current.

1.5.8 Using the MPT

The MPT is a tool to select cells for power applications and to calculate the specifications of batteries using those cells. MPT is not an invitation to operate

TABLE 1.5
Voltage Sag for a Given Specific Current and Maximum Power Time

	Specific Current						
MPT	**0.1 C**	**0.2 C**	**0.5 C**	**1.0 C**	**2.0 C**	**5.0 C**	**10.0 C**
10 s	0.01%	0.03%	0.07%	0.14%	0.28%	0.69%	1.39%
20 s	0.03%	0.06%	0.14%	0.28%	0.56%	1.39%	2.78%
50 s	0.07%	0.14%	0.35%	0.69%	1.39%	3.47%	6.94%
100 s	0.14%	0.28%	0.69%	1.39%	2.78%	6.94%	13.89%
200 s	0.28%	0.56%	1.39%	2.78%	5.56%	13.89%	27.78%

batteries at the maximum power time: MPT does not consider any limitations on the cell current that are imposed by the chemistry and by the interconnections.

1.6 STATES

A variety of measures can represent the state of a battery.

1.6.1 States of Alphabet Soup

There are various measures called "state of (something or other)" to communicate the state of a battery to the device using it and to the user and to help decide how to use the battery and when it is time to change it.

This list is from multiple sources (Table 1.6) [6–15]; I don't claim to understand them all.

Many such states have been proposed, and some are in active use. None of these are defined precisely and consistently: either they are interpreted slightly differently by various people, or they are based on poorly quantifiable measures. State of charge is by far the most common one and the best understood.

Some of these states were introduced by researchers, hoping to expand our understanding of the condition of a battery.[46] Some were introduced by product designers as metrics to maximize battery performance.[47]

1.6.2 State of Charge

There are two interpretations for SoC:

- Physical,[48]
- Operating.

The SoC for a string of cells in series is somewhat different than the SoC for a single cell, and for an array it is also different. While all of these are measured in percent [%], they are slightly different measures.

State of charge and depth of discharge may use relative units (%) or absolute units (Ah).

46. While also providing a subject for a thesis or a newly published paper.
47. While also providing a feature that distinguishes their product from the competition.
48. More precisely, thermoelectric. Chemists may call this residual capacity.

Table 1.6
State of Alphabet Soup

Initials	Name	Definition
SoA	State of abuse	(Not well-defined.)
SoB	State of balance	Difference in SoC between most and least charged cell in a series string [%].
SoC	State of charge	Ranges from 100% for a full battery to 0% for an empty one.
SoD	State of discharge	Same as depth of discharge (DoD) if using % units: 100% - SoC.
SoE	State of energy	Similar to SoC, but instead of keeping track of remaining charge, it keeps track of remaining energy. Ranges from 100% for a full battery to 0% for an empty one.
SoF	State of function	Multiple definitions: Ratio of effective capacity over nominal capacity [%] Product of SoC and SoH For starter battery: ratio of (voltage headroom now) over (voltage headroom with new battery) Performance as is, considering reduction due to reversible effects
SoF1	State of function 1	Indication of the ability to deliver a constant discharge.
SoF2	State of function 2	Indication of the ability to accept a constant charge.
SoH	State of health	Starts at 100% for a new cell or battery, drops as it degrades. Other than that, everyone defines this state differently.
SoL	State of life	Estimate of how much longer a battery may be usable (aka remaining useful life (RUL)).
SoP	State of power	Ratio of peak power over nominal power [%].
SoQ	State of coulometric load	Partial delta SoC starting from a given point.
SoR	State of resistance	100% for a new battery, 200% for a battery that has twice the resistance.
SoS	State of safety	Inverse of state of abuse: 0% to 79%, unsafe; 80% to 100%, safe.
SoV	State of voltage	During discharge, how far away from the low-voltage cutoff: 100% when the battery is full; 0% at the time of low-voltage cutoff.

1.6.2.1 Physical Cell SoC

Physical SoC is the concentration of ions in the negative electrode. You could theoretically take a cell apart, count the lithium ions on either side of the separator, and determine its SoC. Physical SoC is quite useful in the lab. It could be defined in many ways including this extreme definition::

- *0%:* When the cell current through a short circuit drops to 0, the cell is damaged;
- *100%:* When the charging current no longer stores charge in the cell, the cell voltage is excessively high (e.g., 10V) and the cell is damaged.

1.6.2.2 Operating Cell SoC

For the sake of the cell, it is operated in a narrower range than the physical SoC (see Section 2.4.3) to avoid the damaging extremes; yet, this *operating cell SoC* range is large enough to achieve a respectable value for capacity.

1.6.2.3 *String SoC*

A string of cells in series has a *string SoC*, which may be different from the SoC of the cells in the string. The SoC of a string is 100% when any cell within it is at 100% operating SoC. Similarly, the string SoC is 0% when any cell within it is at 0% operating SoC (see Section 3.2.4).

1.6.2.4 *Array SoC*

State of charge is a misleading concept in an array of batteries, each of which may or may not be connected to a bus at a given time. An array of 10 batteries in parallel can deliver 10 times more charge if the batteries remain connected until the end of discharge than if only one battery remains connected (for example, because the other nine batteries overheated and disconnected). For such applications, *charge acceptance* and *discharge availability* are more useful than SoC (see Section 1.6.7).

1.6.3 State of Energy

As the product using the battery it is more likely to operate at a given power rather than a given current, it may prefer knowing the battery's *state of energy* (SoE)[49] rather than its SoC. During discharge, as the voltage drops, the current must increase to deliver a given power, so knowing how much energy is left is more helpful to the load than knowing how much charge is left.

When we say that the SoC is 50%, we're saying that there is still half the charge left in the battery. Yet, because the voltage drops and the internal resistance increases as the battery is discharged, the energy that we can still extract from the battery is less than the energy we already extracted. On the contrary, when we say that the SoE is 50%, we're saying that half the available energy remains.

Neither SoC nor SoE is ideal because both the available charge and the available energy are reduced at high load current. Still, the charge is affected less than the energy. For example, a 300 Ah, 133V battery can deliver about 40 kWh at low current, but only 20 kWh if operated at the maximum power point (which is a factor of 2). Yet it can deliver 300 Ah at low current and about 270 Ah if operated at the maximum power point (which is only 10% off) (Figure 1.18(a)). So, SoC may be more accurate than SoE.

At constant discharge power, an accurately evaluated SoE drops linearly from 100% to 0% regardless of the power level (Figure 1.18(b)). Conversely, SoC is nonlinear (because the voltage drops over time) and depends on the load (at high power, it never reaches 0% because the battery shuts down before it's empty). Note that the 50% SoE point occurs first (when 50% of the energy is used up) when more than 50% of the charge is still in the battery. The 50% SoC points occurs later (when 50% of the charge is used up) by which time less than half the energy remains.

The SoE evaluation assumes that the power is constant, and must take into consideration the battery inefficiency at high current and how the efficiency is reduced as the SoC drops.

If the power is not constant, then the SoE value can be quite far off. It's debatable whether SoE (directly applicable but inaccurate) is more or less valuable than the SoC (accurate but not directly applicable).

49. State of energy is technically an incorrect term because energy is not a state function. Instead, it is a path function.

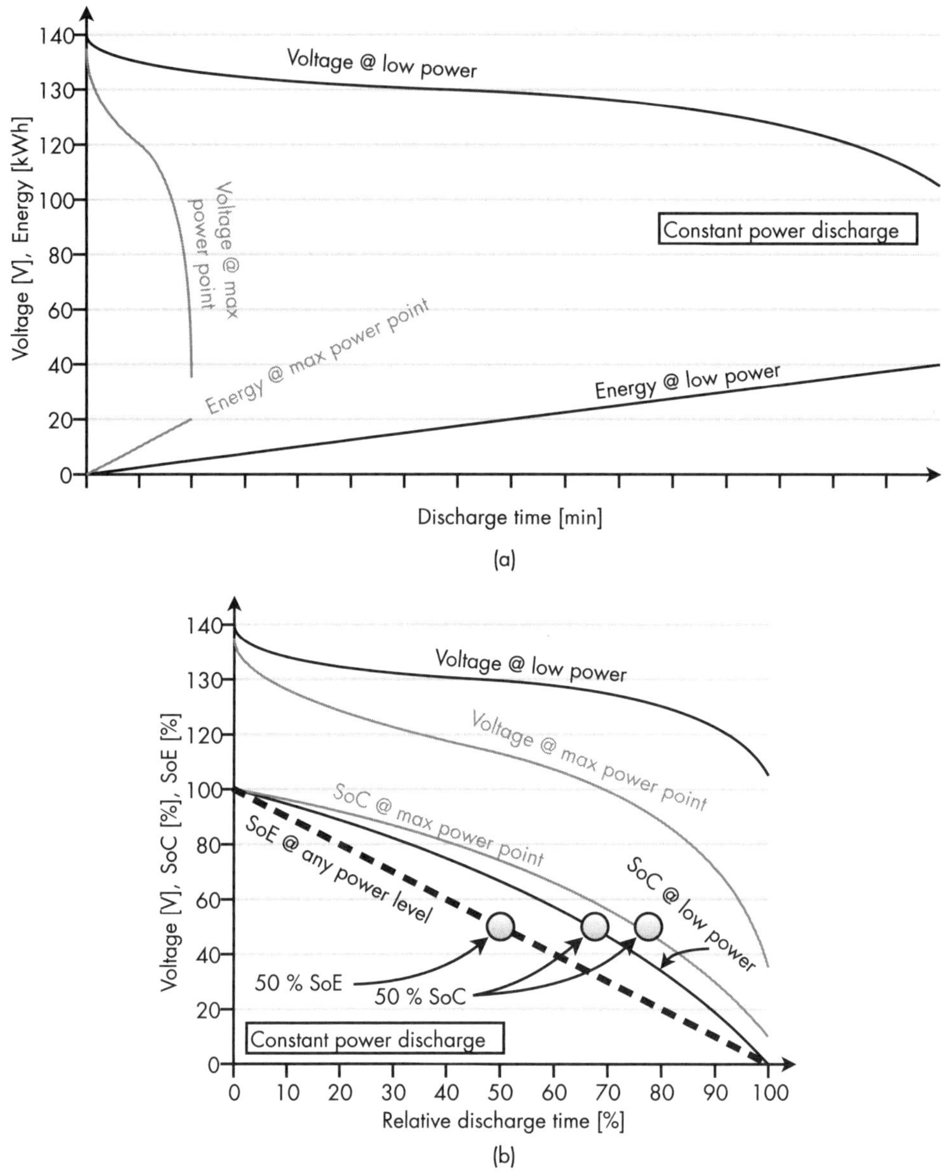

FIGURE 1.18
Constant power discharge plots: (a) versus time, and (b) versus relative time.

1.6.4 State of Health

State of health (SoH) is a useful concept. The idea is that a new cell or battery has an SoH of 100% and that the SoH goes down as the battery ages and degrades. When the SoH drops down to a certain level, it's time to replace the battery.

The problem is that everyone defines SoH differently; plus, there is no definition of the SoH level below which it's time to replace a battery (0%? 50%? 90%?).

Some implementations start reducing the SOH as soon as the battery is in use. Others keep the SoH at 100% for a long time and only start dropping it when the battery has degraded past a threshold. Others base the SoH on the number of cycles.[50]

If genuinely based on the conditions of a given cell, SoH should be based on parameters that are difficult to measure (some extremely so), especially in applications that do not use the entire SoC range. These parameters are:

50. Ignoring the issue of what defines a cycle in real-world applications.

- Actual capacity;
- Actual resistance;
- Actual self-discharge current.

All of these parameters are difficult to measure (some extremely so), especially in applications that do not use the entire SoC range.

For a string of cells in series, SoH should be based on

- The actual capacity of the string (affected by balance as well as the capacity of all of its cells);
- The actual resistance (based on the resistance of all of its cells);
- The actual self-discharge current of the worst cell.

I define SoH thus: If placing two degraded batteries in parallel results in a performance that matches a new battery, then the degraded batteries have an SoH of 50% (Figure 1.19). The degraded batteries have half the capacity and twice the resistance of new batteries.

1.6.5 Other "State of"

A few more states are compelling.

1.6.5.1 State of Function

The *state of function* (SoF) promises to offer the advantages of SoC and SoH in a single value. However, the fact that it is so variously defined keeps it from delivering this promise.

1.6.5.2 State of Voltage

The *state of voltage* (SoV) is meant to replace SoC as an indicator of when the battery may shut down (which under heavy loads occurs before the SoC reaches 0%). However, voltage is variable because the load can be unpredictable, so SoV is of limited use. A well-evaluated SoE would be more useful.

1.6.5.3 State of Power

The *state of power* (SoP) should tell the load the maximum power it may take from the battery in the next few seconds. This information is quite useful in variable, high-power applications. However, it is quite hard to evaluate (it doesn't correspond to any particular parameter), and many different definitions and equations have been proposed for it.

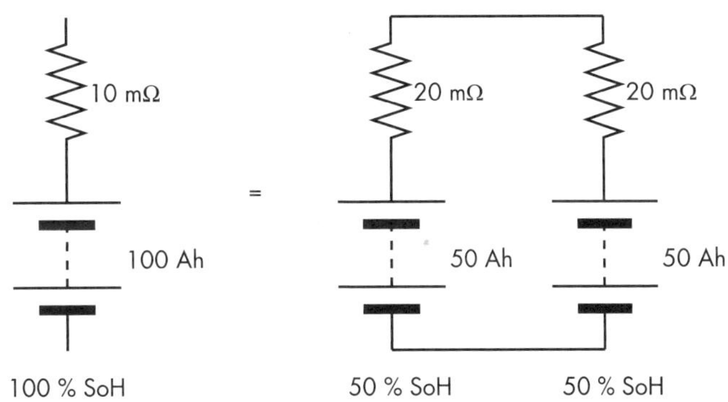

FIGURE 1.19
Two batteries at 50% SoH connected in parallel match the performance of a single battery at 100% SoH.

1.6.5.4 State of Life

The *state of life* (SoL) should tell the user when to replace a battery. However, like SoH, it's based on vaguely defined measures and it depends on the application.

1.6.5.5 State of Balance

The *state of balance* (SoB) ranges from 100% (for a string whose cells are all at the same SoC) to less if the string is imbalanced (see Section 3.2.5.5). There are different interpretations of this measure (see Section 3.2.6):

- At the balance point—a measure of how well a string is balanced, regardless of its present SoC level;
- At any given time—a measure of the variation of cell SoC levels right now, regardless of how well the string is balanced.

1.6.6 Depth of Discharge

Depth of discharge (DoD) is the opposite of SoC; it starts at 0 when the battery is full, and it increases as the battery is discharged (Figure 1.20). It usually uses Ah units (0 Ah when full, up to the battery capacity when empty), although it may use relative units (from 0% when full to 100% when empty).

1.6.7 Charge Acceptance, Discharge Availability

State of charge is a useful measure when discussing the state of a single battery, yet it's of limited use in an array of batteries, each of which may or may not be connected to a bus, because the capacity of an array is not a fixed value.

For example, consider an array of ten 100 Ah batteries connected in parallel to a bus. It matters greatly to the user how many are turned on because if only one is connected, the energy available is less than 10% compared to if all 10 are connected. If the BMS reports that the SoC is at 50%, that's of limited use to the user. In contrast, it is far more valuable if the BMS reports that the array can deliver 50 Ah (one battery connected to the bus), 100 Ah (two batteries), or 500 Ah (all batteries).

The following two states are more general than SoC and should be more useful for array applications.[51]

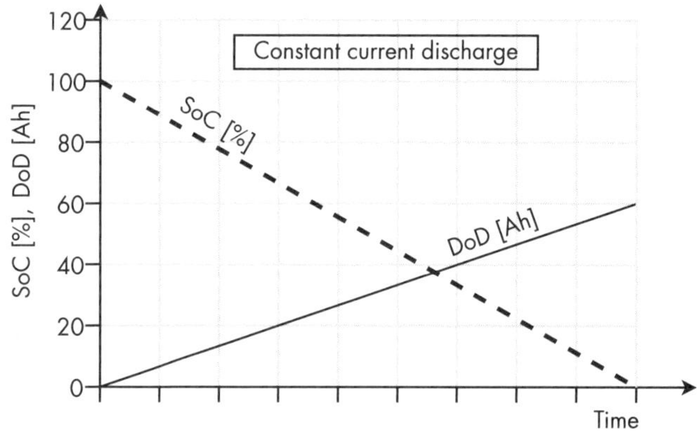

FIGURE 1.20
Depth of discharge for a 60 Ah battery.

51. I may have invented these terms, but the concept is not new.

- *Charge acceptance*:[52] How much charge [Ah] a battery or battery array can accept;
- *Discharge availability*: How much charge [Ah] a battery or battery array can deliver.

That gives us a way to define capacity and SoC, not just for an array, but also for batteries and cells:

$$\text{Capacity [Ah]} = \text{Charge Acceptance [Ah]} + \text{Discharge Availability [Ah]} \tag{1.39}$$

and

$$\text{SoC [\%]} = 100 \times \text{Discharge Availability [Ah]/Capacity [Ah]} \tag{1.40}$$

For a single cell or battery:

- Charge acceptance is simply the DoD [Ah];
- Discharge availability is simply the SoC if expressed in [Ah].

For an array of batteries in parallel:

- Effective capacity is the sum of the capacity of each battery that is connected to the bus;
- Discharge availability is the sum of the discharge availability reported by each battery that is connected to the bus;
- Charge acceptance is the sum of the charge acceptance reported by each battery that is connected to the bus.

If all batteries remain connected to the bus, then the total capacity remains fixed, discharge availability corresponds to SoC, while charge acceptance corresponds to DoD [Ah] (Figure 1.21(a)). If, however, a battery is suddenly disconnected from the bus (Figure 1.21(b)), the following happens:

- The array capacity drops (from 120 to 100 Ah);
- The discharge availability drops accordingly, indicating that the charge that can be extracted from the array is reduced;
- The charge acceptance drops as well, indicating that the charge that can be received by the array is also reduced.

1.7 CHARTS

Various charts are used to represent the properties of an electrical storage technology. They are particularly useful when comparing multiple technologies because they show at a glance the advantages and disadvantages of each technology.

1.7.1 Radar Chart

Radar charts[53] present multiple parameters in two dimensions. They consist of three or more radii emanating from a center, each radius representing a parameter using

52. Note that some people use the term charge acceptance in place of coulombic efficiency, but charge acceptance is how much charge [Ah] a battery can accept, which is a significantly different parameter. See Section 1.4.1.3.

53. Also known as *spider charts*, *web charts*, and other names.

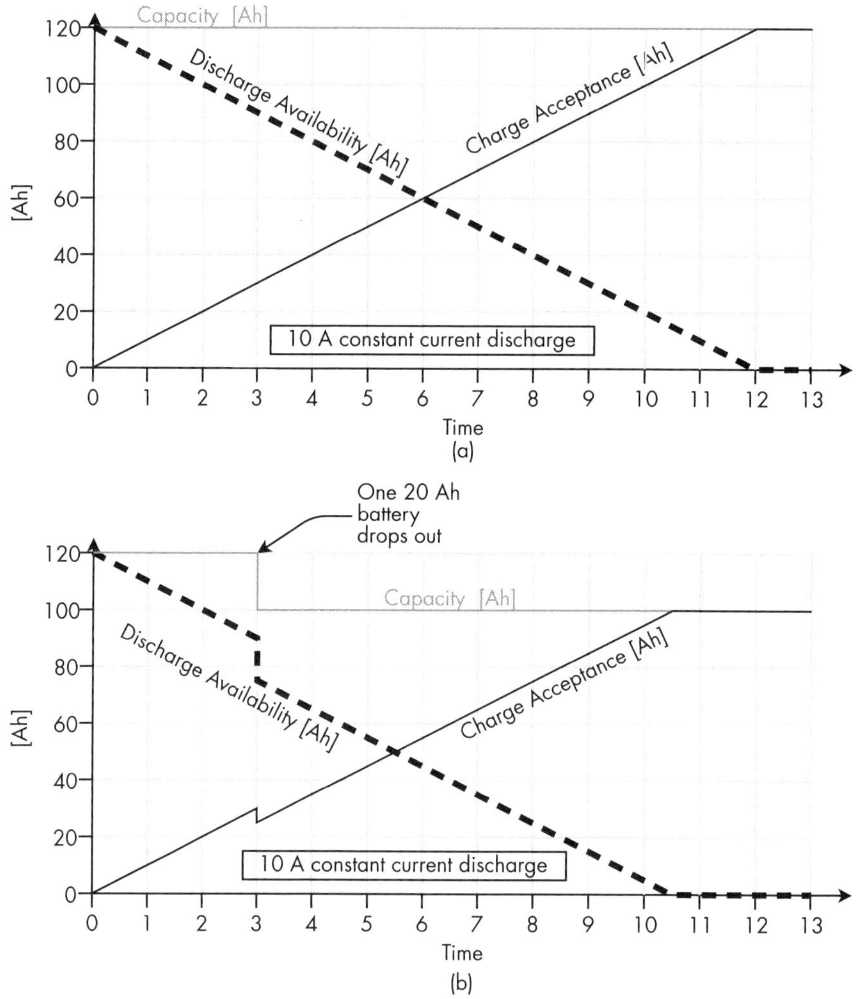

FIGURE 1.21
Plot of charge acceptance, discharge availability, and capacity in an array of six 20 Ah batteries in parallel: (a) all batteries remain connected, and (b) one battery drops out after 3 hours.

arbitrary units, with worst toward the center and best toward the perimeter. The units along each axis are relative, as long as the units are consistent and "better" is outward (for example, a high price is plotted closer to the center).

Radar charts are particularly useful when comparing various battery technologies (Figure 1.22(a)) or determining if a technology is appropriate to the application's requirements (Figure 1.22(b)).

Multiple parameters may be compared, such as (Figure 1.23):

- Specific price (relative to energy or power) and availability;
- Energy density or specific energy;
- Power density or specific power or maximum power time;
- Lifetime (cycle or calendar);
- Operating temperature (charging and discharging);
- Safety (such as thermal runaway temperature);
- Self-discharge current;
- Environmental (benign materials, recyclability).

FIGURE 1.22
Radar charts: (a) comparing two technologies, and (b) comparing an available cell to the design requirements.

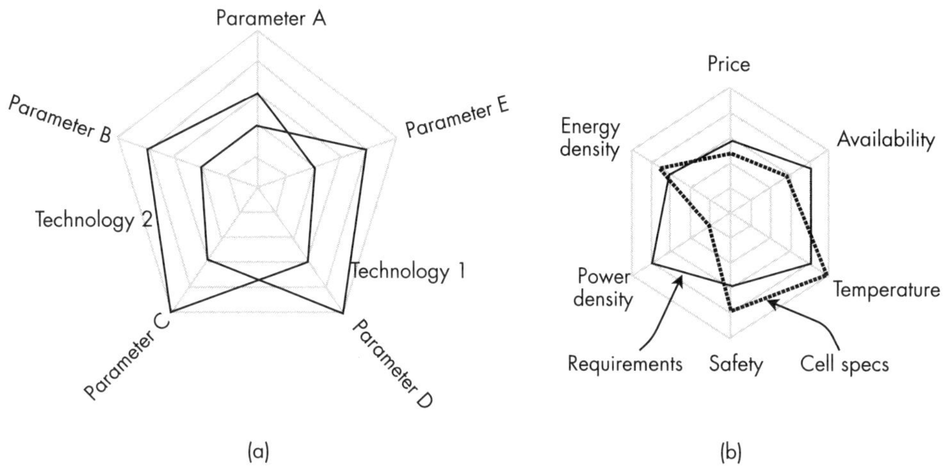

FIGURE 1.23
Radar chart examples: (a) comparing three Li-ion chemistries, (b) comparing various battery technologies, and (c) comparing the same batteries, but for a different application—the relative importance of certain parameters changed.

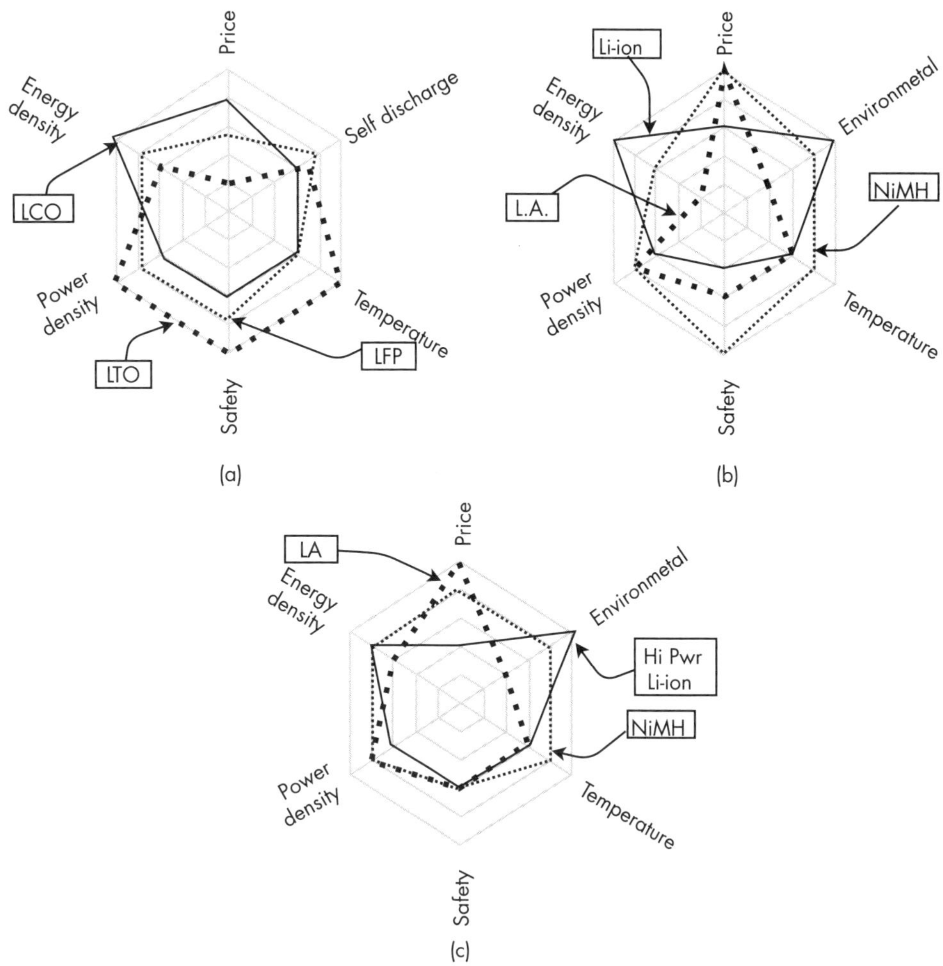

For clarity, no more than four items and no more than eight parameters should be compared. If just for a specific application, only parameters relevant to that application should be compared. For example, for a stationary application, energy density may be more important than operating temperature.

1.7.2 Ragone Plot

Ragone[54] *plots* are useful to simultaneously represent the energy and power of a given storage technology—batteries, capacitors, pumped hydro, flywheels, and more. David V. Ragone devised these plots 50 years ago in a paper [16] that compared batteries for EVs.

This results in four permutations (Figure 1.24):

- *Gravimetric* (for a given mass) or *volumetric* (for a given size);
- Power on the X- or Y-axis.

Regardless, the best devices (highest energy and power) are in the top right corner. The units for energy are either joule or watt-hour, the units for mass are usually kilogram, and the units for volume are liter. These charts normally use a log-log scale.

1.7.2.1 Typical Points

The proper way to represent the capabilities of a storage device onto a Ragone plot is as a curve (Figure 1.25(a)) that ranges from low power at one end (top left corner, where energy is maximum) to the maximum power point at the other end (bottom right corner, where energy is minimum). The curve is not straight. It has a hump toward the top right, at the operating point (the *typical point*) where both the power and the energy are one half of their respective maximums.

When comparing multiple storage devices, the plot can be overcrowded (Figure 1.25(b)). If so, rather than showing the complete curve, we often use only the typical point on each curve (Figure 1.25(c)). Then we can simplify the plot by showing only the typical points (Figure 1.25(d)). Though detailed information is lost, the comparison is clearer.

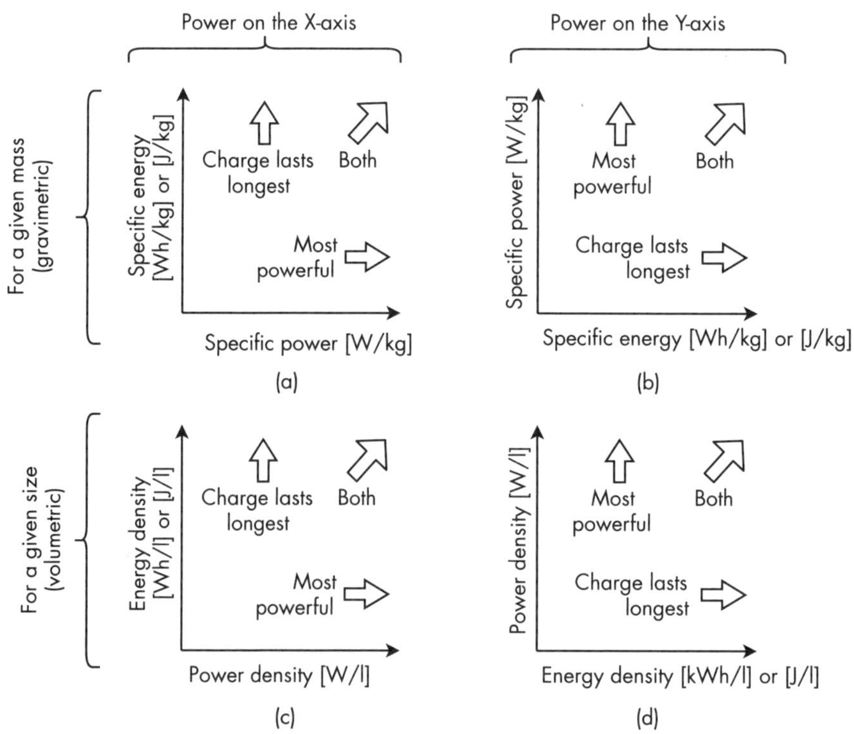

FIGURE 1.24
Ragone plots:
(a) gravimetric, power on the X-axis, (b) gravimetric, power on the Y-axis, (c) volumetric, power on the X-axis, and (d) volumetric, power on the Y-axis.

54. Pronounced "rah GO neh" (Italian), though Case Western Reserve University (of which he was president) says it's "ru GO nee."

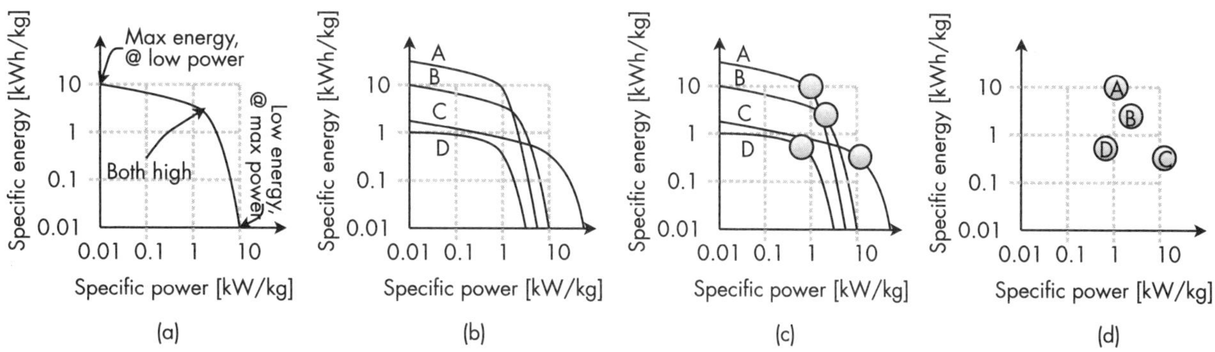

FIGURE 1.25 Ragone plots with typical points: (a) single storage device using complete curve, (b) four devices using complete curves, (c) the same comparison, adding the typical points, and (d) the same comparison, using only the typical points.

1.7.2.2 Using Bubbles

When comparing multiple classes of storage devices using individual curves, the plot can get rather messy (Figure 1.26(a)). Again, we can use just the typical points instead of the entire curve (Figure 1.26(b)), but all those dots still crowd the plot (Figure 1.26(c)).

Instead, we can group all the MPP dots of a given class of power sources into a bubble (Figure 1.26(d)). By showing only the bubbles, the plot clearly emphasizes the differences between classes of power sources (Figure 1.26(e)).

Once more, by showing only bubbles, information is lost.

1.7.2.3 Using Wide Bands

To retain information about operation over the entire range of power and energy, instead of using the typical points, we may use a wide band that encompasses all the

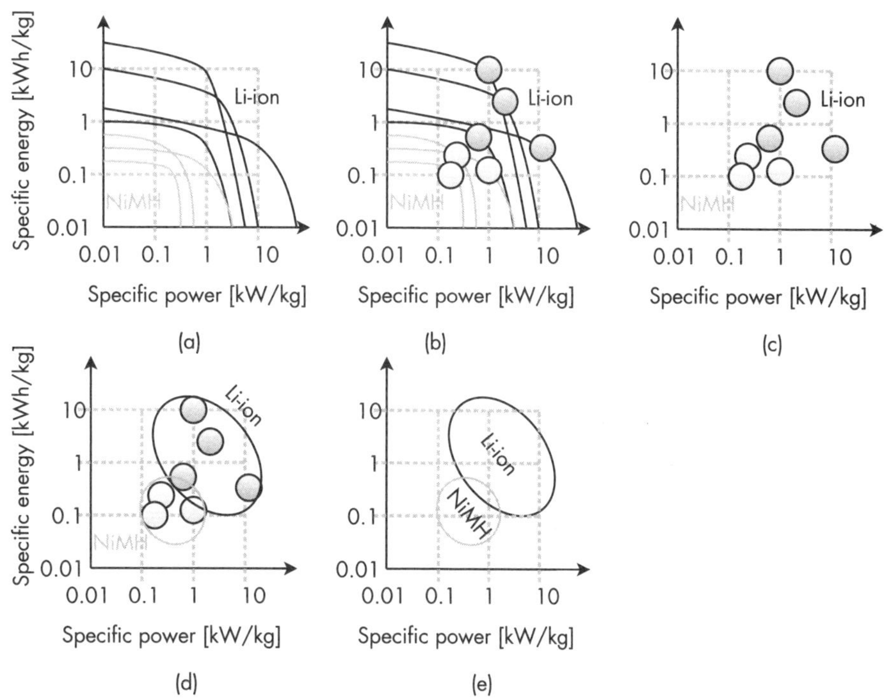

FIGURE 1.26 Comparing classes of storage devices on a Ragone plot: (a) using complete curves, (b) adding typical points, (c) just the typical points, (d) grouping typical points, and (e) bubble for each group.

curves for a given class of power sources (Figure 1.27(a)). Showing just the bands emphasizes the differences between classes of storage devices, including operation in the entire range of power and energy (Figure 1.27(b)). Using bubbles is better when comparing many classes (more than seven classes), while using wide bands is better when you need a comprehensive picture of what each class of storage devices can provide.

1.7.2.4 Time Diagonal Lines

Ragone plots may include diagonal lines to indicate the ratio of the energy and power (Figure 1.28(a, b)). They have units of time because the ratio of energy (Wh) and power (W) has units of hours. This time is the duration of a discharge at the given, constant power. For example, this particular storage device (Figure 1.28(a)) can deliver (1) 0.07 kW/kg for 4 days, or (2) 2 kW/kg for 1 hour, or (3) 7 kW/kg for 26 seconds.

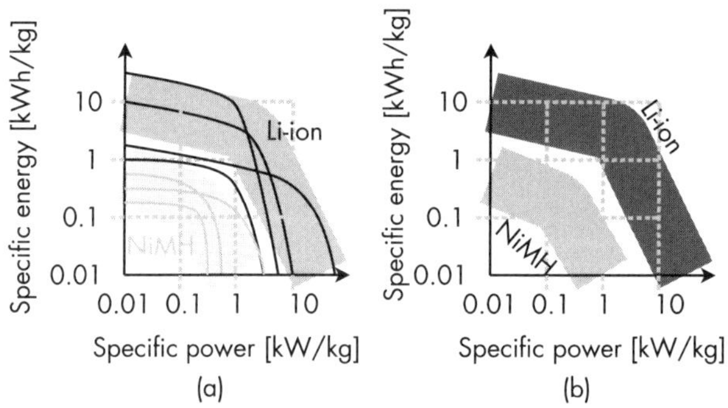

FIGURE 1.27
Comparing classes of storage devices on a Ragone plot:
(a) grouping with wide bands, and (b) showing only the wide bands.

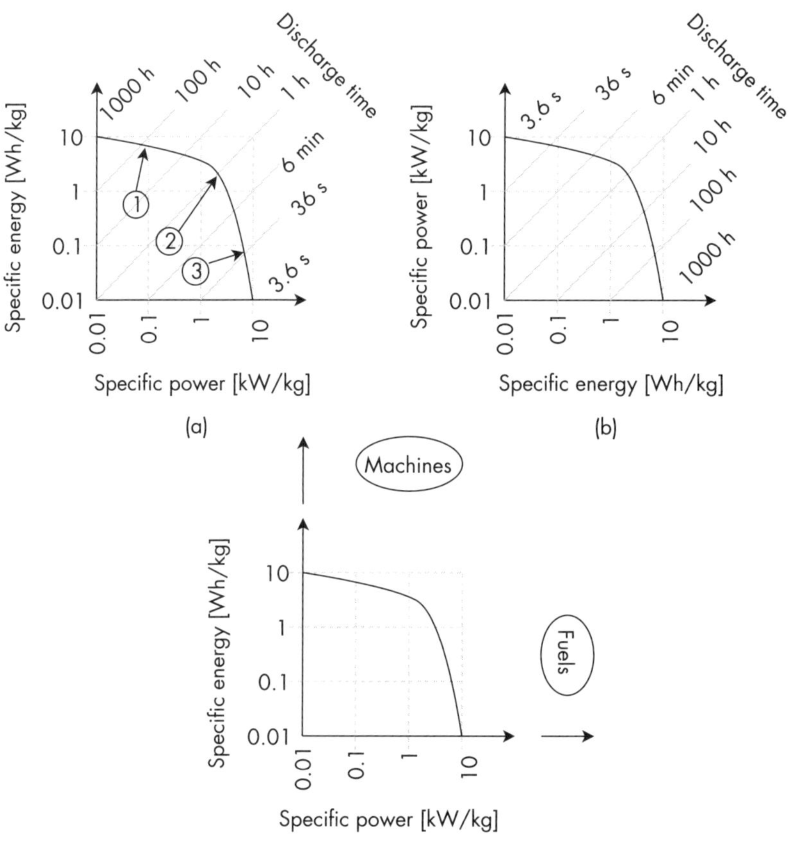

FIGURE 1.28
Time lines in Ragone plots: (a) power in the X-axis, and (b) power on the Y-axis.
(c) Machines and fuels are literally off the chart.

1.7.2.5 Machines and Fuels

Occasionally you may see a plot that also lists machines (e.g., fuel cells, petrol engines) or fuels (e.g., petrol, hydrogen, methanol) (Figure 1.28(c)). This is nonsensical:

- Machines have practically infinite energy (given an unlimited fuel supply and no wear and tear), which places their dot an infinite distance above a Ragone plot (with power on the X-axis);
- Fuels have practically infinite power (limited only by the size of an explosion), which places their dot an infinite distance to the right of that Ragone plot.

1.7.2.6 Examples of Ragone Plots

Let's populate these plots with some real data [17] in the four possible formats (Figures 1.29 and 1.30). The plots show most electrical storage devices. Li-ion cells are subdivided into these groups:

- Power cells (high power, decent energy)
- Energy cells (high energy, decent power)
- Both (a compromise between energy and power)
- Mediocre (general purpose)
- Poor (not worth it)

There are additional battery and capacitor types:

- LIC = Li-ion capacitor (a hybrid of a super-capacitor and a Li-ion cell)
- LMP = lithium metal polymer (not a Li-ion cell)
- NiFe = nickel-iron
- NiCD = nickel-cadmium
- NiMH = nickel-metal hydride
- LA = lead-acid
- Electrolytic capacitors are just outside the plot area (*Cap*)

Note how the relative spacing of the various technologies is not consistent in the volumetric and gravimetric plots. For example, lead-acid and NiCD overlap in the volumetric plots and not in the gravimetric ones. This is due to the difference in the specific energy of the various technologies.

1.7.3 MPT-Based Plots

The Ragone plot is effective when comparing storage devices just based on their size or mass. But what if what all you care about is how efficient cells are at high power? If so, it's more useful to plot energy versus maximum power time.

This plot has MPT on the X-axis and energy on the Y-axis. Unlike in a Ragone plot, "best" is on the top left corner. The plot can be gravimetric (Figure 1.31(a)) or volumetric (Figure 1.31(b)).

In this plot, the diagonal lines represent power when operated at the maximum power point (which is well-defined). For a gravimetric plot, the diagonal lines are specific power [W/kg] (Figure 1.32(a)). For a volumetric plot, they are power density [W/l] (Figure 1.32(b)).

Of course, one could put power on the Y-axis, in which case the diagonal lines would be energy. Or one could put the maximum power time on the Y-axis and put

FIGURE 1.29
Ragone plot, power on the X-axis: (a) gravimetric, and (b) volumetric.

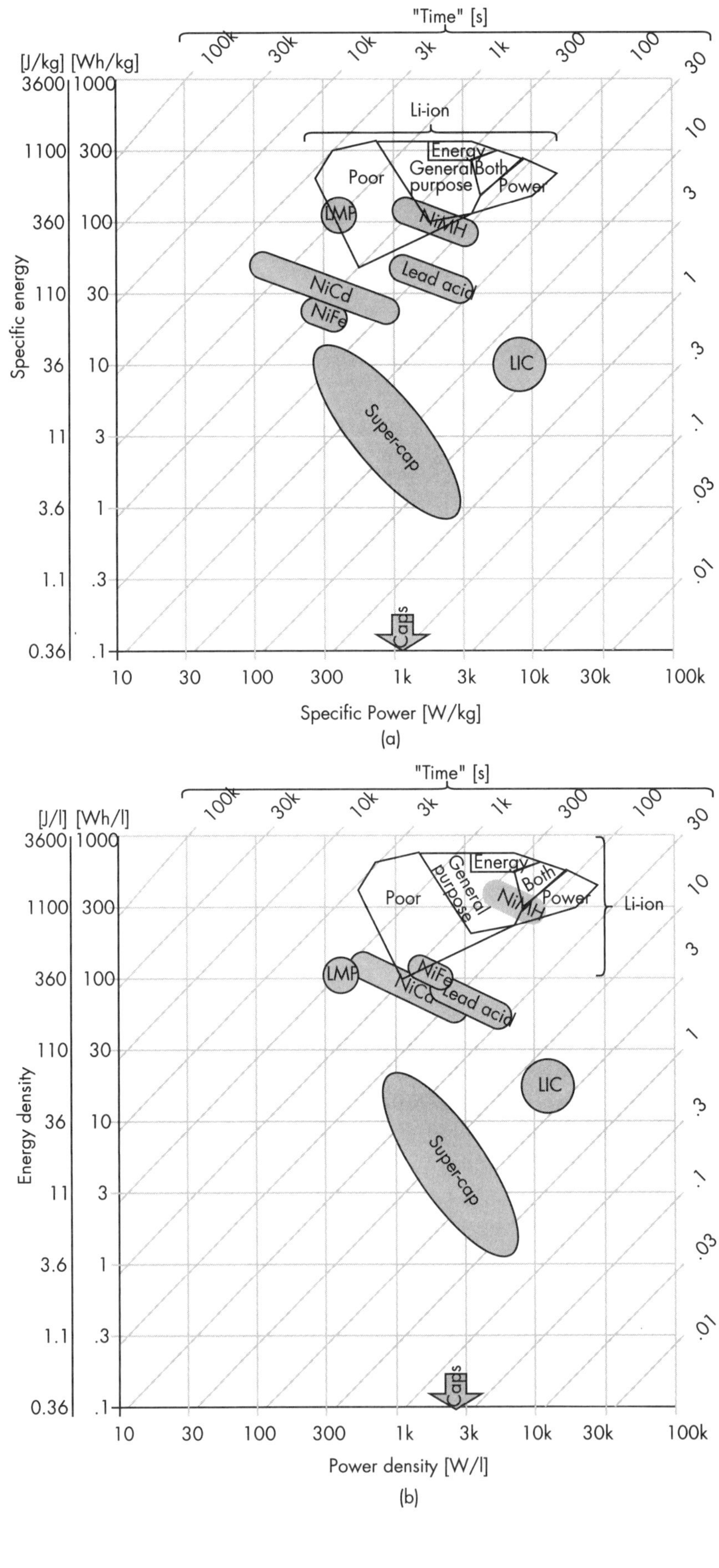

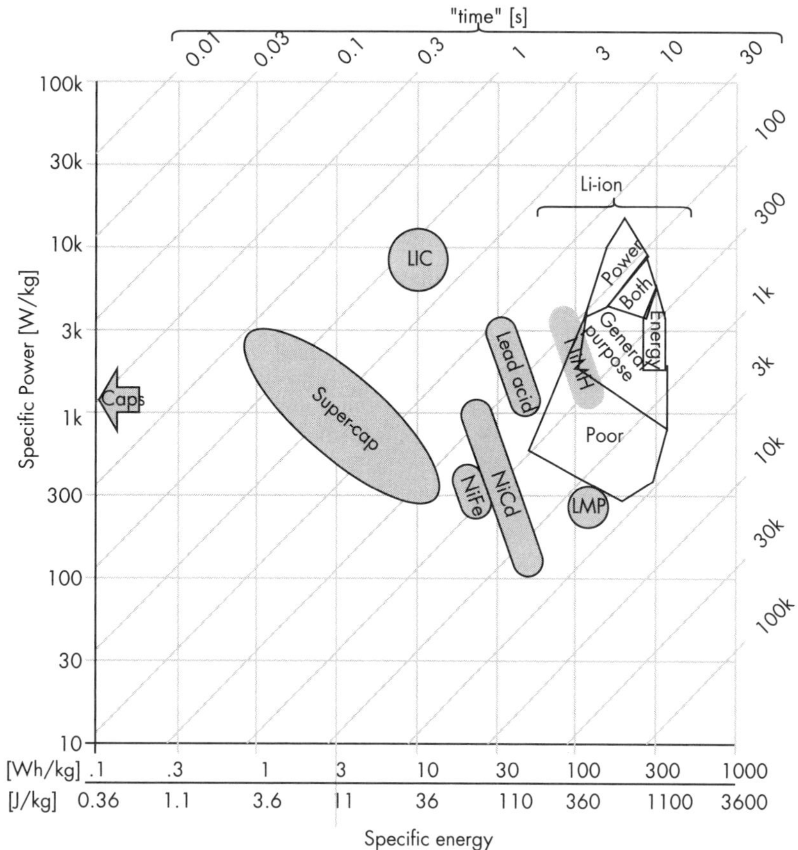

(a)

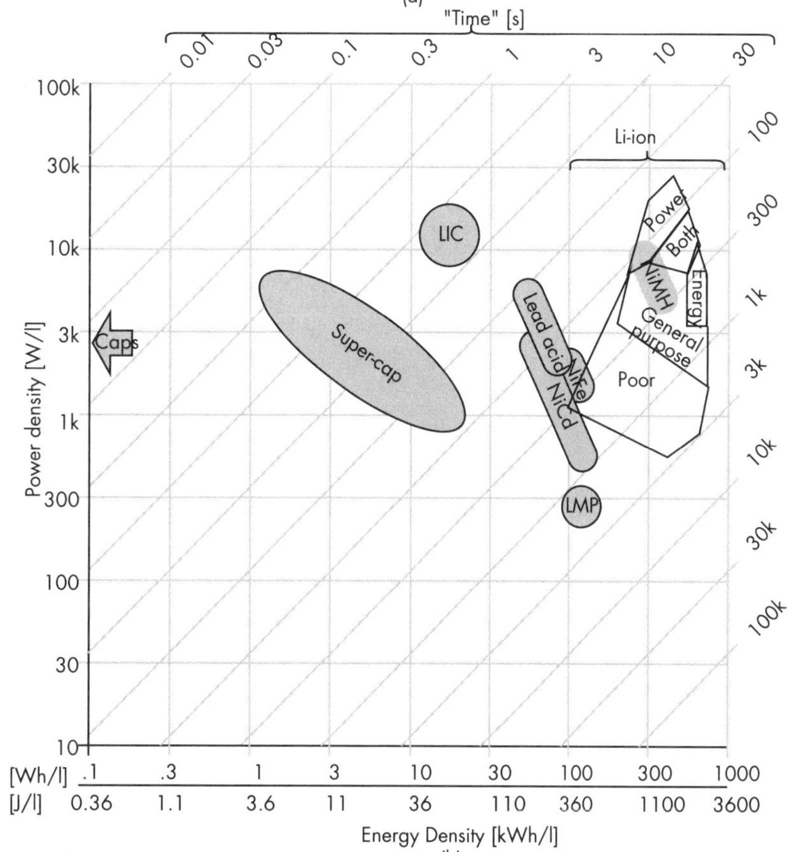

(b)

FIGURE 1.30
Ragone plot, power on the Y-axis: (a) gravimetric, and (b) volumetric.

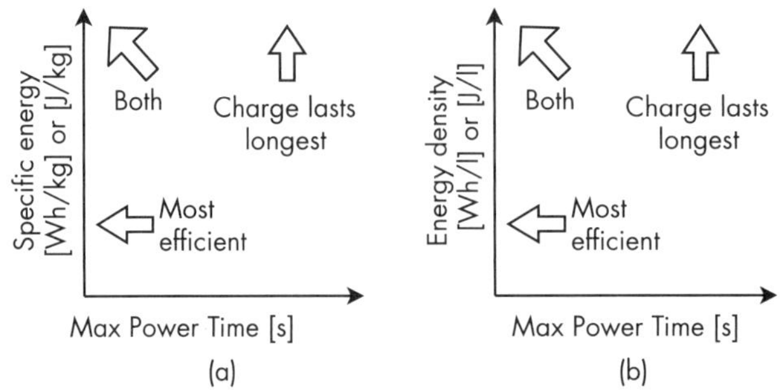

FIGURE 1.31
MPT-based plots: (a) gravimetric, and (b) volumetric.

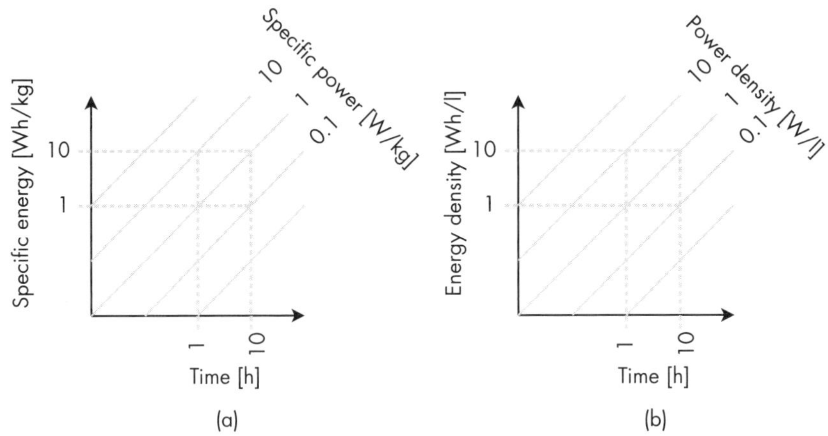

FIGURE 1.32
Function of diagonal lines (a) specific power, and (b) power density.

either energy or power on the X-axis. Having tested all three possibilities, I feel that the version I show here is the most effective.

1.7.3.1 Examples of MPT-Based Plots

Once more, let's plot actual data to see the effect of replacing power with maximum power time (Figure 1.33). We see that compared to the Ragone plot, the areas are mirrored horizontally and skewed 45 degrees counterclockwise.

1.8 DEVICES USED WITH BATTERIES

A battery is commonly connected to the following devices.

1.8.1 Power Converters

Power converters convert between AC and DC, within DC, or within AC. They may convert in one direction, the other direction, or both (Figure 1.34). Converters can be classified as AC to AC, AC to DC, DC to AC, DC to DC or any direction:

AC to AC:

- *Transformer*: Converts an AC voltage to another AC voltage in either direction at the same frequency, usually different voltages. Normally it provides isolation (e.g., a *wall wart*[55] with AC output, or a 110 to 220V travel transformer).
- *Variable Frequency Drive (VFD)*: A driver for AC motors for industrial applications; converts the AC line voltage to another AC voltage at a different and variable frequency to power the AC motor.

55. A module that is plugged directly into an AC power outlet and has a low-voltage output cable.

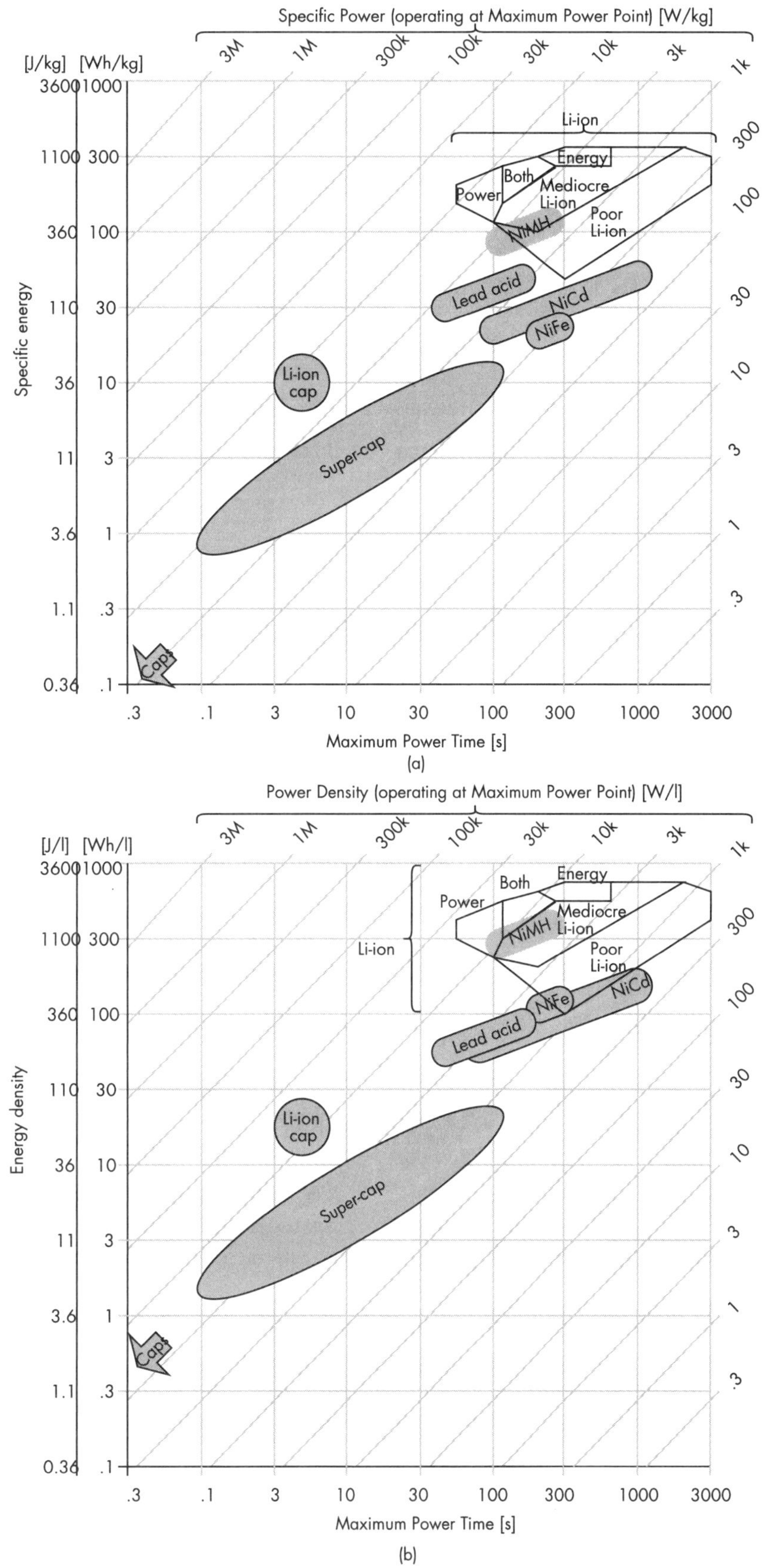

FIGURE 1.33
MPT-based plot:
(a) gravimetric, and
(b) volumetric.

FIGURE 1.34
Power conversion chart.

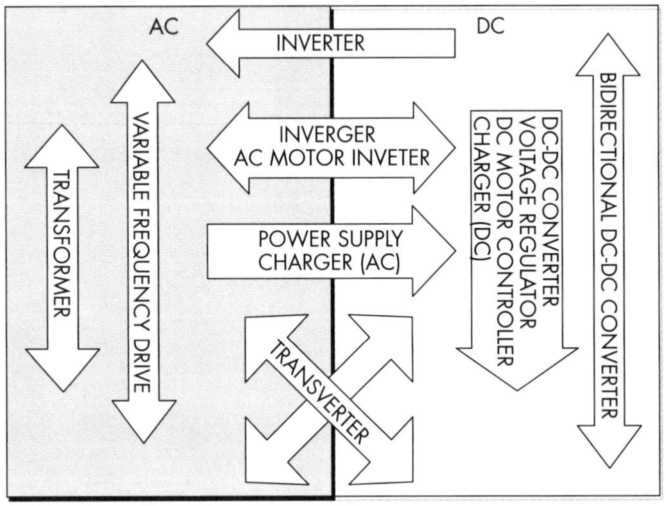

AC to DC:

- *Power supply*: Converts AC to DC (e.g., an AC adapter for consumer electronics); may be linear (quiet) or switching (more efficient).[56]
- *Charger*: Same as above, but current-limited, suitable for charging a battery (e.g., a charger for NiMH cells).

DC to AC, possibly bidirectional (only one direction at a given time):

- *Inverter*: Converts DC to AC; for example:
  - Inverter to power 110Vac loads from a car's cigarette lighter outlet
  - Solar inverter to feed solar power to power an off-grid house
  - Inverger between a 48V battery and the grid.
  - Bidirectional AC motor driver for vehicle traction; supports regenerative braking (e.g., the *inverter* in an electric vehicle).

DC to DC:

- *DC-DC converter*: This switching regulator[57] converts a DC voltage to another DC voltage, usually at a different voltage, only in one direction (it has an input and an output), and may or may not be isolated (e.g., a USB adapter that plugs into a car's cigarette lighter outlet).
- *Bidirectional DC-DC converter:* Same as above, but power may flow in either direction (only one direction at a given time).
- *Voltage regulator:* Same as a DC-DC converter, but linear (not switching), less efficient, and not isolated.
- *Charger with DC input*: Like any of the above, but current-limited, suitable for charging a battery (e.g., in a portable power bank to charge a car battery).

Any combination:

- *Transverter*: From anything to anything, in any direction.

56. Also known as switch-mode power supply (SMPS).
57. Also known as switch mode.

Let's look at some of these converters in more detail.

1.8.2 AC to DC: Chargers and Power Supplies

A *power supply* converts AC to DC. A power supply may be:

- Standard power supply[58] (Figure 1.35(a)): Only operates in constant voltage mode, up to a maximum current; what happens if you attempt to draw more current than the maximum is unclear (Blown fuse? Shutdown? Overheat? Flames?); operation at a voltage other than the constant voltage (CV) is not possible, therefore, it is not suitable for direct connection to a battery.[59]
- Power supply with fold-back (Figure 1.35(b)): Same as above, but operation past maximum current is specified: the voltage drops and the current folds back to a lower value; if used as a charger it is not damaged, though the charging current would be disappointingly low.
- Charger: A current-limited power supply (Figure 1.35(c)); its output is constant current/constant voltage (CCCV), suitable to charge a battery; the supply switches between a CV and a constant current (CC) mode, depending on the current drawn by the load, or the voltage set by the load:
  - In CC mode, it tries to maintain the current to the CC value, as long as the load voltage is less than the CV value (vertical line).
  - In CV mode, it tries to maintain the voltage to the CV value as long as the load current is less than the CC value (horizontal line).

1.8.2.1 Charger

A *charger* is a power supply with a CCCV output (Figure 1.36). It operates in two modes:

1. It starts in the CC mode, *bulk charging* the battery at constant current, letting the battery set the voltage;
2. When the battery voltage reaches the CV setting, the charger switches to CV mode, maintaining the battery voltage and letting the battery set the current.

Specialized chargers use a *profile*, varying the CV setting over time for fast yet safe charging through three or more stages.

For example, for a 12V lead-acid battery:

- Bulk: CC mode at 0.2C current;
- Topping (absorption): CV mode at 14.5V, for a preset time;

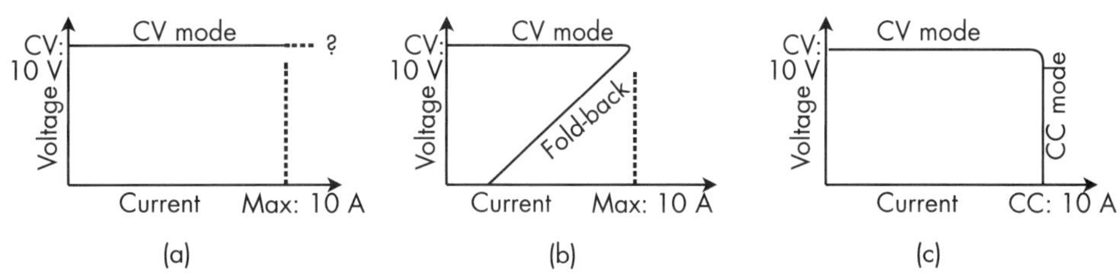

FIGURE 1.35
Power supply output: (a) standard CV, (b) standard with fold-back, and (c) CCCV (charger).

58. An AC adapter is a low-power power supply. It is incorrectly called a "charger" by consumers. The actual charger is in the phone or the laptop.

59. Attempting to charge a battery with a CV supply may result in the supply shutting down (at best) or damage to either device or both (at worst).

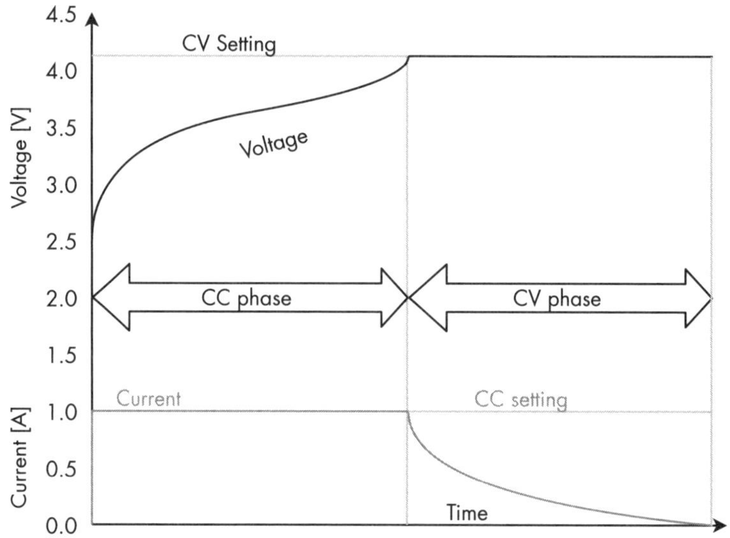

FIGURE 1.36
Charger profile.

- Float: CV mode at 13.5V, indefinitely.

The CV setting is temperature-compensated to match the thermal coefficient of lead-acid batteries: –3 mV/°C.

1.8.3 DC to DC: DC Chargers and DC-DC Converters

A *DC-DC converter* converts from one DC voltage to a different one. Just like a power supply, its output could be CCCV, CV only, or fold-back. It may or may not be isolated.

A nonisolated, unidirectional DC-DC converter can have many topologies, including, most commonly, these:

- Step down (buck) (Figure 1.37(a)): The output voltage is lower than the input; the output current is higher than the input,[60]
- Step-up (boost) (Figure 1.37(b)): The output voltage is higher than the input, the output current is lower than the input;
- Buck-boost (Figure 1.37(c)): The output voltage may be higher or lower than the input.

In all three cases, the voltages are positive and the ground is common between the input and the output.

The output of an isolated DC-DC converter (Figure 1.37(d)) can be higher or lower than the input, or the same. The input and output grounds are isolated, allowing many circuit configurations.

A bidirectional converter (Figure 1.37(e)) may transfer power in either direction, depending on the status of a control line.

1.8.4 DC to AC: Inverters, Invergers, and AC Motor Drivers

Inverters, invergers, and AC motor drivers convert DC to AC; some are bidirectional.

60. The output power is nearly as high as the input power, which is why, as the output voltage goes down, the output current must go up (Power = Voltage × current).

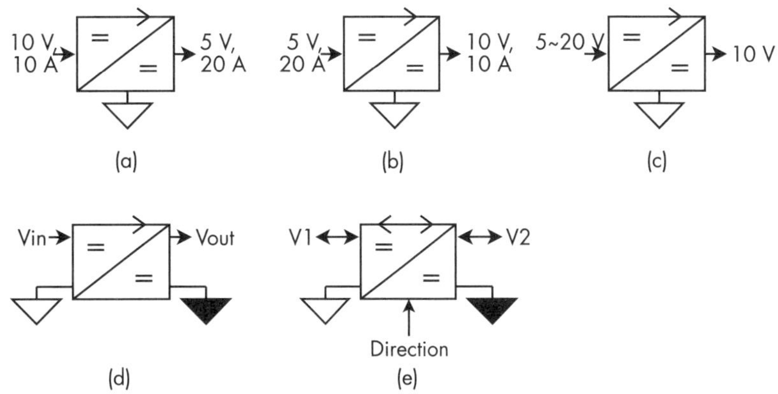

FIGURE 1.37
DC-DC converters: (a) buck (not isolated), (b) boost (not isolated), (c) buck-boost (not isolated), (d) isolated, and (e) bidirectional.

1.8.4.1 Inverter

An *inverter* converts DC to AC, either at a fixed frequency (e.g., 50, 60, or 400 Hz) or at a variable frequency to drive a motor at a variable speed possibly with regenerative braking.

1.8.4.2 Inverger: Charger/Inverter, Combi

An *inverger* is placed between a battery and an AC line (at a fixed frequency). Sometimes it operates as a charger, sometimes as an inverter (Figure 1.38(a)) (see Volume 2, Section 2.2.3.4, and Section 2.5).

1.8.4.3 AC Motor Driver

An *AC motor driver* converts the DC from a battery to variable frequency AC for a motor, specifically for traction. It is typically called an *inverter.* However, an inverter is unidirectional while an AC motor driver is bidirectional.

It operates in four quadrants, which are the four permutations of two directions (Figure 1.38(b)) (See Volume 2, Section 3.3.4.7).

- Power: Drive or brake (this means that power can flow in either direction);
- Rotation: forward and reverse.

An inverger and an AC motor driver are different in a few respects:

- An inverger operates at a fixed AC line frequency; conversely, for an AC motor driver, the frequency to the motor varies with the speed of the motor and with the torque;
- An inverger has two distinct operation modes (either as a charger or as an inverter) and may have to pause between them; conversely, an AC motor can shift rapidly and seamlessly between driving, coasting, and braking.
- The DC side of an inverger is designed to charge a battery; a motor driver may not be current limited, in which case it should be connected to a battery.

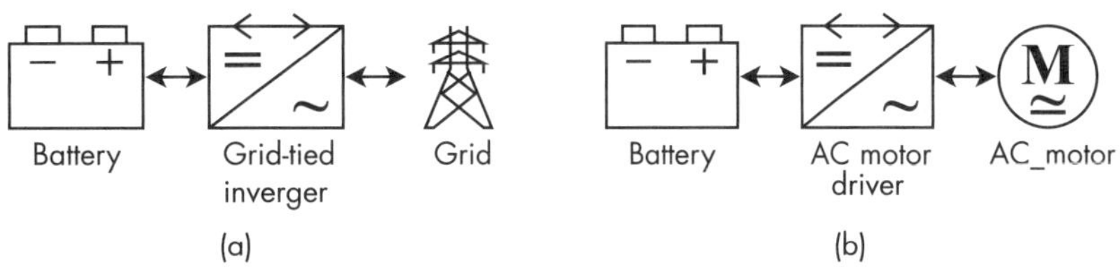

FIGURE 1.38
Bidirectional AC and DC: (a) Inverger, and (b) AC motor driver.

1.8.5 AC to AC: Transformers and Variable Frequency Drives

Both *transformers* and *VFDs* convert AC to AC:

- Transformers: Change the voltage, provide isolation, or both; the frequency remains the same (Figure 1.39(a));
- VFDs: Drive an AC motor at variable speed; they change both the voltage and the frequency; they receive AC power at 50 or 60 Hz, and generate three-phase power for an induction motor (Figure 1.39(b)).

1.8.6 Any Direction: Transverters

In this context,[61] a *transverter* is a universal power converter. It combines and converts power from various sources—solar panels, batteries, the grid, generators—to power both DC and AC loads. Specifically, a transverter reduces the number of power devices, simplifying a system by integrating many functions, such as charger, inverter, and solar conversion (Figure 1.39(c)) (see Volume 2, Section 2.2.3.5).

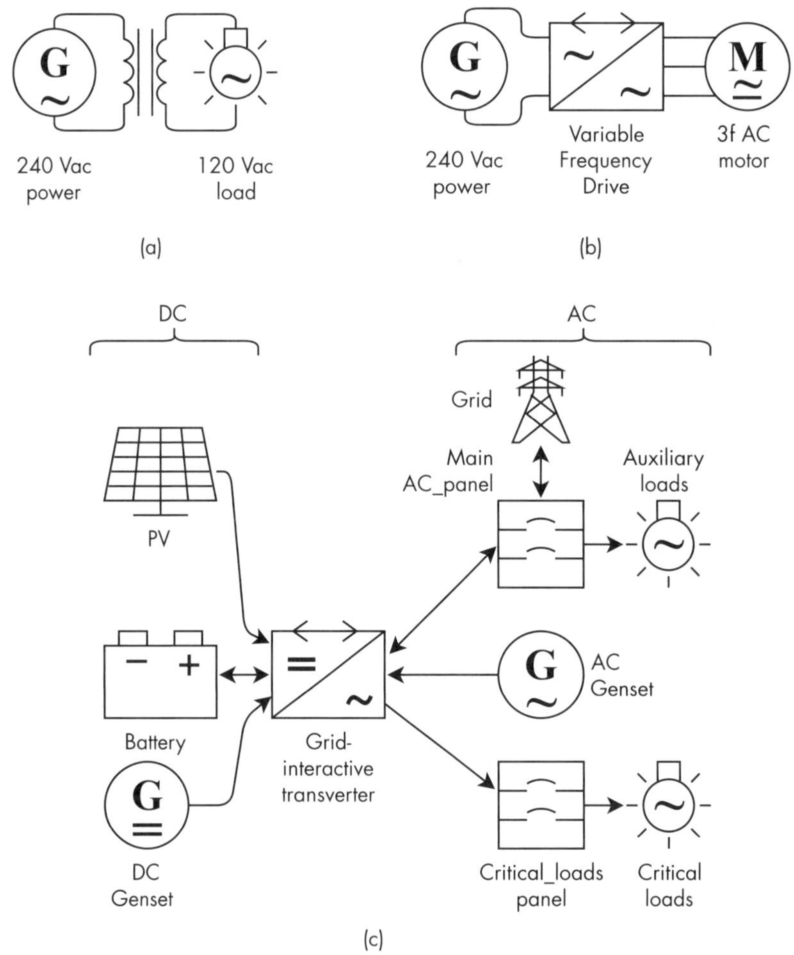

FIGURE 1.39
(a) Transformer,
(b) VFD, and
(c) transverter converting from any source to any load.

61. In other contexts, the term transverter is more widely understood to be a radio frequency device that consists of an upconverter and a downconverter in one unit.

References

[1] Buchmann, I., "Lithium-Polymer: Substance or Hype?" Battery University, batteryuniversity.com/learn/article/the_li_polymer_battery_substance_or_hype.

[2] en.wikipedia.org/w/index.php?title=Talk:Lithium_polymer_battery/Archive_1&oldid=615892015#Article_should_be_deleted.2Fmerged.

[3] Andrea, D., "I know this is controversial, but I would like to point out that Ohm's law only applies to purely resistive loads. The vast majority of loads aren't, so Ohm's law does not apply," https://redd.it/dru5bv.

[4] Dell, R. M., D. A. J. Rand, and R. Bailey, Jr, *Understanding Batteries*, Cambridge, UK: Royal Society of Chemistry, 2001, p. 30.

[5] Pickering, J., "Understanding Coulombic Efficiency Limitations in an Acid-Base Energy Storage System: Mass Transport Through Nafion," M.S. dissertation, Department of Chemical Engineering Case Western Reserve University, 2018.

[6] Zhang, C., and J. Jiang, *Fundamentals and Applications of Lithium-ion Batteries in Electric Drive*, Singapore: Wiley, 2015.

[7] Zhang, L., H. Peng, Z. Ning, Z. Mu, and C. Sun, "Comparative Research on RC Equivalent Circuit Models for Lithium-Ion Batteries of Electric Vehicles," *Applied Sciences,* Vol. 7, No. 10, 2017, p. 1002.

[8] Mamadou, K., E. Lemaire, A. Delaille, D. Riu, S. E. Hing, and Y. Bultel, "Definition of a State-of-Energy Indicator (SoE) for Electrochemical Storage Devices: Application for Energetic Availability Forecasting," *Journal of the Electrochemical Society,* Vol. 159, No. 8, 2012, pp. A1298–A1307, doi: 10.1149/2.075208jes.

[9] S. Xiang, G. Hu, R. Huang, F. Guo, and P. Zhou, "Lithium-Ion Battery Online Rapid State-of-Power Estimation under Multiple Constraints," *Energies,* Vol. 11, 2018, p. 283.

[10] Wang, S., L. Shang, Z. Li, H. Deng, and J. Li, "Online Dynamic Equalization Adjustment of High-Power Lithium-Ion Battery Packs Based on the State of Balance Estimation," *Applied Energy*, Vol. 166, 2016, pp. 44–58, 10.1016/j.apenergy.2016.01.013.

[11] Fuller, T. F., and J. N. Harb, *Electrochemical Engineering*, Hoboken, NJ: John Wiley & Sons, 2018.

[12] Mottier, M., Patent WO2006058970A1, Renault, France, 2004

[13] Cabrera-Castillo, E., F. Niedermeier, and A. Jossen, "Calculation of the State of Safety (SOS) for Lithium Ion Batteries," *Journal of Power Sources*, Vol. 324, 2016, pp. 509–520, doi.org/10.1016/j.jpowsour.2016.05.068.

[14] Protogeropoulos, C., R. H. Marshall, and B. J. Brinkworth, "Battery State of Voltage Modelling and an Algorithm Describing Dynamic Conditions for Long-Term Storage Simulation in a Renewable System," *Solar Energy*, Vol. 53, No. 6, 1994, pp. 517–527.

[15] Corigliano, S., and A. Cortazzi, "Battery Energy Storage Systems Modeling for Robust Design of Micro Grids in Developing Countries," thesis, Politecnico di Milano.

[16] Ragone, D., "Review of Battery Systems for Electrically Powered Vehicles," SAE Technical Paper 680453, 1968, https://doi.org/10.4271/680453.

[17] Wikipedia, "Comparison of Commercial Battery Types," en.wikipedia.org/wiki/Comparison_of_commercial_battery_types; Green Car Congress, "Bolloré Group Introduces BlueCar Lithium-Metal-Polymer EV Concept," http://www.greencarcongress.com/2005/03/bolloreacute_gr.html.

CHAPTER 2

LI-ION CELL

2.1 INTRODUCTION

Li-ion technology has advanced battery technology fundamentally, earning its inventors, John B. Goodenough, M. Stanley Whittingham, and Akira Yoshino, the Nobel Prize in chemistry.

This chapter[1] discusses an individual Li-ion cell, while the following chapter discusses how multiple cells are interconnected.

2.1.1 Tidbits

Some of the more interesting items in this chapter include:

- They are *all* Li-ion cells (2.2);
- At high current, a 2 Ah power cell can outlast a 3 Ah energy cell (2.6.2.2);
- A cell may no longer work in an application because its internal resistance went up, not because its capacity went down (2.4.2.1);
- The higher current the less batteries are charged (2.4.4.4).
- When the low-voltage cutout is reached, there's still charge in the cell—you just can't use it (2.4.4.2);
- Li-ion cells are swell...until they swell (2.4.6);
- If you need 100 Ah, use 150 Ah as it will be cheaper in the long run (2.5.3.1);
- Heat is bad for cells, yet a hot cell performs better (2.5.3.2);
- Beware of claims from cell vendors (2.8.1);
- A "safer" cell chemistry from a questionable source is more dangerous than a standard one from a reliable source (2.2.1.8).

2.1.2 Orientation

This chapter starts by defining a Li-ion cell, discussing cell chemistry and format. Then it goes through all the characteristics of cells. It then discusses how a cell may be used (charged and discharged). Finally, it covers buying and testing cells.

1. For guidance with this chapter, I thank Dr. Gregory Plett, professor at the University of Colorado, author of the book series Battery Management Systems, and director of the GATE Center of Excellence in Innovative Drivetrains in Electric Automotive Technology Education. Thank you to Byron Azarm for copy editing.

2.1.3 Li-Ion Cell Definition

A *Li-ion cell* is a single, rechargeable electrical storage device based on the transfer of lithium ions between two electrodes (see Volume 2, Section A.3.1.3).

There is no bulk lithium metal in a good Li-ion cell. There are only individual lithium ions. If you were to open a good cell (which is not advisable), you would see no solid lithium metal.[2] The lithium ions represent only about 2% of the mass of the cell.[3]

A healthy Li-ion cell:

- Is *secondary* (rechargeable);
- Relies on the transport of positive lithium ions (Li+) and their intercalation in both electrodes (they remain Li+ throughout the process and are not converted to any other compound);
- Contains no bulk lithium metal and no elemental lithium;
- Uses a liquid or gel electrolyte.

While other cells also use lithium, they are not Li-ion cells as they do not meet the above definition (see Section 2.1.3.3).

For an in-depth discussion of the inner workings of a Li-ion cell, let me refer you to other publications [1–3]. This book considers a Li-ion cell as just a black box that can store energy.

Note that a cell, by itself, is not a battery; generally, a battery is a collection of cells (see Sections 1.2.2.2 and 5.1.3)

A *Li-ion capacitor* is neither a Li-ion cell nor a supercapacitor. It is a hybrid of the two whose specifications exceed those of either a Li-ion cell or a supercapacitor (see Volume 2, Section A.3.4).

I assume that you already know about the wonders of Li-ion and its impact on our society, for better or worse.[4] Therefore, I will skip all that and jump right into the heart of the matter.

2.1.3.1 Cell Failure

You may worry that a cell may suddenly become an open circuit or short circuit. In reality, a cell (that is not abused) does not suddenly fail open or shorted. Instead, a cell slowly degrades over time: the capacity decreases, the series resistance increases, the self-discharge current increases, and the coulombic efficiency decreases. Any sudden open or short circuit is normally due to the failure of a connection in the battery, not within the cells.

Resources should be spent on a sound mechanical design of the battery, and on quality control during battery assembly, rather than on awkward solutions to the perceived problem of cell failure (see Section 3.6.2).

Although quite rare, even a properly protected battery may experience *spontaneous lithium combustion* as a result of manufacturing defects. The battery designer should prepare for this eventuality by designing ways to mitigate the consequences, such as by keeping the fire from spreading.

2. A bad cell, however, may have some lithium plating on the negative electrode.
3. Lithium is advantageous for two main reasons: (1) It is the most electropositive element, meaning that it can most easily donate electrons to produce positive ions, and (2) it is the lightest metal. This is not as much as an advantage as one would think, though, because lithium is a small portion of a Li-ion cell's mass.
4. Tens of millions of people use addictive e-cigarettes, which were only made possible by the advent of Li-ion cells.

Earlier I said, "if not abused." If, on the contrary, a cell *is* abused (e.g., not protected by a properly installed BMS, operated at excessive current, overheated externally, or subjected to physical abuse), then, yes, it will fail.

- It may fail as a *soft short* circuit (see Section 1.2.2.9).
  - If charged at below freezing, or at high current
- It may fail as a *hard short* circuit:
  - If over discharged and reversed
  - If punctured or otherwise mechanically abused
- It may fail as an open circuit:
  - If overheated to the point that its internal protection mechanisms open
  - If it experiences a "rapid unintentional disassembly" due to being overcharged, exposed to flames, mechanically abused, or perforated

2.1.3.2 Weak Cell

This book defines a *weak cell* as a cell in a worse state than a new cell or its neighboring cells:

- Lower capacity;
- Higher series resistance;
- Higher self-discharge current;
- Lower coulombic efficiency.

Later, whenever I will mention a weak cell, I will specify which of these parameters I mean.

2.1.3.3 Not Li-Ion Cells

Here are a few lithium-using devices than should not be mistaken for Li-ion cells:

- *Primary cells* (not rechargeable):
  - Energizer's *Ultimate Lithium*™—a family of primary cells compatible with alkaline battery cells that boast impressive power density, especially at low temperatures;
  - Lithium cells—used in hearing aids and for memory backup in desktop computers (see Section 1.2.2.4);
- *Capacitors* (charge is stored electrostatically, not electrochemically);
  - Li-ion capacitors (LIC) (see Volume 2, Section A.3.4)—a high-voltage double-layer capacitor;
  - TDK *CeraCharge*™—presented as a "the world's first solid-state, SMT-compatible Li-ion battery" [4] ; it actually appears to be a double-layer capacitor that uses lithium.[5]
- Not quite Li-ion cells:

5. The capacitance of 0.24 F is heavily dependent on the voltage, with a maximum of 1.5V. Lithium is present in the ceramic electrolyte. A thin layer of lithium metal forms one of the plates of the capacitor; a copper layer forms the other plate. The capacitor is not yet formed when delivered. The user forms it by charging it for the first time.

- Lithium-metal/solid-state cells (see Volume 2, Section A.3.3)—very similar to Li-ion cells, but use lithium atoms, not ions, or use a solid electrolyte; includes lithium-sulfur (Li-S) cells

2.2 TYPES OF CELLS

Li-ion cells can be classified by chemistry and format (Figure 2.1).

All too often people compare Li-ion with LiPo or Li-ion with LiFePO4. This doesn't make sense because all of these cells are Li-ion (see Section 1.2.2.7).

2.2.1 Cell Chemistry

Successfully charging or discharging a Li-ion cell does not involve any chemical reaction.[6] (see Volume 2, Section A.3.2) This is partly why Li-ion cells have such longevity and a near 100% coulombic efficiency.[7]

The *Li-ion cell chemistry* (LFP, LCO, and others) refers to the chemical composition of the cell's positive electrode.[8]

As of this writing, approximately six Li-ion electrode chemistries are commercially available. Table 2.1 lists them in order of nominal voltage.[9] Some chemistries still in development, and some non-Li-ion rechargeable lithium chemistries are included.

In practice, only a few of these are readily available to the general public: LCO, NMC, and LFP; LTO is somewhat available.

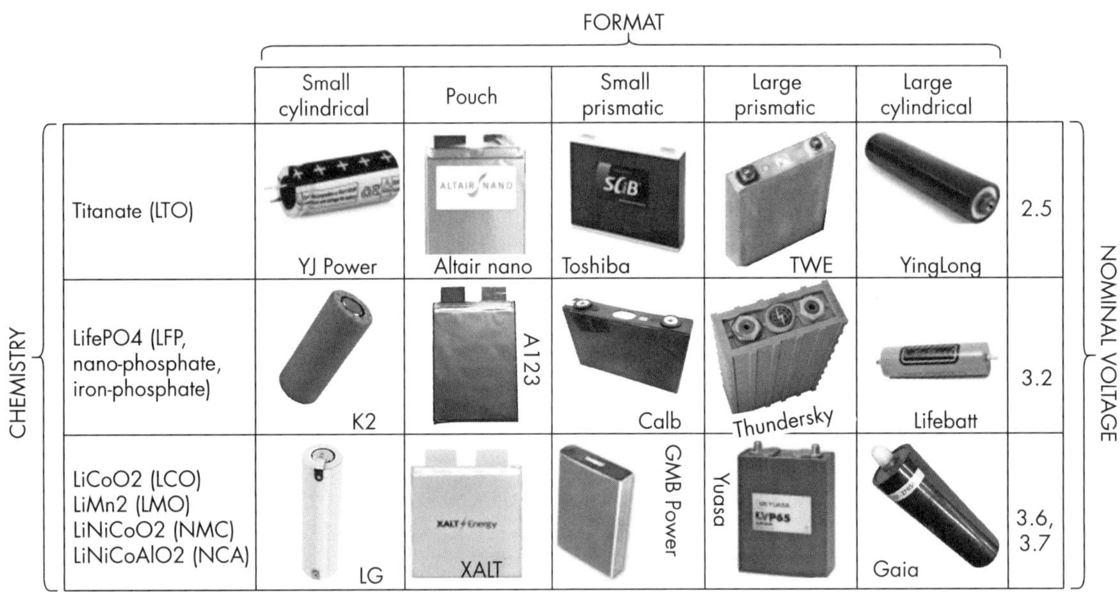

FIGURE 2.1 Types of Li-ion cells.

6. In the neophyte's narrow understanding that a chemical reaction changes one material into another.
7. On the contrary, some other types of cells do rely on chemical reactions, which are not 100 % reversible, leading to lower Coulombic efficiency, and faster loss of capacity.
8. Except for LTO, which is the chemistry of the negative electrode.
9. Slightly higher voltages are achieved with materials such as Silicone-graphene.

Chemistry	Code	Name	Voltage [V]	Energy Density	Power Density	Temperature	Safety	Life	Low Cost
Li-S		Lithium-sulfur*	1.7						
$Li_4Ti_5O_{12}$	**LTO**	Spinel lithium titanate (titanate)	2.3		✓	✓✓	✓✓	✓✓	
$LiFePO_4$ **$LiFeYPO_4$**	**LFP**	Lithium-iron-phosphate (nanophosphate) Lithium-iron-yttrium-phosphate	3.25		✓✓	✓	✓✓	✓✓	✓✓
$LiCoO_2$	**LCO**	Lithium cobalt oxide	3.6	✓✓		✓			✓✓
$LiNiCoAlO_2$	**NCA**	Lithium nickel cobalt aluminum oxide	3.6	✓✓	✓	✓		✓	✓
$LiNiO_2$	**LNO**	Lithium nickel oxide	3.6	✓✓					
$LiNiMnCoO_2$ **$LiMnCoO_2$**	**NMC**	Lithium nickel manganese cobalt oxide Lithium manganese cobalt oxide	3.65	✓✓	✓	✓	✓	✓	✓✓
$LiMn_2O_4$	**LMO**	Lithium manganese oxide	3.75	✓	✓		✓		✓✓
$LiCoPO_4$	**LCP**	Lithium cobalt phosphate †	4.7						

*Not Li-ion. †Not yet available. ✓ = good; ✓✓ = optimal.

TABLE 2.1 Cell Chemistries

Figure 2.2 is a radar chart that compares the benefits of six Li-ion cell chemistries [5, 6]. The chart compares:

- Safety, based on high self-ignition temperature;
- Maximum power time;
- Specific energy;
- Cycle life;
- Cost/energy.

Let's look in more detail at some currently available Li-ion chemistries (again, in order of voltage).

2.2.1.1 LTO

Titanate (*LTO*) cells are inherently safer and longer-lasting[10] than any other Li-ion chemistry. They have the best performance at low temperatures and the fastest charging time (i.e., 10 minutes to 80% SoC). However, they have a low energy density and are expensive.

Some high-performance applications (e.g., vehicles that require an extremely fast charge or continuous operation below freezing temperatures) may require LTO cells, despite their higher mass and cost.

10. LTO cells experience little stress due to swelling when charged, which improves the cycle life. Additionally, their low voltage reduces disassociation of the electrolyte.

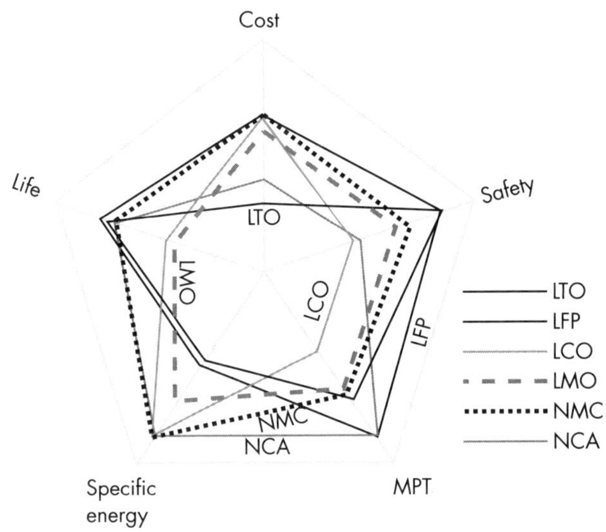

FIGURE 2.2 Comparison of Li-ion cell chemistries.

2.2.1.2 LFP

Iron phosphate (*LFP, LiFePO4*) cells are inherently safer since they have a higher self-ignition temperature. They last longer since the electrolyte doesn't break down as much at their lower voltage. They have higher power density (they are used in power tools) and flatter voltage than previously available chemistries. However, they suffer from lower energy density, mostly due to the lower nominal voltage.

Newer chemistries claim to match the performance of LFP without the energy density penalty.

LFP cells are used in sea-going vessels because safety at sea is essential. They are also used in 12V batteries because four cells in series closely match the voltage of a lead-acid battery.

2.2.1.3 LCO

LCO is the original Li-ion chemistry. It has a high energy density, and it is low cost. Other than that, it is not that great. These cells are used in consumer products.

2.2.1.4 NCA

Nickel cobalt aluminum (*NCA*) cells have an even higher energy density but are expensive and somewhat unsafe. Today they are used in the traction battery of Tesla cars.

2.2.1.5 LNO

Lithium nickel oxide (*LNO*) cells have higher energy density than LCO but are even less safe. This material is often blended with others to achieve a performance that is a compromise between the two. The Nissan Leaf traction battery uses an LNO/LMO blend.

2.2.1.6 NMC

Today *nickel manganese cobalt* (*NMC*) is the standard chemistry used for high-performance batteries, as it provides high energy density and decent power density, as well as an acceptable level of safety. It may be used in most applications.

2.2.1.7 LMO

Lithium manganese oxide (*LMO*) is a lower-cost chemistry. The spinel version has high power and slightly higher voltage. LMO cells have a relatively short cycle life. Auto manufacturers combine it with NMC to achieve a compromise of long cycle life and low internal resistance. These are used in power tools and medical products.

2.2.1.8 Chemistry Safety

Much has been said about the safety of LFP and LTO cells compared to other Li-ion chemistries. While it is true that those two chemistries have a higher self-ignition temperature, that is not the only criterion for safety. Other criteria include

- Internal resistance: a low resistance cell can release more energy more quickly, which can more readily overheat a conductor or a load and possibly ignite nearby materials.
- Internal protection: a safer cell includes a PTC resettable fuse or a separator that shuts off ion flow when it melts.
- Manufacturing quality: a cell rolled by hand in a factory with open windows is more likely to include impurities that may create a short circuit and ignite the cell.
- Response to abuse: an abused cell is dangerous no matter how safe its chemistry may be. A properly designed and installed BMS minimizes electrical sources of abuse, but some chemistries are particularly dangerous when abused; for example, NMC cells may self-ignite during a charge after an overdischarge.
- Arguably, an LCO cell from a reputable cell manufacturer is safer than an LFP cell from a shady one.

Therefore, using LFP or LTO cells is not, by itself, a guarantee of a safe battery.

2.2.2 Cell Formats

Li-ion cells are commonly available in five formats (see Table 2.2).

Some notes:

- Each cell format covers a range of capacity—from 0.1 to 1000 Ah—and maximum power time—from 10s (power) to 200s (energy) and higher (inadequate) (Figure 2.3).
- Cylindrical cells tend to offer lower energy density[11] because dead space remains when packed together.
- Cells with light cases (small prismatic) or no hard case (pouch) tend to offer the highest specific energy, though that is reduced once the required mechanical constraint is added to limit expansion. Cylindrical cells are inherently constrained and do not require any additional constraint.
- Large prismatic cells offer the highest ease of assembly and pack design, whereas pouch cells require the greatest design and manufacturing effort.
- Small cylindrical cells are produced in the millions, making them less expensive. They are also more mature and more reliable.

11. Volumetric.

	Small Cylindrical	**Large Cylindrical**	**Small Prismatic**	**Large Prismatic**	**Pouch**
Image					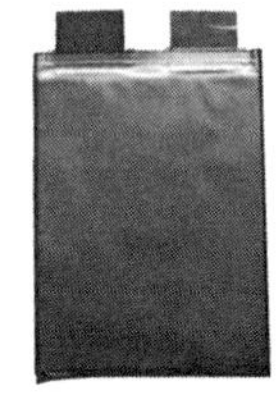
Size	Small	Medium-large	Small~medium	Large	Small~medium
Capacity*	1~3.0 Ah	8~40 Ah	1~20 Ah	20~1000 Ah	0.1~100 Ah
Case	Metal	Hard plastic	Light metal	Plastic	Soft Mylar
Cross section	Round†	Round	Rectangular	Rectangular	Flat
Terminals	None	Threaded studs	Tabs or studs	Threaded holes	Tabs
Expansion Retention	Inherent	Inherent	Good	Some: requires external retention	None: requires external retention
Typical Application	Personal EVs Laptop computers, other consumer products	(Not many)	Now starting to be used in stationary batteries and traction batteries	Hobby EVs Large stationary batteries Marine house	Cell phones Race EVs Production EVs Model airplanes Stationary batteries
Assembly	Hard	Easy	Hard	Quite easy	Quite hard

*As of this writing
†The cross section of Boston Power cells is oblong, but they are basically small cylindrical cells.

TABLE 2.2 Cell Formats, Characteristics, and Applications

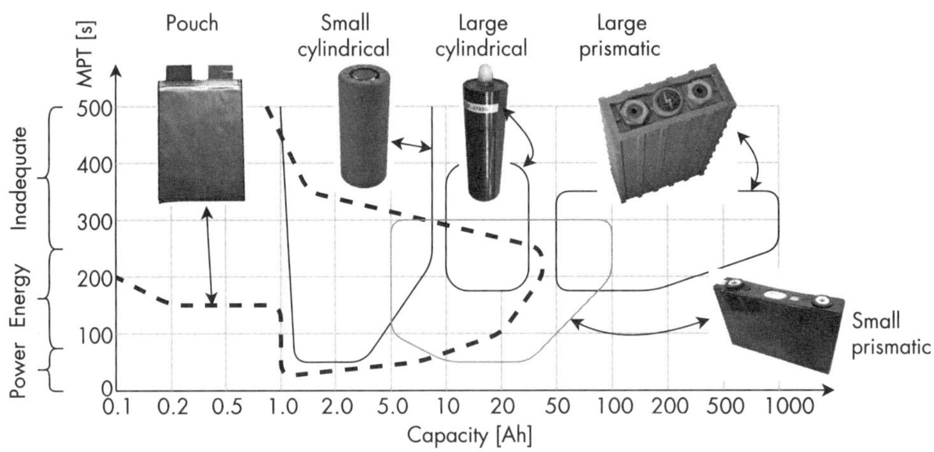

FIGURE 2.3 Presently available ranges of capacity and maximum power time for the five cell formats.

Some call a pouch cell "prismatic" instead. This confusion was worsened when, unfortunately, cell manufacturer A123 advertised its 20 Ah pouch cell as a "prismatic" cell. Here is the difference:

- The body of a pouch cell (including the 20 Ah cell from A123) is a thin, soft bag;

- The body of a prismatic cell is hard (or semihard) plastic or metal;
- The thickness of a prismatic cell is about 1/3 of its width, while pouch cells are far thinner.

Each cell format has unique requirements for its connection and mounting (see Section 5.2.3).

2.2.2.1 *Small Cylindrical*

Small cylindrical cells come in some 20 different sizes, of which the following are the most notable:

- CR123A: Tiny, replacement for nonrechargeable cell for consumer electronics;
- 14500: Slightly smaller than a 18650, used in e-cigarettes;
- 18650: By far the most common;
- 20700: Slightly larger;
- 21700: Even slightly larger, used by Tesla;
- 26650: Twice the volume, the same height as a 18650, typical of LFP cells;

The size code denotes the size (e.g., 18650 = 18 × 65.0 mm). The cell is inherently constrained against expansion (see Section 2.4.6).

The terminals of small cylindrical cells are relatively flat and do not include any fastening. As such, a cell may be connected by installing it in a holder. For a permanent connection, tabs are welded to the cell's terminals.

Usually (but not always!), the cap terminal is the positive terminal. The can and the flush terminal are the negative terminal. The vent is on the cap terminal. Care must be taken not to obstruct it when mounting or interconnecting the cell.

As of this writing, the capacity ranges up to about 3.0 Ah. Unscrupulous vendors falsely advertise a capacity that far exceeds this value.

Note the difference between a 18650 cell, which is not protected, and a 18650 battery, which includes a protector BMS (see Section 4.3.4).

2.2.2.2 *Large Cylindrical*

Compared to small cylindrical cells, *large cylindrical* cells are much larger and have higher capacity, but have a lower power density.[12] Their capacity can range up to 40 Ah. Most large cylindrical cells have two threaded stud terminals on either end.

Only a few companies manufacture large cylindrical cells, and few batteries use them.

2.2.2.3 *Large Prismatic*

Large prismatic cells are brick-shaped in a semirigid plastic case, which doesn't provide sufficient constraint against expansion, so additional constraints are necessary. The capacity ranges from 20 Ah to 1,000 Ah.

Most large prismatic cells have two terminals on the top surface, each with a large threaded hole, sized M8 through M16. There are exceptions:

- Each terminal in a cell made by GBS has four small holes with M4.5 threads;

12. It's harder to remove heat generated deep within the cell compared to a small cylindrical cell.

- Some 200 Ah GBS cells have one terminal at the top and one at the bottom of the cell;
- Some 700 Ah and larger cells have four redundant terminals along the top surface—two are positive and two are negative;
- Some 1,000 Ah and larger cells have six redundant terminals, side by side, clustered in a corner, which is angled—three are positive and three are negative.

A red circle and a + symbol mark the positive terminal. The round cap at the top of the cells is not just a pretty logo. It is a critical safety vent. The vent is not sealed and may release some gas during regular operation.

2.2.2.4 Small Prismatic

Small prismatic cells are shaped like a small box; most cells have two threaded stud terminals on the top surface and are enclosed in a thin metal case (Figure 2.4(a)). These cells are not as common as large prismatic ones but are becoming more popular.

As the case of a small prismatic is metal, we wish to know whether it is electrically floating. In some cells, the case is connected directly to the positive terminal.[13] Curiously, in some LFP cells,[14] the case is electrically referenced to the electrolyte (!): its voltage is about 0.6V below the positive terminal (Figure 2.4(b)). This would not be an issue because the case is coated with an insulating layer. However, the vent (between the two terminals) usually is not coated, and therefore it's electrically hot. An accidental connection between a conductor (a bus bar or ring terminal) and the vent may at best discharge the cell; at worst, it may result in an uncontrolled current.

Cell phone batteries may look like small prismatic cells but are in reality complete batteries that include a pouch cell.

2.2.2.5 Pouch

Pouch cells are in a soft bag and come in a wide variety of nonstandard sizes. Pouch cells require the highest battery design effort; in return, they can provide optimized performance for a given application.

Practically all pouch cells are rectangular and have two tabs exiting the bag at one end; rarely, power cells have one tab on one end and one on the opposite end. Although both tabs look the same due to their plating, they are different: the negative

(a) (b)

FIGURE 2.4
Small prismatic cell: (a) before installation, and (b) the metal case may be hot (courtesy Richard Gleeson).

13. CALB. (e.g., cells from CALB)
14. Vision and Champion. Thank you to Rushikesh Mahajan of Rocket Batteries for this discovery.

terminal tab is normally copper or nickel, while the positive terminal tab is aluminum, which not ideal for soldering.

Specialty cells are available with unusual shapes to accommodate the form factor of small products (Figure 2.5(a)).

Pouch cells are quite delicate and are easily susceptible to damage from nicks, local mechanical pressure, or stress on the terminals. They provide no constraint against swelling and thus require external constraint. There is no vent; in case of overpressure, the bag ruptures.

Pouch cells are infamous for being housed in bags that are not fully isolated. Consequently, once placed in contact with metal cases or metal cooling plates, they are slowly discharged; leaks in the seal may also result in gassing, corrosion, and short circuits.

2.2.3 Energy Versus Power Cells

Cells may be optimized for energy or power:

- *Energy cells* have higher capacity;
- *Power cells* have lower internal resistance, and therefore are more efficient.

I plotted the energy density and the maximum power time (see Section 1.5.2) of 500 different 18650 cells (Figure 2.6).[15] I arbitrarily divided them into these sets:

- Power: MPT of 110s or better;
- Energy: Energy density (ED) of 600 Wh/l or better, with an MPT between 240 and 500s;
- Both: ED/MPT ratio of 2.5 or better, but not included in previous sets;
- General-purpose: ED/MPT ratio of 1.0 or better, but not included in previous sets;
- Inadequate: Not in previous sets.

This plot indicates which cells would be best for a given class (see Section 5.1.4).

- Power: For a power (e.g., a power tool) or buffer battery (e.g., a hybrid car);
- Energy: For an energy battery (e.g., a standard EV);
- Both: For a dual-duty battery (e.g., a plug-in hybrid car).

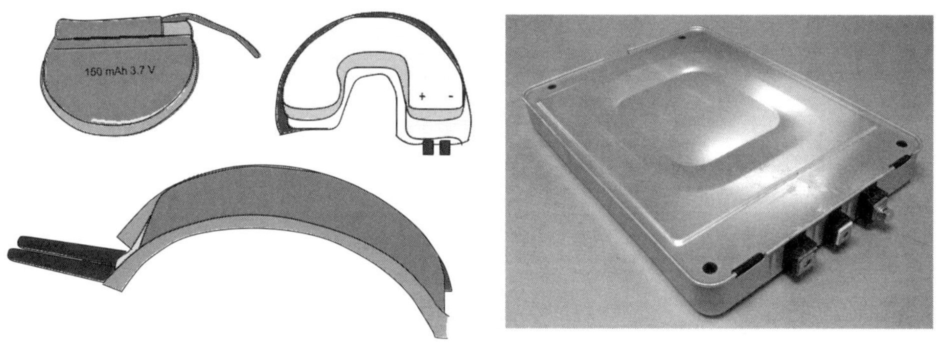

(a) (b)

FIGURE 2.5
Pouch cells: (a) examples of unusually shaped pouch cells, and (b) leaf module.

15. Data came from cell spec sheets, testing done in house, and Henrik K. Jensen's research published in lygte-info.dk.

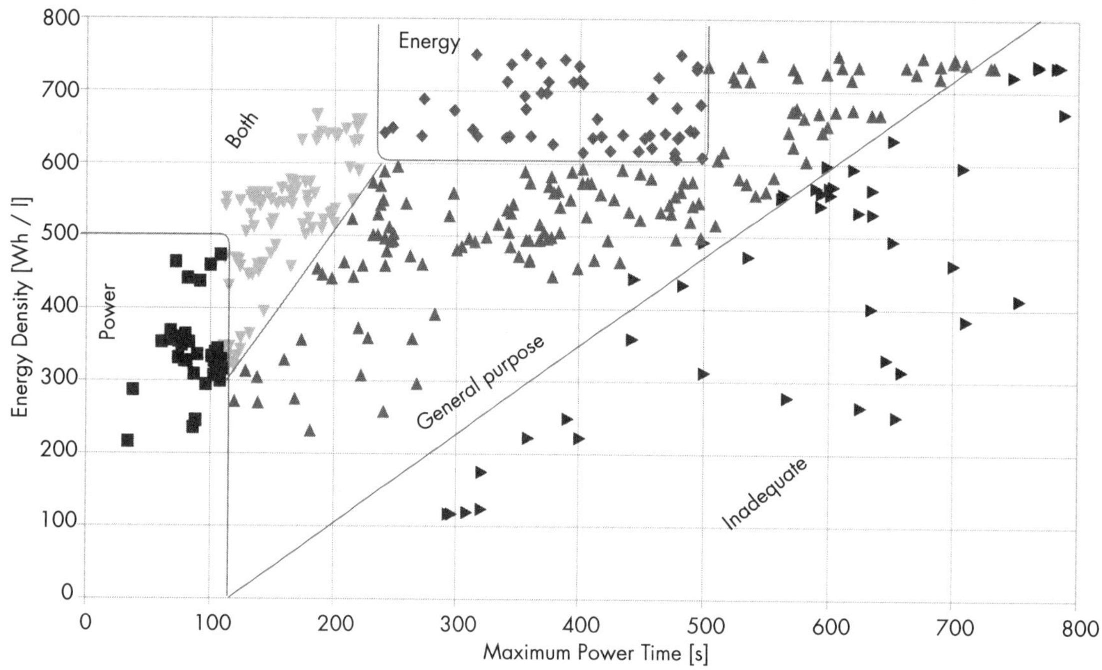

FIGURE 2.6 Classification of 500 18650 cells for power, energy applications.

The general-purpose cells would be OK for any nondemanding application. The inadequate cells include, for example, those that go by the generic name of *UltraFire*. Their poor performance contradicts the outlandish claims in their specs.[16] Cell formats other than 18650 may be classified in the same way.

Today, power cells are only available as small cylindrical and pouch formats. Despite misleading claims from vendors, power cells are not available in large prismatic or large cylindrical formats.

In a power application, a power cell may outlast an energy cell, despite its lower capacity (see Section 2.6.2.2).

2.2.4 Cell Modules

A cell manufacturer may permanently mate a set of Li-ion cells into an indivisible module. Although electrically it is a set of cells, from a mechanical standpoint, it is a single unit and must be used as such.

Electric vehicles often use such modules (see Section 2.8.5). The Nissan Leaf module consists of four pouch cells, in a 2P2S arrangement (Figure 2.5(b)); note that the module comes in two opposite polarizations for ease of battery design. The Chevy Volt uses blocks of three pouch cells in parallel, which are then placed into modules of 6S or 12S.

2.3 CELL CHARACTERIZATION

Note that this chapter provides graphs and limits typical of Li-ion cells in general. Since your cells differ from mine, please don't assume that these graphs apply specifically to your cells.

16. You can buy UltraFire shrink-wrap sleeves online, and convert any 18650 cell to an UltraFire cell!

2.3.1 Perspectives for Characterization

The following sections characterize the Li-ion cell from six perspectives (Table 2.3), each covering specific parameters (see Section 1.4).

2.3.2 Safe Operating Area

For safety and longevity, a Li-ion cell must be kept within its *safe operating area*[17] (*SOA*), delimited by the ranges of these parameters:

- Voltage (OCV and terminal voltage);
- Temperature (charging versus discharging);
- Current (directly, and its effect on temperature).

For example, the SOA of a typical LCO cell may be shown on a three-axis plot (Figure 2.7). The cell must be operated inside this funny-looking envelope.

- VH: Maximum voltage (4.2V);
- VL: Minimum voltage (2.9V);

Table 2.3
Cell Characterization Perspectives

Perspective	Description	Applies to	Parameters
Typical characteristics	What to look for in Li-ion cells	Li-ion cells in general	All
Cell model specifications	Nominal characteristic specified by the cell manufacturer	A specific cell model (e.g., IMR18650PE)	Capacity Min and max voltage Min and max temperature Nominal and peak currents Discharge curves Cycle life
Cell family characteristics	Actually measured characteristics for any cell that uses the same technology, regardless of size	All cells in that family (e.g., a 18650 and a 26650, both EV type)	Energy density, specific energy OCV vs. SoC MPT Cycle and calendar life
Operational characteristics	Operating range when used in a specific application	A cell model or its family	SoC and voltage range Effective capacity, energy Temperature range Current range Cycle and calendar life
Individual cell state	Temporary and reversible characteristics at a given moment in time during use	A particular, individual cell	Voltage and SoC Temperature Instantaneous resistance
Individual cell conditions	Long-term and irreversible characteristics of the cell's or battery's health	A particular, individual cell	Actual capacity Actual MPT Actual self-discharge current State of health

17. The word "area" implies two dimensions, but there can be more. In this case, there are three dimensions, so technically we should call it safe operating volume.

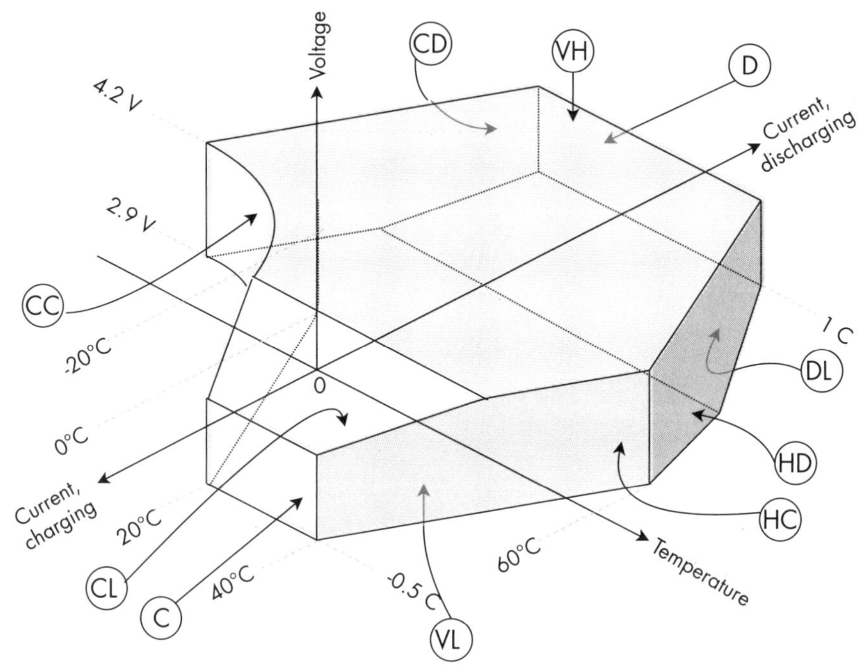

FIGURE 2.7
Example of SOA for an LCO cell.

- C: Charging current limit (0.5C);
- D: Discharging current limit (1C);
- CL: Reduced charging current limit at high cell voltages;
- DL: Reduced discharging current limit at low cell voltages;
- HC: Reduced charging current limit at hot temperatures (up to 60°C);
- HD: Reduced discharging current limit at hot temperatures;
- CC: Reduced charging current limit at cold temperatures (limited charging below 0°C);
- CD: Reduced discharging current limit at cold temperatures (down to -20°C).

Actually, the SOA envelope is not as well-defined as this graph implies—it's not true that inside this envelope the cell is perfectly happy, and that the moment the cell steps outside this shape it is dead. In other words, it's not like a fuse, which is fine until the current exceeds its rating, at which point the fuse blows. Instead, the walls of the envelope are fuzzy, both in space and in time. By this I mean that each wall is actually a continuum, and the cell is degraded more and more as it approaches this wall, and more and more the farther outside this wall it goes and the longer it stays there. For example, the highest temperature may be 60°C. This does not mean that the cell can be stored at 59°C indefinitely with no ill effect and that it melts at 61°C (Figure 2.8(a)). Instead, the cell is slowly degraded while at a high temperature, even at 30°C, a bit more at 40°C, more at 60°C, and a lot at 80°C. The cell degrades just a bit if it's kept at high temperature for 1 minute, a bit more if it's kept there for 1 hour, and a lot if it's kept there for 1 month (Figure 2.8(b)).

Some walls are fuzzier than others. For example, the minimum cell voltage is well-defined because damage occurs rapidly when crossed. On the other hand, the maximum temperature is fuzzier because operating at high temperature will degrade the cells slowly.

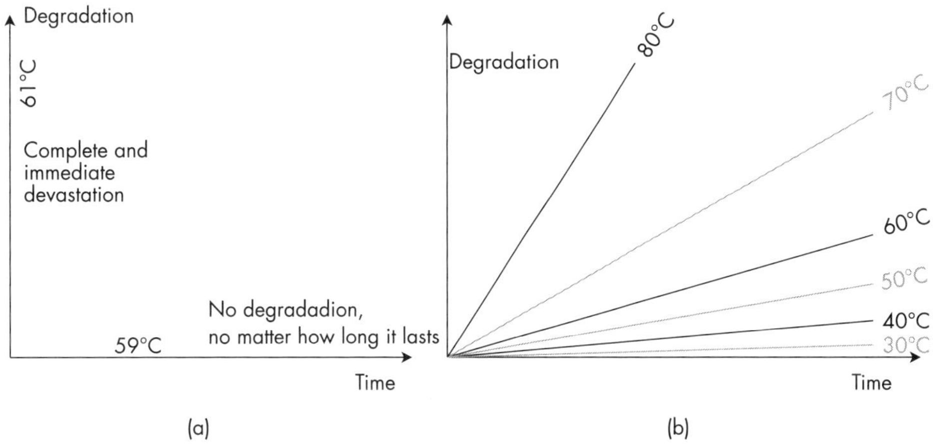

FIGURE 2.8
SOA walls: (a) we'd like to think of it as a well-defined wall, but (b) in reality, it's a continuum.

To maximize the cell's life and performance, a sophisticated BMS may vary the SOA envelope as SoC changes and as the cell ages. For example, a vehicle BMS may use a more conservative envelope initially to maximize battery life; later, the BMS may expand the envelope and go for broke to squeeze more performance from the tired cells at a time when there is little to lose. For another example, a BMS in a power battery may allow more discharge current at high SoC when the cell's internal resistance is low, and then reduce it as the cell gets discharged.

2.3.3 Abuse

If abused, a Li-ion cell may initiate a thermal runaway and self-destruct violently (see Section 8.2.1.5). This abuse could be electrical, mechanical, or thermal. We already discussed electrical abuse.

Mechanical abuse includes

- Localized mechanical pressure (e.g., a thermistor pressed against the side of a pouch cell);
- Deformation (e.g., folding a pouch cell);
- Penetration (e.g., a nail in a cell, in case of an accident);
- Excessive vibration (e.g., the *jelly roll* inside a cylindrical cell may slide and contact the case [7]).

Thermal abuse includes

- Exposing the cell to high temperatures from external sources, from self-generated heat during use, or due to an internal short;
- Charging below 0°C.

2.3.4 Equivalent Model

When considering the behavior of a cell, it helps to visualize it by its equivalent electrical circuit.

The simplest *model* has only a voltage source (the OCV), whose voltage depends on SoC (Figure 2.9(a)). This model is only helpful in low current applications in which the cell voltage does not sag under load.

From a purely DC standpoint, a cell behaves like two components in series (Figure 2.9(b)):

FIGURE 2.9
Cell equivalent model:
(a) plain voltage source,
(b) plus DC resistance, and
(c) plus AC impedance.

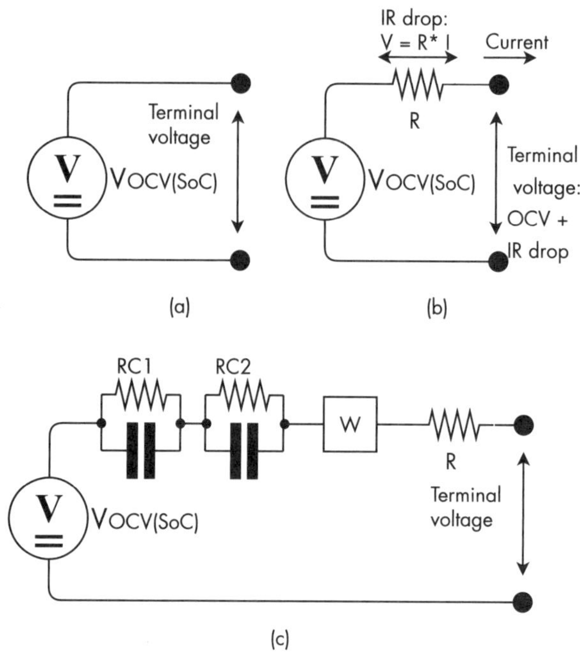

- V: A voltage source (the OCV), whose voltage depends on SoC;
- R: A resistor (the internal resistance), whose resistance varies with the state of the cell.

The DC model is sufficient for most of our discussion because it explains how the cell's terminal voltage changes with SoC and with continuous load current and direction.

However, the DC model does not account for dynamic effects such as the relaxation of the voltage. To account for them, we use a more complex model, one of which uses four impedance elements in series (Figure 2.9(c)) (for a detailed discussion on these impedance elements see Volume 2, Section A.3.5). For our purposes, we can assume that these impedance elements are sufficient to emulate the behavior of a real cell as the value of the current varies.[18]

2.3.5 Cell Life

The life of a Li-ion cell is limited in two ways:

1. *Cycle life*: Degradation for every charge and discharge cycle;
2. *Calendar life*: Constant degradation over months, whether in use or not.

A cell in storage experiences only calendar life. A cell in actual use is degraded by both calendar life and cycle life effects. Indeed, a cell may degrade more while at rest (calendar life) than while cycled (cycle life). Specifically, the battery in a laptop computer degrades over a few years even though it may be cycled only 30 times in its life because it sits at full voltage when the laptop is plugged in.

Degradation affects four parameters. Most people only consider the first, rarely do they consider the second one, and hardly ever the last two [8–10].

18. One cause of these effects is that the lithium ions take some time to diffuse and to cross the separator.

- Decrease in capacity;
- Increase in internal series resistance;
- Increase in self-discharge current;
- Decrease in coulombic efficiency.

Each of these parameters is discussed in more detail in the rest of this chapter.

2.3.5.1 *Cycle Life*

To a great extent,[19] when a cell is cycled (Figure 2.10(a)):

- The coulombic efficiency remains constant at just below 100% and is not affected until the end of life, at which point it drops,[20]
- To a certain extent, as a consequence of this, the capacity drops slowly and linearly, until the end of life, when it starts dropping rapidly;
- The resistance increases with use—typically it changes faster than the capacity does, and the resistance increases rapidly at the end of life;
- The increase in self-discharge current is tied to fast charging or cold charging—cycle count does not affect self-discharge current (unless cycling stresses the cell).

2.3.5.2 *Calendar Life*

Generally speaking, when a cell is stored at room temperature and 50% SoC (Figure 2.10(b)):

- Coulombic efficiency and self-discharge current do not degrade appreciably with age;
- Capacity fades slowly (see Section 2.5.3);
- Resistance increases at a constant rate.

These processes are accelerated at high temperatures or high SoC levels.

2.4 VOLTAGE AND STATE OF CHARGE

The voltage and SoC of a cell are closely related, so they might as well be considered together.

2.4.1 Voltage Ranges

The safe operating voltage range of a cell depends on its chemistry (Figure 2.11).

For most Li-ion cells, if the voltage drops below a specific level, damage starts occurring immediately. In the worst case, if the cell voltage is reversed, there is immediate and catastrophic damage to the cell.

Similarly, all Li-ion cells slowly degrade at higher voltages. In the worst case, if the cell voltage goes well above its rated maximum voltage, especially while being charged at high currents, the cell will go into thermal runaway, with potentially catastrophic results for itself and everything around it (see Section 8.2.1.5).

---

19. Various studies, based on different cells, report a range of results.
20. In the end of life region, coulombic efficiency (and therefore capacity) drops because even normal charging results in lithium plating rather than storage.

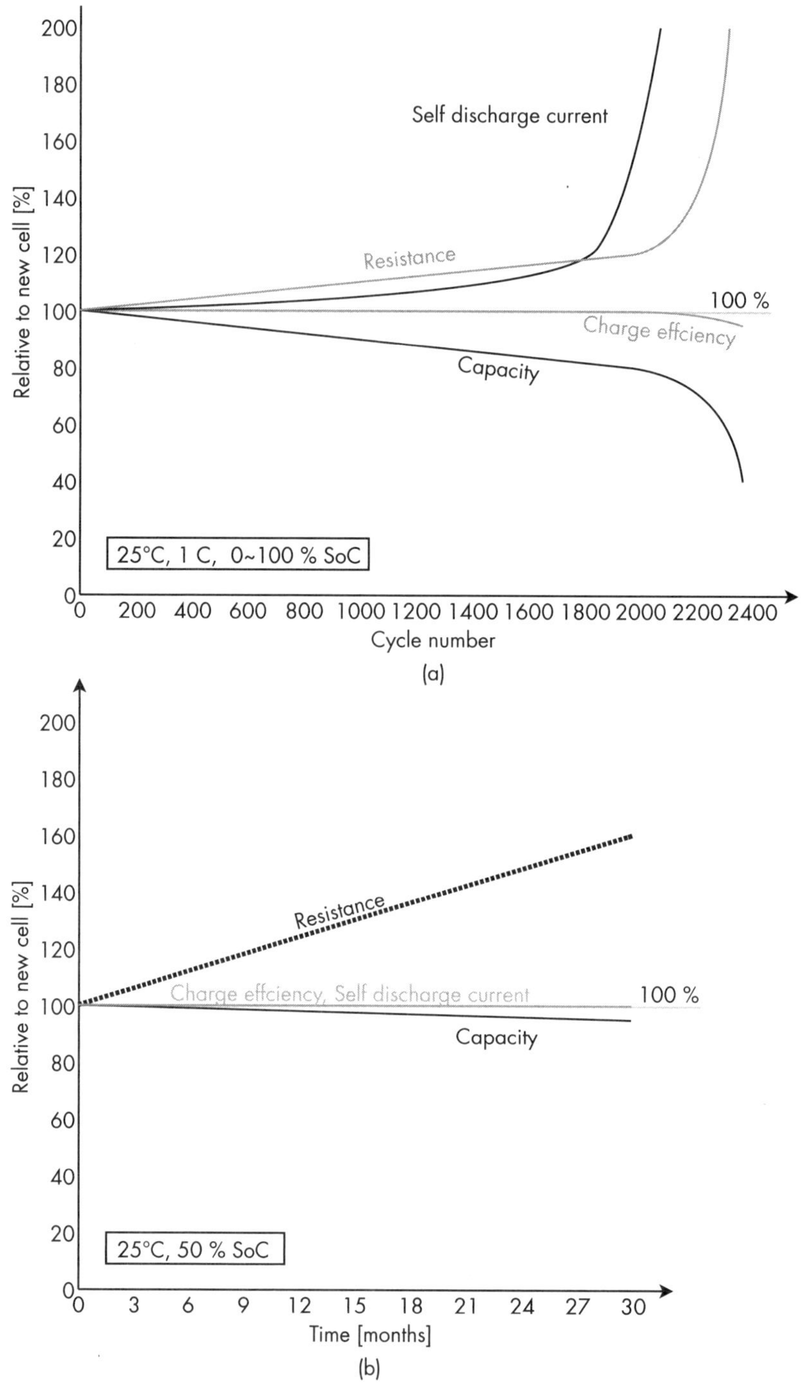

FIGURE 2.10
Cell life: (a) cycle life, and (b) calendar life.

The specifications sheet for a cell specifies its safe voltage range. The application may operate the cell in a narrower voltage range than specified by the cell manufacturer to prolong its life (see Section 2.5.3.2).

2.4.1.1 *Battery on DC bus*

In some energy and power applications, the charger is not turned off. Instead, the battery is on a DC bus, a power supply powers the bus and charges the battery, and loads are powered by the DC bus. The resting voltage of the cells is crucial because,

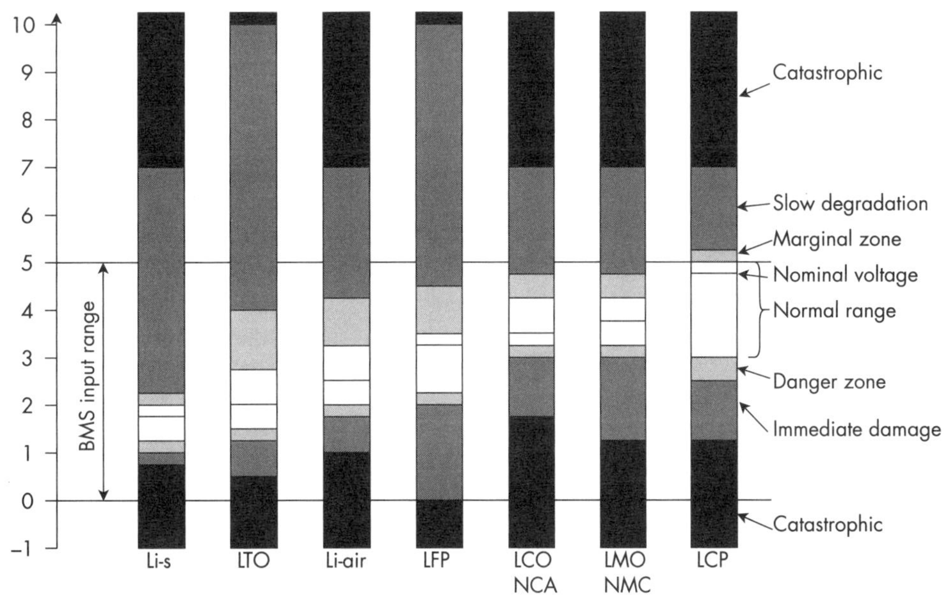

FIGURE 2.11 Approximate voltage ranges for various Li-ion and other lithium-based chemistries.

if a charger keeps the voltage high (even if it is within the cell specs), the cells will degrade rapidly.

Reducing the bus voltage minimizes this effect. Consequently, the battery is never fully charged. This is a disadvantage, but the longer life of the cells is a more significant advantage.

Three-state charging (see Section 5.15.3.2) could be used to reduce charging time (e.g., after AC power returns). The higher voltage in the *absorption stage* shortens the CV charging phase. However, its duration must be limited, and it must be followed by a *float stage* with reduced voltage (e.g., 3.4V/cell for LFP batteries).

2.4.2 Terminal Voltage and Open-Circuit Voltage

When dealing with cells, we refer to two different voltages:

- *Open-circuit voltage;*
- *Terminal voltage.*[21]

While both are measured in volts, they are different parameters (Figure 2.12).

The terminal voltage is the voltage that we can measure at the cell's terminals at any time. The OCV is the internal voltage of the cell, to which we don't have direct access. The only time it can be evaluated is after the cell has been resting for some time at zero current, when the terminal voltage is nearly equal to the OCV. Otherwise, the terminal voltage is different from the OCV due to the voltage across the cell's impedance.

The OCV of a Li-ion cell is affected exclusively[22] by its SoC.

The terminal voltage is affected by the SoC, the load current, and the previous history:

- With steady discharge current, the terminal voltage is below the OCV;

21. Sometimes called closed-circuit voltage (CCV), though inaccurately because the circuit may be open (the cell is unloaded), yet the terminal voltage is still relaxing, and therefore, it is different from the OCV.
22. Actually, temperature and aging do have a slight effect. (See Sections 2.4.2.4, and 2.4.4.5.)

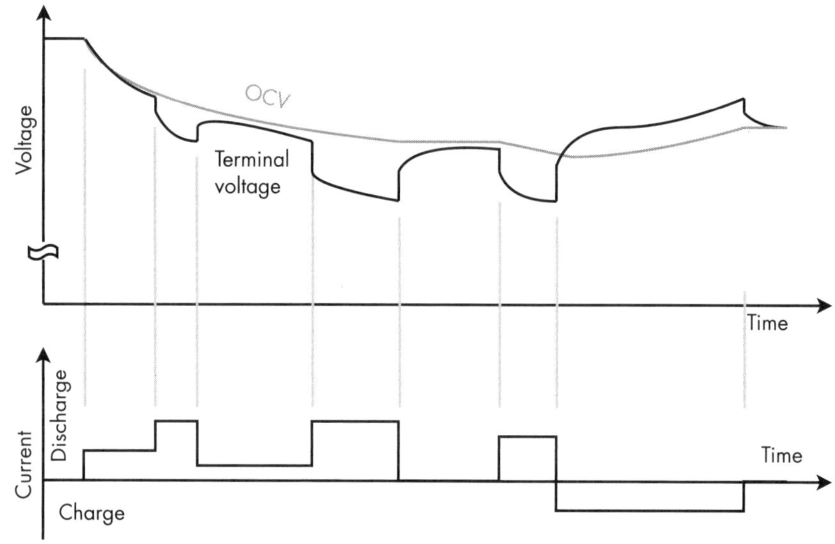

FIGURE 2.12
Plot of OCV and terminal voltage of a cell.

- With a steady charge current, the terminal voltage is above the OCV;
- With no current, the terminal voltage slowly approaches the OCV, but never quite reaches it (see Section 2.4.2.3).

2.4.2.1 *IR Drop, Voltage Sag*[23]

At constant current, the terminal voltage differs from the OCV by the voltage that electrical engineers call the *IR drop*.[24] During discharge, the voltage is decreased (it sags) and, during charge, it increases. The IR drop is proportional to current and resistance:

$$V_{IR\text{-}drop}\,[V] = -\,\text{current}\,[A] \times \text{cell\_internal\_resistance}\,[\Omega] \tag{2.1}$$

The minus sign signifies that when the current is positive (that is, the cell is discharging) this voltage is negative, which *reduces* the terminal voltage relative to the OCV. Conversely, during charging, the current is negative, which makes this voltage *positive*, which in turn *increases* the terminal voltage relative to the OCV.

Again, this is for a *continuous* current. We will discuss a variable current next.

2.4.2.2 *Relaxation*

For a step increase in discharge current, the drop in terminal voltage is not sudden, as predicted by the simple DC equivalent model. Instead, there is an initial, sudden drop, followed by a slow drop, as explained by the complex AC equivalent model (Figure 2.13(a)). While the current is constant, the terminal voltage relaxes towards a final value. Specifically, while the current is zero, the terminal voltage *relaxes* towards the OCV.

There are two different definitions for IR drop, both shown in detail in Figure 2.13(a):

23. Researches may call this polarization. (See Volume 2, Section A.3.5).
24. Electrical engineers use I for current (from intensity), and R for resistance; IR means I × R, current times resistance, which is voltage; that is, the voltage drop across a resistor when a current flows through it.

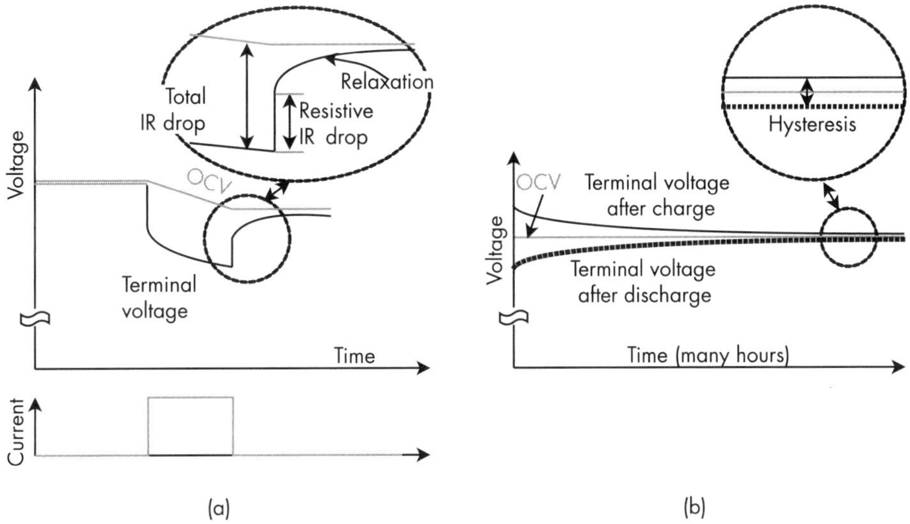

FIGURE 2.13
OCV and terminal voltage: (a) IR drop and relaxation, and (b) hysteresis.

- *Total IR drop*: The entire change in voltage, including relaxation (which considers all the resistive elements in the cell impedance);
- *Resistive IR drop*: The height of the initial step when the current changes by a sudden step; this considers just one element in the cell impedance—the stand-alone resistor—and ignores all the resistors in the RC circuits (see Section 2.3.4).

In this book, we use the first definition: total IR drop.

2.4.2.3 *Hysteresis*

The terminal voltage never reaches the OCV, even after a long relaxation time at zero current (Figure 2.13(b)):

- When charging ends, the terminal voltage slowly drops toward the OCV, but never quite reaches it—it stops right above it;
- When discharging ends, the terminal voltage slowly climbs toward the OCV, but never quite reaches it—it stops right below it.

This effect is due to the poor kinetics of the ions — the lithium ions bunch up near an electrode when being added to that electrode, or become sparse near an electrode when being taken from that electrode.[25]

An analogy may help visualize this effect: consider an empty 55-gallon oil barrel, with a hole in the top cover near the circumference, and a tap near the bottom of the sidewall.

- If you pour water in the barrel, it diffuses evenly (because of the good kinetics of water) (Figure 2.14(a)). That means that the water level is the same throughout; if you look down the hole, the water level you measure there is the true water level.

25. This explanation is not entirely technically correct, but close enough to envision the effect. Hysteresis is also due to the activation overpotential.

FIGURE 2.14
Barrel analogy: (a) water diffuses evenly, (b) sand bunches-up under entry hole, and (c) sand dips above exit tap.

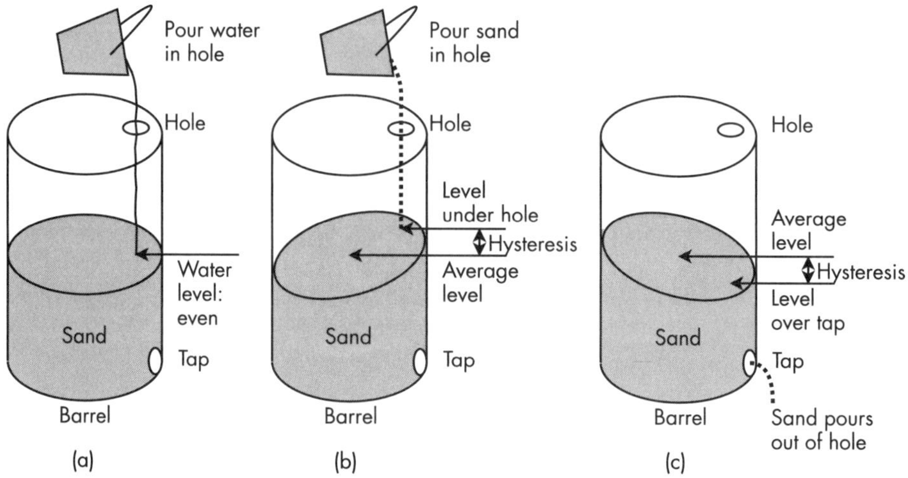

- If you pour sand in the barrel through the top hole, the sand builds up unevenly (because of poor kinetics of sand): it bunches up right under the hole (it does not diffuse equally throughout the barrel, the way water would) (Figure 2.14(b)). That means that if you look down the hole and measure the sand level, it is higher than the average level (when considering the entire barrel).
- If you let sand pour out of the barrel through the bottom tap, the sand sinks unevenly: it sinks deeper right under the hole (Figure 2.14(c)). That means that if you look down the hole and measure the sand level, it is lower than the average level.

The difference between the sand level at the point of entry or exit relative to the average level is analogous to the difference between the terminal voltage and the OCV. The inability of sand to diffuse evenly in the barrel is analogous to the inability of lithium ions to diffuse evenly in a cell terminal.

The difference between the final terminal voltage after charging and the final terminal voltage after discharging is the *hysteresis* in the terminal voltage. This hysteresis is quite small but can be significant when we wish to use the cell voltage to evaluate the SoC in regions where the OCV versus SoC curve is nearly flat. Between 50% and 70% SoC, the OCV of an LFP cell changes by 10 mV, yet the hysteresis may be as high as 10 mV. Therefore, such a level of hysteresis may result in a 20% error in SoC evaluation!

Hysteresis depends on

- Cell chemistry—less hysteresis with technologies with high ion mobility[26];
- Temperature—less hysteresis at high temperatures,[27]
- Current—less hysteresis if the current had been high.

A researcher can remove this effect by quickly charging and discharging a cell at high current and measuring the voltages. The average of the two measurements is close to the OCV. However, a real-world application has no way of doing this because

26. I wish I could give you a sense of hysteresis for the various types of cells, but researchers are still working this out.
27. Ion mobility increases at high temperatures, therefore better diffusion.

the battery is in active use. Attempting this trick would probably interfere with the system.

2.4.2.4 Temperature Coefficient

The OCV of a Li-ion cell is barely affected by temperature.[28] Yes, the *terminal* voltage under load is strongly affected due to variation on the internal resistance, but not the OCV.

There *may be* a small increase in OCV as the temperature increases, of on the order of 100 μV/°C, plus a slope at SoC levels below 40% of on the order of -20~-50 μV/°C/% [11]. It's so small that it's really hard to quantify. It is never specified.

2.4.3 Cell SoC

Cell SoC (see Section 1.6.2) is one of the states of a cell (see Table 2.3). The range of cell SoC is not precisely defined, as it depends on who defines it (Figure 2.15(a)):

- A scientist may define it as the physical SoC (see Section 1.6.2.1).
- For the sake of the cell, its manufacturer specifies a narrower range of SoC, small enough to avoid operation in the damaging extremes, yet large enough to achieve a respectable value for capacity. A manufacturer promoting high capacity may specify a wider range than a manufacturer concerned with cell longevity.
- Because it results in higher effective capacity, the cell vendor may specify a wider range for marketing reasons.
- The user may shrink the range to extend life, or expand it to extend performance; the range over which the cell is actually used is the operating SoC (see Section 1.6.2.2).

Psychology and marketing have more to do with SoC range than physics and engineering.

2.4.4 Voltage vs. SoC

The voltage of a cell is related to its state of charge. We represent this with various curves:

- *OCV versus SoC curve*: the cell's OCV is mostly determined by its SoC;
- *Discharge curves*: while discharging, the terminal voltage is lower due to the voltage drop through the cell's resistance;
- *CCCV charge curves*: while charged by a CCCV charger, the terminal voltage is higher than the OCV, then drops toward the OCV as the current decays during the CV phase;
- *CC charge curves*: when charged by the application, there may not be a CV phase, and charging stops earlier;
- *Charging and discharging curves*: these curves show the discharge curves and the CC charge curves in a single set of curves.

The spec sheet for a cell is likely to include only the discharge curves and CCCV charge curves. However, these may be of limited use due to low resolution and possibly

28. Unlike some other battery technologies. For example, the voltage of a lead-acid cell has a thermal coefficient of -3 mV/°C.

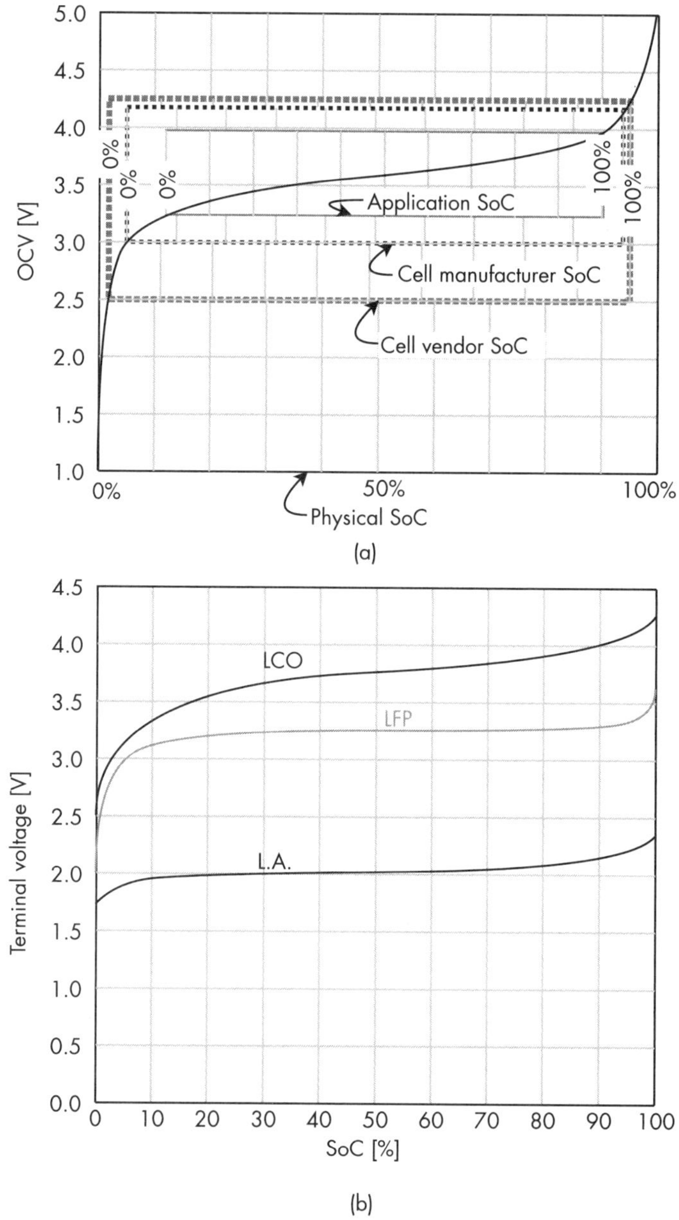

FIGURE 2.15
SoC: (a) cell OCV versus SoC for various cells, and (b) different organizations define the SoC range differently.

because they may be unrelated to the actual cell if, for example, the manufacturer changed the cell design but did not update the spec sheet.

Your BMS may require you to enter the OCV versus SoC (or SoC versus OCV) as a table. If you are a valuable customer, a reputable cell manufacturer may provide you with an accurate table of OCV versus SoC (possibly with a resolution of 0.1 mV for OCV, and 0.002% for SoC). Otherwise, you may hire a lab to create such a table (see Section 2.9.1). This is something you may end up doing, regardless, as a double-check. A more sophisticated BMS may be able to derive this table on its own, through many full cycles.

2.4.4.1 OCV Versus SoC Curve

The *OCV versus SoC curve* of a cell (Figure 2.15(b)) indicates how its OCV is related to its SoC. This curve depends on the cell chemistry.

This curve may be used to evaluate a cell's SoC roughly given its OCV (see Section 4.8.1.2). Note how the curve gets steeper at the empty and full ends. This means that converting cell voltage to SoC is more reliable when the cell is full or empty.

2.4.4.2 *Discharge Curves*

The *discharge curves* give a more detailed picture of how a cell will behave in an application. These graphs plot the cell voltage as the cell is discharged at various levels of specific current[29] at room temperature.

The cell voltage drops more deeply at higher current. Toward the end of charge, this higher drop in voltage means that the cell reaches the lowest voltage limit sooner, at which point discharge must be stopped, even though charge remains in the cell. Therefore, discharge at high current reduces the effective capacity.

This effect is more significant in an energy cell than in a power cell.

See how, at a given current, the voltage drops more for an energy cell (Figure 2.16(a)) than for a power cell (Figure 2.16(b)) (both are 18650 small cylindrical cells from the same manufacturer).

Note also how a power cell has a lower actual capacity, but at high current, the effective capacity is similar since an energy cell stops discharging sooner.

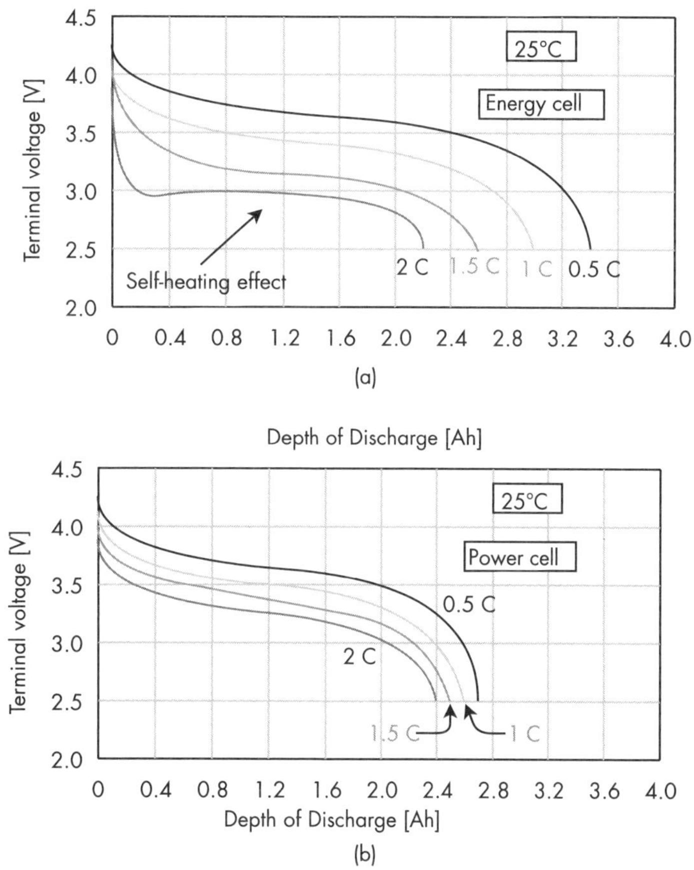

FIGURE 2.16 Discharge curves at various levels of specific current: (a) energy cell, and (b) power cell.

29. Yet battery applications are more likely to discharge at a given power, not at a given current.

Finally, in the energy cell at the highest current, note how self-heating decreases the internal resistance after a while, therefore reducing the voltage drop. It indicates that the curves were taken during a continuous discharge cycle. Therefore, these curves don't tell us what the terminal voltage would be during a single high-power pulse because the cell would still be cold.

The internal resistance of a cell increases at cold temperatures, causing the voltage drop at a given current to increase when the cell is cold. See how an energy cell (Figure 2.17(a)) is more affected by temperature than a power cell (Figure 2.17(b)). For an energy cell at the coldest temperature, we see that self-heating decreases the internal resistance after a while, reducing the voltage drop (Figure 2.17(a)).

Temperatures above 30°C are not shown because those curves would overlap the curve for 30°C.

2.4.4.3 CCCV Charging Curves

Charge curves show the typical charging process. Charge curves are taken using a CCCV charger, in two distinct phases:

- Constant current (CC, solid line)—until the cell voltage reaches the CV setting for the charger (various levels of specific current may be shown);
- Constant voltage (CV, dotted line)—as the current decreases asymptotically toward zero (charging is stopped after a while).

The X-axis may show either time (Figure 2.18(a)) or charge (related to SoC) (Figure 2.18(b)).

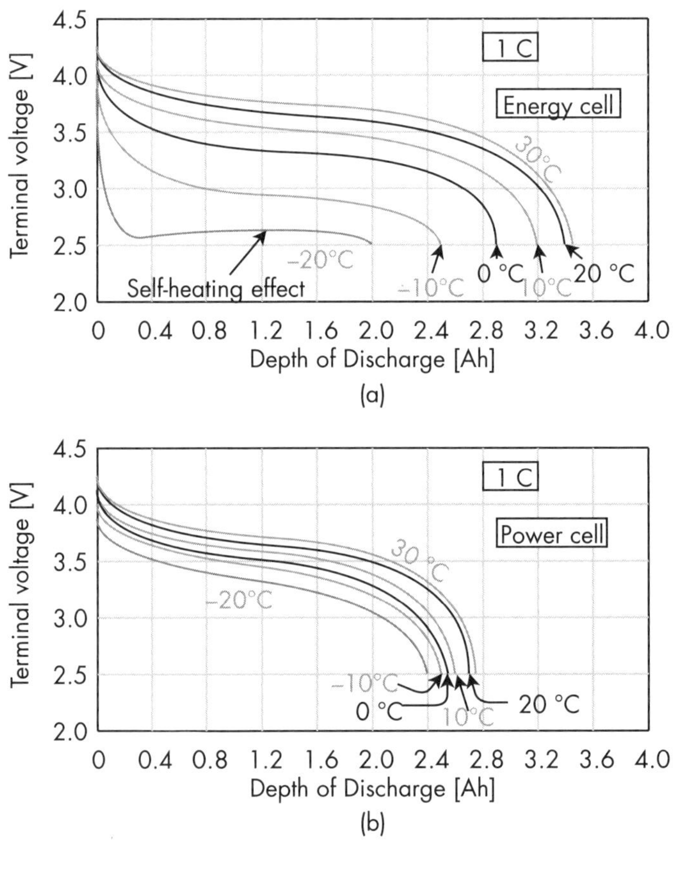

FIGURE 2.17
Discharge curves at various temperatures: (a) energy cell, and (b) power cell.

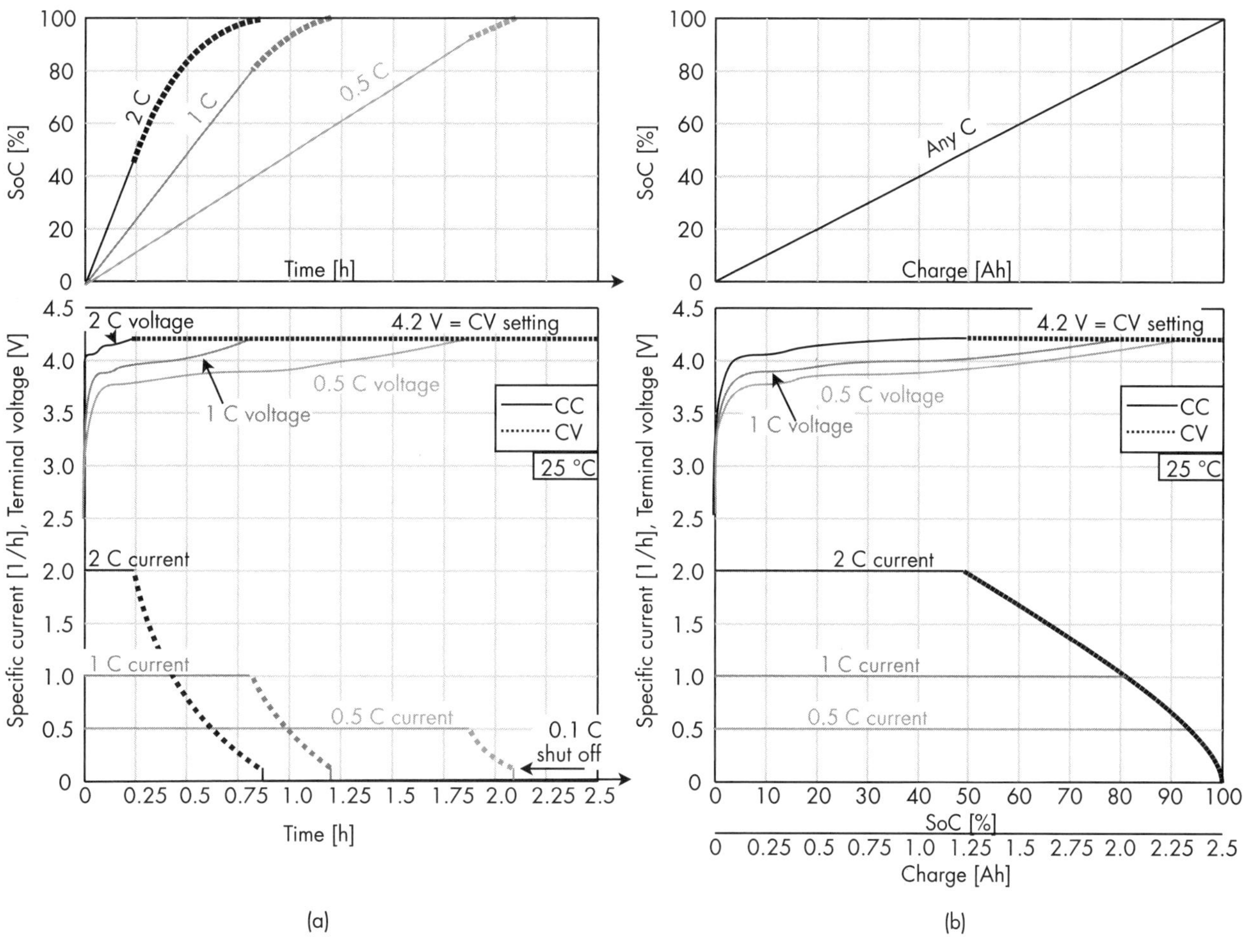

FIGURE 2.18 CCCV charging curves at various levels of specific current: (a) versus time, and (b) versus SoC (or charge).

These curves are for an 18650 NMC cell with a capacity of 2.5 Ah and an internal resistance of 80 mΩ. The charger has a CV setting of 4.2V and a shut-off current of 0.25 A (corresponding to 0.1C). We note that

- *At high current (2C):* Little time is spent in the CC phase because the cell voltage reaches the CV setting quite fast, with the SoC at only 50%. Then a long time is spent in the CV phase, as the current drops towards zero and as the SoC reaches 100%. The total charging time is 0.8 hours.
- *At medium current (1C):* More time is spent in the CC phase. At 85% SoC, the cell voltage reaches the CV setting, so the CV phase starts. Total charging time is 1.2 hours.
- *At low current (0.5C):* Most of the time is spent in the CC phase until the SoC reaches 93%. It is followed by a short CV phase to finish charging the cell. Total charging time is 2.1 hours.

Fast charging time depends significantly on the cell MPT—when charging a cell with low MPT, the charger remains in the CC phase much longer and spend little time in the CV phase. For example, the cell in the previous example (with a resistance of 80 mΩ) charged in 0.8 hours at 2C. However, if it had a resistance of 10 mΩ, it would have charged in 0.5 hours at 2C, 1 hour at 1C, or 2 hours at 2C.

A plot of total charging time versus MPT for various CC settings shows that power cells can be charged in close to the theoretical minimum charging time thanks to their low value of MPT:

$$\text{Minimum\_charging\_time [h]} = 1/\text{Specific\_current [1/h]} \tag{2.2}$$

Conversely, energy cells (with a high MPT) take almost 1 hour to charge no matter how high the CC setting is (Figure 2.19(a)).

With a charger's CC setting of 1.4 C, for a battery with an MPT of 400s, half the time is spent in the CC phase and half in the CV phase (Figure 2.19(b)). Of course, the total charging time is lower at high current, but it is not directly proportional to the current because at high current the CC phase lasts a smaller portion of the charging time. The result is that charging takes a certain amount of time no matter how high the current (e.g., 49 minutes for a cell with an MPT of 400s).

Why damage the cell by charging it at high current if the total charge time is nearly the same? If charging fully, then fast charging is not worth the reduction in the cell lifetime. However, if charging only part way (such as for power tools or public transit vehicles), then it may be worth it.

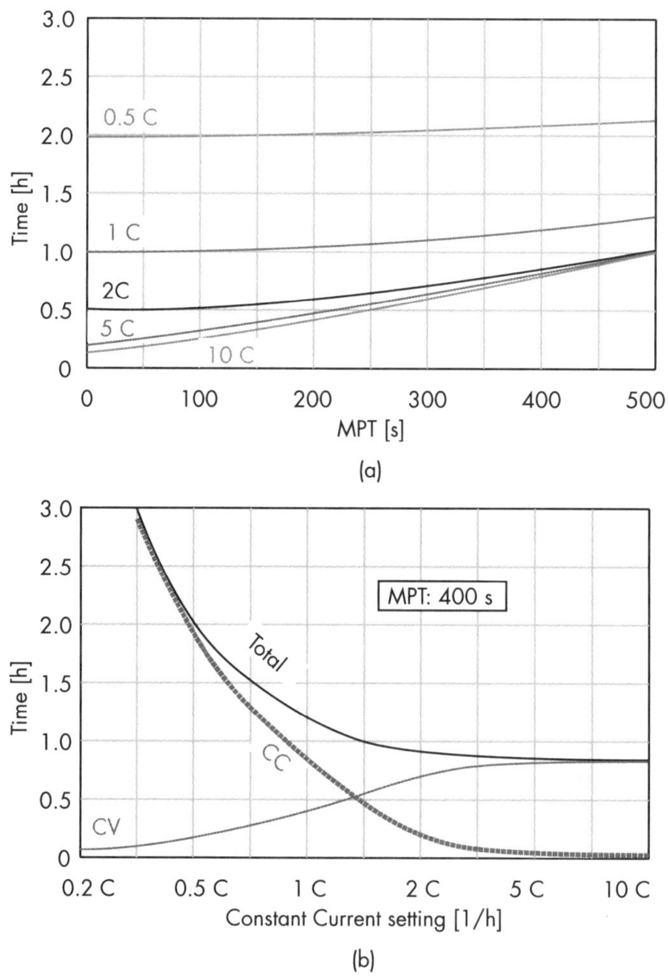

FIGURE 2.19
Charging time: (a) versus cell MPT at various currents in the CC phase, and (b) versus current in the CC phase.

The battery MPT and the constant current setting determine the proportion of the time spent in the CC phase and the CV phase (Figure 2.20(a)). The total charging time is minimum when most of the time is spent in the CC phase.

For energy cells with a high MPT, a constant current setting of 0.3~0.5C is ideal, because the CC phase lasts about 90% of the time. Charging at a higher current is not worthwhile because it degrades the cells, and in any case, it doesn't expedite charging by much.

For power cells, charging at high current within the limits specified by the cell manufacturer does speed up the total charging time.

The fastest charging time depends on the chemistry and is shorter for power cells than energy cells (Figure 2.20(b)). High-power LTO cells may be charged to 80% in 10 minutes at 6C and completely over a total time of about 20 minutes. This graph is for general reference only. It is not a strict specification for these cell chemistries.

2.4.4.4 CC Charging Curves

The CCCV charging curves in the previous section are applicable when cells are fully charged with a CCCV charger. In some applications, the battery is charged by something other than a charger, probably at constant power. In such cases, curves based on constant current are more applicable (Figure 2.21). These curves are a subset of the CCCV charging curves.

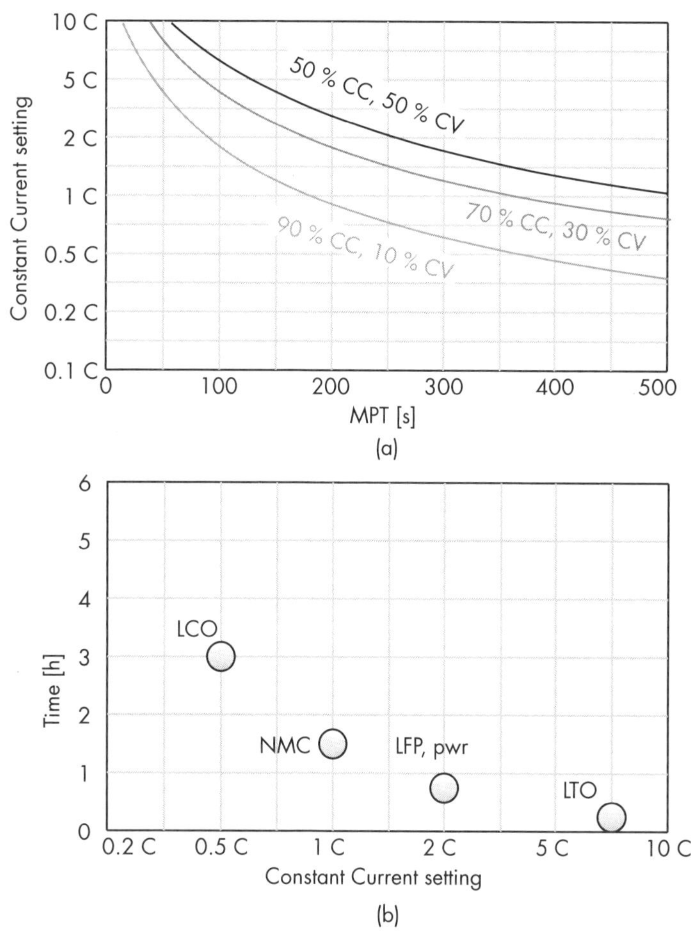

FIGURE 2.20
(a) Constant current setting versus maximum power time that results in a given time in each charging phase, and (b) fastest charging time and constant current for certain Li-ion chemistries.

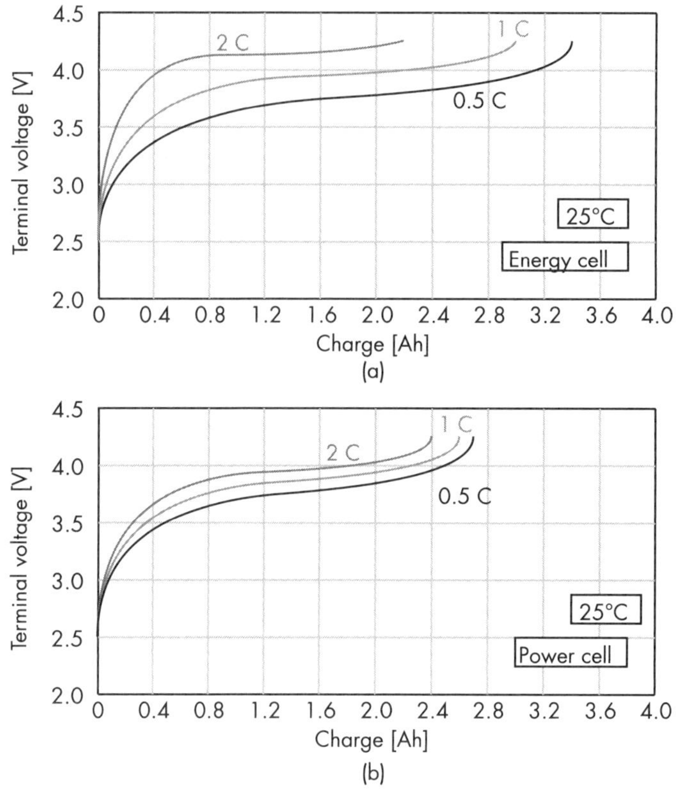

FIGURE 2.21
CC charging curves at various levels of specific current: (a) energy cell, and (b) power cell.

Note how more current results in a higher terminal voltage. At a given current, for an energy cell (Figure 2.21(a)) the voltage is higher than for a power cell (Figure 2.21(b)).

The cell cannot be fully charged at constant current: these curves stop at 4.2V regardless of current, even if the cell is not yet full (Figure 2.22). In this example, for an energy cell charged at 2 C specific current, the final charge is only 2.1 Ah (62% SoC), well short of the cell capacity of 3.4 Ah. Counterintuitively, the higher the current, the less the cell is charged.

The curves at various levels of specific current are displaced from the OCV versus SoC curve due to cell resistance—the terminal voltage is lower than the OCV while discharging and higher while charging (Figure 2.23(a)).

The same information can be presented in terms of specific current at various SoC levels (Figure 2.23(b)). Although this is a useful curve, it is never provided by the cell manufacturer. The flatness of the lines is due to a relatively constant internal resistance of a cell at a given SoC and temperature. As the discharge current increases (toward the right), the terminal voltage drops. Similarly, as the charge current increases (toward the left) the terminal voltage increases.

Note how the slope of the lines is nearly the same between 25% and 90% SoC. This means that the MPT of the cell is nearly constant in this range (480 s in this example). Yet, the slope becomes steeper—that is, the MPT increases—at low and high SoC levels; as high as 730s at 0% SoC.

Note also how the 0% line is only on the left side of the graph. This is because an empty cell can only be charged. Similarly, the 100% is only in the right side of the

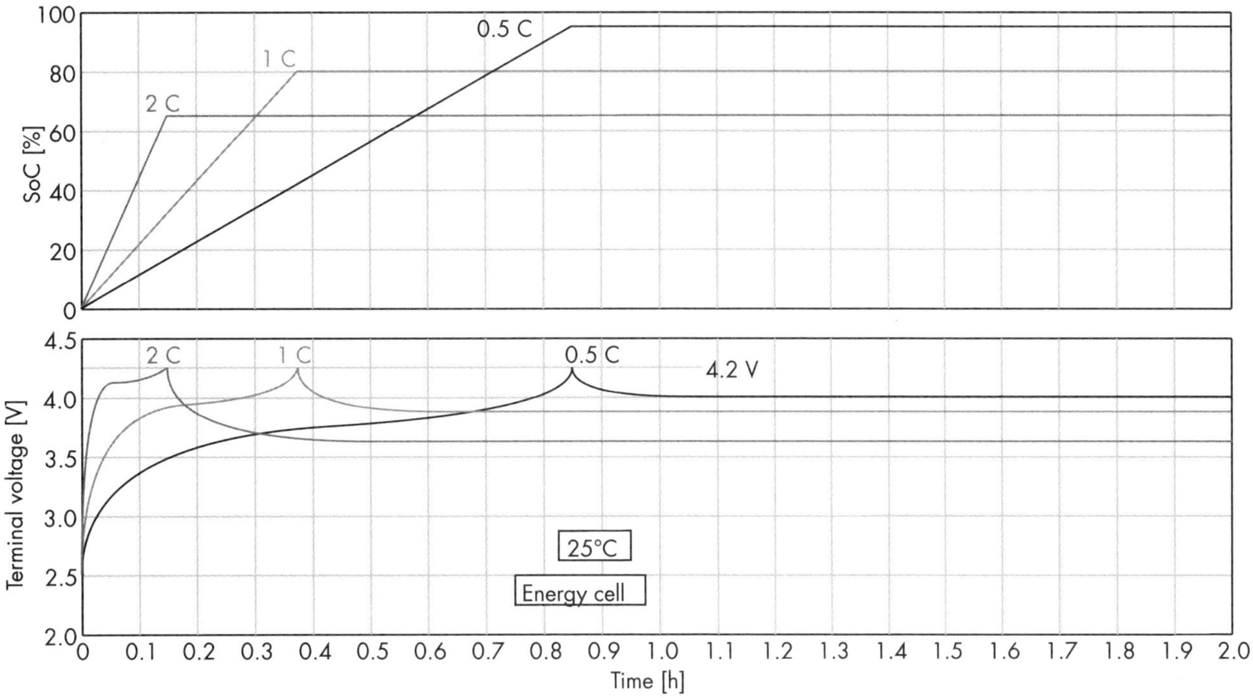

FIGURE 2.22 CC charging at various levels of current.

graph because a full cell can only be discharged. Lines above 50% SoC are shorter on the left because the maximum cell voltage is 4.2V.

These graphs suggest that we should plot MPT versus SoC (Figure 2.23(c)). The MPT is flat from 25% to 90% and increases when the cell is nearly empty or full, as expected.

2.4.4.5 Evolution of OCV Curves

The shape of the OCV versus the SoC curve changes as the cell degrades (Figure 2.24).[30] This has a strong effect on the differential curves as well [12, 13]. Monitoring these curves can be used to estimate the conditions of a cell.

2.4.5 Differential OCV versus SoC

A valuable though often overlooked parameter is the rate of change of OCV and charge—it reveals interesting points in the cell's state [14].

There are two curves, one the reciprocal of the other:

- *Differential voltage,* $\delta V/\delta Q$[31] versus SoC: Change in voltage for a given change in charge over the full SoC range (Figure 2.25(a));
- *Differential charge,*[32] $\delta Q/\delta V$ versus OCV: Change in charge for a given change in voltage over the full OCV range (Figure 2.25(b)).

30. Especially due to lithium plating.
31. δ is the lowercase Greek letter Delta. It indicates a difference, and therefore a change.
32. Scientists call this differential capacity, which is a misnomer because what is differentiated is the charge, not the capacity. The capacity is constant, so the differential capacity is zero.

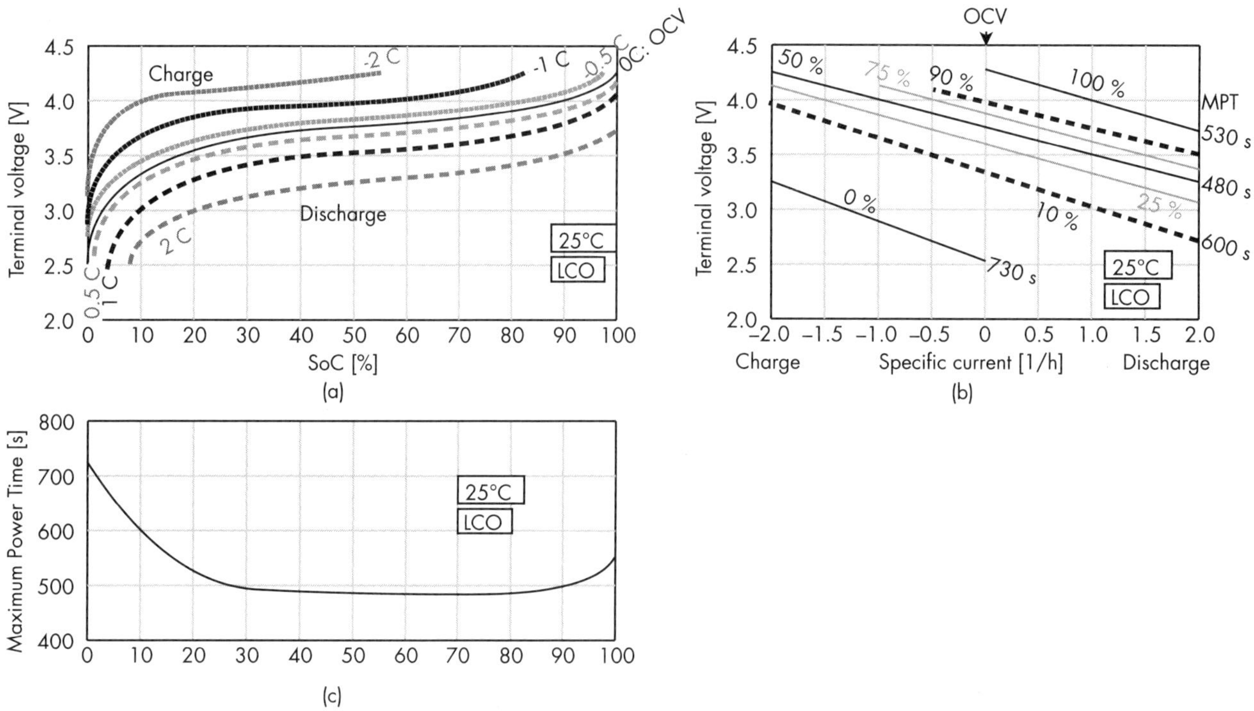

FIGURE 2.23 Charge and discharge curves: (a) terminal voltage versus SoC, (b) terminal voltage versus specific current at various SoC levels, and (c) maximum power time versus SoC.

FIGURE 2.24 Evolution of OCV versus SoC curve as cell degrades.

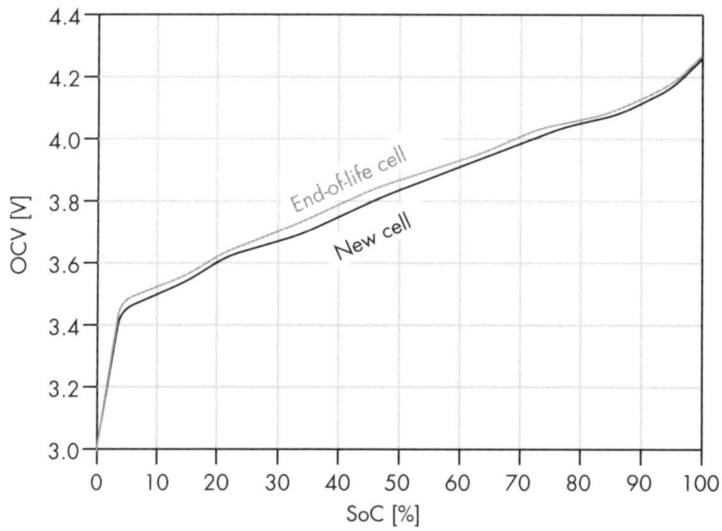

The first set of curves is for LiCo. Notice the small bump around 57% SoC,[33] which is all but invisible in the OCV curve but pops out in the $\delta V/\delta Q$ curve. This bump is as close as we can get to having a direct indicator of the absolute cell SoC.[34] Careful though—the position of the bumps is fixed only when the charge

33. Each bump corresponds to a phase transition in the lithium intercalation/deintercalation. These bumps vary with cell chemistry, but also among cells of the same chemistry but of a different design.

34. An even more sensitive measurement is the rate of cell swelling versus SoC. (See Section 2.4.6.)

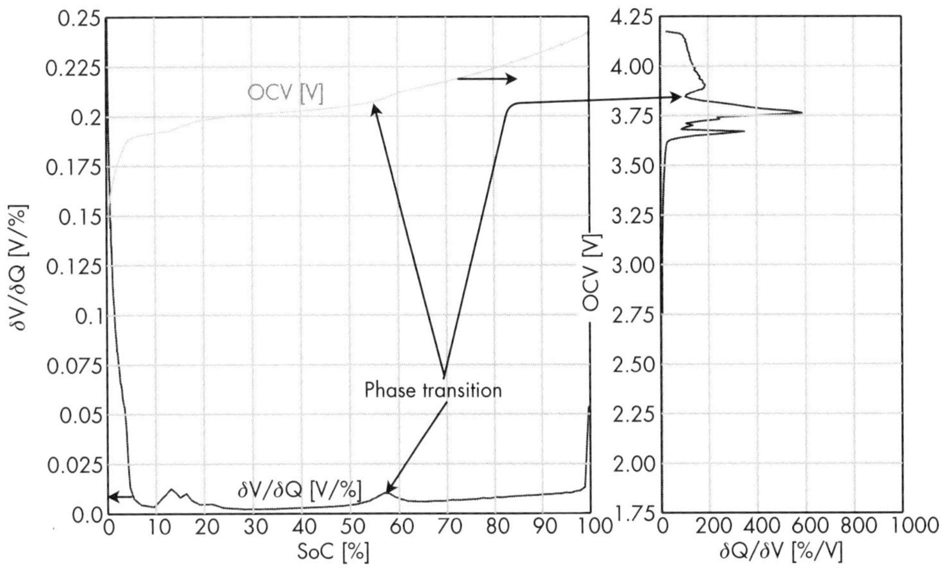

FIGURE 2.25 Characteristics for an LCO cell (Sanyo 18650): (a) OCV and differential voltage, and (b) differential charge.

and discharge current is extremely slow. At significant current, the bumps shift toward higher terminal voltage during charging, and the other way when discharging.

Note that the $\delta V/\delta Q$ curve is always positive, which means that the OCV versus SoC curve is monotonic—the OCV always increases as the SoC increases.[35]

Let's compare these curves for three Li-ion chemistries: LCO (Figure 2.25), LFP (Figure 2.26), and LTO (Figure 2.27). The graphs in these three figures use the same scale. We already saw the curves for LCO; let's now look at LFP.

An LFP cell has a flatter curve. Consequently, the differential voltage (Figure 2.26(a)) is much closer to zero and has much flatter bumps, such as the bump at 75% SoC. It also means that the differential charge (Figure 2.26(b)) has much higher peaks, which are closer to each other.

An LTO cell has a rather smooth curve [15]. Consequently, the differential voltage curve has no bumps (Figure 2.27(a)), and the differential charge curve has a single, smooth, low peak (Figure 2.27(b)).

2.4.5.1 Voltage Monotonicity

The differential charge curves in the previous section show that the $\delta Q/\delta V$ versus OCV is always positive. This means that the OCV of a cell is monotonic—when discharging, it always decreases, it never increases.[36]

35. Earlier we saw how, at rather cold temperatures, the terminal voltage (not the OCV) may increase as the cell discharges due to self-heating and, therefore, reduction of internal resistance. See Section 2.4.4.
36. Some researchers report that under certain circumstances, the open-circuit voltage of an electrode with respect to lithium metal may not be monotonic—during very slow discharge, during a phase transition at specific SoC levels, that voltage may actually increase slightly. However, the IR drop counteracts that effect, so the terminal voltage shows a plateau, rather than an increase. I have found no indication that this is the case in any commercially available Li-ion cell that I examined.

FIGURE 2.26
Characteristics for an LFP cell (K2 26650): (a) OCV and differential voltage, and (b) differential charge.

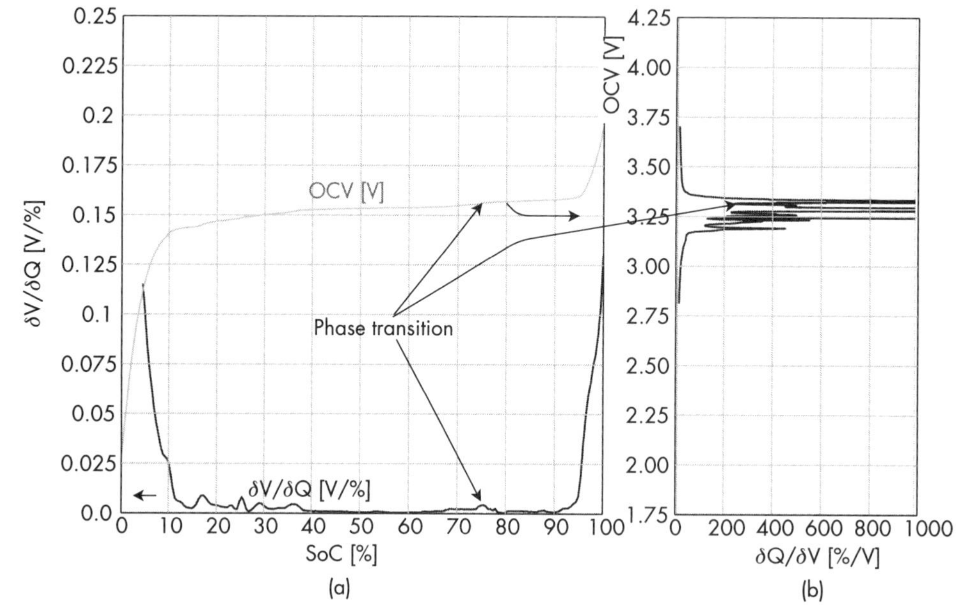

FIGURE 2.27
Characteristics for an LTO cell: (a) OCV versus SoC and differential voltage ($\delta V/\delta Q$ versus SoC), and (b) differential charge ($\delta Q/\delta V$ versus OCV).

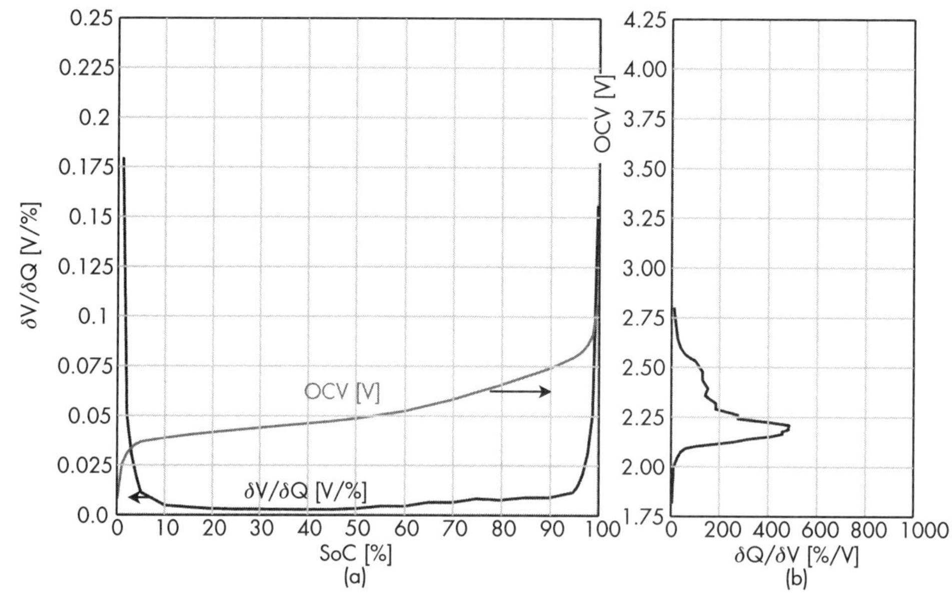

2.4.6 Expansion and Contraction

A pouch or prismatic Li-ion cell swells at either high SoC[37] or high temperature.[38] The swelling is highest in the center of the face.[39] The cell contracts back when the SoC is lower and the cell is cooler [16].

The swelling of an NMC pouch cell may be plotted during charge, then during discharge at various rates, and finally through a relaxation period (Figure 2.28). Note

37. While charging, the negative electrode swells as it's being filled with lithium ions, while the positive one contracts as it loses ions, but not as much. In an LFP cell, the negative electrode swells about 10% and the positive one contracts about 3%. This difference results in more total swelling when the cell's SoC is high. LTO cells do not experience much swelling due to their internal structure.
38. In some cells, gases may be formed when the cell is at the highest voltage, which may contribute to the swelling.
39. For two reasons: (1) the edges are more constrained by the shape of the cell, and (2) at high current the cell heats, and the temperature rise is higher in the center (away from the edges).

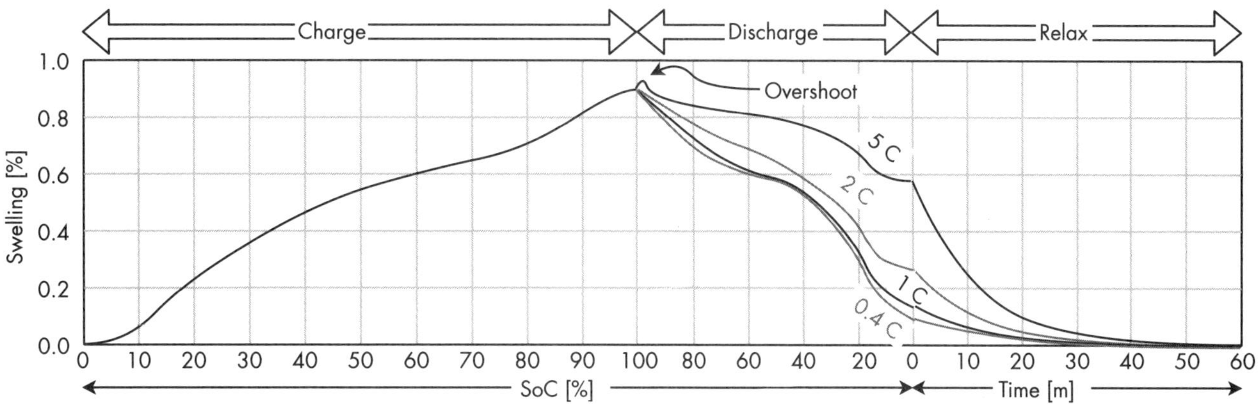

FIGURE 2.28 Expansion, contraction, and relaxation of an NMC cell at various discharge rates.

that the amount of swelling depends on its *state* (the SoC), not on the *act* of being charged.

Note how the swelling during charge is independent of charging rate, but that the contraction is delayed at high discharge rates (due to internal heating) and is completed during a relaxation time as the cell cools, eventually returning to the original size. The overshoot at 5C is believed to be due to heat still flowing even after charging is stopped.

The swelling depends on the cell and may be as high as 3%. Also, cells experience an irreversible swelling as they age, up to 6% over the life of the cell [17].

Expansion may be tested by varying the SoC of a cell while compressed between two plates, using a force gauge (Figure 2.29(a)).

A cell that is charged fully and then left unconstrained may expand irreversibly, well beyond the 1%~3% shown the previous graph (Figure 2.29(b)).

2.4.6.1 Managing Expansion

If a pouch or prismatic cell[40] is allowed to expand unrestrained, it degrades. Therefore, the swelling must be contained to a certain extent. While the case of small or large prismatic cells offers some restraint, it is not sufficient—additional restraint is required (see Section 5.5). Pouch cells are soft, and as such require external restraint. A resilient material such as compressible foam is placed between the cell and the restraint. It is compressed as the cell swells and is allowed to expand as the cell contracts.

Pouch cells in consumer products are rarely restrained. Therefore, their life tends to be short.

2.5 CAPACITY, COULOMBIC EFFICIENCY, AND ENERGY

Because the cell voltage is nearly constant, the energy in a cell is closely related to its capacity; both may be discussed together. From one point of view, gradual loss of capacity during cycling may be tied to Coulombic efficiency, so we'll discuss this subject here as well.

40. The case of cylindrical cells (small or large) limits the swelling inherently.

(a) (b)

FIGURE 2.29
Large prismatic cell expansion: (a) testing, and (b) expanded cell.

2.5.1 Capacity

Cell *capacity* (see Section 1.4.2) may be viewed from three perspectives (see Table 2.3.):

- Specification: Nominal capacity;
- Operational: Effective capacity;
- Condition: Actual capacity (often used as one of the factors when evaluating state of health or state of function).

The actual capacity fades

- Over time, whether in use or not, which affects the calendar life of the cell;
- During use, which affects the cycle life of the cell.

2.5.2 Capacity Fade over Time: Calendar Life

Whether or not a cell is used, its capacity degrades over time; faster at first, then more slowly[41] (Figure 2.30(a)).

The rate at which capacity is reduced is on the order of 0.01% to 1 %/month, depending on the cell. It is greatly affected by temperature and SoC (Figure 2.30(b)) [18]. Capacity fades faster

- At high SoC levels;
- At high temperatures.

To minimize capacity fade, keep cells cool and store them at mid to low SoC levels.

2.5.3 Capacity Fade During Use: Cycle Life

The capacity of a cell may increase during the first few cycles (even beyond the rated capacity) and then decreases during the remaining cycles (Figure 2.31(a)).

In the long term, capacity fades faster if the cells are used at high current (Figure 2.31(b)). Fade is nearly linear[42] until the end of life, at which point the capacity starts

41. The fade is proportional to the square root of time. It is due to loss of lithium (consumed by undesired side reactions, mostly while charging, growing the SEI layer on the negative electrode) and loss of storage sites in the electrodes.
42. More precisely, it drops a bit faster at first and slows down afterward.

FIGURE 2.30
Capacity fade: (a) capacity over time, and (b) capacity loss rate versus SoC.

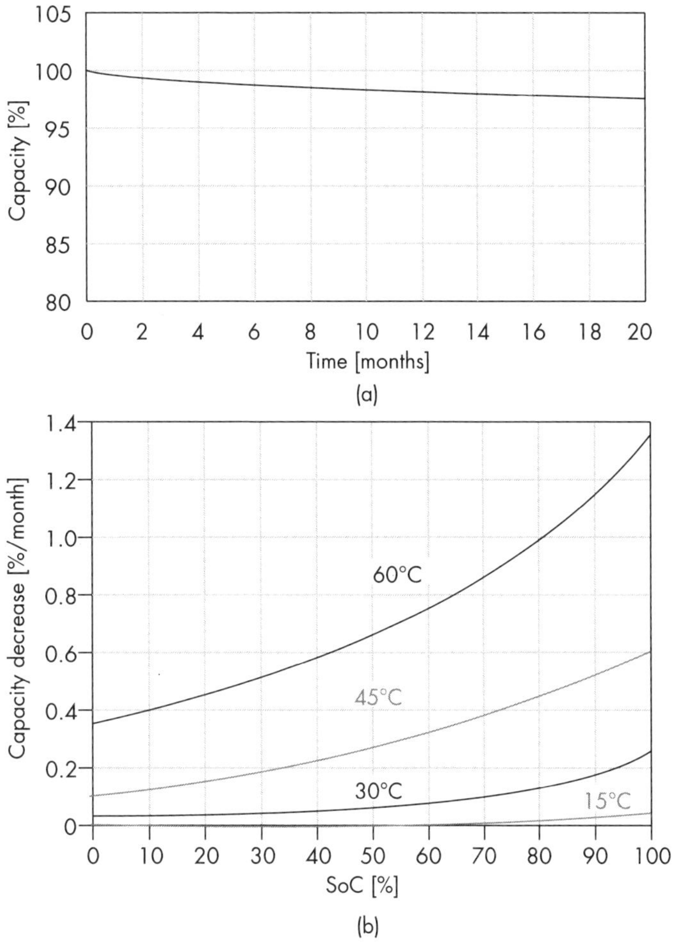

dropping precipitously. Also, in the long term, capacity fades when cycled at high temperatures (Figure 2.31(c)).

2.5.3.1 *Minimizing Capacity Fade, Maximizing Cell Use*

The capacity fade curves in the spec sheet (if any) are not directly applicable to your application because cell manufacturers measure capacity fade under carefully controlled laboratory conditions. In actual applications, in which a cycle may be shallower, temperatures vary, and current is not constant, the profile of capacity fade is significantly different (it may be better or worse).

Cycle life is increased by avoiding operation in the areas at the two ends (low and high SoC), where cells degrade more rapidly (Figure 2.32(a)).

To discover the ideal operating range for a specific cell, we use contour plots to graph the number of cycles achieved when operated between a given bottom SoC and a given top SoC (Figure 2.32(b)). For example, when operated from 20% to 90% SoC, after 2,000 cycles, the capacity is reduced by 10%. However, if we limit the operating range to 30%~70%, then the cell life increases to 6,300 cycles.

As the following study is for a specific cell operated under stressful conditions, the number of cycles may appear lower than expected. The curve for your cell, used in your application, is different, although the general principles remain the same.

This contour plot (Figure 2.32(b)) shows contour lines for the number of cycles until the cell capacity drops to 90% of its original value.

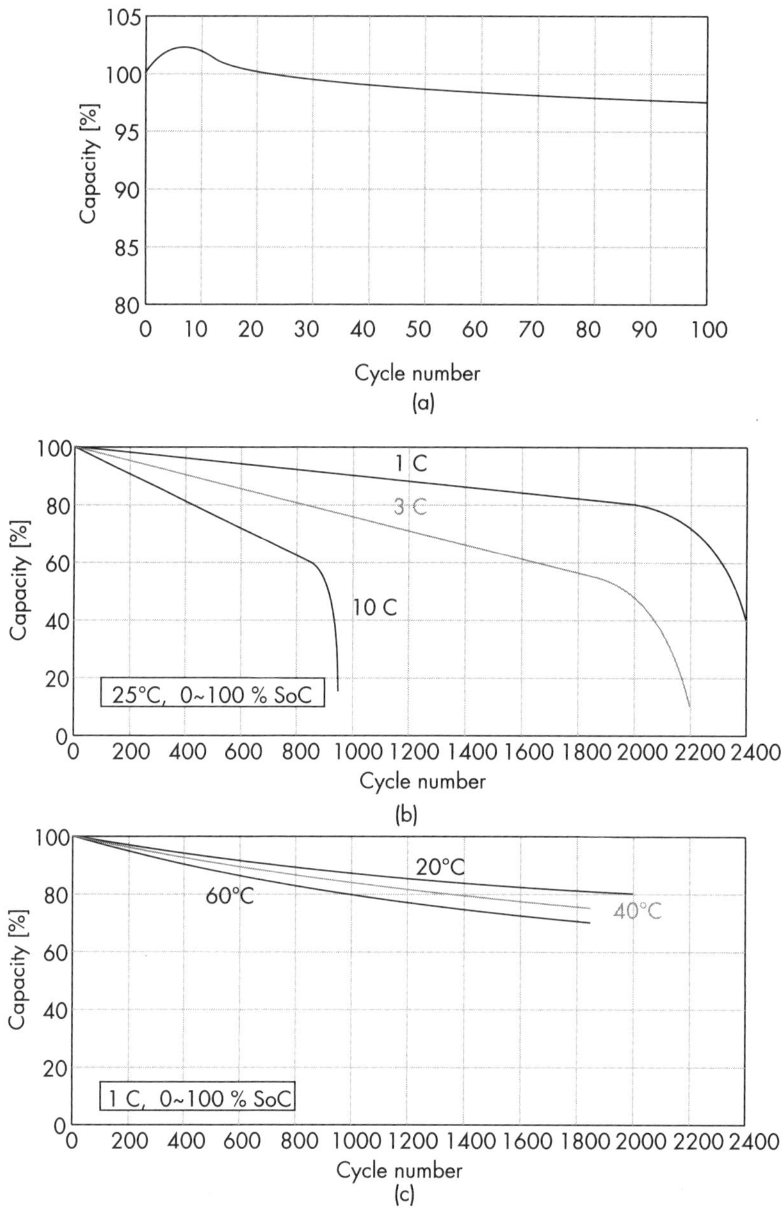

FIGURE 2.31 Capacity versus cycle number: (a) short term, (b) long term at various levels of current, and (c) long term at various temperatures.

Note that

- The diagonal lines indicate the SoC range, which is the difference between the top SoC and the bottom SoC. For example, the 50% SoC line includes operation from 0% to 50%, from 20% to 70%, and from 50% to 100%.
- Because the top SoC must be higher than the bottom SoC, the lines are all contained on the top-left section of the graph, where the SoC range is greater than zero.
- The number of cycles is less than 200 if the SoC ranges from 100% to 0% ("0~100%" label in the top-left corner).
- The maximum number of cycles (more than 9,000) occurs if the SoC stays close to 30% ("Max cycles" label toward the bottom).

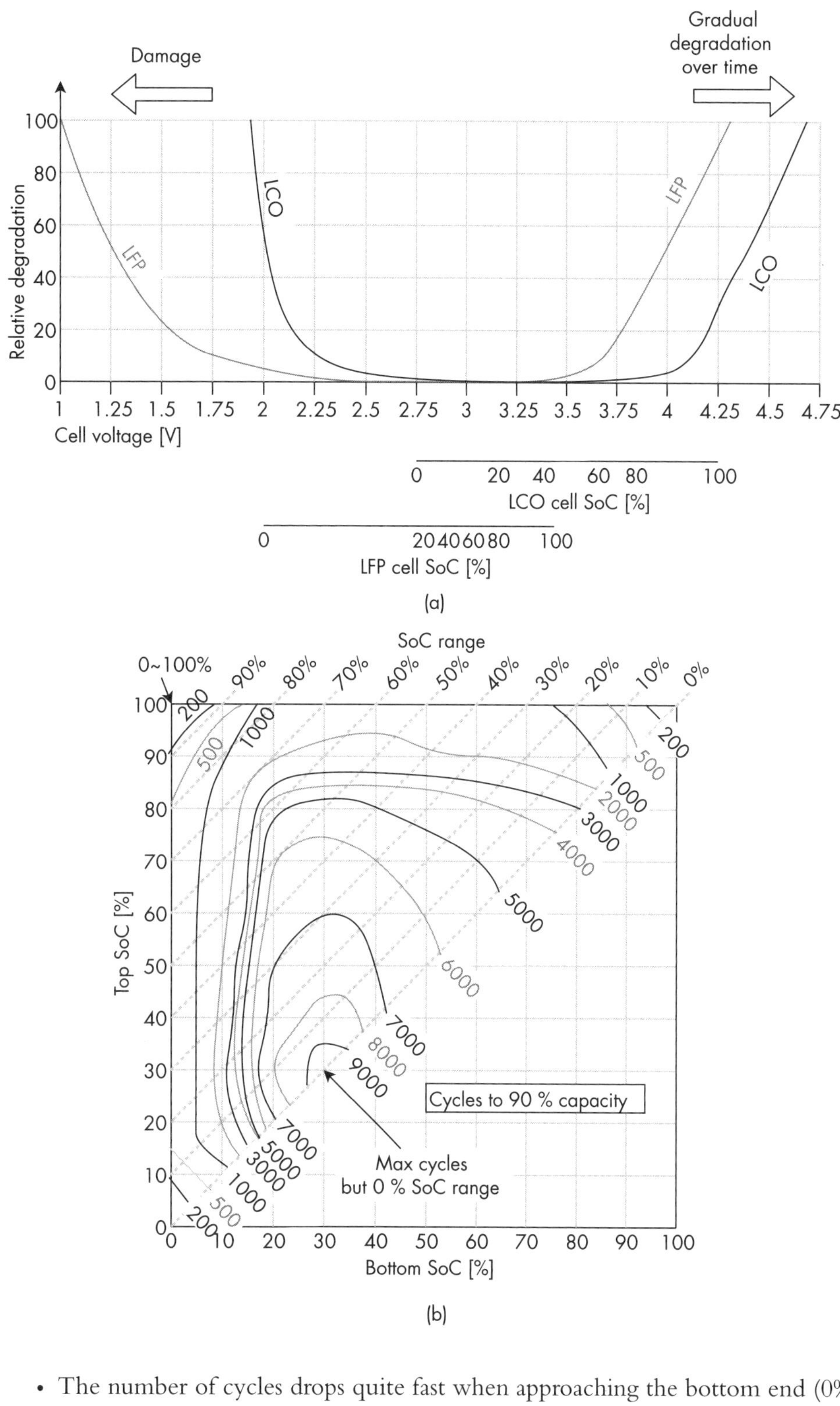

FIGURE 2.32
Cell cycle life: (a) relative degradation versus SoC for LCO (black) and LFP (gray), and (b) number of cycles when operated over a limited SoC range.

- The number of cycles drops quite fast when approaching the bottom end (0% SoC at the left edge of the graph) because the cell resistance is higher and the cell works harder.
- The number of cycles also drops when approaching the top end (100% SoC) (top edge of graph), but not as fast.

Conclusion: for a buffer battery (see Section 5.1.4), this particular cell should be operated around 30% SoC. For an energy or power battery, a good compromise range would be 25%~85 % SoC.

Reducing the SoC range increases cycle life but also reduces the charge transferred per cycle; yet we are also interested in the *total* charge transferred by a cell throughout its useful life. How is this total charge related to the operating range?

To answer this, let's use the TSCT, with its units of emery (see Section 1.4.1.2). For now, let's define the end of life of a cell when its capacity drops by 10%. The second contour graph (Figure 2.33(a)) shows the TSCT when the cell is used in a given SoC range.

Note that

- When the SoC ranges from 100% to 0% ("0~100%" label in the top-left corner), the lifetime TSCT is less than 200 emery.
- The maximum TSCT (more than 3,000 emery) occurs if the SoC ranges from 80% to 25% ("Max TSCT" label).[43]
- The number of cycles drops quite fast when approaching the diagonal line because the SoC range is small, meaning that each cycle has a small charge; indeed, the TSCT right on the diagonal line is 0 emery.

Operating over a smaller SoC range maximizes the TSCT. Doing so requires cells whose capacity exceeds the capacity required by the application. For example, if the application needs a 10 Ah charge, and we operate a cell between 80% and 25% (which is an SoC range of 55%), then we need an 18 Ah cell, of which we only use 10 Ah.

After 3.000 cycles of use within this range, the capacity will drop by 10% (from 18 Ah to 16.2 Ah), which is still a lot more than the 10 Ah we need. This means that we can keep on using this cell for many more cycles—until its capacity is down to 10 Ah. Therefore, let's now redefine "end of life" for the cell: instead of using a cell until its capacity has dropped by 10%, let's use it until its capacity is as low as the application needs.

Assuming that the capacity degrades linearly (an assumption that we will address shortly), we can ask how much total charge we can get out of an oversized cell until its capacity drops to the required capacity for the application.

The third contour graph (Figure 2.33(b)) shows the lifetime TSCT until the capacity becomes the limiting factor.

Note that

- The maximum lifetime TSCT (more than 15,000 emery) occurs if the SoC ranges from 73% to 23% ("Max TSCT" label), which is an SoC range of 50%, and therefore, the cell capacity must be twice the desired capacity (e.g., 20 Ah cell to get 10 Ah of charge);
- The dotted curve follows the maximum SoC range for each level of TSCT—the cell should be operated somewhere along this line to maximize the lifetime TSCT.

Maybe a lifetime TSCT of 15,000 emery is more than the application needs, and using a cell that is twice as big is too expensive.

43. This point is for an SoC range between 20% and 80%; that is, a 60% range. In the previous graph, we can see that if using a 20%~80% SoC range, the lifetime is 5,000 cycles. Therefore, the lifetime TSCT at this point is 5,000 cycles × 60% = 3000 emery.

FIGURE 2.33
Lifetime TSCT when operated over a limited range of SoC: (a) to 90% capacity, and (b) until capacity becomes the limiting factor.

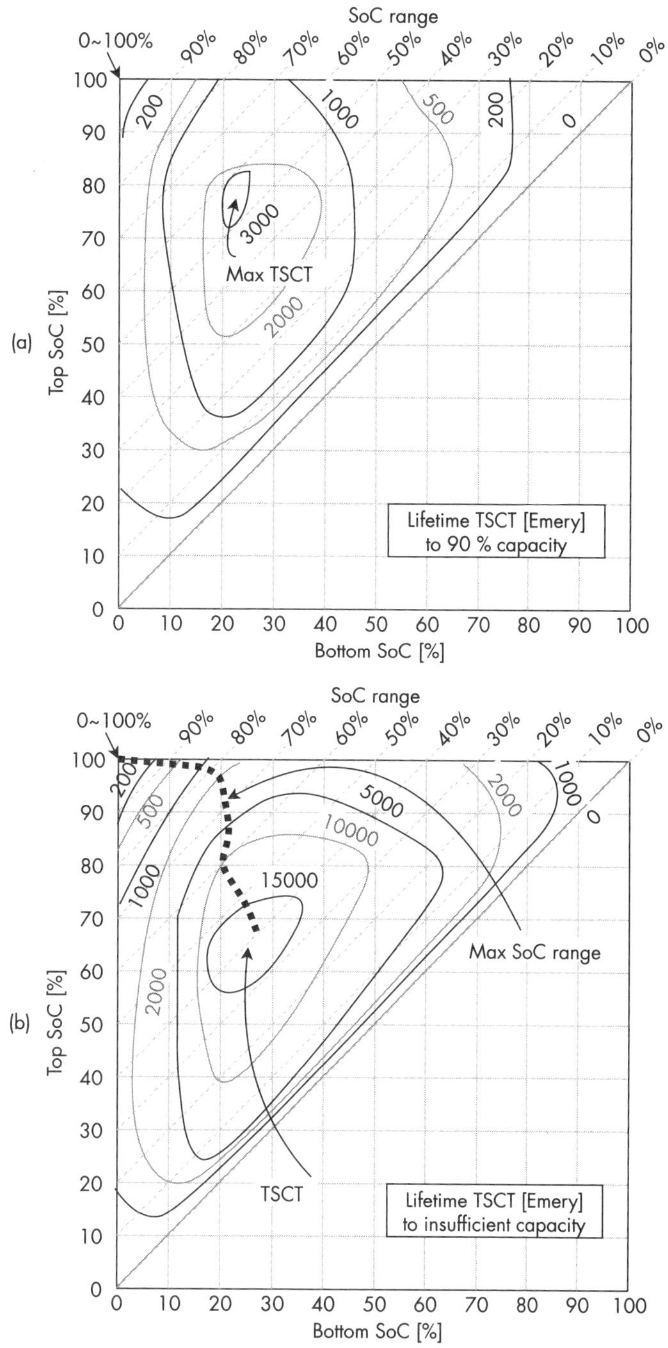

Let's plot the extra capacity required, versus lifetime TSCT, to establish how much larger a cell needs to be to achieve the desired lifetime TSCT. Let's first assume that the capacity degrades linearly (the black line in Figure 2.34(a)).

Actually, the cell degrades more and more rapidly due to a positive feedback mechanism (the gray line in Figure 2.34(a)). As the cell capacity drops, it is operated in a wider and wider SoC range, which includes more and more of the two degrading areas at either end, which in turns makes the cell degrade more rapidly. Therefore, let's derate the capacity by 30% to accommodate this effect (gray line).

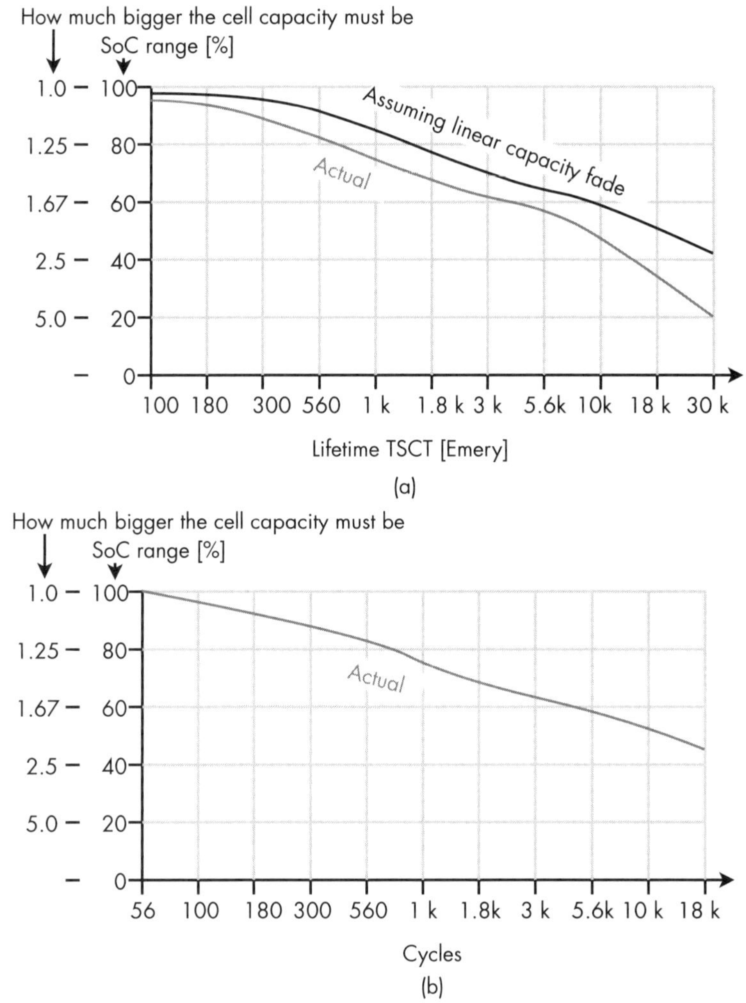

FIGURE 2.34
Capacity required to achieve a desired: (a) lifetime TSCT, and (b) number of cycles.

Rather than having to convert the number of cycles and capacity to TSCT manually, it is more convenient to plot the extra capacity versus the number of cycles (Figure 2.34(b)).

For example, if we want to get 3,000 cycles, we need to operate a new cell in a SoC range of 60%, which means that the initial cell capacity needs to be 1/60% = 1.67. That is, we need a cell that has 67% more capacity (e.g., use a 33 Ah cell to power a 20 Ah application for 3,000 cycles). At around the 3,000th cycle, the cell capacity has dropped to the desired capacity (20 Ah), and the cell is operated in the range of 0% to 100% SoC. After that, the capacity drops below the application's requirement and it must be replaced.

Let's plot the cost of a larger cell versus the higher number of cycles it gives us (Figure 2.35(a)); note how this cost drops rapidly with as little as 20% extra capacity, and settles to a somewhat constant cost, when the cell is at least twice as large as the needed capacity.

The conclusion is that, for an energy application, we can get a cell anywhere between these two points:

- 10% more capacity, and operate it initially over a 91% range of SoC, to get 200 cycles;

- Twice the capacity, initially operate it over a 50% range of SoC and get 10,000 cycles.

Again, this study was for a specific cell and application; a study for your cell and application will result in different conclusions.

2.5.3.2 Cycle Life Prolongation

To get the longest life from a cell (see Section 5.1.4):

- For a power battery:
  - Do not charge below freezing;
  - Do not float charge at the maximum voltage; either stop charging after the battery is full or reduce the CV setting of the charger (e.g., 3.4V/cell for LFP instead of 3.6V/cell) (see Section 2.4.1.1);
  - If possible, during long periods of inactivity, allow the SoC to drop close to 50%.
- For an energy battery, the same as above, plus:
  - Avoid high current pulses, especially charging.
- For a buffer battery, the same as above, plus:
  - Center operation on an SoC level that maximizes cycle life, hopefully in an area where the maximum power is the same whether charging or discharging (see Section 2.7.2);
  - Avoid high current pulses, especially while charging;
  - Do not use below freezing.

2.5.4 Coulombic Efficiency

For all practical purposes, the *coulombic efficiency* (*CE*) (see Section 1.4.1.3) of a Li-ion cell is 100%,[44] meaning that every electron that goes into it during charge is recovered during discharge. If a cell starts at 50% SoC and we take 1 Ah from it and then put 1 Ah back into it, its SoC returns to 50%, precisely.

2.5.4.1 Coulombic Efficiency Is Less than Ideal

Careful scientific analysis reveals that coulombic efficiency is actually slightly less than 100%.[45] Some sources report that the coulombic efficiency is "80%~90 %" [19], others that it is greater"greater than 95%," or "99%" [20]. In reality, for a cell that has a lifetime of thousands of cycles, it's better than that: it is on the order of 99.9%, meaning that for every 1.001 Ah we put in a cell, we can recover 1.0000 Ah.[46] In other words, for every 1,001 lithium ions that leave one electrode, 1,000 make it to the other electrode, and one gets lost along the way and ends up in the wrong place.

Coulombic efficiency is reportedly rather low for LMO cells—as low as 98% [21]. It is far better in an LTO cell, on the order of 99.998%!

Research [22] shows that coulombic efficiency drops significantly at cold temperatures and is somewhat worse at high discharge current (Figure 2.35(b)).

44. Just to be clear, we are not talking about energy efficiency, which is noticeably less than 100%.
45. Mostly due to irreversible loss of active material that occurs at each full charge and discharge cycle (irreversible capacity); also due to side reactions, mostly growth of the SEI layer.
46. For the derivation, see Volume 2, Section A.3.5, "Coulombic Efficiency."

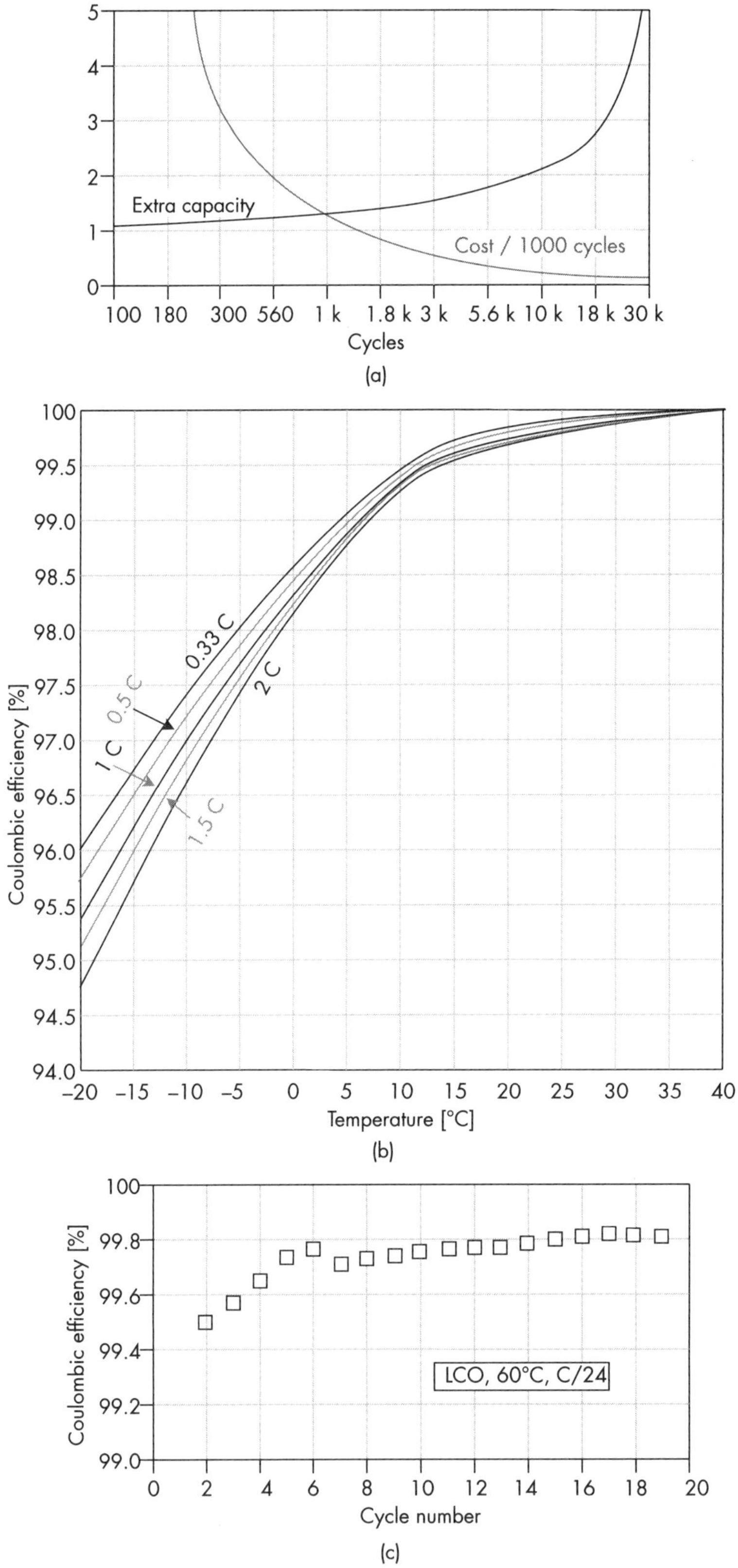

FIGURE 2.35
(a) Extra capacity and cost of larger cell relative for 1,000 cycles versus number of cycles,
(b) effect of discharge current and temperature on coulombic efficiency,
(c) improvement of coulombic efficiency after a few cycles.

Other research [23] shows that the coulombic efficiency of a brand new cell starts slightly low and quickly increases after a few cycles at high temperature (Figure

2.35(c).[47] As the cell ages, the coulombic efficiency remains pretty much constant until the end of life, at which point it starts dropping (Figure 2.10(a)).

2.5.4.2 Coulombic Efficiency and Capacity Loss

There appears to be a direct connection between coulombic efficiency and capacity loss during cycling.[48] As a cell is cycled, a few ions block sites in the electrodes instead of transferring charge between electrodes. This results in two related effects:

- During each cycle, blocked sites reduce the capacity;
- The fact that less than 100% of the ions transfer charge means that the coulombic efficiency is less than 100%.

2.5.4.3 Coulombic Efficiency during Charging and Discharging

Coulombic efficiency is better during discharge, though various sources report different results:

- Some researchers report that during discharge the coulombic efficiency is 100%[24, 25];
- Others say that at a cold temperature of -20°C, the coulombic efficiency during discharge can be more than 100% (!), while during charge, it can be as low as 97% [26].

Regardless, all researchers tell us that the coulumbic efficiency is lower during charging than during discharging.

2.5.4.4 Drop in SoC Due to Capacity Loss

If an ideal cell is charged and then discharged by exactly the same amount of charge, its final SoC is exactly the same as its starting SoC (Figure 2.36(a)). In reality, CE is less than 100%—a cell loses capacity during a cycle. As a consequence, the final SoC during this test is less than the initial one (Figure 2.36(b)):

- *During charge:* The SoC ends-up higher than it would have for an ideal cell because the capacity is slowly reduced during charge (83% instead of 80% in the example);
- *During discharge:* the SoC ends up lower than it would have for an ideal cell because the capacity is lower throughout the discharge (16% instead of 20% in the example).

The increase of SoC during charge (3%) is less than the drop of SoC during discharge (7%) because the capacity is slowly reduced during charge, yet it is lower during the entire discharge. Therefore, the SoC loss during discharge is preponderant, resulting in the final SoC being lower than the initial one.

The example above uses an exaggerated low coulombic efficiency of 90% during charging, resulting in a drop if SoC of 3.5% after a single cycle from 20% to 80% SoC. As the coulombic efficiency is closer to 99.5%, the drop in SoC after a full cycle is actually on the order of 0.2%. A plot of SoC loss versus coulombic efficiency during

47. The researchers conclude from these data "the time of exposure is really the bad actor here in the failure of these cells at elevated temperature." Yet these data shows that the coulombic efficiency improves over the first few cycles at high temperatures.

48. Normally called capacity fade; Professor Jeff Dahn of Dalhousie University has written extensively on this. See Volume 2, Section A.3.6,

FIGURE 2.36
Effect of coulombic efficiency and capacity loss during a cycle:
(a) ideal cell, and
(b) cell with exaggerated coulombic efficiency of 90% during charge.

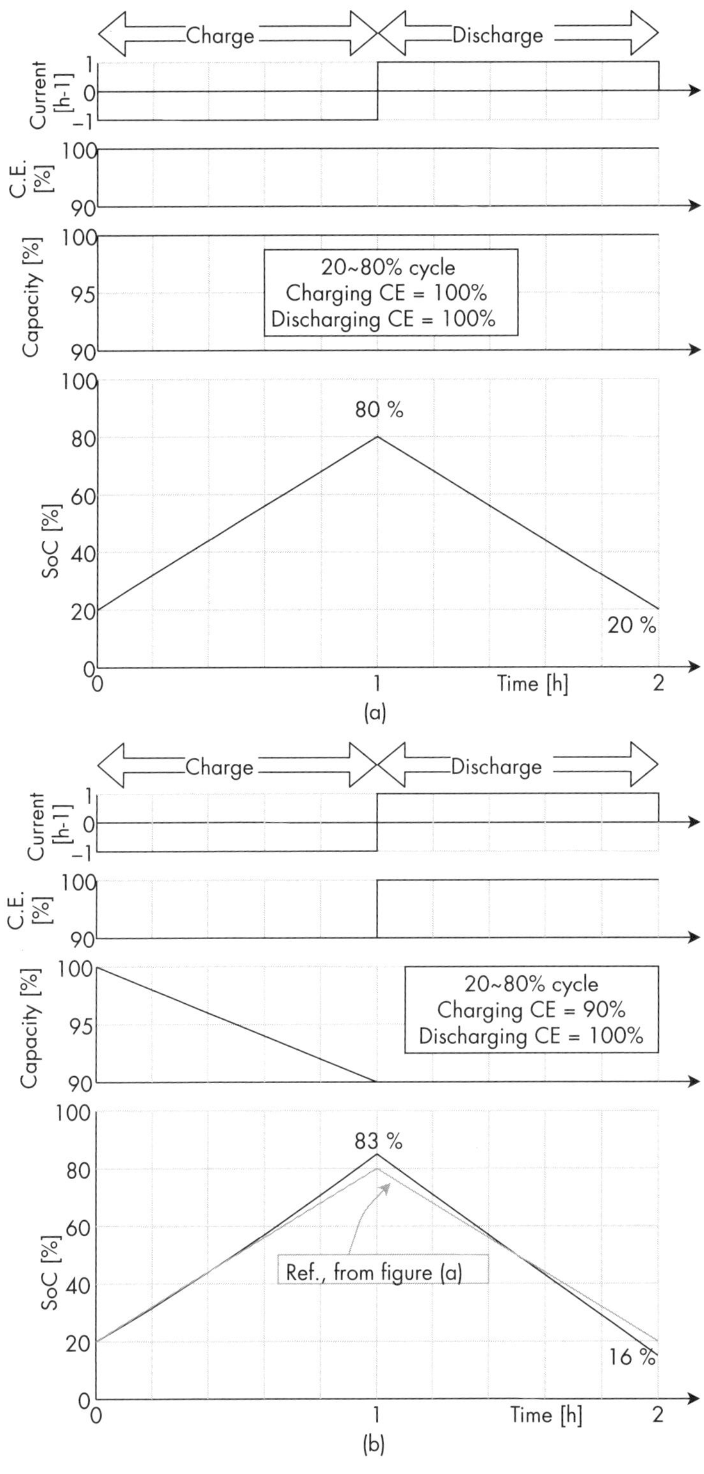

discharge shows that the SoC loss is linear (Figure 2.37(a)). The plot assumes a 0% to 100% cycle and a 100% coulombic efficiency during discharge.

2.5.4.5 *Effects of Coulombic Efficiency*

Coulombic efficiency and this change in the final SoC level have little practical consequence to a single Li-ion cell; so what if it takes an extra 0.001% more time to

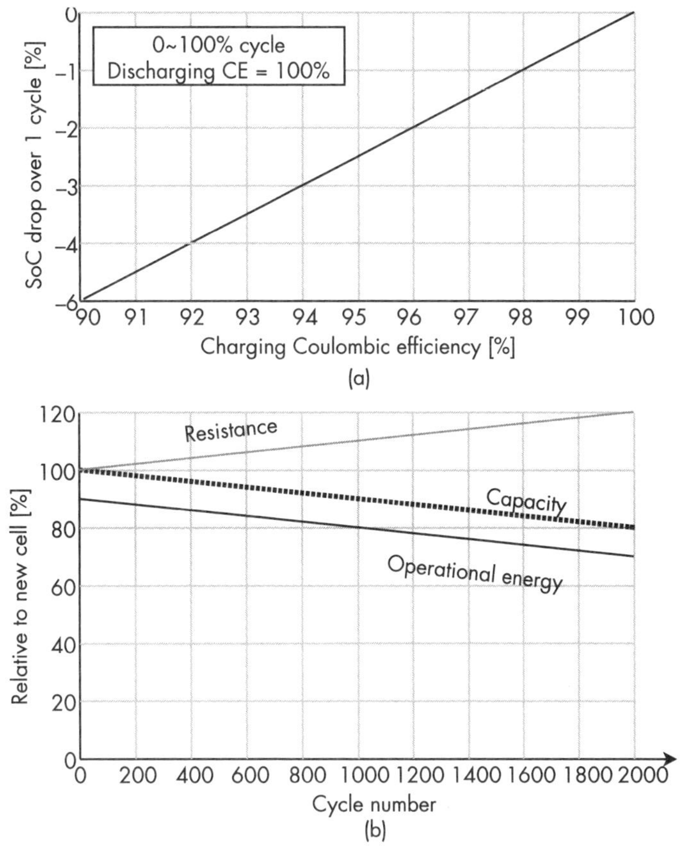

FIGURE 2.37
(a) Drop in SoC over one cycle versus Coulombic efficiency, and (b) energy reduction as cell degrades.

charge a cell? This change in the final SoC level only matters with a string of cells in series, each with a different coulombic efficiency—the string will become more unbalanced at every cycle (see Section 3.2.8).

2.5.4.6 *Coulombic Efficiency Measurement*

Direct measurement of coulombic efficiency is challenging because the measurement error in a typical current sensor is larger than the effect being measured, and because it is hard to measure the SoC precisely to confirm that the cell has returned to the initial SoC level.

Researchers can measure coulombic efficiency over a single cycle in the laboratory.[49] Otherwise, the effect may only be noticeable over many, many cycles.

2.5.5 Energy

Energy (see Section 1.4.4) may be viewed from four perspectives (see Table 2.3).

- Specifications: Nominal energy;
- Family characteristics: Energy density and specific energy;
- Operational: Effective energy;
- Condition: Actual energy

49. Dalhousie University researchers under the guidance of Jeff Dahn.

As the cell degrades, the energy that can be extracted from a cell is reduced (Figure 2.37(b)):

- Directly, due to the drop in actual capacity;
- Indirectly, due to the increase in resistance:
  - A higher resistance reduces the cell's efficiency; therefore, more charge is required to deliver a given power to load;
  - A higher resistance results in a deeper voltage sag—the low-voltage cutoff is reached sooner.

2.5.6 Energy Density and Specific Energy

The *specific energy* of lithium by itself (not the complete cell) is theoretically close to the one for gasoline (46 MJ/kg). Yet, Li-ion cells do not come close to this value (0.3~0.9 MJ/kg). One reason is that the lithium ions comprise a small portion of the mass of the cell (on the order of 1.7%).[50] The other components in a cell account for the rest of the mass—the enclosure, the current collectors, the electrodes, the separator, all impregnated in the electrolyte and conductive additives.

2.5.7 Energy Efficiency

Energy efficiency is <100% because some power goes into heating the cell as it's used.[51] The energy efficiency is reduced at high current and by high internal resistance. It is different during charging and discharging.

Energy efficiency for a given discharge time can be easily derived from the MPT of a cell and the duration of a discharge (see Section 1.5.7.2). This value applies not only to the cell but also to any battery built from those cells.

2.6 RESISTANCE, IMPEDANCE, MAXIMUM POWER TIME

Maximum power time (see Section 1.5.2) is a more general way to look at cell resistance (see Section 1.4.6), so let's discuss the two of them together.

2.6.1 Resistance

When we talk about *cell resistance*[52] we are assuming the simple DC equivalent model, which has just a voltage source and a single resistor (Figure 2.9(a)). The cell resistance determines the voltage sag under load and the voltage rise when charged. Therefore, it affects the energy efficiency.

Contrary to some common misunderstandings, cell series resistance does not affect coulombic efficiency. It does not directly affect the balance of cells in a string.

Resistance may be viewed from four perspectives (see Table 2.3).

- Specifications: Nominal DC resistance;
- State: Instantaneous DC resistance;

50. It does represent a larger portion of the volume because lithium is very light.
51. Li-ion charging is endothermic, meaning that the cell is slightly cooled as it's charged. However, the heating due to the current flowing into the battery is a stronger effect.
52. This is not the ohmic resistance, measured at high frequency in EIS. This is the total resistance of all the series elements. See Volume 2, Section A.3.4, "EIS and Nyquist Plots."

- Condition: Actual DC resistance;
- Family characteristics: Nominal MPT.

2.6.1.1 *Nominal DC Resistance*

Nominal DC resistance is the DC resistance specified by the manufacturer, at room temperature, and 50% SoC. Under these conditions, new cells should have a resistance that is less than this value.

2.6.1.2 *Instantaneous DC Resistance*

At a given time, the actual DC resistance of a specific, individual cell is affected by three of its states [24]:

- *Temperature:* resistance decreases with temperature; LFP resistance increases rapidly below −15°C[53], and LTO cells are much less affected by temperature (Figure 2.38(a)) [28].
- *Current:* The resistance of a new cell at room temperature is hardly affected by current; on the other hand, if the cell resistance is high (e.g., due o low temperature or degradation), then it increases further at low current 103 whether charging or discharging (Figure 2.38(b)).
- *SoC:* Resistance is pretty constant over a wide range of SoC, except that it becomes significant at high and low SoC levels; LTO cells are barely affected by SoC (Figure 2.38(c)) [29].[54]

The DC resistance of a cell is a state that is short-term and is reversible. The short-term resistance affects the maximum current and power that the cell can handle at a given SoC (see Section 2.7.1).

2.6.1.3 *Actual resistance*

Over the long term, the resistance of an individual cell increases irreversibly with age, use, and especially stress. Specifically:

- *With use*: The resistance increases rapidly during the first few cycles of use, then increases slowly during the life of the cell, and finally increases rapidly at the end of life. Deeper discharge cycles degrade the resistance more than shallow ones (Figure 2.39(a))[30].
- *With time:* The resistance increases at a relatively constant rate, on the order of 1% per month (again, depending on the cell). This is significantly worsened at high temperatures and high SoC (Figure 2.39(b)) [31].

This increase in resistance is the cause of what is commonly called *power fade* as an indication of how the maximum power that a cell can deliver over time decreases.

Since high temperatures and high SoC levels accelerate degradation, cells should be stored at cool temperatures (such as 20°C) and about 50% SoC (higher than this, they degrade faster, lower than this, there is less headroom for self-discharge).

53. Headway cell, measured by Elektromotus.
54. Specifically, ohmic resistance remains relatively constant and typically dominates, while polarization resistance increases dramatically at the two ends, and becomes dominant. (See Volume 2, Section A.3.5,)

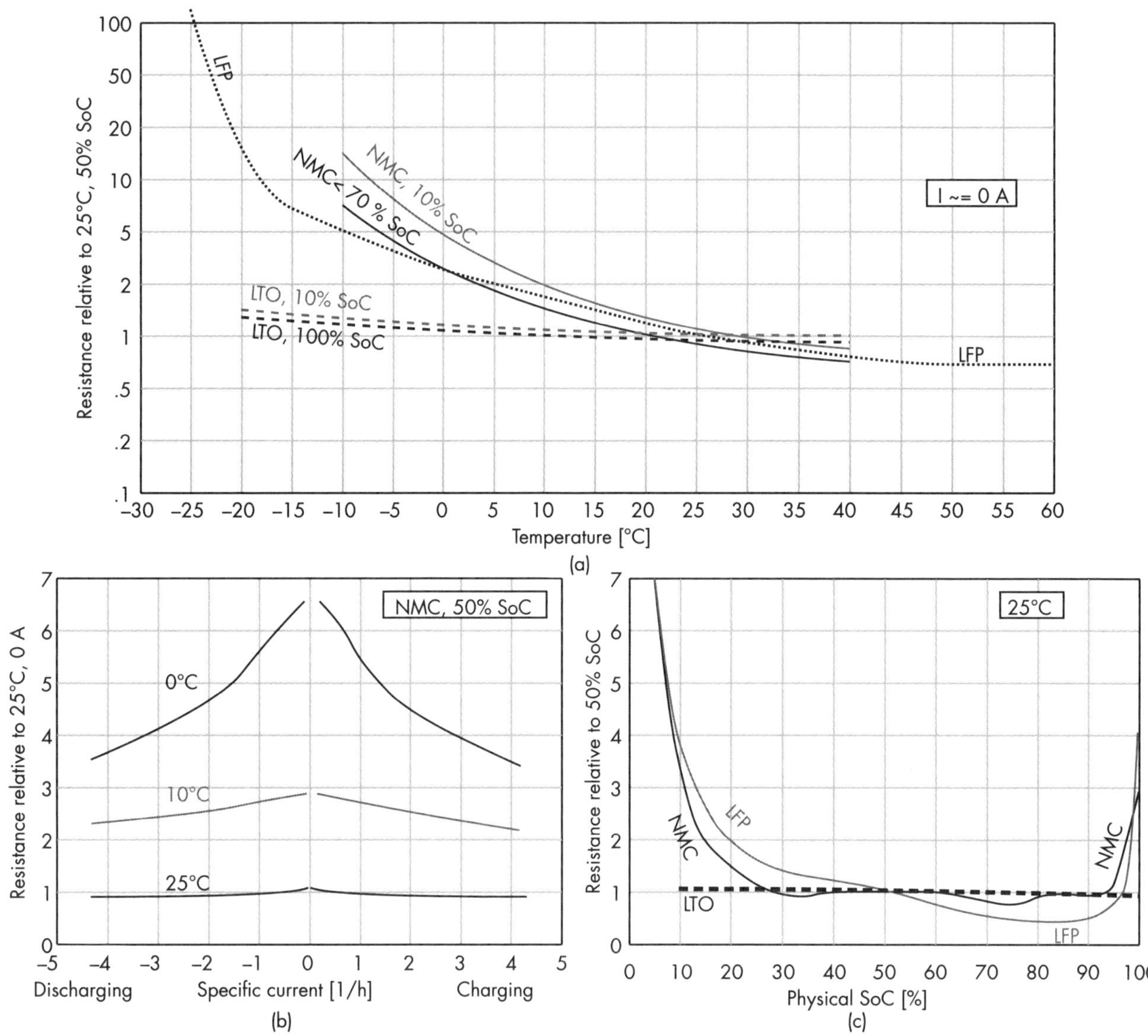

FIGURE 2.38 Instantaneous resistance: (a) versus temperature, (b) versus current, and (c) versus SoC.

2.6.2 Maximum Power Time

For power applications, knowing the nominal MPT of a family of cells is more useful than knowing the DC resistance of a specific cell within that family. MPT allows the battery designer to quickly compare families of cells and estimate characteristics of a battery constructed with those cells—energy efficiency, voltage sag, and total battery resistance (see Section 1.5.7).

2.6.2.1 Nominal MPT

The MPT of Li-ion cells varies considerably. It can range from 16 s for an excellent power cell to more than 1,000s for a nearly worthless cell. Earlier, we saw a plot for 18650 cells (Figure 2.5). Table 2.4 lists cells for energy applications.[55] Table 2.5 lists

55. Data came from cell spec sheets, testing done in house, and Henrik K. Jensen's research published in lygte-info.dk.

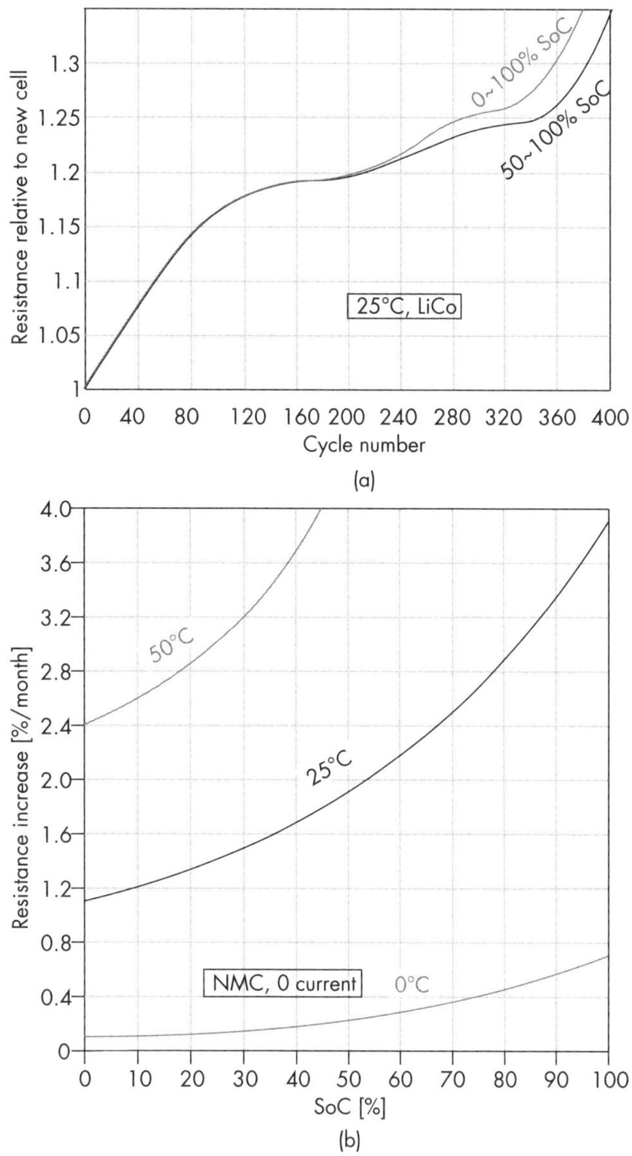

FIGURE 2.39
Degradation of resistance: (a) versus cycle number and (b) versus time.

cells for energy applications. Some cells in this table are dishonestly advertised for high discharge currents, even though their high MPT makes them particularly unsuitable for power applications.[56] Designers of batteries for race vehicles have been badly disappointed after using some of these cells, having been mislead by such false claims.

2.6.2.2 *Power Cells Outlasting Energy Cells*

For power applications, a power cell may be able to power a given application longer than an energy cell despite the lower capacity. While the energy cell stores more energy, the power cell is more energy-efficient. Therefore, it can deliver more of the stored power to the load. Also, the power cell has lower resistance and therefore, it has lower voltage sag, which delays the time when the low-voltage cutoff is activated.

56. Headway, GBS.

Table 2.4
Nominal MPT of Particular Li-Ion Cell Families Up to 200 s

Brand	**Model**	**Format**	**MPT [s]**
GEB	6619140 [LiFePO4]	Small prism	16
EIG	F007 pouch	Pouch	25
Toshiba	SCiB small prismatic	Small prism	26
Kokam	SLPB....H5, 4, 4.5, and 5 Ah pouch	Pouch	30
Enerdel	PHEV 16 Ah pouch	Pouch	42
Kokam	SLPB98188216P 30 Ah pouch	Pouch	60
Sony	US18650VTC3	Small cylindrical	61
Sony	US18650VTC4	Small cylindrical	72
Molicell	IHR18650C	Small cylindrical	82
GS Yuasa	LVP [LiCoO2]	Metal prismatic	82
XALT	75Ah HP pouch	Pouch	82
A123	M1 26650 grade A [LiFePO4]	Small cylindrical	86
Altair Nano:	Titanate	Pouch	86
K2	LFP26650P [LiFePO4]	Small cylindrical	88
Valence	IFR26650-Power [LiFePO4]	Small cylindrical	90
LG	INR18650HE2	Small cylindrical	93
Samsung	INR18650-25R	Small cylindrical	94
Amperex	35 Ah pouch	Prismatic	99
Enerdel	EV 17.5 Ah pouch	Pouch	102
Sony	US18650VTC5	Small cylindrical	104
Molicell	INR18650A	Small cylindrical	113
Panasonic	UR18650RX	Small cylindrical	115
Amperex	66 Ah pouch	Pouch	122
XALT	75Ah HE pouch	Pouch	125
Chaoyang Liyuan	Hybrid supercap [LiFePO4]	Prismatic	131
K2	LFP26650E [LiFePO4]	Small cylindrical	162
RealForce	Prism [LiFePO4]	Prismatic	168
A123	20-Ah pouch [LiFePO4]	Pouch	169
LG	INR18650E1	Pouch	180
Boston Power	Swing 5300	Small cylindrical	181
EIG	F014 pouch	Pouch	189
RealForce	Pouch	Pouch	196
EIG	C020	Pouch	198
Gold Peak	E10, 10-Ah pouch	Pouch	200

For example, K2 Energy makes two versions of its 26650 cell—power and energy. When discharged at constant power, below 95W, the energy cell lasts longer; above 95W, the power cell lasts longer.

In the 50W discharge graph (Figure 2.40(a)), the energy cell powers the load longer, as expected. In the 150W discharge graph (Figure 2.40(b)), the power cell lasts longer.

TABLE 2.5
Nominal MPT of Particular Li-Ion Cell Families Above 200 s

Brand	**Model**	**Format**	**MPT [s]**
Sinopoly	SP prism [LiFePO4]	Prismatic	253
Panasonic	NCR18650PF	Small cylindrical	255
LG	HD2	Small cylindrical	268
CALB	CA [LiFePO4]	Prismatic	286
Headway	HW [LiFePO4]	Large cylindrical	288
Winston	Prism [LiFeYPO4]	Prismatic	293
CALB	SE [LiFePO4]	Prismatic	326
Samsung	ICR18650-22	Small cylindrical	343
LG	INR18650MH1	Small cylindrical	367
Boston Power	Sonata 5300	Small cylindrical	375
Panasonic	UR18650ZY	Small cylindrical	410
Panasonic	UR18650ZT	Small cylindrical	425
Samsung	ICR18650-20Q (M)	Small cylindrical	444
Molicell	ICR18650M	Small cylindrical	447
Molicell	ICR18650J	Small cylindrical	448
Molicell	IHR18650B	Small cylindrical	477
GBS	Prism [LiFePO4]	Prismatic	518
Molicell	ICR18650K	Small cylindrical	533
Panasonic	NCR18650BF	Small cylindrical	546

Note how even though the power cell starts at a lower charge (it has a smaller capacity), it doesn't discharge as fast because it is more efficient, so the current required to deliver 150W to the load is lower than for the energy cell. At around 50s, the charge remaining in the two cells is the same, and after that, the power cell has more charge than the energy cell.

Note also how the voltage sag in the energy cell is so high that it reaches the low voltage cutoff sooner, even though it still has charge left. Conversely, the power cell is discharged completely.

2.6.3 Impedance

A researcher can measure the *complex AC impedance* (see Section 1.4.6) of a cell over a wide range of frequencies and plot it on a Nyquist plot (see Volume 2, Section A.3.5). The curve on this plot changes slightly with the state (short term, temporary) and the conditions (long term, irreversible) of the cell; by looking at this plot, a scientist may be able to derive the cell's state and condition.

Cell manufacturers measure just the real part of a cell's nominal impedance and just at 1 kHz. This value helps the manufacturer spot any outlier cells during manufacture. This value has little impact on the typical application, which principally operates at DC and certainly not at 1 kHz.

However, some applications draw power in narrow pulses (e.g., a cell phone). Normally, a filter capacitor powers the application during this short pulse. The battery provides only the long-term power. Yet, if the application relies on the battery rather than on a capacitor, the AC impedance of the cell becomes important.

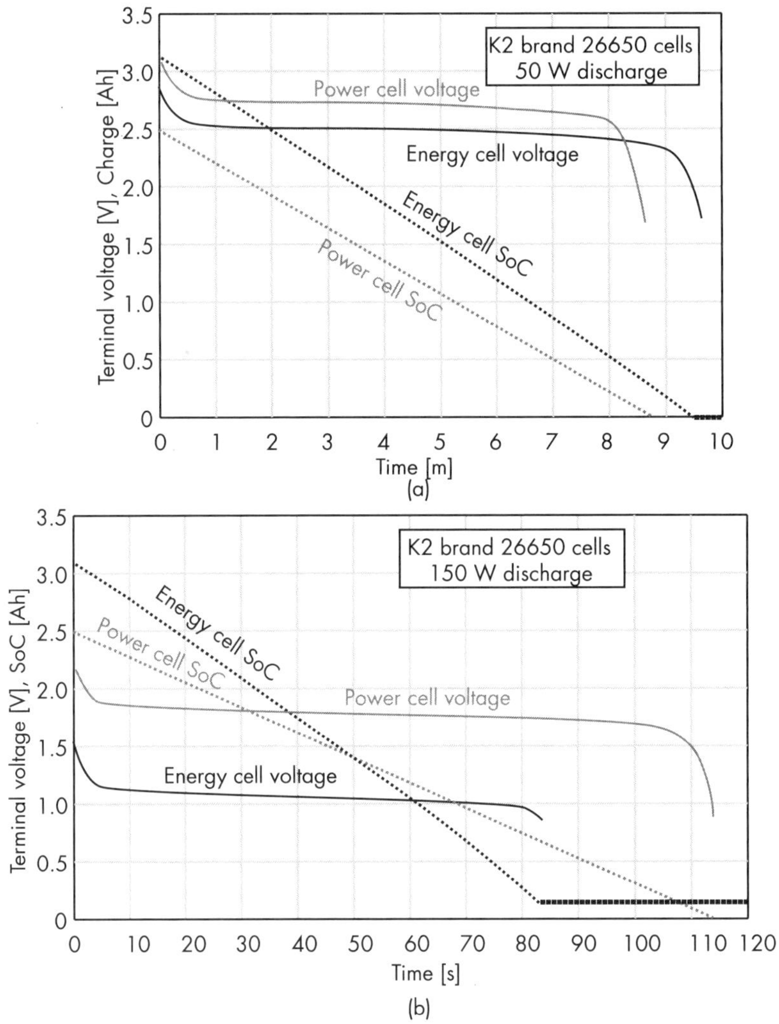

FIGURE 2.40
Discharge duration for power cell and energy cell: (a) 50W discharge, energy cell lasts longer, and (b) 150W discharge, power cell lasts longer.

While the impedance of a capacitor is reliably constant, the impedance of a cell varies considerably with SoC, temperature, and age. Therefore, relying on the cell's impedance instead of on a capacitor would be a questionable design choice.

Although of great value to the researcher, analysis of the impedance is not yet practical in today's products, which do not yet include the sophisticated spectrographic hardware required for the measurement of complex impedance over frequency, nor the software to analyze it. Some smart algorithms claim to come close to doing this analysis without the need for additional hardware and with minimal disruption to the operation of the application.

2.7 CURRENT, POWER, AND SELF-DISCHARGE

Because the cell voltage is nearly constant, cell power is closely related to its current. Therefore, we can cover them together.

2.7.1 Current

Current (see Section 1.4.3) may be viewed from these perspectives (see Table 2.3).

- Specifications: Nominal and peak charging and discharging current;

- Cell family characteristics: maximum power current;
- Operational: the actual range of current in the given application, peak and continuous (see Section 1.4.3.3).

2.7.1.1 *Specifications*

Cell manufacturers specify a standard charging current and a standard discharging current.[57] The continuous current limit is such that the cell doesn't overheat and does not degrade too quickly:

- In a buffer application, the cell is hot, but still below the maximum temperature;
- In an energy application, the temperature does increase during discharge, but it is still within range at the end.

Cell manufacturers may also specify peak current for a given duration (such as 10 seconds). Some may even specify current limits for high-power applications, regardless of how operation at that level may degrade the cell.

Operating at peak current heats and degrades the cell faster. Therefore its duration and number of events should be kept at a minimum; roughly speaking, the effect of a short, higher current peak may be similar to the effect of a peak that is half as high and last twice as long.

The current rating of a cell is ill-defined:

- On the one hand, cell manufacturers would like to specify a small number because it results in the longest cycle life;
- On the other hand, they would like to specify a high value because it is more impressive.

The marketing department may have more of a say in this than the cell designers. The vendors may inflate these limits further to sell more cells.

When comparing cells for power applications, do not use the C-rating because it is bogus (see Section 1.2.2.8). Instead, use their MPT (see Section 1.5.2), which is based on measurable characteristics.

2.7.1.2 *Cycle Life*

The current limit for a cell is a continuum, not a well-defined wall. It is a trade-off—the higher the current, the shorter the cycle life.

As a rule of thumb, if cell life is more important, design the application so that the cell discharge is 0.5C or less and the charge is 0.2C or less. If performance is more important, operate the cell up to the maximum power point, but within its maximum temperature, knowing that the cell will degrade rapidly.

In this example (Figure 2.41(a)), if operated at the maximum power point, the cell would lose 10% of its capacity after only five cycles. If at 2C, after 75 cycles; at 1C, 200 cycles; at 0.5C, nearly indefinitely (limited only by calendar life).

Cycle life is affected by both current and temperature while charging (Figure 2.41(b)). It is maximized by charging the cell at 1C or less, and between 15°C and 45°C. This study [32] indicates that it *is* possible to charge below 0°C, as long as it's done at a lower current. This conclusion contradicts the researchers and cell manufacturers who state that cells may not be charged below freezing. It also contradicts the cell

57. The discharging current is usually higher than the charging current.

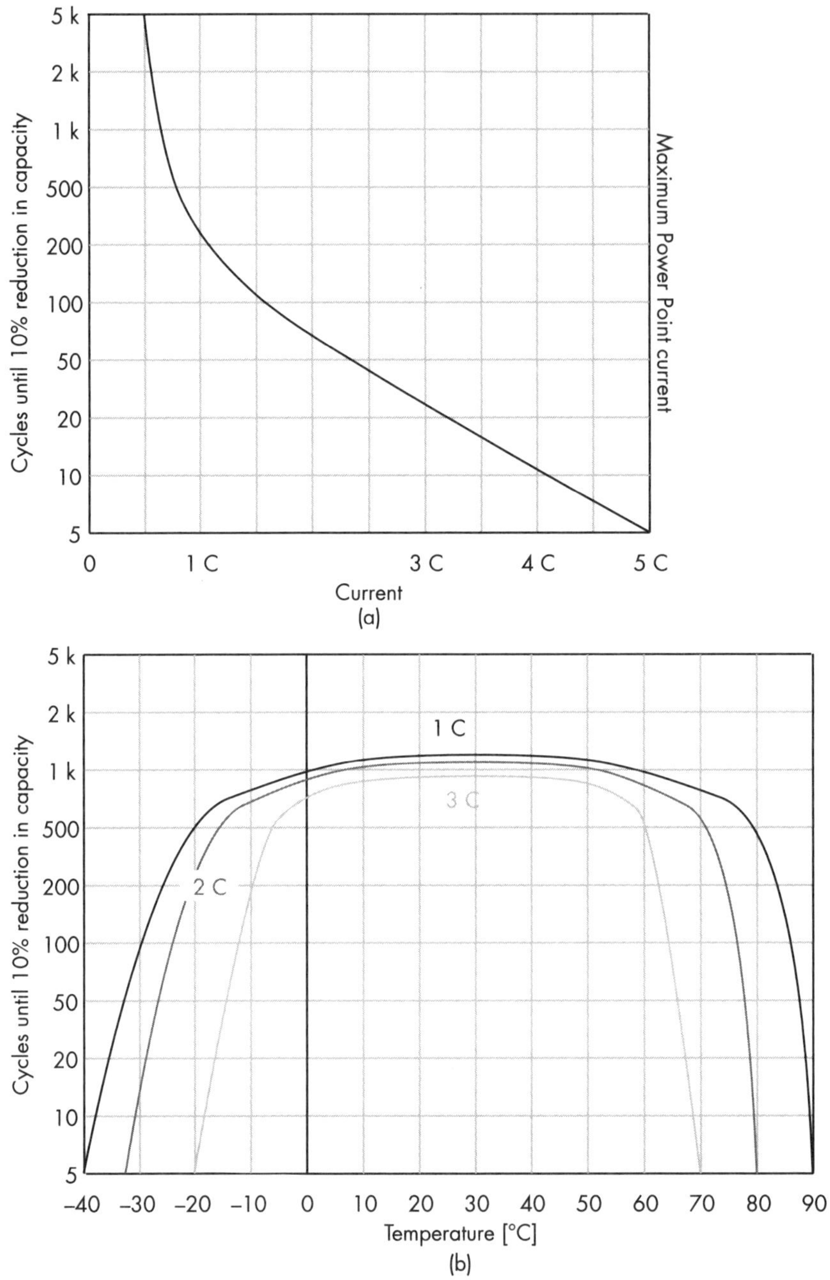

FIGURE 2.41
(a) Cycle life versus current, and (b) cycle life versus charging current and temperature.

manufacturers who state that charging between -30°C and 0°C should be done at most at 0.1 C.

2.7.2 Power

This section discusses the maximum power of a cell and the power wasted in heat.

2.7.2.1 Power Delivered

The maximum power (see Section 1.4.5) that a cell can deliver is limited by its temperature and by its resistance.

For a given application, the maximum cell power may be

- Chosen to be a fixed value;
- Computed by the BMS in real time based on the present conditions;
- Set indirectly by monitoring the cell temperature and keeping it below a maximum.

Occasionally, manufacturers of power cells specify the maximum pulse power. This chart compares the specific pulse power curves for charging and discharging for three cells from manufacturers that do provide such data (Figure 2.42).[58]

The different levels of specific power may have more to do with each manufacturer's definition of maximum power than on the cell's actual maximum power time. Notice how power decreases towards the end of charge. It also does so toward the end of discharge. Note also the SoC level where power is the same for both charging and discharging: 32% for A123, 43% for Enerdel, 75% for EiG. In a power application, you may want to operate the cell around that SoC level.

A cell can deliver more power when it is full because its voltage is highest and its DC resistance is lowest. It can also deliver more power when it is hot because cell

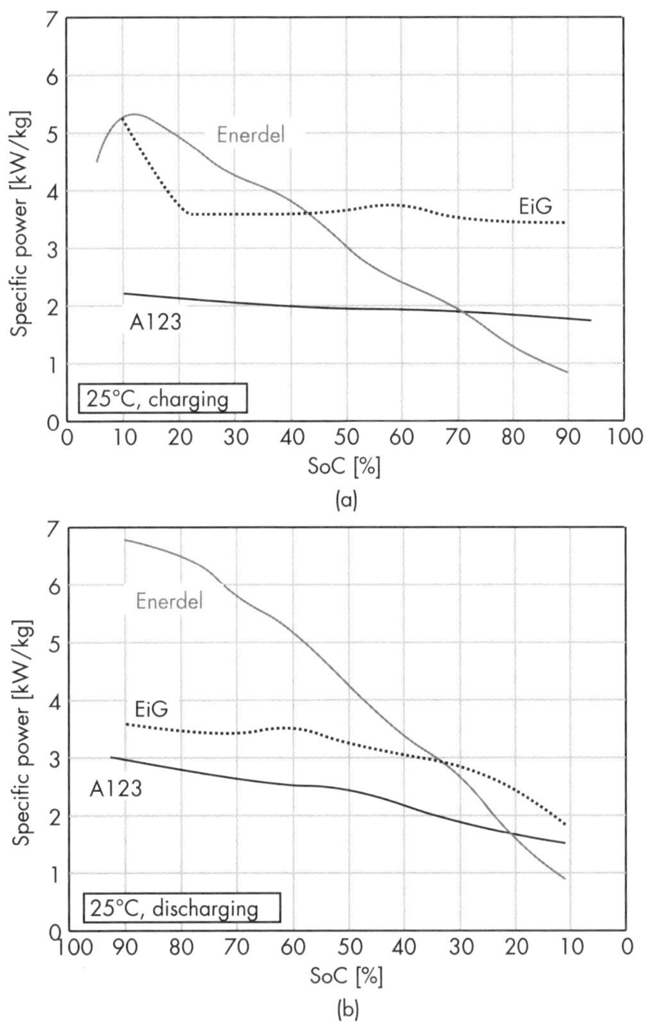

FIGURE 2.42
Pulse power versus SoC for Enerdel's HEV cell, EiG's F007, and A123's AMP20M1HD-A cells: (a) charging, and (b) discharging.

58. EiG and A123 used the FreedomCAR HPPC test; Enerdel does not say so.

resistance decreases with temperature. For instance, a racecar may heat the cells to 60°C for maximum performance during a race.

2.7.2.2 *Self-Heating Power*

While in use, a cell wastes power through heat, mostly due to the current flowing through its internal resistance.[59] The lower the current and the maximum power time (see Section 1.5.2) of the cell, the less heat is wasted, and the higher the efficiency (see Section 1.5.7.2).

2.7.3 Self-Discharge Current

A Li-ion cell is discharged quite slowly by a tiny current flowing internally from the positive terminal to the negative one. While the cell is not in use, this self-discharge current slowly reduces the cell's SoC.

Self-discharge current increases considerably with temperature and may increase as it degrades.[60] Since the self-discharge current occurs inside the cell, it's impossible to measure it directly. The self-discharge current can be evaluated indirectly (see Volume 2, Section A.3.7), though the evaluation is time-consuming and suffers from low accuracy.

Cell manufacturers do not specify the self-discharge current because it is hard to evaluate and because it varies considerably with SoC and temperature. The approximate self-discharge current may be deduced from the number of months for total self-discharge (if specified).

Table 2.6 gives the approximate self-discharge current, self-discharge rate, and time for a complete discharge for three cell chemistries. This table assumes that the SoC starts at about 50% (self-discharge current is higher at higher SoC levels). The full discharge time is how long it would take for a full cell to be discharged from 100% to 0% with the indicated self-discharge current (if it were constant, which it isn't). Use this table only as a general guide because the data used to create them [33, 34] are imprecise.

The table gives the approximate specific self-discharge current at 20°C. To approximate the current for other temperatures, use this equation[61]

$$\text{Specific current at a given temperature}\left[1/\text{h}\right] = \text{Specific current at }20°\text{C}\left[1/\text{h}\right] \times e^{\left(\left(temperature\left[°\text{C}\right]-20°\text{C}\right)/43480\right)} \tag{2.3}$$

Figure 2.43(a)) shows approximate specific self-discharge current versus temperature for LCO and LFP cells, as well as a general range of current for a bad cell with high self-discharge current, above which the cell becomes dangerous.[62]

As I stated earlier, the self-discharge current is strongly affected by SoC and temperature (Figure 2.43(b)) [35].

Some points to note:

59. There are also some minor entropic effects, which heat the cell during discharge (exothermic) and cool the cell during charging (endothermic). However, they are small compared to the heat generated by the internal resistance.
60. While the separator inside a cell offers high resistance to electron flow, it is not an ideal insulator. If a dendrite is formed within the cell, reaching across between the two terminals, it offers a path for a high self-discharge current.
61. Which I derived empirically.
62. I picked a somewhat arbitrary limit of 0.001 C.

TABLE 2.6
Approximate Self-Discharge Current of Li-Ion Cells at about 50% SoC, 20°C

	Self-Discharge Specific Current	Rate/ Month	Full Discharge Time
LTO cell	0.000005 C (5 μC)?	0.4%?	20 years?
LFP cell	0.000008 C (8 μC)	0.57%	15 years
LCO cell	0.000023 C (23 μC)	1.7%	5 years
Bad cell	0.001 C (1 mC)	72%	6 weeks

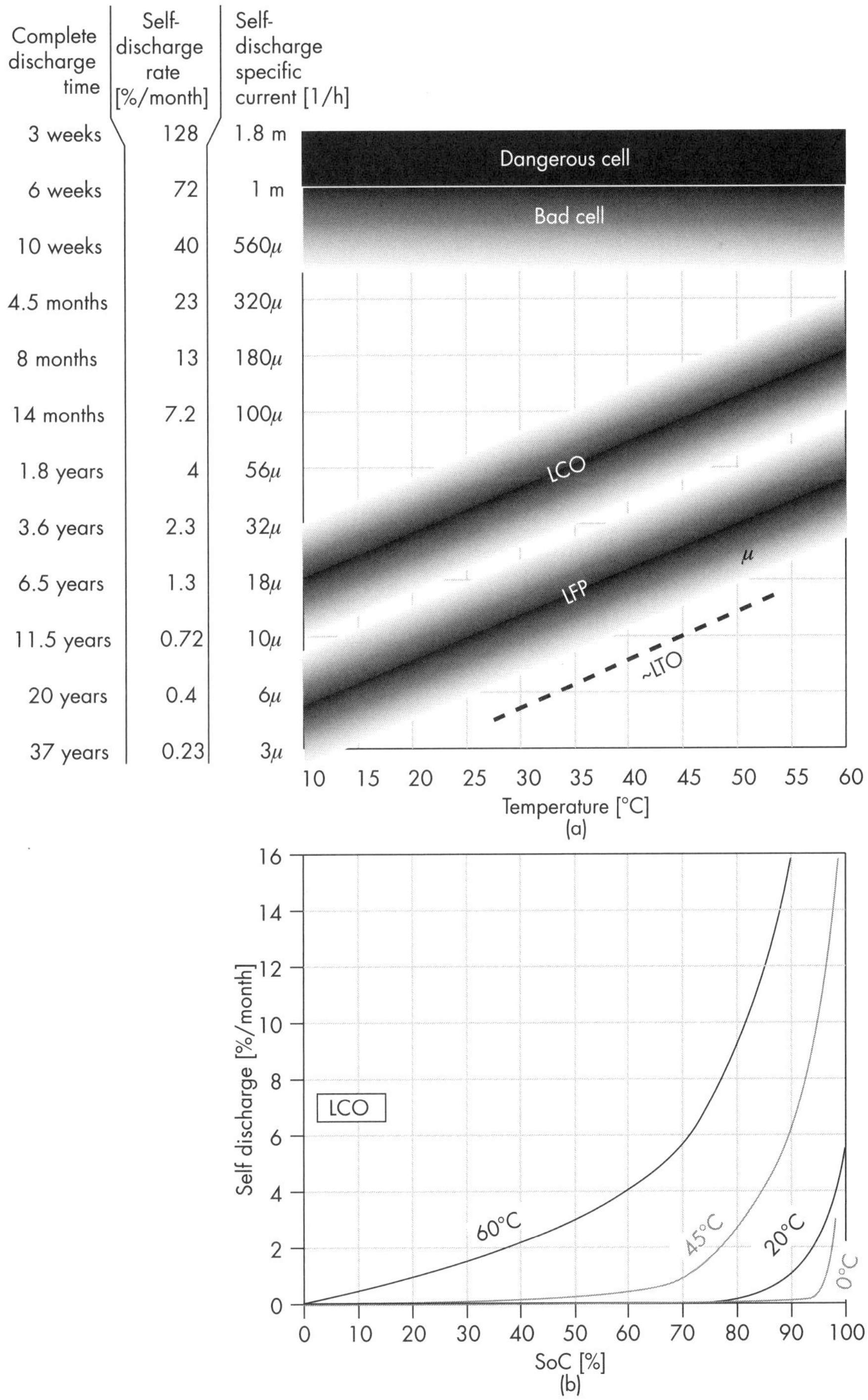

FIGURE 2.43
Self-discharge:
(a) approximate self-discharge of Li-ion cells at about 50% SoC, and
(b) approximate self-discharge current versus SoC at various temperatures.

- LFP has lower self-discharge than LCO[63]; there's little data on LTO, though it appears to be even less;
- The self-discharge current increases logarithmically with temperature;
- A still-usable cell that is suffering from internal dendrites may have a self-discharge current as high as 0.001C (1 mC, which would discharge a full cell in 6 weeks); with a higher self-discharge current than that, the cell may go into thermal runaway and should be replaced.

At a given SoC and temperature, the self-discharge current seems to remain relatively constant over time.

Self-discharging itself is not an issue of calendar life because the effect is reversible—the cell may be charged again.

A significant increase in self-discharge current is an indication of dendrite growth.[64] Dendrites may puncture the separator and create a soft short circuit (see Section 1.2.2.9) between the electrodes, discharging the cell. In the worst case, the discharge may be so fast that the cell could self-ignite (see Section 8.2.1.5).

In a properly designed cell, dendrites form only if the cell is abused, especially by

- Charging below freezing temperatures;
- Recharging a cell that was allowed to over-discharge.

When looking at a single Li-ion cell in regular use, self-discharge is of practically no consequence, just as coulombic efficiency is of little consequence (see Section 2.5.4). In both cases, if it takes an extra 0.0001C to charge a cell it is of little consequence. The only reason to care is if cells are in a series string—if they have different self-discharge currents, the string will become more imbalanced over time (see Section 3.2.8).

2.8 CELL SELECTION AND PROCUREMENT

Selecting just the right cell can be tricky; buying them is even more so.

2.8.1 Liars, Damn Liars, and "Battery Suppliars"

Battery manufacturers are perpetually fighting off a poor reputation for trustworthiness. In all fairness, that reputation is sometimes undeserved, as vendors are more to blame for embellished specifications than cell manufacturers. Possibly my favorite specification sheet is from a "royal" vendor to the "hobby" market: "Unfortunately with other big brands; numbers, ratings, and graphs can be fudged. Rest assured, [XYZ] nano-techs are the real deal, delivering unparalleled performance!"[65]

I tested one of those cells myself: their 70C rating is wildly hyperbolic.[66] Ironically, this particular vendor is correct to say that "numbers, ratings, and graphs can be fudged"—about its own cells.

63. Some sources state the opposite: see Batteries in a Portable World by Isidor Buchmann.
64. Tiny trees of lithium metal, starting from the negative electrode (on top of the SEI layer), reaching toward the positive electrode.
65. Uh huh, yeah.
66. Plus the fact that this particular cell manufacturer has no website. Its name, a bit too coincidentally, sounds like Tenergy, a reputable cell manufacturer.

2.8.2 Types of Specification Sheets

Unfortunately, not all spec sheets are useful. The amount of detail in specification sheets is quite variable, as is the style and level trustworthiness:

- Skeletal;
- Useful;
- Bureaucratic.

None of them specify every one of the parameters the battery designer truly needs (see Volume 2, Section A.3.9).

2.8.2.1 Skeletal Spec Sheet

The skeletal spec sheet is straight and to the point:

> 3.2 V, 2200 mAh, 58g, 99 × 34 × 8 mm, 40 C

That's it; no charging or discharging limits, no temperature rating, no curves, no life specs.

The skeletal spec sheet is in plain text on a web page, created by a vendor that may present exaggerated specs in response to exaggerations from competing vendors.

The skeletal spec sheet is sufficient for the hobbyist who doesn't want to be bothered with details and does not understand that the 2,200 mAh capacity is measured at 0.2C and not at a "C-rating" of 40C.

2.8.2.2 Useful Spec Sheet

The useful spec sheet contains two pages with all the essential data:

- The model number;
- A table of parameters and values (12 to 24 items);
- Between one and four curves;
- A dimensioned mechanical drawing.

The scientists and engineers who designed the cell generate this pdf file. It bears the mark of the manufacturer, which increases its trustworthiness.

This spec sheet is appropriate for the engineer who needs to select a cell and who needs to calculate the expected performance of a battery, configure a BMS, and set up useful incoming QA[67] parameters.

2.8.2.3 Bureaucratic Spec Sheet

The bureaucratic spec sheet is many pages long, formatted as a table, and written in the kind of language that you'd expect from a lawyer:

> Cell shall be visually inspected from no less than six directions to verify that no stains or discolorations are visible.

This style is tailored to the large, incoming QA department (see Section 2.9) familiar with inspecting screws. It gives the technicians in receiving something to

67. Quality assurance.

measure with a micrometer and the chance to reject a pouch cell because its tabs are 1% thicker than specified.[68] Yet they pass cells with excessive resistance because this parameter is not listed in the spec sheet and in any case they don't have the equipment to test cell resistance.

2.8.3 Reading Specification Sheets

When selecting a cell, it is crucial to know how to read and interpret the data in its spec sheet. Professional bettery designers try to verify the cell characteristics themselves (see Section 2.9), though only some of the parameters in a spec sheet may be verified.

- A few of the parameters can be measured directly and are not open to interpretation, such as nominal cell voltage, weight, and size;
- Some other parameters are completely unverifiable, such as calendar life, which may be specified to be longer than the cell has been in production;
- Other parameters are hard to define and are open to interpretation.

2.8.3.1 Cell Part Number

A part number is essential so that a cell may be precisely identified. Yet some manufacturers (even major ones[69]) don't bother to devise a part number. Occasionally, a given part number may be used for two different cells with different dimensions[70] or even different chemistry[71]!

Some model numbers are rather long and hard to read, such as those that include the three dimensions of a pouch cell,[72] making it all too easy to confuse two different cells with similar part numbers.

2.8.3.2 Voltage

The voltage listed in a spec sheet is just a rough classification of the cell:

- 2.3V for LTO;
- 3.2V for LFP;
- 3.6V or 3.7V for all other chemistries.

2.8.3.3 Capacity

The spec sheet lists the nominal capacity of a new cell at a low discharge current. Occasionally, it specifies the test current (such as 1C). Better cell manufacturers declare the minimum capacity. The actual capacity for new cells should always be higher than this value. A 10% headroom in the rated capacity allows the cells to reach the nominal capacity after they are cycled for the rated number of life cycles at standard conditions. Vendors and questionable manufacturers may exaggerate the capacity for marketing reasons.[73]

68. Just wait and see what happens when they measures the distance between the terminals with their metal calipers!
69. Altair Nano sells a "50 Amp Hour Cell" instead of coming up with a simple part number, such as AN-50.
70. Thundersky changed the mechanical dimensions of some of its cell multiple times, all using the same part numbers!
71. Gregory Plett reports seeing two mobile-phone batteries with the same part number yet different chemistry.
72. Dow Kokam's SLPB160460330H.
73. In some cases, cells have been tested to have a capacity that is 20% of the specified capacity!

2.8.3.4 Charging Instructions

The cell manufacturer specifies how the cell was charged during its testing and recommends you do the same. These charging instructions result in reasonably good cycle life and reasonable charge time. The spec sheet may also include fast charging instructions, though charging under those conditions will degrade the cell more rapidly.

The charging instructions include at least two parameters: CC and CV. They may also include termination current and temperature range.

The CC rating specifies the continuous current to be used to charge the cell initially (see Section 1.8.2.1). This is usually expressed as a specific current, such as 0.5C (recommended) or 2C (fast). During this phase, the cell voltage increases until it reaches the CV.

The CV rating specifies the voltage to be used to complete charging. During this phase, the cell current decreases naturally and exponentially towards 0A.

The CV rating is the terminal voltage, not the OCV. For example, the specs for an LFP cell specify a CV of 3.6V. This means that the charger should be set at a CV of 3.6V and must be turned off once the cell is charged. This does not mean that it is acceptable to keep a cell at an OCV of 3.6V—doing so would reduce its life.

Some vendors specify two values for CV to cater to hobbyists who do not use a BMS:

- One for cells by themselves (e.g., 3.6V for an LFP cell);
- One for cells used in a series string (e.g., 3.4V/cell for a string of LFP cells).

By lowering the CV, these manufacturers are telling those users to set their charger for the equivalent of 3.4V per cell, in the hope that fewer cells will be overcharged.

During the CV phase, charging should be stopped once the current drops below a *termination current*. The spec sheet often lists this limit (typically 0.1C). However, when charging a large battery, the charger may not be capable of generating 0.1C current. For example, a 1,500W charger connected to a 20 kWh battery charges it at a current of 0.075C, which is less than 0.1C. If following the directions in the spec sheet, charging would be stopped immediately, which would be absurd. In practice, when charging below 0.1C, there is no CV phase—the CC phase lasts until the cell voltage reaches the maximum (e.g., 4.2V) and then charging stops.

The minimum charging temperature is normally 0°C, while the maximum one may vary from 40°C to 60°C. Charging below freezing degrades the cell far more than charging at high temperatures.

Some spec sheets specify a peak charging current as well. This is the maximum charging current that can be accepted for a short time (such as 10s). It is useful when designing traction batteries that need to absorb current from regenerative braking. The peak charging current is different from the fast charging current, which is continuous.

2.8.3.5 Discharging Limits

The discharging limits are specified like the charging limits, with a few differences:

- Instead of a CV phase, there is a termination voltage;
- There is no termination current;
- The temperature range is wider, especially at the cold end;
- The maximum discharge current is larger.

A termination voltage is specified because typical loads do not have a constant voltage phase (unlike CCCV chargers). Instead, the cell manufacturers specify the terminal voltage at which discharging must stop. Compensating for the IR drop, an application may extend this point further, allowing the terminal voltage to drop below this limit, as long as the OCV remains above the limit.

As there is no constant voltage phase, a termination current is not applicable.

Cells can handle a wider temperature range when discharging than when charging. Both ends of the discharging temperature range are not walls but fuzzy continua. Therefore, both the minimum and maximum specified temperatures reflect marketing concerns as well as technical ones.

Be aware that the maximum discharge current is open to wild exaggerations.

2.8.3.6 Energy

Spec sheets rarely specify the energy stored in a cell (see Section 2.5.5). If they do, they specify the energy delivered in a slow discharge. Energy parameters are measurable. Therefore, these data are relatively reliable.

Some spec sheets specify energy density [Wh/l] and specific energy [Wh/kg] even though they can easily be derived from other specifications (capacity, voltage, volume, and mass). The tabs in pouch cells are not included in the energy density calculation because it is assumed that they will be cut or bent. Similarly, the studs in large cylindrical cells and large prismatic cells are not included.

Remember that the energy density and specific energy of a battery are lower than those of the cells within it, due to the additional mass and volume of the battery enclosure and other components.

2.8.3.7 Power

Maximum power is rarely specified. When it is, the conditions are unclear (see Section 2.7.2). Power cannot be precisely quantified, as it depends on

- Where "maximum" is defined;
- The SoC level;
- Whether charging or discharging;
- The duration (limited by heating);
- The recent history (was the cell heated by a previous cycle?);
- The ambient temperature.

For example, the specifications could include

- A set of curves showing maximum power versus SoC. These are for a pulse of a given duration, separate curves for charging and for discharging, and different curves for various temperatures.
- A curve showing maximum power throughout a discharge cycle when operated at the maximum power point, but limited by the maximum temperature.

Power density [Wh/l] and specific power [Wh/kg] may be given or can be approximated from other parameters (power, volume, and mass). Since power is ill-defined, power density and specific power are also ill-defined.

2.8.3.8 *Temperature*

Besides the charging and discharging temperature range, the specs may list a storage temperature range. Again, temperature limits are not hard walls, but fuzzy continua.

2.8.3.9 *AC Impedance and DC Resistance*

Knowing the DC resistance is quite useful as long as you understand that it varies considerably with SoC, temperature, and age; yet few manufacturers specify it.

Often, what's listed as "resistance" is actually the real part of the AC impedance at 1 kHz, which is of limited use to the battery designer (see Volume 2, Section A.3.5 and Section 2.6.3). On rare occasions, the specs also list a "DC impedance," which is really the DC resistance.

If DC resistance is specified, a charging and discharging resistance may be listed separately, though they tend to be quite close to each other.

2.8.3.10 *Cycle Life*

Cell manufacturers cycle their cells to measure the loss of capacity over time, then translate this to the number of cycles during which the cell is useful. There are multiple problems with this parameter:

- Tests are conducted under carefully controlled lab conditions that are quite likely to differ from the conditions in the application.
- Often the final threshold is not well-defined: it could be a capacity loss of 10% or 20%. A lower capacity threshold results in an apparent increase in the cycle life.
- The loss of capacity could be specified relative to either the initial capacity or the nominal capacity. If the nominal capacity is purposely understated, it takes many more cycles to reach a 10 % loss, resulting in an apparent increase in cycle life.
- In a power application, the true limitation is likely to be the increase in cell resistance, not the drop in capacity. A discharge cycle terminates sooner due to excessive voltage sag, even though the actual capacity is still high and the cells still have charge left in them.

Therefore, the cycle life parameter is of limited use to the battery designer. Yes, it can be used to compare different cells, but only if

- They all define capacity the same way;
- They use the same threshold for a drop in capacity;
- The application uses low current.

A cycle life curve, if provided, is more useful (see Section 2.8.3.13).

2.8.3.11 *Calendar Life*

Calendar life is an even less defined parameter than cycle life because it cannot be tested directly. It can only be estimated through accelerated testing by subjecting the cell to high temperatures and voltages.

2.8.3.12 *Mass (Weight) and Size*

Specs typically list the mass (in grams) and dimensions of the cell. Regrettably, specs for large prismatic cells rarely include the thread size and terminal spacing, which is required to plan connections to a BMS, especially a distributed one (see Section 4.2.2.7, "Distributed").

2.8.3.13 *Curves*

Any respectable spec sheet includes at least one curve.

A plot of full charge/discharge cycle over time (Figure 2.44(a)) illustrates the recommended charging and discharging instructions. This plot is taken at a single level for current and therefore is not terribly useful.

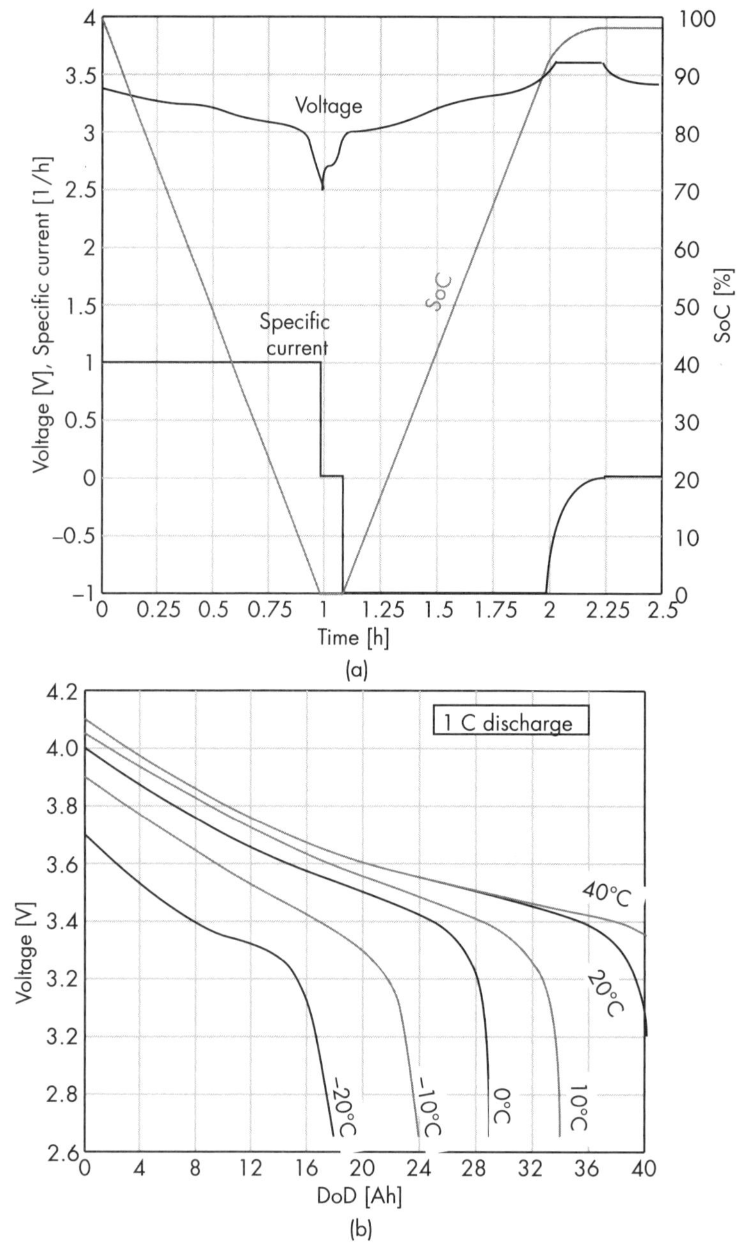

FIGURE 2.44
Curves: (a) discharge and charge curve, and (b) discharge curves at various temperatures.

A chart of discharge curves shows the terminal voltage versus SoC at multiple levels of specific current (Figure 2.45(a)). At times, the curves show actual current or constant power instead of specific current. This chart contains a wealth of information, more than people may realize. From it, one can extract the maximum power time, the resistance, the nominal capacity, the effective capacity at various levels of current, and the stored energy (Figure 2.45(b)).

For example, from this curve, we can derive that

- The minimum and maximum cell voltages are 2.7 and 4.2V;
- The MPT is 103s (see Section 1.5.5),[74]
- The resistance at 50% SoC is 1.7 mΩ,[75]
- The nominal capacity is 31.5 Ah, but at 7C the effective capacity is down to 28.4 Ah;

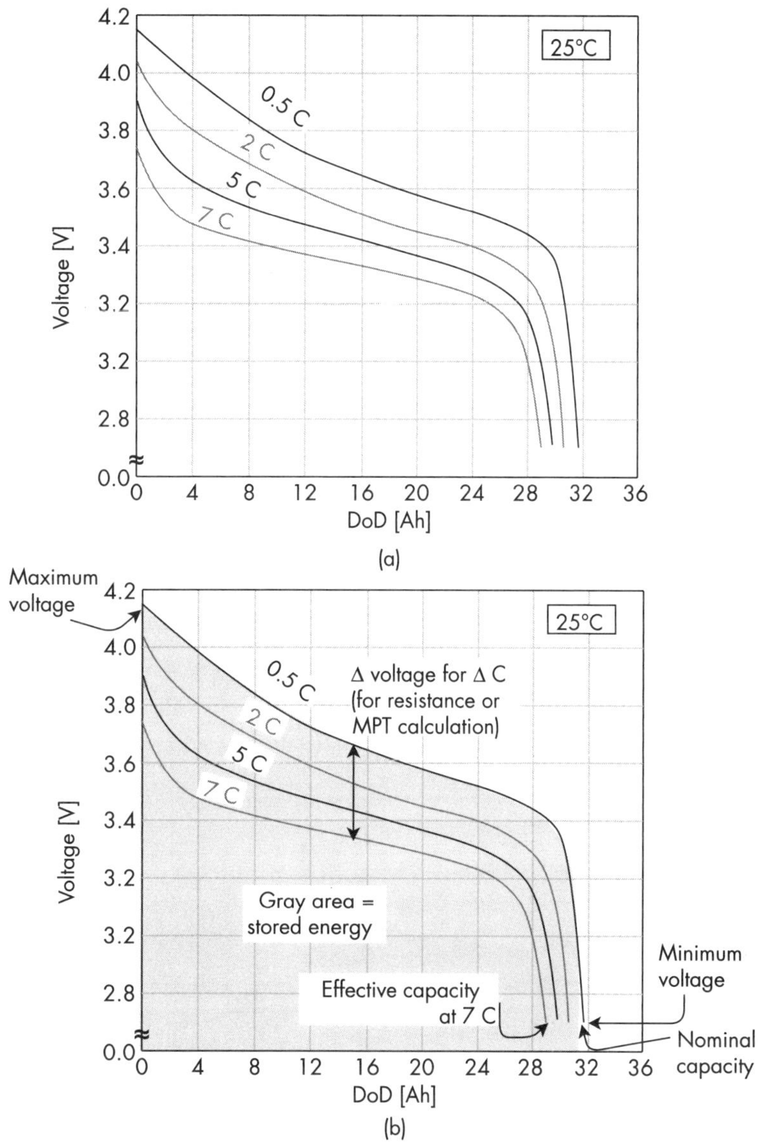

FIGURE 2.45 Discharge curves: (a) in the spec sheet, and (b) parameters that may be derived from it.

74. MPT [s] = 7200 × 0.34V/6.5 C/3.65V = 103 s. .

75. Resistance [Ω] = ΔV/ΔI = 0.34 V/(6.5 C × 32 Ah) = 1.7 mΩ.

- The stored energy is 114 Wh.[76]

The spec sheet may also include a set of discharge curves at different temperatures (Figure. 2.44(b)). These curves indicate the low-temperature performance of the cells at a given current (such as 1C). One may derive from them an approximate curve of DC resistance versus temperature.

Occasionally, in addition to discharge curves, one may also find charge curves (at multiple levels of charging current).

Another chart often found on spec sheets is a plot of capacity versus cycle number (Figure 2.46(a)). Note that the capacity decays slowly and linearly, dropping by 20% after 2,400 cycles.

This chart stops at 2,400 cycles, giving the impression that the capacity will continue to decrease linearly down toward 0 Ah after many thousands of cycles. In reality, past a certain point, the capacity crashes rapidly. From this chart, we don't know if the manufacturer stopped taking data at 2,400 cycles or if it cropped off unflattering data past this point.

A few manufacturers of high-power cells provide a chart of maximum charging and discharging power versus state of charge at various temperatures (Figure 2.46(b)).

The y-axis units may be power [W], power density [W/l], or specific power [W/kg]. The x-axis units may be SoC [%] or DoD [Ah].

This kind of graph is helpful when designing a power battery or buffer battery (see Section 5.1.4).

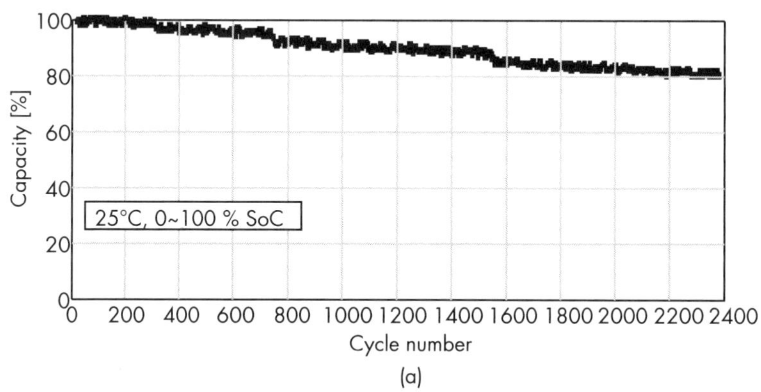

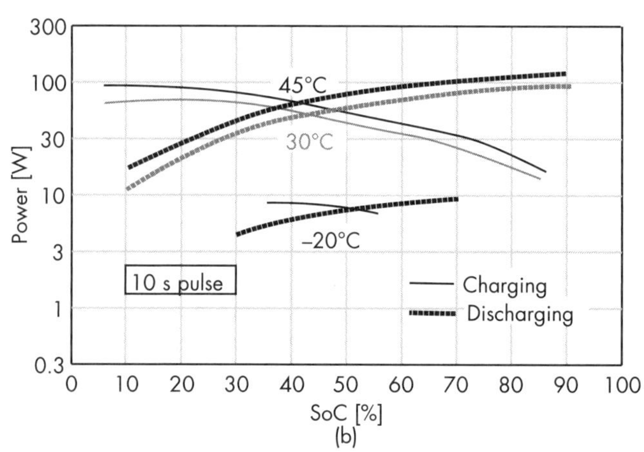

FIGURE 2.46
Spec sheet curves: (a) cycle life curve, and (b) power curves versus SoC at various temperatures.

76. From integrating the gray area under the 0.5 C curve.

2.8.4 Cell Sourcing

While technical considerations are critical in cell selection, ultimately, cost and availability are as important.

Major cell manufacturers (e.g., BAK, Panasonic, Samsung, Sanyo, Sony) court large automotive companies. The two companies work closely to ensure cell availability at the desired price.

These cell manufacturers are unlikely to offer such attention to medium-sized companies. These companies are likely to be served better by a smaller manufacturer (in the United States: K2 Peak Power,[77] and Enerdel come to mind), or manufacturers with a distribution network (for example, Chinese CALB has a warehouse and rep in California). You may need to start with cells bought from Alibaba, and, only when in steady production, you may approach the cell manufacturer directly, with a demonstrable history or production.

Hobbyists are likely to buy small cells from companies that serve the hobby market.[78] Prototype designers are likely to buy large cells from companies that serve the EV conversion market.[79]

I maintain a list of 18650 cells, sorted by energy density and maximum power time,[80] and a general list of Li-ion cell manufacturers.[81]

2.8.5 Second Use

By now, the world is full of Li-ion cells that are still viable but no longer in use. They may be found in

- Laptop batteries (Figure 2.47);
- Traction batteries of salvaged electric vehicles;
- Consumer products that were returned to the manufacturer and not refurbished.

The value of these cells is offset by the cost of retrieving them. Often, the labor to take apart a consumer battery costs more than a new set of cells. On the other hand, the 3–4 hours required to open a Nissan Leaf battery pack and remove all of its modules may be worth the effort. Add to this the cost of testing the cells to ensure they are still OK and of sorting them into bins according to capacity and resistance.

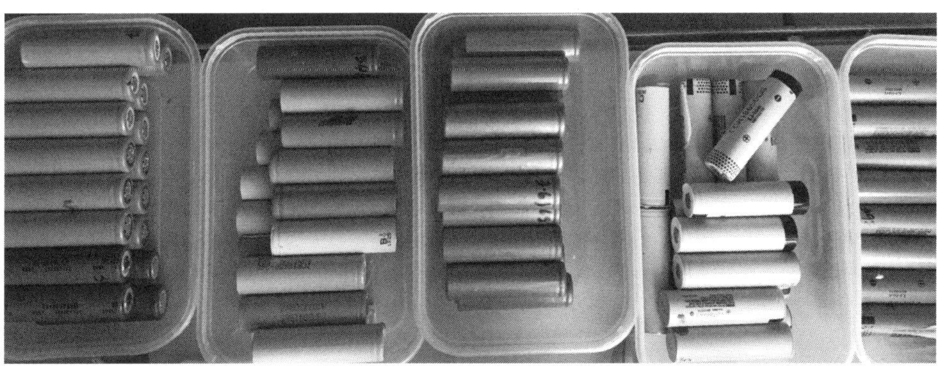

FIGURE 2.47
18650 cells salvaged from laptop batteries.

77. Boston Power failed during the writing of this book.
78. Battery Space, Hobby King.
79. Such as Evolve Electrics and Electric Car Parts in the United States and E-transportation and GWL Power in the European Union.
80. http://liionbms.com/php/small_cylindrical_cells.php.
81. http://liionbms.com/php/cells.php.

Compared to using new cells, using older cells carries a higher risk of premature battery failure. There are warranty costs associated with this risk.

For all those reasons, cells are rarely reused. When they are, it is generally because the labor costs are low:

- Hobbyists extract laptop cells to build battery banks (see Volume 2, Section 1.5.1) (generally a bad idea, but it happens all the time);
- In developing countries, cheap labor is used to extract and test cells from consumer products before reusing them in large, low voltage, stationary batteries.

2.8.5.1 Case Study: Laptop Cells

In Puerto Rico, thousands of small cylindrical cells were harvested from laptop batteries and then reassembled into five large 48V modules, all connected in parallel (Figure 2.48) for use with a solar system. A distributed digital BMS from Batrium protects the cells and communicates with the inverger.

The low cost of local labor and the value of off-grid power on an island that experiences frequent and long-lasting blackouts justified the effort.[82]

2.8.5.2 Case Study: Nissan Leaf Modules

There is a desire to find a second use for the tens of thousands Li-ion traction batteries for the Nissan Leaf EV after they may no longer be suitable for EV use. At this point, they may still have enough capacity for other uses.

FIGURE 2.48
A 48V, 70 kWh battery in Puerto Rico, made of second-use small cylindrical cells; five strings in parallel, 2,100 cells each in a 150P14S arrangement.

82. The most recent one, after Hurricane Maria, lasted 7 months!

People wishing to convert a used 350V traction battery to a 48V battery for residential solar applications contact me often. This conversion is not practical as it's either too messy or too expensive see Volume 2, Section 2.12.6).

2.8.5.3 Case Study: Chevy Volt Modules

Traction batteries from crashed Chevy Volt HEVs are now becoming available. Again, these are high-voltage batteries (355V) and are best reused as is. Yet vendors disassemble these batteries to harvest the modules (two each are 6S, and seven each are 12S). Each of the six or 12 blocks in a module consists of three pouch cells, bonded directly in parallel.

Each 12S module has a nominal voltage of 44V, which is not quite 48V, but close enough. When fully charged, the voltage is 50.4V, which is well within the range of an inverger or solar charge controller (see Volume 2, Section 2.5). Therefore, people are starting to buy these modules to create a ~48V battery for their solar system. For a BMS, they need a wired BMS—either a centralized BMS or master/slave BMS (see Section 4.2.2). Yet they have a problem finding a BMS that can handle strings in parallel, and, if they do, it costs on the order of $3,000, which in most circumstances is far more than the value of the harvested modules.

Again, the traction battery should be reused as is—as a 350V battery—as reusing its modules is economically nonviable.

2.9 CELL TESTING

Most end users accept the data in a cell's spec sheet at face value. A diligent company using a large volume of cells also tests a large sample of cells, not only to verify the data in the spec sheet but especially to gather valuable data that are not specified.

A set of new cells would be tested for actual characteristics:

- Charge and discharge curves, OCV versus SoC;
- Electrochemical impedance spectroscopy (EIS);
- Mechanical test.

Simultaneously, a different set of cells would be subjected to long-term testing:

- Calendar life over time and temperature;
- Cycle life over time and temperature;
- Self-discharge;
- Coulombic efficiency.

2.9.1 OCV versus SoC curves and Table

An accurate OCV versus SoC table (see Section 2.4.4) is required to evaluate SoC through voltage translation. Cell manufacturers do not typically offer such a table, and may not offer an OCV versus SoC curve, so you may have to generate your own.

The IR drop must be minimized when generating an OCV versus SoC curve. To achieve that, an extremely low test current is used, such as 0.01C. This means that a full charge and discharge cycle takes a total of 8 days and 8 hours.

This test produces two closely spaced curves, one for charging and one for discharging. A line drawn midway between these curves matches the OCV versus SoC curve to a great approximation (Figure 2.49).

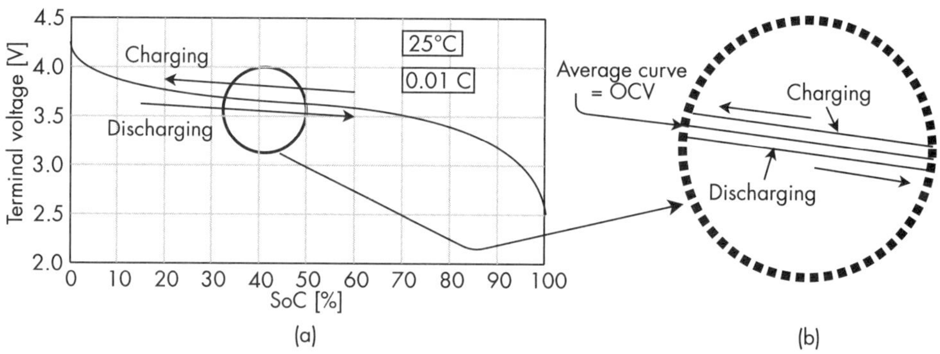

FIGURE 2.49
OCV/SoC curve: (a) derivation, and (b) detail showing the closely spaced discharging and charging curves and the OCV line between them.

Alternatively, the OCV versus SoC curve may be obtained faster (at a somewhat higher current), using *titration* (Figure 2.50) [36]:

Start:

- Charge the cell fully to the voltage that is defined as 100% SoC;
- Clear the integral used to measure the total charge.

Discharge:

- Discharge the cell slowly (C/10) for 10 minutes; measure the current *accurately,*[83] and integrate it to get the charge extracted from the cell during this time;
- Stop, wait another 10 minutes for the voltage to relax, measure it;
- Repeat the above two steps;
- Stop when the cell voltage drops to the voltage that is defined as 0% SoC.

Charge:

- Same as above but in the opposite direction (charging);
- Stop when the cell voltage reaches the voltage that is defined as 100% SoC; make sure to stop charging at the same exact SoC points as during discharge.

This test produces a stepped curve with some 60 steps[84] in each direction.
At this point, there are two sets of data:

- Terminal voltage versus SoC during charge;
- Terminal voltage versus. SoC during discharge.

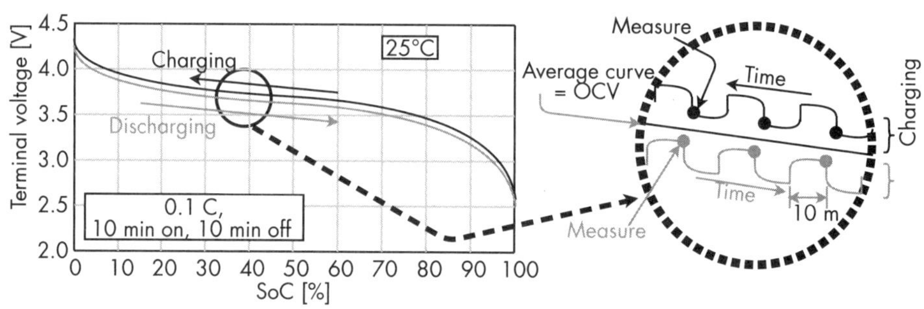

FIGURE 2.50
Determining OCV versus SoC through titration.

83. Better than 0.001 C.
84. C/10 means 10 hours or 600 minutes. If we stop every 10 minutes, that's 60 steps.

The OCV for a given SoC is obtained by averaging the measurements taken at this SoC level while charging and while discharging to remove the effects of relaxation and hysteresis. Use interpolation to derive the intermediate data points.

Now you may prepare a curve and a table of OCV versus SoC. The table should have about 200 points.

This table could use either SoC or OCV as the index, each having disadvantages:

- A table indexed by a constant step in SoC lacks precious detail at the two ends, where the OCV changes rapidly (Figure 2.51(a));
- A table indexed by a constant step in OCV lacks precious detail in the middle, where the OCV changes slowly (Figure 2.51(b));
- A table that picks data points judiciously to give detail where needed won't be evenly spaced, which may be a problem if the BMS requires you to enter the table with a constant step size (Figure 2.51(c)).

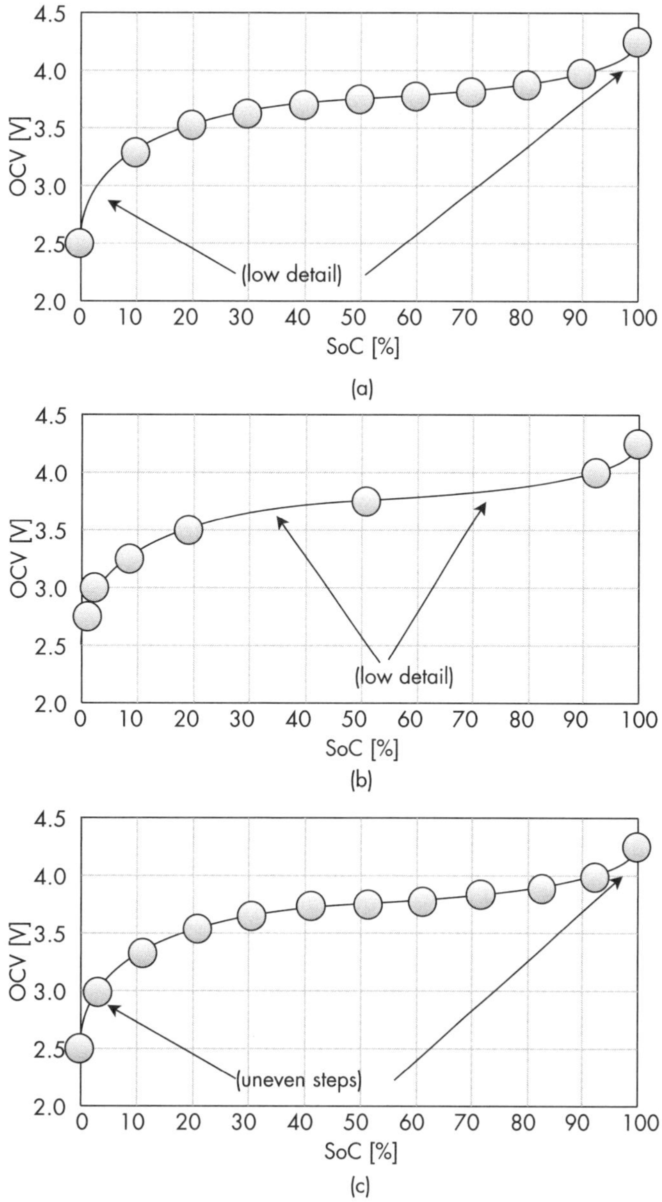

FIGURE 2.51
Order of OCV versus SoC table: (a) constant SoC step, (b) constant OCV step, and (c) constant travel along the curve.

From this data, one may also calculate the capacity ass the total charge extracted from the cell.

2.9.2 Charge and Discharge Curves

Charge and discharge tests characterize the cell under various conditions:

- At various levels of specific current (such as 0.2C, 0.5C, 1C, 2C, 5C, 10C), various levels of actual current or power, or both;
- At various temperatures (such as –20°C,[85] 0°C, 20°C, 40°C, and 60°C).

Cycle the cells, starting from a full cell:

- Discharge the cell under the selected conditions:
  - Measure the current *accurately* and integrate it to get the charge extracted from the cell; convert this to SoC using the capacity calculated in the previous tests:
  - Measure the terminal voltage
  - Measure the cell temperature
- Recharge the cell under the same conditions, taking the same measurements.

These tests produce a useful set of curves:

- Charge and discharge voltage versus SoC at various currents and temperatures;
- Plots of temperature versus time, showing how the cell heats up during use.

From those, you can calculate new curves:

- Cell resistance versus temperature at various SoC levels;
- Alternatively, cell resistance versus SoC at various temperatures;
- Cell resistance versus current (charging and discharging) at various temperature and SoC levels;
- Maximum power versus SoC at various temperatures, for charging and for discharging.

You can also calculate the nominal maximum power time at 50% SoC and room temperature. If you're brave, you can measure the MPT directly, by discharging a full cell completely when powering a maximum power point tracking load and seeing how long the charge lasts.

2.9.3 Electrochemical Impedance Spectroscopy

Electromechanical impedance spectroscopy (EIS) generates a *Nyquist plot* (see Volume 2, Section A.3.5) of the impedance of the cell over a wide range of frequencies. Doing EIS of several new cells allows spotting cell-to-cell variations (which may be an indication of manufacturing quality issues). It also provides a baseline to be used later when comparing the impedance for used cells to the one for new cells.

2.9.4 Mechanical Test

The standard cell-integrity tests are nail penetration and crushing. Other mechanical tests include

85. Except for charging: do not charge below freezing.

- For a pouch cell, measurement of the expansion versus SoC level and versus temperature;
- Effect of vibration on cell capacity, resistance, self-discharge;
- Effect of impact on cell capacity, resistance, and self-discharge;
- Effect on terminal integrity of vibration between the cell body and the terminals.

2.9.5 Calendar Life

Calendar life tests measure the change in cell characteristics over time. The cell is not used, except that the capacity measurement requires occasional cycling.

2.9.5.1 Capacity and Resistance Versus Time

The following tests determine how capacity decreases and resistance increases over time at various SoC levels and various temperatures.

To measure capacity:

- Charge the cell fully, using the manufacturer's recommended charge profile;
- Discharge the cell at C/20, measuring the current with high accuracy and integrating it;
- Stop when the voltage drops down to the manufacturer's recommended low-voltage cut-out;
- Note that the value of the integral is the total charge and, therefore, the capacity.

To measure the discharge resistance:

- Measure the cell voltage at rest;
- Connect the cell to a resistor that draws on the order of 1C;
- After 10s, measure the voltage and the current.
- Calculate the resistance as R [Ω] = (V-rest – V-loaded) [V]/Current [A].

To measure charge resistance, do the same but with a charger instead of a resistor.

Take this measurement with multiple cells (e.g., 10 cells) to spot variations from cell-to-cell.

To see how capacity and resistance change over time, take these measurements periodically (e.g., once a month) starting with a new cell, until the capacity starts dropping precipitously. Between tests, keep different cells at different temperatures (such as –20°C, 0°C, 20°C, 40°C, and 60°C) and SoC Levels (such as 0%[86], 25%, 50%, 75%, and 100%[87]). However, perform the tests themselves at room temperature. Note that when using the above recommendations, these tests require a total of 250 cells.[88]

2.9.5.2 OCV versus Time, Self-Discharge

The following test detects cells with excessive self-discharge current, indicating the consequences of leaving a product unused for an extended period.

Start from a full cell and measure its voltage periodically (e.g., once a month). Initially, the voltage drops faster; later it will drop more slowly for two reasons:

86. Of course, you cannot test discharge at 0% SoC.
87. Of course, you cannot test charge at 100%SoC.
88. 10 cell × 5 temperatures × 5 SoC levels = 250 cells.

- Self-discharge is reduced at lower SoC levels;
- The OCV versus SoC curve is flatter at medium SoC levels.

End the test when the cell voltage drops to the minimum voltage for the cell, when the voltage stabilizes, or when management defunds your lab, whichever occurs first.

Take this measurement with multiple cells (e.g., 10 cells) because there are variations from cell to cell. Keep different cells at different temperatures (such as -20°C, 0°C, 20°C, 40°C, and 60°C). There is no need to bring the cell back to room temperature for the measurement because the voltage of a Li-ion cell is hardly affected by temperature. Note that when using these recommendations, these tests require a total of 50 cells.[89]

2.9.6 Cycle Life

Cycle life tests measure the change in cell characteristics during use. Ideally, you would use a cycle that mimics your application. If this is unknown, you could use a standard cycle (as defined by the cell manufacturer) or an industry-standard cycle (such as those used in automotive testing.[90])

2.9.6.1 Capacity, Resistance and Coulombic Efficiency versus Cycle Number

The following tests reveal how capacity, resistance, and Coulombic efficiency change during use.

Start from a cell at the specified top SoC level and clear the count of cycles. Then cycle the cell:

- Discharge the cell until its SoC drops to the specified SoC level:
  - Discharge using the selected discharging profile
  - Measure the current *accurately*, and integrate it
  - Vary the current during discharge to enable calculation of the resistance:

$$R\ [\Omega] = (V1 - V2)\ [V] / (I1 - I2)\ [A]$$

  - Stop when the voltage drops *precisely* to the specified bottom level; at this point, the value of the integral is the total charge and therefore the effective capacity in that application
- Recharge the cell back to the specified top SoC level:
  - Use the selected charging profile
  - Measure the current *accurately*, and integrate it
  - Stop when the voltage reaches *precisely* the specified top level
  - The value of the integral is the total accepted charge
  - Try to calculate the coulombic efficiency as the ratio of the capacity over the accepted charge (hard to do)
  - Increment the count of cycles

Repeat the points above until the capacity starts dropping precipitously.

89. 10 cell/test × 5 temperatures = 50 cells.
90. Such as the USABC/FreedomCAR profile for EVs.

This test produces curves specific to your application:

- Capacity versus cycle number;
- Resistance versus cycle number and resistance versus SoC (multiple curves, at various SoC levels);
- Coulombic efficiency vs. cycle number (apply averaging to remove the noise in this measurement, which suffers from the fact that the efficiency is nearly 100%).

2.9.6.2 Self-Discharge versus Cycle Number

The following test reveals how cycling affects self-discharge:

- Run 100 cycles (no measurements);
- Measure the self-discharge (as described above) for a long time (such as one week);
- Repeat.

Do this test at various temperatures using different cells.

This test produces a set of curves of self-discharge versus cycle number and versus temperature.

References

[1] Gholam-Abbas, N., and G. Pistoia (eds.), Lithium Batteries: Science and Technology, New York: Springer; January 14, 2009.

[2] Yuan, X., H. Liu, and J. Zhang (eds.), Lithium-Ion Batteries: Advanced Materials and Technologies, Boca Raton: CRC Press, 2011.

[3] Ozawa, K., Lithium Ion Rechargeable Batteries: Materials, Technology, and New Applications, Weinheim, Germany: Wiley-VCH, November 23, 2009.

[4] TDK, "CeraCharge Rechargeable Multilayer Ceramic Battery," https://www.tdk-electronics.tdk.com/download/2427656/b08f30417762baafa9bd0358f7333716/dl---cera-charge-dbl.pdf.

[5] Buchmann, I., "BU-205: Types of Lithium-ion," Battery University, https://batteryuniversity.com/learn/article/types_of_lithium_ion.

[6] Weicker, P., A Systems Approach to Lithium-Ion Battery Management, Norwood, MA: Artech House, p. 31.

[7] Zhang, L., Z. Ning, H. Peng, Z. Mu, and C. Sun, "Effects of Vibration on the Electrical Performance of Lithium-Ion Cells Based on Mathematical Statistics," Applied Sciences, Vol. 7, No. 8, August 2017.

[8] Buchmann, I., "BU-808: How to Prolong Lithium-based Batteries," Battery University, http://batteryuniversity.com/learn/article/how_to_prolong_lithium_based_batteries.

[9] Electric Vehicle Wiki, "Battery Capacity Loss," http://www.electricvehiclewiki.com/Battery_Capacity_Loss.

[10] Electropaedia, "Battery Life (and Death),", https://www.mpoweruk.com/life.htm.

[11] Plett, G., Battery Management Systems, Volume I, Norwood, MA: Artech House, Figure 2.18.

[12] Dubarry M., et al., "Evaluation of Commercial Lithium-Ion Cells Based on Composite Positive Electrode for Plug-In Hybrid Electric Vehicle Applications III. Effect of Thermal Excursions without Prolonged Thermal Aging," Journal of the Electrochemical Society, Vol. 160, January 2013.

[13] Stevens, M., "Hybrid Fuel Cell Vehicle Powertrain Development Considering Power Source Degradation," 2009, UWSpace. http://hdl.handle.net/10012/4244.

[14] Zhang, C., J. Jiang, L. Zhang, S. Liu, L. Wang, and P. Loh, "A Generalized SOC-OCV Model for Lithium-Ion Batteries and the SOC Estimation for LNMCO Battery," Energies, http://www.mdpi.com/1996-1073/9/11/900/htm.

[15] Stroe, A.- I., J. Meng, D. I. Stroe, M. Swierczynski, R. Teodorescu, and S. Kær, "Influence of Battery Parametric Uncertainties on the State-of-Charge Estimation of Lithium Titanate Oxide-Based Batteries," Energies, March 2018.

[16] Ki-Yong Oh et al., "Rate Dependence of Swelling in Lithium-Ion Cells," Journal of Power Sources, Vol. 267, 2014.

[17] Dow Kokam, Pack Integration Handbook.

[18] Delaille, A., et al., "SIMCAL Project: Calendar Aging Results Obtained on a Panel of 6 Commercial Li-Ion Cells," 224th ESC Meeting, San Francisco, CA, October 27–November 3, 2013.

[19] Wikipedia article on Li-ion: https://en.wikipedia.org/wiki/Lithium-ion_battery.

[20] Buchmann, I., "BU-409: Charging Lithium-ion," Battery University, http://batteryuniversity.com/learn/article/charging_lithium_ion_batteries.

[21] Qiao, Y., et al., "Visualizing Ion Diffusion in Battery Systems by Fluorescence Microscopy: A Case Study on the Dissolution of LiMn2O4," Nano Energy, Vol. 45, 2018.

[22] Feng, F., R. Lu, and C. Zhu., "A Combined State of Charge Estimation Method for Lithium-Ion Batteries Used in a Wide Ambient Temperature Range," Energies, Vol. 7, No. 5, 2014.

[23] Dahn, J., WIN Seminar Series, https://www.youtube.com/watch?v=9qi03QawZEk.

[24] Plett, G., Battery Management Systems, Volume II, Norwood, MA: Artech House, p. 239.

[25] Weicker, P., A Systems Approach to Lithium-Ion Battery Management, Norwood, MA: Artech House, p. 153.

[26] Feng, F., R. Lu, and C. Zhu, "A Combined State of Charge Estimation Method for Lithium-Ion Batteries Used in a Wide Ambient Temperature Range," Energies, Vol. 7, No. 5, 2014, Figure 8.

[27] Waag, W., S. Käbitz, and D. U. Sauer, "Experimental Investigation of the Lithium-Ion Battery Impedance Characteristic at Various Conditions and Aging States and Its Influence on the Application," Applied Energy, Vol. 102, 2013.

[28] Farmann, A., W. Waag, and D. U. Sauer, "Application-Specific Electrical Characterization of High Power Batteries with Lithium Titanate Anodes for Electric Vehicles," Energy, Vol. 112 (C), pp. 294–306.

[29] Lu, L., X. Han, J. Li, J. Hua, and M. Ouyang. "A Review on the Key Issues for Lithium-Ion Battery Management in Electric Vehicles," Journal of Power Sources, Vol. 226, February 2013.

[30] Prof. Kung, Chung-Chun, 306 control lab.

[31] Schmalstieg, J., S. Käbitz, M. Ecker, and D. U. Sauer, "A Holistic Aging Model for Li (NiMnCo)O2 Based 18650 Lithium-Ion Batteries," Journal of Power Sources, Vol. 257, June 2014.

[32] Asakura, K., M. Shimomura, and T. Shodai. "Study of Life Evaluation Methods for Li-ion Batteries for Backup Applications," Journal of Power Sources, Vol. 119, June 2003.

[33] Khan, K. A., M. H. Bakshi, and A. A. Mahmud, "BryophyllumPinnatum Leaf (BPL) is an Eternal Source of Renewable Electrical Energy for Future World," American Journal of Physical Chemistry, Vol. 3, No. 5, 2014.

[34] Buchmann, I., "BU-802b: What Does Elevated Self-Discharge Do?" Battery University, https://batteryuniversity.com/learn/article/elevating_self_discharge.

[35] Jeff VanZwol, Micro Power Electronics, "Designing Battery Packs for Thermal Extremes," Power Electronics, Vol. 32, No. 7, 2006.

[36] Metrohm Autolab BV, "Galvanostatic Intermittent Titration Technique," Autolab Application Note BAT03.

CHAPTER 3

CELL ARRANGEMENT

3.1 INTRODUCTION

This chapter[1] discusses how two or more cells are connected in series, in parallel, or both to achieve the desired capacity and voltage.

3.1.1 Tidbits

Some interesting items in this chapter include:

- Save money and make a better battery by connecting cells in parallel first (3.4);
- Making a battery out of standard modules sounds great, but there are dangers and costs (3.5.2);
- Knowing the total voltage of a battery tells you little about the voltage of each cell (3.2.2.1);
- It's a myth that variations in cell resistance result in an unbalanced string (3.2.8);
- Balancing a good battery requires only 0.0003 C of specific current (3.2.9.1);
- Balance is optional: it's a performance function, not a safety function (3.2.9);
- Without a BMS, a cell voltage can be reversed during discharge (3.2.11);
- Balancing a string at around 50% SoC is hard, especially for an LFP battery (3.2.6.2);
- Adding a fuse on each cell in a parallel block is controversial (3.3.12);
- Connecting a battery directly to a load can blow up your BMS (3.2.12.1);
- Connecting two cells or batteries directly in parallel can be nasty (3.3.6);
- Yet, it may be not that bad if it's just two of them (3.3.6.1);
- As long as there's a BMS, it's OK to use mismatched cells in a battery (3.2.3);
- Connecting Li-ion *directly* to lead-acid or NiMH is a bad idea (3.5.6.2).

3.1.2 Orientation

This chapter starts by listing the four basic arrangements. It discusses issues specific to each of these arrangements: series, parallel, parallel-first, and series-first. Finally, it touches on arrangements other than the four basic ones.

1. Thank you to Byron Azarm for copy editing.

3.1.3 Basic Cell Arrangements

Inside a battery (or a single battery module) cells are connected in one of these basic arrangements:

- *Series string*: Two or more cells are connected in series—for example, a 12V lead-acid battery, a 9V alkaline battery, a traction battery for a hybrid car (Figure 3.1(a));
- *Parallel block*: Two or more cells are connected in parallel—for example, a 3.2V, 90 Ah battery from K2 which contains many 26650 small cylindrical cells in parallel (Figure 3.1(b));
- *Parallel-first*: Two or more cells are connected in parallel to form a block[2], then two or more such blocks are connected in series—for example, laptop batteries (Figure 3.1(c);)
- *Series-first*: Two or more cells are connected in series to form a string,[3] then two or more such strings are connected in parallel—for example, a product that is available in multiple options, such as small, medium, or large capacity (Figure 3.1(d)).

In practice, we primarily use only two of these arrangements: series and parallel-first.

Cells are often connected both in series and in parallel to achieve the desired voltage and capacity:

- Cells are added in parallel to increase the battery capacity, which conveniently also reduces the internal resistance;
- Cells are added in series to increase the battery voltage, though at the expense of higher internal resistance.

Most people see no difference between series-first and parallel-first since, with ideal cells in either arrangement, the complete battery appears to have the same specifications. Yet a parallel-first battery has additional wires connecting adjacent cells, which make all the difference: they improve performance if the cells are not identical, and reduce cost by allowing a simpler BMS.

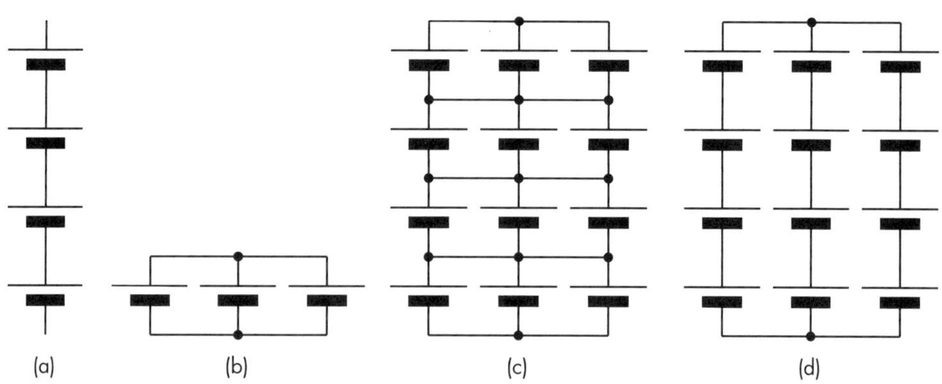

FIGURE 3.1 Arrangements: (a) series, (b) parallel, (c) parallel-first, and (d) series-first.

2. Gregory Plett calls this a parallel cell module (PCM), not to be confused with PCM for protector circuit module. [1]
3. Gregory Plett calls this a series cell module (SCM) [1].

Series-first and parallel-first *are* physically different and perform differently, despite the arguments of the naïve designer. This crucial point is often lost on many designers, even experts in this field, and deserves emphasis:

Connection order *does* matter

More complex arrangements are possible, expanding on the parallel-first or series-first arrangements. For example, a modular battery may consist of several modules connected in series, while inside each module, the cells may be connected in parallel-first.

3.1.4 Cell Arrangement Notation

With two or more cells, we use a standard notation to specify the cell arrangement, such as 4S for four cells in series (Figure 3.2(a)), or 3P for three cells in parallel (Figure 3.2(b).[4]

In a battery with cells both in series and in parallel, the notation specifies the complete cell arrangement. Conveniently, the order of S and P in the notation for the arrangement indicates whether cells are first connected in series or parallel.[5]

- 3P4S (parallel-first) means that three cells are connected directly in parallel-first to form a block; later, four such blocks are connected in series (Figure 3.2(c).
- 4S3P (series-first) means that four cells are connected in series-first, to form a string; later. three such strings are connected in parallel (Figure 3.2(d)).

Misunderstandings and time-consuming requests for clarification occur when people do not use this notation to specify the arrangement. The price for a BMS for a 100S4P battery is astronomical. After it becomes clear that the arrangement is actually 4P100S, a much cheaper BMS may be used.[6]

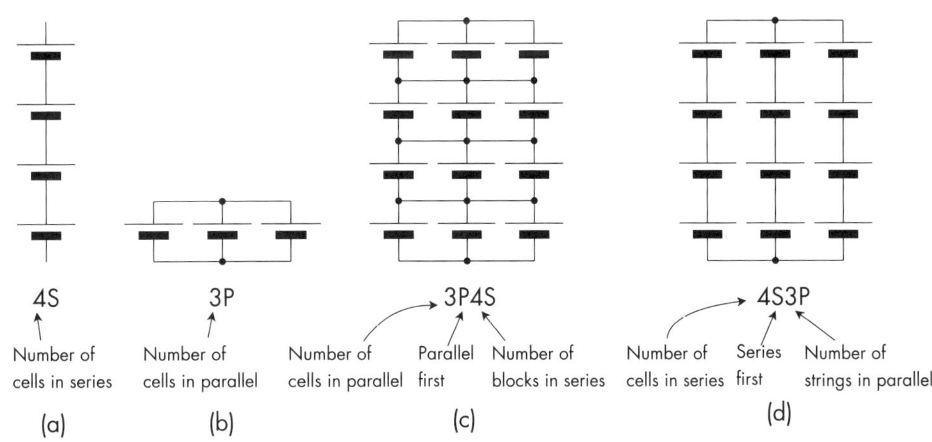

FIGURE 3.2
Cell arrangement notation: (a) series, (b) parallel, (c) parallel-first, and (d) series-first.

4. Sometimes, the notation 1S is used with a single cell, which is technically imprecise, since a single cell is no more 1S than it is 1P. Some people add 1P to the notation for a series string (e.g., 13S1P), or 1S to a parallel block. While this is unnecessary, technically it is not incorrect.
5. It has been suggested that I may be the originator and disseminator of this notation. While it's true that I came up with it independently, others have as well. For example, I have seen it in a NASA presentation.
6. A BMS for a 100S4P battery costs about $8000. For a 4P100S battery, it costs as low as $30. Why such a wide spread? Because the former is a specialized BMS that can handle strings in series and monitor 4,000 cells, while the latter is a Chinese protector BMS manufactured in high volume that only needs to handle a single string of 4 cells.

Different groups of people have different points of view about this notation:

- Experts in small batteries, who wouldn't even consider using a series-first arrangement have no need to distinguish the two arrangements; they have always placed the S first and insist that is the *only* correct notation. They have never heard of placing the P first for parallel-first and have no incentive to start doing so now.
- Experts in batteries of all sizes, who deal with both parallel-first and occasionally with series-first arrangements. They have a critical need to distinguish the two arrangements; once they are made aware of this notation, they welcome it.
- Everyone else. They may not even realize that the two arrangements are different, much less understand the need for such a notation.

For the sake of the second group, I encourage you to start using this notation and to make others aware of it as well.

I prepared an online utility to display the notation of the various arrangements and to help visualize their differences.[7]

3.1.5 Module Arrangement

Large batteries may be subdivided into modules for flexibility and ease of handling and placement. Regardless of how cells are arranged inside a module, several modules may be connected in one of the arrangements described above:

- Series (Figure 3.3(a));
- Parallel (Figure 3.3(b));
- Parallel-first (Figure 3.3(c));
- Series-first (Figure 3.3(d)).

Describing both the arrangement of the modules and the arrangement of the cells inside the modules can get complicated: "Two modules in series, each module with six cells in a 2P3S arrangement" (Figure 3.4(a)). It usually takes three sets of verbal back-and-forth to communicate such a complex arrangement. A drawing is unambiguous.

A notation such as 2P3S-2S is helpful. In more complex cases, a sentence such as "six modules in a 3P2S arrangement, each module with six cells in a 2P3S arrangement" (Figure 3.4(b)) may be more digestible than 2P3S-3P2S or other notation.

3.1.6 Cell Arrangement Characteristics

Table 3.1 compares the characteristics of each arrangement for a balanced battery (see Section 3.2.5) that uses identical cells.

In reality, cells are not identical. Table 3.2 compares the characteristics of each arrangement, given the actual characteristics of the cells. It still assumes the battery is balanced.

If the battery is unbalanced, its capacity is reduced (as is its energy); all other parameters are not affected (Table 3.3).

7. http://liionbms.com/php/wp_series_parallel.php.

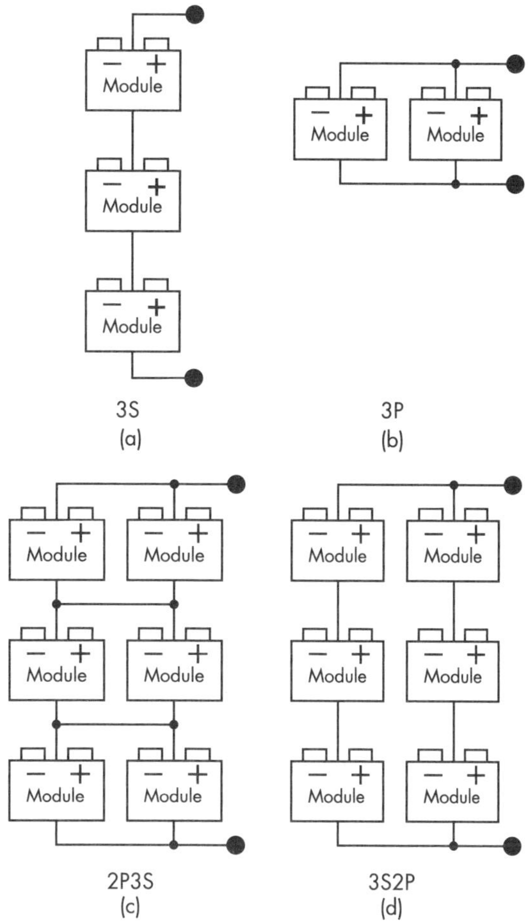

FIGURE 3.3
Module arrangements: (a) series, (b) parallel, (c) parallel-first, and (d) series-first.

TABLE 3.1
Specifications for Each Arrangement, Balanced Battery Using Identical Cells

	Series String	**Parallel Block**	**Parallel-First**	**Series-First**
Capacity	Cell capacity	Cell capacity × Pc	Cell capacity × Pc	Cell capacity × Ps
Resistance	Cell resistance × Sc	Cell resistance/Pc	Cell resistance × Sb/Pc	Cell resistance × Sc / Ps
Voltage	Cell voltage × Sc	Cell voltage	Cell voltage × Sb	Cell voltage × Sc
Current	Current in any one cell	Cell current × Pc	Cell current × Pc	Cell current × Ps
SoC	SoC of any cell			
Energy	~Total capacity x nominal battery voltage			
Power	Total current × battery voltage			
Max power	N × maximum power for a cell			
BMS taps	N +1	2	Sb +1	N - Ps +2

N = total number of cells, Sc = number of cells in series, Sb = number of blocks in series, Pc = number of cells in parallel, Ps = number of strings in parallel.

There are extensive considerations for each of the four arrangements, as explained in the following four main sections.

TABLE 3.2
Specifications for Each Arrangement, Balanced Battery Using Real-World Cells

	Series String	Parallel Block	Parallel-First	Series-First
Capacity	Capacity of the cell with the lowest capacity	Sum of capacities of each cell	Capacity of the parallel block with the lowest capacity	Sum of capacities of each series string
Resistance	Sum of resistance of each cell	1/(sum of (1/ resistance of each cell))	Sum of resistance of each parallel block	1/(sum of (1/ resistance of each string))
Voltage	Sum of voltage on each cell	Voltage of any cell	Sum of voltage on each block	Sum of voltage on each cell in any one string
Current	Current in any one cell	Sum of current in each cell	Current in any one parallel block	Sum of current in each string
SoC	SoC of the cell with the lowest capacity	SoC of any cell	SoC of the parallel block with the lowest capacity	~SoC of any string

TABLE 3.3
Specifications for Each Arrangement, Unbalanced Battery Using Real-World Cells

	Series String	Parallel Block	Parallel-First	Series-First
Capacity	Limited at one end by the cell that becomes fully charged first; limited at the other end by the cell that becomes fully discharged first	Sum of capacities of each cell (not reduced: a parallel block cannot be unbalanced)	Limited at one end by the block that becomes fully charged first; limited at the other end by the block that becomes fully discharged first	Limited at one end by the cell that becomes fully charged first; limited at the other end by the cell that becomes fully discharged first
All others	Same as for a balanced battery			

3.2 SERIES STRINGS

Cells are connected in a simple series in a string, to achieve the desired voltage (Figure 3.1(a)). Most multicell batteries use this topology.

3.2.1 Current in Series Strings

In a string of cells in series, the same current flows from one cell to the next one. However, this may be modified while balancing the string.

3.2.1.1 Charging and Discharging

In general, the string is bulk charged (Figure 3.5(a)), meaning that the current flows equally into each cell's positive terminal. Over a given period, each cell receives the same amount of charge, regardless of its SoC. Once any one cell is full,[8] there's no way to use just the bulk charger to finish charging the other cells.

Similarly, in general, the string is bulk discharged (Figure 3.5(b)), meaning that the current flows equally out of each cell's positive terminal. Consequently, once any one cell is empty, the load has no way to use any of the remaining charge in the other cells. Discharging must be stopped to avoid overdischarging the empty cell.

8. I use "full" as a shorthand for "at 100% SoC" and "empty" as a shorthand for "at 0 % SoC.or thereabouts"

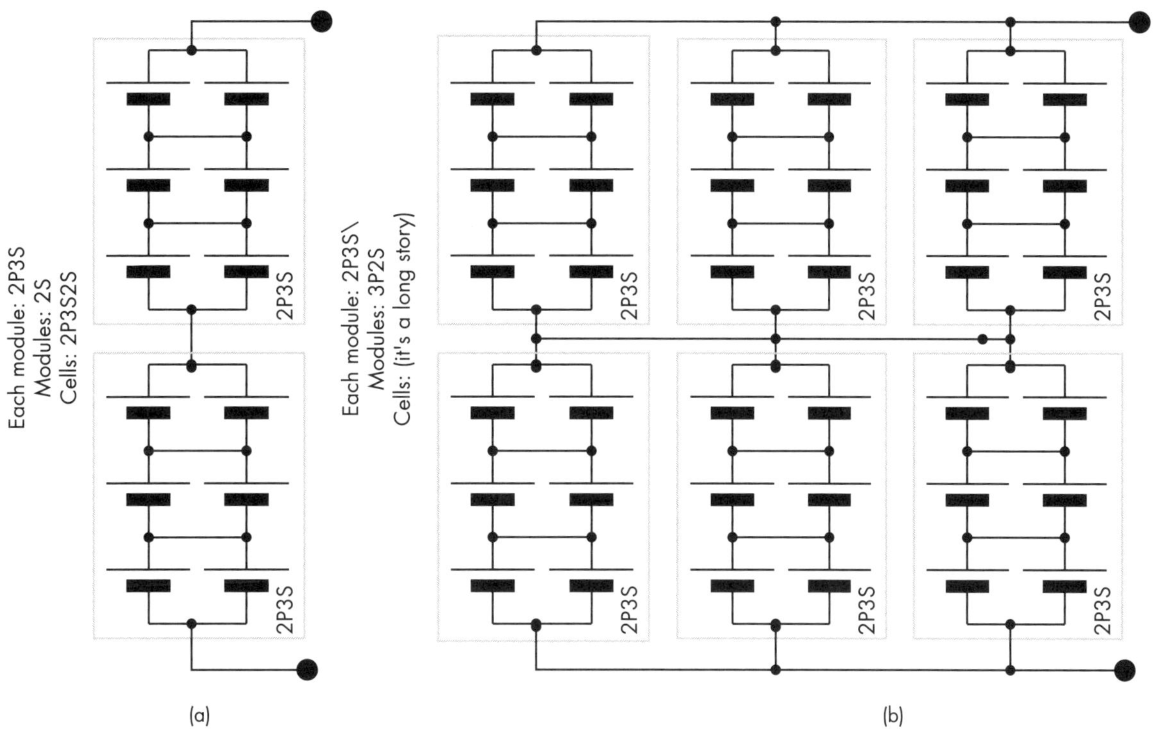

FIGURE 3.4 Notation for battery modules and cells inside the modules: (a) 3P2S modules, each 3P4S (2P3S-2S), and (b) 2P3S-3P2S.

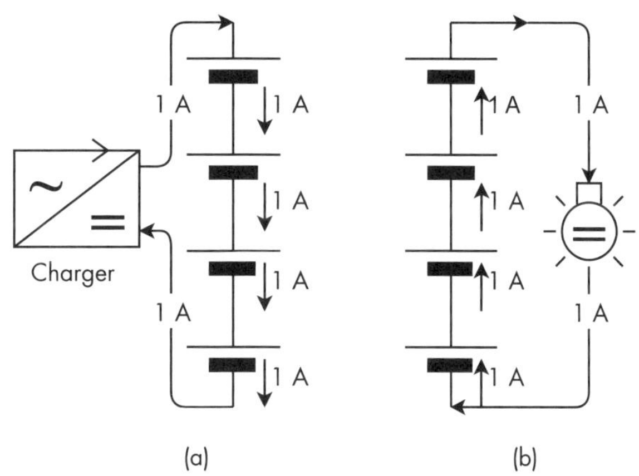

FIGURE 3.5
Current in a series string:
(a) bulk charging, and
(b) bulk discharging.

The only way to control the current in individual cells is through tap wires connected to the junctions between adjacent cells. These solutions control the current in individual cells in a beneficial way:

- Distributed charging (such as with a *balance charger* (see Section 6.5.4) uses a *balance connector* (see Volume 2, Section 1.3.3) to charge each cell individually (Figure 3.6(a));
- Bypass balancing removes some charge from an individual cell (Figure 3.6(b)) (see Section 3.2.9);
- Charge-transfer balancing may add some charge to an individual cell (Figure 3.6(c)) or remove some charge from it (see Section 4.7.2).

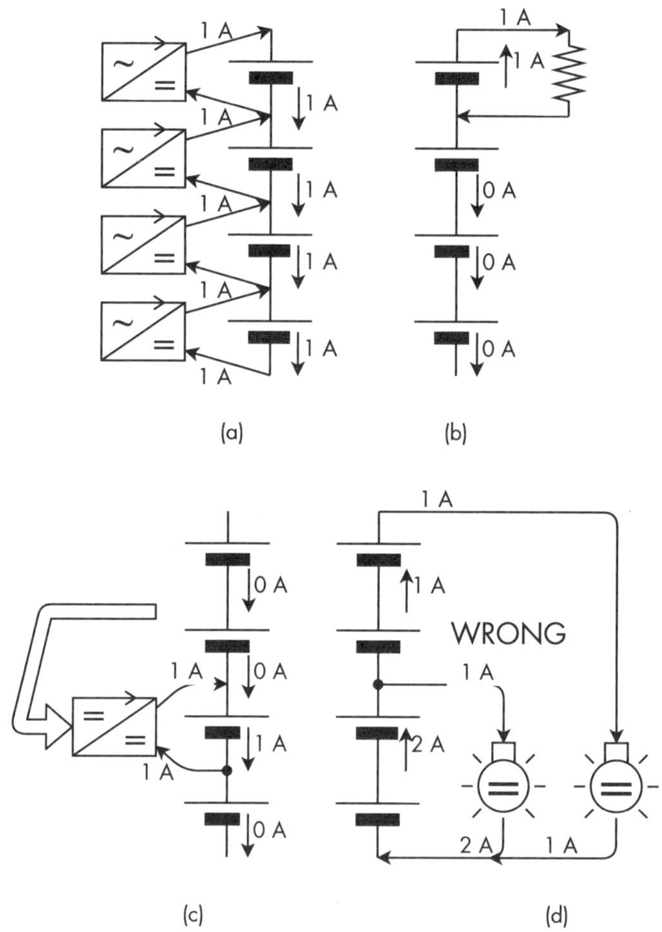

FIGURE 3.6
Current to individual cells in a series string: (a) distributed charging, (b) balancing by removing charge, (c) balancing by adding charge, and (d) load powered by tap.

Any other intermediate connections are not beneficial. A load connected to only half the string draws current from just some cells (Figure 3.6(d)), resulting in a severely unbalanced string (see Volume 2, Section 2.6.2.4).

3.2.1.2 *Stopping Charging and Discharging*

When any cell reaches the maximum voltage, charging must stop, even if other cells in the string are not yet full (Figure 3.7(a)).[9] Otherwise, the most charged cell will be overcharged. Charging may be restarted after some charge is removed from the full cells, through balancing (see Section 3.2.9).

Similarly, when any cell reaches the minimum voltage, discharging must stop, even if other cells in the string are not yet empty (Figure 3.7(b)). Otherwise, the most discharged cell will be overdischarged.

3.2.1.3 *Main Fuse*

A fuse should be placed in series with the cells to protect them in case of a short circuit. The fuse also provides secondary protection should the BMS fail to react to an overcurrent.

9. After the fullest cell reaches its maximum voltage, charging may be continued a while longer by reducing the charging current to maintain a constant voltage across the fullest cell until the current drops to zero. In practice, this is rarely done, as balancing allows charging all the cells equally.

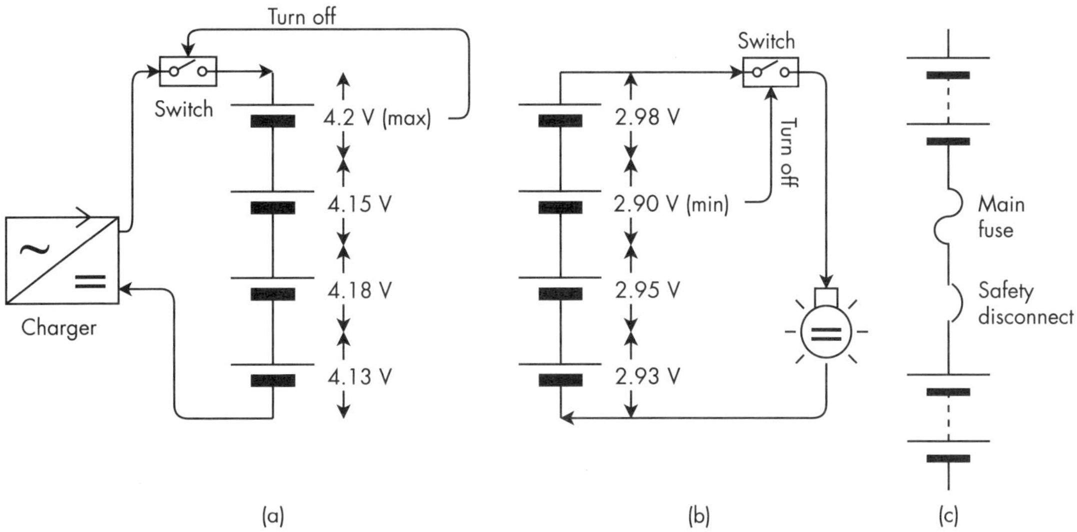

FIGURE 3.7 Protecting cells in a series string: (a) charging stops when any cell voltage is max, (b) discharging stops when any cell voltage is min, and (c) placement of main fuse and safety disconnect.

The ideal placement is midpack, as this is where it is statistically most likely to blow in case of a short circuit between any two points inside the string (Figure 3.7(c)). However, placing the fuse midpack requires a BMS with two or more banks, so that different banks can handle the cells on either side of the fuse (see Section 5.9.1.5). If the BMS has a single bank, the fuse must be placed at one end of the string of cells.

In many applications, a circuit breaker may be used instead of a main fuse.

3.2.1.4 Safety Disconnect

In large batteries, a safety disconnect allows the user to disable the battery. For instance, this is useful for transportation, service, or in case of an emergency (Figure 3.7((c)). The safety disconnect is critically important. It is carefully regulated in automotive applications.

The disconnect could be

- A high-power switch;
- A connector into which a shorted shunt is plugged in (a.k.a. *loop key*);
- A connector that contains the main fuse (e.g., many electric cars) (see Volume 2, Section 3.14.2.3);
- A contactor, controlled by a low-power switch accessible to the user (see Volume 2, Section 3.9.1.6,);
- A circuit breaker (in which case a main fuse may not be needed) (see Volume 2, Section 2.12.1.2);
- The battery's connector itself.

The best placement for a safety disconnect is midpack, the same as for the main fuse.

3.2.2 Voltage in Series Strings

Care is required to manage the voltages of individual cells in a string.

3.2.2.1 Cell Voltage Distribution

You may not assume that the string voltage is divided equally among the cells in a string (Figure 3.8(a)) (see Section 3.2.5). In reality, the string voltage could be distributed unequally (Figure 3.8(b)), even dangerously so, as some cells could have too high a voltage, too low a voltage, or even a reversed voltage (Figure 3.8((c)).

In all these examples, the cell voltages add up to the same total string voltage, yet the cells are safe only in the first case. Knowing just the string voltage tells us nothing about the individual cell voltages. This is why a BMS must monitor the voltage of each Li-ion cell in series. Contrast this with a 12V lead-acid battery, in which the cell voltages are not monitored (and, indeed, are not even accessible) because a lead-acid battery self-balances during overcharging.

3.2.2.2 Maximum String Voltage

In theory, any battery voltage may be achieved with enough cells in series. Each cell only sees the voltage across it and has no idea whether it's part of a 12V battery or a 1,200V battery. In practice, there is a limit, not because of the cells, but because of what's next to them (e.g., a metal enclosure) or connected to it (i.e., the BMS). Special precautions are required in high-voltage batteries (see Volume 2, Section 4.7).

3.2.3 Mismatched Cells in Series Strings

The cells in a properly designed and assembled string should all be well matched; but should they not be, it is helpful to understand the effect:

- Connecting otherwise identical cells in series but at different SoC reduces the capacity of the string because when the most charged cell is full, a bulk charger cannot finish charging the other cells, and when the least charged cell is empty, the load cannot finish discharging the other cells (Figure 3.9(a)).
- Connecting cells of different capacity in a series string is possible, though it is disadvantageous because the capacity of the string is (at best) the capacity of the cell with the lowest capacity (Figure 3.9(b)).
- With a properly installed BMS, connecting cells of different internal resistance in series is not a safety problem; under high current, the BMS prevents overvoltage or undervoltage of the cell with the highest resistance (Figure 3.9(c)). However, the performance of the string is limited. Sensing the temperature of each cell is important to prevent overheating of the cell with the highest resistance.

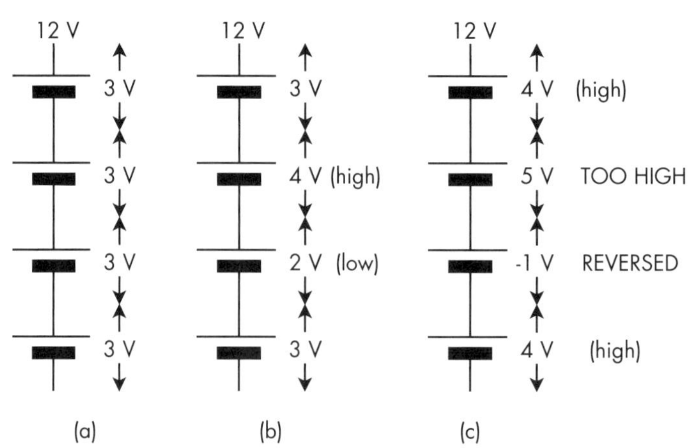

FIGURE 3.8
Voltage sharing: (a) equal, (b) unequal, and (c) dangerously unequal.

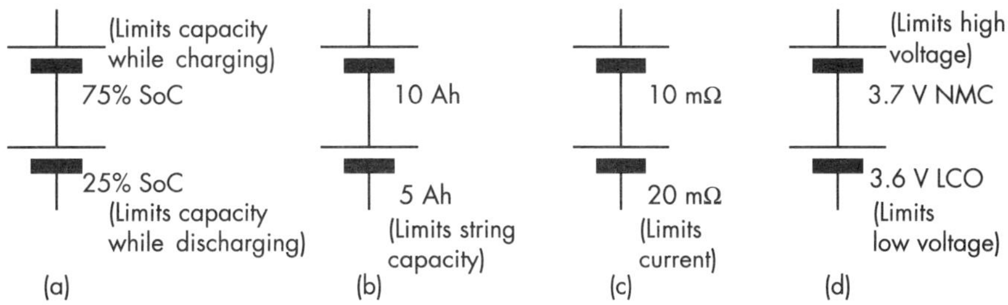

FIGURE 3.9
Mismatched cells in series:(a) different SoC, (b) different capacity, (c) different resistance, and (d) different chemistry.

- Connecting cells of different chemistry in series (e.g., 3.6V LCO cells and 3.7V NMC cells), or cells with different history (as a hobbyist may do with reclaimed cells), is not inherently a problem (as long as a BMS is properly included), but results in suboptimal performance (Figure 3.9(d)).

A string with mismatched cells may perform poorly but is not dangerous because a properly configured and installed BMS ensures that each cell is operated within its safe operating area (see Section 2.3.2).

3.2.4 String Capacity, String SoC

The capacity of a single cell is directly related to the condition of the cell itself and is independent of its state. In contrast, the capacity of a series string is an indirect consequence of the state and condition of each of its cells. The capacity of a string is related to how much charge the string can accept and how much charge it can deliver:

$$\text{string\_capacity [Ah]} = \text{DA [Ah]} + \text{CA [Ah]} \tag{3.1}$$

Where

- DA = discharge availability—how much charge the string can deliver, which is set by the cell with the least remaining charge [Ah];
- CA = charge acceptance—how much charge the string can accept, which is set by the cell with the least space available to receive more charge [Ah].

Similarly, the SoC of a single cell is directly related to the state of the cell itself and is independent of its condition. In contrast, the SoC of a series string is an indirect consequence of the state and condition of each of its cells. The SoC of a string is related to how much charge the string can accept and how much it can deliver:

$$\text{string\_SoC [\%]} = 100 \times \text{DA [Ah]}/\text{string\_capacity [Ah]} \tag{3.2}$$

If the string is balanced (see Section 3.2.5), the cell with the lowest capacity sets the capacity and SoC of the string. For example, with two cells in series, one 5 Ah and one 10 Ah, the string capacity is 5 Ah (Figure 3.9(b)). If the string is not balanced, its capacity is lower.

String SoC becomes less and less meaningful as a string is more and more imbalanced. In the extreme case that one cell is at 100% SoC and another cell is at 0% SoC, the DA and CA are both 0 (we can neither charge nor discharge), so the string capacity is 0. Then the equation for string SoC has a 0 in the denominator. Dividing by 0 is a no-no. In other words, string SoC ceases to have any meaning when a string is 100% imbalanced.

3.2.5 String Imbalance

In this context, *imbalance* refers to the mismatch in the SoC, and therefore voltage, among the cells in a series string. *Balancing* refers to the act of reducing this mismatch by transferring charge away from fuller cells or by transferring charge to emptier cells.

This section discusses the state of balance of a string; how a string behaves when it's balanced or unbalanced.

3.2.5.1 Balanced String

In an ideal string, all the cells have the same capacity and are at the same SoC at all times, which is also the SoC of the string (Figure 3.10(a))

- When the string is at 100% SoC, all the cells are also at 100% SoC, and they all limit any further charging;
- When the string is at 50% SoC, all the cells are also at 50% SoC;
- When the string is at 0% SoC, all the cells are also at 0% SoC, and they all limit any further discharging.

3.2.5.2 Unbalanced String

Even if all the cells have the same capacity, if the cells are at different SoC levels, the string is imbalanced (Figure 3.10(b)), because

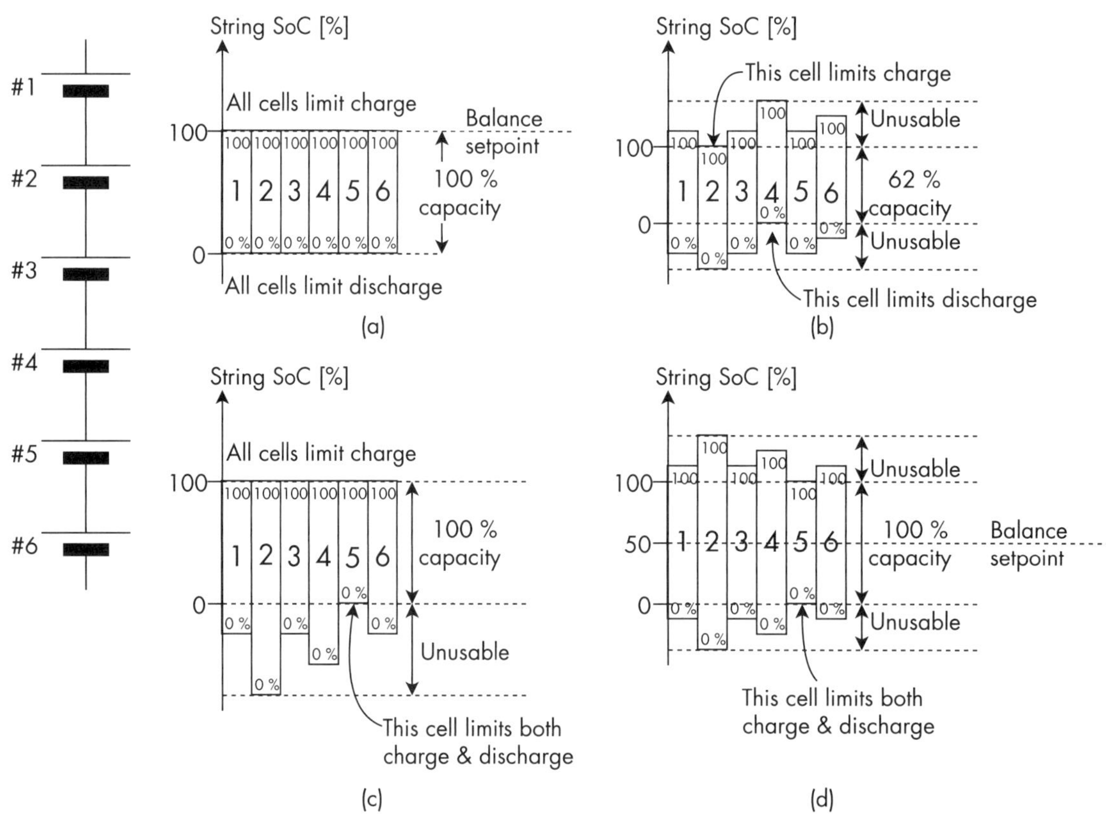

FIGURE 3.10 Balance of a string of six cells in series. Each rectangle represents one of the cells, the height represents its capacity, and the vertical position represents its SoC. (a) Same capacity, balanced, (b) same capacity, unbalanced, (c) different capacity, top-balanced, and (d) different capacity, midbalanced.

- When cell #2 is at 100% SoC, it limits any further charging. Therefore, the string is at 100% SoC; all other cells still have some more space, but a bulk charger cannot charge them further because doing so would overcharge cell #2.
- When the string is at 50% SoC, the SoC of each cell ranges from about 30% to 70%.
- When cell #4 is at 0% SoC, it limits any further discharging. Therefore, the string is at 0% SoC; all other cells still have some more charge left in them, but a load cannot discharge them further because doing so would overdischarge cell #4.

Note that two different cells limit the capacity of the string, #2 at the top, and #4 at the bottom. This is a sign of an unbalanced string.

3.2.5.3 Variable Capacity, Top-Balanced

If the capacity varies from cell to cell, the string for an energy or power battery (see Section 5.1.4) can be top-balanced (Figure 3.10(c)). The limits are

- When the string is at 100% SoC, all the cells are also at 100% SoC, and they all limit any further charging.
- When cell with the lowest capacity (#5) is empty it limits any further discharging. Therefore, the string is at 0% SoC; all other cells still have some more charge left in them, but a load cannot discharge them further because doing so would overdischarge cell #5.

Note that a single cell (#5) limits the capacity of the string for both charging and discharging; this is the sign of a balanced string.

3.2.5.4 Midbalanced String

The string can also be midbalanced[10] (Figure 3.10(d)). The limits are

- When cell with the lowest capacity (#5) is full it limits any further charging;
- When the string is at 50% SoC, all the cells are also at 50% SoC;
- When cell #5 is at 0% SoC, it limits any further discharging.

Again, note that a single cell (#5) limits the capacity of the string for both charging and discharging: once more, this is the sign of a balanced string.

3.2.5.5 State of Balance, State of Imbalance

At a given time, the *state of imbalance* (*SoI*) of a string is the difference between the SoC of most charged cell in the string and the SoC of the least charged cell. For example, if the SoC of the most charged cell is 75%, and the SoC of the least charged cell is 70%, then the SoI of this string is 5%. For a string whose cells are all at the same SoC, the SoI is 0%.

The *state of balance* (*SoB*) of a string is 100% minus the SoI. If the SoI of a string is 5%, its SoB is 95%.

In the discussion above, I used boxes to represent the range of SoC for each cell. If this representation does not work well for you, I hope that the following plots of cell voltages and SoC levels versus string SoC may convey the concept more clearly.

10. For buffer batteries.

3.2.5.6 Same Capacity, Balanced

Ideally, all the cells in a series string have the same capacity, and at any given time, the same SoC (Figure 3.11). As the string is charged or discharged, all the cell voltages remain in lockstep. The SoC of each cell is the same as the SoC of the string (e.g., when the string is at 50% SoC, each cell is also at 50% SoC).

You may recognize the top curve as simply the cell OCV versus SoC for an NMC Li-ion cell. All three cells follow this curve. The bottom curve is the cell SoC versus string SoC, which, in this case, is 1:1—a straight line from 0% to 100%.

The schematic diagram of three cells in series at 50% SoC corresponds to the vertical line midway through both plots.

All cells are at 100% SoC when the string is full; no single cell limits charging—they all do so at the same time. Similarly, all the cells are at 0% SoC when the string is empty; no single cell limits discharging—they all do so at the same time. The capacity of the series string is the same as the capacity of each cell (10 Ah in the example). That is, the entire capacity of each cell is accessible. Therefore, the state of balance of the string is 100% (meaning that the string is fully balanced), and its state of imbalance is 0% (i.e., there is no imbalance).

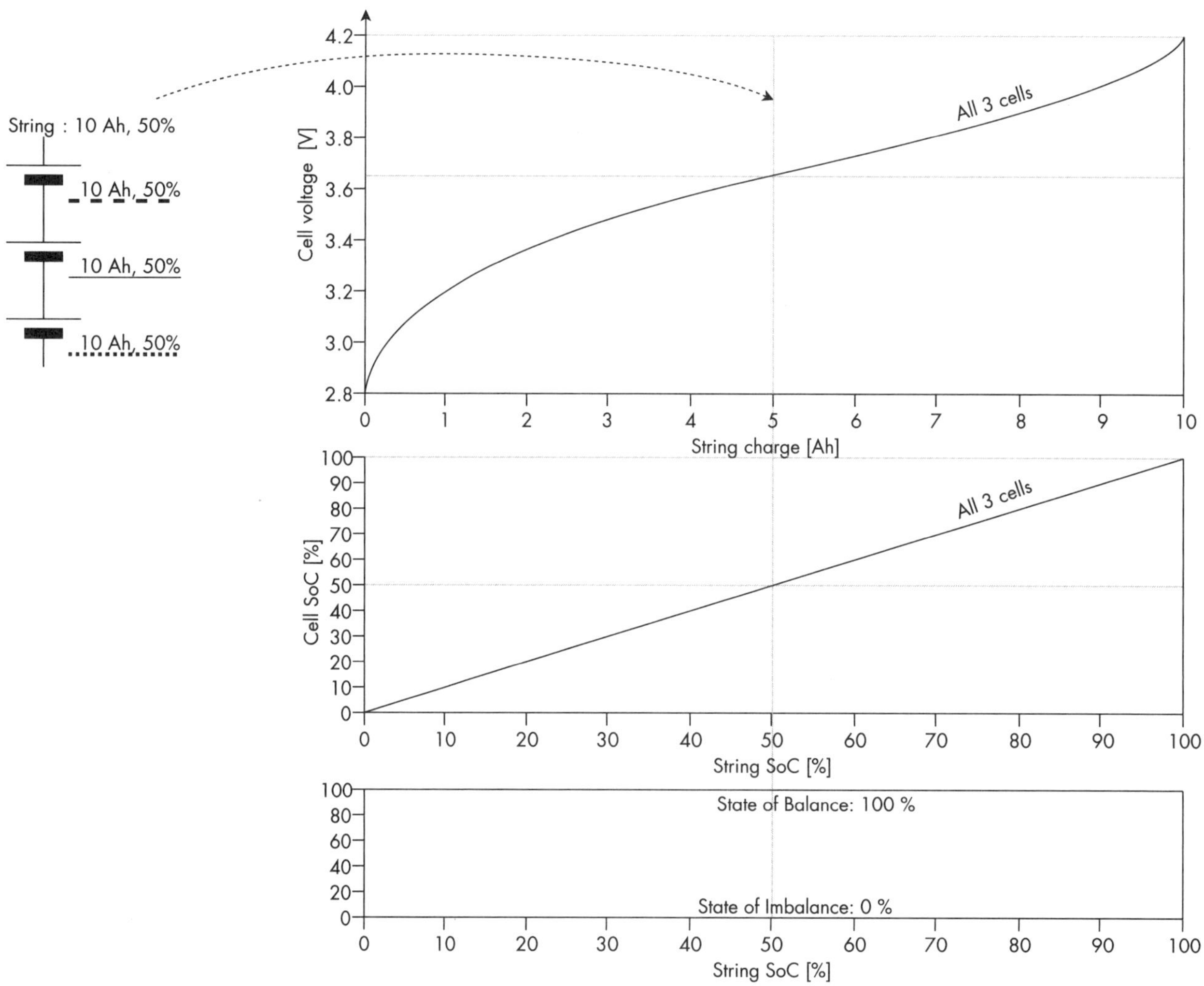

FIGURE 3.11 String of three cells of equal capacity, balanced.

3.2.5.7 Same Capacity, Unbalanced

If the cells have the same capacity but do not start at the same SoC (Figure 3.12), then one cell limits charging (cell #1 in this example), and a different cell limits discharging (cell #3). As a result, the capacity of the string is reduced.

The top curves are the "voltage versus charge" for the three Li-ion cells. Note that the shape of the curves is the same. They are just shifted left (cell #1 with higher SoC) or right (cell #3 with lower SoC).

The bottom curves are the "cell SoC vs. string SoC" for the three cells (which in this case is not 1:1). Note that the slopes are the same because all three cells have the same capacity. They are just shifted left or right. Specifically:

- When the string is at 0% SoC (left end of the plot), cell #3 is empty and discharging is disabled. The other two cells still have a higher voltage (cell #2 is at 3.38V, cell #1 is at 3.6V) and still have charge in them (cell #2 is at 20% SoC, cell #1 is at 40% SoC). However, this charge is not available because a load cannot access it without overdischarging cell #3.
- When the string is at 50% SoC (vertical line in the middle of the plot), cell #1 is at 70% SoC, cell #2 is at 50% SoC, and cell #3 is at 30% SoC.

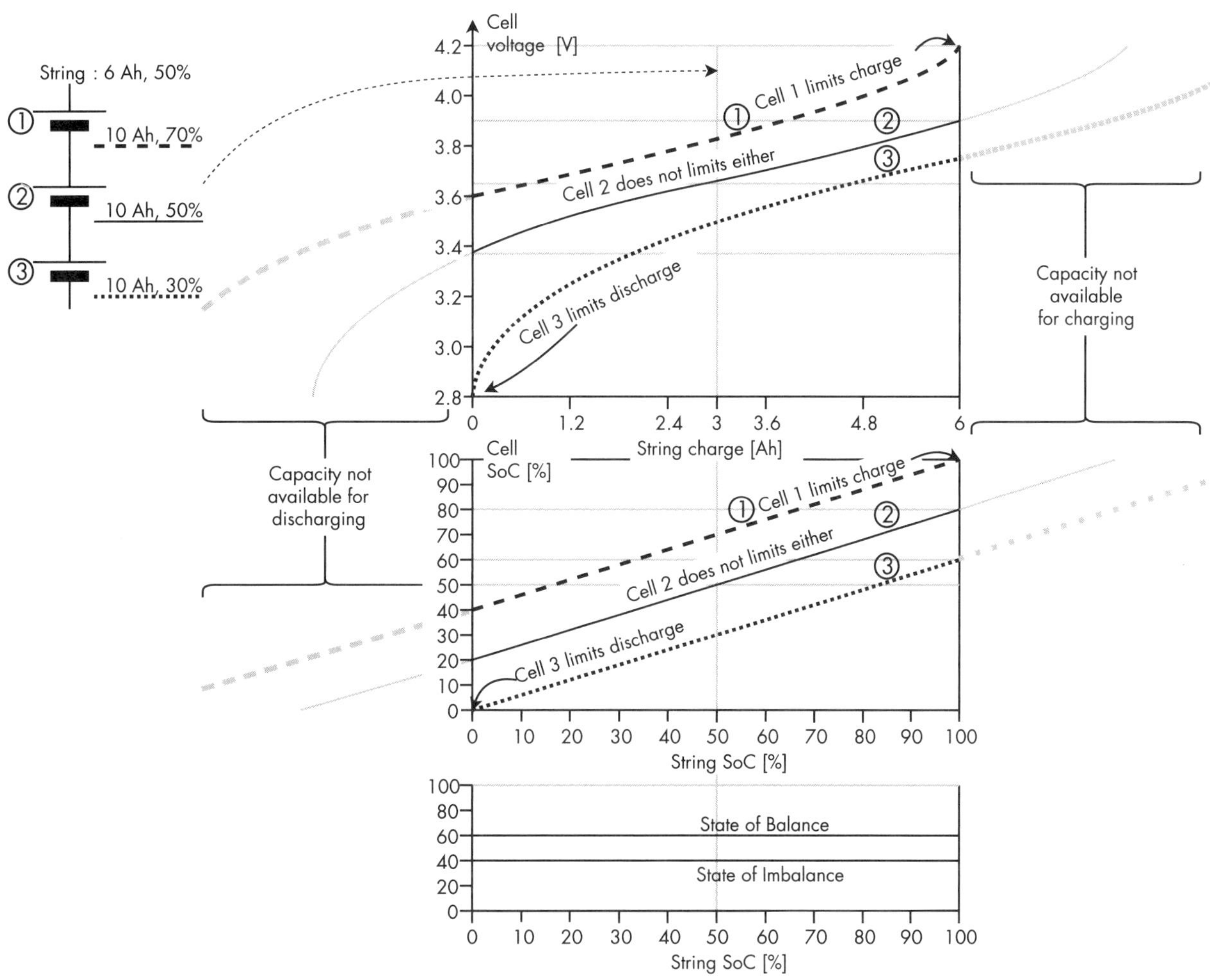

FIGURE 3.12 String of three cells of equal capacity, unbalanced.

- When the string is at 100% SoC (right end of the plot), cell #1 is full, and charging is disabled; the voltage of the other two cells is lower (cell #2 is at 3.87V, cell #3 is at 3.75V) and could still accept more charge (cell #2 is at 80% SoC, cell #3 is at 60% SoC). However, this space is not available because a bulk charger cannot access it without overcharging cell #1.

Each cell operates in a limited range of SoC:

- The SoC of cell #1, which limits charging, ranges from 40% to 100%;
- The SoC of cell #2 ranges from 20% to 80%;
- The SoC of cell #3, which limits discharging, ranges from 0% to 60%.

Throughout the range, the delta in the SoC range is 60% for all the cells. That is, only 60% of the cell capacity is available to the string, and 40% is not available due to imbalance — the state of balance of the string is 60%, and its state of imbalance is 40%. The capacity of each cell is 10 Ah, but the capacity of the string is only 6 Ah (60 % of 10 Ah).

Balancing the string would give access to that previously inaccessible charge and restores the full capacity of the string (see Section 3.2.9).

3.2.5.8 Different Capacity, Midbalanced

When cells in a string have different capacities (and in practice, they do) the string can only be balanced at a single SoC level. Everywhere else, the cells have different SoC levels. When the string is charged or discharged from this point, the SoC of the cells diverges. Then, when the string is brought back to the original SoC level, the SoC of all the cells reconverge.

The top curves in Figure 3.13 plot the voltage versus charge for the three Li-ion cells. Note that the curves have the same general shape. Only their width changes: the curve is wider if the capacity is higher (cell #1) or narrower if the capacity is lower (cell #3). The bottom curves are the "cell SoC versus string SoC" for the three cells. Note that the slopes are different because the three cells have different capacities: for a given current flowing into or out of a string of cells, the SoC of a low capacity cell changes faster than the SoC of a high capacity cell.

In this example:

- When the string is at 0% SoC (left end of the plot), cell #3 is empty and discharging is disabled; the other two cells still have charge in them (cell #1 is at 40% SoC, cell #2 is at 20% SoC) and their voltage is still higher than the minimum (cell #1 is at 3.5V, cell #2 is at 3.38V). The difference in SoC between cell #1 and #3 is 40%, meaning that the string's state of imbalance is 40%.
- When the string is at 50% SoC (vertical line in the middle of the plot), it is balanced: all three cells are at 50% SoC, and therefore at the same voltage (3.65V). The SoB is 100%, and the SoI is 0%.
- When the string is at 100% SoC (right end of the plot), cell #3 is full and charging is disabled; the other two cells could still accept more charge (cell #1 is at 60% SoC, cell #2 is at 80% SoC) and their voltage is still lower than the maximum (cell #1 is at 3.82V, cell #2 is at 3.9V). The difference in SoC between cell #1 and #3 is 40%, meaning that the string's state of imbalance is 40%.

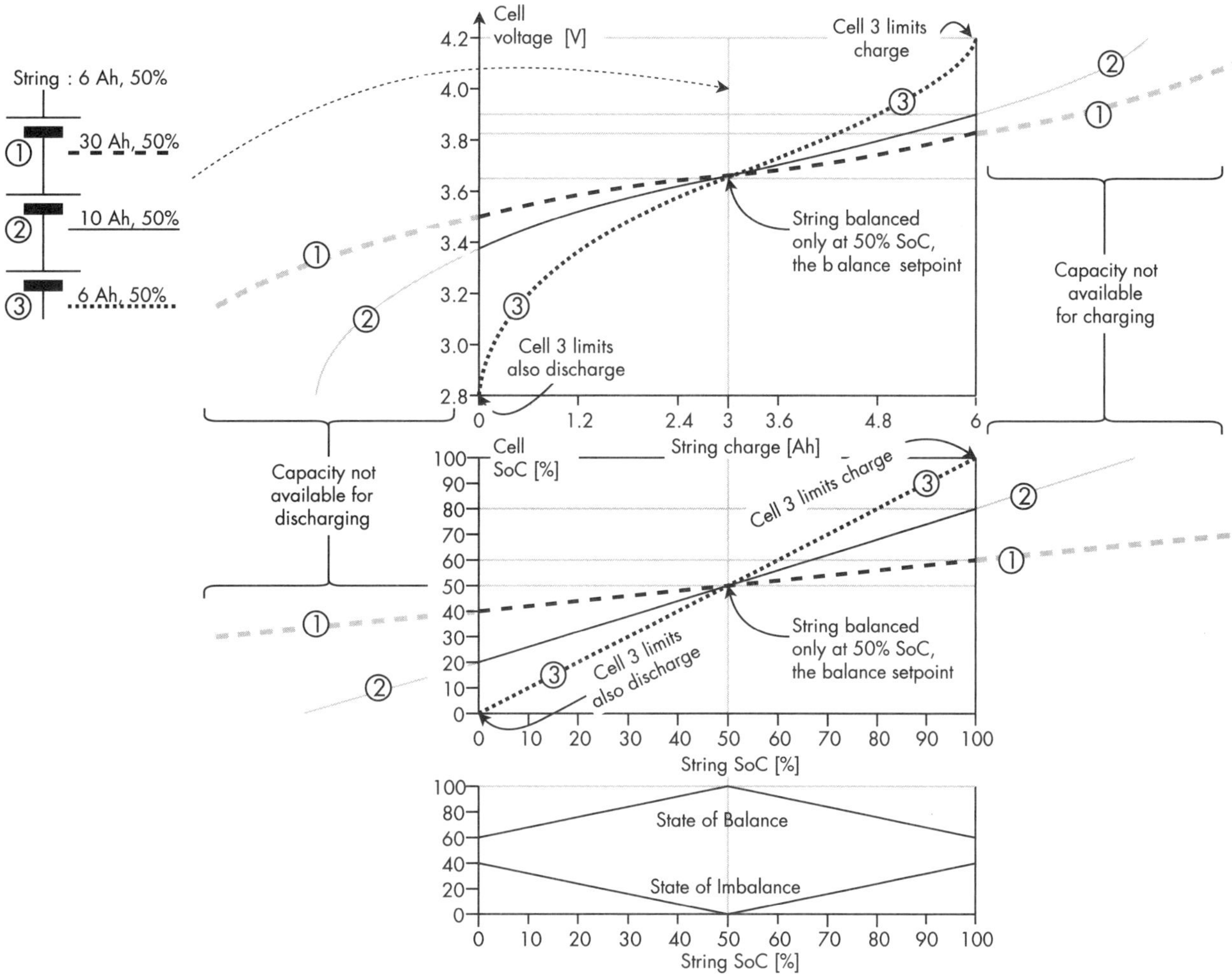

FIGURE 3.13 String of three cells with unequal capacity, midbalanced.

3.2.6 Optimal Balance Setpoint

A string is balanced if a single cell limits both charging and discharging. If so, its capacity is maximized. Since the string could be balanced at any SoC between 0% SoC and 100% SoC, the question is: where *should* the string be balanced? (For bottom balancing, see Volume 2, Section A.5.5.)

It depends on the type of battery's operating SoC range.[11]

3.2.6.1 Energy or Power Battery: Top-Balanced

An energy or power battery (see Section 5.1.4) is charged to 100% SoC regularly, and operates from 100% SoC down (possibly as low as 0% SoC).

This battery is *top-balanced*—the BMS sets the balance setpoint at 100% SoC. The battery can store maximum energy because, when fully charged, each cell is at the maximum voltage, which means that it stores the maximum energy.[12]

11. Note that the application sets the operating range, while the BMS sets the balance setpoint.
12. Energy is capacity times voltage. The capacity is already maximum because the string is balanced. However, to also maximize the voltage, we balance the string at the top.

When a top-balanced string is discharged toward the end of discharge, the cell voltages start diverging, and the SoI increases (Figure 3.14). The cell with the lowest capacity drops the fastest. When recharged, the cell voltages reconverge and end up balanced at 100% SoC, meaning that the SoI returns to 0%.

3.2.6.2 *Buffer Battery: Midbalanced*

A buffer battery is operated around an average SoC that is somewhere in the middle. That battery is *midbalanced*. The balancing setpoint SoC is also somewhere in the middle, but not necessarily at the same as the average operating SoC.

If the operating range is selected for maximum SoC range (so that there is just as much headroom on the charged side as on the discharged side), then the balance setpoint is the same as the reference SoC (e.g., 50 SoC). The lowest capacity cell is the limiting factor at both the charged and discharged ends.

When the string is charged or discharged, the cell voltages diverge and the SoI increases (Figure 3.15). The cell with the lowest capacity diverges the most. When the

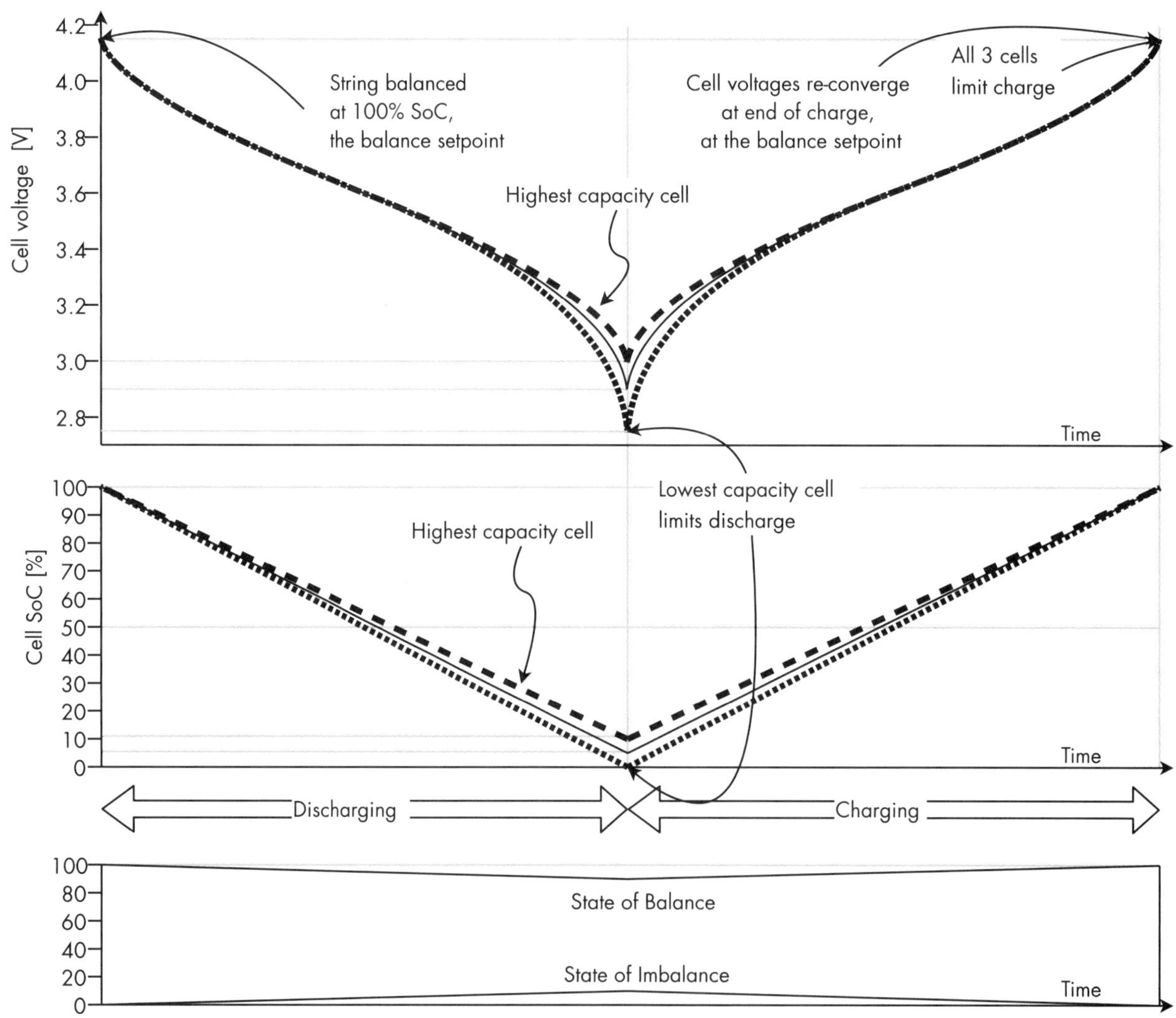

FIGURE 3.14 Discharging and recharging a top-balanced string of three cells with unequal capacity.

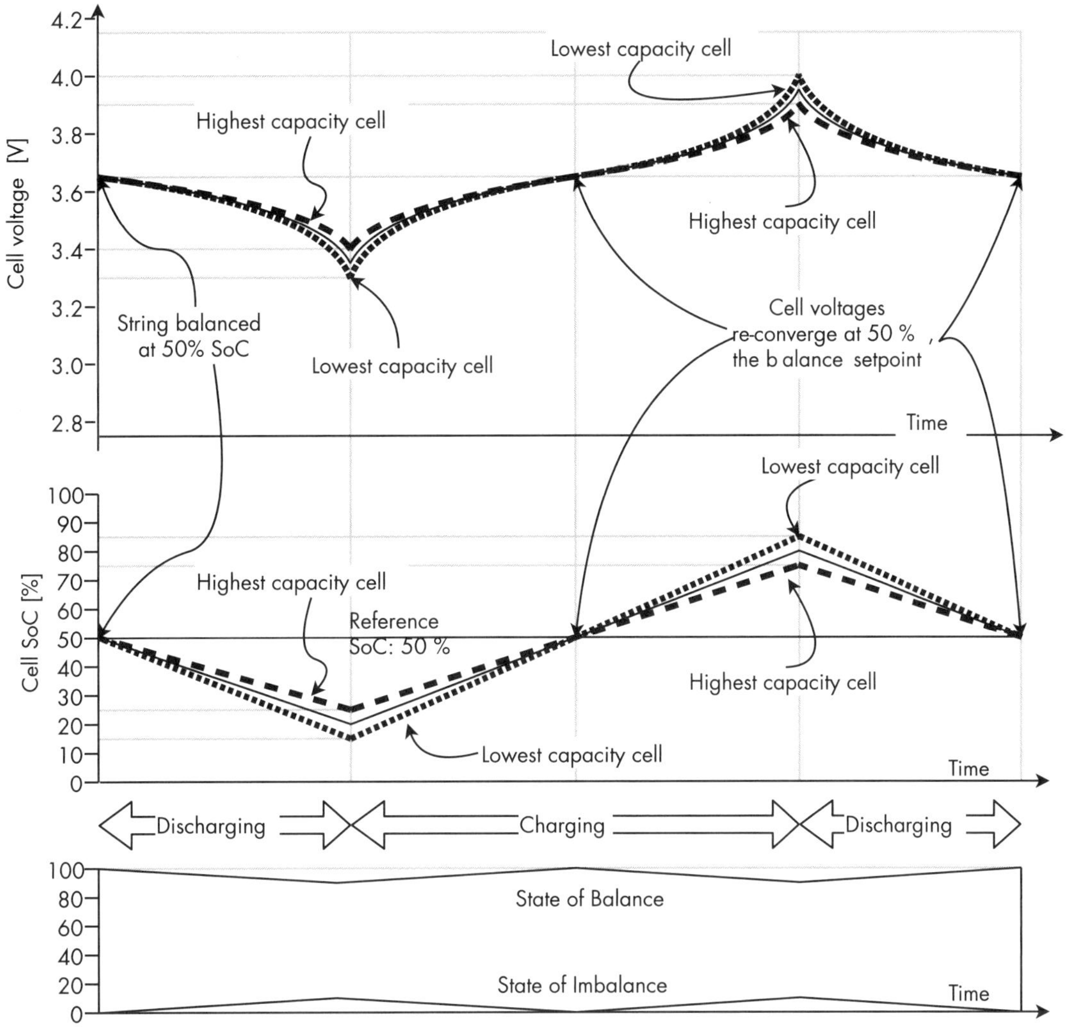

FIGURE 3.15 Charging and discharging a midbalanced string of three cells with unequal capacity, with a reference SoC and a balance setpoint of 50%.

string is brought back to 50% SoC, the cell voltages reconverge and the SoI returns to 0%.

The balance setpoint is usually selected to be the same as the reference SoC, regardless of the reference SoC, such as 50% SoC (Figure 3.16(a)), 30% SoC (Figure 3.16(b)), or 70% SoC (Figure 3.16(c)).

However, it may make more sense to set the balance setpoint somewhere other than the average SoC to keep the cell with the lowest capacity from operating in an undesirable region. For example, a buffer battery may be operated in a range centered around 30% because doing so results in the longest lifetime (see Section 2.5.2). At the same time, we want to keep the cell with the lowest capacity from dipping too low into the 0% SoC area because it degrades it further. We do this by biasing the balance setpoint at a higher SoC level, such as 40% (Figure 3.17).

3.2.6.3 Energy/Buffer Battery

For a battery that is used both as an energy battery and as a buffer battery (e.g., a plug-in hybrid vehicle or a stationary battery for a building), top-balance is best

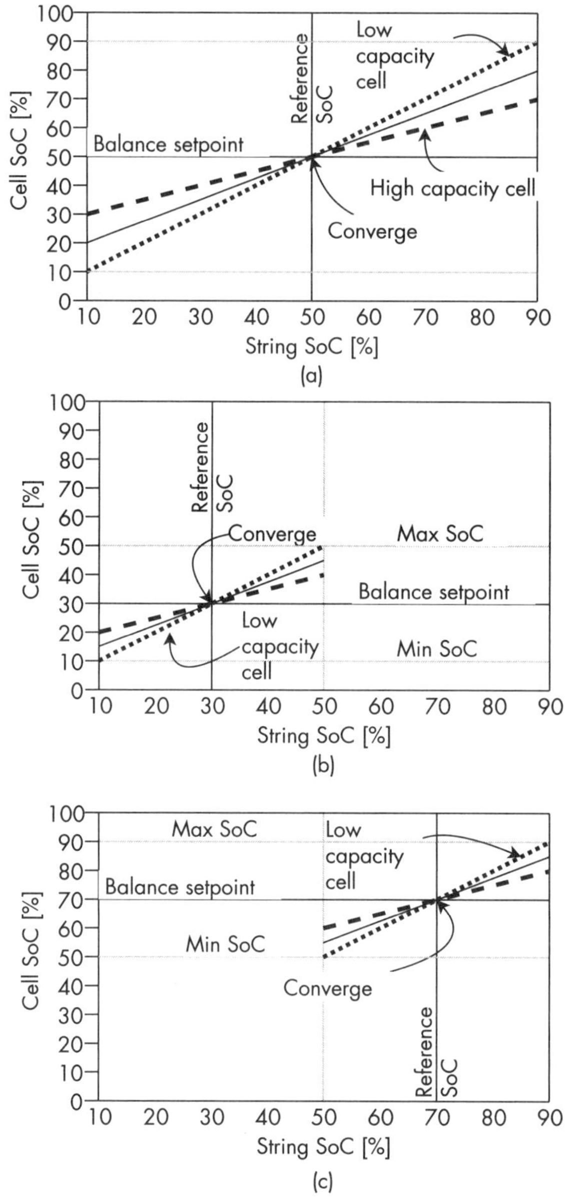

FIGURE 3.16
Buffer battery, balance setpoint same as reference SoC: (a) both at 50% SoC, (b) 30% SoC, and (c) 70% SoC.

because the balancing point is more important when the battery is used as an energy battery. The balancing point is less critical when the battery is used as a buffer battery because the battery capacity is so large, and the SoC never reaches the extreme SoC levels.

3.2.7 Imbalance Detection

To balance a string, the BMS looks at the cell voltages to decide which cells are out of balance. However, because the OCV versus SoC curve is not a straight line, the difference in voltage between two cells is not directly proportional to the difference in their charge. For a given imbalance, the difference in the voltages of two cells in series depends on multiple factors:

- Amount of imbalance: More imbalance increases the voltage difference;

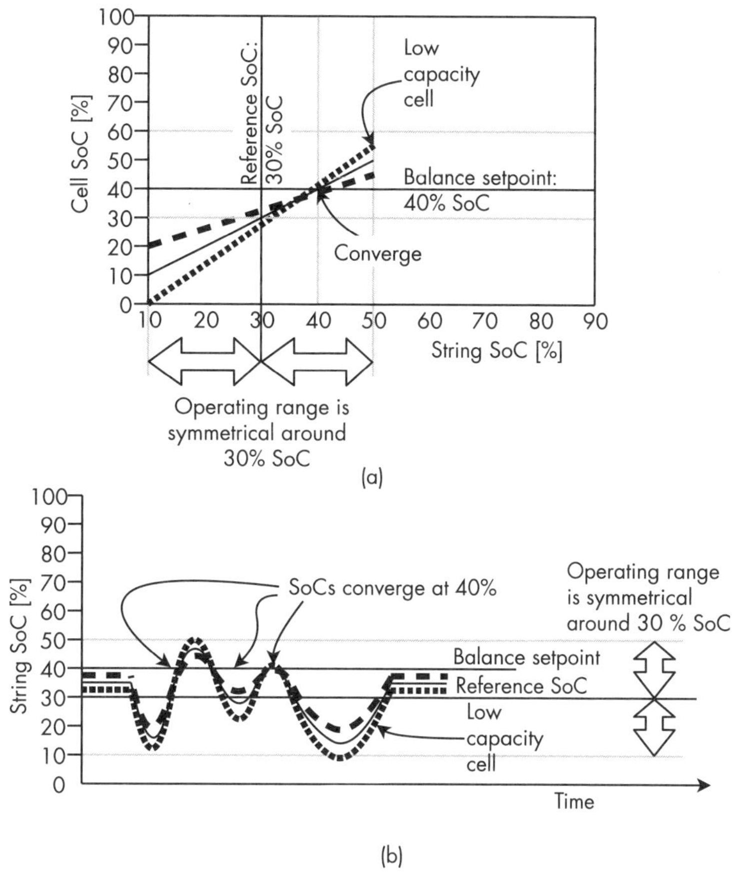

FIGURE 3.17
Balance setpoint at 40% and SoC reference at 30%: (a) graph, and (b) plot.

- SoC of the string: The voltage difference is more pronounced at low and high SoC levels due to the shape of the OCV versus SoC curve;
- Cell chemistry: With LFP cells, at mid-SoC levels the cell voltage are nearly the same regardless of SoI; a voltage difference becomes evident only at low and high SoC levels.

These graphs plot the difference in voltage between two cells versus the string SoC for different amounts of SoC imbalance between them. For example, the 10% curve is for when the SoC of one cell is 10% higher than the SoC of the other cell.

For NMC cells, it's hard to detect a 1% imbalance, except below 5% SoC (Figure 3.18(a)); a 10% imbalance is easily detectable, except around 35% SoC.

For LFP cells, a 1% imbalance is easily detected below 10% SoC and above 95% (Figure. 3.18(b)). Everywhere else, it's pretty much undetectable; at 50% SoC, only a terrible imbalance of 30% can be detected. Therefore, while LFP cells are excellent for power applications, a buffer battery with LFP cells is practically impossible to midbalance.

Attempting to balance a string in a region where the delta voltage is too low to be useful is not only pointless but is likely to be counterproductive, as other factors affect the difference in cell terminal voltage more than imbalance.

3.2.8 Imbalance Causes

String imbalance may be due to a variety of reasons.

FIGURE 3.18
Delta OCV versus SoC for various levels of imbalance: (a) NMC, and (b) LFP.

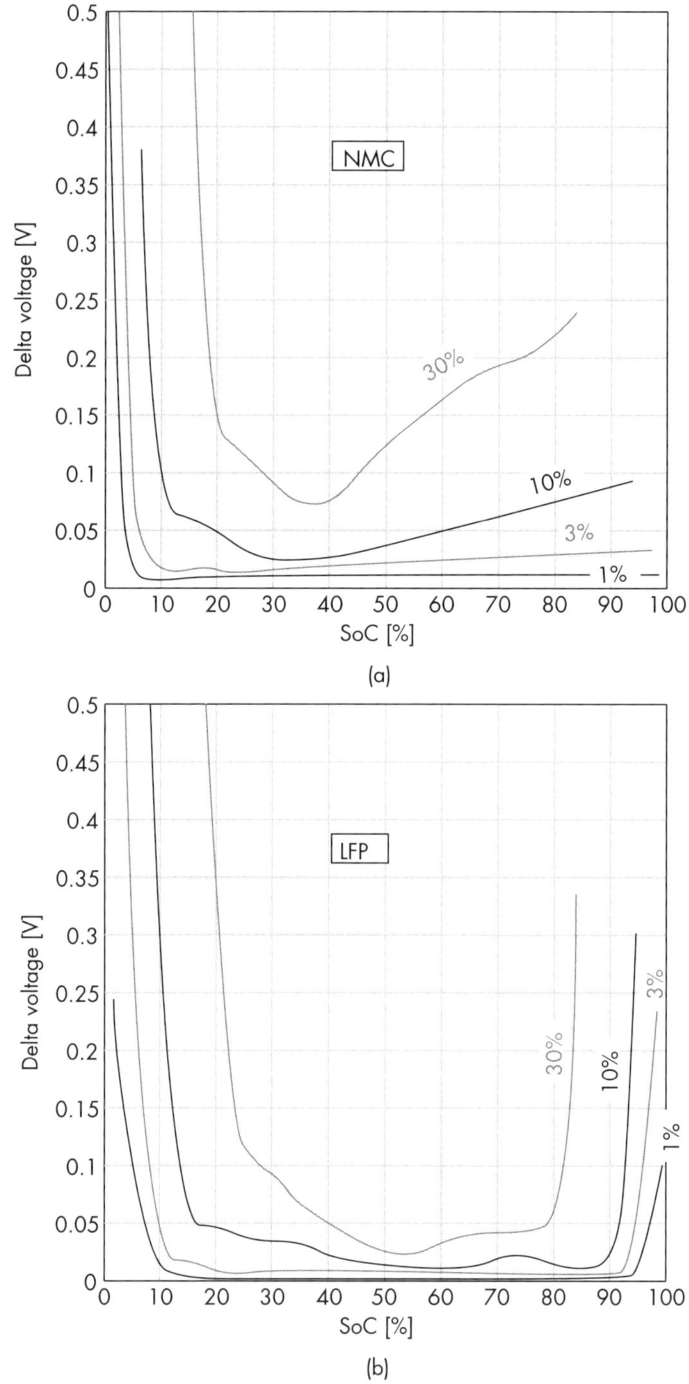

When the string is first assembled, its imbalance is due to using cells at various levels of SoC.

In a string with cells of various capacities, the string is only balanced at the balance setpoint. Everywhere else it's unbalanced. Be aware that if the string is balanced at some point other than the desired balance SoC setpoint, then it will be unbalanced at the SoC setpoint. For example, if the cells are prebalanced at 50% SoC and then used to build an energy battery, the battery will not be top-balanced and will have to be balanced again, but this time at 100% SoC.

The battery assembler performs *gross balancing* to balance the battery the first time (see Section 7.6.1.1).

Normal imbalance is due directly to self-discharge, and indirectly to cycling; I discuss each below. The BMS compensates for such differences in self-discharge through *maintenance balancing*[13]. *Abnormal* imbalance is due to one or more of the following:

- A bad cell with high self-discharge current;
- A BMS that draws different amounts of current from different cells (see Section 4.5.1.4);
- A broken or misconfigured BMS, or design flaws in the BMS;
- A tap in the middle of the string that draws current or supplies current (see Section 3.2.1);
- A cell that was replaced with no regard to its SoC compared to the SoC of other cells.

Any abnormal imbalance is corrected by first fixing the cause (such as repairing the battery or BMS, replacing the bad cell, removing the tap (see Section 8.7)), and then manually balancing the string (see Section 7.6.1.1).

Note that the following are *not* causes for imbalance:

- Variations in cell series resistance;
- Variations in cell capacity.

However, these variations have effects indirectly related to balance:

- Variations in cell series resistance might cause cell-to-cell variations in temperature, which in turn result in differing self-discharge rates, which in turn may cause some imbalance.
- Large variations in internal resistance may confuse some balancing algorithms, which may result in the BMS actually unbalancing the string.
- When a string is not at the balance setpoint, its cells are at different SoC levels; the wider the range in the capacities of the cells, the wider the difference in the SoC levels. This may give the false impression that variations in cell capacity result in imbalance. In reality, the string is still balanced: once back at the balance setpoint, the cells return to being at the same SoC level.

3.2.8.1 Self-Discharge

Differences in self-discharge currents (see Section 2.7.3) are the direct cause of string imbalance. The SoC of cells with higher self-discharge drops faster than the SoC of cells with lower self-discharge. As time goes by, the SoC levels of the various cells diverge, increasing the state of imbalance.

While all the cells self-discharge at a minimal rate, some do so faster than others, especially if they are at different temperatures (Figure 3.19(a)).

The worst case occurs when some cells are at room temperature while others are hot (for example, because they are mounted next to a charger). The coldest cells may have a self-discharge current of 0.000023C, while the hot ones may have 0.00017C,

13. Greg Plett did a simulation that confirms that the causes of imbalance are (1) difference self-discharge between cells and (2) difference in coulombic efficiency between cells (See Section 3.2.9.2) [2].

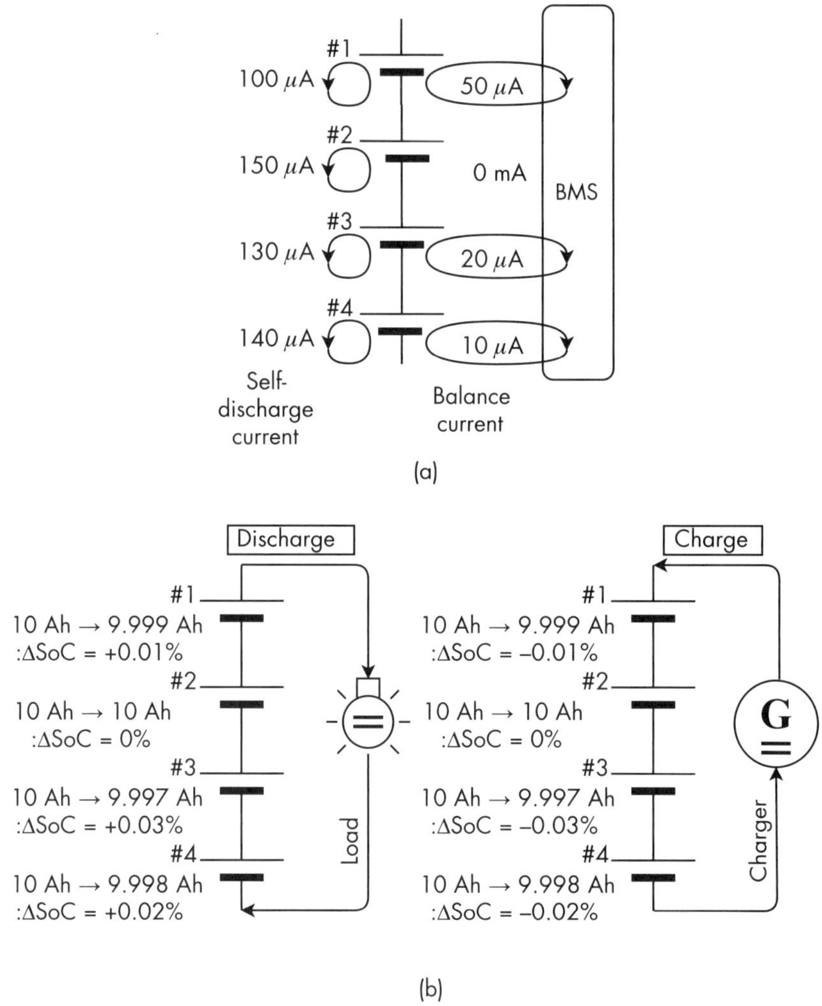

FIGURE 3.19
Normal causes for imbalance: (a) different self-discharge current (the BMS compensates for it), and (b) different capacity fade during cycling.

which is a difference of 0.00015C. The string will become 11% more imbalanced each month due to this difference.[14]

To compensate for the cell-to-cell variation in self-discharge current, the BMS draws a *balance current* from the cells with the lowest self-discharge current. This balance current is exactly equal to the difference in self-discharge current between this cell and the one with the highest self-discharge (see Section 3.2.9.1). In the example in the figure, note how the BMS ensures that each cell is discharged at the same current of 150 μA by providing an additional discharge current to the cells with low self-discharge current. Alternatively, the BMS can charge the cells with the highest self-discharge current to make up for the difference (see Section 4.7.2.4).

3.2.8.2 Cycling

An indirect cause for an increased imbalance in a series string is charging and discharging (cycling) through two mechanisms (Figure 3.19(b)):

- Variations in self-discharge current in cells at different temperatures;
- Variations in coulombic efficiency.

14. 0.00015 C × 24 hours/day × 30 days/month = 0.108 = 11%.

If cycling heats the cells in a string differently, the hotter cells self-discharge faster than the cooler ones. This mechanism is one of the indirect causes of imbalance during cycling:

- Various cells generate a different amount of heat due to variations in internal resistance;
- Various cells heat and cool differently due to their different placement and asymmetry in the thermal management system;
- In any case, the same temperature may affect various cells differently.

We saw that a coulombic efficiency of less than 100% results in the loss of a small charge and a small reduction in capacity during each cycle (see Section 2.5.4). As a consequence, if a cell is charged and then discharged by exactly the same charge, the final SoC is lower than the initial SoC. If this effect were the same in all the cells in a string, all of their SoC levels would drop by the same amount, and the string would remain balanced. However, cell-to-cell variations in coulombic efficiency result in differing drops of SoC over a cycle and therefore cause the string to become slightly unbalanced. This mechanism is the other indirect cause of imbalance during cycling. This reduction of SoC due to capacity fade during cycling has the same effect as if there were an additional self-discharge current during cycling. For a string that is charged and discharged at1C, this apparent self-discharge current is on the order of 0.0001C. Variations in this apparent self-discharge current could be seen as the cause of imbalance during cycling.

3.2.9 Balancing

Large imbalances should be corrected manually during assembly or after a repair through gross balancing (see Section 7.6.1.1). Once this is done, keeping a string in balance is far easier and requires just enough balance current to compensate for the imbalance current. The BMS does maintenance balancing during normal operation (see Section 3.2.9.2).

3.2.9.1 Required Balancing Current

If a series string is built correctly, it starts balanced. Afterward, it can be kept in balance by compensating for the imbalance current. The question is, how much current is required to do so?

We saw that there are two causes of imbalance: self-discharge current and cycling. Let's take a cell that has the worst drop in SoC due to both effects:

- The cell is warm (60°C), and therefore has a self-discharge of about 0.0002C.
- Assume that all the capacity loss happens only during charging with a charging efficiency of 99.99%. Assume that a complete cycle lasts 1 hour (2C charging, 2C discharging); then, for every cycle, the SoC loss is 0.01%, which is equivalent to a 0.0001C self-discharge current.

The total imbalance current is 0.0003C.

If all the cells in the string lose SoC at that rate, then the string remains balanced. However, they don't. If we assume that the ratio between the best and worst affected cells is 1:2, the BMS has to compensate for half of this current; that is, 0.00015C (see Section 3.2.8.1).

The above is valid for a battery with good cells. If a cell is bad (i.e., it has a high self-discharge current or a low coulombic efficiency), then a higher balance current is required to keep the string in balance.

I arbitrarily selected a level of 0.001C as the worst self-discharge of a cell suffering from dendrites, beyond which it becomes dangerous and should be replaced (see Section 2.7.3). This current is an order of magnitude higher than the normal level estimated above. To keep a string balanced, a BMS would have to charge the bad cell with an average current of 0.001C or discharge the other cells by this current. Note that for a 1,000 Ah battery, that is a balance current of 1A; many batteries exceed 1000 Ah, yet few BMSs provide more than 300 mA balance current.

If the BMS is only able to balance part of the time, then it must do so at a higher current to make up for the time when it cannot balance. For example, if an electric car can only balance at the end of a charge, it will do so for only about 1 hour a day. Thus, the required balance current is about 24 times higher (see Section 4.7.1).

3.2.9.2 Balancing Methods and Time

A string may be balanced in different ways at different points in the life of the battery:

- Before battery assembly: Use *pre-balancing* (see Section 7.4.3);
- After battery assembly: Use gross balancing (see Section 7.6.1.1);
- During use: Maintenance balancing, by the BMS (see Section 3.2.9.2);
- After cell replacement in the field: Do manual gross balancing (see Section 8.7).

Each referenced section will discuss how long balancing takes.

3.2.10 Balancing Not Implemented

Balancing is a performance function, not a safety function. Therefore, as long as a BMS protects the string, balancing could be considered to be optional.

As long as a battery is top-balanced at the time of manufacture, it may operate sufficiently long without any maintenance balancing. If the cells have a low self-discharge current and a high coulombic efficiency, the battery's imbalance may remain within acceptable limits by the time the battery is replaced for other reasons.

Applications that may use Li-ion batteries without balancing include

- Cheap consumer products with a short expected life;
- Price-sensitive power tools, in which cell degradation limits the usefulness of the battery faster than the imbalance limits the capacity;
- Homemade EV conversions, due to some hobbyist's inexperience and distrust (see Volume 2, Section A.5.5, "I Don't Need No Stinking BMS");
- Student race vehicles, as they run only a few times and are not reused in next year's race (see Volume 2, Section 3.10.4).

Again, a BMS that monitors each cell voltage is required, whether or not balancing is implemented.

3.2.10.1 Self-Balancing

A battery does not require external balancing if its chemistry is self-balancing (e.g., lead-acid).

A self-balancing battery chemistry would have one or both of these characteristics:

- A coulombic efficiency (see Section 2.5.4) that decreases sharply as the cell is nearly full;
- A self-discharge current (see Section 2.7.3) that increases sharply at the top voltage.

Li-ion is not self-balancing because its coulombic efficiency is practically 100% and its self-discharge current is extremely small. Self-balancing Li-ion cells are a myth (see Volume 2, Section A.4.1).

3.2.11 Overdischarge and Voltage Reversal

Overdischarging a series string without a BMS reverses the voltage in some of its cells. The mechanism is not immediately obvious and requires some thought.

Let me start with an analogy. You have a bank account with overdraft protection, you set up an automatic monthly payment for your electric bill, and your bank will pay those bills from your bank account on the first of the month. If you lose your job and stop depositing paychecks, eventually the account will be drained to zero by those payments. The next time the bank will pay your electric bill, the account balance will go negative. The power company gets paid, but where does this money come from? It comes from the bank's funds, or, more precisely, from the money in other people's accounts.

The sequence of events in a battery with a cell with a low SoC and no BMS to stop the current drain when a cell voltage gets too low is similar (Figure 3.20). The cell voltage of the cell with the lowest SoC (#4) drops toward zero. As the discharge continues, the cell voltage goes negative, driven by the charge still left in the other cells.

You may ask: If this cell's voltage is 0V, how can there still be any more current? Yes, the voltage of *this* particular cell is 0. However, the voltage of the *other* cells is still higher than 0, and *that* is the voltage that drives the discharge current that continues flowing. Since current flows in those other cells, it must also pass through cell #4.[15] As this current continues to flow, it discharges the other cells, but it starts charging the first cell in the reverse direction, which means that its voltage now starts increasing in the opposite direction: the cell voltage is reversed.

Note that there is no reversal in case of complete string discharge due to cell self-discharge (after a few years without being charged): there is no battery current, and the discharge current is internal to each cell. Each cell is drained to 0V and no lower.

3.2.12 Transitional Spikes

The cell terminal voltage is not rock-solid because the cell has a nonzero series impedance. A sudden change in the current results in a spike in the cell terminal voltage. Its polarity is

- *Positive:* During a sudden increase in the charge current or a sudden decrease in the discharge current;
- *Negative:* During a sudden decrease in the charge current or a sudden increase in the discharge current.

Even if the terminal voltage momentarily exceeds the cell's safe operating area, the cell itself may be fine, as the spike may be too short to cause noticeable damage;

15. Because all the cells are in series, and there's no other path for the current to go.

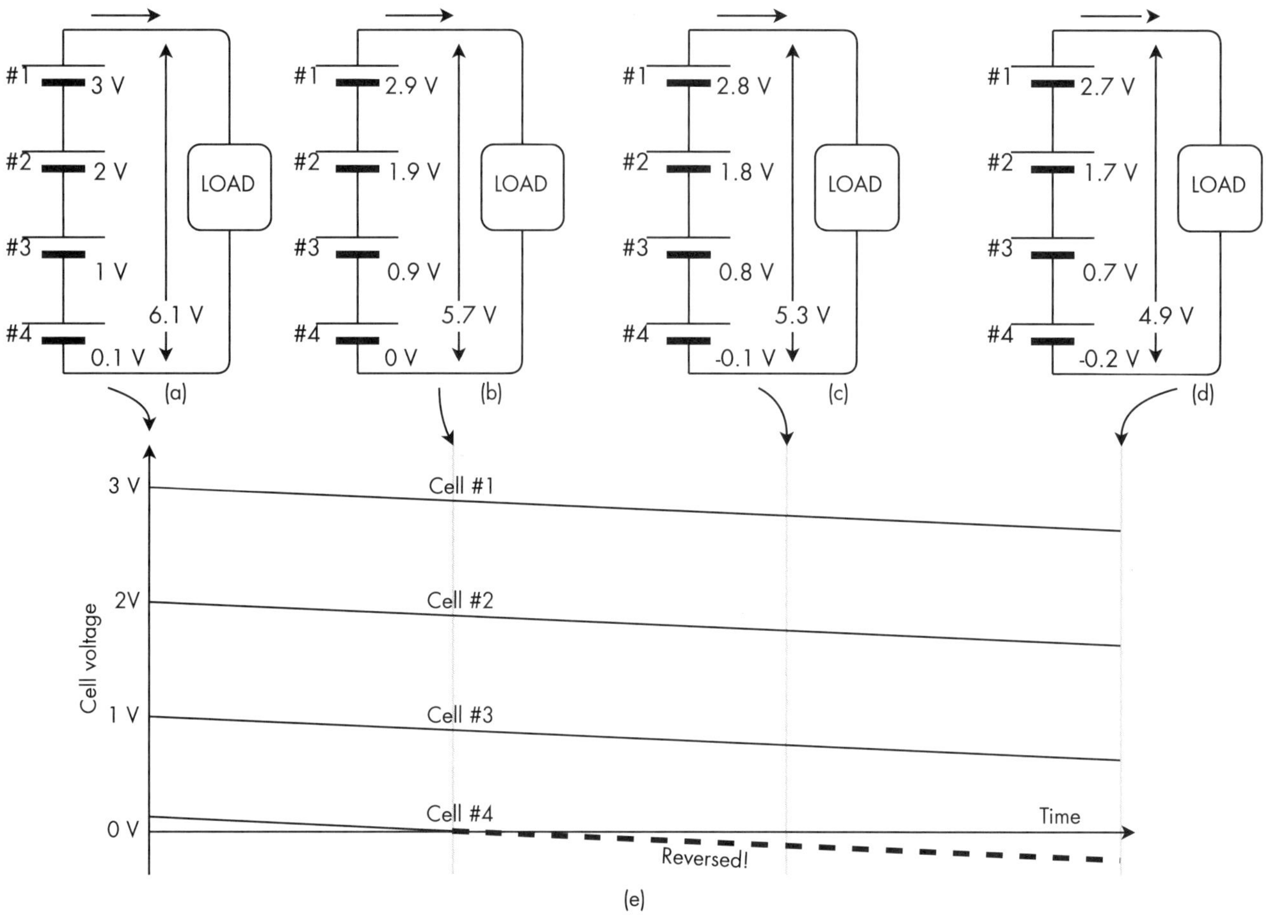

FIGURE 3.20 Cell voltage reversing process: (a) lowest cell nearly empty, (b) at 0V, (c) reversed, (d) even more reversed, and (e) plot showing cell voltage reversal.

however, the terminal voltage may exceed the maximum ratings of the BMS input, damaging it (see Section 8.3.4).

3.2.12.1 Negative Spikes

In particular, the cell terminal voltage can be reversed temporarily at the moment of a direct connection to a discharged capacitive load, which is likely to damage the BMS.

For example, a 12V battery is about to be connected directly to a discharged capacitive load (Figure 3.21(a)). The battery voltage is 12V and the load voltage is 0V. The moment the contactor is turned on, it connects the 0V of the discharged load capacitor directly to the battery, and the terminal voltage of the battery drops immediately down to 0V.

If the four cells are identical (Figure 3.21(b)), the battery voltage is distributed equally to each cell: the terminal voltage of each cell is 0V. Inside the cell, the OCV is unchanged, still at 3V, but the voltage across the series resistance is –3V. These two voltages add up to the 0V of the terminal voltage.

In a new, fully charged, top-balanced battery, with all the cells at the same temperature, the cells' internal resistances are going to be pretty even and relatively low. The cells can survive this for an instant, and the BMS is fine because it can handle a 0V input.

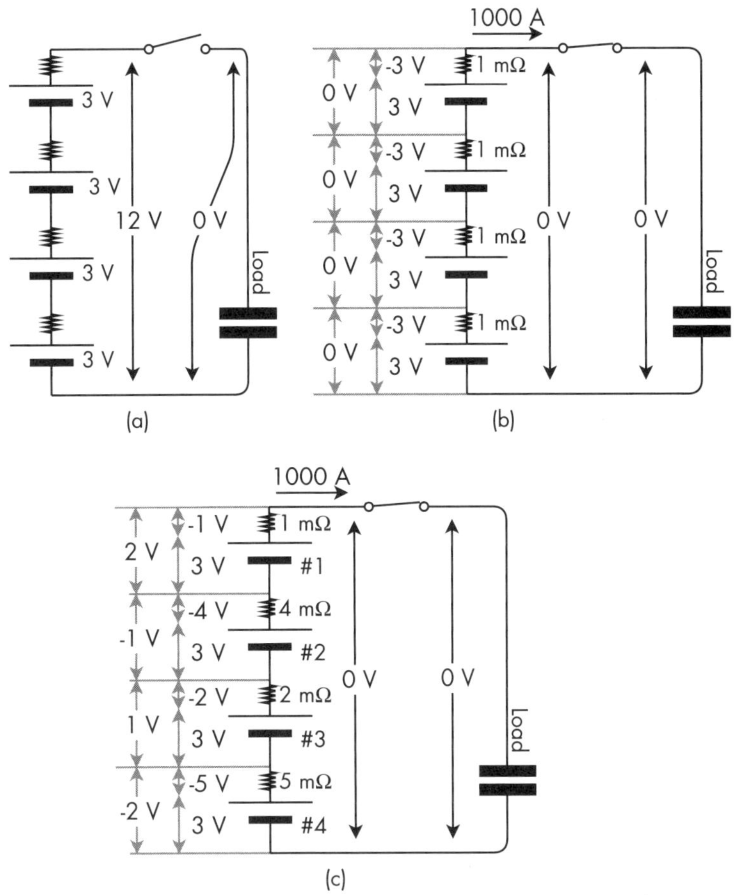

FIGURE 3.21
Initial effects of direct connection to capacitive load: (a) before connection, (b) with identical cells, and (c) with cells of differing internal resistance.

In reality, the resistance of each cell varies:

- If the battery is severely unbalanced, the cells with the lowest SoC has the highest resistance;
- If the battery is close to empty, the cell with the lowest capacity is nearly empty and has high resistance;
- If the cells are at different temperatures, the coldest ones have higher resistances;
- The battery may include a weak cell with a high resistance.

Given that the internal resistances of the cells are different under load, the battery voltage is not be distributed equally (Figure 3.21(c)). Specifically, the voltage of the lowest resistance cell (#1) does not drop too much, while the voltages of the highest resistance cells (#2, #4) become negative (so that the total battery voltage adds up to 0V). This reverse voltage, whose duration is too short to degrade the cells, is still long enough to damage a BMS connected to the string.

3.2.12.2 Positive Spikes

We saw how connecting a battery to a discharged capacitive load results in negative spikes on some cell voltages. Similarly, connection to a higher voltage results in positive spikes due to the same mechanism. Specifically, a charger that is turned on and not connected to any load has an output voltage equal to its CV setting. The moment this a charger is connected to a battery with a lower voltage, it brings the battery voltage up to its CV voltage temporarily until its output capacitor discharges into the battery.

Eventually, the voltage drops to the previous battery voltage and the charger switches to the CC mode.

The voltage at the string terminals is not distributed equally to all the cells—the voltage of low impedance cells hardly goes up, while the voltage of high impedance cells jumps up to make up the difference. Again, this voltage spike, whose duration is too short to degrade the cells, may still be long enough to damage the BMS.

3.2.13 Charging a Series String

The way a series string is charged depends on on whether it's top-balanced or mid-balanced.

3.2.13.1 Top-Balanced Battery

Let us first look at a battery that uses top-balancing.

If a string that is already balanced is connected to a CCCV charger set for an appropriate CV setting, at the end of charge, all the cells reach the top voltage simultaneously. At this point, the charger switches to the CV mode. Afterward, all the cells are topped up as the current decays to 0 A (Figure 3.22(a)).

On the other hand, if the string is not yet balanced, things get complicated.

If the charger has fixed CC and CV settings, the BMS is only able to turn off the charger when the most charged cell reaches its top voltage, wait for balancing to do its job, and then turn balancing on again (Figure 3.22(b)). The BMS repeats this until the string is top-balanced. At this point, the total battery voltage reaches the CV setting of the charger and the charger remains on in the CV mode, topping off the cells, as the current slowly drops to 0 A.

With a remotely programmable charger, the BMS can instruct it to act as a CCCV charger for the most charged cell (Figure 3.22(c)). This allows the charger to remain on while charging the remaining cells as some charge is removed from the most charged cell. Assuming 4.2V cells:

1. Initially, the BMS configures the charger for the final CV setting (N × the maximum cell voltage). This doesn't affect charging because the charger is in the CC mode.
2. Once the voltage of the most charged cell reaches its CV voltage, the BMS starts reconfiguring the charger's CV setting to the sum of the voltages of the cells that are not charged, plus 4.2V for each cell that is already full. This keeps the voltage across all fully charged cells at 4.2V.
3. In the meantime, balancing removes some charge from the most charged cells.
4. The cell voltages converge towards 4.2V, as the BMS slowly increases the CV setting of the charger. The current in the most charged cells naturally drops towards 0 A because those cells are at a constant 4.2V; therefore, the string current also drops towards 0 A.
5. Finally, all the cells are at 4.2V, as the string current has dropped to nearly 0 A.

3.2.13.2 Midbalanced Battery

Usually, one would not use a CCCV charger to charge a midbalanced string (i.e., in a buffer battery). Instead, the application would charge this battery at a constant current or a constant power. For example, in a hybrid electric vehicle, the battery would be charged by the electric motor and may be charged to only 70% SoC.

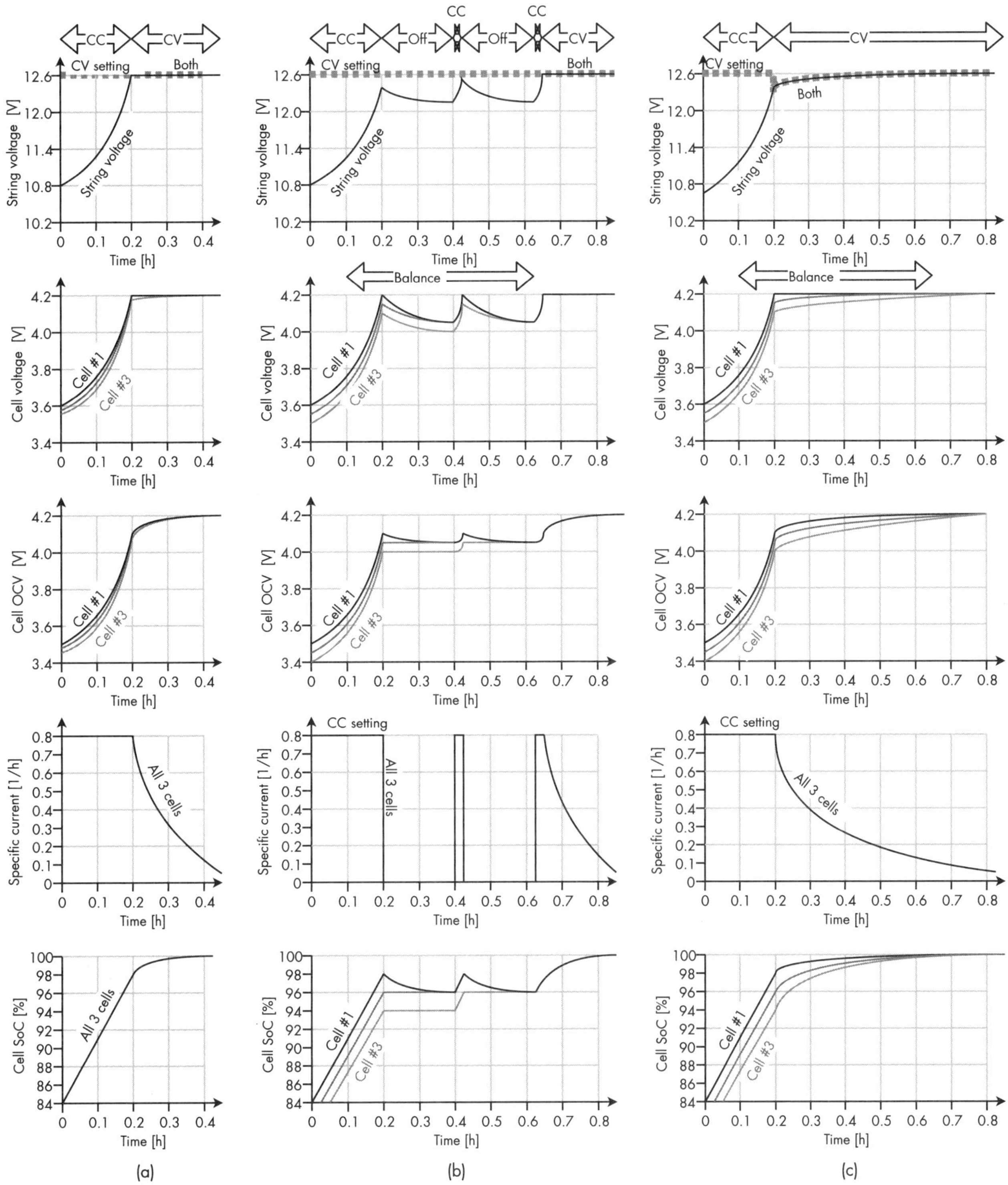

FIGURE 3.22 Charging a series string of three cells: (a) balanced string, (b) imbalanced string, charger with fixed CV, and (c) imbalanced string, charger with CV configurable by the BMS.

3.3 PARALLEL BLOCKS

Cells are connected directly in parallel to achieve a parallel block with the desired capacity (Figure 3.23(a)).

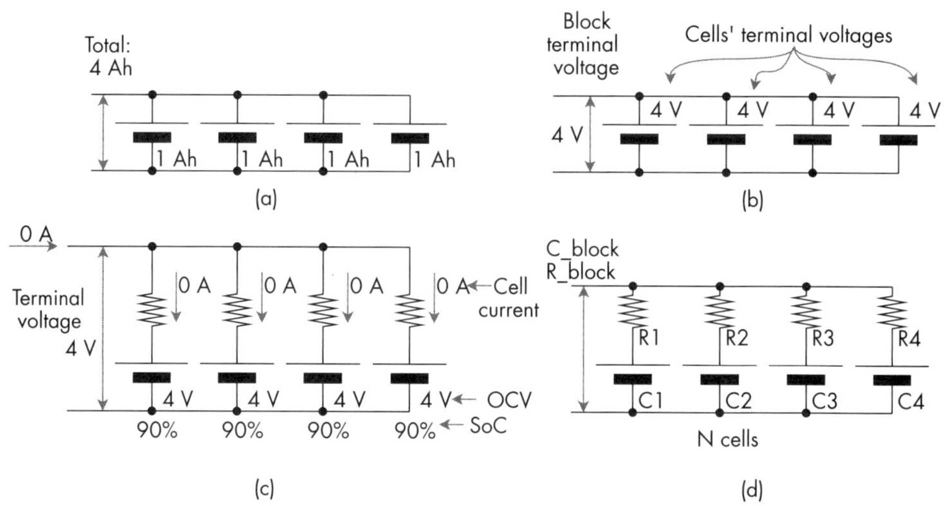

FIGURE 3.23
Parallel block: (a) blocks are used to increase capacity, (b) terminal voltage, (c) OCV, and (d) block capacity and resistance.

Li-ion power banks with a 5V output may use this arrangement. Other than that, hardly any applications use this arrangement. Our primary interest in parallel blocks stems from the fact that they are used within a parallel-first arrangement, which is quite common (see Section 3.4).

As all the cells in a parallel block are connected in parallel, they all have the same terminal voltage (Figure 3.23(b)).

After a long rest at zero current, all the cells in a parallel block have the same OCV, which is equal to the block's terminal voltage because there is no current, and therefore no voltage across the cells' internal impedance (Figure 3.23(c)).

Given that all the cells have the same OCV, they also have practically the same SoC.[16]

Electrically, a parallel block is equivalent to a single cell that is as big as the total of the cells within it (see Section 3.3.5) (Figure 3.23(d)). In general terms, the capacity of the block is higher than the capacity of the cells within it; its resistance is lower than the resistance of the cells within it.

Current flowing into a parallel block is shared by its cells, though not necessarily equally. The specifics depend on whether or not the cells are identical, as discussed in the next sections.

For simplicity, the following discussion considers only the DC resistance and glosses over the complete picture of cell impedance.

3.3.1 Parallel Block with Identical Cells

Ideally, all the cells in a block are identical—they all have the same capacity and internal resistance. If so, with N cells:

- The capacity of the block is proportional to the number of cells:

$$C\_block\ [Ah] = N \times C\_cell\ [Ah]$$

- The resistance of the block is inversely proportional to the number of cells:

16. Except for secondary effects due to hysteresis (See Section 2.4.2.3).

R_block [Ω] = R_cell [Ω] / N

- The battery current is divided equally among all the cells:

I_cell [A] = I_block [A] / N

- Consequently, the SoC of all the cells changes equally:

SoC_block [%] = SoC_cell (N) [%] = 100 × C_block [Ah] × ∫ I_block [A] di/dt = 100 × C_cell [Ah] × ∫ I_cell [A] di/dt

- Therefore, all the cells have the same OCV.

For example, if a block is composed of four cells, each with a capacity of 100 Ah and an internal resistance of 10 mΩ, the block has a capacity of 400 Ah and a resistance of 2.5 mΩ (Figure 3.24(a)).

If this block is charged at 40A, each cell receives 10A.

As each cell sees the same current, the SoC of each cell increases by 10% every hour (Figure 3.24(b)). Given that all the cells have the same SoC, they also have practically the same OCV (Figure 3.24(c)).

In most applications, it is acceptable to assume that the cells in a block are identical and that, therefore, the general points in this section are acceptable. If you do not wish to make this assumption, you must perform a more careful analysis, as discussed in the next section.

3.3.2 Parallel Block with Dissimilar Cells

Because real-world cells differ in capacity and resistance, the block current is not divided equally. Therefore, during use or soon afterward, each cell has a slightly different SoC and OCV. After the block has been at rest for some time without

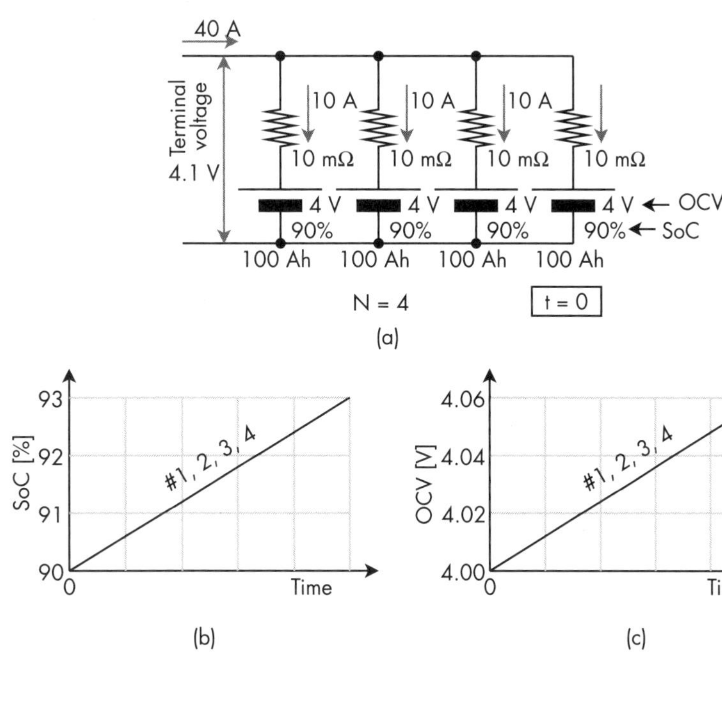

FIGURE 3.24 Parallel block with identical cells: (a) initial state, (b) SoC versus time, and (c) OCVs versus time.

current, all the cells settle at the same SoC and OCV; these are also the SoC and the terminal voltage of the block.

3.3.2.1 *Block Characteristics*

The capacity of a block is equal to the sum of the capacities of its cells:

$$\text{C\_block [Ah]} = \text{C1 [Ah]} + \text{C2 [Ah]} + \text{C3 [Ah]} + \text{C4 [Ah]}\ldots \quad (3.3)$$

Note how, if all the cells have the same capacity, this reduces to C_block [Ah] = N × C_cell [Ah], as we saw in the previous section.

In electrical engineering, we rarely use conductance (the inverse of resistance, measured in siemens). This is one case where using conductance would be convenient. The conductance of a block is equal to the sum of the conductance of each of its cells:

$$\text{G\_block [S]} = \text{G1 [S]} + \text{G2 [S]} + \text{G3 [S]} + \text{G4 [S]}\ldots \quad (3.4)$$

Considering resistance instead of inductance complicates the calculation. The resistance of a block is equal to the parallel of the resistances of its cells:

$$\text{R\_block } [\Omega] = 1 / (1/\text{R1 } [\Omega] + 1/\text{R2 } [\Omega] + 1/\text{R3 } [\Omega] + 1/\text{R4 } [\Omega] \ldots) \quad (3.5)$$

Note how, if all the cells have the same resistance, this reduces to R_block [Ω] = R_cell [Ω] / N, as we saw in the previous section.

3.3.2.2 *Initial Current Sharing*

As the resistances differ among the cells in a parallel block, the current is not shared equally among them. Rather, the current is divided in a convoluted way (see Volume 2, Section A.4.2).

Initially, while charging, low-resistance cells receive more than their share of current while high-resistance cells receive less (Figure 3.25(a)).

Therefore, the SoC of low-resistance cells increases faster than for high-resistance cells (Figure 3.25(b)). Also, for a given current, the SoC of low capacity cells increases faster—compare cells #1 and #4, both of which receive 5A, yet the SoC of cell #1 rises faster because its capacity is lower.

The cells' OCVs diverge according to their SoC levels (Figure 3.25(c)), yet all the cells still have the same terminal voltage because they are all connected in parallel.

Again, this is only initially. In the long run, the evolution of the state of each cell is convoluted. A self-correcting mechanism tends to keep the SoC of each cell close to the average SoC. This scenario is typical:

1. First, the block is at rest, there is no current, and all the cells are at the same SoC and OCV;
2. A constant charging current is applied to the block of cells;
3. The current is shared unevenly among the cells—more current goes into the cell with the lowest resistance, less into the cell with the highest resistance;
4. Consequently, the low-resistance cell charges faster than the other cells—its SoC and OCV rise faster;
5. As the OCV of the low-resistance cell increases, the voltage across its internal resistance decreases: V_series_resistance [V] = V_block [V] – OCV [V];

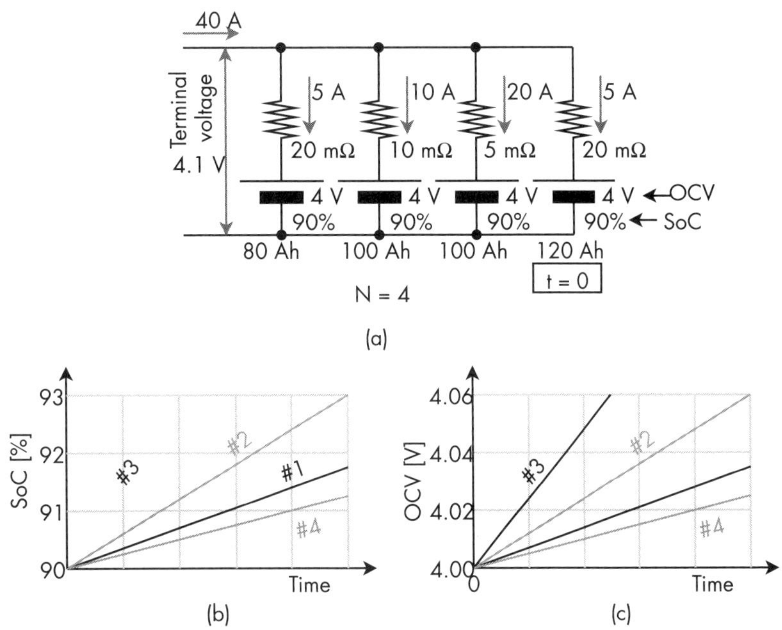

FIGURE 3.25
Parallel block with dissimilar cells: (a) initial state, (b) SoC versus time, and (c) OCVs versus time.

6. Therefore, its current is also reduced: I_cell [A] = (V_block [V] – OCV [V]) /R_cell [Ω];
7. Consequently, more current is available to charge the other cells, allowing them to catch up with the first cell.

Additionally, for a given current, the cell with the lowest capacity charges faster. That is, its SoC increases faster as well.

As a result, the SoC of the cells with low resistance of high capacity leads slightly ahead of the SoC of the block. The SoC of the high-resistance and the high-capacity cells lags slightly behind the SoC of the block (Figure 3.26). The higher the cell resistance or capacity, the more its SoC lags behind the block SoC.

So far, we discussed the cell SoC levels in general terms. In the next section, we will consider them in a specific example.

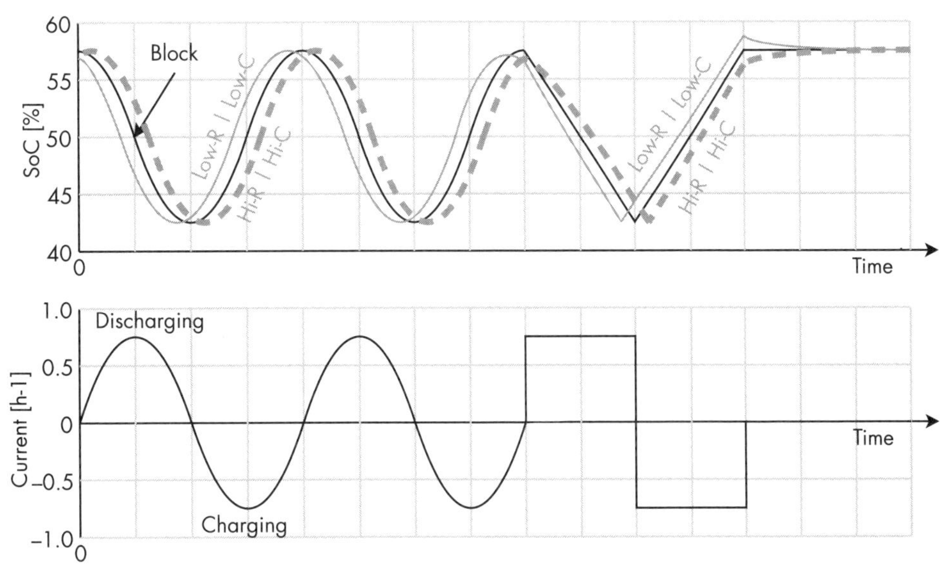

FIGURE 3.26
Plot of cell SoC levels in a parallel block with dissimilar cells.

3.3.2.3 Constant Current Charge Cycle

The SoC levels in a block of parallel cells with dissimilar cells behave strangely, following unexpected paths. As a complete analysis of this effect is rather complicated, it is helpful to start by considering the simple case of a block of cells of the same capacity charged at a constant current for a given period. Specifically, let's consider a block of four NMC cells with various resistances, charged for 250s at a constant current of 10A (Figures 3.27, 3.28).

Please bear with me, as it is rather convoluted:

$t < 0$ — at rest (Figures 3.27(a), 3.28 before $t = 0$):

- There is no current.
- All cells are at the same SoC and OCV.

$t = 0$ — Charging starts (Figure 3.27(b), 3.28t = 0s):

- The terminal voltage jumps up due to the IR drop (cell resistance times charging current).
- Initially, the current is shared unequally: the lowest resistance cell receives the most current and the highest resistance cell receives the least current.

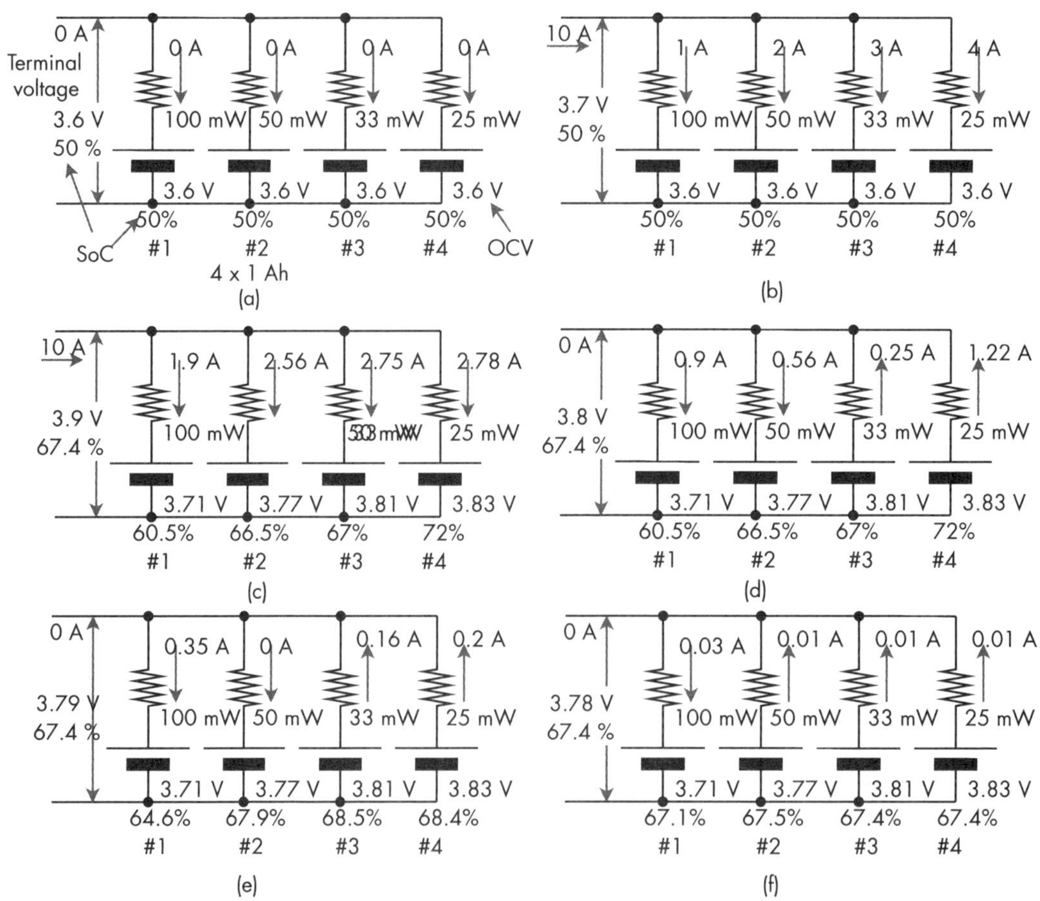

FIGURE 3.27 Block of cells with different resistances: (a) At rest, (b) start of charging, (c) end of charging, (d) start of equalization, (e) during equalization, and (f) end of equalization.

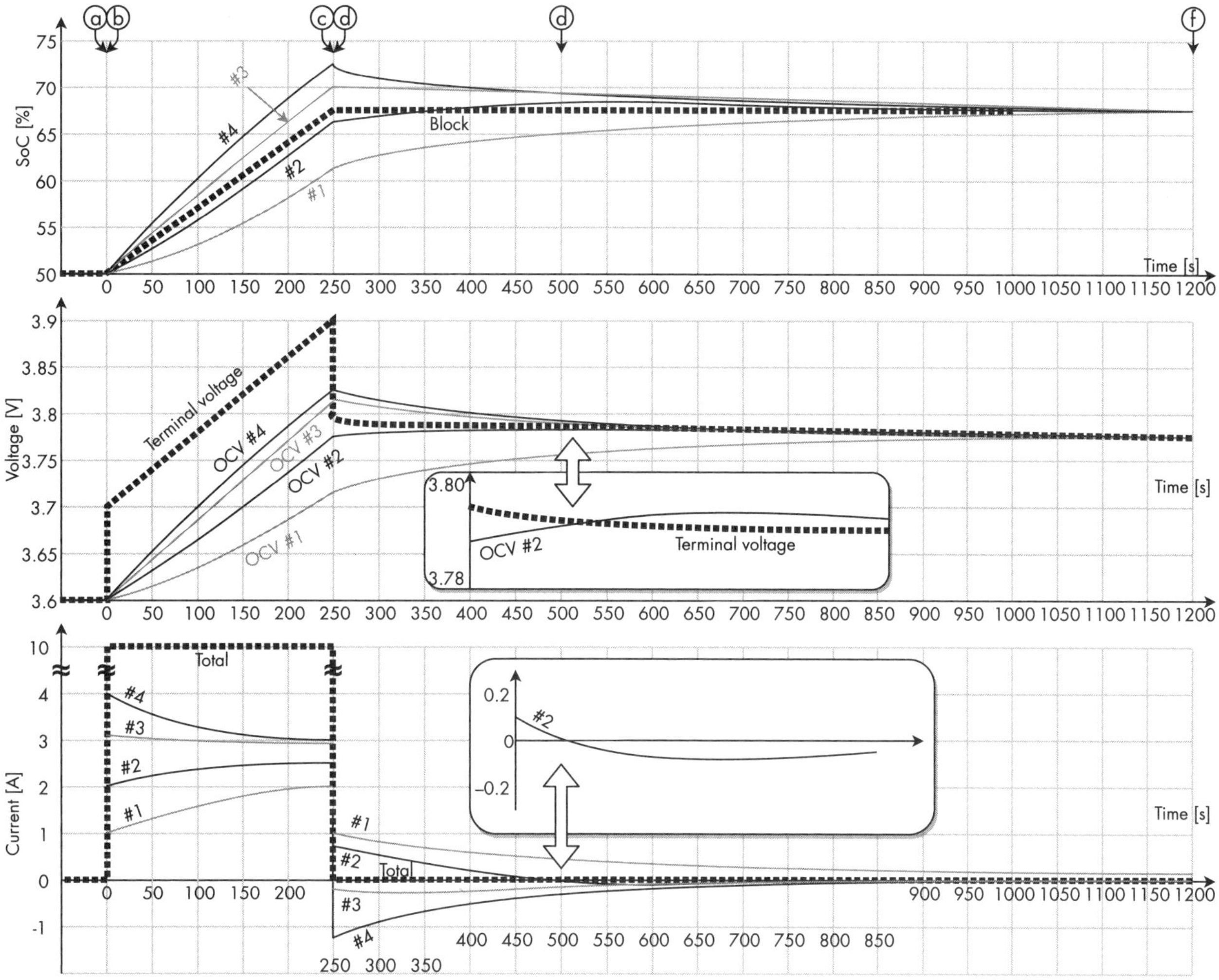

FIGURE 3.28 Plot of block of cells with different resistances. The letters a through f at the top correspond to a through f in Figure 3.27.

$0 < t <$ ~100s — Charging (Figure 3.28t = 0–100s):

- The SoC of the lowest-resistance cell increases the fastest because it receives more current than the other cells.
- Consequently, its OCV increases the fastest.
- The resulting negative feedback begins to equalize the current in each cell—as the lowest resistance cell's OCV increases the fastest, the voltage across its series resistance decreases the fastest, decreasing its current as well, leaving more current available to the other cells, which soon start catching up:
    - The clockwise bend in the OCV of the low-resistance cells and the counterclockwise bend in the OCV of the high-resistance cells show this effect. All these curves approach parallel paths.
    - Similarly, the bend of the curves for the cell current show this effect; these curves become progressively more horizontal as the block current is shared progressively more equally among the cells.

~100 $< t <$ 250s — charging (Figure 3.28t = 100–250s):

- The current in each cell approaches the average current because the block current is shared progressively more equally among the cells.
- The SoC and OCV of all the cells increase nearly at the same rate, though cells #1 and #2 lag behind because of their late start. The cell SoC levels increase nearly linearly because the block current is constant and the terminal voltages increases nonlinearly due to the nonlinearity of the OCV versus SoC curve.

t = 249s—just before the end of charging (Figures 3.27(c), 3.28t = 249s):

- The current in all the cells is about 2A.
- The block has an SoC of 67.4%; the SoC of the cells ranges from 60.5% for the highest-resistance cell to 72% for the lowest -resistance cell.

t = 250s—charging is stopped (Figures 3.27(d), 3.28t = 250s):

- The terminal voltage jumps down because there is no longer an IR drop due to the charging current.
- The SoC levels and the OCVs remain the same.
- An equalizing current starts flowing inside the block, from the cells with the highest SoC (#3, 4) to the ones with the lowest SoC (#1, 2).
- Of course, all the cell currents add up to zero because the block current is zero.
- Charge is conserved—the total charge in all the cells remains constant because it has nowhere else to go.

250 < t < 500s—equalization start (Figure 3.28t = 250~500s):

- The SoC of all the cells slowly converge towards the block's SoC.
- Similarly, the OCV of all the cells converge toward the block's terminal voltage.
- These are not simple asymptotical curves; instead, they are complex curves[17] that behave strangely, as seen below.

t = 500s—reversal (Figures 3.27(e)), 3.28t = 500s):

- The OCV of cell #2 crosses the terminal voltage.
- The current for cell #2 is zero as it reverses direction (!)
- Cells #3 and 4 are still charging cell #1.

500 < t < 1200s—equalization completion (Figure 3.28t = 500~1200s):

- The curves finally behave more normally, as they converge toward their final values.

t = 1200s—end of plot (Figures 3.27(f), 3.28t = 1200s):

- All cells are practically at the same SoC.
- There is just a little bit of equalizing current into cell #1 from the other three cells.

17. Despite the outward similarity, these curves are not the result of relaxation. These curves are purely the result of the equalization currents between cells.

3.3.2.4 Conservation of Charge and Energy

During equalization, one may view the block as a closed system because it's disconnected from anything else. At first glance, one may conclude that the total charge and energy stored in the cells must be conserved. Are they?

- *Conservation of charge*[18]: Every electron that leaves one cell enters another cell,[19] and therefore, the total charge *is* conserved.[20]
- *Conservation of energy:* There are equalizing currents flowing inside the block, generating heat. The energy for this heat comes from the cells, and therefore, the total energy *is not* conserved.

Take a closer look at the curves and note the slight drop in the block's voltage during equalization (Figure 3.28, middle curve, t = 250~500s). Given that energy is equal to charge multiplied by voltage and that the total charge is constant, this tells us that the energy stored in the block decreases slightly during equalization. That decrease in energy corresponds exactly to the heat generated as current flows through the cells' internal resistance.

Look at it this way: during equalization, as cell #4 charges cell #1, the voltage of cell #4 goes down and the voltage of cell #1 goes up, yet the voltage of cell #4 goes down more than the voltage of cell #1 goes up, so the average voltage goes slightly down. However, then you go look at the curves and see that, actually, the voltage of cell #1 changes *more* than the voltage of cell #4! How can this be? The reason is that cell #4 has a much lower resistance than cell #1. Hence, even though the voltage of cell #1 changes more, it has a smaller effect on the block's voltage.

A plot of the energy inside each cell and the total energy in the system is fascinating (Figure 3.29):

$0 < t < 250$s—charging:

- Initially, cell #4 (lowest resistance, highest current) wastes the most power while cell # 1 (highest resistance, lowest current) wastes the least power, as expected;
- At around 100 s, the power of all four cells converges;
- Toward the end of charging, the current in all four cells is nearly the same and the power of the four cells is reversed, with cell #1 wasting the most power and cell #4 the least;
- If charging had continued past 250s, the four curves would have become parallel to each other, illustrated by the thin dotted curves in the top plot.

$t > 250$s—Equalizing:

- The energy of cells #3 and 4 decrease as they charge cell #1 and 2, whose energy increases;
- The ever-so-slight decrease in the total energy in the block corresponds exactly to the heat released by the cells' internal resistances as they equalize.

---

18. In physics, the principle of charge conservation means something different; partially, it means that for every positive charge that is created, an equal and opposite negative charge is also created, leaving the total net charge in the universe unchanged.
19. Assuming a coulombic efficiency of 100%.
20. Because this closed system includes only cells. If the system also included a load, charge would not be conserved—it would be dissipated in the load. In the extreme case, the load could be a short circuit; even though the short circuit generates no power, it does dissipate the entire charge from the cell.

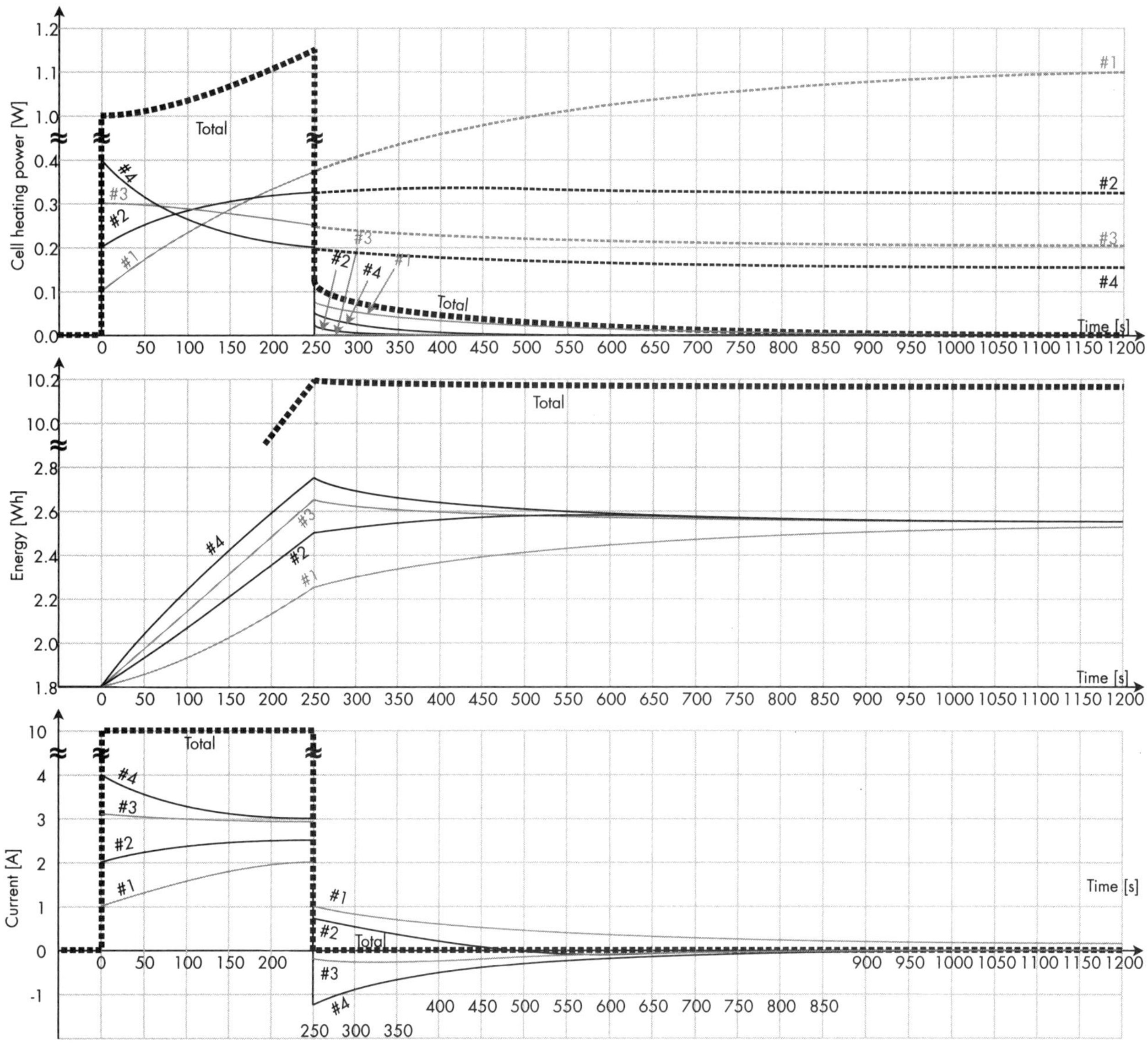

FIGURE 3.29 Plot of block of cells with unequal resistance, power, and energy. The thin dotted curves in the top plot would be the power if charging had not stopped at 250s.

Note that these curves occur in a region where the OCV versus SoC curve has a slope of about 10 mV/% (see Section 2.4.5). For an LFP cell, which has a much flatter curve, equalization would take much longer because the OCV difference between cells is much lower, and hence the equalization currents would be much lower. Note also be that the MPT of these cells is on the order of 500s; better cells would equalize faster because they would have lower internal resistance, and therefore the equalization currents would be higher.

3.3.2.5 Degradation of Low-Resistance Cells

The cell with the lowest total resistance (internal cell resistance plus external resistance of the interconnections) does most of the work in applications in which the current varies continuously.

Over the short run, this is a runaway problem due to a positive feedback mechanism: the lowest resistance cell does most of the work and gets hotter, which reduces its resistance further and makes it grab even more of the workload.

Over the long run, this is a self-correcting problem due to a negative feedback mechanism: cells with the lowest total series resistance work harder and experience a faster increase in their internal resistance. Eventually, they will have a total resistance (internal plus external) that is as high as the total resistance of the other cells. At that point, the current will divide equally among the cells in parallel, and all the cells will degrade at a similar rate.

Counterintuitively, the cell with the lowest external resistance ends up with the highest internal resistance. If, in the end, all the cells have the same total resistance, then

- A cell with high external resistance doesn't work hard, and its internal resistance won't change much;
- A cell with low external resistance works hard, and its internal resistance increases until its total resistance (low external resistance plus high internal resistance) equals the other cell's total resistance (high external resistance plus low internal resistance).

3.3.3 Temperature in Parallel Blocks

All cells in a parallel block have the same voltage, but not necessarily the same temperature:

- Some cells may be closer to a heat source;
- Cooling may be uneven, which may be caused by constricted airflow around particular cells;
- Cells have different resistance; a counter-intuitively low-resistance cell heats more than a high-resistance cell:
  - We saw that, when the current varies, low-resistance cells carry more than their fair share of current
  - On the one hand, power increases as current increases; therefore, low-resistance cells should produce *more* heat
  - On the other hand, power is inversely proportional to resistance; therefore, low-resistance cells should produce *less* heat
  - You may think that the two effects would cancel out; however, power in a resistor is proportional to the *square* of the current, so more current trumps higher resistance

Some applications (e.g., student race teams) require measuring the temperature on multiple cells in a parallel block (see Volume 2, Section 3.10.4).

3.3.4 Mismatched Cells in Parallel Blocks

Although the cells in a properly designed and manufactured parallel block should be well-matched, it is helpful to understand the effect mismatched cells would have on a parallel block:

- Placing cells with widely different SoC levels in parallel is not a good idea (Figure 3.30(a)) because, upon connection, an inrush current flows between the

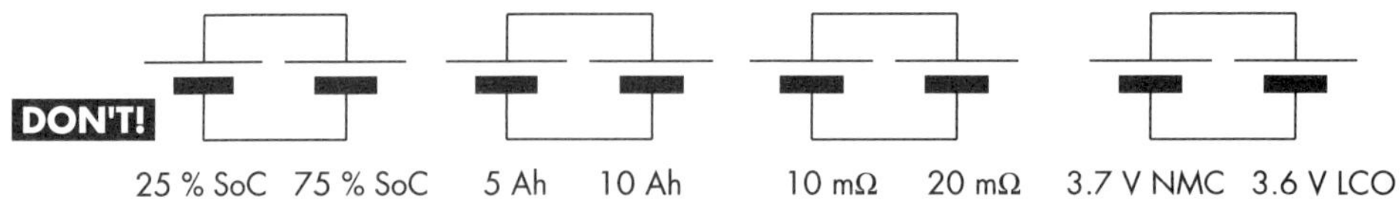

FIGURE 3.30 Mismatched cells in parallel: (a) different SoC, (b) different capacity, (c) different resistance, and (d) different chemistry.

cells to equalize the SoC and because half of the transferred energy is wasted in heat (see Section 3.3.6). To avoid this, the voltages should be matched before connection in parallel.

- Placing cells of different capacity in parallel is fine (Figure 3.30(b)), though unusual.[21] It's best if the cells use the same technology.
- Placing cells with different internal resistances in parallel is not a significant problem; the strong cells support the weak cells (Figure 3.30(c)) (see Section 3.3.2).
- Placing Li-ion cells with different but similar chemistries in parallel (e.g., 3.6V LCO cells and 3.7V NMC cells) or histories, as a hobbyist may do with reclaimed cells, is not intrinsically a problem, as long as the voltages are the same before connection. The BMS must be configured for the limits of the cell with the most stringent requirements (Figure 3.30(d)).
- Placing Li-ion cells with significantly different chemistries in parallel (e.g., 3.6V LiCo cells and 3.2V LFP cells) is of little use because the BMS limits must be set to protect both cells, leaving a narrow range of operation. As a result, a big portion of the charge in one cell or the other is unavailable.

3.3.5 Many Small Cells in Parallel versus One Large Cell

The battery designer may have a choice between using several small cells in parallel or a single large cell of the same total capacity. Electrically, they are equivalent (Figure 3.31).

Advantages of using a single large cell:

- Fewer connections;
- A large-format cell is easier to connect;
- Fewer hassles for students competing in races;[22]

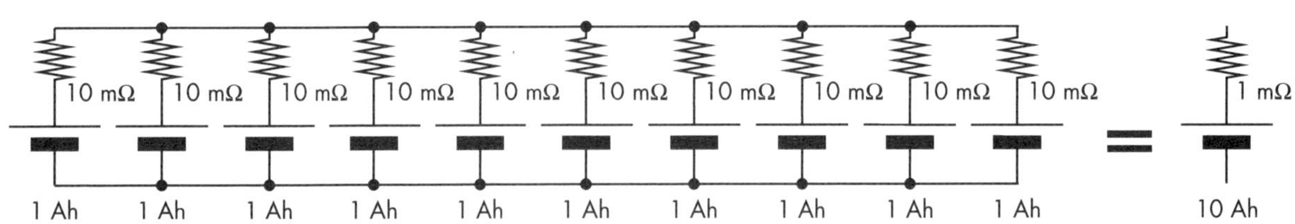

FIGURE 3.31 Parallel block equivalence.

21. Reusing reclaimed cells may be one reason to use cells of different capacities in a single battery. Also, I suppose one might want to achieve a 75 Ah capacity by adding a 45 Ah cell and a 30 Ah cell in parallel, though most people would use a single 80 Ah cell or two identical 40 Ah cells.

22. Designers of vehicles for student races are advised to use one large cell rather than many small cells in parallel, even though the two are equivalent from most perspectives. The race rules place excessive demands on the user of multiple cells in parallel (e.g., one fuse for each cell, a temperature sensor for every three cells (see Volume 2, Section 3.10.4.

- Requires less development effort, which is more appropriate for prototypes or low volume production.

Advantages of using many small cells in parallel:

- Better utilization of available space: easier to form the battery into a convenient shape;
- Better thermal management: in a small cell, internally generated heat travels a shorter distance to the case; also, it results in multiple paths for coolant to flow between cells;
- Versatility: if the initial design requirement is for a 100 Ah battery, but its changes to 110 Ah, it is easier to add one more small 10 Ah cell than to find a 110 Ah cell;
- It is cheaper to replace just a single cell in a block than to replace a large cell.
- In particular, small cylindrical 18650 cells are produced in huge quantities, making them cheaper and more reliable;
- If a cell goes into thermal runaway (see Section 5.18.3), there's hope that the other cells in the block won't.

3.3.5.1 Reliability

Using multiple cells may or may not improve reliability compared to using a single cell.

Let us compare a block of ten 18650 LFP cells in parallel and a single large cylindrical cell with ten times as much capacity using the same technology.

From a statistical standpoint, reliability is not affected.

Let us assume that a 1 cm^3 of a given LFP technology has a one in a million chance of becoming unusable during one year of use. If so:

- On the one hand, a single 18650 cell (whose volume is 16 cm^3) will lose 1/16 of its capacity, and therefore, the block of ten cells will lose 1/160 of its capacity;
- On the other hand, the single large cylindrical cell (whose volume is 160 cm^3) will lose 1/160 of its capacity.

In both cases, the total capacity loss is the same.

Note that the statistical distribution did not affect these results because it is relative to volume and therefore is the same for all cells, regardless of their size.

Considering that good sections support a weak section in parallel, reliability is not affected either:

- On the one hand, a weak cell is supported by the other cells in parallel;
- On the other hand, a weak section in a large cell is supported by the rest of the cell.

From this point of view, many cells in parallel appear to be equivalent to a single large cell.

In case of a thermal runaway, reliability is affected:

- On the one hand, if a single 18650 cell goes into thermal runaway, it conceivable that the battery design prevents propagation to other cells;
- On the other hand, if any portion of a large cell goes into thermal runaway, there is no way of stopping propagation to the rest of the cell.

From this point of view, many cells in parallel appear to be better than a single large cell.

Beyond the internal reliability of the cells, one must consider the external reliability of the connection between cells:

- On the one hand, each additional connection reduces reliability, so having a single large cell should be more reliable than having many small cells;
- On the other hand, if the single connection to a large cell opens, the capacity drops to 0; with multiple small cells in parallel, if one connection opens, depending on where the connection is, the capacity may either drop to zero or may just be reduced.

From this point of view, it is unclear which solution is more reliable.

3.3.6 Inrush Current upon Parallel Connection

When cells with different SoC levels are connected in parallel, current rushes between them to equalize their SoC levels.

3.3.6.1 Inrush Current between Two Cells in Parallel

When two cells with different SoC levels are connected in parallel, current rushes from the most charged cell to the least charged one. In the end, both cells end up at an SoC level that is somewhere between the two starting levels. Specifically, if both cells have the same capacity, the final SoC will be the average of the two initial SoC levels.

This process is inefficient because part of the energy is transferred and the rest is wasted in heat. The efficiency is worsened

- When the difference in SoC between the two cells is wider;
- At both ends of the SoC, where the slope of the OCV vs. SoC curve is steepest;
- If cells have a low MPT; counter-intuitively, the equilibrium process is least efficient for power cells, which are the most efficient cells during normal operation.

The peak current depends on the SoC of the two cells and their maximum power time.

Figure 3.32(a) shows the peak specific current when two LFP cells (with a maximum power time of 100s) are connected in parallel. For other values of MPT, the peak specific current is inversely proportional to the MPT (e.g., if the MPT is 50s, the specific current is twice the level shown in this graph).

The specific current is not too high (less than 2C) because

- If both cells are at middle SoC levels, the difference in voltage is low; therefore, the current is also low.
- If only one cell has a low or high SoC, its resistance is high; hence, even though the voltage difference is high, the current is low.
- If one cell is at low SoC and the other at high SoC, both of their resistances are high; consequently, even though the voltage difference is even higher, the current is still low.

NMC cells behave differently (Figure 3.32(b)). The curve for 30% SoC exceeds the one for 0% (because the relationship of resistance versus SoC, and the OCV versus SoC, don't match as nicely as they do for LFP cells).

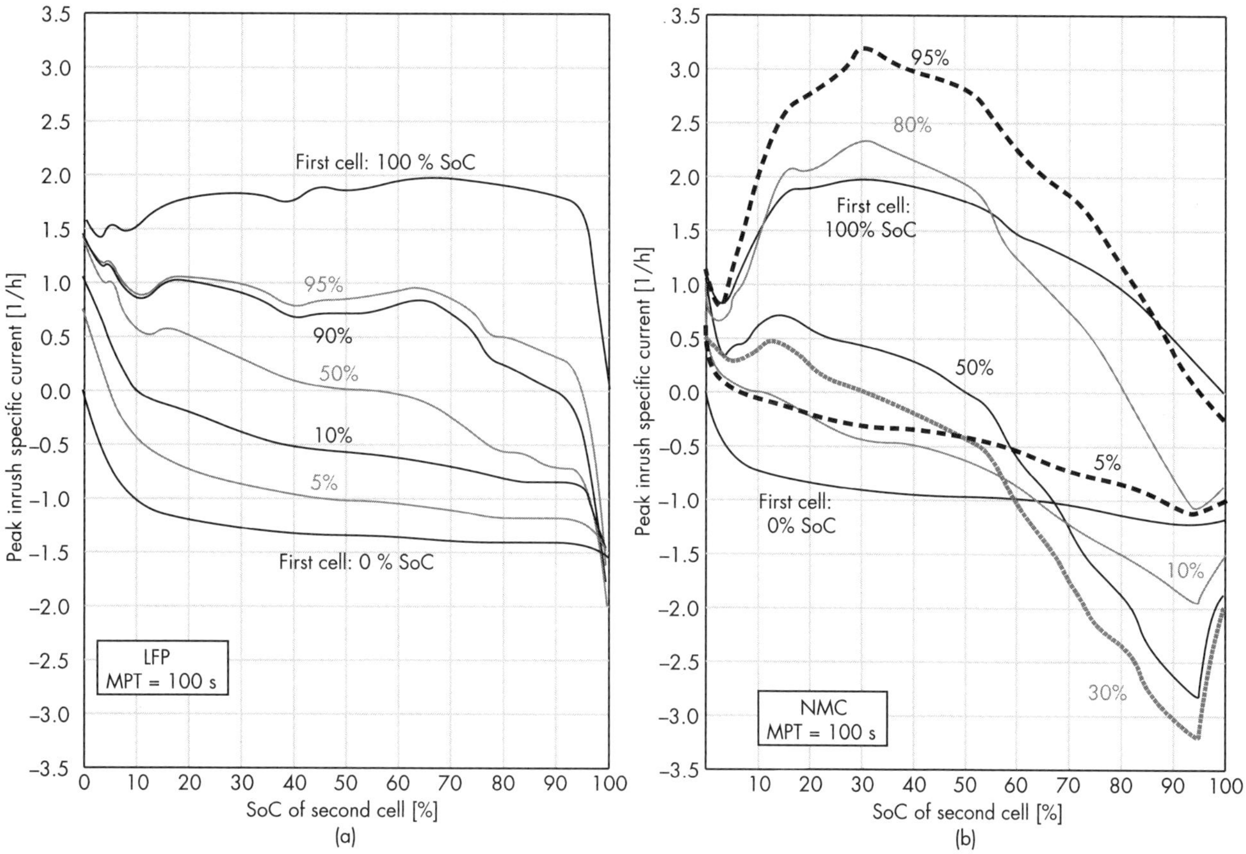

FIGURE 3.32 Peak inrush specific current from one cell upon parallel connection to another cell of different SoC versus SoC: (a) LFP, and (b) NMC.

The following graphs present the same information in a different format, as surface maps. The diagonal line has zero current because it corresponds to connecting cells at the same SoC. With only two cells, the curves are symmetrical.

For LPF cells, the maximum current of 2 C occurs when the cells are at 55% and 100% SoC (Figure 3.33(a)). For NMC cells, the maximum current of 3C occurs when they are 30% and 95% SoC (Figure 3.33(b)).

NMC cells experience a higher current than LFP cells, mostly because NMC cells have a wider voltage swing than LFP cells.

A specific current of 3C may not sound too bad until you consider that the maximum charging current for standard LCO cells is on the order of 0.5C and that at 3C, the inrush lasts about 20 minutes. This current is not good for the health of the cell with the lower SoC.

3.3.6.2 *Inrush Current Between a Single Cell and a Block of Cells*

The situation is worse when connecting one cell to a block of cells already in parallel: the resistance of the parallel block is much lower, so the current is limited only by the resistance of the additional cell. The initial inrush current is much higher and depends on the number of cells already connected in parallel. Under some conditions, the inrush current can be extremely high.

Figure 3.34 shows the peak inrush current when connecting a single cell across a large block of cells already connected in parallel. Note how the peak inrush current

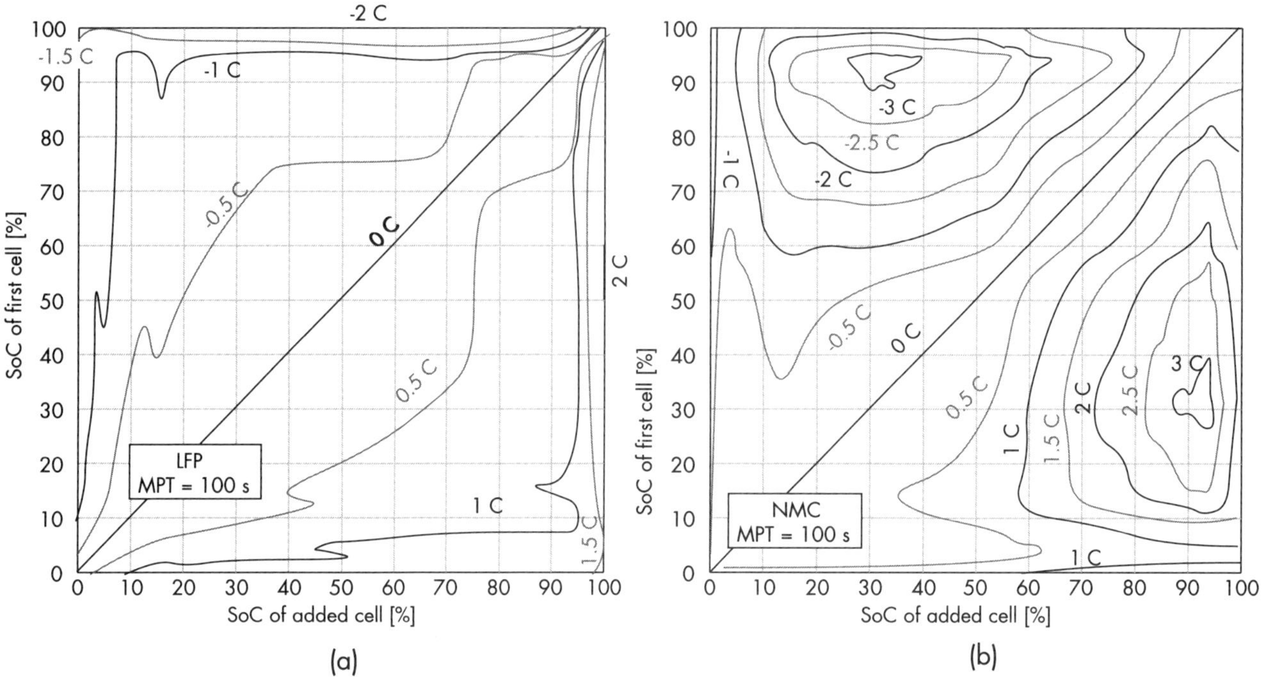

FIGURE 3.33 Peak inrush specific current from one cell upon parallel connection to another cell of different SoC, surface maps: (a) LFP, and (b) NMC.

can be as high as 67 C for LFP and 30 C for NMC! Also, note that it's far worse to connect an empty cell to a full bank rather than the other way around. This time, the curves are not symmetrical: connecting a cell at 30% SoC to a block at 70% SoC is not the same as connecting a cell at 70% SoC to a block at 30% SoC.

Figure 3.35 shows the worst-case specific current versus the number of cells already in parallel.

The current increases rapidly as the number of cells in the block increases, then settles to a maximum, corresponding to the block having effectively zero resistance; only the resistance of the added cell limits the current. The current is much higher if the added cell has a lower SoC than the block.

3.3.7 Safe Parallel Connection of Cells

Great care must be taken to minimize inrush current when assembling a parallel block (see Section 7.4.3):

- Check that the voltage difference is not too great;
- Connect cells in a binary fashion: first pairs, then quads, and so forth.

3.3.8 Charging a Parallel Block

Charging a parallel block is the same as charging a single cell: use a CCCV charger and mind the temperature of the cells.

3.3.9 Matching Resistance in All Connections

While not much can be done about variations in cell resistance, variations in the external circuit can be minimized by presenting the same external resistance to each

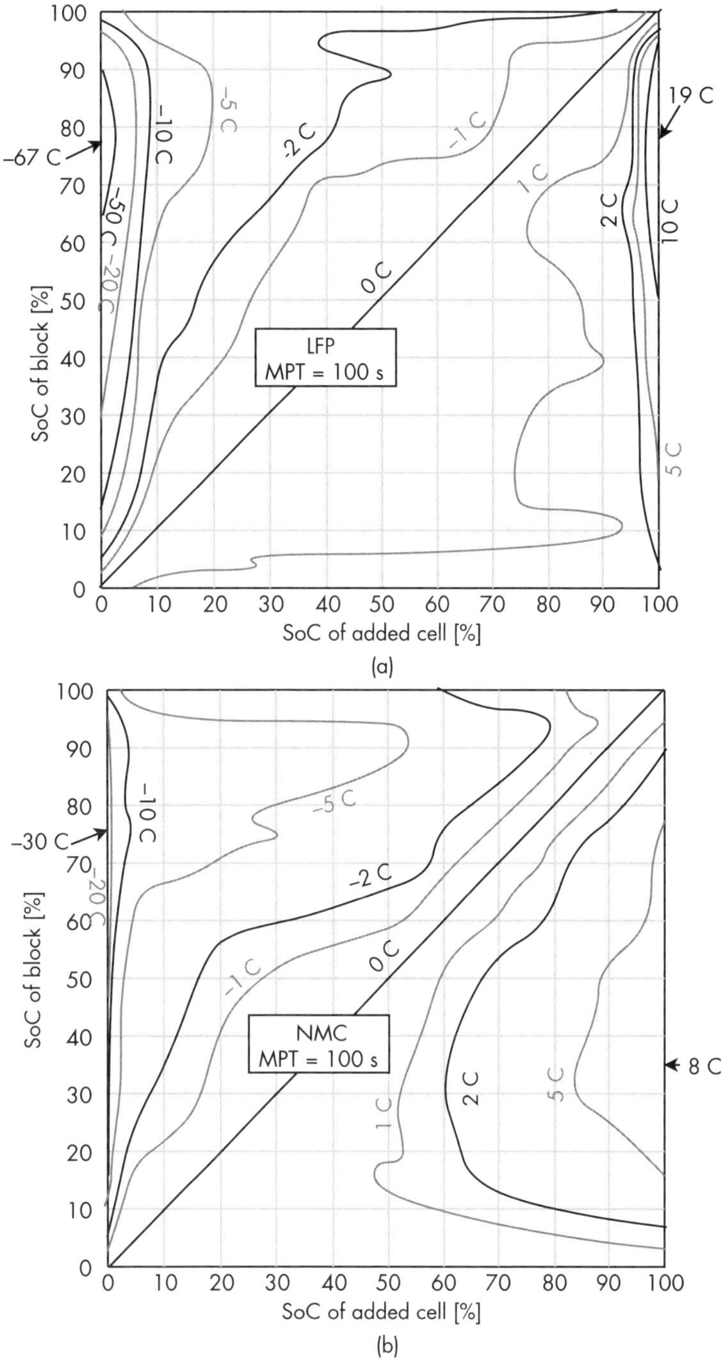

FIGURE 3.34
Peak inrush specific current of a cell upon connection to large block of cells of different SoC with a MPT of 100 s: (a) LFP, and (b) NMC.

cell, using one of the following (These solutions are analog to the solutions to equalize airflow for battery cooling. See Section 5.17).

- Paths of extremely low resistance, so that length does not matter (Figure 3.36(a));
- Equally long paths to all the cells on the positive side along with equally long paths to all the cells on the negative side (Figure 3.36(b));
- Different path lengths on the positive side of the cells with inversely long paths on the negative side, such that the total path length is the same for each cell (Figure 3.36(c)).

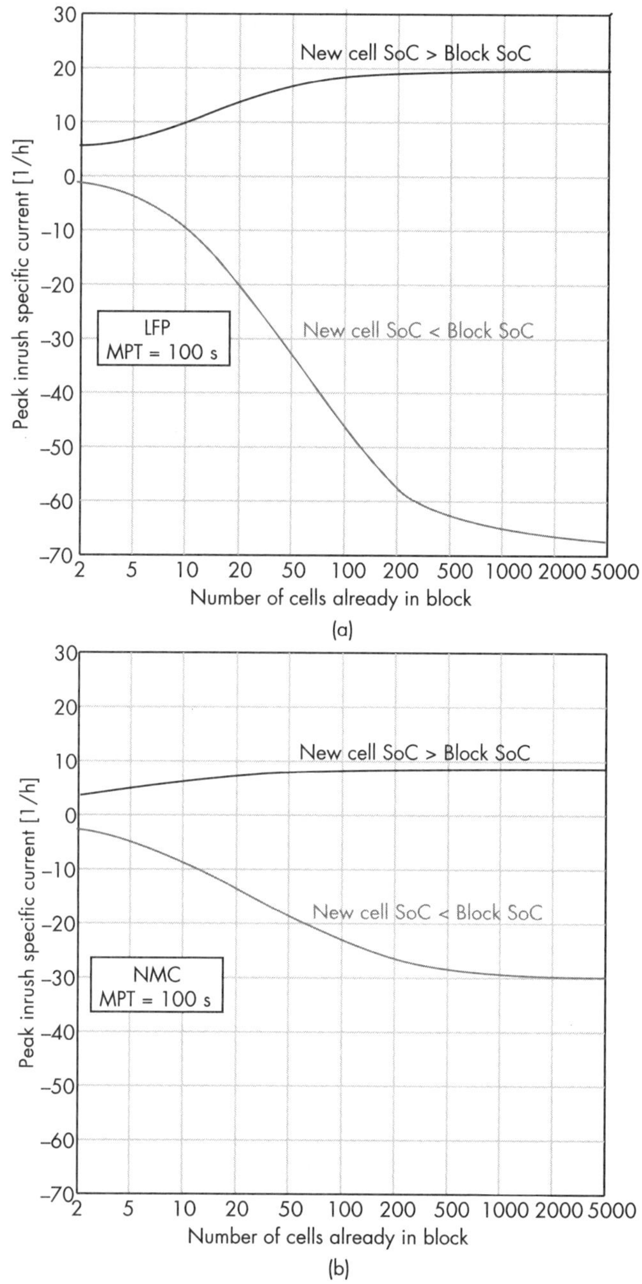

FIGURE 3.35
Worst-case inrush specific current for a cell upon connection across a block of cells of different SoC versus number of cells in the block: (a) LFP, and (b) NMC.

The above assumes that all the cells have the same internal resistance. In reality, they do not, and the lowest resistance cell handles more than its fair share of the current. Over the long run, this is somewhat of a self-correcting problem due to cell degradation (see Section 3.3.2.5).

3.3.10 Shorted Cell

Should a cell develop a short (see Section 1.2.2.9), the other cells in the parallel block are discharged through this cell.

3.3.10.1 Hard Short

A cell may develop a hard short in these cases:

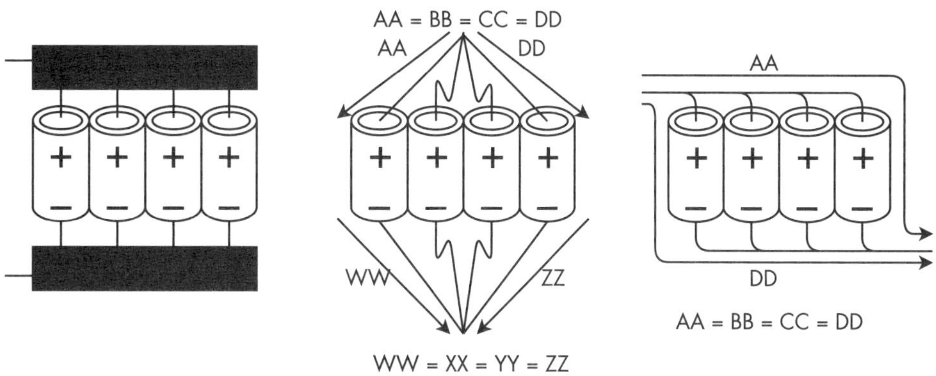

FIGURE 3.36
Equal paths to all parallel cells: (a) large conductors, (b) equally long paths on positive side and equally long paths on negative side, and (c) equally long total paths on both sides.

- Overdischarged, then recharged (NMC chemistry);
- Reverse voltage: discharged below 0V and then charged backward (see Section 3.2.11);
- Thermal runaway (see Section 8.2.1.5);
- Accidental penetration or deformation.

The first one and the second one are prevented by a properly installed BMS, in which case the likelihood is zero. There is only so much that a BMS can do to prevent the last two.

3.3.10.2 Soft Short

It is more likely that a cell develops a soft short as it ages. This soft short manifests itself as an significant self-discharging current. A parallel block is advantageous, as the other cells in the block will supply that current, reducing the rate at which the block SoC drops.

3.3.11 Cell Installed Backward

While the likelihood that a cell suddenly develops a hard short circuit is rather low, the likelihood that a cell is installed backward in a parallel block is much higher.

It can happen in the field (a hobbyist designs a parallel block using cell holders, and someone inserts a cell backward) or while a battery is being assembled.

It happened at one of my companies—a 26650 cell among a group of 10 was placed upside-down. Once welded, it discharged the other cells in its row. The assembler had no indication of it (no smoke, no pops, no fire). It was only discovered because its row had a low voltage.

Yes, a fuse per cell would have helped in this case: it would have saved the other cells in the row. But this happened at a factory, and it happened only once. It's just as easy to replace a whole row of cells as it is to replace a single cell. And the battery failed testing,[23] so it could not have left the factory in this condition. That company could not justify the complexities and potential disadvantages of adding a fuse per cell just to handle the one time that a cell was placed upside down at the factory.

3.3.12 Fuse-per-Cell

As we saw, if a cell develops a hard short, current from the other cells in the parallel block flows into the shorted cell (Figure 3.37(a)). If a fuse is placed in series with

23. Quality assurance testing.

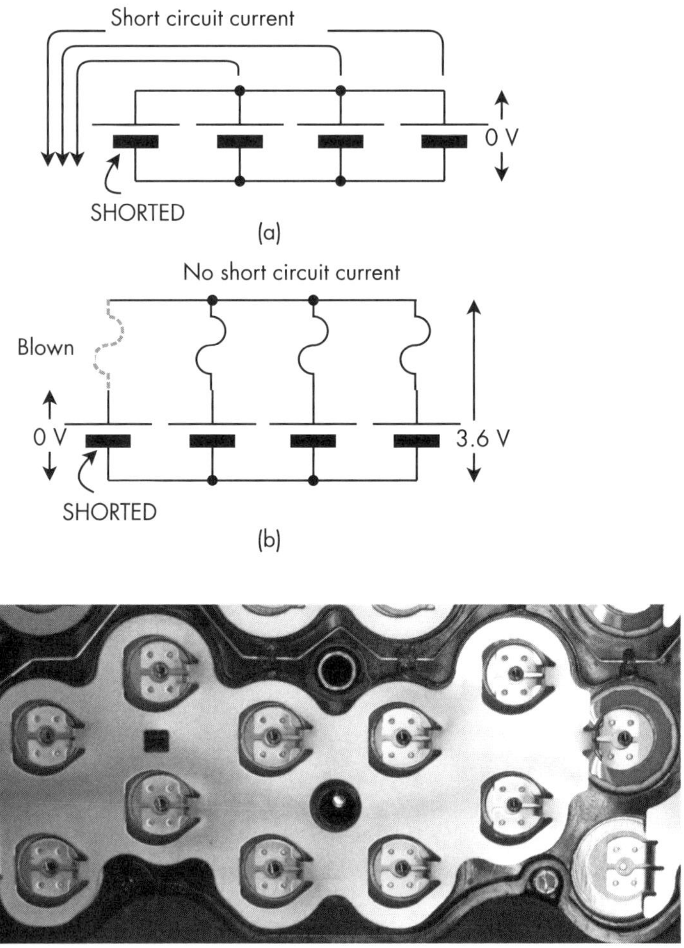

(c)

FIGURE 3.37
Parallel first: (a) shorted cell in a parallel block with no fuses, (b) with fuses, and (c) fuse per cell in a Hymotion PHEV traction battery.

the shorted cell, it blows, preventing it from discharging the other cells (Figure 3.37(b)).

The fuse can be implemented simply by forming a fusible link into the interconnecting plate, making the cost insignificant (Figure 3.37(c)).

Placing fuses in series with each cell in a parallel block is a hotly debated topic in the battery industry:

- Some battery designers do it;
- Other battery designers argue that fuses cause more problems than they solve;
- Others argue that fuses are a solution in search of a problem.

I will do my best to present all three points of view objectively.

3.3.12.1 Individual Fuses Are Required

Many battery designers [3] wish to address the possibility that a cell in a parallel group develops an internal hard short circuit.

The total current through the shorted cell is the short-circuit current of each cell times the number of cells in parallel. This current could be 10 kA. This is of particular concern when the power capability of other cells is maximum—100% SoC, hot cells, large bus bars.

A few cell manufacturers may be able to specify the maximum number of cells that can be safely connected directly in parallel. This is the maximum number of cells that can be connected in parallel that will not go into thermal runaway if connected to a hard short.

For example, if the cell manufacturer specifies 10 cells, then a parallel block of up to 10 such cells should not require fuses. However, in general, cell manufacturers do not perform this analysis. If they do, they do not want the responsibility resulting from specifying this parameter. In any case, they may give this datum only to a large manufacturer. Therefore, most battery designers must perform this analysis on their own.

In the particular case of a cell undergoing thermal runaway, other cells on the same block are doubly harmed by the heat received from the cell in thermal runaway and by the self-heating due to the short-circuit current.

These designers make the case that fuses in series with each cell are required to protect against this event. A fuse blows if its cell experiences an internal short circuit. This prevents a dangerous temperature rise and avoids discharging of the other cells.

Stopping the current in the other cells in parallel with a cell in thermal runaway removes a source of heat in these cells, therefore reducing the chance for thermal runaway propagation. Also, the fuses reduce the thermal conductivity between cells slightly, therefore reducing the chance for thermal runaway propagation.

Therefore, this argument goes:

- There is no reason not to implement this solution:
  - It has no downsides;
  - It prevents a cell that experiences an internal short from initiating a thermal runaway;
  - It reduces propagation of thermal runaway to other cells in the same parallel block;
  - If does not discharge the other cells in the same parallel block.
- The concerns about fusible links are unfounded:
  - Breakdown voltage of the fusible link: It's practically impossible that every single fuse in a block blows; therefore, there is never the chance that the full battery voltage appears across an open fuse;
  - Fire danger from arcing in the fuse: No flammable materials are placed near the fusible link.

3.3.12.2 Individual Fuses Create More Problems than They Solve

Other battery designers say that, yes, a fuse does reduce issues in case of a shorted cell, but it does so by adding a separate fire danger and at a disproportionate cost.

Fusible links integrated into the interconnecting plate are not reliable fuses:

- Low maximum breaking current: They may blow at 10 times the rated current but won't clear at 10 kA, which is precisely the current we want to protect from, because that current exceeds the fuse's breaking current rating.
- Low DC rating: Yes, the fuses can withstand the 3.5V or so of a single cell voltage. However, when the last fusible link in a parallel block blows,[24] the entire

24. When the first fuse blows, the other fuses carry the entire current, bringing them closer to their fusing current. The remaining fuses may start blowing one after another, like firecrackers. This is particularly a problem with only two or three cells in parallel.

battery voltage appears across this link (see Section 8.3.2.4). In a high-voltage battery, the fusible link may arc, which in itself is a fire danger, given that the link is not enclosed.

The fuses must be rated for the full battery voltage, DC rated, and for an extremely high breaking current to be able to break the short circuit current successfully. An off-the-shelf fuse that is rated for 400 Vdc and a breaking current of 50 kA costs about $100 and is larger than a 18650 cell. Imagine one such a fuse for each $2 cell!

This solution

- Is uneconomical—the fuses are more expensive than the cells;
- Has lower energy density—the fuses double the volume of the battery;
- Is unreliable—the additional components reduce reliability.

Therefore, this argument goes:

- Fuses may work, but the low risk of a fire from a shorted cell is preferable to the higher risk of fire from a fusible link;
- Actual fuses are too expensive.

3.3.12.3 Individual Fuses Are a Solution in Search of a Problem

Some designers state that cells do not fail as a hard short. Even if they did, fuses could not protect against such eventuality.

These designers pose and answer five questions:

1. Do cells fail as a hard short? Yes and no.
   - Yes. A cell that is reversed (charged in the wrong direction) may fail as a complete short, depending on the chemistry; however, a properly designed and installed BMS prevents a cell from going below a minimum voltage, let alone reverse the voltage. In any case, if reverse voltage damages a cell, the cells in parallel with it are also damaged; therefore, fuses won't protect those cells.
   - Yes. A cell undergoing thermal runaway becomes a hard short; however, at this point, discharging the other cells in the block is the last of your concerns.
   - No. Internal dendrites that manage to cross the separator and reach the opposite electrode create a soft short that has a resistance on the order of 1 kΩ for a small cell such as an 18650.
2. Will the fuse blow? No.
   - A soft short discharges a cell at a current of around 0.001C, far less than the approximately 10C that is required to blow a fuse in series with it.
   - If a cell does develop a hard short, it most likely did because its voltage was reversed. If so, the voltage was reversed on all the cells in the block, not just the one cell. Therefore, all the cells are discharged and cannot generate the current required to blow a fuse.
3. Will the other cells in the parallel block catch fire? Not likely.
   - An energy cell is designed to maximize the energy density at the expense of a high internal resistance. This high resistance limits the discharge current;

the cell discharges over 1 to 15 minutes,[25] which is too slow to heat the cell to its self-igniting temperature.[26]

- On the other hand, power cells are optimized for maximum power density and low internal resistance; they discharge into a short circuit in only 20 seconds or so.[27] Such cells may reach the self-igniting temperature when shorted, depending on the chemistry.[28]
- Regardless, today's Li-ion cells usually include a protection mechanism that disables the current before the cell becomes overheated[29]; a short circuit doesn't cause even a power cell to ignite.
- Given that a complete short is unlikely (thanks to the BMS preventing voltage reversal), and that a high-resistance path failure is more likely, the current through the bad cell is limited by the high resistance in the bad cell rather than by the internal resistance of the good cells. Therefore, the worst-case scenario considered by the fuse proponents is not realistic.
- Spacing between cells, whether through the air or heat-absorbing material,[30] may prevent propagation should one cell undergo a thermal runaway.

4. Will a cell that shorts out catch fire? That is unlikely.
   - A cell with a 0 Ω short circuit is not be heated by the current from adjacent cells because power is proportional to resistance, the resistance of the short circuit is zero, so power across the short circuit is zero. Therefore, the short circuit doesn't generate heat. However, the cell's resistance is not zero; the short circuit current does generate heat inside the cell; the arguments above indicate that this is unlikely to start a fire.
   - A cell that develops a soft short *does* dissipate heat because the resistance is nonzero. Therefore, the power is nonzero, yet the current is limited by this resistance, so the power is small—on the order of 10 mW to 1 W for an 18650 cell.
5. Do fuses improve reliability? No.
   - They won't blow;
   - The presence of the additional components, the fuses, actually reduces the reliability.

Consider also the perceptions of the battery designer:

- On the one hand, the designer sees ten 18650 cells in parallel forming a 25 Ah block and worries about one shorting out, discharging the other ones.

25. Energy cells have a maximum power time of 120~1000 s. The discharge time through a short circuit is half of the MTP: 1 to 15 minutes.
26. Some cell chemistries are safer than others (notably, LFP has a higher self-ignition temperature).
27. Power cells have a maximum power time of 20~120 s, for a discharge time through a short circuit of 10 to 60s.
28. Note that the number of cells in parallel is irrelevant from an electrical standpoint. This is because it's the power wasted in heat generated inside each cell that matters, and when discharged through a complete short circuit, this power is independent of how many other cells may also be discharging through this same short circuit. This is from an electrical standpoint. On the other hand, from a thermal standpoint, the number of cells does matter: a large number of cells in close proximity all heating up at the same time results in a higher localized temperature than a single cell by itself.
29. The separator is engineered to melt and stop ion flow at a given temperature.
30. AllCell promotes its phase change material as a safety mechanism, as it absorbs heat from one cell, drastically reducing the heat from reaching the adjacent cells (see Section 5.17.7.2).

- On the other hand, the designer doesn't worry about a 25 Ah large prismatic cell, which internally is composed of several Li-ion bags connected directly in parallel.

Both have the same amount of active material; relative to electrode area, both have the same risk of containing impurities or of developing dendrites. Yet, in the first case, the paralleling paths are external and visible (worrying the battery designer), while in the second case, they are internal and not visible (the battery designer is oblivious). Even though the risk is virtually identical, the battery designer is only concerned about the first case.

3.3.12.4 Deciding Whether Fuses Are Required

I do see the validity of all three points of view.

My recommendation is that you should consider fuses only if all of the points below apply:

- There is a mechanism that may allow a cell to fail as a hard short circuit;
  - It may be subjected to a reverse voltage;
  - Or it may be charged after a deep discharge;
  - Or it may undergo thermal runaway.
- The cells do not include a thermal cutout mechanism;
- The MPT of the cells is less than 120s;
- The cell chemistry has a low threshold for self-ignition temperature;
- The cells are in physical contact with each other.

3.3.12.5 Fuse Size

How should the fuses be rated? Should they be rated for the maximum current that a cell should handle? Or should they be rated for the maximum total current of all the cells in a block?

If the cell resistances are not equal, one cell does most of the work when there is a sudden change in the current (see Section 3.3.2.2). If the fuse is rated for the maximum current for one cell, a spike in the current would blow the fuse in series with the low-resistance cell.

3.3.12.6 Detecting a Blown Fuse

It is OK if the BMS cannot detect a blown fuse; the battery's capacity is reduced, but it won't be unsafe because the blown fuse isolates the shorted cell.

Still, a BMS may be able to detect that a fuse has blown (without the need for a tap for each cell in parallel) by noticing that the capacity of the parallel block has decreased suddenly. This doesn't work if the block uses a large number of cells in parallel because the effect is too small to see.

3.4 PARALLEL-FIRST

In the parallel-first arrangement,[31] cells are connected in parallel to form a block, and then blocks are connected in series (Figure 3.1(c)). This is the standard arrangement

31. Sometimes called PCM (parallel cell module), not to be confused with protection circuit module or phase change material.

for batteries that use cells of lower capacity than the desired total capacity. For example, it is often used in laptop computer batteries (Figure 3.38(a)).

The advantages and disadvantages of parallel-first are the same as for parallel blocks (see Section 3.3, "Parallel Blocks"). Compared to series-first, the parallel-first arrangement is advantageous because a weak cell is supported by the cells in parallel with it:

- A low-capacity cell doesn't reduce the capacity or the battery as much;
- Similarly, the effect of a high-resistance cell is reduced;
- The voltage of a cell with a high self-discharge current doesn't drop as fast, as the other cells furnish that current.

If a fuse-per-cell is required, each cell in each row has a fuse (Figure 3.38(b)).

Charging a parallel-first arrangement is the same as charging a series string. If the cells are imbalanced, the same concerns apply (see Section 3.2.13).

3.4.1 Paralleling Wire Size

In an ideal battery, the wires that parallel cells in a block can be thin, as they only carry the sense voltage to the BMS and the balance current; 24-AWG wires are suited to carry 3A of balancing current.

In reality, there may be a high current between adjacent cells in a block due to variations in cell impedance and especially if a cell is weak (high resistance). In the worst case,[32] the paralleling wire may need to carry the entire battery current.

Using large conductors between cells in parallel is not a problem for new battery designs. It is of great concern for people who buy two stacks of pouch cells from a hobby store and attempt to connect them in parallel through their balance connectors (see Volume 2, Section 1.3.3.2)—those wires and contacts are woefully inadequate to carry the full battery current. It is also of great concern to people who buy a traction

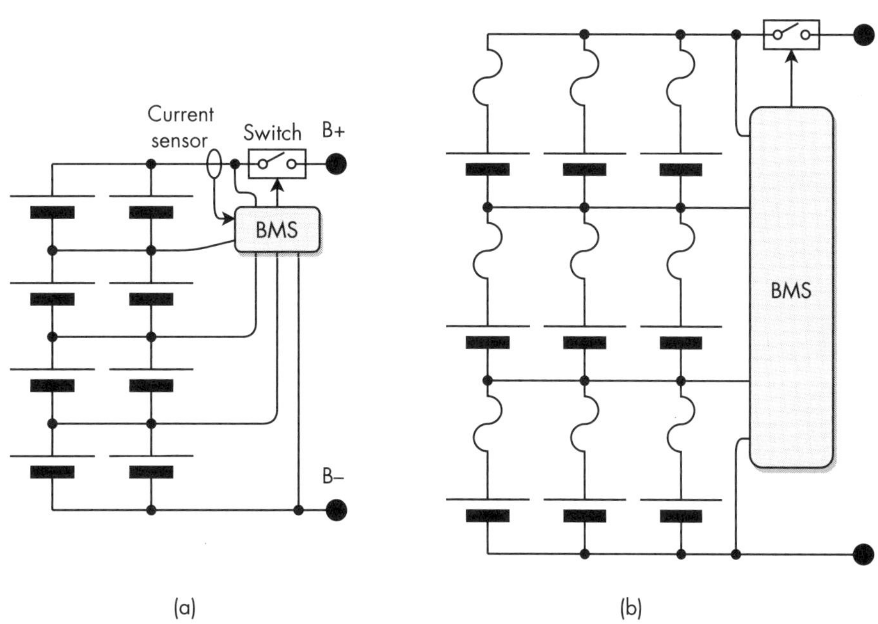

FIGURE 3.38
Parallel first: (a) typical battery, and (b) battery with fuse-per-cell.

32. The worst case is when there are two cells in parallel and one cell becomes disconnected.

battery from a crashed car and rearrange the cells in parallel to drop the voltage to 48V (see Volume 2, Section 2.12.6).

3.5 SERIES-FIRST

In the series-first arrangement,[33] cells are connected in series to form a string. Then, two or more strings are connected in parallel (Figure 3.1(d)).

Even though the series-first arrangement is more expensive and less reliable than parallel-first, it is often considered by

- The hobbyists who wish to reuse reclaimed batteries;
- The inexperienced designer who thinks it is advantageous.

Sometimes, the series-first arrangement is necessary for professionally designed batteries due to unavoidable mechanical constraints. Obviously, the number of cells in each string must be the same. Otherwise, the strings cannot be connected in parallel because their voltages are different.

3.5.1 Disadvantages of SERIES-FIRST

The series-first arrangement is problematic due to its many disadvantages.

3.5.1.1 Higher Cost

Often, people new to battery design envision a battery with a large number of strings in parallel. After they recover from the shock of the price and complexity of a BMS for this arrangement, usually these designers discover the advantages of the parallel-first arrangement and reconsider their design. In the end, they are pleasantly surprised by the significantly lower cost for a BMS for a parallel-first arrangement.[34]

The series-first arrangement is more expensive than parallel-first because it uses a more complex BMS. For example, a 3P4S battery requires a BMS with five taps (Figure 3.39(a)), while a 4S3P battery requires a BMS with eleven taps (Figure 3.39(b)). This disadvantage grows as the number of strings increases. For example, a 10P10S battery requires a BMS with 11 taps, while a 10S10P battery requires a BMS with 92 taps!

A BMS that can handle strings in parallel is rare. Using 10 separate BMSs is not practical because they cannot be easily coordinated to work as a single BMS.

It may be tempting to combine the tap wires from all the cells in a row through resistors that give the BMS the average voltage of all the cells in one row (Figure 3.39(c)). While this would reduce the number of taps, it does not work:

- With 10 cells in parallel, if one is down to 0V and the others are at 3.5 V, the average voltage is 3.15 V; the BMS accepts this voltage because it has no idea that a cell is in such a bad shape;
- With two cells, one at 5V and one at 2 V, the average is 3.5V; the BMS has no idea that one cell is way overcharged and the other is way overdischarged;
- Balancing is impractical because of these resistors;

33. Sometimes called series cell module (SCM).
34. As I write this, I am quoting a BMS for a 10S60P battery at $9,000; rearranging the same cells to 60P10S brought the price down to $800.

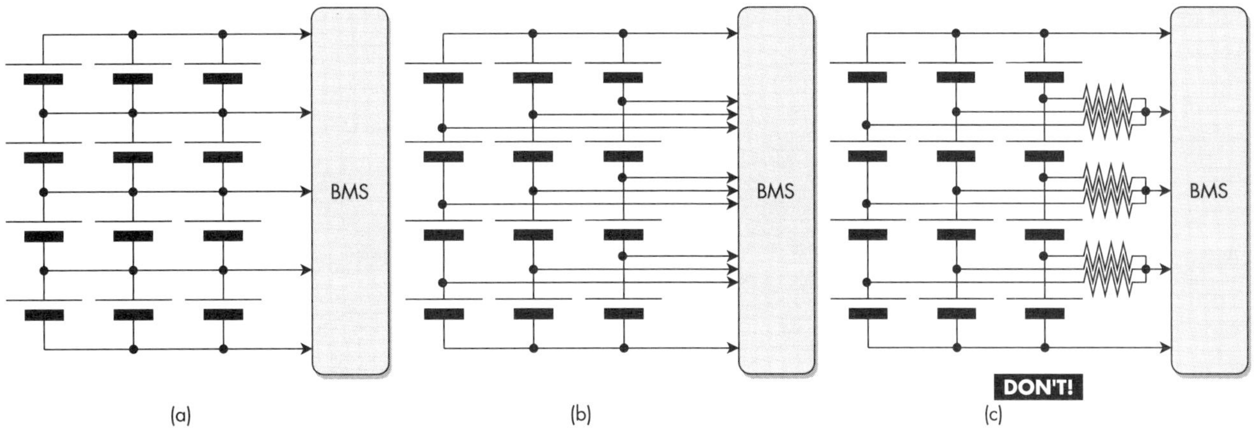

FIGURE 3.39 Number of taps: (a) Parallel-first, five taps, (b) series-first, 11 taps, and (c) cheating: five taps (DON'T DO THIS!).

- The BMS reads the wrong voltage because the small current into the BMS sensing input results in a small voltage drop across the resistors.

3.5.1.2 Worse Performance

In the real world, cells vary. A few cells may be weak—they have a lower capacity or a higher resistance. By connecting cells directly in parallel, the cells in parallel with the weak cell help carry its weight and reduce its effect on battery performance.

If low-capacity cells are distributed randomly in a battery, the battery capacity is higher with parallel-first (Figure 3.40(b)) than with series-first (Figure 3.40(c)). Similarly, if high-resistance cells are distributed randomly in a battery, the resistance is lower with parallel-first (Figure 3.41(b)) than with series-first (Figure 3.41(c)). Granted, a single high-resistance cell won't affect the total resistance of the battery as much as a low-capacity cell would.

The voltage of a high-resistance, weak cell swings significantly under load. The BMS shuts down the entire battery when a weak cell's voltage exceeds the limits, even though the other strings are OK:

- Without a load current, all the cells are at 4V, and the total voltage is 16 V (Figure 3.42(a)).
- Under 100 A load current, the BMS shuts down the battery (Figure 3.42(b)):

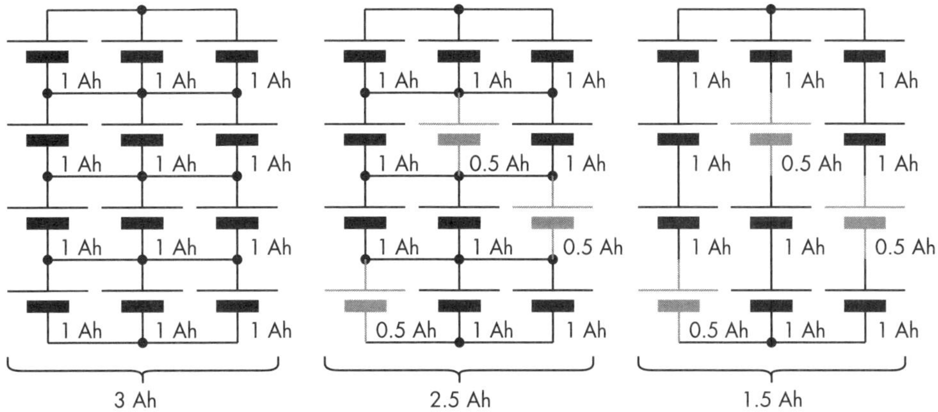

FIGURE 3.40
Effect of weak cell in a parallel block:
(a) capacity, (b) low capacity, parallel-first, and (c) low capacity, series-first.

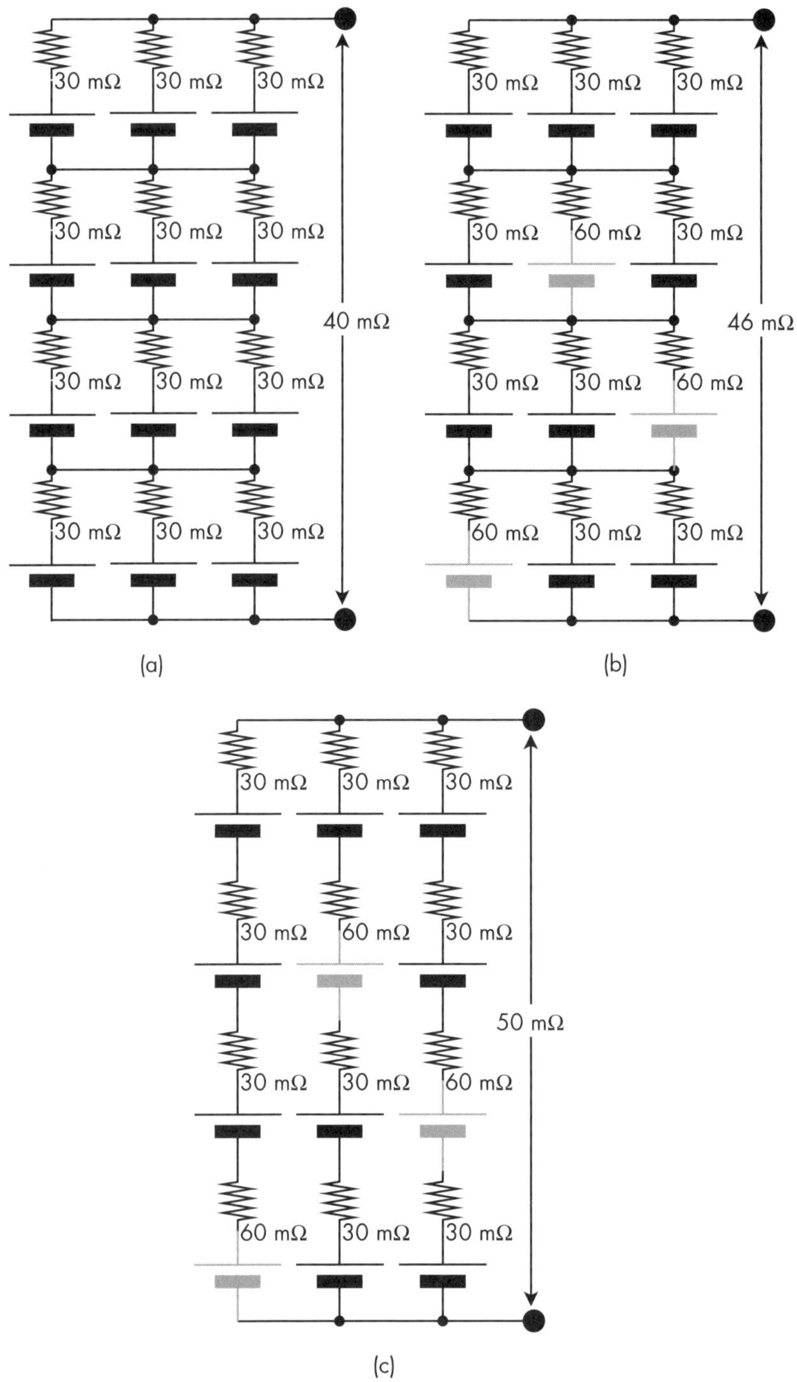

FIGURE 3.41
Effect of weak cell in a parallel block: (a) normal resistance, (b) high resistance, parallel-first, and (c) high resistance, series-first.

- The center and right strings carry practically all of the current (50A each), and the voltage of their cells drops down to 3V; the total voltage drops to 12V.
- The leftmost string carries little current because of the weak cell. Because they are carrying practically no current, the good cells in this string remain at 4V, for a total of 12V; this leaves 0V across the weak cell's terminals, which it is unable to overcome through its high resistance.
- The BMS sees that the voltage on the high-resistance cell dropped to 0V, and shuts down the battery.

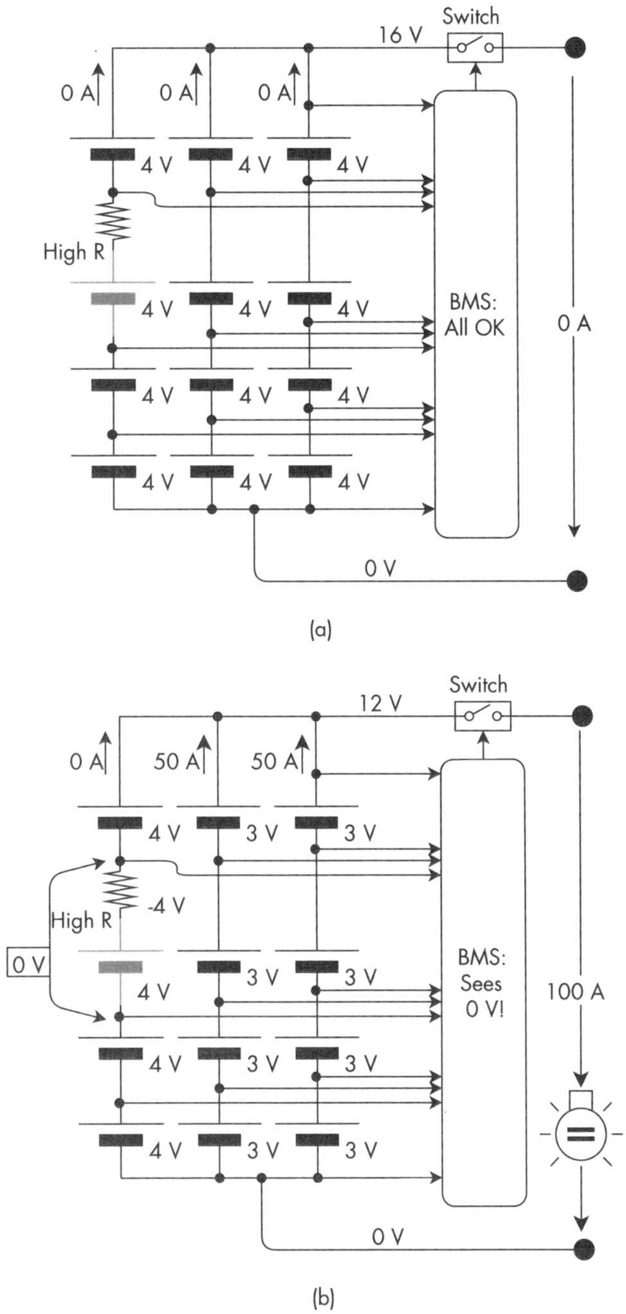

FIGURE 3.42
High-resistance cell in a series-first arrangement: (a) no load current, and (b) high load current.

Finally, a cell with high self-discharge current forces the BMS to balance the battery by discharging all the other cells at high current.

3.5.1.3 *Equalizing Inrush Current*

If the strings are not at the same voltage, when they are first connected in parallel, an inrush current attempts to equalize their voltages. When two strings of the same capacity are connected in parallel, the specific current of this inrush is the same as when two identical cells are connected in parallel (see Section 3.3.6.1). As we saw, the intensity of the inrush current may or may not be acceptable, depending on the respective SoC levels and the MPT of the batteries.

However, when a string is connected to two or more strings already connected in parallel, the inrush current can be quite large and damaging (see Section 3.3.6.2).

To minimize the danger, connect the strings in a binary fashion: start by connect the strings in pairs, then connect two pairs, then two quads, and so forth. This ensures that a set of strings is connected to a set of strings of the same size (therefore of similar internal resistance), limiting the inrush current in each string.

3.5.1.4 Poorly Defined Evaluation

The BMS only measures the total battery current. It does know the current in each string. Therefore, evaluation of the state of the battery (SoC, resistance, SoH) is more difficult.

3.5.1.5 Unequal Balancing

Because the current does not divide equally in the strings, hysteresis in the cell voltages of one string won't match the hysteresis in the other strings. Therefore, a BMS may not balance the cells in one string to the same SoC as other strings (see Section 4.7.4).

For these reasons, the series-first arrangement provides worse performance than the parallel-first arrangement.

3.5.2 Perceived Advantages of Series-First

At first glance, a battery designer may assume that the series-first arrangement is preferable due to

- Flexibility—capacity may be increased on-demand by adding more strings in parallel;
- Redundancy—if a string fails, it can be removed;
- Modularity—a single series string may be housed in a single case, and multiple identical cases can be connected in parallel.

Yet, after careful consideration of the disadvantages, it becomes clear that the series-first arrangement does not have these perceived advantages because

- There is the potential for damage when connecting strings at different voltages (see Section 3.3.6);
- The BMS must monitor the cells in each string, increasing the complexity of the BMS, which requires a sensing connector for each string;
- Bad things are bound to happen if the user is allowed to connect strings in parallel without ensuring that the strings have the same voltage;
- If a string is in a module, the string voltage appears on the contacts unprotected, exposing it to a potential short circuit.

3.5.2.1 Flexibility

A battery designer may consider using a modular battery for a small vehicle or a portable device. The goal is to allow the user to decide each day how many battery modules to install, so as not to carry more energy that will be required during the day. On a typical day, the user may use no extra batteries (Figure 3.43(a)). If the user expects heavy use, they may add an extra battery (Figure 3.43(b)). Before going on a long trip, the user may add three extra batteries (Figure 3.43(c)).

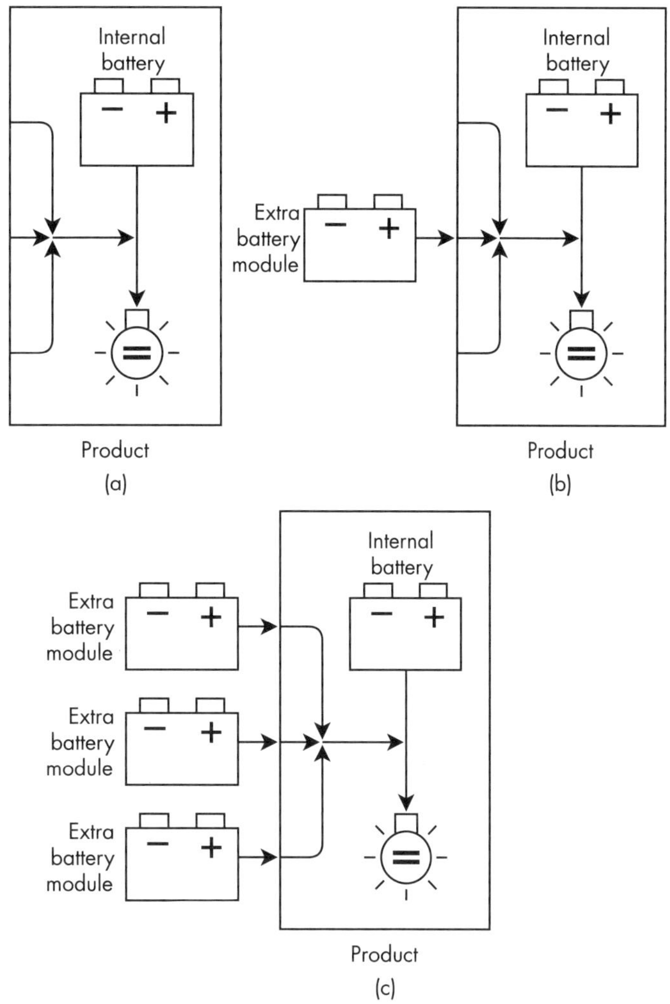

FIGURE 3.43
Product with modular battery: (a) no extra modules, (b) one extra module, and (c) three extra modules.

Carrying as many as are needed works for water bottles, gas tanks, or lunch sandwiches. For the reasons listed above, it doesn't work as well for battery modules.

3.5.2.2 Redundancy

Should a string fail, one would assume that it would be taken out of service. The rest of the strings can continue powering the product. This is correct, except that

- The user must not be allowed to replace the string or to reconnect a string after it has been repaired because it could be at a different voltage;
- An automated system must not be allowed to reconnect a string unless it has a way to check the voltages.

3.5.2.3 Modularity

In a modular battery, several identical modules are assembled to achieve the desired battery size (see Section 6.2). The modules are permanently connected in parallel at the factory. In the field, they are connected by a qualified technician who knows how to check voltages before making a parallel connection. If the technician forgets to check, the battery could be damaged.

3.5.3 Actual Advantages of Series-First

Despite all its disadvantages, the series-first arrangement is used when unavoidable:

- When there is a concern about too many cells in parallel (see Section 3.3.12);
- When reusing a battery from a different product (e.g., reusing modules from a traction battery);
- When regulations require fuses between parallel battery connections (see Section 3.5.5).

3.5.4 Voltage and Current in Series-First

The voltage of each string in a series-first arrangement is the same as the voltage of all the other strings since they are all connected in parallel. As I said, before connecting a string to other strings, ensure that their voltages are the same.

Current is not be shared equally by all the strings in a series-first arrangement for the same reasons that it is not shared equally in a parallel block (see Section 3.3.2).

3.5.5 Fuse per String

In a series-first arrangement, placing a fuse in each string increases safety (Figure 3.44(a)) compared to using a single fuse for the entire battery (Figure 3.44(b)).

Should the fuses be rated for the maximum current that a string should handle? Or for the maximum total current of all the strings? If the string resistances are not equal, when the current changes suddenly, one string does most of the work. If the fuse is rated for the maximum current for one string, the string with the lowest resistance with do most of the work and carry most of the current. A peak in current will blow the fuse in series with that string.

3.5.6 Mismatched Strings, Mixing Battery Types

Ideally, all the strings in parallel should be identical. This section explores what happens when they aren't.

3.5.6.1 Mismatched Strings

Connecting strings in parallel with different SoC levels, capacities, or resistances presents the same issues as when connecting cells in parallel (see Section 3.3.4).

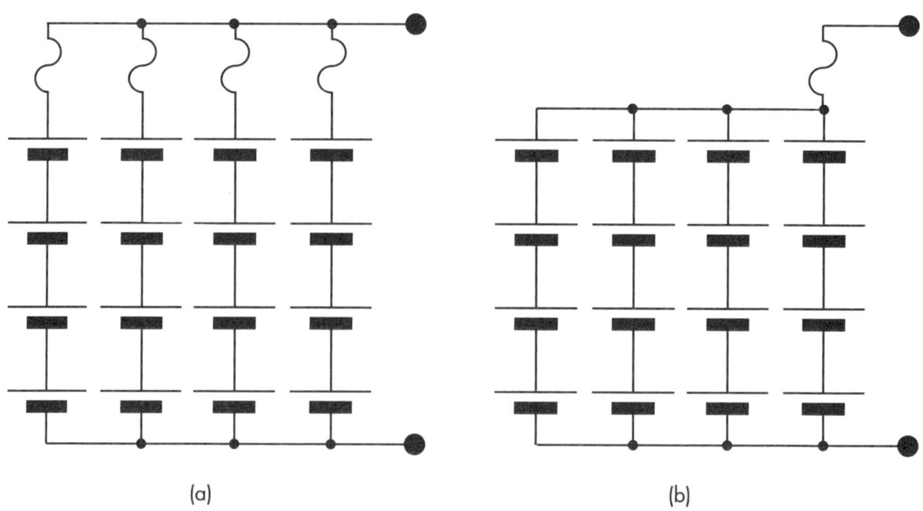

FIGURE 3.44 Series-first fuses: (a) fuse per string, and (b) single fuse.

3.5.6.2 *Different Types of Batteries*

Connecting completely different types of batteries is not ideal and can be dangerous.

For example, connecting a string of four LFP cells[35] across a 12V lead-acid battery may be OK at first, since the voltages can be the same (see Section 6.6). However:

- Initially, the LFP string carries most of the current because it has lower internal resistance than the lead-acid battery; eventually, the two strings will equalize.
- The OCV of a lead-acid battery has a significant temperature coefficient (unlike Li-ion); as temperature increases, the OCV of the lead-acid battery drops, resulting in an equalization current between the two strings. Thus, if there is a sudden change in temperature, then the equalizing current between strings may be quite large; a fuse in each string would help in case of excessive equalization current.

Connecting a NiMH to a Li-ion battery is a problem because they use different criteria to terminate charging:

- Charging NiMH stops when cells get warm or the voltage stops increasing and actually starts *dropping* (!)
- Charging Li-ion stops when the current drops below a certain level.

3.5.7 Charging a Series-First Arrangement

Charging a series-first arrangement is the same as charging a series string, with the same concerns for imbalance (see Section 3.2.13). Also, there may be some concern if the strings have a different series resistance because the current is not divided equally, resulting in one string having cells with a lower SoC than the other strings. After the charger is turned off, current flows from the other strings to the one string, eventually balancing all the strings. However, it takes a long time (see Section 3.3.2.3).

3.6 OTHER ARRANGEMENTS

Cells may be arranged in more ways than the four basic arrangements discussed above.

3.6.1 Complex Arrangements

The four basic arrangements may be combined for more complex arrangements. This is typically done to overcome limitations in the available space or to allow reuse of premade modules.

For example, if you are given 24V, 10 Ah modules, composed of seven cells in series (Figure 3.45(a)), you may want to make a 48V, 50 Ah battery by arranging the modules as 5P2S (Figure 3.45(b)). Finding an off-the-shelf BMS for such arrangements can be a challenge.

For another example, a 24V house power battery for a boat is composed of cells that are stashed in the few available spaces around the vessel. Some spots can hold only a few cells, other more cells (Figure 3.45(c)). The notation for the resulting arrangement is rather unwieldy.[36]

---

35. Four cell at 3V equals 12V.
36. Possibly 2S2P+3S2P+2S2P.

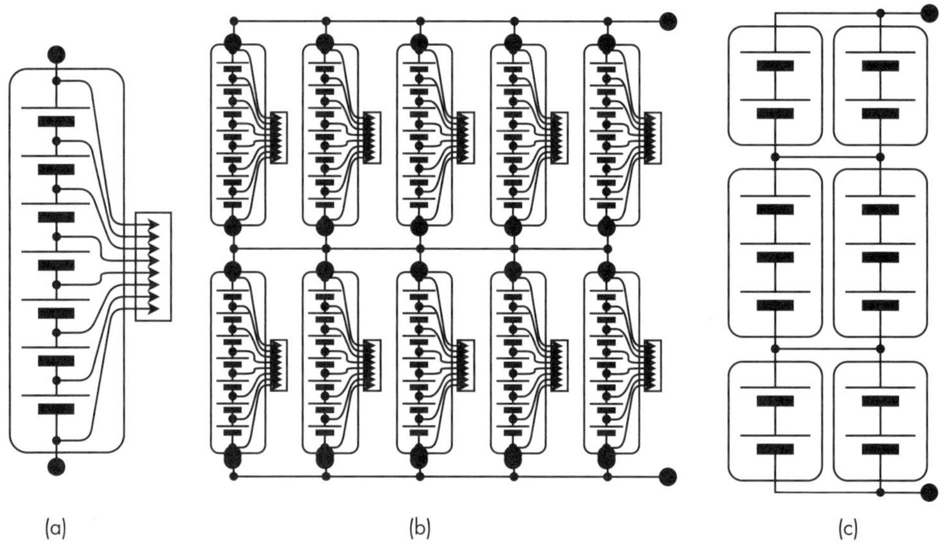

FIGURE 3.45
Complex arrangements: (a) 7S module, (b) modules arranged for a 48V battery, and (c) uneven module arrangement.

3.6.2 Dynamic Arrangements

I regularly encounter people reinventing ways to rearrange cells on the fly. These are all actual examples:

- Bypass a bad cell, reducing the number of cells in series by one (Figure 3.46(a));
- Start a race car with two halves of a battery in parallel for higher torque, then, above a certain speed, reconnect them in series for higher top speed[37] (Figure 3.46(b));
- Connect five cells in parallel for use inside a USB power bank; some other time, connect them all in series to power an 18V laptop directly;
- Balance a battery by connecting all its cells in parallel for a while, then restore the normal series string during use (Figure 3.46(c));
- Change the arrangement of a power tool battery from 15S to 5P3S automatically, depending on the voltage of the tool on which the battery is installed (Figure 3.46(d)).

This last example is a commercially available product—a large slide switch in the DeWalt FlexVolt battery is operated by the shape of the tool it's inserted in. It reconnects the cells for either 20V or 60V. This product violates the rule that strings should not be connected in parallel without first making sure they are at the same voltage. The designers of this battery assume that the cells are all superbly matched and that their voltages are always close to each other regardless of their history. I would not make this assumption. This battery degrades each time it is transferred from a 60V tool to a 20V tool as equalizing currents rush between strings.

The race car example is also real. It uses contactors to rearrange two halves of its traction battery. While connecting two battery halves in parallel without first checking their voltage is cringeworthy; winning a race is more important than the health of the cells.

37. Parallel = low resistance = high current = high torque; series = high voltage = high top speed.

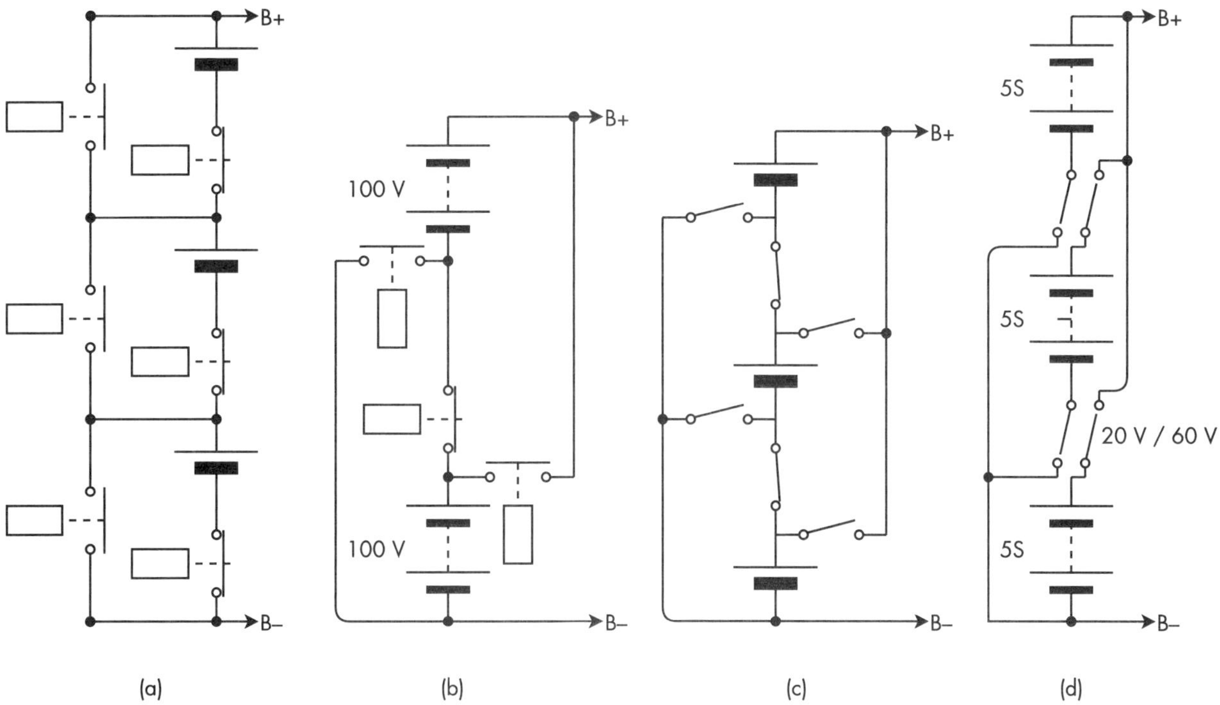

FIGURE 3.46 Rearranging cells: (a) bypass a cell, (b) parallel for high torque, series for high speed, (c) parallel to balance, series to use, and (d) DeWalt FlexVolt.

All other examples are of dubious merit. Some are solutions in search of a problem; regardless, one must consider the challenges of technologies that allow rearranging cells on the fly:

- Using a mechanical switch (as DeWalt does) is not so bad;
- Contactors and electronic circuits that can rearrange cells on the fly do exist, but they are inefficient, inelegant, expensive, and prone to spectacular failure when the conditions are nonideal.

While it is fun to theorize about dynamic arrangements, engineering resources are better spent on improving the design of fixed-arrangement batteries because they are safer, more effective, and more economical.

3.6.3 Series Strings with Fuses or Resistors

At times alternative arrangements are suggested as an attempt to overcome limitations of the standard arrangements. The series-first arrangement has the disadvantage of requiring a BMS with many taps: $N = S \times P$ taps. The parallel-first arrangement has the perceived disadvantage of the risk should one cell suddenly turn into a complete short circuit.

A group of respected researchers proposed one of the weirdest and potentially most dangerous arrangements I have seen to try to overcome the real disadvantage of series-first and the perceived disadvantage of parallel-first. The circuit is like a parallel-first arrangement, except that it uses resistors or fuses between rows of cells (Figure 3.47). Though these circuits are presented as fail-safe, they can actually be downright dangerous, as explained in the appendix (see Volume 2, Section A.4.3,).

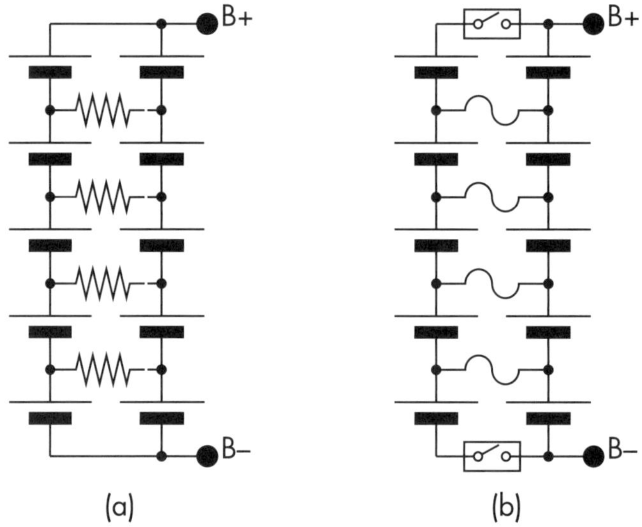

FIGURE 3.47
Kim arrangements:
(a) with resistors, and
(b) with fuses.

References

[1] Plett, G., Battery Management Systems, Volume 2, Norwood, MA: Artech House, 2016.

[2] Ibid., p. 260.

[3] Pesaran, A. A., G.- H. Kim, J. Neubauer, K. Smith, and S. Santhanagopalan, Design and Analysis of Large Lithium-ion Battery Systems, Norwood, MA: Artech House, 2014, p. 168.

CHAPTER 4

LI-ION BMS

4.1 INTRODUCTION

This chapter discusses the full range of BMSs for Li-ion batteries, to guide the selection of the type, functionality, and topology that best fits a given application. If you have already selected a BMS, it helps you understand how your BMS works, and why it does what it does.

While this chapter is not a guide to designing a BMS (see Volume 2, Section A.5.4), it may be useful when initially defining its specifications.

While any decent BMS can protect your battery, a better BMSs may extend battery life or its performance. It may also be more helpful during troubleshooting. A BMS of a given topology may be easier to install in a particular battery.

When discussing BMSs, I strive to remain impartial, despite my association with Elithion. My only interest is to help you find the BMS that best fits your application.

4.1.1 Tidbits

Some interesting items in this chapter include

- You are not a BMS; any product that requires your intervention is not a BMS (4.1.3.1);
- You may think that your BMS is "crap," because it's just not well matched to your battery—or to you (4.13.4);
- Charge transfer balancing is sexy, but I bet you don't need it (4.7.2);
- Be skeptical of a BMS specification of accuracy in the SoC measurement (4.8.1);
- It's normal for the charger to go off and on as the BMS balances the battery (4.7.5.3).

4.1.2 Orientation

This chapter starts by defining a BMS, listing its types and topologies. It talks about PCMs, digital protectors, and digital BMUs. It discusses topics common to various types of BMSs, including measurement, protection, balancing, evaluation, and wiring the BMS. Finally, it considers BMS reliability and gives suggestions on how to buy a BMS.

4.1.3 BMS Definition

The main job of a battery management system is to protect each cell in a battery by keeping it within its safe operating area (see Section 2.3.2). At a minimum, it does so by stopping the battery current when required. Optionally, a BMS may also optimize the battery performance, evaluate its state, and report it.

Specifically, a BMS performs some or all of these functions:

Protects the cells:

- By stopping the battery current, controlling a protection switch directly (e.g., relays, contactors, MOSFETS);
- By asking the system to reduce the battery current when required;
- By managing the temperature, controlling heating and cooling directly.

Maximizes the battery performance by

- Balancing strings of cells in series to maximize capacity;
- Dynamically adjusting the operating area to extend cell life by slowing down their degradation.

Evaluates and reports the state of the battery, including

- State of charge;
- Warnings and faults;
- Maximum available discharge current, maximum acceptable charge current (or power);
- State of health.

Communicates

- To the external system;
- To a charger.

4.1.3.1 Not a BMS

Products and solutions that do not meet the BMS definition (above) are not worthy of the name:

- *Monitors only:* Report cell voltages, do nothing about it;
- *Balancers/equalizers/regulators:* Clamp the cell voltage by bypassing current;
- *State of charge displays:* Do not monitor individual cell voltages, do not control the current.

Any "BMS" that relies on a human being is not a BMS. A monitor in a car that beeps when a cell voltage gets low, to tell the driver to go easy, is not a viable solution. The first time that the stereo is too loud, or the driver is a bit deaf or is a teenager who just doesn't care, the battery is damaged.

4.1.3.2 A BMS Is Not a Charger

A BMS's protector switch is either on or off. It does not change the voltage or limit the current. Some BMS can *request* a gradual reduction, but they cannot reduce the current by themselves.

For example, between a 12V lead-acid battery (not current-limited) and a string of three Li-ion cells in series, you need both (Figure 4.1):

- A charger (to limit the current and to drop the voltage from 12V to whatever the string voltage is at the time);

FIGURE 4.1
A charger and a BMS are both required between a voltage source and a Li-ion battery.

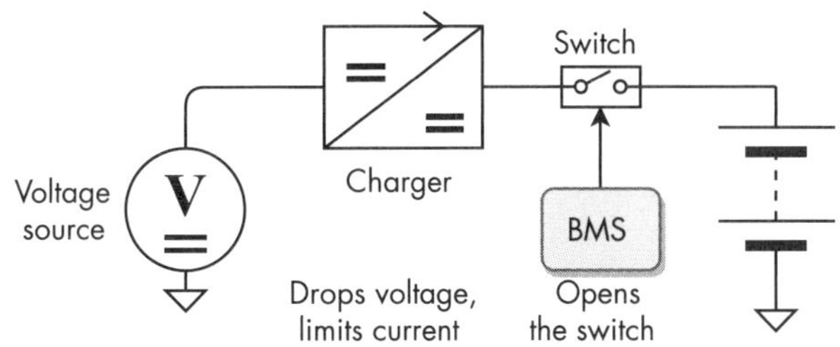

- A BMS (to interrupt the current if any cell voltage is too high, or for some other reason).

Some products for small batteries include a charger and a BMS (see Section 4.3.5).

4.1.4 A BMS Is Not Optional

Any question that starts with "Do I need a BMS if..." can only be answered with an emphatic "Yes, you do!": a Li-ion battery must always a BMS of some form.

For a single cell, the BMS may consist simply of using the correct charger and adding a low-voltage cutoff on the load. For an artificial satellite, it may involve testing 100,000 cells and selecting 10 cells whose parameters are matched to 0.1% to guarantee that they all behave exactly the same. In all cases, the Li-ion cells are managed in some way.

4.2 TYPES OF BMS

The BMS type can be classified based on

- Technology;
- Topology;
- Format.

4.2.1 BMS Technology

BMSs can be classified based on the technology they use (Table 4.1):

TABLE 4.1
Comparison between Analog and Digital BMSs

	Analog	**Digital**
Sophistication	Dumb: knows there is a problem, but not what, where, or by how much	Smart: knows what the problem is, where it is, and how bad it is
Sensing	Cell voltage (current, few temperatures*)	Cell voltage, current, temperatures (battery voltage and isolation, supply voltage*)
Limits	On/Off	Requests gradual reduction, then off
Communications	None, or On/Off	Reports data
Troubleshooting	Unhelpful	Quite helpful
Deployment	Easy enough for a hobbyist	May require careful setup
Cost	Cheap	More expensive

*Optional.

- *Analog:* Knows there's a problem, but not what, where, how bad;
- *Digital:* Also knows what the problem is, where it is, how bad it is, and can tell you.

BMSs can also be classified based on whether they include a protector switch:

- *Protector:* Turns off the current using an included power switch;
- *Battery management unit/battery monitor unit (BMU)*[1]: Controls external power switches or tells the system to turn off the current.

This results in four permutations (Figure 4.2). Figure 4.3 shows three examples. Table 4.2 compares the four permutations.

4.2.2 BMS Topologies

BMSs can be classified based on how the subdivision of the electronic assemblies, and on how it connects to the cells.

4.2.2.1 By Assembly Subdivision

The BMS topology can be classified based on how its electronics are subdivided (Table 4.3):

- Single board: All in one;
- Single BMU: The same, except for the protector switch;
- Modular: Multiple assemblies.

The modular topology is quite flexible and is great for large, modular batteries. Each battery module includes a slave. A small cable runs from slave to slave, all the

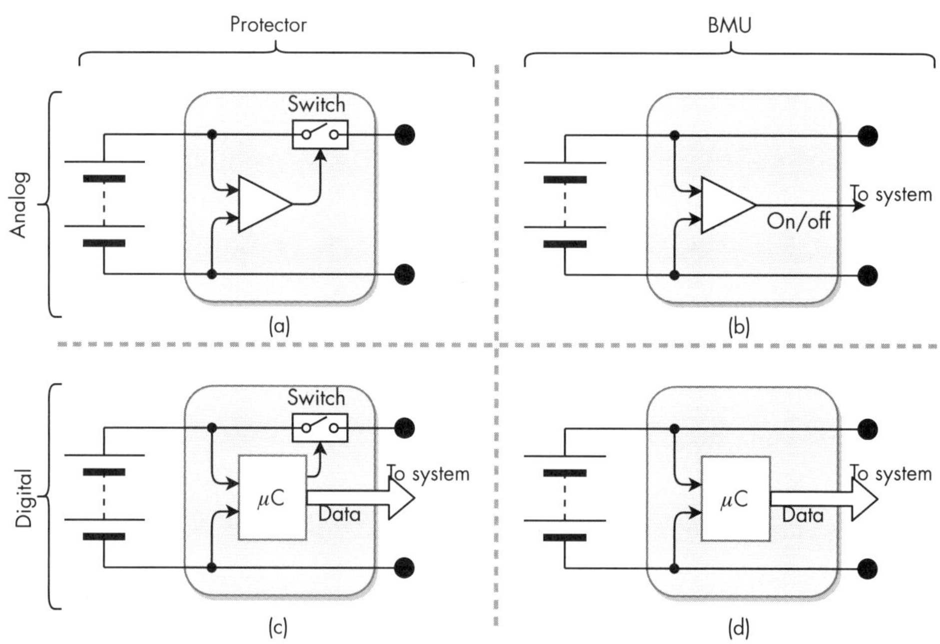

FIGURE 4.2
Four permutations for BMS types: (a) analog protector, (b) analog BMU, (c) digital protector, and (d) digital BMU.

1. Also known as battery control module (BCM) or battery electronic control module (BECM).

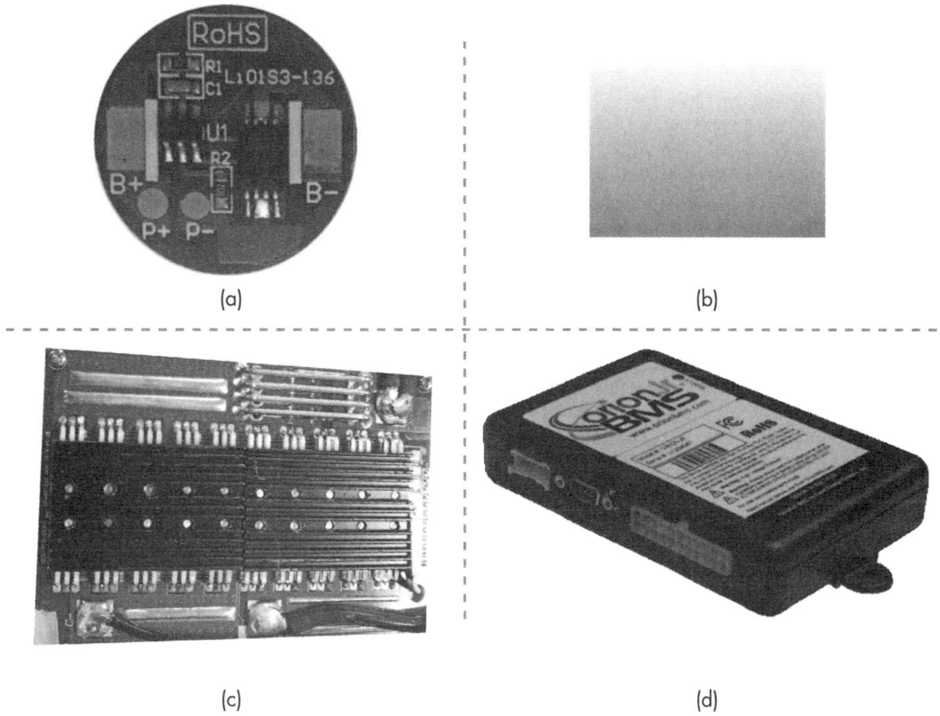

FIGURE 4.3
Examples of BMS types:
(a) analog protector,
(b) analog BMU (no example found),
(c) digital protector, and
(d) digital BMU (courtesy of Ewert Energy).

TABLE 4.2
Four Permutations for BMS Types with Typical Applications

	Protector	**BMU**
Analog	Analog protector (Figures 4.2(a), 4.3(a)) Most common type for small batteries	Analog BMU (Figures 4.2(b), 4.3(b)) (Uncommon) engine starter battery
Digital	Digital protector (Figures 4.2(c), 4.3(c)) Small batteries that require monitoring	Digital BMU (Figures 4.2(d), 4.3(b)) Most common for large batteries

TABLE 4.3
Comparison of BMS Topologies Based on Its Subdivision

	Single-Board	**Single-BMU**	**Modular**
Description	The entire BMS is on one electronic assembly	The BMU is a single electronic module, the protector is external	The BMU consists of multiple modules, one of which is a master
Versatility	None	Quite good	Excellent
Cost	Lowest	Higher	Highest

way back to the master.[2] The master may include a slave function as well, so it could be used by itself or be expanded by adding slaves.

4.2.2.2 *By Connection to Cells*

The BMS topology can also be classified based on how it connects to the cells (Figure 4.4).

A wired BMS is easier for the new user to understand. In large quantities, it can be less expensive. However, a wired BMS requires high-voltage wires and thermistors

2. If this term offends you, I apologize; none of the proposed alternatives express the concept as clearly. In engineering, master/slave has only technical connotations; I acknowledge that, in other contexts, this term is offensive.

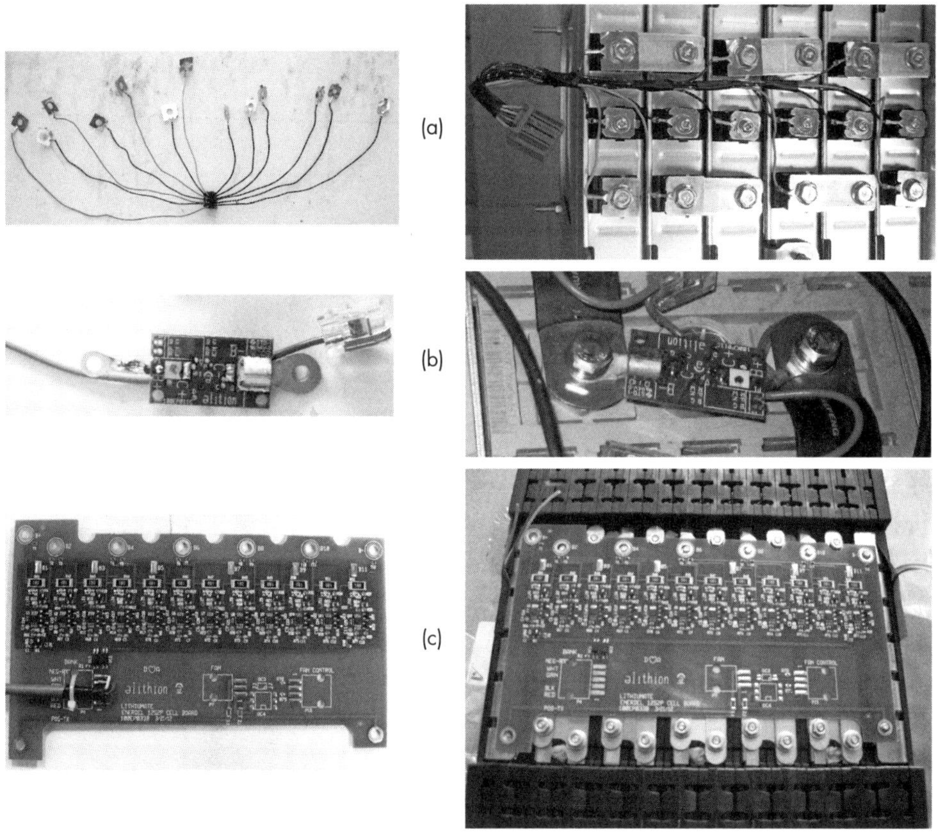

FIGURE 4.4
Connection methods to the cells: (a) wired, (b) cell board, and (c) bank board.

throughout the battery, which can lead to short circuits; a mated BMS (cell boards and bank boards) does not, which makes it safer. A wired BMS is cheaper than a mated one, but the cost of preparing a wire harness reduces this difference.

It's easy to damage a wired BMS by wiring to the wrong cell, in which case the whole unit must be replaced. It's also easy to damage a cell board by connecting it backward, but at least only that one cell board needs to be replaced. Bank boards (for multiple cells) are unlikely to be installed incorrectly because they fit only one way. Table 4.4 compares the three ways of connecting cells.

4.2.2.3 Permutations of BMS Topologies

I listed three ways to subdivide the electronics and three ways of connecting the cells. This results in nine permutations (Table. 4.5). Of those, one is not possible, leaving only eight topologies that are practical.

Figure 4.5 shows the block diagrams of the eight BMS topologies and Figure 4.6 shows examples. Table 4.6 lists the eight topologies, Table 4.7 compares them, and Table 4.8 lists sources of off-the-shelf BMSs for each topology.

4.2.2.4 Wired PCM

A *wired PCM* is a single, open PCB assembly, mounted next to the cells. Tap wires from the cells are soldered to pads on the board; alternatively, a harness from the cells is mated to a connector on the board (Figures 4.5(a) and 4.6(a)).

This topology is typical of off-the-shelf PCMs for small, low-volume production batteries.

Table 4.4
Comparison of BMS Topologies Based on Connection to the Cells

	Wired	**Cell boards**	**Bank boards**
Connection to Cells	N + 1 tap wires to N cells	Direct connection: a cell board is mounted onto each cell	Direct connection: a bank board is mounted to a block of cells
Wiring to Master	Same wires as above	Small, low-voltage cables to the two cell boards at either end of a bank	Small, low-voltage cable to the bank board (if applicable)
Accuracy	Great	Good	Good to great
Noise Immunity	Great	Good	Good to great
Overvoltage, Reverse Voltage Immunity	None or some	None	None
Temperature Sensing	Wired thermistor, typically one sensor for six cells	One sensor per cell Integrated into the board	Varies
Installation Labor	Average Miswiring prone	Average Backward prone	Easy Error unlikely
Short Circuits	Risky	Low risk	Low risk
Versatility	Excellent	Great	Low
Cost	Medium	Highest	Higher

Table 4.5
Nine Permutations for BMS Topology, of Which Eight Are Possible

	Single-Board	**Single-BMU**	**Modular**
Wired	Wired PCM	Centralized BMS	Master/Slave BMS
Cell boards	N.A.	Distributed BMS	Distributed master/slave BMS
Bank boards	Mounted PCM	Banked BMS	Banked master/slave BMS

4.2.2.5 Mounted PCM

A *mounted PCM* is a single, open PCB assembly, mounted directly on the cells (Figures 4.5(f), 4.6(f) and 4.7).

This topology is typical of custom PCMs for small, high-volume production batteries.

The difference between the wired PCM topology and this one is that that one has wires to the cells and this one does not. Instead, nickel tabs from the cells are connected directly to the PCB.

4.2.2.6 Centralized

In a *centralized BMS* (a.k.a. *monolithic BMS*), all the electronics are contained in a single assembly, and wire harnesses go to the cells (Figures 4.5(b), 4.6(b) and 4.8).

The difference between the wired PCM topology and this one is that that one is mounted on the cells, and this one is not. Instead, it is some distance away from the cells or even outside the enclosure that contains the cells.

Table 4.6
Eight Possible BMS Topologies

Topology			**Example**	**Applications**
Wired	**Single board**	Wired PCM (Figure 4.5(a))	PCM BMSs from China (Figure 4.6(a))	Low-volume small batteries
	Single BMU	Centralized BMS (Figure 4.5(b))	Orion BMS (Figure 4.6(b))	Traction batteries, medium batteries
	Modular	Master/slave BMS (Figure 4.5(c))	Nuvation (Figure 4.6(c))	Large, modular batteries
Cell boards	**Single BMU**	Distributed BMS (Figure 4.5(d))	Elektromotus EMUS BMS (Figure 4.6(d))	Traction batteries, large, low voltage, high voltage
	Modular	Distributed master/slave BMS (Figure 4.5(e))	Elithion Vinci HV (Figure 4.6(e))	Large-traction batteries High-voltage batteries
Bank boards	**Single board**	Mounted PCM (Figure 4.5(f))	Laptop batteries (Figure 4.6(f))	High-volume small batteries
	Single BMU	Banked BMS (Figure 4.5(g))	Lithiumate for Enerdel (Figure 4.6(g))	Medium batteries with cell blocks
	Modular	Banked master/slave BMS (Figure 4.5(h))	Elithion Vinci EV (Figure 4.6(h))	High-voltage batteries

This is the most common topology for a large battery in a single enclosure for up to 256 cells in series.

An off-the-shelf centralized BMS doesn't scale well because it supports a fixed number of cells (e.g., 12, 24, or 36 cells), which means that some of its capabilities are unused when used with a battery with a lower number of cells in series. This makes the BMS slightly more expensive than it needs to be. On the other hand, the cost of a custom centralized BMS is lower because it supports exactly the required number of cells.

A centralized BMS is often enclosed in a plastic or metal case. It may even be sealed, which would make it more appropriate for marine and industrial applications.

4.2.2.7 Distributed

In a *distributed BMS* (Figure 4.5(d) and 4.6(d)), a cell board (Figure 4.9) is mounted on each cell (Figure 4.10).

A cell board should use ring terminals (Figure 4.9(b, g)) rather than bolting the PCB directly to the cell:

- A ring terminal is made of metal and can be tightened to the correct torque;
- In contrast, a PCB is resilient and cannot be torqued correctly; additionally, its thermal coefficient is different from the one for the bolt, and constant thermal cycling during use may loosen the bolt.

For the same reason of incompatible thermal coefficient, a cell board should include a way to handle changes in the terminal spacing due to thermal cycling during use, such as having one terminal on a wire (Figure 4.9(b)). A PCB with two mounting holes is stressed as the cell expands at a different rate compared to the PCB.

Cells may be divided into banks, each one with between 2 and about 25 cells in series. Two small cables run from a master to the most positive and the most negative cell board of a bank.

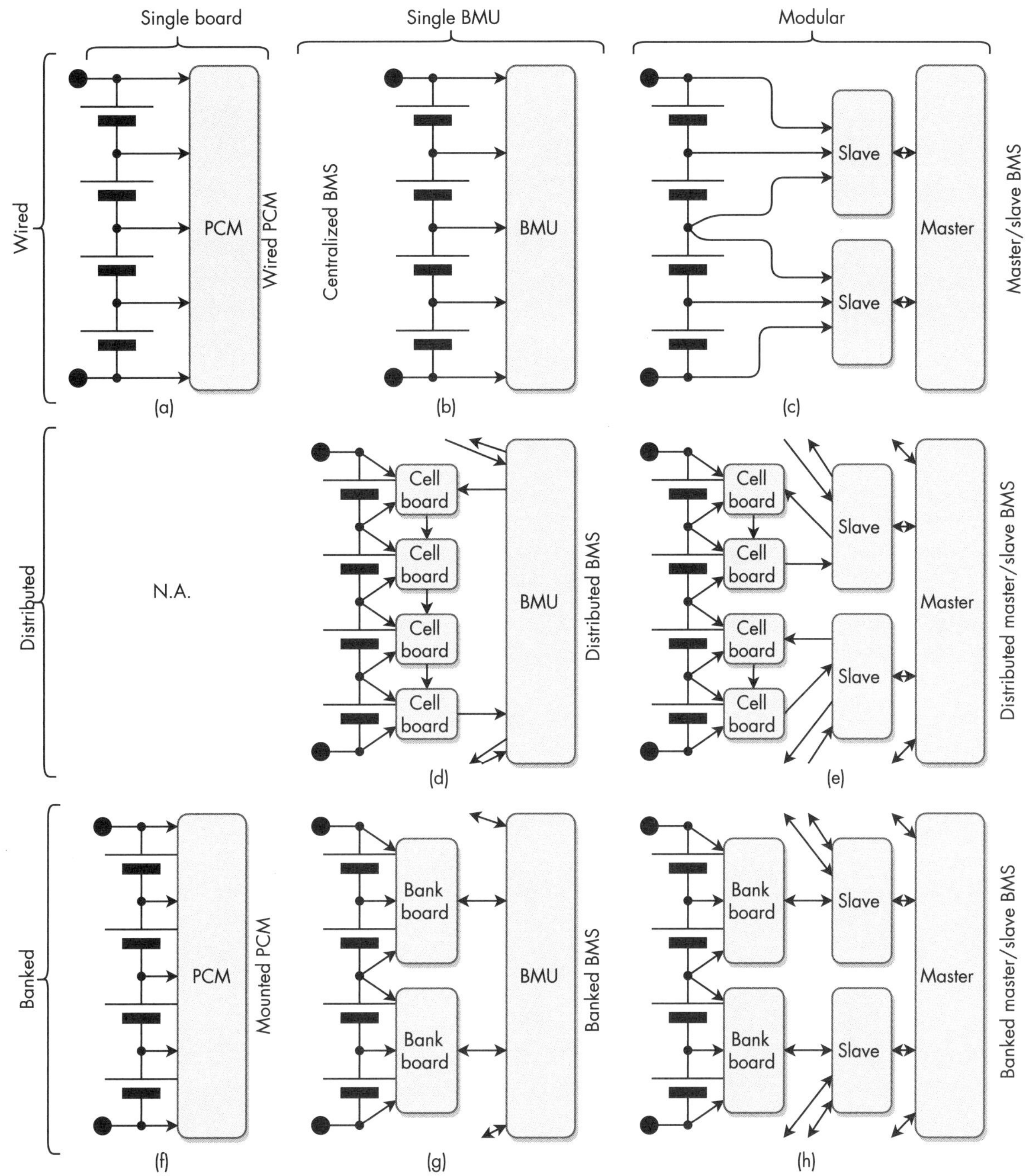

FIGURE 4.5 Block diagrams of BMS topologies: (a) wired PCM, (b) centralized BMS, (c) master/slave BMS, (d) distributed BMS, (e) distributed master/slave BMS, (f) mounted PCM, (g) banked BMS, and (h) banked master/slave BMS.

Each cell board is powered by its cell. An advantage is that the cell board can continue monitoring its cell even while the BMS is off, for example, to record if the cell is abused. A disadvantage is that it stops operating when the cell voltage is too low—typically less than 1.8V—which can happen in a race car application where the cells are used close to the MPP (see Section 1.5). The voltage of LTO cells is too low for a distributed BMS.

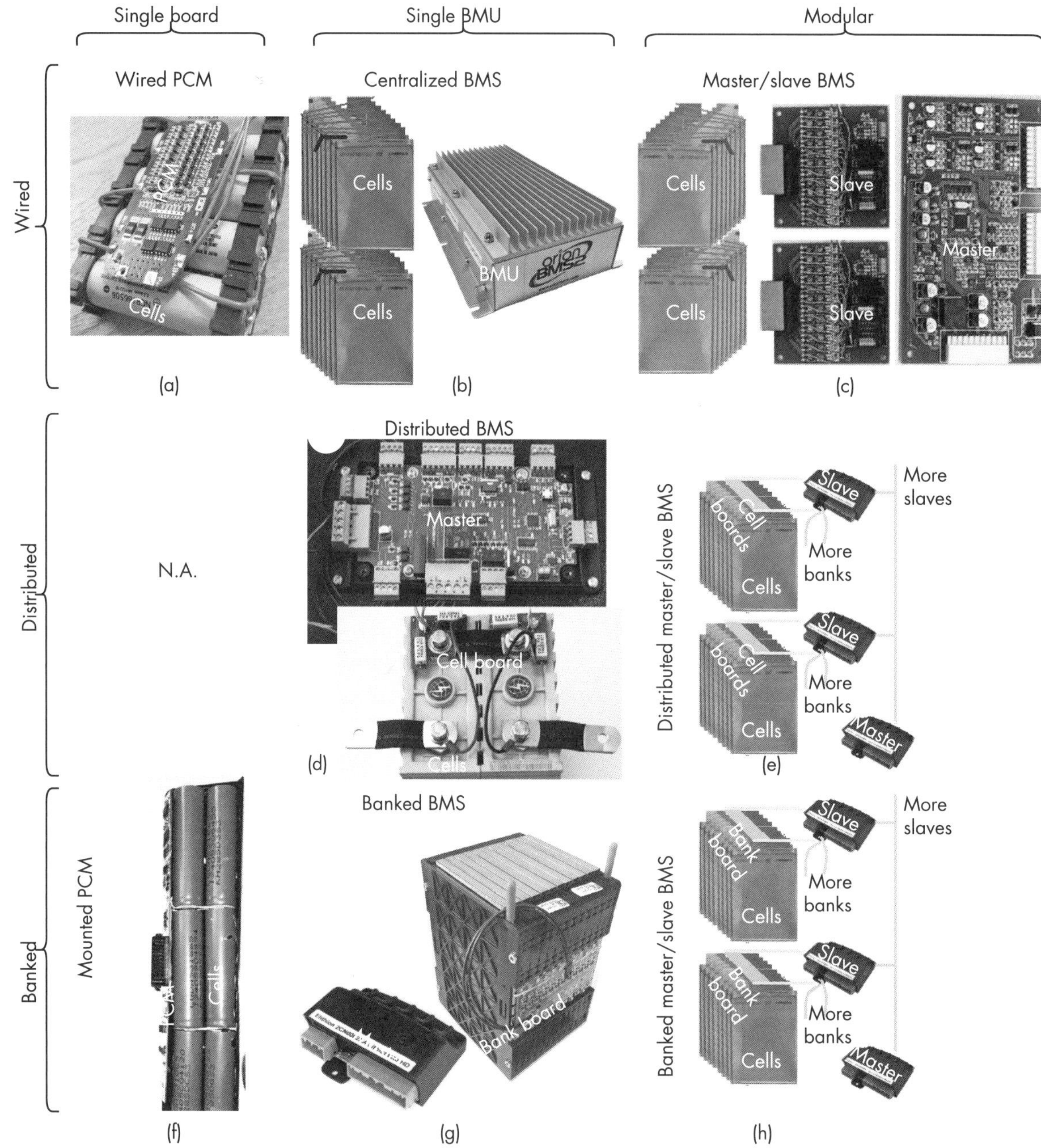

Figure 4.6 Examples of BMS topologies: (a) wired PCM—Chinese PCM, (b) centralized BMS—Orion BMS (courtesy of Ewert Energy), (c) master/slave BMS, (d) distributed BMS, (e) distributed master/slave BMS—Vinci HV, (f) mounted PCM—laptop battery, (g) banked BMS—Lithiumate for Enerdel Moxie™ block, and (h) banked master/slave bms—Vinci EV.

The most noticeable difference between this topology and the centralized one is that this one has lots of electronic assemblies, and the centralized one has lots of wires. Unlike in a centralized BMS, there are no wasted channels in a distributed BMS because only the required number of cell boards needs to be used. If a cell board is damaged, only that cell board needs to be replaced, not the entire BMS (as is the case for a centralized BMS).

Topology			Figure	Battery size	Battery case	Cost	Cells (Typical)	Banks (Typical)	Protector
Wired	**Single board**	**Wired PCM**	4.6a	Small	Single	Low	1~48	1	Y
	Single BMU	**Centralized**	4.6b	Large	Single	Medium	4~256	1~16	N
	Modular	**Master/slave**	4.6c	Large	Modular	High	6~256	1,600	N
Cell boards	**Single BMU**	**Distributed**	4.6d	Large	Either	Medium	8~256	1~16	N
	Modular	**Distributed Master/slave**	4.6e	Large	Modular	High	200~ 40,000	1,600	N
Bank boards	**Single board**	**Mounted PCM**	4.6f	Small	Single	Low	1~4	1	Y
	Single BMU	**Banked**	4.6g	Large	Either	Medium	2~192	2~16	N
	Modular	**Banked M/S**	4.6h	Large	Modular	High	200~ 40,000	1600	N

Table 4.7 Comparison of BMS Topologies

Table 4.8 Sources of BMSs of Each Topology

Topology			Source
Wired	**Single board**	**Wired PCM**	Many companies in China
	Single BMU	**Mounted PCM**	None*
	Modular	**Centralized**	E-pow, Ewert, JTT, Nuvation, REC, Volrad
Cell boards	**Single BMU**	**Distributed**	123 Electric, Batrium, Elektromotus, Elithion, Jon Elis
	Modular	**Banked**	None*
Bank boards	**Single board**	**Master/slave**	E-pow, EVPST, Freemens, I+ME Actia, JTT, Ligoo, Lithium Balance, Nuvation, REC, KLClear, Volrad, Zeva
	Single BMU	**Distributed M/S**	Elithion
	Modular	**Banked M/S**	None*

(*)= Custom-made for the particular battery.

Each bank uses

- A positive cell board, mounted on the most positive cell in the bank, with a link to the master;
- A negative cell board, mounted on the most negative cell in the bank, with a link to the master;
- Any number of midboards mounted on all the other cells.

This topology is used in larger batteries or modular batteries. Typically, it supports up to 256 cells in series.

In some respects, this topology is safer than a wired topology because it avoids high-voltage sense wires and thermistors throughout the battery.

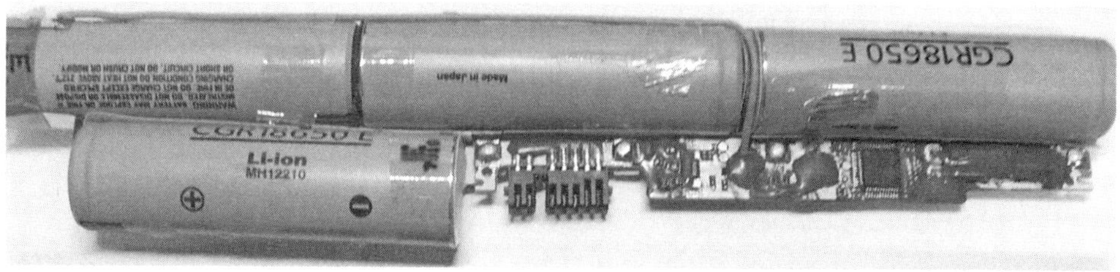

FIGURE 4.7
PCM mounted on the cells of a laptop battery. (Courtesy of karosium.com.)

FIGURE 4.8
Centralized BMS with wires going to the cells.

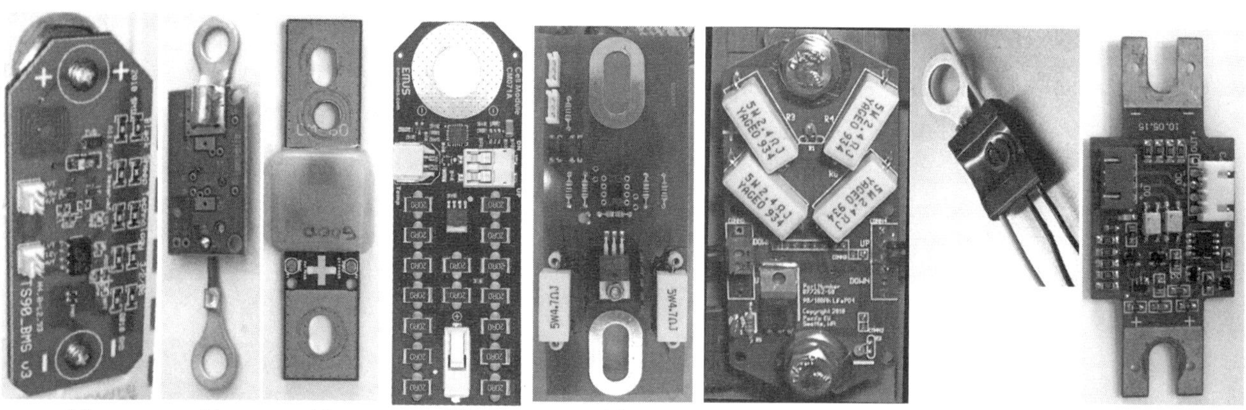

(a) (b) (c) (d) (e) (f) (g) (h)

FIGURE 4.9 Cell boards for a distributed BMS for large prismatic cells: (a) Blacksheep, (b) Elithion, (c) EV power, (d) EMUS (courtesy of EMUS), (e) Blade, (f) PEV, (g) Vinci, and (h) unknown.

The master sends a request to the first cell board. This board reads its cell voltage and temperature, and passes the information to the second cell board. The second cell board reads its cell voltage and temperature, appends those data to the data from the first cell board, and passes them all to the third cell board. This process is repeated with the remaining cell boards. The last cell board in the bank sends the data from all the cells in that bank back to the master.

For ease of assembly and lower cost, the link between adjacent cell boards may consist of a single wire. The bus bar between the cells doubles as the return wire.

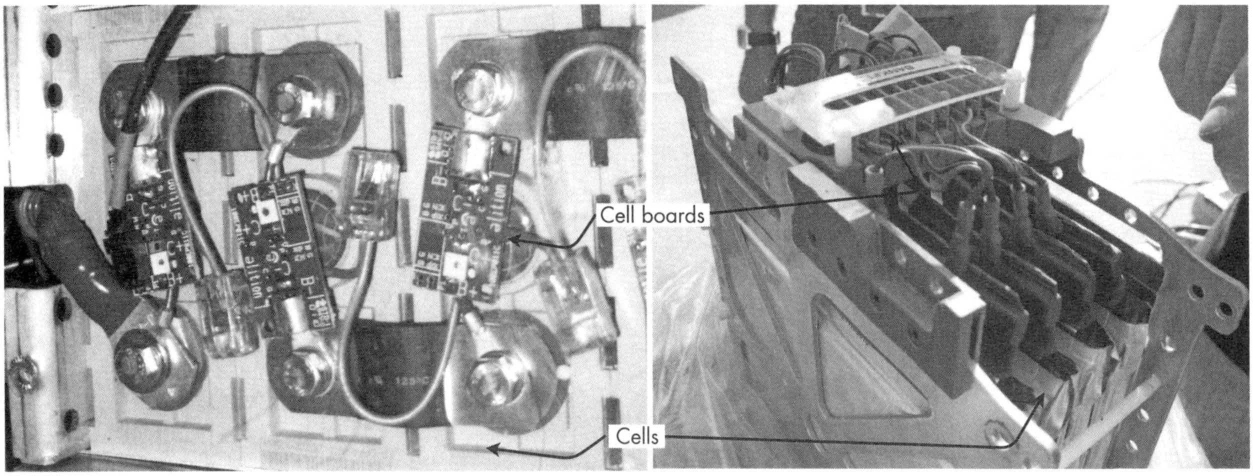

FIGURE 4.10 Cell boards mounted on cells: (a) large prismatic cells, and (b) pouch cells.

If the battery is divided into multiple banks (for example, because it is physically divided into multiple modules), the master queries each bank and receives the data back from all the banks (Figure 4.11(a)).

The isolation between the high-voltage and the low-voltage sections is implemented inside the two cell boards at either end of a bank. In some BMSs, each cell board is isolated from the next cell board, which is better for noise rejection, but is more expensive and requires at least two wires between cell boards.

The distributed topology moves some of the electronics out of the BMU and onto the cells. This means that now there are electronic assemblies in the same environment of the cells—exposed to vibrations, dust, oils, moisture. Most cell boards are, at best, conformally coated, which offers some environmental protection, but still not as much as a sealed, potted module.

4.2.2.8 Banked

In a *banked BMS*, a bank board is mounted directly on a group of cells that forms a bank (Figure 4.5(g), 4.6(g), and 4.12). A small cable runs to the master. One or more bank boards handle all the cells in the battery. This is an effective solution for blocks of cells, such as pouch cells contained in frames. A bank board provides a fast and elegant BMS installation.

The difference between this topology and the distributed one is that in the banked topology, a single board handles multiple cells. This topology scales nicely: to increase the battery size, more cell blocks are added, each with a bank board (Figure 4.11(b)). On the other hand, this topology is not flexible because a bank board designed to mate to a specific cell block of a specific size cannot be used with a different cell model or a different number of cells.

Electrically, a banked board can be implemented in two ways:

- One electronic circuit for each cell: Electrically, this is the same as a distributed BMS, with all its limitations (including a minimum operating voltage of 2V);
- A single electronic circuit for the bank: Electrically, this is the same circuit used in a centralized BMS, but there is no need for thermistors hanging on wires because the thermistors can be on the bank board.

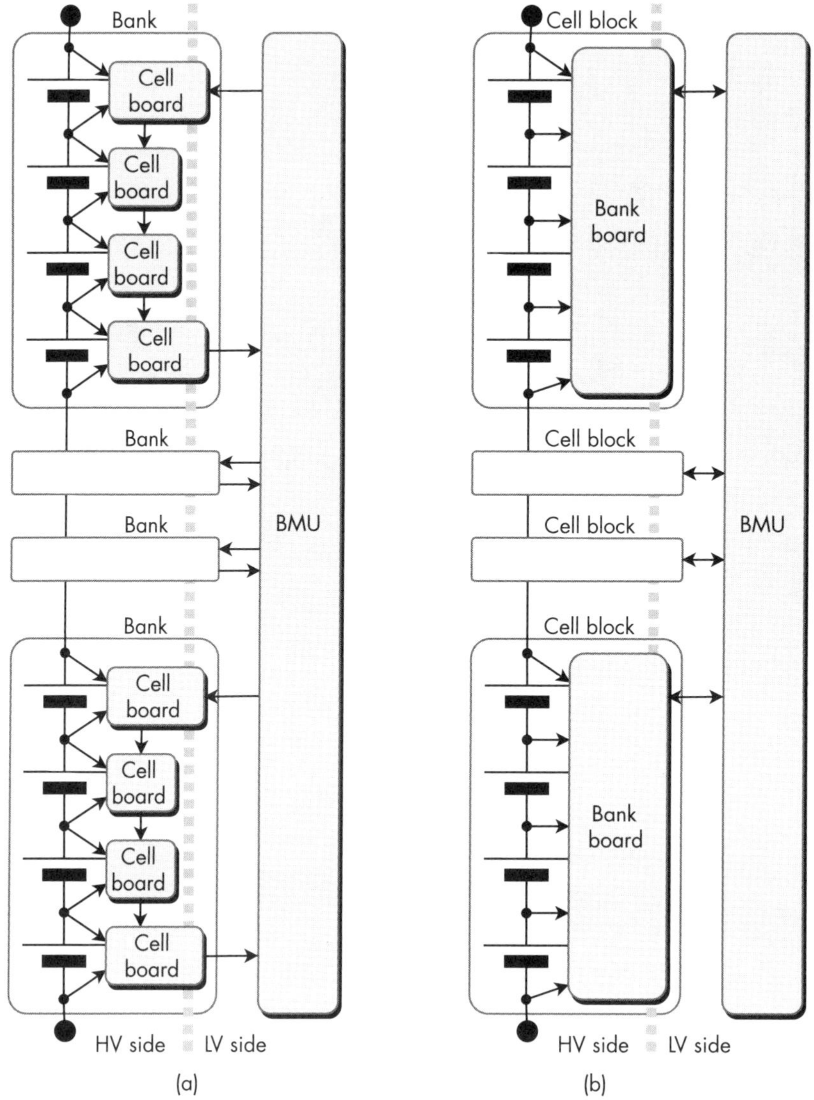

FIGURE 4.11
High-voltage and low-voltage sides in a mouted BMS: (a) distributed BMS, and (b) banked BMS.

FIGURE 4.12
Bank board for a banked BMS mounted on an Enerdel Moxie™ module.

Typically, given that a bank probably doesn't have more than 12 cells in series, and that a master probably doesn't support more than 16 banks, this topology is limited to 192 cells in series.

The isolation between the high voltage (at the cells) and the low voltage occurs inside each bank board.

In terms of serviceability, a banked BMS is somewhere between the centralized and distributed BMS. If a single bank board is damaged, the entire board must be replaced, which is better than having to replace the entire BMS (as in a centralized BMS), but not as good as being able to replace just a single cell board (in a distributed BMS). Banked BMSs are custom-designed for each particular battery module.

4.2.2.9 Master/Slave

A *master/slave BMS* is divided into several modules—a master and one or more slaves.

A slave module handles several cells; a wire harness connects it to its cells. A small cable runs to the master (Figure 4.6c).

The difference between this topology and the banked topology is that in this one the slave has wires to the cells, while in the banked topology the bank board is mounted directly to a block of cells. Therefore, this topology is more flexible while the banked topology is easier to install.

This topology should more accurately be called a *wired master/slave BMS*, to distinguish it from the two other master/slave topologies (see below). The difference between this topology and those other two is that in this one, a slave is wired to its cells with wires; in those other two topologies, the slave communicates with cell boards or bank boards mounted on the cells.

This topology is ideal for use with battery modules. A slave inside the module handles all the cells in the module. The module only needs a small cable to the master. It also scales nicely: to increase the battery size, more modules are added, each with a slave.

Typically, given that a slave probably doesn't support more than 12 cells in series, and that a master probably doesn't support more than 16 slaves, this topology is limited to 192 cells in series. In some BMSs, the slaves support more than 12 cells, but the master supports fewer than 16 slaves. Therefore, all off-the-shelf master/slave BMSs are limited to about 192 cells, with one exception that can handle up to 256 cells.

The master may communicate with the slaves with a daisy chain (Figure 4.13(a)), a bus (Figure 4.13(b)), or to each individually (Figure 4.13(c)). Note how the isolation between the high voltage (at the cells) and the low voltage occurs inside each slave.

For example, Enerdel uses a custom master/slave BMS.

4.2.2.10 Distributed Master/Slave

The *distributed master/slave BMS* topology is the same as above, except that a cell board[3] is mounted on each cell. Two small cables run from a slave to the most positive and most negative cell board of a bank of cells. A small cable runs from the slave to the master. A single slave can handle several banks; one or more slaves handle all the cells in the battery (Figure 4.6(e)).

3. A.k.a. cell supervisory circuit (CSC).

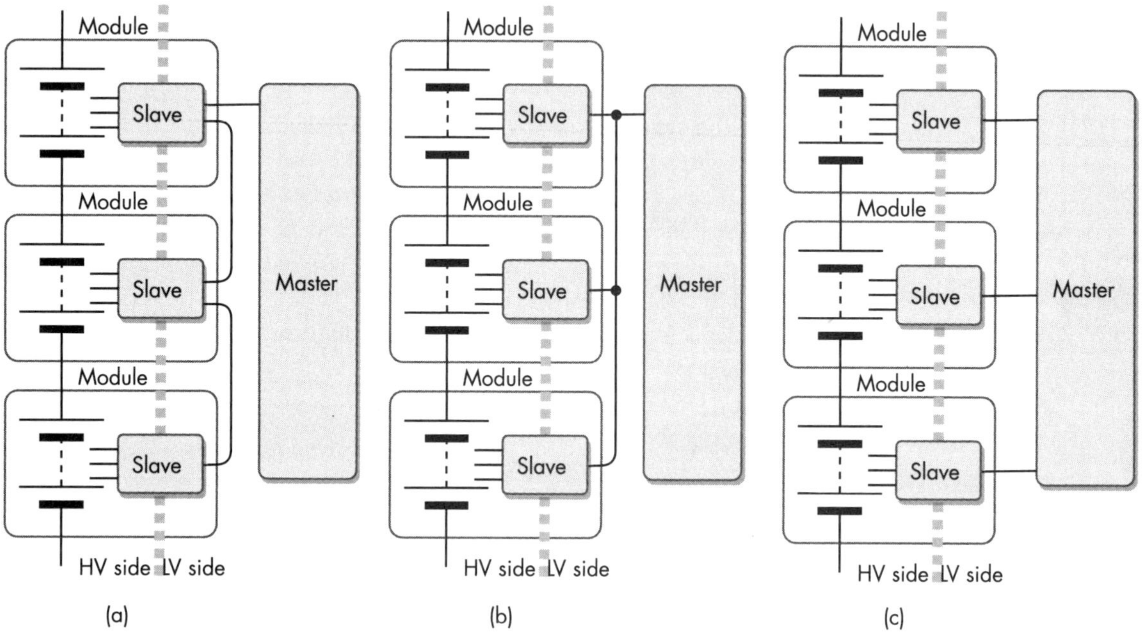

FIGURE 4.13 Master/slave communications: (a) daisy chain, (b) bus, and (c) individual links.

The difference between this topology and the distributed one is that, in this one, a slave is between the cell boards and the master (Figure 4.14); in the distributed topology, the cell boards communicate directly with the master.

The difference between this topology and the master/slave one is that this one uses cell boards and the master/slave topology uses wires. Also, in this topology, a slave can handle several banks of cells; in the master/slave topology, it only handles one bank. For example:

- A slave in a distributed master/slave BMS has four cables, for up to four banks, and each bank has 24 cells, so the slave handles up to 96 cells;
- A slave in a master/slave BMS has 13 wires for only 12 cells.

This topology is also ideal for battery modules. It has the added advantage that there are no high-voltage wires run through the module. It is ideal for sizable batteries since it can support tens of thousands of cells. For example:

$$25 \text{ cells/bank} \times 16 \text{ banks/slave} \times 100 \text{ slaves/master} = 40{,}000 \text{ cells} \quad (4.1)$$

4.2.2.11 Banked Master/Slave

The topology of a *banked master/slave BMS* is the same as the previous one, except that it uses bank boards instead of cell boards (Figure 4.6(h)). This topology is ideal for a huge battery that uses hundreds of pouch cell blocks (Figure 4.15).

Since the typical bank board handles only 12 cells, the maximum number of cells in this topology is about half as the distributed master/slave topology:

$$12 \text{ cells/bank} \times 16 \text{ banks/slave} \times 100 \text{ slaves/master} = 19{,}200 \text{ cells} \quad (4.2)$$

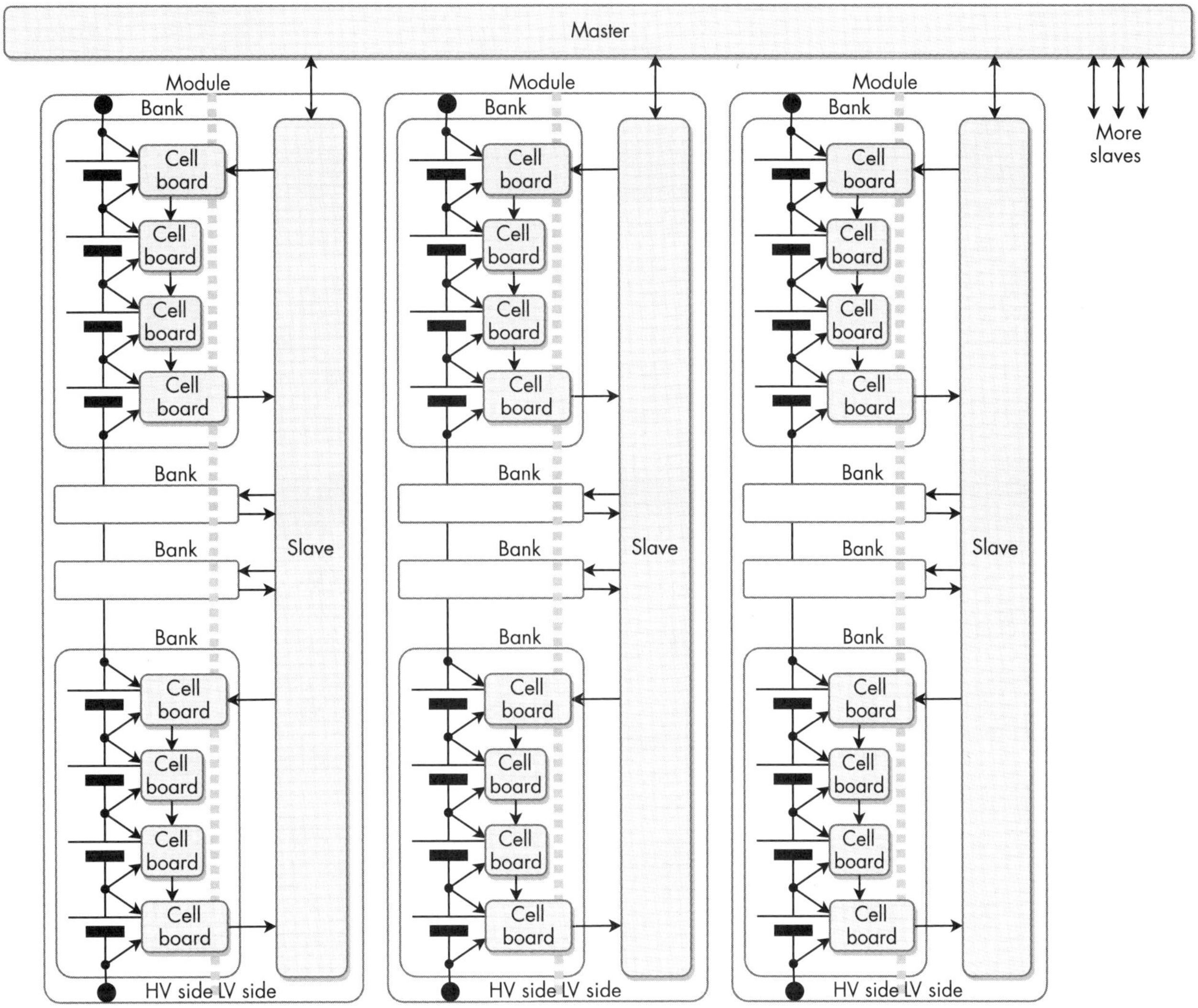

FIGURE 4.14 Distributed master/slave topology.

4.2.3 BMS Format

BMS assemblies can be open boards, enclosed or sealed (Figure 4.16). In some cases, there are multiple assemblies, some of which are open boards while others are enclosed.

4.2.4 BMS Cost

Generally, the cost of a BMS is proportional to the number of cells it can handle. For most off-the-shelf BMSs, this number is either fixed or it increases in discrete steps (e.g., 12, 24, 36, or 48 cells in series). Therefore, the cost increases in large steps (you're paying for channels that you don't use). On the other hand, a distributed BMS contains only as many cell boards as there are cells, so the cost does increase granularly and linearly with the number of cells handled. A custom BMS may be designed optimally to include only the number of channels required.

There are some significant, nonrecurring costs associated with the deployment of a BMS into the first article of a new product line. These costs are for planning,

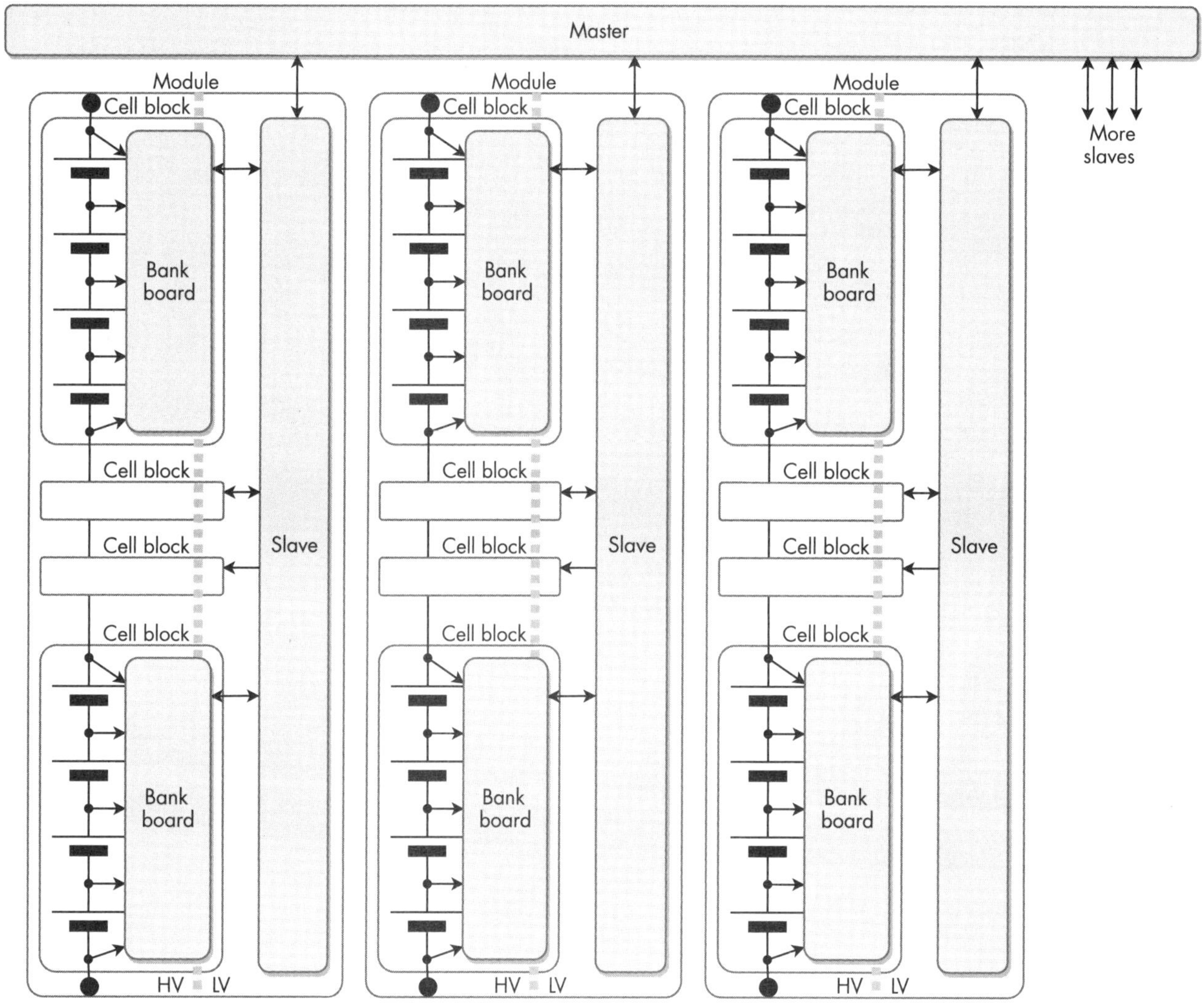

FIGURE 4.15 Banked master/slave topology.

FIGURE 4.16
BMS formats: (a) open board, (b) enclosed, and (c) sealed.

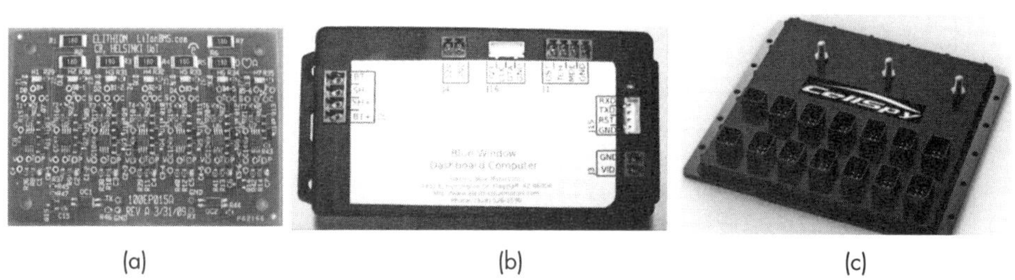

configuration, some software development, troubleshooting, and the learning curve. The battery designer pays for most of these costs; if using an off-the-shelf BMS, its manufacturer may incur some of these costs as well. The BMS manufacturer may absorb its costs during the initial deployment (hoping to recover them with future sales), may charge a one-time fee,[4] or may charge an hourly consulting fee.

4. I understand that this is what Lithium Balance does.

For a small battery, a PCM is by far the most economical solution. A digital PCM adds just a little to the cost. Chinese companies offer PCMs at less than the total cost of the individual components on the board.[5] For large batteries, a custom BMS, board level, using the centralized topology, is the most cost-effective solution. For all other options, the BMS cost per cell is an order of magnitude higher.

4.3 ANALOG PROTECTOR BMS (PCM)

An analog protector BMS (a.k.a. PCM or PCB[6]) is a simple yet complete Li-ion BMS on a single-board (Figure 4.17). PCMs are ubiquitous and ideal for low cost, self-contained batteries. They perform well given their amazingly low price from China.

4.3.1 PCM Placement

The way a PCM is connected in the circuit depends on its type:

- *Load-side:* Between the cell(s) and the load;
- *Single-port:* Between the cell(s) and everything else (charger, loads);
- *Dual-port* (see Section 5.12.2): Between the cell(s), the charger on one side, and the load on the other side.

A load-side solution is only possible if the charger can be trusted not to overcharge the cell(s).

A single-port PCM is the safest, as it can reliably protect against overcharge and overdischarge. A dual-port PCM is more efficient, but it can't prevent discharge through the charge port or charge through the discharge port (see Section 4.3.3.4).

Unlike a single-cell PCM, a multicell PCM is also connected to the taps between adjacent cells; it uses them to monitor the voltage of each cell. This results in six permutations for the placement of a PCM (Table 4.9). Also, the PCM could interrupt B+ (Figure 4.18) or B- (Figure 4.19).

All of these placements are in actual use in real-world applications (Table 4.10).

Physically, the topology of these PCMs is either wired PCM (through a balance connector and tap wires) (Figures 4.5(a), 4.6(a)) or mounted PCM (Figures 4.5(f), 4.6(f)).

FIGURE 4.17 Analog PCM.

5. How? It's Chinese magic.
6. Protection circuit module (not to be confused with phase change material or parallel cell module (also PCM)), protector circuit board.

Table 4.9
Six Permutations for PCM Placement

	Load-Side	Single-Port	Dual-Port
Since-cell	Single-cell, load-side (Figures 4.18(a), 4.19(a))	Single-cell, single-port (Figures 4.18(b), 4.19(b))	Single-cell, dual-port (Figures 4.18(c), 4.19(c))
Multicell	Multicell, load-side (Figures 4.18(d), 4.19(d))	Multi-cell, single-port (Figures 4.18(e), 4.19(e))	Multi-cell, dual-port (Figures 4.18(f), 4.19(f))

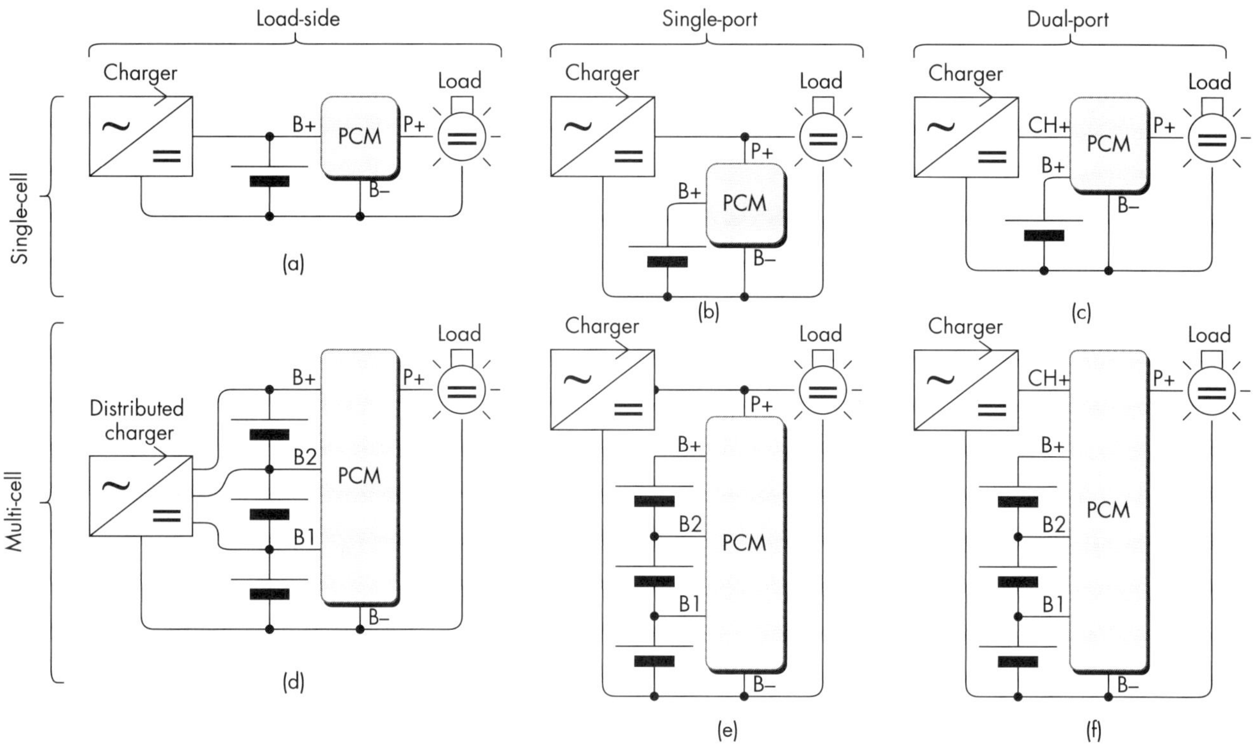

Figure 4.18 Placement of PCM that opens B+: (a) single-cell, load-side, (b) single-cell, single-port, (c) single-cell, dual-port, (d) multicell, load-side, (e) multicell, single-port, and (f) multicell, dual-port.

4.3.2 PCM Terminal Labels

The PCM is connected through terminals to the charger, the load, and the cells. The labels on the terminals follow a convention depending on the PCM topology. These are shown in the block diagrams in Figure 4.19.

For multicell PCMs, the voltage sense taps are named B#[7] (Figure 4.20(a)):

- B0 or B-: Negative of the most negative cell;
- B1: Positive of the most negative cell and negative of next cell up;
- B2: Positive of the second most negative cell and negative of next cell up;
- ... (unknown number of intermediate steps);
- Bn or B+: Positive of the most positive cell (n is the number of cells in series).

For example, with four cells in series, they are named B0, B1, B2, and B3.

7. I regret that the letter B was chosen for the cell taps, instead of C for cell.

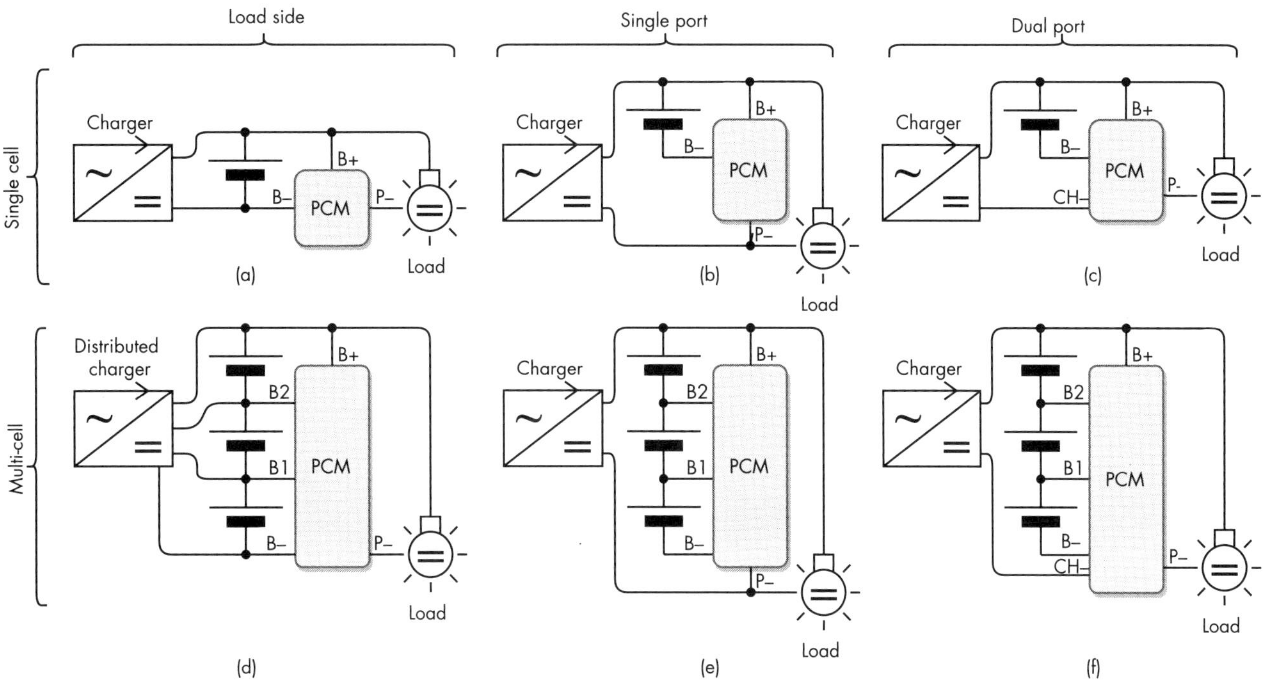

FIGURE 4.19 Placement of PCM that opens B-: (a) single-cell, load-side, (b) single-cell, single-port, (c) single-cell, dual-port, (d) multicell, load-side, (e) multicell, single-port, and (f) multicell, dual-port.

TABLE 4.10
Example Applications of the Six Permutations of Protector BMS Placement

Placement		**Figure**	**Applications**	**Notes**
Single-cell	**Load-side**	4.18(a) 4.19(a)	Toys, cheap consumer products, 3.6V	The 'PCM' is simply a low voltage cutout in the load
	Single-port	4.18(b) 4.19(b)	Cheap cell phones, self-contained, small 3.6V batteries	
	Dual-port	4.18(c) 4.19(c)	Better consumer products, 3.6V	PCM offers extra protection if the charger fails
Multi-cell	**Load-side**	4.18(d) 4.19(d)	Small UAVs (drones)	Only the discharge side is protected within the vehicle; a balancing charger protects the cells during charge
	Single-port	4.18(e) 4.19(e)	Self-contained, small batteries > 3.6V	
	Dual-port	4.18(f) 4.19(f)	Traction batteries for personal EVs	The charging port is always powered and could discharge the battery, uncontrolled by PCM

Depending on the PCM, there may be either B- or B0, or both. Similarly, there may be either B+, Bn, or both. The power terminals use the labels[8] listed in Table 4.11.

8. I have no idea what P stands for.

Table 4.11
Labels of Power Terminals in a PCM

	Switched Positive	Switched Negative
Load-Side	B+: Positive of most positive cell and charger P+: Positive of load B–: Negative of most negative cell, charger, and load	B+: Positive of most positive cell, charger, and load B–: Negative of most negative cell and charger P–: Negative of load
Single-Port	B+: Positive of most positive cell P+: Positive of charger and of load B–: Negative of most negative cell, charger, and load	B+: Positive of most positive cell, charger, and load B–: Negative of most negative cell P–: Negative of charger and of load
Dual-Port	B+: Positive of most positive cell CH+: Positive of charger P+: Positive of load B–: Negative of most negative cell, charger, and load	B+: Positive of most positive cell charger, and load B–: Negative of most negative cell CH-: Negative of charger P–: Negative of load

4.3.3 PCM Functionality

Internally, a PCM is rather basic (Figure 4.20(b)).

4.3.3.1 Voltage Protection

The PCM is connected to each cell (or the single cell). It compares the voltage of each cell to fixed thresholds (set in the hardware) and generates a single signal that is "OK" if all the cell voltages are within those limits.

4.3.3.2 Current Protection

The PCM may not deal at all with battery current, or it may have a simple circuit to shut off the battery in case the current exceeds a threshold set in the hardware. There may be two thresholds, one for charging and one for discharging.

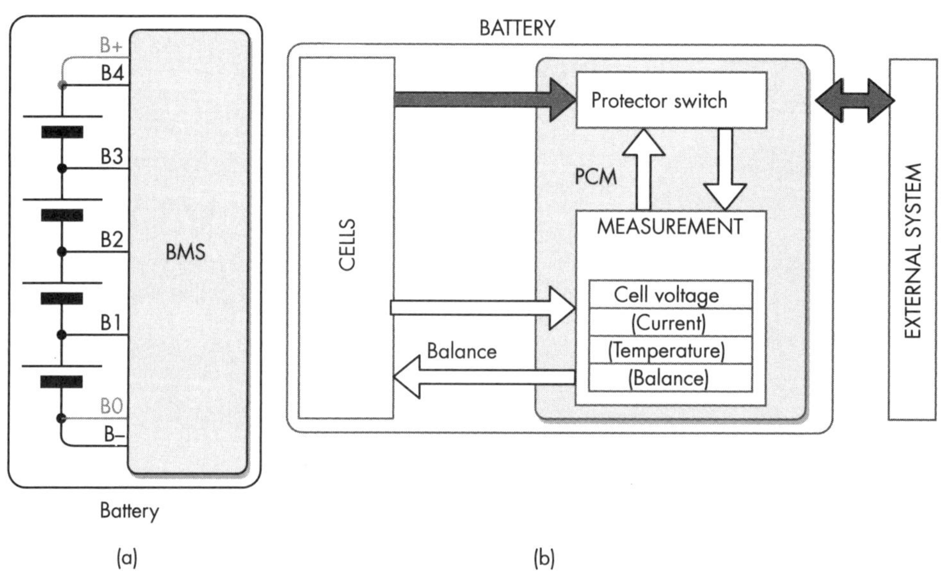

Figure 4.20
PCM: (a) voltage taps, and (b) protector BMS block diagram.

The PCM senses the battery current by routing it through a resistor and measuring the resulting voltage drop (see Section 4.5.4).

4.3.3.3 Temperature Protection

Optionally, a thermistor is placed on one of the cells to sense its temperature (see Section 4.5.3). The PCM shuts down the battery if the temperature is above a fixed threshold (set in hardware). The PCM should also sense for cold temperature to prevent charging below freezing, but this is rarely implemented.

A high-power protector BMS also monitors the temperature of its protector switches.

4.3.3.4 Protector Switch

The PCM controls two solid-state transistor switches (MOSFETs), one to control charging, one to control discharging.[9] MOSFETs are transistors that are quite useful as switches but can turn on and off current only in one direction. When attempting to run current through them in the opposite direction, the current flows unimpeded. Therefore, to control both charging and discharging, two MOSFETs are required, one to control charging and one to control discharging.

Two topologies are possible (see Section 5.12.2):

- Single-port (Figure 4.21(a, c));
- Dual-port (Figure 4.21(b, d)).

In a single-port PCM, the charging or discharging current flows through the same input/output. The PCM places the two back-to-back MOSFET switches in series between the cells and the port. One MOSFET controls charging, the other one controls discharging. This allows the BMS to be always in control, as it can

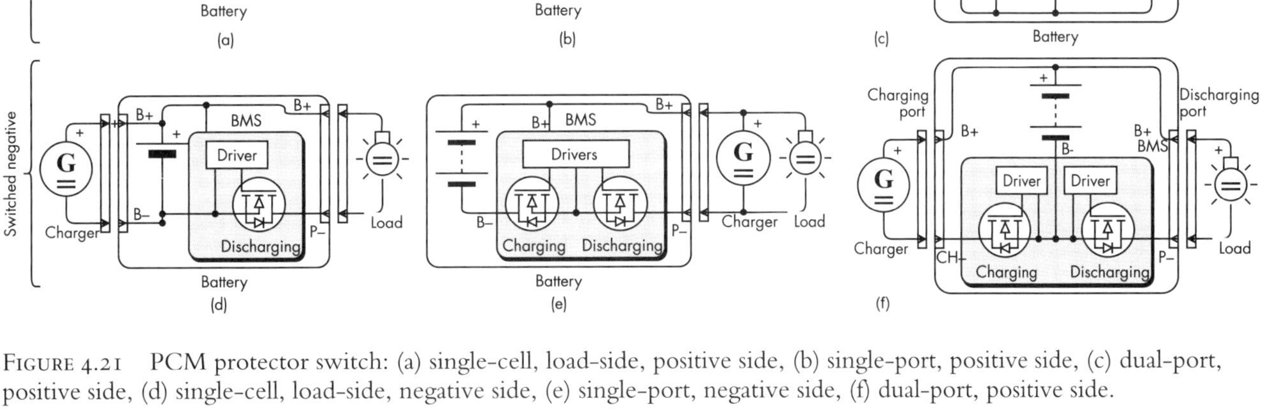

FIGURE 4.21 PCM protector switch: (a) single-cell, load-side, positive side, (b) single-port, positive side, (c) dual-port, positive side, (d) single-cell, load-side, negative side, (e) single-port, negative side, (f) dual-port, positive side.

9. A load-side PCM only needs one MOSFET to control discharging because the charger controls charging.

always disconnect the battery regardless of the direction of the current. Single-port protectors are required in conjunction with bidirectional devices, such as motor drivers that include regen, or with invergers.

In a dual-port PCM, one port is for charging and one for discharging. There is a single MOSFET switch between the battery and the charging port to control charging, and another MOSFET switch between the battery and the discharging port to control discharging. This allows the BMS to control charging as long as the charger's voltage is higher than the battery voltage, and to control discharging as long as the load voltage is lower than the battery voltage.

Dual-port protectors are more efficient because the current goes through a single MOSFET switch, and therefore their voltage drop is lower.

However, dual-port protectors cannot control the current flowing in the opposite direction. For example, if the charging port is wired to an exterior connector, the user could draw current from the battery through this connector, discharge the battery dangerously, and the BMS would not be able to do anything about it. Conversely, if the discharging port is wired to an exterior connector, the user could charge and overcharge the battery through this connector, and the BMS would not be able to do anything about it. Therefore, dual-port PCMs are generally less safe than single-port ones and require some mechanical means to prevent abuse of the charging or discharging port.

A dual-port BMS with a bidirectional switch of each port (for a total of four MOSFETs) would not suffer from this limitation. Such a BMS is technically possible, though I haven't seen any.

Note that you cannot convert a dual-port protector to a single one by connecting the charging and discharging terminals. If you do, you'll place the MOSFETs in parallel (they should be in series), and the protector won't be able to control either charging or discharging.

The protector switch could be placed in series with the positive or the negative output:

- In the positive side, P-channel MOSFETs are used (Figure 4.21(a, b, c));
- In the negative side, N-channel MOSFETs are used (Figure 4.21(d, e, f)).

For a single-port battery, this makes no difference to the user: in both cases, the battery has a single, 2-pin connector. It does make a difference to the user or a dual-port battery because, in one case, the negative is common between the charger and the load, and in the other case, the positive is common.

4.3.3.5 Fuse

Some PCMs include a *chemical fuse*, which the BMS can blow to shut off the battery current if it can't control the power switches (see Section 5.12.7).

4.3.3.6 Balancing

Many PCMs include a top-balancing (see Section 4.7.4.1), bypass balancing (see Section 4.7.2) function.

4.3.4 Protected 18650 Batteries

Protected 18650 batteries[10] include a cell and protector BMS inside the same package (Figure 4.22). The PCM is circular and placed against the negative terminal.

These batteries must be used on their own:

- No more than two should be connected in parallel, as the user is likely to install a charged battery in parallel with a discharged one, which would result in an inrush equalizing current (see Section 3.3.6),
- They should not be used as part of a series string of more than a few batteries because the protector switch in each 18650 battery may not be rated for the total string voltage.[11]

If an 18650 measures 0V, is it a dead 18650 cell (unprotected) or an 18650 battery (protected) that disabled discharging? There are a few ways to tell:

- By the part number: Research the part number to see if it's an 18650 cell or an 18650 battery;
- With a meter in the diode range: A dead cell shows infinite resistance. a shorted cell show zero resistance. and a battery may show a diode when measured with the red probe to the negative battery terminal;
- With a CCCV power supply set for a low-charging current: A cell starts at less than 1V and a battery start at about 3V; after a few minutes, remove the CCCV supply: A cell still measures 0V, and a battery measures about 3V;
- By looking at the negative terminal and comparing it to pictures of an 18650 cell and an 18650 battery;
- By checking the length: If it is slightly longer than 65 mm, it's a battery (the extra space is for the protector PCM and an additional cap);
- By noting that the heat shrink is stepped at the negative end, with a narrower diameter, where the PCM is placed: If so, it's a battery;
- By looking for a thin wire along the body of the cell: If present, it's a battery;
- By checking if the negative terminal is magnetic: If so, it's a cell;
- By checking if the color of the negative terminal is something other than silver: If so, it's a battery.

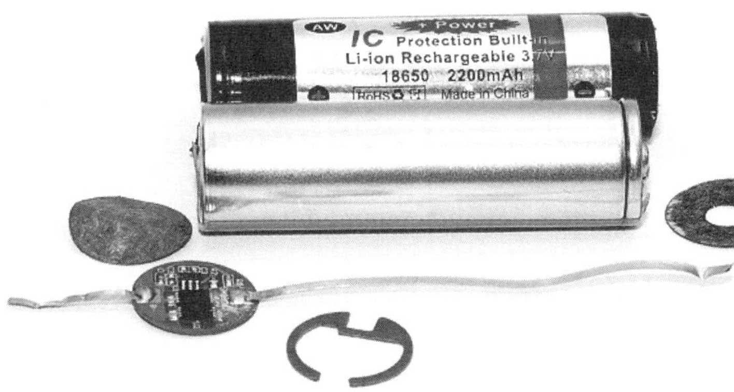

FIGURE 4.22 Components in a protected 18650 battery. (Courtesy of Henrik K. Jensen, lygte-info.dk.)

10. While 18650s on their own are just cells, protected 18650 batteries are "batteries" because they are more than just a cell since they include a protector BMS.
11. When the protector switch of one battery opens, the entire string voltage appears across its switch.

4.3.5 Charger/PCM Combo

Nifty little products include a charger and a PCM on a single board. They charge a single 3.6V Li-ion cell from a USB port, and they protect the cell (Figure 4.23(a)). They are often based on the TP4056 IC[12] and therefore the whole board may be called by this generic name.

The board has three ports (Figure 4.23(b)):

- USB, for charging;
- Bat+ and Bat– for the cell;
- Out+ and Out– for the load.

Note that the three ports are not isolated from each other—using multiple boards causes a short circuit. Despite what a web search shows you, these boards may not be used to create a multicell battery:

- These boards are not isolated, so they may not be used for a series string; any attempt to charge the cells simultaneously results in a short circuit across all the cells, except for the most positive one (Figure 4.23(c));

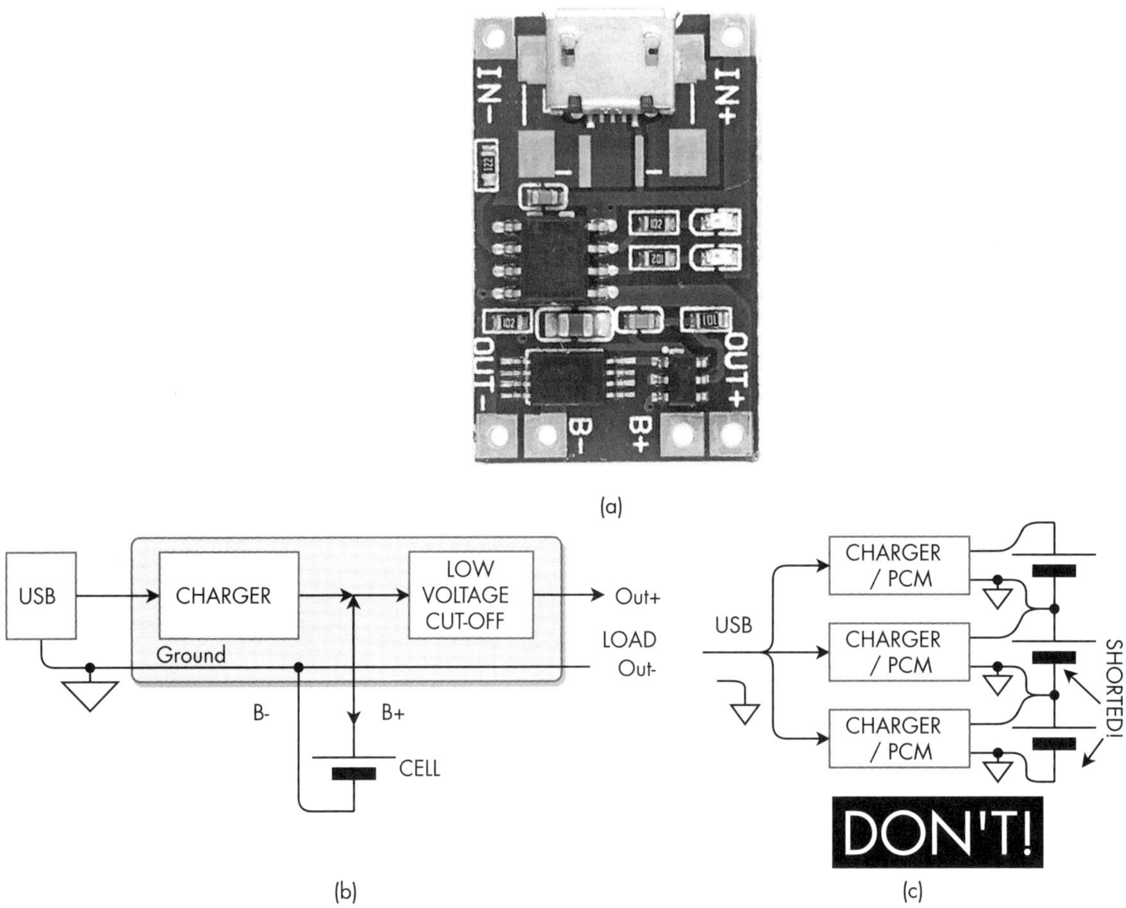

FIGURE 4.23 TP4056 board: (a) image, (b) circuit, and (c) do not use with a series string!

12. NanJing Top Power ASIC Corp.

- These boards should not be connected in parallel due to the equalizing inrush current between cells at different SoC (see Section 3.3.6).

Note also that USB-powered boards that only have two contacts (*B*–, *B*+) are just chargers, not BMSs.

4.4 DIGITAL BMS

Unlike an analog BMS, a digital BMS knows the state of a battery, evaluates its conditions, and reports them.

A digital BMS may be

- A digital protector—includes a protector switch;
- A digital BMU—a protector switch must be added externally.

4.4.1 Digital Protector

A digital protector is a digital BMS (see Section 4.4.2) that includes a protector switch, just as an analog protector does (see Section 4.3) (Figure 4.24).

4.4.1.1 Small Batteries

Digital protectors are used in batteries for larger consumer electronics, such as laptop computers.

In some instances, the BMS function is split between the battery and the product. For example, a cell phone battery may contain little more than a thermistor and protection switches. The phone itself performs the actual BMS function (see Volume 2, Section 1.3).

Recently, some Chinese manufacturers of off-the-shelf, open-assembly BMSs have graduated from analog BMSs to digital ones. They may include a Bluetooth interface to a smart-phone app to monitor the battery (see Volume 2, Section 1.6.2.1,).

4.4.1.2 Medium Batteries

Protector BMSs are used extensively in the traction batteries for electric bikes and other small EVs (see Volume 2, Section 3.6).

4.4.1.3 Large Batteries

There aren't many off-the-shelf digital protectors for large batteries, mainly because of the constraints of the protector switch: either the switch is not able to handle the

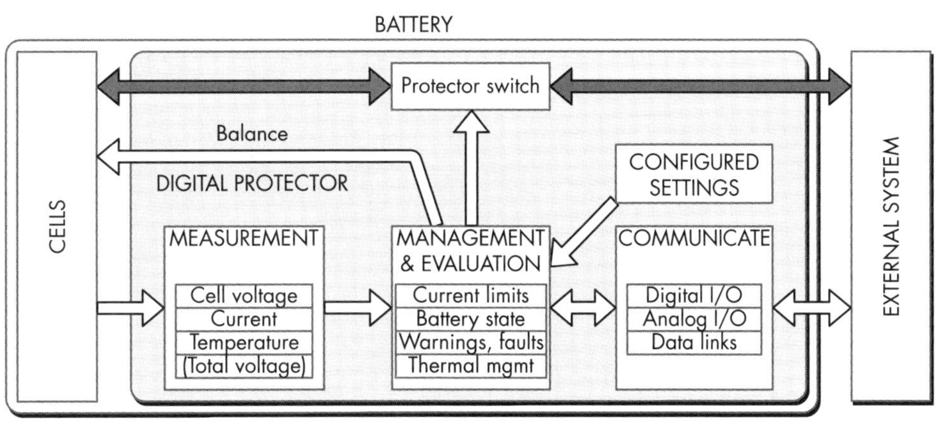

FIGURE 4.24 Digital protector block diagram.

current for a given application, or it's too large for that application, making it too expensive. Also, solid-state switches (typically used in protector BMSs) are unable to handle high currents. Contactors may be used instead. For large batteries, a digital BMU is typically used instead.

4.4.2 Digital BMU

A digital BMU (battery management unit) is the heart of a BMS. It is completed by adding an external protector switch, connecting to an external system that obeys the BMU, or both (Figure 4.25).

Unlike an analog BMS, a digital BMU knows the state of a battery, evaluates its conditions, and reports them. Unlike a protector, a digital BMU does not include a protector switch.

4.4.3 Digital BMS Accessories

The manufacturer of a digital BMU may offer additional modules and accessories to augment the BMS functionality (Figure 4.26):

- A thermistor sense module for a wired BMS;
- A high-voltage front end (HVFE) that operates directly at the full battery voltage:
  - Powers the BMS from the battery voltage through a DC-DC converter;
  - Senses the battery current;
  - Drives the contactors;
  - Senses the precharge;
  - Senses the battery voltage (before and after the protector switch);
  - Tests battery isolation and ground faults;
  - Interfaces to a charger;
  - Interfaces to a charging station.
- A display to show the state of the battery:

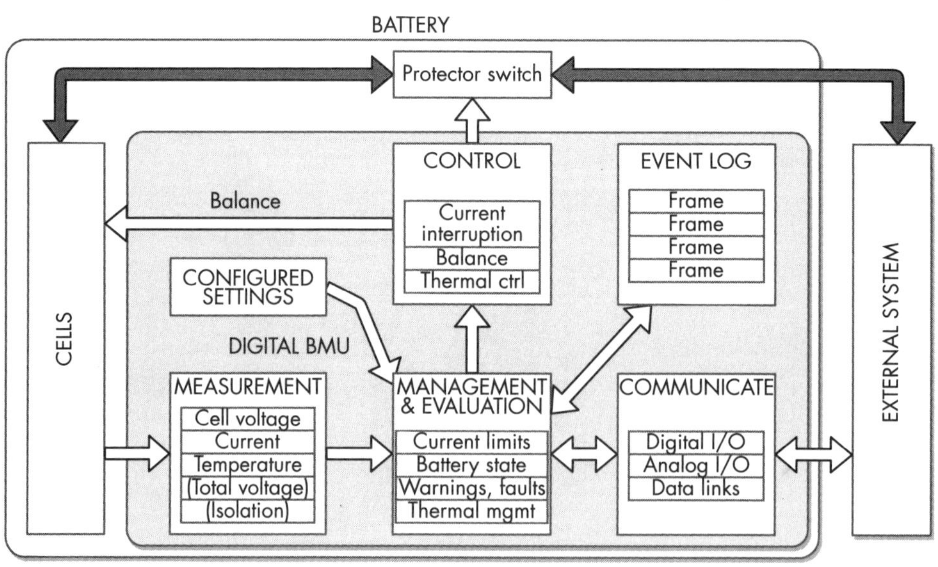

FIGURE 4.25 Digital BMU block diagram.

(a) (b) (c)

FIGURE 4.26
Accessories: (a) SoC display, (b) HVFE, and (c) current sensor.

- Analog meters: Current, SoC, voltage, temperature;
- LED bar: To display the SoC;
- LCD: A display with two or four lines of alphanumeric characters;
- Touch-screen display: To display a wide range of parameters.

4.4.4 Digital BMS States

Most BMSs have only one power state—powered on.

A BMS for a product that has distinct operating modes may have a state machine with two or more BMS states. It may also have a *sleep* state, used during shipping,[13] or in the off season.[14]

For example, for an electric vehicle, a BMS may have five power states (Figure 4.27(a)):

- Charging: Gets the value of the current from the charger; is allowed to balance; uses a slow measurement and evaluation rate; uses a current sensor for a small current.
- Ignition: Measures the value of the current to the load; balancing is disabled; measures and evaluates more rapidly; uses a bidirectional current sensor able to sense the high current during acceleration and regenerative braking.
- Balancing: When the vehicle is off and not charging, but the battery is imbalanced.

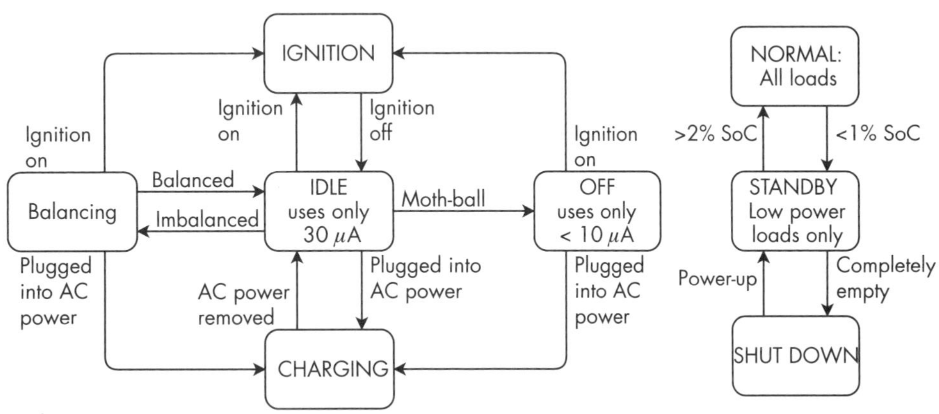

(a) (b)

FIGURE 4.27
BMS power states: (a) EV, and (b) low-voltage stationary.

13. For example, an EV built in Japan and transported to Europe by ship.
14. For example, a boat docked during the winter.

- Idle: When the vehicle is off and not charging, and the battery is balanced; during this time, the BMS may sleep for hours, and then wake up for a short time to measure the fully relaxed voltage of cells.
- Off: Completely off, during transportation or *mothballing.*[15]

For another example, a BMS for a large, low-voltage stationary battery, may have three states (Figure 4.27(b)):

- Normal: Charging and discharging enabled;
- Stand-by: Charging enabled, discharging only to power itself and a solar charge controller;
- Shut-down: When a cell voltage is too low even to power the BMS.

4.4.4.1 Protection State

The BMS may also have a protection state machine (Figure 4.28(a)).

Note that once in the fault state, clearing the fault and restoring normal operation requires manual intervention (such as powering down the battery), which the service technician may do only after addressing the cause for the fault.

4.4.4.2 Contactor State

If a BMU drives contactors, it would have a contactor state machine (Figure 4.28(b)). This is a simplified state machine for a battery with only one contactor. The state

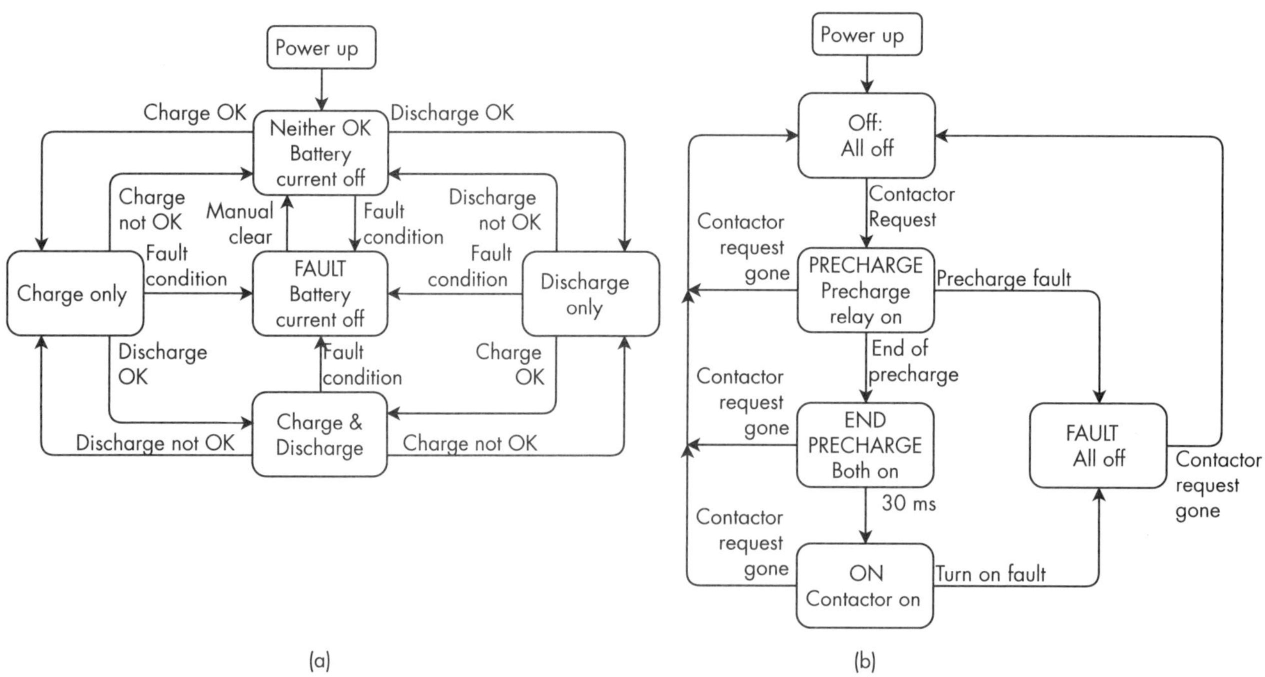

FIGURE 4.28 More BMS states: (a) protection state, and (b) contactor state.

15. Long period without use.

machine for a battery with a positive and negative contactor that also tests the contactors and the battery isolation testing is far more complex.

4.4.5 Digital BMS Functions

The following seven sections describe each function of a digital BMS in detail:

- Measurement;
- External system control;
- Balancing;
- Evaluation;
- Battery device control;
- Inputs and outputs;
- Communications and logging.

4.5 MEASUREMENT

A digital BMS measures cell voltages and temperatures, as well as battery voltage, temperature, and current.

4.5.1 Cell Voltage Measurement

The BMS measures the voltages of each cell.

4.5.1.1 Range

A wired BMS measures cell voltages from 0 to 5V, which covers all chemistries and operating conditions (Figure 4.29). A distributed BMS is limited to a smaller range (2~5V) because it's powered by the cell itself. This limits its applications:

- It does not support LTO[16] cells, whose minimum voltage dips below 2V;
- It's inappropriate for race vehicles, whose cells may be operated at the maximum power point, bringing the terminal voltage down to one half the OCV (therefore as low as 1V for LFP cells).

A banked BMS may or may not be limited in this regard, depending on whether it uses the circuits of a wired or distributed BMS.

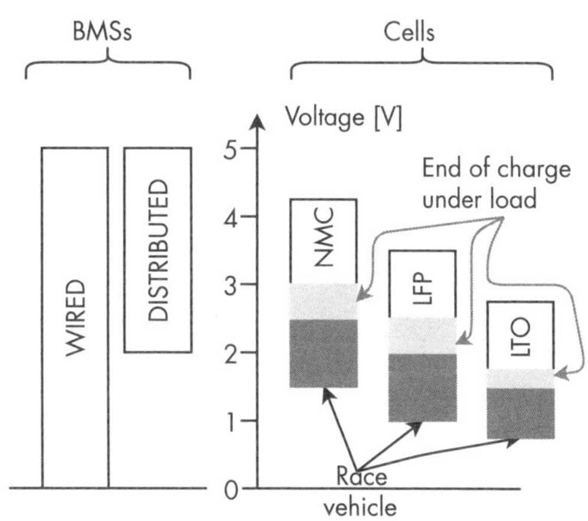

FIGURE 4.29 Cell voltage range and ability of a BMS type to handle it.

16. Titanate.

4.5.1.2 Measurement Accuracy and Resolution

A digital BMS converts the cell voltage to a reading using an analog-to-digital converter with a resolution[17] that ranges from 8 to 16 bits (Table 4.12).

The accuracy of an 8-bit measurement (± 10 mV) is sufficient to protect against under and overvoltage and for top-balancing an LFP battery.

A 12-bit measurement (±600 μV) is required if using the cell voltage to evaluate SoC, and for mid-balancing.

Due to errors and noise, a 16-bit measurement (±38 μV) doesn't have a 16-bit accuracy. In practice, a BMS cannot reliably measure more accurately than ±300 μV.

A laboratory-grade data acquisition system can achieve higher accuracy when measuring a single cell in a controlled scientific experiment. Long-term averaging can improve the resolution; that is possible when the OCV of a cell that is at rest is measured for many hours. It is not possible when measuring a cell voltage that varies rapidly in actual use.

4.5.1.3 Measurement Rate

A measurement rate of one reading per second is typical and sufficient for most applications. In high-power applications, in which the voltage may change rapidly, a rate of 10 readings per second may be required. A faster rate than this is probably unnecessary because Li-ion cells should be able to handle short pulses of under or overvoltage, which end before the BMS would have a chance to notice and react.

4.5.1.4 Current into Cell Voltage Sense Inputs

A cell voltage sense input must draw negligible current so as not to discharge the cell whose voltage it is measuring.

As a rule of thumb, this parasitic current must not be higher than the self-discharge current of the cells. A self-discharge current of 0.00001 C (see Section 2.7.3), that is 10 mA for a 1,000 Ah cell and 10 μA for a 1 Ah cell. Given that an off-the-shelf BMS doesn't know beforehand whether it will be used to protect a small battery or large one, it should be designed for the worst case—a small battery. The BMS's parasitic current should be less than 30 μA in case it is used to protect a battery with a single string of 18650 cells (which have a capacity of about 2~3 Ah). If the battery has a

TABLE 4.12
Analog-to-Digital Conversion Bits and Resolution

	Resolution			Appropriate for			
Bits	**Ratio**	**Assuming 0~5V**	**Accuracy**	**Protection**	**Top-Balance**	**Mid-Balance**	**Mid-SoC Evaluation**
8	1:256	±9.8 mV	~±16 mV	✓	✓	—	—
10	1:1024	±2.44 mV	~±4 mV	✓	✓	—	—
12	1:4096	±610 μV	~±1 mV	✓	✓	✓*	—
14	1:16384	±152 μV	~±500 μV	✓	✓	✓	✓*
16	1:65536	±38 μV	~±300 μV	✓	✓	✓	✓

*Except for LFP.

17. Note the difference between resolution and accuracy; resolution is the smallest changes that a measurement can see, while accuracy is how close the measurement is to the actual value. Accuracy may be better or worse than resolution. If I estimate the number of sheep in a flock to be 98, my resolution is 1, but my accuracy may be more like ±10. If I tell you that there are 365 days in an average year, my resolution is 1, but my accuracy is -0.25.

standby and an *operating* mode, then it may be acceptable for the BMS to draw a higher parasitic current while operating.

When a BMS monitors a series string, all the sense inputs must draw exactly the same current. Otherwise, the BMS would contribute to the imbalance of the string (see Section 3.2.8). This is particularly a problem for a distributed BMS because the two cell boards at the end of a bank use a different circuit from the others. As the slave in a master/slave BMS is powered by its cells, it must draw the exact same current from each cell to avoid imbalancing the battery.

An often-overlooked aspect of the parasitic sense current is its effect on the cell voltages when waiting for them to relax to the OCV. For however small the sense current may be, it is still sufficient to affect the cell voltage. The final voltage that a cell reaches when it is connected to a BMS is different from the voltage that it would reach if it were truly disconnected from everything. It's not just that the sense current discharges the cells slightly; it's also because the hysteresis in the cell voltage depends on the direction of the current and is worse at low levels of current (see Section 2.4.2.3).

4.5.1.5 Open Tap Detection

In a wired BMS, if the wire from the voltage sense tap is disconnected, the BMS is no longer able to monitor the voltage of two cells in series.

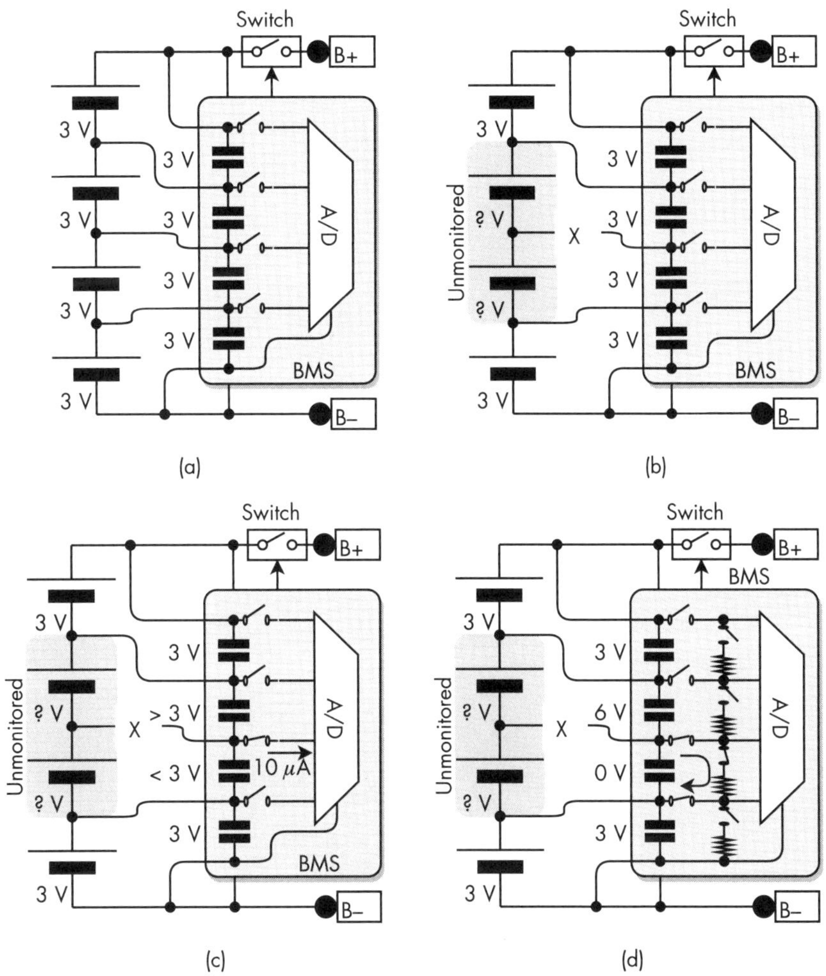

FIGURE 4.30
Open tap detection: (a) normal, (b) open tap, sense voltage remains unchanged, (b) sense current makes voltage drift, and (c) advanced BMS detects an open tap.

The cell voltage sense inputs in a BMS include capacitors that filter noise in the cell voltage (Figure 4.30(a)). The moment a sense wire is disconnected, the capacitor retains the voltage. Therefore, the BMS assumes incorrectly that the cell voltage has not changed (Figure 4.30(b)). Yet, the BMS draws a bit of current while measuring a cell voltage. This current may slowly discharge the capacitor; eventually, the BMS may see a low voltage and issue a "low cell voltage" fault (Figure 4.30(c). This is not a reliable method to detect an open sense line because at the same time that one sense current discharges the capacitor, the sense current for the next cell recharges it.

An advanced BMS runs a periodic test to detect an open sense wire reliably and rapidly. For example, it can do so by turning on the bypass balance load, which drops the sensed voltage immediately (Figure 4.30(d)).

This is not an issue for a distributed BMS—if a wire to a cell board is disconnected, it is no longer powered, and it does not report, which the BMU detects readily.

4.5.1.6 Fault Protector

A BMS may also include a fault protector for redundancy. A fault protector is an analog BMS added to the digital BMS as a back-up; it acts in case the digital BMS fails to act (Figure 4.31(a)).

This analog BMS compares each cell voltage to fixed limits. If any cell voltage exits the range set by the fault protector, the BMS shuts down the battery, immediately and directly, regardless of what the digital BMS may think. The protector switch is closed only if both the digital BMS and the analog fault protector agree that the battery is OK.

The fixed limits of the fault protector are beyond the configured limits used by the digital BMS, so that the digital BMS trips first, and the fault protector only trips if the digital BMS has failed to respond (Figure 4.31(b)).

4.5.1.7 Redundant BMSs

For critical applications, two independent BMS may be used on the same cells. Then the question is, which one to believe? For some applications, it may be acceptable to leave the battery on if at least one BMS says that is OK. For other applications, it may be preferable to shut down the battery if either BMS says things are not OK. With three BMSs, the majority wins.

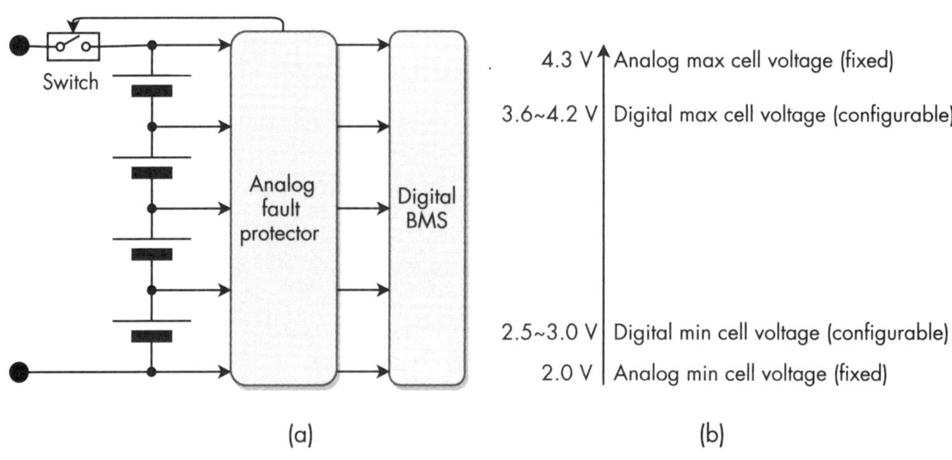

FIGURE 4.31
Fault protector: (a) block diagram, and (b) limits.

4.5.1.8 Banking

A BMS for large batteries divides the battery into several banks:

- Each bank in a wired BMS may handle up to 12 or 16 cells (Figure 4.32(a));
- Theoretically, the size of a bank in a distributed BMS is not limited; in practice, it is limited to about 25 cells[18] (Figure 4.32(b)).

Within a bank, cell voltages must be applied to the BMS sense inputs in order and in the correct direction. The BMS is easily damaged if wired to the cells in the wrong order (see Section 8.3.2.2).

Banks are electrically isolated from each other. Therefore, they can handle any voltage between them. When designing a battery, take advantage of this isolation between banks. That is a good place to locate any item that might open (a fuse, a connector, a safety disconnect, or a contactor), or a high-resistance cable (between modules).

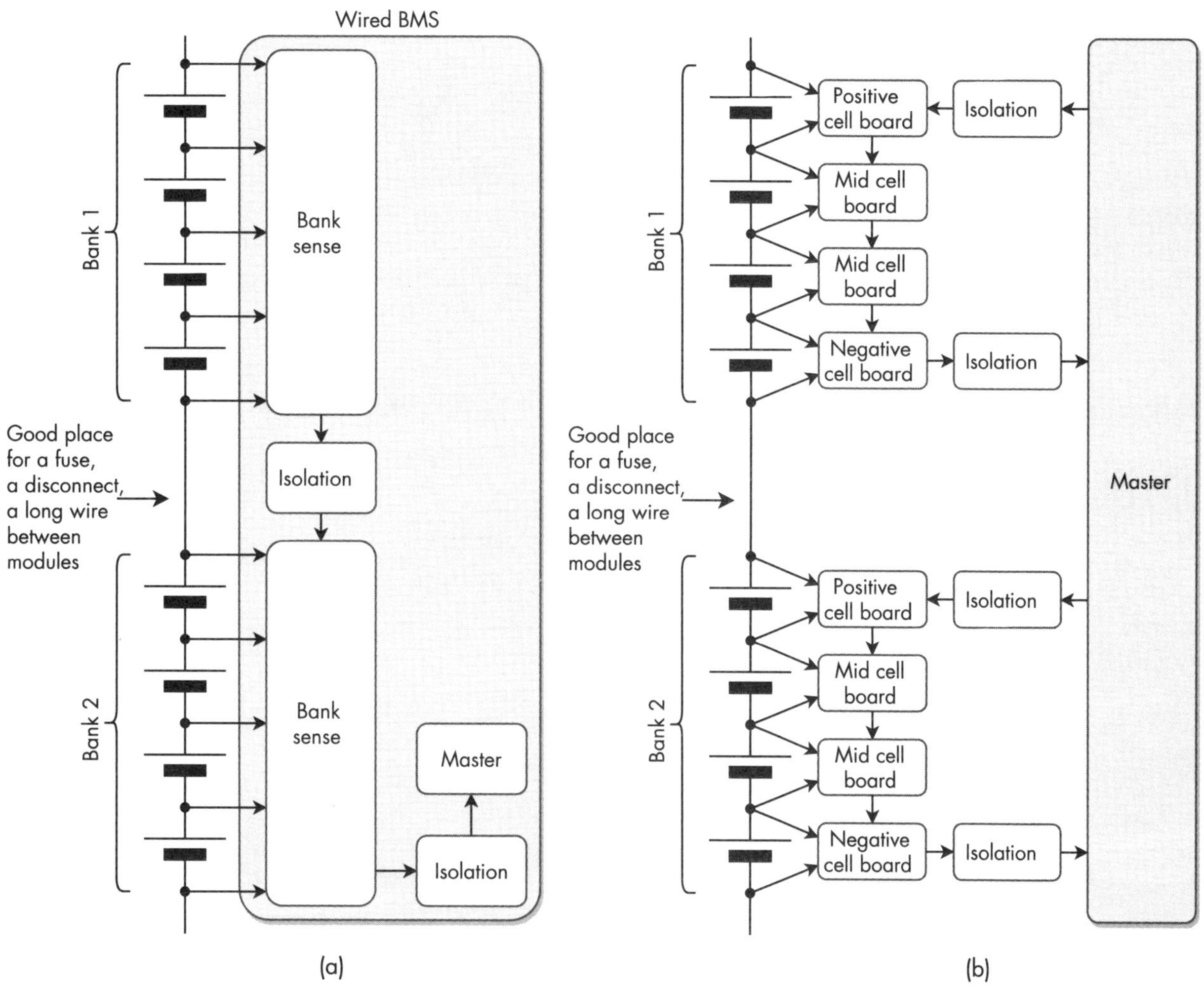

FIGURE 4.32 Banking: (a) wired, and (b) distributed.

18. Because each cell board degrades the signal slightly; by the time the signal from the first cell board passes through all the cell boards and reaches the last cell board, it may be distorted.

4.5.1.9 *Numbering*

Note that a BMS may number the banks either from the most positive one or the most negative one, and it may consider the first bank to be either number 0 or number 1. Similarly, the BMS may number cells from the most positive or the most negative one, and it may consider the first cell to be either number 0 or number 1. Probably, the BMS uses the same direction and the same starting number for both the cells and the banks. The BMS may restart numbering the cells for each bank or may continue the numbering across banks. This results in 32 different ways the BMS may number the cells and the banks. Figure 4.33 shows just four of these permutations.

A BMS may allow you to select how the cells and banks are numbered (actually, just how they are displayed to the user; internally, the BMS keeps on using its original numbering scheme).

The BMS's ordering may be different from the ordering preferred by the battery designer. When working on a battery, be careful not to confuse the BMS's numbering with the battery designer's numbering, as they may be different.

4.5.2 Additional Voltage Measurements

Besides the cell voltages, the BMS may measure other voltages (Figure 4.34), such as

- *Total voltage:* The BMS adds the voltages of cells in series to calculate the total battery voltage; optionally, the BMS may also measure the total battery voltage directly. A noticeable discrepancy between the two values would indicate a problem with either measurement.
- *Bank voltage:* Optionally, the BMS may also measure the total voltage of a bank directly (again, as a double-check);
- *Battery terminal voltage:* Monitoring the terminal voltage of the battery (which is different from the voltage of the cells when the protector switch is open) may

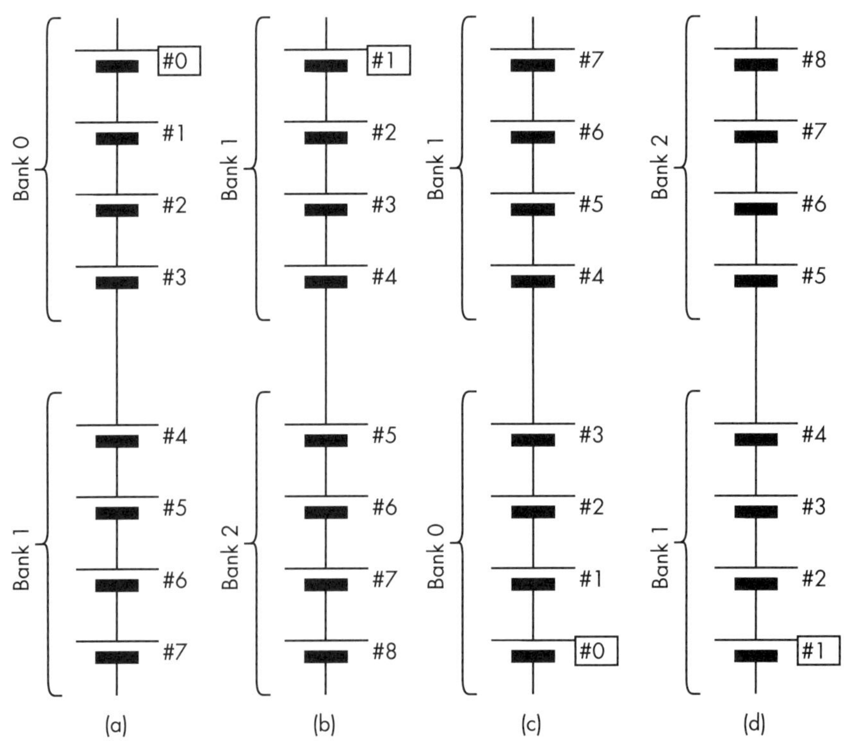

FIGURE 4.33
Four of the 32 possible ways to number cells and banks: (a) From positive and from 0, (b) from positive and from 1, (c) from negative and from 0, and (b) from negative and from 1.

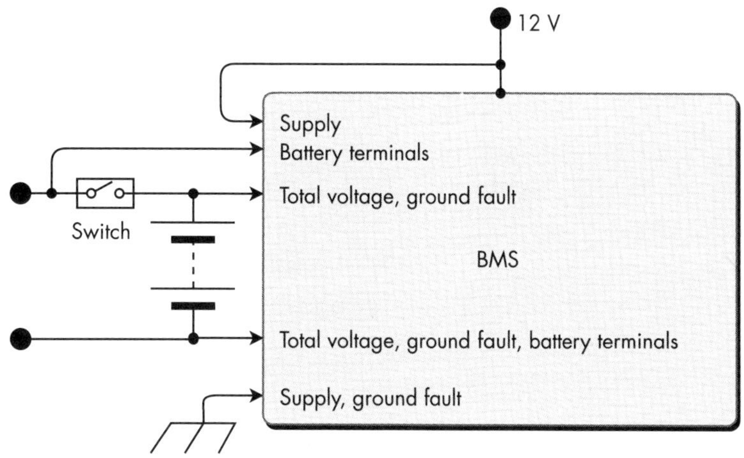

FIGURE 4.34 Additional voltage measurements.

help the BMS test the protector switch, and may let it determine what kind of load is connected to the battery and decide if it is safe to turn on the switch.

- *Ground voltage:* Measuring the ground voltage relative to the battery terminals may be useful when determining if there is a ground fault.
- *Power supply:* The BMS may also measure the voltage of its power supply so it may log it or report it; such information may be useful while troubleshooting.

4.5.3 Temperature Measurement

Various BMSs handle cell temperatures in different ways:

- Not measure them at all;
- Measure just a few temperatures somewhere in the battery;
- Measure a few temperatures for a group of cells (e.g., one temperature every six cells);
- Measure the temperature of each cell.

Nearly universally, temperature measurement uses NTC[19] thermistors. These may look like a grain of rice at the end of two long wires, or a cylindrical probe, or a ring terminal, or a surface mounted component (on a PCB or flex circuit).

If the thermistors are not included with the BMS, they must be bought separately. Typically, these thermistors have a resistance of 10 kΩ at room temperature. The rate at which the resistance changes (the *beta* parameter) is critical; many values for β are possible. It is unlikely that the BMS allows configuration of resistance and β.

The wires of a thermistor are likely to be referenced to the BMS's power supply. In a slave, they are likely to be referenced to the most negative cell in the bank. A short circuit between those wires and a cell terminal causes a ground fault (at best) or a dangerous short circuit (at worst). Therefore, great care must be taken when wiring thermistors to avoid the potential for short circuits.

At a minimum, a BMS uses the temperature information to shut down the battery if it's too hot. The BMS may also use this information to manage the battery more carefully:

- Prevent charging at temperatures below freezing;

19. Negative temperature coefficient.

- Reduce the current limits as the temperature starts getting too high or too low;
- Manage climate control by turning on a fan, a heater or a coolant pump.

Most BMSs consider only the absolute temperature. A more sophisticated BMS also considers the rate of temperature increase because a rapid increase in temperature is of concern even if the absolute temperature is still within limits.

Also, a BMS may monitor other temperatures, such as

- Air intake and exhaust;
- Liquid coolant;
- Electronic assemblies;
- MOSFETs in the protector switch.

The BMS would use different limits for these temperatures than the ones used for cell temperatures.

4.5.4 Current Measurement

The BMS may measure the current directly (using a resistive or Hall effect sensor), or may rely on the external system to measure it and reports it to the BMS (see Section 5.9.2).

4.5.5 Other Measurements

The BMS could also measure other parameters for improved safety:

- Measure pouch cell thickness, for SoC and SoH estimation;
- Sniff the gases from a vented cell, to shut down the battery and send an alarm before the battery catches on fire, and possibly to activate a fire extinguisher;
- Humidity measurement, in case the battery includes some form of a dehumidifier, to avoid condensation;
- High-frequency noise in a cell voltage, which may be an indication of a soft short inside it;
- Airflow, to detect cooling issues before they result in an over-temperature.

These techniques are still in the researchers' labs and have not been implemented in off-the-shelf BMSs.

Perplexing methods of SoC evaluation have been proposed, through light[20] or sound[21]; I don't see how these solutions could be cheaper, more reliable, and more accurate than present technology.

4.6 EXTERNAL SYSTEM CONTROL

A digital BMU controls the external system by evaluating and reporting current limits and issuing warnings and faults.

---

20. Liliana Zdravkova of the University of Waterloo estimated cell SoC by looking at the surface of an electrode. She stuck an optical fiber in a cell and correlated the reflectivity of an electrode with the SoC; The accuracy was approximately 14% after calibrating the sensor to each cell. Other researchers have done similar work. [1]
21. Titan Advanced Energy Solutions, which recently received a $ 10 M investment, is developing the "ionView advanced acoustic BMS" to provide "nonintrusive, providing an in situ real-time measurement of the battery's current and voltage." According to a patent, each cell is ultrasonically shaken and the state of the cell is deduced from analysis of the response. [2]

4.6.1 Current Limits

A digital BMS calculates two current limits:

- Charge current limit;
- Discharge current limit.

It calculates those limits based on the readings:

- As the voltage of the cell with the highest voltage gets close to the maximum threshold, the BMS reduces the CCL; similarly, as the voltage of the cell with the lowest voltage gets close to the minimum threshold, the BMS reduces the DCL (Figure 4.35(a));
- As any temperature gets close to the maximum or the minimum thresholds, the BMS reduces both the CCL and DCL (different thresholds for charging and discharging) (Figure 4.35(b));
- If a current peak lasts too long, the BMS may reduce both the CCL and DCL.

The BMS communicates these current limits to the external system. Ideally, the external system obeys them, so there's no need for the BMS to resort to turning off the battery.

The limits could be expressed in

- Current (e.g., max current of 200A);
- Specific current (e.g., max current of 2C);
- Relative current (e.g., max current of 75% of the "max current" setting).

4.6.1.1 Operating Range

These limits set the operating range of the battery. In an off-the-shelf BMS, these limits may be configurable;[22] but once configured, they usually remain fixed.

The limits are selected as a compromise between these parameters which are somewhat mutually exclusive:

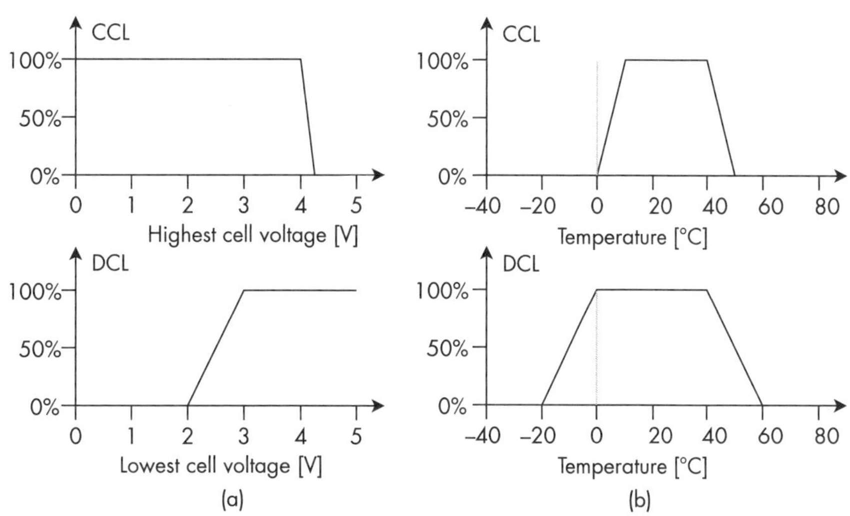

FIGURE 4.35 Current limits: (a) voltage-based, and (b) temperature-based.

22. Usually, in a custom BMS the limits are fixed.

- Battery performance;
- Cell longevity;
- Safety of the battery, product, and user.

For example, lowering the limit for the minimum cell voltage allows for a deeper discharge, and therefore a longer run time. However, this degrades the cells faster, reducing the cycle life.

For another example, allowing an EV to charge below freezing is more convenient for the user, but this degrades the cells in a way that may result in a cell self-igniting.

The person selecting those limits is responsible for those compromises.

4.6.1.2 Adaptive Operating Range

In a more advanced BMS, the software may be able to vary these limits dynamically to adapt to the cell's SoC (allow more performance when the battery is full) and SoH (expand the cells' operating range as they degrade). Such a BMS uses the cells' potential at its fullest, maximizing the value of the battery by optimizing its performance and longevity. In turn, this allows the battery designer to use the minimum number of cells in the smallest battery volume.

4.6.2 Current Turn-Off

If the CCL reaches zero, charging must stop. Similarly, if the DCL reaches zero, discharging must stop.

As already mentioned, a BMU does not include a protector switch. Therefore, it must stop the battery current in one of two ways:

1. If the BMU includes drivers for protector switches and the battery includes such switches, the BMS opens those switches;
2. If the BMU relies on the external system, it tells the system to stop charging or discharging.

4.6.2.1 Avoiding Nuisance Turn-Offs

To avoid nuisance turn-offs, the BMS may delay shutting down the battery (to give the system a chance to recover) by

- Adding a specific time delay;
- Time-averaging the parameter being monitored.

The difference is that

- Time delay is fixed regardless of the severity of the problem, whether deep (Figure 4.36(a)) or shallow (Figure 4.36(b));
- Time-averaging responds faster if the problem is severe (e.g., deep overvoltage) (Figure 4.36(c)), slower if it is mild (e.g., shallow overvoltage) (Figure 4.36(d)).

Time-averaging has the advantage that it is more forgiving of a small problem and reacts quite fast if the problem is severe.

Regardless, this delay could result in a reversed cell if the BMS is configured to allow the battery to operate at the edge of the cells' safe operating area. That is because, when the voltage of the least charged cell drops quite low, it doesn't take long to overdischarge a it, which can happen during this delay and before the BMS reacts.

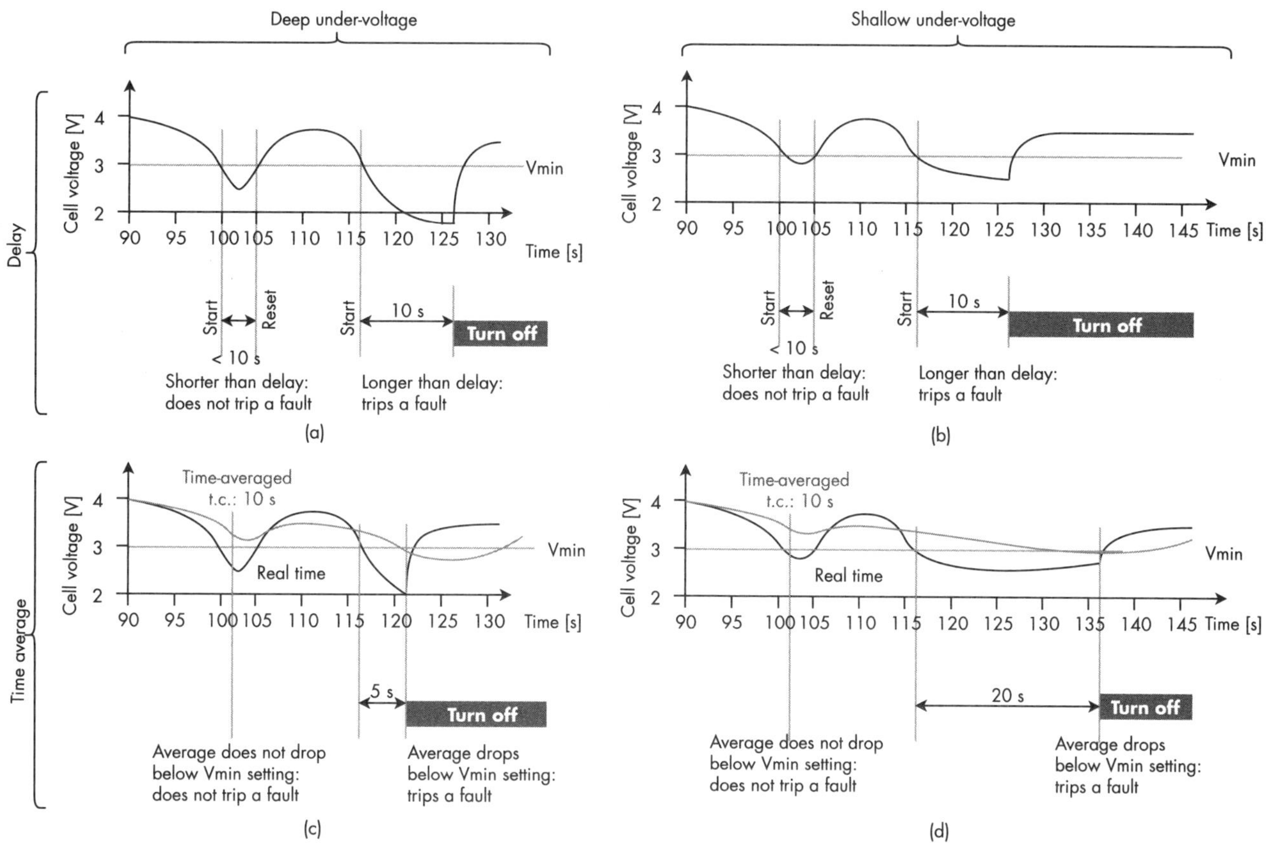

FIGURE 4.36 Nuisance turn-off mitigation: (a) delay, deep undervoltage, (b) delay, shallow undervoltage, (c) time-average, deep undervoltage,(d) time-average, shallow undervoltage.

4.6.3 Warnings and Faults

As a value exceeds a warning threshold, the BMS issues a warning (e.g., low voltage). If then the value continues past the warning threshold and exceeds a fault threshold, the BMS issues a fault (Figure 4.37).

The BMS may delay issuing a fault, to give the external system a chance to recover.

In the previous example, we saw the reaction to a low voltage. That is just one of the various parameters that may result in warnings and faults. The following example uses all the parameters.

4.6.3.1 Cell Voltage, Temperature, Current

The BMS can issue warnings and faults based on cell voltages and temperatures. For example, while charging a battery using LCO cells, the BMS can do so under the following conditions. In Figure 4.38, the left column is for warnings and faults, and the right column is for the current limit:

- Highest cell voltage (Figure 4.38(a)):
  - If the voltage of the most charged cell reaches 4V, the BMS issues a "high cell voltage" warning;
  - At 4.2V, the BMS disables charging;

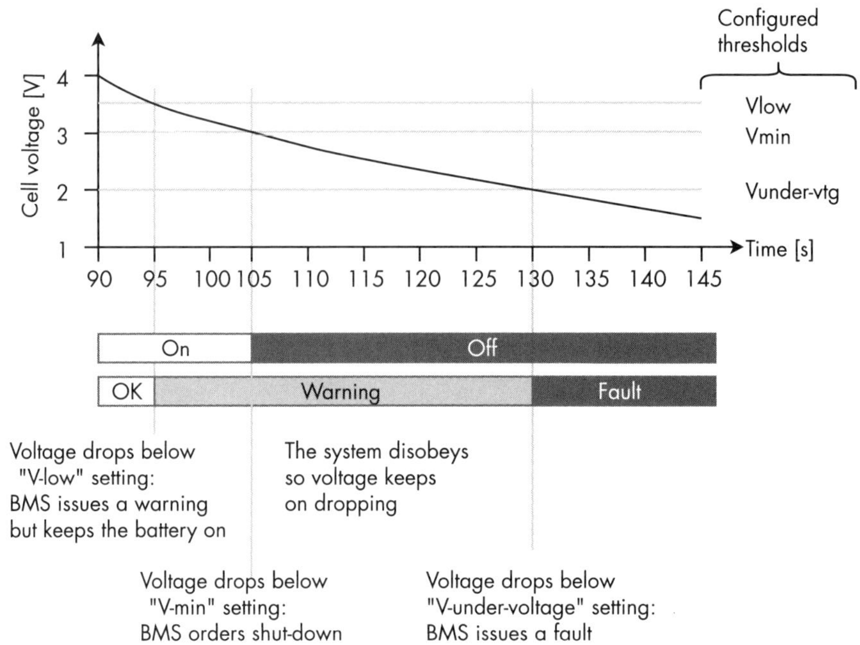

FIGURE 4.37
Low-voltage plot, showing warnings and faults.

    - If, despite this, charging continues, at 4.3V the BMS issues a "high cell voltage" fault.
- Cell temperature (Figure 4.38(b)):
    - If the temperature of the hottest cell exceeds 40°C, the BMS issues a "hot" warning;
    - At 45°C, the BMS disables charging;
    - If, despite this, charging continues, and the temperature continues to rise, at 50°C the BMS issues a "hot" fault;
    - If the temperature of the coldest cell drops below 10°C, the BMS issues a "cold" warning;
    - At 0°C, the BMS disables charging;
    - If, despite this, charging continues, and the temperature continues to drop, at -5°C the BMS issues a "cold" fault.
- Battery current (Figure 4.38(c)):
    - If the charging current exceeds 0.5C, the BMS issues a "charge overcurrent" warning;
    - At 0.6 C, the BMS disables charging;
    - If, despite this, charging continues, and the current increases further, at 0.75C the BMS issues a "charge overcurrent" fault.

Similarly, while discharging, the BMS issues warnings and faults and controls discharging, though using different thresholds levels (Figure 4.38(d–f)).

4.6.3.2 Other Causes

The BMS may issue warnings and faults for other causes as well, such as

- Battery voltage out of range;

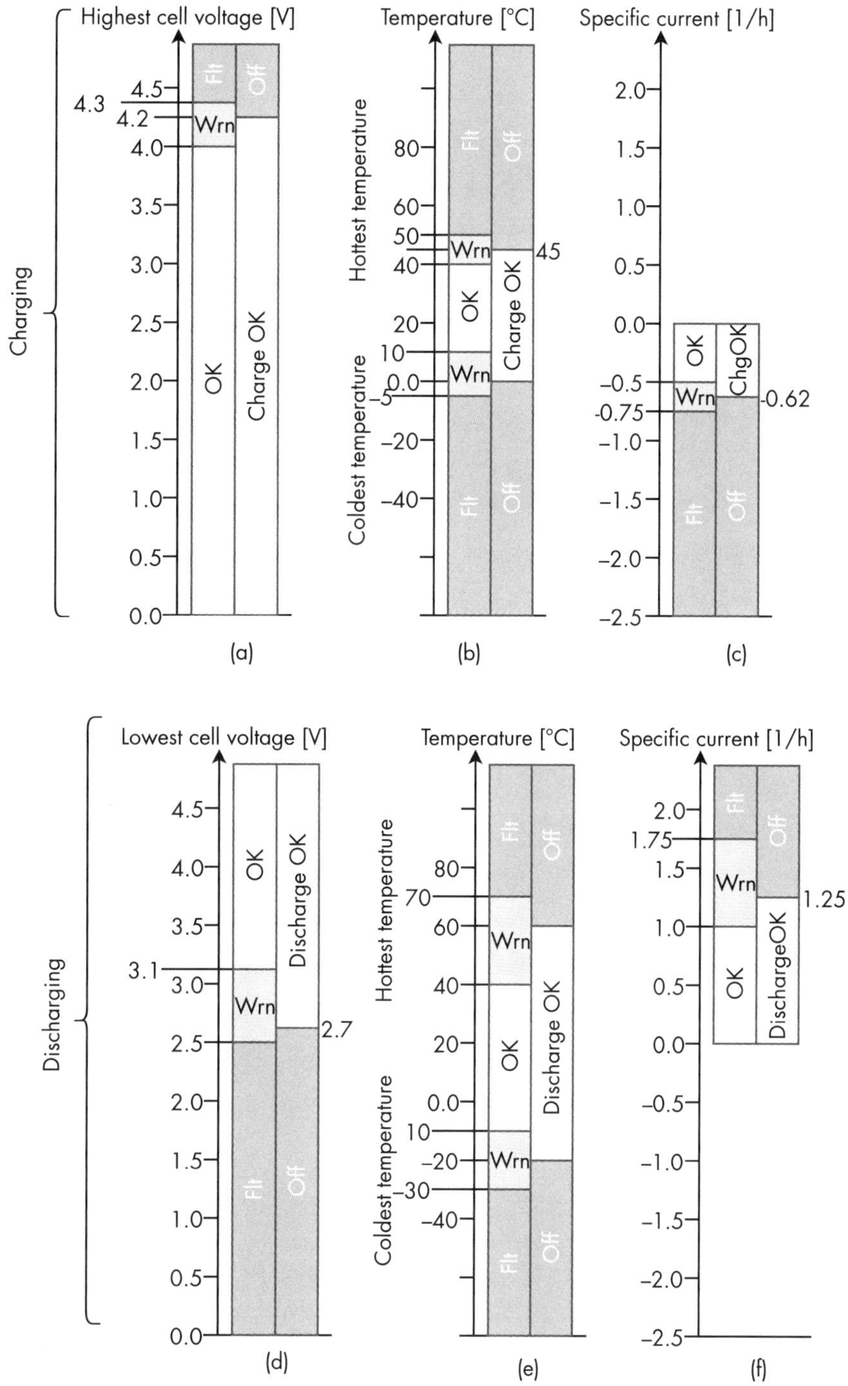

FIGURE 4.38
Warnings and faults:
(a) charging cell voltage,
(b) charging temperature,
(c) charging current,
(d) discharging cell voltage, (e) discharging temperature, and
(f) discharging current.

- Fast temperature rise;
- Temperature while neither charging nor discharging;
- Ground fault;
- Communications errors;
- Hardware faults in the battery (e.g., welded contactor contacts);
- Hardware faults in the BMS (e.g., damaged open-drain driver);
- Software errors internal to the BMS (e.g., stack overflow).

4.6.3.3 *Crawl Home Mode*

Rather than completely shutting down the battery, in some cases, the BMS may reduce the current limits down to a low level for a while longer, under the assumption that doing so won't damage the battery. This level may be sufficient for the application to continue at a low rate—allowing a car to crawl out of traffic and safely pull over.

4.7 BMS BALANCING

Balancing (see Section 3.2.9) maximizes the capacity of a series string (limited only by the cell with the lowest capacity) by correcting any imbalance in the string, which can be due to a variety of causes (see Section 3.2.8). A single cell doesn't require balancing, nor does a simple parallel block of cells.

Balancing is a performance function, not a safety function. While it is optional, it is implemented in all BMSs other than the cheapest analog PCMs.

Balancing is done by charging cells with low SoC or discharging cells with high SoC.

Balancing is done at a single SoC point (the *balance setpoint*) (see Section 3.2.6). Elsewhere, the SoC levels of all the cells differ due to variations in cell capacity (see Section 3.2.5.7).

This chapter discusses balancing performed by the BMS. Balancing may also be performed by a *balancing charger* (see Section 6.5.4) or manually (see Section 7.4.3).

4.7.1 Required BMS Balancing Current

We saw that a good string of Li-ion cells may experience an imbalance current of about 0.00015C (150 μC[23]). We also saw that a bad cell may self-discharge at a specific current as high as 0.001C (1 mC) (see Section 3.2.9.1).

That is not much current at all. Yet, it can become significant for large-capacity strings or if the BMS is not able to balance all the time. In many applications, the BMS can balance only part of the time (yet imbalance current flows in the cells all the time). If so, the balancing current needs to be higher, in inverse proportion to how much time is available for balancing. For example, assume that the string's imbalance current is 1 mA. If the BMS can do balancing continuously, then the required balancing current is simply 1 mA. However, if the BMS is only able to balance one hour every day, the required balancing current is 24 mA to achieve an average of 1 mA. In extreme cases, the BMS won't be able to balance a string (see Section 8.7).

Figure 4.39 plots the balancing current required to balance a string, given the imbalance current and the portion of the time that the BMS can balance.

The *X*-axis in this graph has the amount of imbalance current. It also shows the approximate battery capacity that would experience a given imbalance current, for either a normal battery (assuming 0.00015C of imbalance current) or a battery with a bad cell (assuming 0.001C of imbalance current).

The diagonal lines indicate the proportion of the time that the BMS can do balancing, from 1% to full time. The vertical axis has the minimum balancing current that the BMS must be able to provide.

23. μC is not microcoulomb, it is a specific current equal to one millionth of the battery capacity, which is 1/Mh; that is, one over megahour.

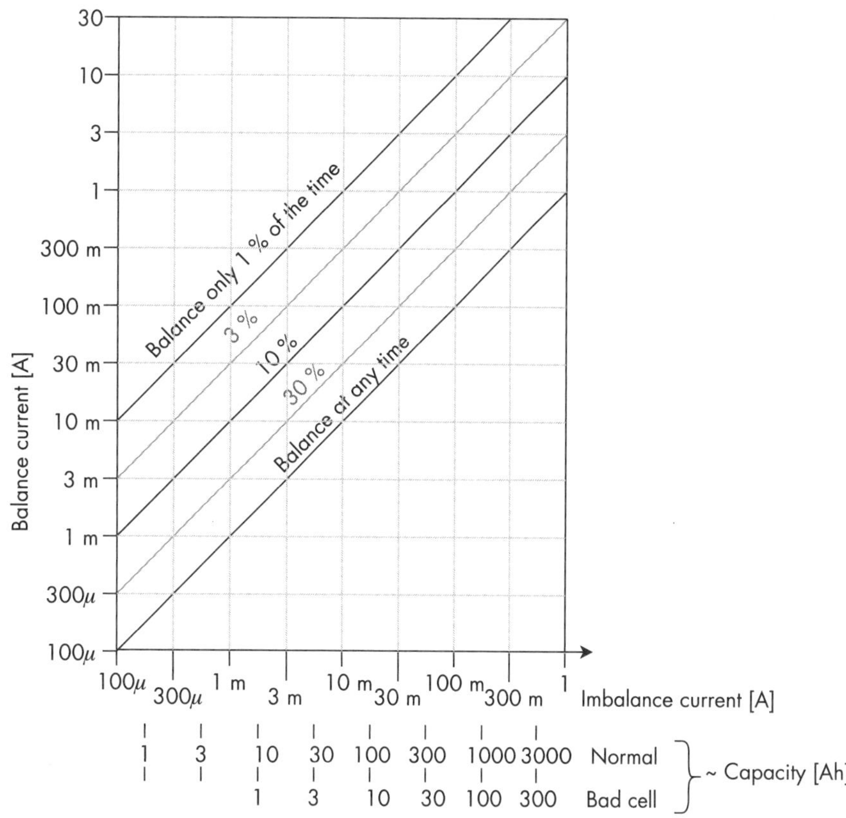

FIGURE 4.39
Balance current required to compensate for an imbalance current.

For example, a BMS that can balance at least 0.15A, and at least 10% of the time,can keep a normal 100 Ah battery balanced. On the other hand, if there's a bad cell, it must balance 100% of the time to keep up with the self-discharge current in the bad cell.

The advantage of a balancing algorithm that allows balancing at all times is that it allows the use of BMS whose hardware provides less balancing current. Such an algorithm can allow a given BMS to balance a battery that is 2~10 times larger.

Normally a BMS can balance at a much higher current than the battery requires. The BMS matches the imbalance current in one of two ways:

- By turning balancing on and off with a duty cycle that reduces the average current;
- By actually reducing the current.[24]

A BMS that was able to keep a battery balanced when new may not be able to do so later, as cells age and their imbalance current increases.

The various "Applications" chapters in Volume 2 will discuss the balancing current requirements for each application.

It makes little economic and engineering sense to specify a BMS capable of an extremely high balancing current. Yes, this BMS may be able to balance a grossly unbalanced battery just the one time in its life it is required. However, this BMS would be more expensive, would be bulkier, and would produce far more heat compared to BMS that is designed just for the job required 99.9% of the life of the battery—maintaining the balance.

24. Usually through PWM, which, if you think about it, is the same solution, except at a higher frequency.

In some cases, the BMS itself contributes to cell imbalance in a series string due to damage, misconfiguration, or a design flaw (see Section 8.7).

4.7.2 Balancing Technologies: Bypass and Charge Transfer

There are two balancing technologies.[25]

- Bypass balancing: Charge is removed from a cell and wasted into heat;
- Charge transfer balancing: Charge is transferred in or out of a cell, wasting much less power.

4.7.2.1 Bypass versus Charge Transfer Balancing

Practically every BMS uses bypass balancing because it's quite simple yet effective. Charge transfer balancing is sexy but often an expensive overkill, so it is rarely seen in high-volume commercial products. However, it is appropriate in batteries that require a balancing current of 3A or more. Table 4.13 compares the two technologies.

4.7.2.2 Passive and Active Balancing

The general public knows bypass balancing as *passive balancing* and charge-transfer balancing as *active balancing*. These terms are ambiguous because, unfortunately, "active balancing" has also been used to mean "all-the-time, SoC-based balancing" (see Section 4.7.4.2) or "redistribution" (see Section 4.7.7); "passive balancing" has been used to mean "end-of-charge, voltage-based balancing" as well.

In contrast, the terms "bypass balancing" and "charge-transfer balancing" are precise and unambiguous. Therefore, other authors and I have taken the lead in introducing the industry to these terms. We encourage you to do the same. You won't see the terms "active balancing" and "passive balancing" anywhere else in this book.

4.7.2.3 Bypass Balancing

Bypass balancing connects a resistor[26] across a cell, to remove the excess charge and turn it into waste heat (Figure 4.40(a)). Bypass balancing is appropriate for most applications.

Table 4.13 Comparison between Bypass and Charge Transfer Balancing, Assuming 1 A Balancing Current

	Bypass	**Charge Transfer**
Balancing Current	10 mA~3 A	10 mA~30 A
Power Wasted in Heat	100%	Cell-to-cell in a long string: 90%~100 % Other topologies: 5%~30 %
Heat Power per Cell	~4 W	0.2~1.2 W
Cost per Cell	~0.5 $	~ 3$
Volume per Cell	0.1 cm^3	10 cm^3
Balancing Component	Resistor	DC-DC converter

25. I believe that Texas Instruments selected the term bypass balancing. I believe that Phil Weicker selected the term charge transfer in his book A Systems Approach to Lithium-Ion Battery Management; by now, a few book authors have adopted these terms as well.
26. Some BMSs use a transistor instead of a resistor. Any load could be used, including a light source or radio transmitter that radiates power away from the cell.

The balancing current depends on the type of BMS:

- Tiny BMSs for small batteries: 1~3 mA;[27]
- Protector BMS: 50~250 mA;
- Centralized BMS: 200~400 mA;
- Distributed BMS: 300 mA ~ 3A.

Above 3 A, bypass balancing is impractical because it generates excessive heat, which is not easy to manage.

Note that some BMSs are not able to balance all the cells simultaneously. Some are limited to a maximum number of balancing loads at a given time. Some can only balance every other cell at a given time: they balance the even-numbered ones for a while, then the odd-numbered ones for a while. This reduces the average balancing current, in the sense that a BMS that states that it balances with 10 mA on the average actually balances at 5 mA.

A BMS may enable balancing only when the battery is connected to a charger because it "feels wrong" to waste energy when the battery is not being charged. In reality, so little energy is wasted during maintenance balancing (see Section 3.2.9.2) that whether or not a charger is connected has a small effect.

4.7.2.4 Charge Transfer Balancing

Charge transfer balancing uses DC-DC converters to either move some energy into cells or out of them. The source or destination of this energy depends on the topology (see Section 4.7.3) (Figure 4.40(b)).

Charge transfer balancing should be considered in applications that truly require a high balancing current:

- High-capacity batteries, especially if balancing can occur for only a short time;
- There is a considerable variation in temperature among cells;
- There is a concern about bad cells with high self-discharge current;

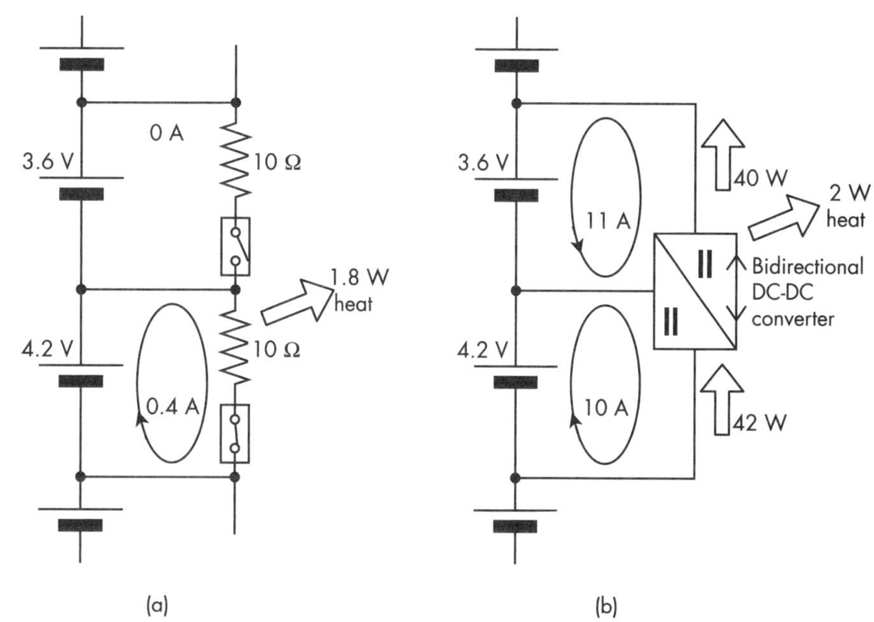

FIGURE 4.40 Balancing technologies: (a) bypass, and (b) one example of charge transfer: cell-to-cell.

27. Wasted in heat inside the BMS integrated circuit.

- A small load or a charger is connected to only some cells, severely unbalancing the battery;
- Cells are likely to be replaced in the field without regard to their SoC.

Otherwise, bypass balancing should be used because it's simpler, cheaper, and more reliable.

The DC-DC converters use magnetics (inductors or transformers) to transfer energy to or from cells. Many have suggested using capacitors or relays for this purpose, but it doesn't work as expected (see Volume 2, Section A.5.1).

4.7.3 Charge Transfer Topologies

There are many ways that charge can be moved in or out of a cell, depending on where power is sent to or received from:

- *Cell-to-cell:* Energy is transferred between adjacent cells (Figure 4.41(a));
- *String:* Energy is transferred between a cell and the string (Figure 4.41(b));
- *Bus:* Energy is transferred between a cell and a bus (not to the string) (Figure 4.41(c)).

The process of transferring charge is not 100% efficient, as the electronics waste some power as heat. If the process were allowed to go on unchecked, it might eventually completely discharge the battery. Therefore, regardless of topology, a converter must shut down if its cell voltage is low. On the same token, a converter must shut down before it overcharges its cell which would start a fire (see Volume 2, Section 5.2.2.3).

4.7.3.1 Cell to Cell

In the *cell-to-cell* topology, energy is transferred between two adjacent cells, in either direction (Figure 4.42(a)).

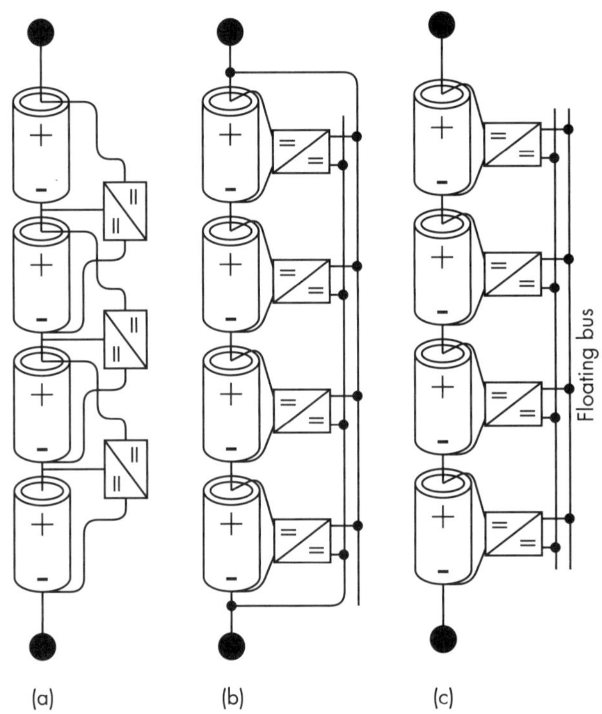

FIGURE 4.41
Charge transfer topologies:
(a) cell-to-cell,
(b) string, and (c) bus.

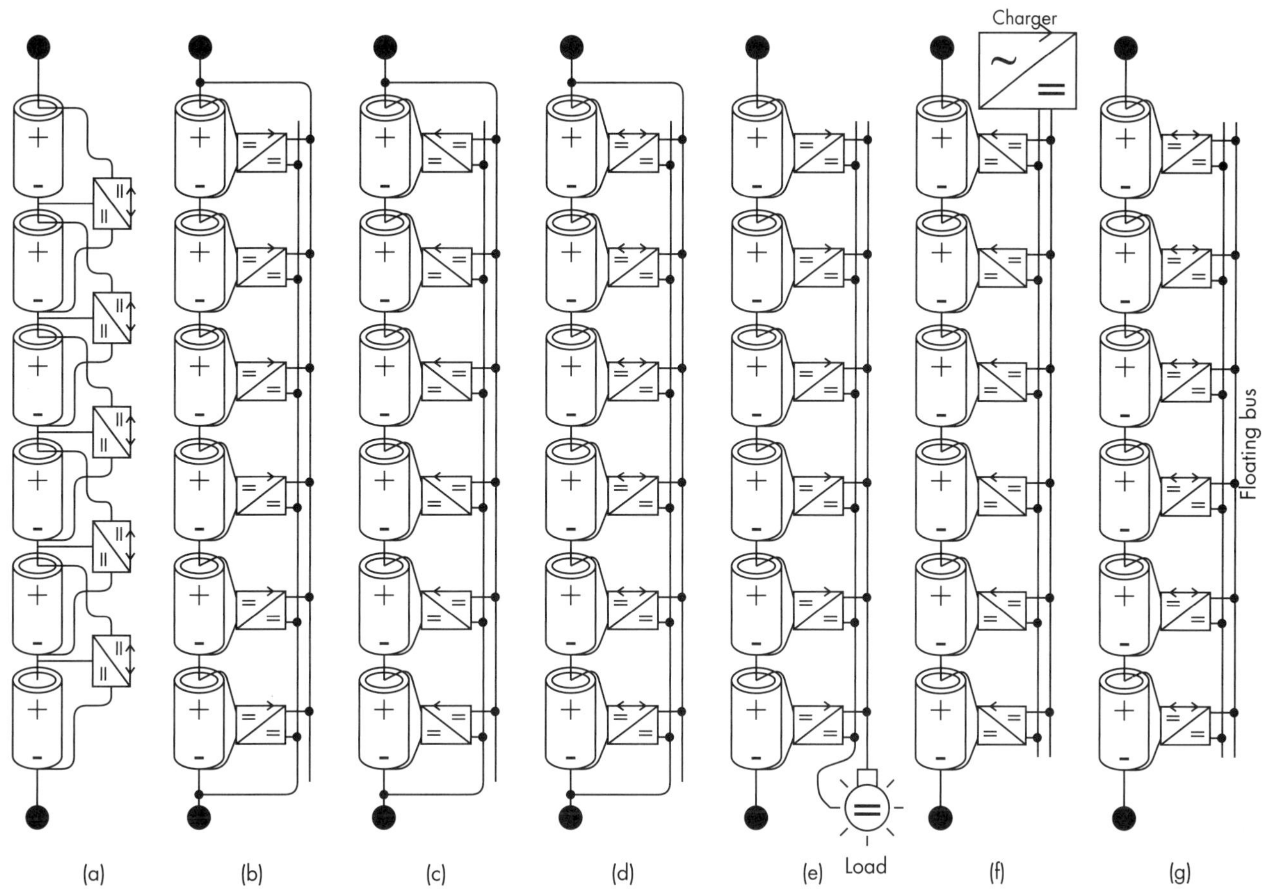

FIGURE 4.42 Balancing topologies: (a) cell-to-cell, (b) cell-to-string, (c) string-to-cell, (d) bidirectional string, (e) loaded bus, (f) powered bus, and (g) floating bus.

For N cells, there are N-1 bidirectional DC-DC converters. The converters are connected to the three terminals of two adjacent cells.

An advantage of the cell-to-cell topology is that it uses low voltage converters because each converter sees only the low voltage of its two cells (e.g., +4.2V and -4.2V). Therefore, the same converters can be used in strings of any length.

This balancer operates autonomously, always trying to balance the voltages of two adjacent cells. This is counterproductive:

- It may try to balance two cells even though they are already balanced because it bases the power flow on relative cell voltages; if one cell has higher resistance, under load its voltage changes more than the other cell's voltage. The balancer assumes that the cells are out of balance and starts transferring energy between them, which actually imbalances them.
- It may balance cells regardless of the SOC, instead of just at the desired balance setpoint, which is counterproductive and wasteful.

If there is a need to transfer energy between two cells that are not adjacent, it is transferred through the intermediate cells, in a *bucket brigade*[28] fashion.

28. In the 18th century, in case of fire, people would form a line between the well and the house, passing water buckets from the well to the first person, to the next person, and down the line eventually to the house. That was called a bucket brigade.

That's fine when the energy has to travel through only a few cells. However, if there are many cells in between the giver and the taker, not much energy may make it all the way because each hop is not 100% efficient; by the time the energy arrives at its destination, little of it may be left.[29] For this reason, cell-to-cell balancing becomes pointless for strings of more than about 16 cells.

4.7.3.2 String

In a string topology, energy is transferred between a string of cells and one of its cells. There are three subtopologies:

1. Cell-to-string (Figure 4.42(b)):
   - Each cell powers a unidirectional DC-DC converter, which powers the string
   - The converters on the most charged cells are turned on to discharge them; this power goes to the entire string
2. String-to-cell (Figure 4.42(c)):(This topology may use a single, shared transformer. See Volume 2, Section A.5.1.1). (This topology may use a bunch of relays to connect a single DC-DC converter to whichever cell needs it. See Volume 2, Section A.5.2.1).
   - The string powers unidirectional DC-DC converters, which charge individual cells
   - The converters on the least charged cells are turned on to charge them; this power comes from the entire string
3. Bidirectional string (Figure 4.42(d)):
   - Bidirectional DC-DC converters are connected between each cell and the string
   - The converters on the most charged cells are turned on in the direction that discharges the cells
   - The converters on the least charged cells are turned on in the direction that charges the cells

The advantage of the string topologies (compared to the cell-to-cell topology) is that the energy from any one cell can be transferred to any other cell regardless of how near or how far the two cells are. The disadvantage is that the DC-DC converters are connected to the string and must be able to handle the full string voltage. For high-voltage strings, DC-DC converters that can handle the high voltage of the string are more expensive and less efficient.

4.7.3.3 Bus

In the bus topology, energy is transferred between a bus (that is isolated from the string) and a cell. There are three subtopologies.[30]

1. Loaded bus (Figure 4.42(e)):

29. Continuing the fire bucket brigade analogy: it's as if each time a bucket is passed to the next person, 10% of the water is spilled. With one person in line, 10% of the water is lost. With two people, 19% is lost. With three people, 27% is lost; ... with seven people, half of the water is lost; ... with 22 people in line, 90% of the water is lost.

30. It should be noted that, strictly speaking, the term charge transfer implies transfer between cells. In the powered bus and floating bus topologies, charge is transferred to or from a bus instead.

- The DC-DC converters are unidirectional, from the cells to the bus
- The bus is powered by the most charged cells (or all the cells if the string is balanced)
- The bus powers some loads (e.g., the 12V bus in a vehicle)[31]
- If the load on the bus is too light, balancing slows down

2. Powered bus (Figure 4.42(f)):
   - The DC-DC converters are unidirectional, from the bus to the cells
   - The bus is powered externally (e.g., from a charger)
   - The bus charges the least charged cells
   - If most cells need some charge, the external power source may not provide enough power, so balancing slows down
3. Floating bus (Figure 4.42(g)):
   - The DC-DC converters are bidirectional
   - The bus is connected just to the DC-DC converters (no charger, no loads)
   - Works on the socialist principle: "From each according to his ability, to each according to his needs"[32]
   - At any time, the power from the most charged cells must be exactly the same as the power to the cells that need charging; otherwise, the bus voltage would drift up or down

The advantage of bus topologies (compared to the string topologies) is that the bus voltage is low and does not depend on the number of cells in series. A lower voltage allows the use of cheaper DC-DC converters. The same DC-DC converters can be used regardless of the number of cells in series (up to a maximum string voltage due to the isolation in the DC-DC converters). A disadvantage of the two unidirectional topologies is that the balancing speed is limited by the power of the load or charger connected to the bus.

4.7.3.4 Multilevel Balancing

For high-voltage batteries, rather than balancing the entire string within one level, it may be better to do it in two levels:

- The string is divided into banks; each bank is balanced on its own;
- The banks are then balanced relative to each other, using high-voltage DC-DC converters.

Doing so is advantageous in the first level:

- For the cell-to-cell topology, having fewer cells in each bank increases the efficiency;
- For the other topologies, low-voltage DC-DC converters are used, which are more efficient and cost less.

On the other hand, having a second level reduces the efficiency and increases complexity.

31. It could power the 12V bus even if the battery's main protector switch is off.
32. Popularized by Karl Marx in 1875.

Any of the seven balancing topologies may be used to balance the banks:

- *Bank-to-bank:* Energy is transferred between two adjacent banks in either direction (Figure 4.43(a));
- *String:* Energy is transferred between the string one of its banks:
  - Bank-to-string (Figure 4.43(b));
  - String-to-bank (Figure 4.43(c));
  - Bidirectional string (Figure 4.43(d)).
- *Bus:* Energy is transferred between a bus (that is isolated from the string) and a bank:
  - Loaded bus (Figure 4.43(e));
  - Powered bus (Figure 4.43(f));
  - Floating bus (Figure 4.43(g)).

Figure 4.44(a) shows bidirectional DC-DC converters used both within a bank and between banks. It is possible to implement a hybrid solution, one that uses bypass balancing for the first level and charge transfer balancing for the second level (Figure 4.44(b)). In this example, the second level uses the bank-to-bank (bucket

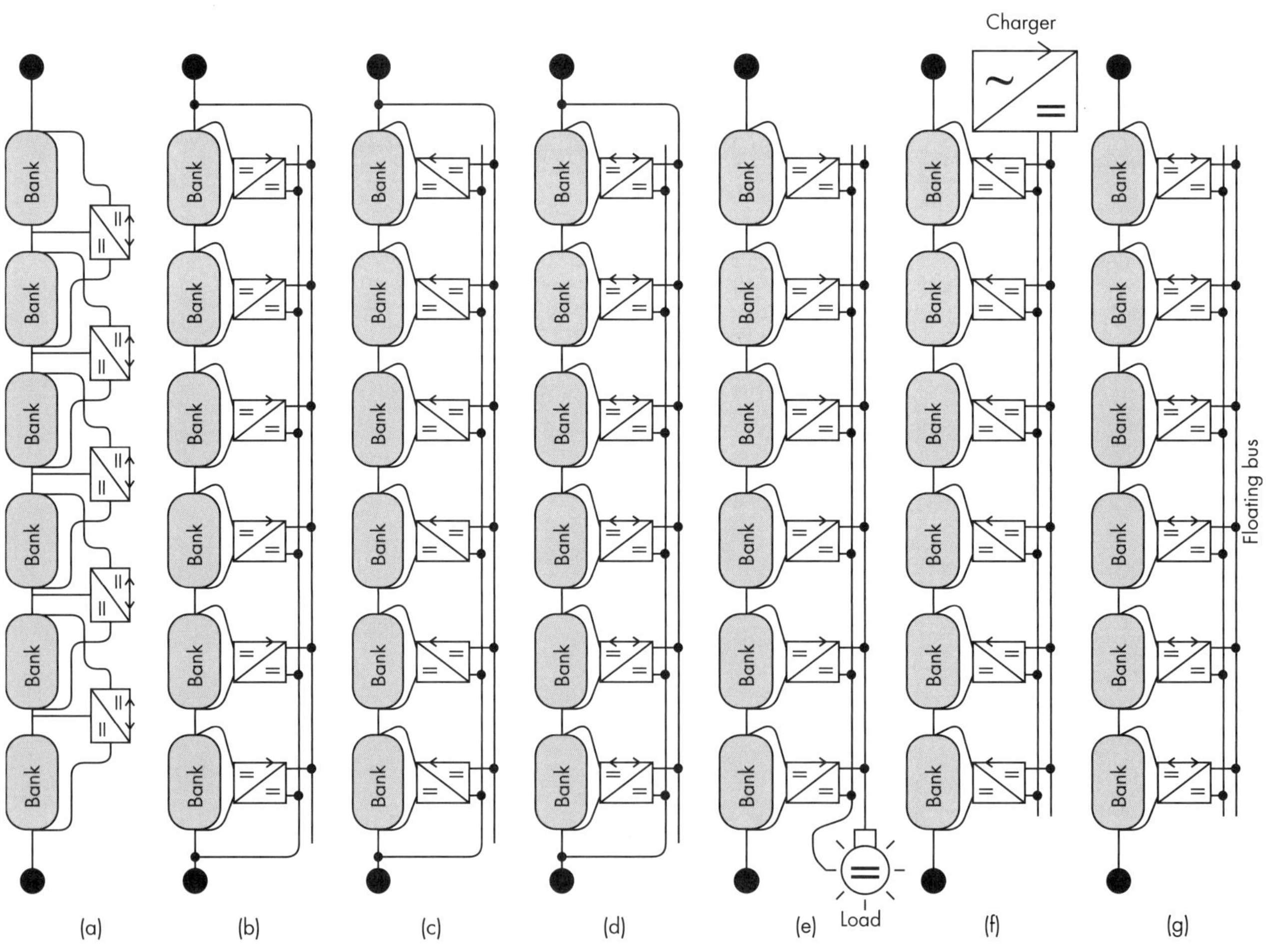

FIGURE 4.43 Bank balancing topology: (a) bank-to-bank, (b) bank-to-string, (c) string-to-bank, (d) bidirectional string, (e) loaded bus, (f) powered bus, and (g) floating bus.

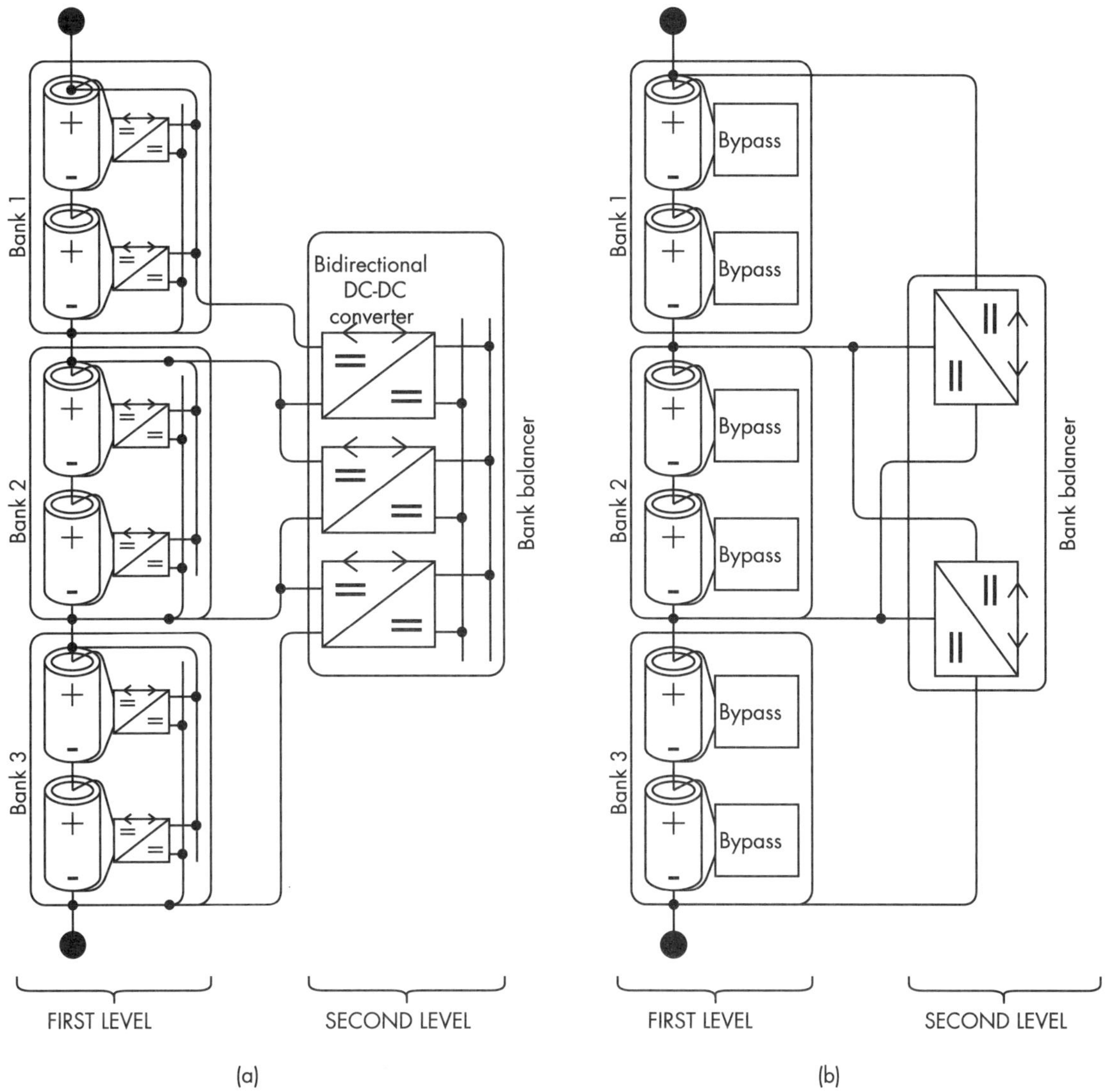

FIGURE 4.44 Two-level balance—cell level and bank level: (a) charge transfer both levels, and (b) hybrid—bypass and bank-to-bank charge transfer.

brigade) topology, though any charge transfer topology may be used. However, this solution is only helpful if the cells in a given bank are already balanced. In practice, this is unlikely. Therefore, this solution offers no advantages compared to plain bypass balancing.

4.7.4 Balancing Algorithms

Balancing algorithms[33] decide how and when to balance a string. Algorithms and topologies are unrelated: Any of the following algorithms may be used with any of the previously described topologies.

The BMS balances a string at the balance setpoint (see Section 3.2.6):

33. No, not named after Al Gore. And no, he never claimed to be the inventor of the internet.

- For an energy or power battery (see Section 5.1.4), the string should be top-balanced;
- For a buffer battery, it should be mid-balanced.

The two main algorithms for balancing are

- *Voltage-based:* Balances only when the battery is at the balance setpoint, based on the cell voltage readings;
- *SoC-based (predictive):* Can balance at any time, based on the estimated cell SoC levels.

Table 4.14 compares these two algorithms.

A few products use the simpler algorithm of always trying to match the voltages, regardless of voltage. These products are counterproductive because during discharge, they remove charge from the cells with the lowest resistance, not from the cells with the highest charge. They are also energy inefficient because they balance all the time, no matter what.

Balancing algorithms are only as good as the measurement of the cell voltage or the SoC evaluation. These situations are particularly challenging:

- Mid-balanced LFP cells because the OCV versus SoC curve is quite flat around 50% SoC;
- Top balance of NMC cells with a rather flat OCV around 100%, while in use;
- If the voltage measurement has low resolution.

If balancing within 3% SoC is desired, the accuracy of the cell voltage measurements can be critical; roughly speaking, the accuracy required is shown in Table 4.15 (see Section 3.2.7).

Only a BMS using state-of-the-art ICs[34] can reliably achieve a 14-bit accuracy.

Regardless of the algorithm, attempting to correct the imbalance too much is counterproductive because errors in the SoC estimation may lead the BMS to move

TABLE 4.14
Comparison between Voltage-Based and SoC-Based Balancing Algorithms

	Voltage-Based	SoC-Based (Predictive)
At the balance setpoint	Remove charge from cells with the highest voltage (or add charge to the cells with the lowest voltage)	Evaluate the SoC of each cell
Elsewhere	Do nothing	Remove charge from the cells with the highest SoC (or add charge to the cells with the lowest SoC)
Stop condition	When the cell voltages are close to each other, or if the cell voltages are too low	When enough charge has been moved, or if the cell voltages are too low
Balancing charge	Limited by the balancing current and the time at the balancing point	Limited only by the balancing current
Applications	Top balanced batteries, especially LFP, especially if usually fully charged	All batteries, any SoC

34. Such as the Linear Technology LTC6804.

TABLE 4.15 Resolution in the Cell Voltage Measurement Required for Balancing within 3% SoC

Balancing	Chemistry	Resolution	Bits
Top	**NMC**	10 mV	9
Mid	**NMC**	2 mV	12
Mid	**LFP**	0.5 mV	14

charge based on readings that are not as accurate as expected. Therefore, there is a lower limit to the delta SoC between cells, below which the BMS must stop balancing.

4.7.4.1 *Voltage-Based Top Balancing*

Voltage-based top balancing is the simpler and by far the most commonly used algorithm (Figure 4.45(a)). It works well for top-balanced batteries that at the end of charge dwell some time at the balance setpoint (which is common for energy and power batteries). This algorithm is based on the idea that bringing all the cells to the same voltage brings them to the same SoC.

When the battery is full, some charge is removed from the cells with the highest voltages. The voltage must be higher than a certain minimum (such as 3.4V for a LFP cell). Alternatively, in some charge transfer topologies, some charge may be added to the cells with the lowest voltage.

For LFP cells, the OCV versus SoC curve is steep only at the fully charged end, meaning that when the cell is nearly full, as the voltage increases rapidly, the cell is charged further. A change of 100 mV may indicate a change in SoC of about 1%~3%. If all the cell voltages are within 100 mV of each other, all the SoC levels should be within 1%~3%. This limits the usefulness of this algorithm to top-balanced LFP

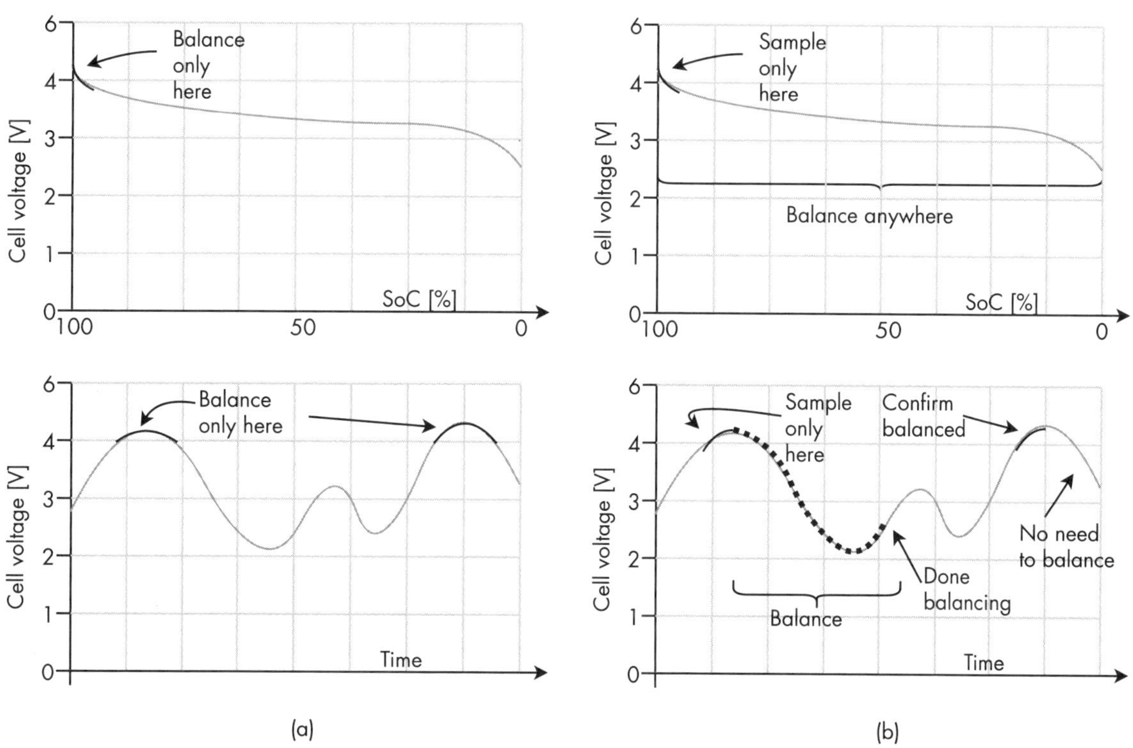

FIGURE 4.45 Balance algorithms: (a) voltage-based, and (b) SoC-based.

strings. This approach is either ineffective or counterproductive for mid-balanced strings, where the OCV versus SoC curve is flat.

For LCO and NMC cells, the OCV versus. SoC curve is not as sharp at the full end, making this algorithm a bit less effective. On the other hand, the curve is not as flat in the middle, allowing this algorithm to work for mid-balanced strings.

A problem with this algorithm is that, while charging, it is confused by the fact that each cell's terminal voltage depends on its internal resistance.[35] Even if charging is interrupted, the terminal voltage is higher than the OCV until the voltage relaxes, which takes a long time.

Normally, the charging current is much higher than the balancing current, so the BMS turns on the charger for short bursts, and then off for long periods. During those long periods when the charger is off, the cell voltage settles towards the OCV, allowing the BMS to sample each cell's voltage. The BMS uses this sample to determine which cells require balancing, with limited accuracy. In any case, the internal resistance varies from cell to cell, so basing balancing of cell voltage during high current balancing may actually increase the imbalance, as charge would be removed from the highest resistance cell rather than from the cell with the highest SoC. If the BMS knows the resistance of each cell, it can calculate its OCV by removing the IR drop from its terminal voltage.

Another problem with this algorithm is that the time available for balancing is limited to the time that starts when a cell voltage first goes above the balancing threshold and ends when charging ends. For example, an EV that is plugged in only part of the day may have two hours for charging, leaving only 10 minutes for balancing at the end of charge. A BMS with a much higher balancing current may be required to compensate for the short time available.

Hysteresis in the cell voltage is not much of a problem with this algorithm because all the cells see the same battery current and therefore experience similar hysteresis. We assume that hysteresis doesn't change much from cell to cell and that all the cells are somewhat close to the same SoC. However, in a series-first arrangement, the current is not the same in all the strings, so the hysteresis in one string is different from another string. This hinders the BMS as it tries to balance the cells in all the strings.

4.7.4.2 SoC-Based Balancing

The SoC-based balancing algorithm is based on an accurate estimate of the SoC of each cell, which allows the BMS to determine how much charge must be removed from individual cells. With some charge transfer topologies, some charge may be added to the cells with the lowest SoC.

This algorithm allows balancing to occur as long as the BMS is turned on. This maximizes the utilization of the balancing hardware. A 100 mA balancing current can balance 2.4 Ah a day instead of the 0.017 Ah that the same hardware can balance if voltage-based balancing is used for 10 minutes at the end of charge (Figure 4.45(b)).

Here is one implementation of this algorithm, assuming bypass balancing or charge transfer balancing that can only take charge from a cell (Figure 4.46).

- The BMS waits until the string SoC happens to be at the balance setpoint, and then remains there at no current, while the cell voltages relax to their respective OCV levels;

35. A cell's terminal voltage is higher than its OCV due to the cell's IR drop: the voltage drop across the cell's internal resistance times the charging current. If the internal resistance varies among cells, cells with higher cell resistance have a higher terminal voltage.

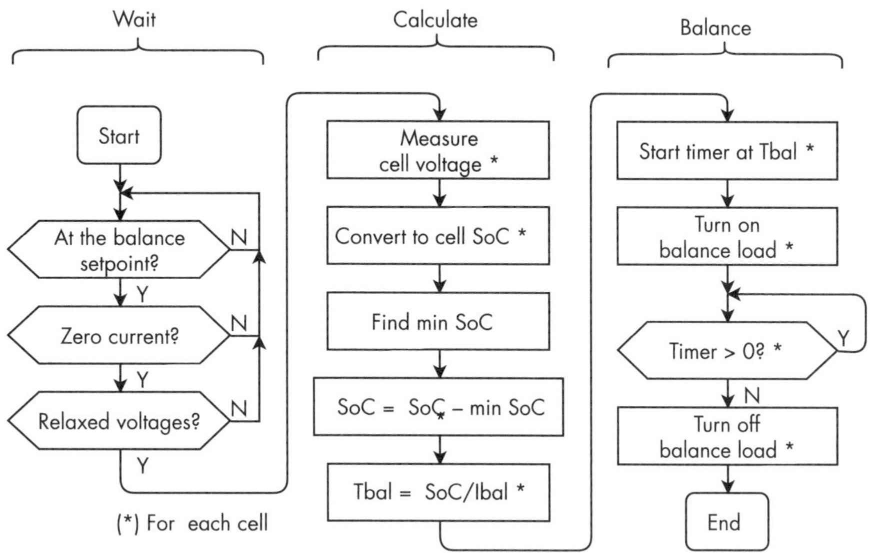

FIGURE 4.46
Flowchart for SoC-based balancing algorithm.

- The BMS calculates the balance time required by each cell:
  - It measures each cell voltage and converts this to its SoC.
  - It calculates how much extra charge each cell contains compared to the cell with the least SoC.
- Knowing the balancing current, the BMS calculates the balance time for each cell—how long it will take for the balancing current to remove the extra charge in a given cell.
- The BMS balances;
  - For each cell, it sets a timer to its balance time;
  - It turns on the balance load for each cell with excess charge;
  - It decrements the timers for each cell;
  - When a cell's timer expires, the BMS turns off the load across it.

More sophisticated versions of this algorithm can determine the extra charge in cells even if the string is not at the SoC point or even under the presence of battery current. Some can compensate for variations in the balancing current. If the BMS uses a charge transfer topology that can add energy to a cell, the algorithm is slightly different.

4.7.5 Charging during Top Balancing

Toward the end of charge, the most charged cell must receive no net charging current or it will be overcharged; balancing achieves this by removing from the cell just as much charge as the charger places in it.[36] Since the balancing current is normally much less than the charging current, this can be done in three ways:

- Providing a high balancing current, as high as the maximum charging current (Figure 4.47(a));
- Reducing the charger's current down to the same level as the balancing current (Figure 4.47(b));

---

36. Assuming one of the balancing technologies that can remove charge from the most charged cell.

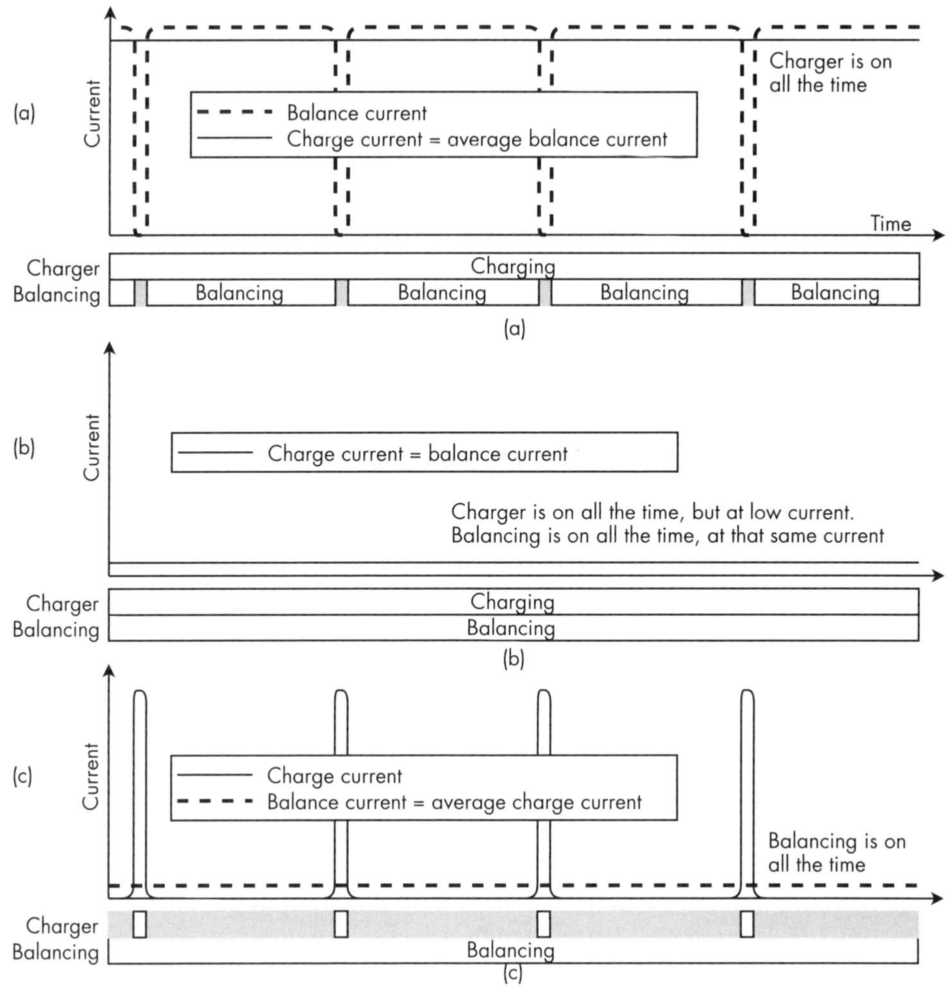

FIGURE 4.47
Avoiding shutdown:
(a) by increasing the balancing current, (b) by decreasing the charging current, and (c) by turning charging off and on.

- Turning the charger off and on; on the average, the charging current equals the balancing current (Figure 4.47(c)).

Regardless of the method, the balancing time is the same because it's set by the balancing current and the delta SoC between the most charged and least charged cells:

balance_time [h] = delta SoC [Ah] / balancing_current [A] (4.3)

Let's look next at each approach.

4.7.5.1 High Balancing Current

The most obvious solution is to use a BMS with a balancing current that is as high as the maximum charging current. When the BMS turns on the load across a cell, all the charging current goes through the load instead of the cell (Figure 4.48). This operates as follows:

1. Initially, the charger is in the CC mode. The charging current flows through all the cells; the voltage of the most charged cell (#1) rises the fastest.

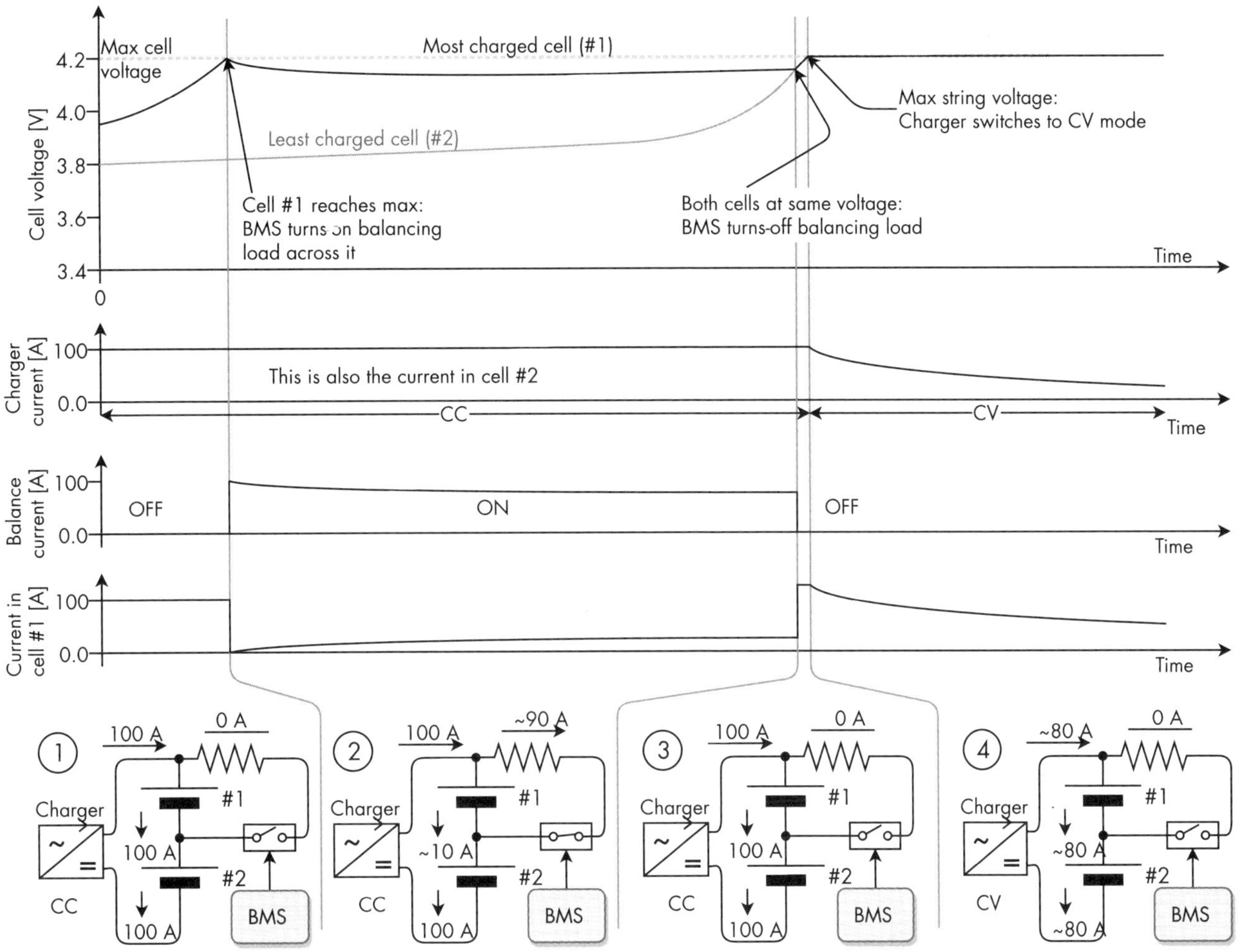

FIGURE 4.48 Balancing at full current. The numbers correspond to the bullet points in the text.

2. When the voltage cell #1 reaches the maximum cell voltage, the BMS turns on the balancing load across it. The charger current flows through the load instead of through cell #1, and it keeps on flowing through the other cell (#2). Note that the string voltage is still below the CV setting of the charger, so the charger remains in CC mode. After a while, the voltage of cell #1 relaxes; consequently, less current flows through the load, and some current starts flowing through cell #1 again.
3. When the voltage of cell #2 reaches the voltage of cell #1, the BMS stops balancing; the charging current goes back to flowing through both cells.
4. When both cells reach the maximum cell voltage, the total string voltage reaches the CV setting of the charger and the charger switches to the CV mode. The charging current starts to decay naturally; this continues as charging is completed, until the current drops to below 0.1C and the charger turns off.

The advantage of this solution is that it results in the fastest possible balancing time. Of course, this is moot with maintenance balancing (see Section 3.2.9.2), as it doesn't require much time. The problem with this solution is that the high current through the balance load produces considerable heat. Charge transfer balancing can minimize this problem.

Finding a BMS that implements this would be challenging; I do not know of any off-the-shelf BMS that can balance by bypassing such a high current.

4.7.5.2 *Reduced Charger Current*

If the charger provides a way for the BMS to reduce its current, then a BMS with a low balancing current will work.

The operation is the same as before, except in step 2, when the BMS tells the charger to drop the current down to the same as the balancing current (Figure 4.49).

The advantage of this solution is that the BMS can use a reasonable level of balancing current. The disadvantages are that you need a charger that can be remotely controlled to adjust its current to a significantly lower level, and that balancing takes longer.

4.7.5.3 *Turn Charger Off and On*

If the previous solutions are not possible, the only remaining solution is for the BMS to cycle the charger off and on, so that the average charging current matches the balancing current.

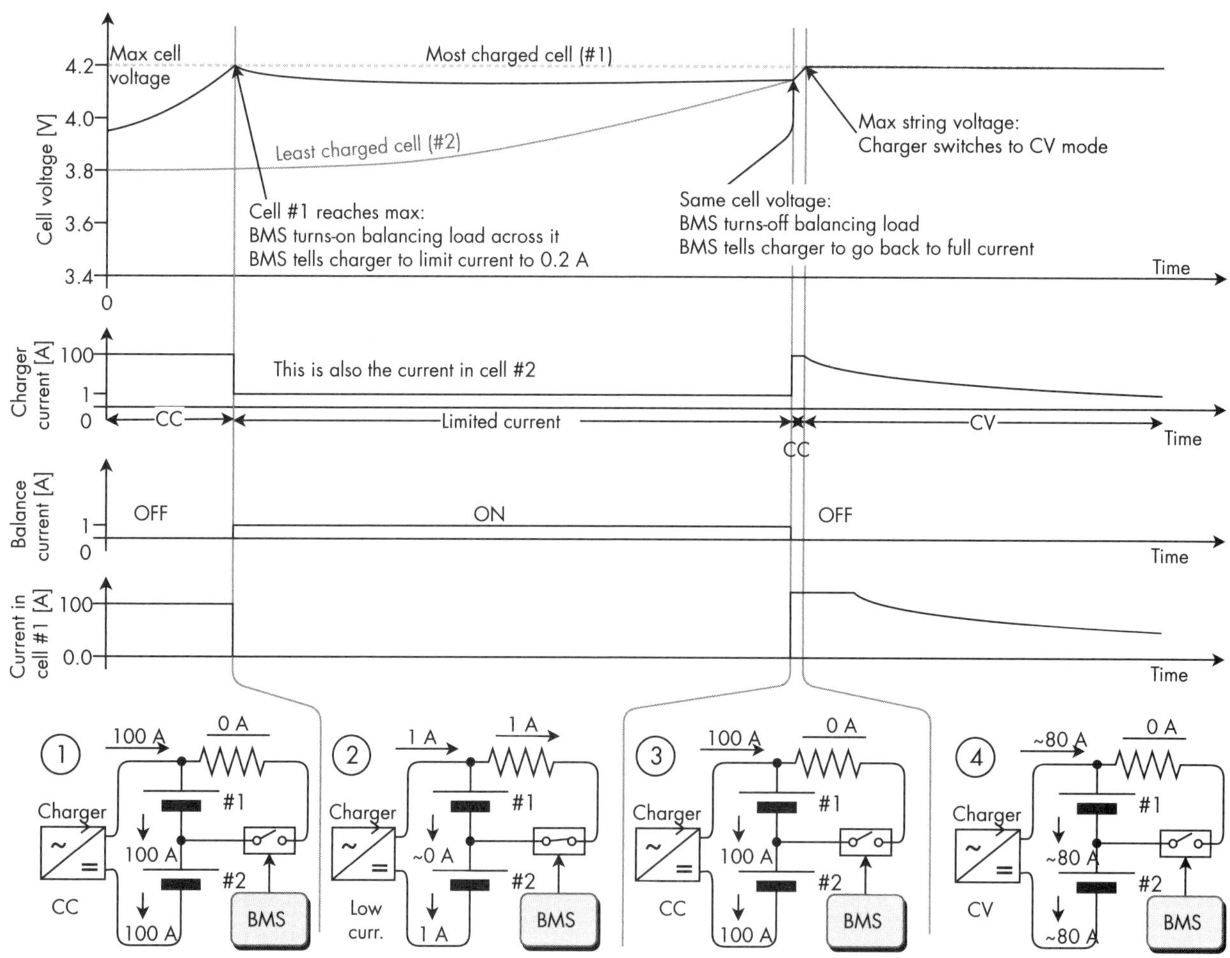

FIGURE 4.49 Balancing with reduced charger current.

Figure 4.50 shows the balancing process for a two-cell string.

1. While the charger is the CC mode, both cell voltages rise.
2. When the voltage of the most charged cell (#1) reaches the maximum cell voltage, the BMS turns off the charger. While the charger is off, the voltage of cell #1 drops due to the balancing current; the voltage of the other cell (#2) also drops, but just a bit, due to relaxation.
3. When the voltage of cell #1 drops to the "High" threshold, the BMS turns the charger back on. The entire charging current flows into cell #2, but cell #1 sees less current because its load is on. The cycle repeats; during each on-off cycle, the voltage of cell #2 climbs a bit, until it reaches the voltage of cell #1, and at this point, the string is balanced.

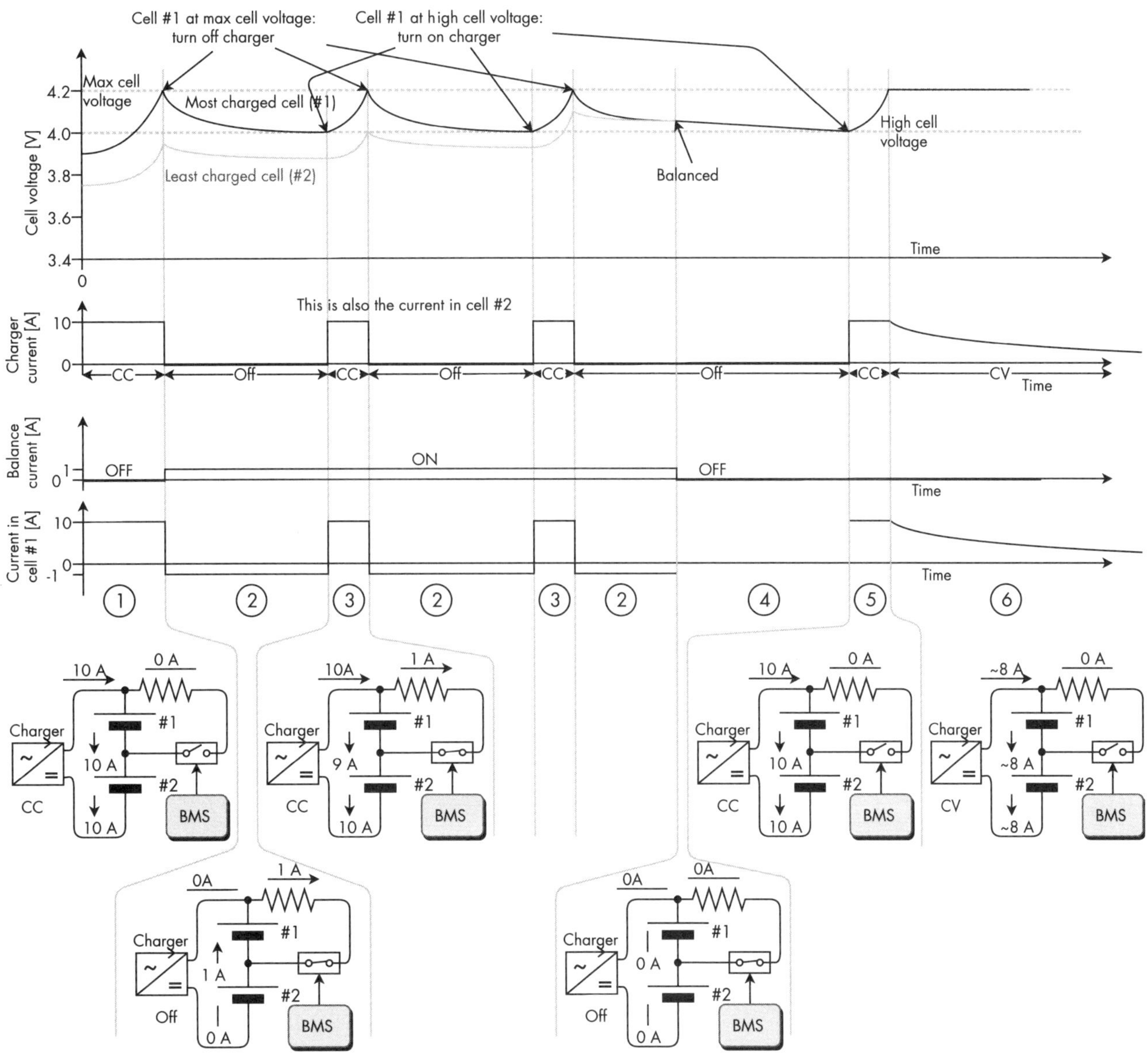

FIGURE 4.50 Balancing by cycling the charger off and on. The numbers correspond to the bullet points in the text.

4. When the string is balanced, the BMS stops balancing, and both cell voltages relax down.
5. When a cell voltage drops to the "High" threshold, the BMS turns the charger back on. The entire charging current flows through both cells, and the cell voltages increase together.
6. When both cells reach the maximum cell voltage, the total string voltage reaches the CV setting of the charger and the charger switches to the CV mode. The charging current starts to decay; this continues as charging is completed, until the current drops to below 0.1C, and the charger shuts down.

Note that the BMS doesn't calculate how long the charger should be on. Instead, the duty cycle is a natural result of the excursion in the voltage of the most charged cell. This duty cycle is such that the average charger current matches the balancing current precisely. Increasing the difference between the "Max cell voltage" and "High cell voltage" settings in the BMS result in increasing the period of the on-off cycle, but doesn't affect its duty cycle.

4.7.6 Generated Heat

Balancing generates heat in the BMS, on the order of 1 W per cell. This is true not just for bypass balancing, but also for charge transfer balancing because it is not 100% efficient. For a large battery, this heat could be a concern because as a battery gets larger, more heat is generated during balancing. In most types of BMSs, this is not a problem because the higher heat is spread out over a larger area (more slaves, more cell boards, or more bank boards), and therefore the temperature does not increase. However, a centralized BMS doesn't scale well because all the heat is dissipated within a single box. As the battery voltage increases, more heat is dissipated within its enclosure, increasing its temperature. A larger BMS enclosure or better cooling may be required.

4.7.6.1 Balancing to Heat Cells

Since bypass balancing generates heat, a BMS may try to use it to heat a battery. If attempting to charge below freezing, the BMS postpones charging and turns on all the balancing loads. Once the battery temperature reaches 0°C, charging can start.

In theory.

In reality, the heat generated by balancing, which is on the order of 1 W per cell, is not enough to overcome the heat simultaneously lost into a cold ambient. It may work for a small battery that uses insulation. Other than that, it's not likely to work. The solution is to use a heating pad (see Section 5.17).

4.7.7 Redistribution

Some of the capacity in a real-world battery remains unused because the capacities of its cells vary. At best, balancing ensures that a single cell limits both charging and discharging. The extra capacity in the other cells remains unused.

A different technique, redistribution,[37] allows the full use of the capacity in each cell. The battery capacity is not limited by any one cell, and the charge in every cell is accessible and used.

37. Some people use the term active balancing for the redistribution technology, which is one reason why we now use the term charge transfer balancing instead of active balancing for the balancing technology.

Redistribution shuttles energy in the battery continuously to keep all the cells always at the same SoC. While charging, redistribution gives additional energy to the cells with the highest capacity or takes additional energy from the cells with the lowest capacity. Similarly, while discharging, redistribution takes additional energy from the cells with the highest capacity or gives additional energy to the cells with the lowest capacity. At any given time, all the cells are at the same SoC level as the string. During discharge, all the cells start at 100% SoC; at the end of discharge, they all end at 0% SoC.

Redistribution uses the same hardware as charge transfer balancing, except that the DC-DC converters are more powerful. The redistribution algorithms are more complex than for balancing.

For example, consider a top-balanced, two-cell string. The capacity of cell #1 is 20% above nominal, while the capacity of cell #2 is 20% below nominal. Then

- Without redistribution (Figure 4.51(a)), after 40 minutes, cell #2 is empty and therefore the battery is empty. There is still charge left in cell #1, but it's inaccessible.
- With cell-to-string redistribution (Figure 4.51(b)), a DC-DC converter takes extra energy from cell #1 to make up for the low capacity of cell #2, so that they will both discharge fully. After 60 minutes, both cells are empty.
- With string-to-cell redistribution (Figure 4.51(c)), a DC-DC converter takes extra energy from the string, charging cell #2 to make up for its low capacity, so that they will both discharge fully. After 60 minutes, both cells are empty.
- If using bidirectional converters, some take extra energy from the high-capacity cells, while others send extra energy to the low-capacity cells.

Without redistribution, the battery capacity is 80 Ah because it's limited by cell #2. With redistribution, the battery capacity is 100 Ah because both cells limit the capacity equally.

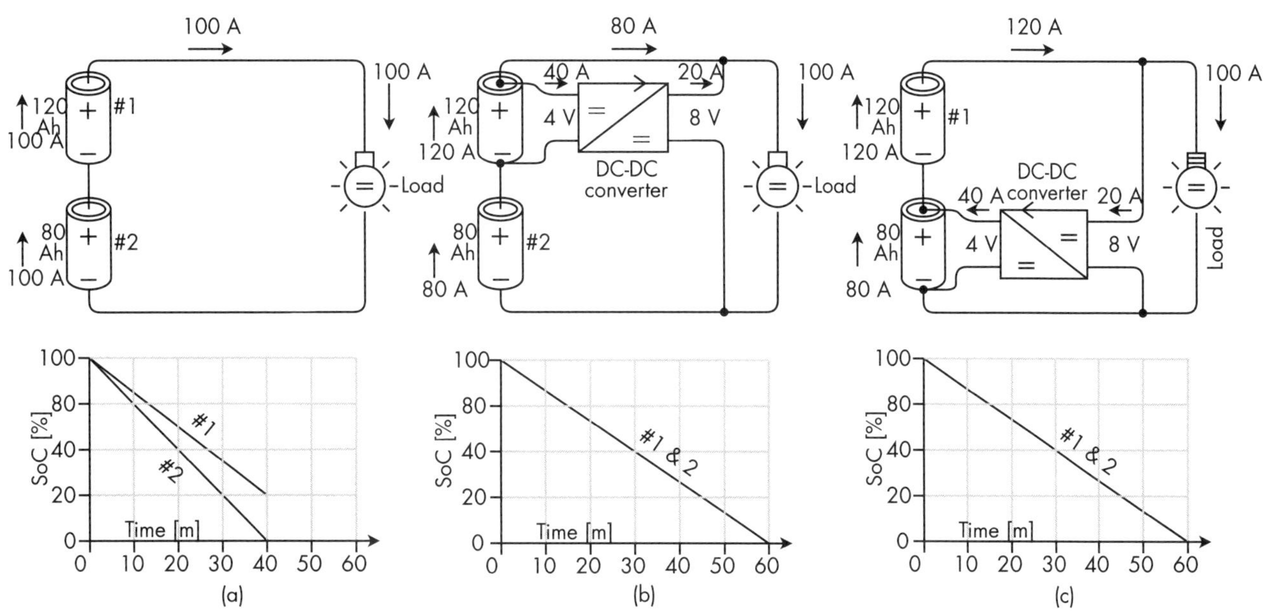

FIGURE 4.51 Redistribution: (a) no redistribution, (b) cell-to-string redistribution, and (c) string-to-cell redistribution.

Table 4.16 compares top balancing and redistribution. Today, no available off-the-shelf BMS offers redistribution.

4.7.7.1 Converter Power

Redistribution requires high-power DC-DC converters. Roughly speaking, the power required for each converter is

$$\text{DCDC\_power}[\text{W}] = \text{average\_load\_power}[\text{W}] \times \text{cell\_capacity\_delta} / \text{number\_of\_cells\_in\_series} \quad (4.4)$$

For example, for a 10 kW load, +0%/−10% variation in cell capacity, and 100 cells in series:

$$\text{DCDC\_power}[\text{W}] = 10 \text{ kW} \times 10\% / 100 = 10 \text{ W} \quad (4.5)$$

You may be surprised to see that 10 W DC-DC converters can power a 10 kW load until you note that

- There are 100 converters. At most 99 of them can be on,[38] so, in total, they may convert up to 990 W.
- Converters can work throughout the discharge period to transfer only 10% of the charge, giving an additional 10:1 advantage.

Therefore, the total power from the DC-DC converters can be as high as 10 × 99 × 10 W = 9.9 kW.

The BMS controls each converter. The converters for the least capacity cells (or for the most capacity cells, depending on the topology) remain powered for the entire discharge period. The other ones operate at a lower current (if they allow it) or are turned on and off with a duty cycle that reduces their average current.

4.7.7.2 Redistribution Pros and Cons

At first brush, redistribution may seem advantageous. Careful consideration may show otherwise.

Table 4.16
Comparison between Top Balancing and Redistribution

	Top Balancing	**Redistribution**
Charge Utilization	Typically 90%~100%	Almost 100%
String Capacity	Equal to minimum cell capacity	Equal to the average cell capacity
String SoC	Equal to SoC of least capacity cell	Equal to the SoC of all the cells
Technology	Bypass or charge transfer	Charge transfer
Operating Time	Stops once the string is balanced	Operates whenever the battery is in use
DC-DC Current (Cell Side)	Low: 10 mA to 3 A	High: 10 to 100 A

38. At least one converter remains off because its cell doesn't require help.

If the only goal is to maximize the use of a battery, redistribution is probably not worthwhile. A careful analysis[39] may show that redistribution is more expensive than simply using a larger battery. My analysis shows that if the discharge time is less than 20 minutes, or if the variation in capacity is less than 20%, a larger battery is cheaper than implementing redistribution [3]. Since a larger battery improves the lifetime of the battery (see Section 2.5.2), this solution is better than redistribution.

Redistribution does have a benefit: it prolongs the life of the battery by supporting the weakest cells [4].

In the extreme case, redistribution may save the day if a cell becomes incapacitated. The DC-DC converter of the failed cell takes over, and the battery continues operating. This requires a powerful DC-DC converter for each cell, one capable of handling the total battery current.[40] The tremendous increase in the complexity, cost, and size of the battery makes this solution prohibitive.

4.7.7.3 Multilevel Redistribution

Typically, redistribution uses one DC-DC converter for each cell (Figure 4.52(a)).

It is possible to divide the battery into banks and use one DC-DC converter for each bank (Figure 4.52(b)). This topology uses fewer DC-DC converters, but each converter must handle more power. For example, if the battery is divided into banks of 10 cells each, one-tenth as many converters are used, but each converter must handle ten times as much power.

The disadvantage of a bank-level redistribution topology is that it doesn't use all the charge in all the cells. That is because it doesn't compensate for an individual cell with low capacity; it only compensates for a bank with low capacity. As a rule of thumb, it only compensates for about 30% of the cell-to-cell variation in capacitance. Compare this to the standard redistribution topology, which compensates for 100% of the cell-to-cell variation in capacitance.

A two-level topology solves this limitation (Figure 4.52(c)) (see Section 4.7.3.4). Low-voltage converters can be used at the slave level, and only a few converters are required at the battery level. Each level can use a different type of converter. However, a two-level topology uses more DC-DC converters than a single-level topology.

4.8 EVALUATION

Optionally, the BMS may evaluate the state of the battery, primarily so it may report it, though the BMS may use these evaluations internally to affect its operation, such as dynamically reducing the operating area as the SoC goes down.

As a user of a BMS, you probably have little to do with the actual methods used to evaluate the state, but knowing a bit about them helps you work within their limitations.

4.8.1 State of Charge Evaluation

The user wants to know the SoC of the battery accurately, especially the driver of an electric vehicle.[41] However, SoC cannot be measured directly, and evaluating it accurately is challenging.

39. This includes a statistical analysis of cell capacity distribution, consideration of the losses in DC-DC converters, and a cost analysis of the full system.
40. Or a single converter plus a bunch of contactors (2 × N, where N is the number of cells in series) to connect the converter to whichever cell needs it. This may actually be more expensive than N converters.
41. Car drivers are accustomed to the inaccuracy and nonlinearity of the fuel gauge in gas-fueled vehicles yet don't accept errors in the SoC evaluation in an EV, probably because of range anxiety.

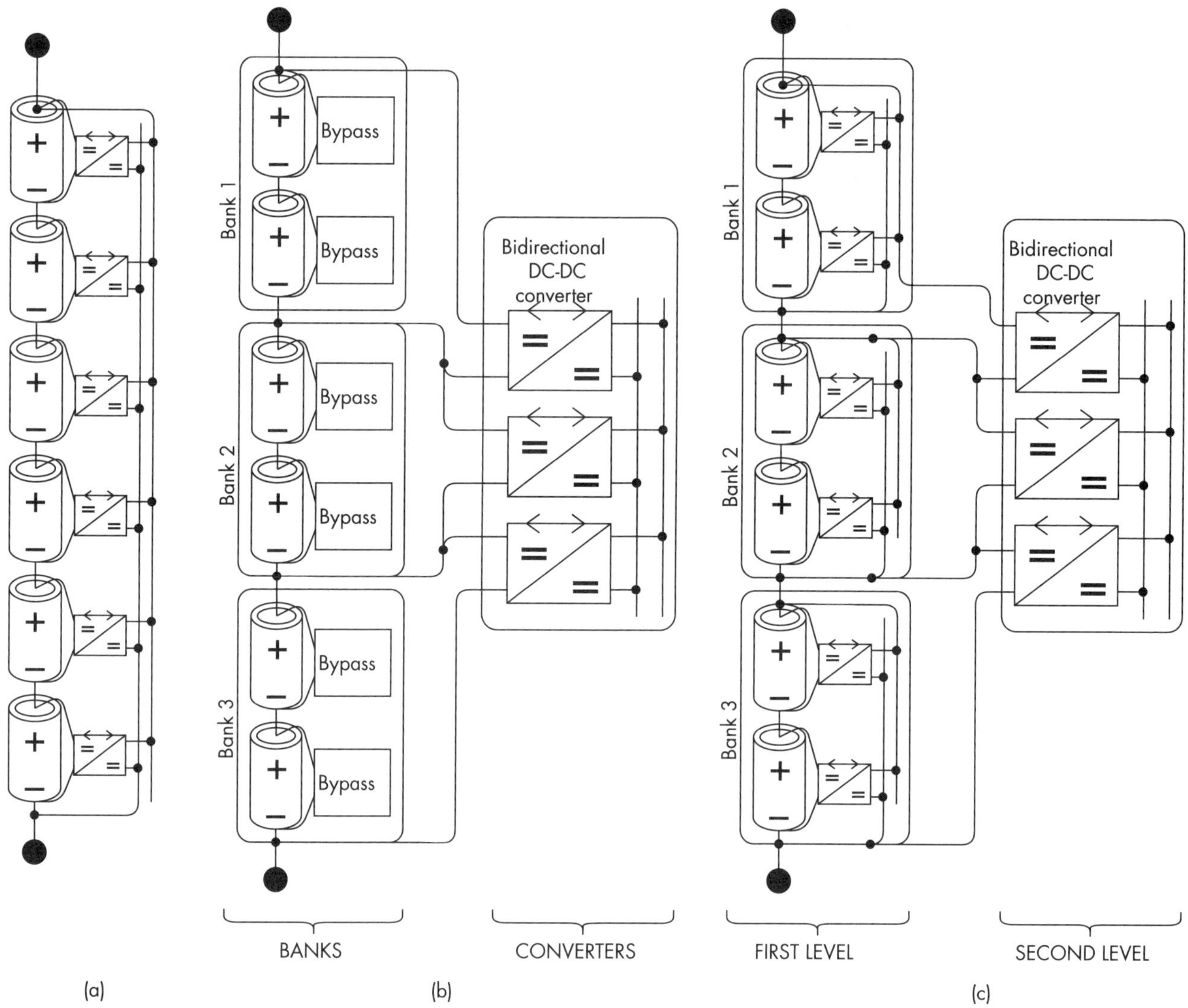

FIGURE 4.52 Redistribution topologies: (a) standard, (b) bank-level, and (c) two-level.

4.8.1.1 *SoC Evaluation Accuracy*

Any BMS that specifies a good SoC accuracy (e.g., 1%) must also specify under what conditions that accuracy holds because there are always some conditions under which the BMS has only a vague idea of the SoC. For example, with an LFP battery that has operated for weeks around 50% SoC, the actual value cannot be known any more accurately than ±10%, no matter how good the BMS is. This is because

- The voltage of LFP cells is a poor indicator of SoC around 50% SoC;
- Errors in the current sensing, integrated over weeks, will make the evaluated SoC drift far more than 10%.

Also:

- Battery SoC is indirectly related to cell SoC, and few BMSs know the SoC levels of each cell.

- Cell SoC is poorly defined because the point at which a cell is considered full or empty is really up to the cell manufacturer first, and to the battery designer second; what I call 96%, you may call 93 % (see Section 2.4.3).

When selecting a BMS, rather than requiring a certain SoC accuracy (which forces the BMS company to lie for the sake of a sale), you should work with the BMS manufacturer to establish reasonable expectations of SoC evaluation reliability, in your particular application, under various operating conditions.

4.8.1.2 Coulomb Counting and Voltage Translation

The most common way to evaluate the SoC of a Li-ion battery is by combining two techniques, each of which would be quite inaccurate on its own:

- Coulomb counting;
- Voltage translation.

Coulomb counting integrates the battery current to get the change in SoC relative to what it was when the integration first started (Figure 4.53a)). This method has many limitations:

- The value is relative to the initial SoC, which is unknown;
- The unavoidable offset in the current sensor results in a slow drift in the evaluated SoC;
- The effective capacity (used to convert from charge [Ah] to SoC [%]) depends on how the battery is used and changes as the cells degrade and as imbalance increases;
- Nonlinearity in some current sensors results in slight errors.

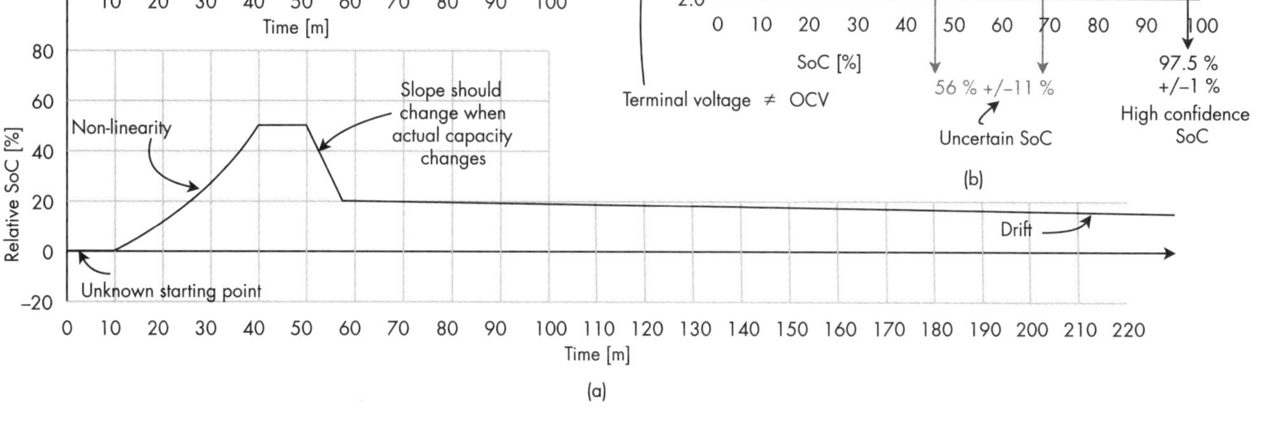

FIGURE 4.53 Inaccuracies in SoC evaluation: (a) coulomb counting, and (b) voltage translation.

Voltage translation converts a cell OCV to the corresponding absolute SoC, using the OCV versus SoC curve for the cells (Figure 4.53(b)). This method also has many limitations:

- OCV cannot be measured directly; terminal voltage is affected by the current into the battery and by its previous history;
- The OCV versus SoC curve is somewhat flat, especially for LFP at mid-SoC levels, meaning that a small error in voltage results in a significant error in SoC;
- Various SoC levels may produce the same cell voltage due to hysteresis (see Section 2.4.2.3).

Combining the two methods may reduce the limitations of each technique.

For example, for an energy battery (Figure 4.54):

- Using voltage translation after the end of charge (when the current is zero, the cell voltage is relaxed and the cell voltage changes rapidly with SoC);
- Using coulomb counting elsewhere (because the initial value has been established with some accuracy, and the time since the next full charge is less than one day).

This method is useful but somewhat inaccurate.

To visualize the effect of combining of these two techniques, let us use the analogy of evaluating the volume of water in a cave, knowing only the water level and the flow of water in and out of the cave's entrance (Figure 4.55):

- Integrating the flow of water is analogous to coulomb counting and tells us the relative amount of water added to or removed from the cave. However, it does not tell us the absolute quantity of water inside the cave. Additionally, errors in the flow measurement result in a drift over time between the estimate and the actual value.
- Measuring the water level is analogous to voltage translation and, as such, it is a poor indicator of the quantity of water because of the irregular shape of the cave. However, when the cave is dry, we know it's empty.

By combining these two techniques, we can calibrate the evaluation when we see that the cave is dry, and then integrate the flow of water into the cave to estimate its quantity. Over time, this estimate diverges from the actual quantity because of errors

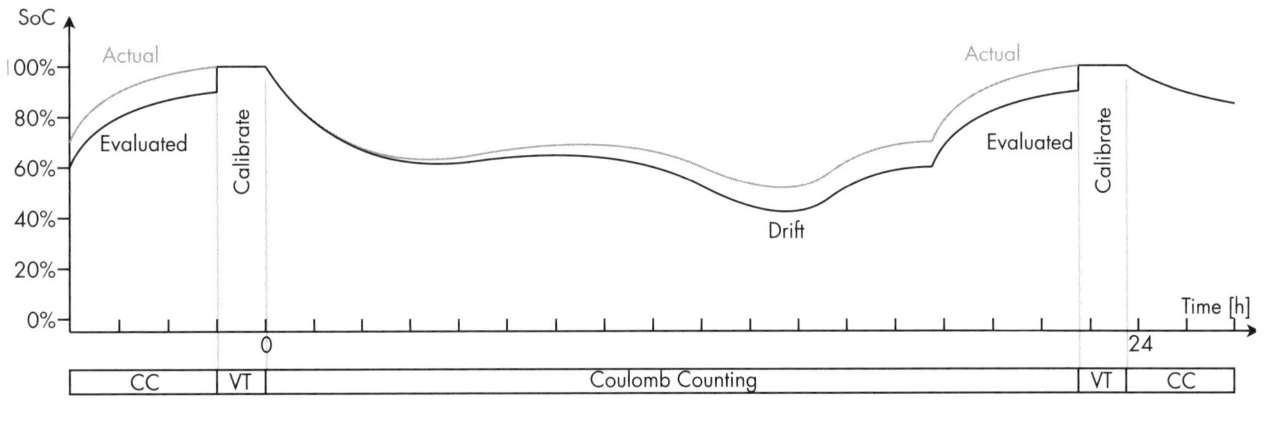

FIGURE 4.54 SoC evaluation: combining voltage translation and coulomb counting.

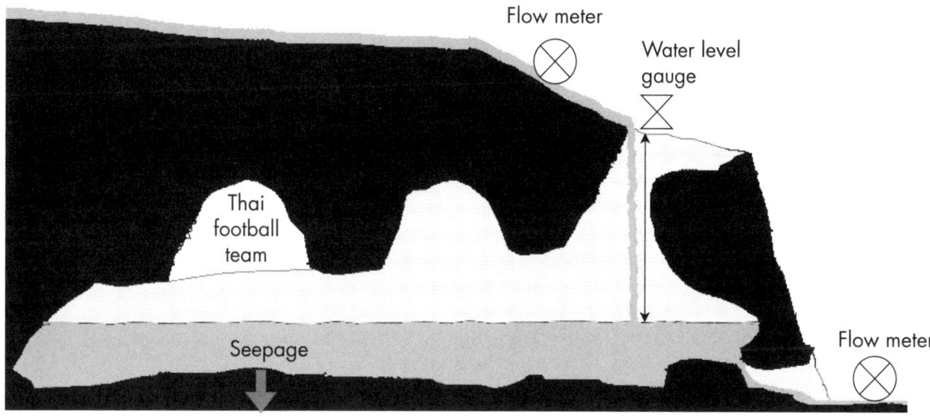

FIGURE 4.55 Cave analogy to SoC evaluation.

in the flow measurement (which is analogous to errors in the current sensor), and because some water seeps out through the walls of the cave (which is analogous to self-discharge current).

4.8.1.3 Advanced SoC Evaluation Methods

Far more sophisticated methods are used to evaluate SoC. These methods require an accurate model of the cells, which an off-the-shelf BMS doesn't have, at least not out of the box. Either the model must be configured into the BMS, or the BMS must learn about the cells on its own.

One such method is Kalman filtering [5]. Roughly speaking, this technique employs an accurate model for the cell and measurements of the cell voltage and current; it then applies the readings to its model, to calculate the expected cell voltage. It compares this result to the actual voltage, and it uses the error to correct its assumptions and therefore its estimate of the cell's SoC (Figure 4.56(a)).

Another such method is Impedance Track™, developed by Texas Instruments [6]. Roughly speaking, Impedance Track profiles the internal impedance of a cell over SoC. The BMS measures the impedance of a cell and uses the profile to derive its SoC. As the cell ages, the BMS updates this profile, to maintain accuracy (Figure 4.56(b)). This method is effective in products that place the BMS in the product rather than in the battery. If the user replaces the battery, the BMS can estimate the SoC of the new battery quickly.

Many other methods exist. Scientists in a lab can evaluate a pouch cell's SoC level from its thickness or a cylindrical cell from its internal pressure. They can do so from the shape of the curve in a Nyquist plot (see Volume 2, Section A.3.5).

A BMS that is designed for a specific application may use a method that is particularly adept at evaluating the SoC under the peculiarities of its operating conditions. For example, A BMS may be designed for prolonged operation at mid-SoC levels,[42] or operation as either a power battery or an energy battery (see Section 5.1.4), or managing a PHEV[43] battery. An off-the-shelf BMS does not have this luxury, and therefore must be flexible enough to be able to handle a variety of conditions.

42. Such a BMS may use a deadband to ignore any current readings below a specific low level; it assumes that any reading within the deadband is just due to offset from the current sensor.
43. Plug-in hybrid electric vehicle.

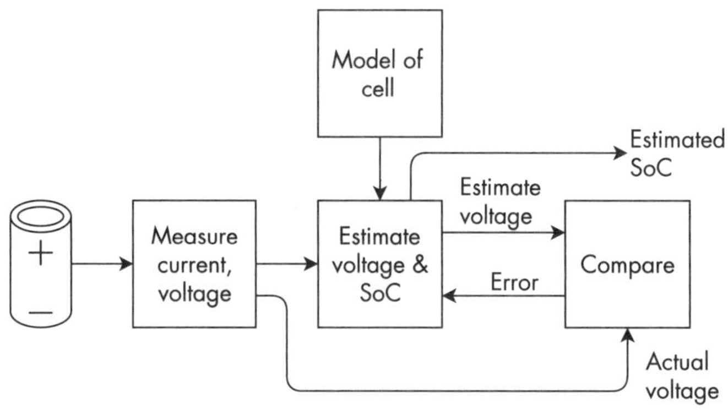

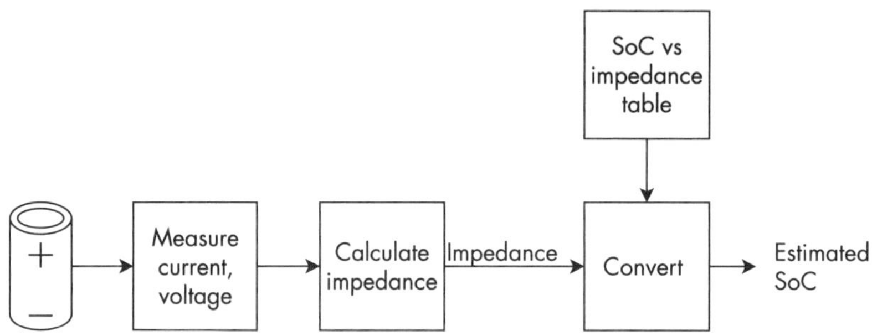

FIGURE 4.56
Other SoC evaluation methods, rough block diagrams: (a) Kalman filtering, and (b) Impedance Track.

Consequently, its SoC evaluation is not as accurate as for a BMS designed for a specific application.

Regardless of the method, be aware that SoC evaluation is challenging under these conditions:

- At zero or minimal current, as the evaluation drifts due to the offset in the current sensor;
- During high-current pulses due to nonlinearities in the sensor;
- If the battery is never or rarely fully charged or fully discharged;
- If the operating conditions change significantly.

4.8.2 Effective Capacity Evaluation

The effective capacity (see Section 1.4.2.3) of a battery varies over time and with load current (see Section 2.5.3 and Section 3.2.5).

If the battery is cycled from full to when the battery shuts off, the BMS can evaluate the effective capacity by integrating the current, which is the charge that the battery released during this discharge cycle. That is the effective capacity (Figure 4.57).

If a full discharge cycle is not available, the effective capacity can be evaluated indirectly by extrapolating two widely spaced points in the battery's operation:

- The SoC at the most charged and least charged points;
- The charge transferred into or out of the battery between those points.

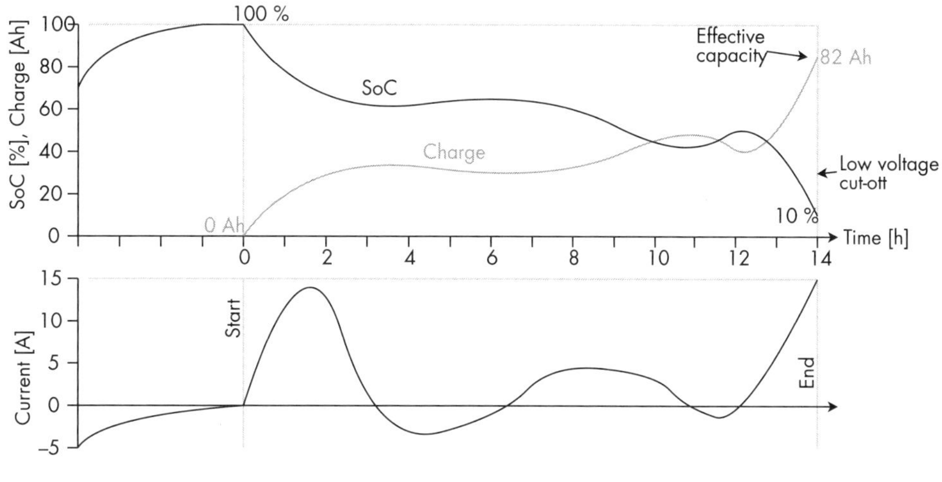

FIGURE 4.57 Effective capacity evaluation.

The effective capacity can be evaluated as

$$\text{Effective\_capacity [Ah]} = \text{Delta\_charge [Ah]} \times 100\ [\%]/\text{Delta\_SoC}\ [\%] \tag{4.6}$$

In this example, the battery delivers 40 Ah during a partial cycle, during which time the SoC drops from 75% to 25% (Figure 4.58(a)). We can use the equation to extrapolate the capacity:

$$\text{Effective\_capacity [Ah]} = 40\ [\text{Ah}] \times 100\ /(75\% - 25\%) = 80\ \text{Ah} \tag{4.7}$$

Alternatively, we can extrapolate the capacity graphically (Figure 4.58(b)) using a graph with the relative charge on the vertical axis and SoC on the horizontal axis. When we plot the charge versus Soc, the curve is a mostly straight line that retraces

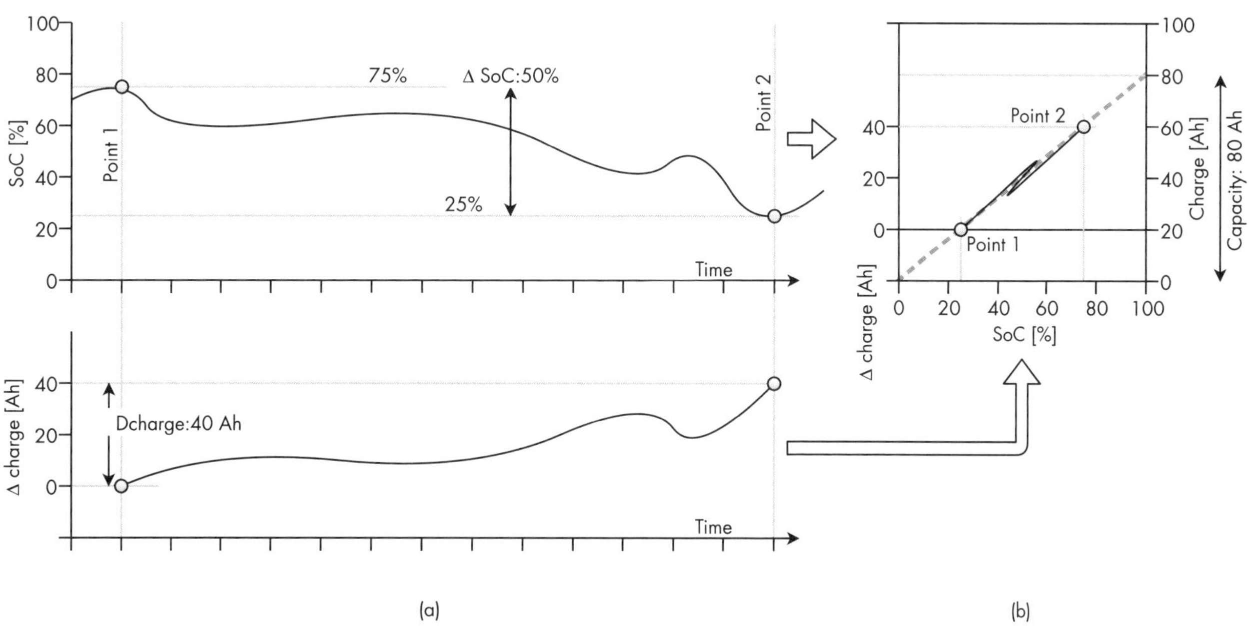

FIGURE 4.58 Capacity extrapolation: (a) calculated, and (b) graphically.

itself as the battery is recharged and discharged. When done, we extrapolate this curve by extending the line to the 0% and 100% SoC points. The vertical extent of the extrapolated line spans the total charge in the battery; that is, the battery capacity.

This extrapolated capacity evaluation is inaccurate because the SoC evaluation is less accurate at mid-SoC levels and because it does not consider the cell resistance.[44]

4.8.3 OCV Evaluation

Assuming that the BMS knows the internal series resistance of a cell, it can try to evaluate its OCV using the simplified model with just a voltage source and a resistor (Figure 4.59(a)) (see Section 2.3.4) as

$$\text{OCV [V]} = \text{Terminal\_voltage [V]} - \text{Current [A]} \times \text{Cell\_resistance } [\Omega] \quad (4.8)$$

During discharge, the terminal voltage is lower than the OCV of the cell, due to the *IR drop* (that is, the voltage across its internal resistance, which is proportional to the resistance and the current).

This evaluated OCV won't be accurate in the short term because the simple model neglects the dynamic effects inside the cell. If the BMS knows the value of each component in the cell impedance, then it can evaluate the OCV of the cell more accurately, using a more complex model (Figure 4.59(b)).

4.8.4 Resistance Evaluation

To evaluate the OCV of a cell, a BMS requires knowledge of the cell's resistance. Also, it may use the battery resistance to evaluate the state of health.

The internal resistance is derived from the voltage and current at two points in time:

$$\text{Cell\_resistance } [\Omega] = (\text{Voltage1} - \text{Voltage2 [V]})/(\text{Current\_1} - \text{Current\_2 [A]}) \quad (4.9)$$

The current and the voltage must vary over time for this method to work (Figure 4.60(a)) because if the current is constant, this equation divides by 0.

If we graph the current versus voltage over time (Figure 4.60(b)), the path forms a wiggly curve. While this path jumps around, it has a distinguishable slope, which is due to the resistance. As the discharge current increases, the voltage decreases. If we

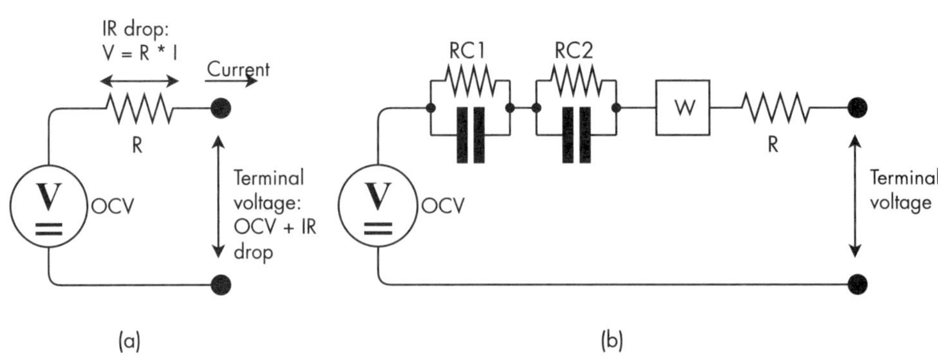

FIGURE 4.59
Cell OCV evaluation models: (a) resistance only, and (b) full impedance.

44. Cell resistance determines the point when the battery is shut down at the end of a high-current discharge.

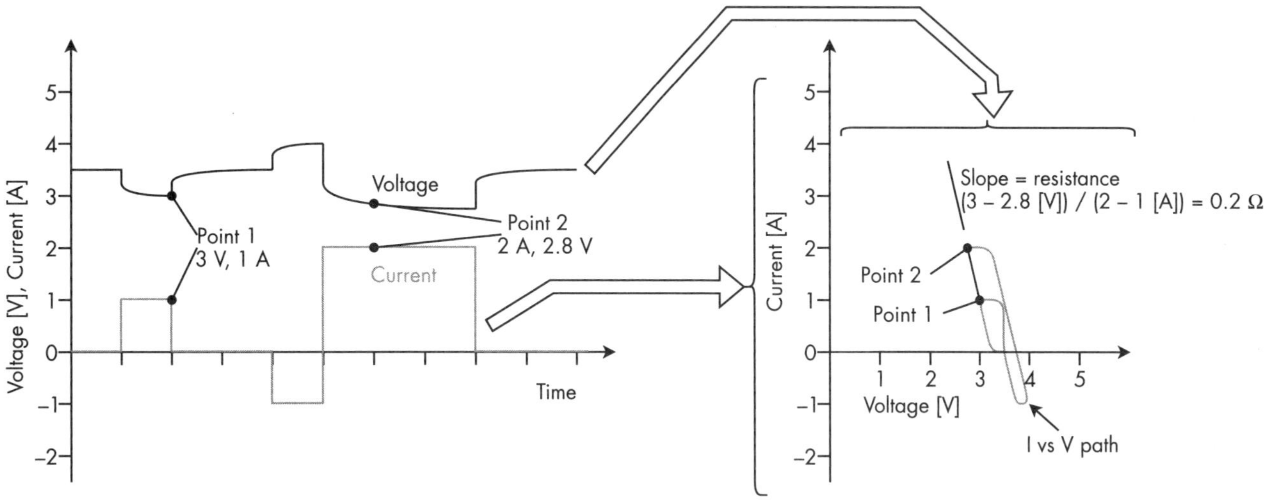

FIGURE 4.60 Cell resistance evaluation: (a) plot of voltage and current, and (b) graph of current versus voltage.

take any two points in this path, two points at two different voltages and currents, we can draw a line through them (black line). The slope of this line is the resistance calculated from those two points.

In practice, this technique is imprecise because the terminal voltage is affected by more than just the current and the resistance. For one, over a long-time discharge, the SoC (and therefore the voltage) drops, making the path drift to the left.

The BMS can use this technique to evaluate battery resistance or individual cell resistance. If the latter, the BMS can calculate the battery resistance from the resistance of each of its cells (see Section 3.1.6).

4.8.5 State of Health Evaluation

The BMS may use the effective capacity (relative to the nominal capacity) and the actual battery resistance (relative to the nominal resistance) as part of its calculation of the battery SoH (see Section 1.6.4). Or not—every BMS designer defines SoH differently.

An accurate SoH evaluation is critical if the user relies on it to decide when to replace a battery. If the BMS reports too low an SoH, a relatively good battery will be replaced too soon. Conversely, if the BMS reports too high an SoH, a product may fail unexpectedly because its battery is no longer good enough to power it. Of course, all this assumes that both the BMS and the user definite SoH in the same way.

4.8.5.1 Failure Prediction

Ideally, the BMS can predict that a cell is about to die (rapidly decreasing capacity, rapidly increasing series resistance, high self-discharge current). The BMS can communicate this prediction by lowering the SoH below a threshold to request service. The service technician can then query the BMS to find out what the problem is, and the BMS can report the number of the cell that is about to die.

4.8.6 State of Power Evaluation

The BMS tells the external system the maximum current that the battery can accept or deliver. The external system could make better decisions if it could also get a sense of how much power it could use in the near future. The BMS can generate a state of power (SoP) for this purpose; unfortunately, there are multiple definitions of SoP (see Section 1.6.5.3).

4.8.7 Ground Fault Evaluation

The BMS may include hardware to check if the battery has a ground fault (see Section 5.14.3).

4.8.7.1 Ground Capacitance Evaluation

A ground fault detector may also test the capacitance between the battery and ground. For a large battery in a metal enclosure, it can be on the order of 10 nF. For a battery that uses pouch cells with cooling plates between the cells, it can be on the order of 1 μF.

Measuring the battery capacitance to ground may detect particular faults:

- A low reading may indicate that a wire is disconnected;
- A high reading may indicate that a load is connected when it shouldn't be.

4.9 BATTERY DEVICES CONTROL

The BMU may implement functions to control devices inside the battery.

4.9.1 Protector Switch and Precharge Control

Most likely, the BMU has a *Charge OK*, *Discharge OK*, and *Fault* state (or some similar name). It may include power drivers (see Section 4.10.7) associated with those states that can drive contactor coils or control MOSFETs. It may have separate drivers for two contactors, one for the positive battery terminal, one for the negative one (see Sections 5.12.2 and 5.13).

The BMU may support precharge by including a driver for a precharge relay, and possibly a precharge sensor (current sensor or load voltage sensor). Logic in the BMS controls precharge timing in conjunction with the timing of the protector switch.

The typical BMS starts the precharge process when initially powered up or when it receives a request to turn on the protector switch. That works for applications in which that request always occurs before a capacitive load is connected. In some applications the BMS has no way of knowing when that will occur; for example, a stand-alone traction battery with no communications to the vehicle doesn't know when the driver will turn on the ignition. An advanced BMS may be able to detect when a capacitive load is applied by monitoring the battery current; if it sees a huge spike in current, it can turn off the protector switch within a ms or so, then start the precharge sequence and finally turn the protector switch back on.

4.9.2 Thermal Management Control

The extent of a BMS's thermal management can range greatly:

- None at all
- Sudden shutdown of the battery is the temperature exceeds some thresholds.
- Reduction of current limits (communicated to the external system, so that it draws less current, therefore slowing down the internal heat generated by the

battery). This doesn't help if the heat source is external; it is of no help against cold temperatures.

- Full climate control through heating and cooling.

A few BMSs may try to bring the battery temperature into the desired range by controlling heating and cooling. They may include drivers for a fan, coolant pump, heating, or cooling:

- Cooling occurs when the maximum cell temperature exceeds a threshold;
- Heating occurs when the minimum cell temperature is below another threshold;
- The BMU may try to heat cells to above freezing before allowing charging.

The BMS may control cooling and heating proportionally (Figure 4.61). For example, cooling may be proportional from 30°C to 50°C, and then at full power above 50°C. Or, heating may be proportional from 0°C to -20°C, and then at full power below -20°C.

4.10 INPUTS AND OUTPUTS

A BMU may have a variety of inputs and outputs to power it, to support devices inside the battery, and to interface to the system outside the battery (Figure 4.62(a)).

A BMS may use dedicated control lines to be connected to devices inside the battery or to communicate with the external system. Dedicated lines have certain advantages:

- Can add reliability and redundancy to a digital link (e.g., a *fault* line can be used to shut down the system, in addition to the *fault* message on a CAN bus);

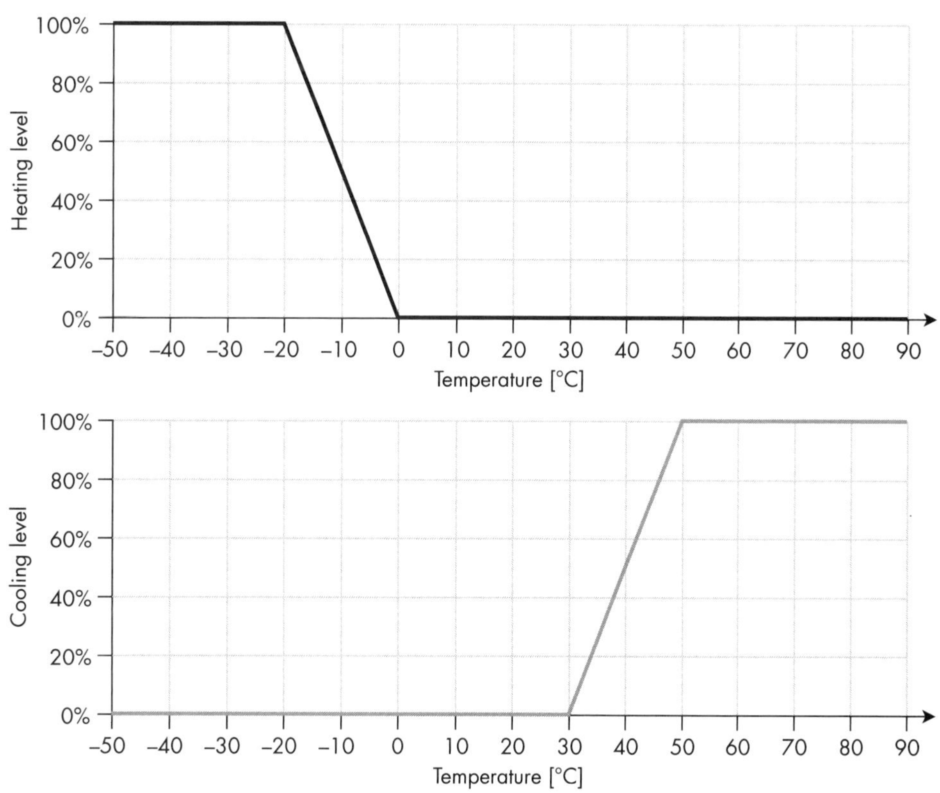

FIGURE 4.61 Thermal management control.

- Are simpler to understand and implement for the new user (e.g., an *SoC* analog output with a 0~5V signal can be sent to the analog fuel gauge of an electric vehicle conversion more easily than converting a CAN message to something that the dashboard instrument panel can understand);
- Can react instantaneously (e.g., removing the contactor request on a line can shut down the battery within a few ms).

4.10.1 Power Supply Inputs

Protector BMSs are powered by the cells and do not require a power supply (Figure 4.62(b)). This is also the case for many BMUs for low-voltage batteries. Powering directly from the cells is risky because the BMU may overdischarge the cells by not shutting itself off in case of undervoltage.

Otherwise, the BMU has one or more power supply inputs (typically expecting 12V). This gives the battery designer the flexibility to devise the best way to power the BMU without the risk of overdischarging the cells. There may be one (Figure 4.62(c)), two (Figure 4.62(d)), or even three (Figure 4.62(e)) separate power supply inputs.

Typically, one power input is enough, though BMUs for automotive use offer as many as three separate power supply inputs:

- Permanent: Used to power the BMU when the battery is off, such as to check the battery once a day or so, and it uses hardly any current;
- Charge: Powered only when a vehicle is plugged in for charging. This may be not just a power supply input, but also a way to instruct the BMS to operate in a *charge mode;*
- Ignition: Powered only when a vehicle is running; similarly, also, this may be a way to instruct the BMS to operate in an *ignition mode.*

While the BMU itself requires little current, the high-power loads it drives (e.g., contactors) may take much more current.

If the BMU is ultimately powered from the cells, to protect them, it must be able to cut off its own supply when any cell voltage is particularly low, even though doing so means that the BMU shuts itself off.

4.10.2 Power Supply Outputs

The BMU may provide power for small devices inside the battery, such as Hall effect current sensors, displays, relays, and contactors.

The supply voltage could be 12V, 5V, or a dual supply for current sensors that require ± 15V. The maximum current is usually quite low.

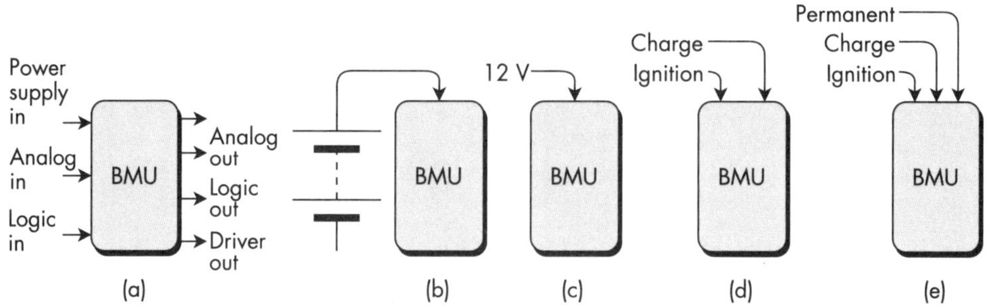

FIGURE 4.62 BMU connections: (a) inputs and outputs (b) power supply from cells, (c) single power supply input, (d) dual power supply input, and (e) triple power supply input.

4.10.3 Analog Inputs

Besides the inputs to measure cell voltages and temperatures, the BMU may include analog inputs to measure:

- The battery current, from a Hall effect sensor or a current shunt;
- The total battery voltage or the battery terminal voltage;
- Battery temperature (e.g., air intake, exhaust, coolant, enclosure).

These inputs typically have a range of 0V to 5V and include a pull-down resistor to ground (Figure 4.63(a)).

4.10.4 Analog Outputs

The BMS may include analog outputs that generate a 0V to 5V signal proportional to the SoC, the CCL, the DCL, or some other parameter.

The SoC signal may be used to drive an analog SoC meter. The DCL signal may be used to limit the range of a throttle in a small EV (see Volume 2, Section 3.3.3.4).

An alternative analog output is a PWM output,[45] with a fixed frequency and a duty cycle that varies linearly from 0% to 100% to indicate the value of a variable.

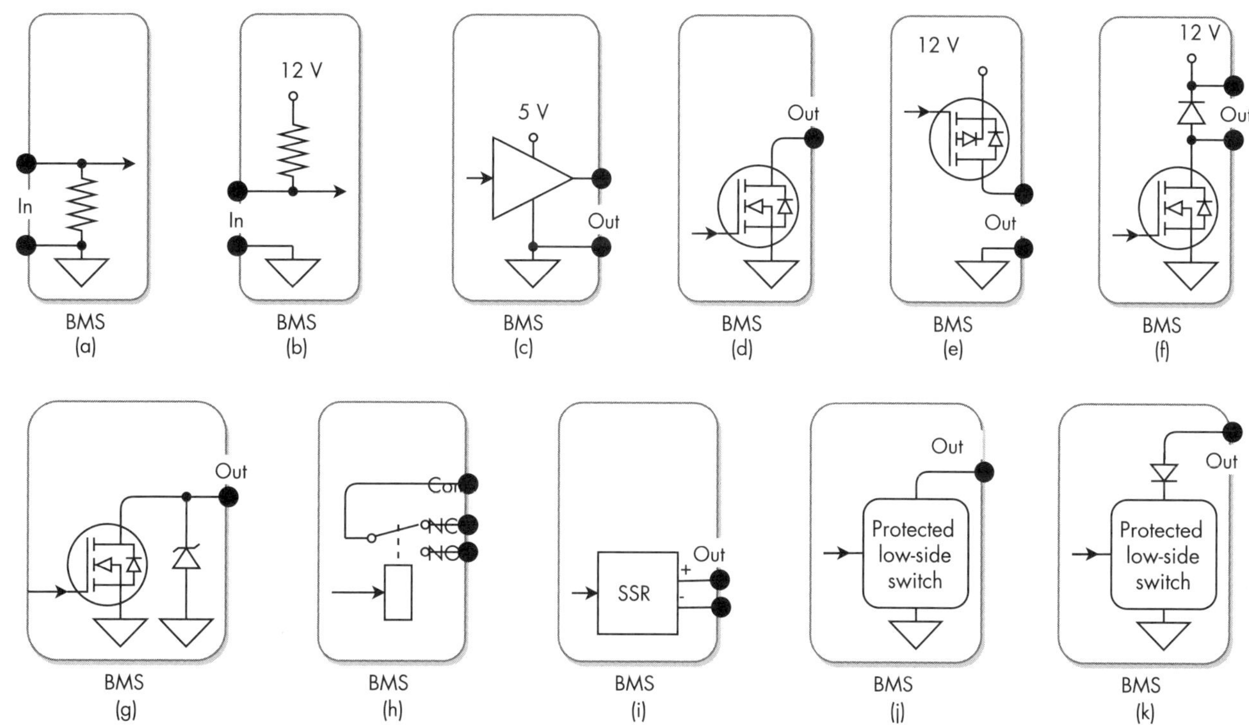

FIGURE 4.63 Inputs and outputs: (a) input with pull-down, (b) input with pull-up, (c) logic output, (d) open-drain output, (e) high side output, (f) open-drain output with flyback diode, (g) open-drain output with TVS, (h) relay dry contact output, (i) DC SSR output, (j) protected low side switch, and (k) with series diode.

45. Pulse width modulation.

4.10.5 Digital Inputs

In general, the BMS tells the outside world what to do, not the other way around. Therefore, the BMS has little use for digital inputs. Some exceptions include

- Operating mode (e.g., on, off, standby, charging mode, normal mode);
- If the BMS includes high power drivers for contactors or thermal management, it may include an input to request those functions;
- Interlock loop, a wire loop that is normally closed, but is opened in case a cable is disconnected or cut (see Volume 2, Section 3.14.6.2);
- The external system may tell the BMS that there is no current at the moment (the battery protector switch is open); the BMS can calibrate the current sensor to cancel out its offset.

Digital inputs may have a pull-down resistor (Figure 4.63(a)). If driven by an open contact, it must be placed between the input and a supply voltage (e.g., 5V or 12V). Or, they may have a pull-up resistor (Figure 4.63(b)). If driven by an open contact, it must be placed between the input and ground (see Section 5.11.1).

4.10.6 Logic Outputs

A BMU may have some logic outputs (that can be either at 0V or 5V) (Figure 4.63(c)):

- *Logic level outputs:* Lines to inform or control the external system, usually with a dedicated function (e.g., Charge OK, or Fault), though the function may be configurable;
- *PWM outputs:* Generate a square wave at a constant frequency (e.g., 20 kHz) and variable duty cycle, such as to drive a fan at a variable speed.

These outputs may be fed to the digital inputs of other devices (see Section 5.11.2).

4.10.7 Open-Drain Outputs

A BMU may have a few power outputs to drive contactors and other power components:

- *Open-collector or open-drain:* A solid-state switch that is either open or connected to ground (Figure 4.63(d));
- *High side:* The same as above, but connected to the positive supply instead of ground (Figure 4.63(e));
- *Dry contact relay:* An isolated switch that is either open or closed (may drive AC or DC loads)[46] (Figure 4.63(h));
- An isolated solid-state relay (SSR) output, typically just for DC loads, rarely for AC loads (Figure 4.63(i)).

An open-collector or open-drain may include protection against the kickback from inductive loads.[47]

46. NC: normally closed; NO: normally open.
47. When the switch driving a contactor opens, current continues to flow through the contactor's coil (due to its inductance) for a while longer. This current needs somewhere to go or it creates a damaging high-voltage pulse. The BMS may provide a path for this current.

- No protection against the kickback from inductive loads (Figure 4.63(d)); the battery designer must include it externally.
- With flyback diode protection (Figure 4.63(f)) with the caveat that the same supply must power the load and the BMU.
- With TVS[48] protection (Figure 4.63(g)); this results in the fastest turn off of the load. The load may be powered by any DC supply, as long as its voltage is less than the maximum rating for the output.
- Instead of a transistor, the switch may be an integrated circuit that includes protection against high current and its own temperature (Figure 4.63(j)).
- Placing a diode in series with the output protects against connection to a negative supply voltage, though this is rare because usually there are no negative power supplies in a battery and the diode wastes power (Figure 4.63(k)).

These outputs may drive relays and contactors (see Section 5.11.3). They usually have a dedicated function (e.g., Charge OK or Fault), though some BMSs allow the battery designer to configure the function of each output.

4.11 COMMUNICATIONS AND LOGGING

The BMS may communicate the state of the battery to the external system and may log it for future reference.

4.11.1 Communication Links

The BMS may use links to communicate internally (e.g., between modules in a master/slave BMS) or externally. The links are usually wired (conductive) but may also be optic or wireless.

Since these communication links occur within and throughout a battery, they are discussed in Section 5.10.

4.11.1.1 Noncopper Communication Links

A BMS may communicate with cell boards, bank boards, or slaves with something other than copper cables:

- Fiber optic:
  - Ideal for high-voltage batteries, as it is inherently isolated;
  - Point-to-point only (cannot be used for buses);
  - Handling the fiber requires some expertise.
- Wireless:
  - Avoids messy wires and cables;
  - Concerns over reliability in noisy environments, and security (hacking).

4.11.2 Data Logging

Data logging is useful for troubleshooting, as it allows going back in time and seeing how a system behaved in the past.

48. Transient voltage suppressor diode.

Logging is a system function (not a battery management function). The external system should do the logging because it has access to not just the state of the battery (as reported by the BMS) but also the state of other devices. A data logger placed on a shared bus sees data from all the devices in the system.

In the automotive world, a common name for this function is the *event data recorder (EDR)*.

The data logger takes a snapshot (a *freeze frame*) regularly, and when special events occur. Each snapshot contains salient information about the system at the time the snapshot was taken.

Each frame could contain

- Battery voltage and current;
- Minimum and maximum cell voltages;
- Temperature;
- CCL and DCL;
- Warnings and faults;
- State of inputs and outputs;
- If due to an event, the name of the event.

Most BMSs do not do any logging because it is a system function. A few BMSs implement this function to some extent by logging just main events such as faults and can retain just a few events in memory. Other BMSs log all events and take periodic snapshots.

As logging many events takes an enormous amount of memory, BMS designers strive to condense the data in each frame down to the bare minimum. These highly condensed data may be hard to interpret. I find that logging is a highly requested feature in a BMS; yet, in actual use, I find it to be an underused resource.

4.12 BMS RELIABILITY

As Li-ion cells become more reliable and last longer, the BMS should not cause battery failure. A battery is only as reliable as its BMS. Therefore

- Its hardware should have longevity;
- Its software should be able to operate continuously for years without shutting down and without bugs due to data overflow.

4.12.1 BMS Hardware Longevity

BMS longevity is not just about hardware reliability. It is also about serviceability and support.

4.12.1.1 Mean Time between Failures

The likelihood of hardware breakdowns is somewhat higher during an initial period of infant mortality and a final period of wear-out. The failure rate is relatively low and constant in the intermediate time, during which we define a mean time between failures (MTBF).

As processes have improved over the years (both for electronic components and electronic assemblies), failure rates have decreased dramatically.

4.12.1.2 Failure Causes

It may be helpful to relate our company's experience with failure rates.

In the early years, failures in our BMS were due to

- ~80%: Incorrect installation in the field;
- ~18%: Poor soldering of wires, terminals ;
- ~2%: Infant mortality in a DC-DC module we bought.

Having corrected the quality issues on our end, today virtually 100% of the hardware failures occur during installation in the field and are due to incorrect installation:

- Misconnection to the cells;
- Working on the battery without disconnecting it;
- Not floating the battery, followed by an accidental short circuit;
- Accidental short circuit between a grounded shield and a battery terminal followed by connecting a laptop to the BMS and a grounded AC adapter.

These can be virtually eliminated by the installer reading the f manual and being attentive.

The vast majority of later failures are due to connector issues (see Section 5.3.3). We haven't seen a single one of our products fail during the flat plateau after infant mortality due to failure in the electronics. It's still too early to see wear-out failures.

4.12.1.3 Serviceability

Should a BMS fail, will you be able to get spare parts in 10 years? Is the BMS repairable? Is there a broad base of users, from which you can look for used products?

It is reassuring to select a BMS that is used widely, sold by a company that has been around for some time, and that readily tells you where to buy replacement components in your own country if you need to repair your BMS. It is good to know that the latest version of the product is always backward-compatible.

4.12.1.4 Warranty

A 1-year warranty on parts and labor on manufacturing defects is standard.This means that if the BMS was built incorrectly or a component within the BMS failed in its own, then the BMS manufacturer will repair or replace the assembly (as it deems appropriate), for free. This doesn't mean that the manufacturer will repair, for free, an assembly that was damaged by the user or that failed after the expiration of the warranty.

Since in-warranty repairs are rare, it's more helpful to look at the manufacturer's out-of-warranty repair policy. How much does the manufacturer charge for repairs? Ideally, it should charge not much more than its costs for the repair or a new board. How fast is the turnaround? Ideally, it should be less than one week.

4.12.1.5 Return Policy

At times, a particular BMS doesn't match the needs of your particular application. If so, it is helpful if the BMS manufacturer allows you to return the product within a specified time, as long as it is in resellable conditions, possibly charging a small restocking fee.

4.12.2 BMS Software Longevity

When selecting a BMS, few of us consider the BMS software and how long it can operate.

4.12.2.1 Software Reliability

As a user, you have no way of knowing if the BMS software is reliable because you don't have access to the code and, even if you did, you don't have the skills to evaluate its reliability.

A certification that the software was developed under a software development standard is the best you can hope for, even though you're just taking the manufacturer's word for it; a bug in certified software is still a bug.[49]

Nothing is as reassuring as extensive testing. However, testing a product 100 times doesn't help the one time the product was used under conditions that were not accounted for in the testing.

The maturity of a BMS, with stable software used successfully in a great variety of applications, is perhaps more reassuring than a certificate.

4.12.2.2 Continuous Operation

Unlike a computer or a phone, a BMS can be used continuously for years on end. Its software must be able to operate without a reset during this time.

Variables must account for 10 years of continuous operation. For example, the *up-time* clock may top one million seconds, which requires more than a 16-bit word. Either the variable size must be sufficiently large, or the software must clamp the value to a maximum. An overflow back to 0 may also be acceptable, but not an overflow to negative.

Some constants (e.g., number of cells in series) are stored in a variable in volatile memory (RAM) for easy access. The software designers must assume that over 10 years, a variable will be corrupted by a cosmic ray or whatever. If the variable is set only once at power-up, only a reset can recover from the consequences of a corrupted value. Therefore, the software must regularly refresh values in volatile memory.

Embedded product such as a BMS use a nonvolatile *EEPROM*[50] to store event logs. These devices are rated for some 100,000 write cycles. Due to their relatively small size, once the log is full, the BMS overwrites old data with recent data.

Excessive logging may result in the EEPROM reaching the maximum number of write cycles, particularly in the location that stores an index to the most recent record since that index is updated each time a frame is stored [7].

The problem is that, as a user, you have no way of knowing whether the software designers addressed these situations. Frankly, the software designers themselves may not even know!

4.12.2.3 Software Upgrades

Some companies offer free software upgrades that add features and fix bugs. Ideally, the upgrade can be performed in the field, with just an internet connection, without the need for special tools.

---

49. As I write this, Boeing's 737 MAX 8 airplanes are grounded due to a software bug. Pundits blame the FAA for not catching it. However, I don't see any way that the FAA could have found that bug, no matter how many resources it had invested, either through testing or by analyzing the "quality" of the code. The pilots discovered the bug while flying the plane.
50. Electrically erasable programmable read only memory.

4.12.3 Electromagnetic Interference Immunity

Batteries are likely to operate in electrically noisy environments, exposed to electromagnetic interference (EMI). The BMS must operate reliably despite this interference.

Of course, as a BMS user, you have little control of how a given BMS is designed, and how sensitive it is to EMI. The point of this discussion is to give you an overview of the issue.

What you *can* control is your willingness to install the BMS as instructed, in a way that doesn't expose it to excessive EMI.

4.12.3.1 Interference Sources

Batteries are exposed to EMI if placed near high-power switching devices or high-power transmitters (see Section 5.16.1).

Various sources impose voltages onto the cell and common-mode voltages onto the battery at various frequencies:

- Differential mode[51] (Figure 4.64(a)):
  - When charging from a single-phase AC-powered charger, the cell voltage bounces by 10~100 mV at twice the line frequency (100 or 120 Hz)
  - When connected to a high-power switching converter, each cell voltage bounces by 10~100 mV at the switching frequency (20~300 kHz)
- Common-mode[52] (Figure 4.64(b)):
  - When connected to a high power switching converter, the entire battery bounces up and down (relative to earth ground), by 10~1000V (up to the battery voltage)

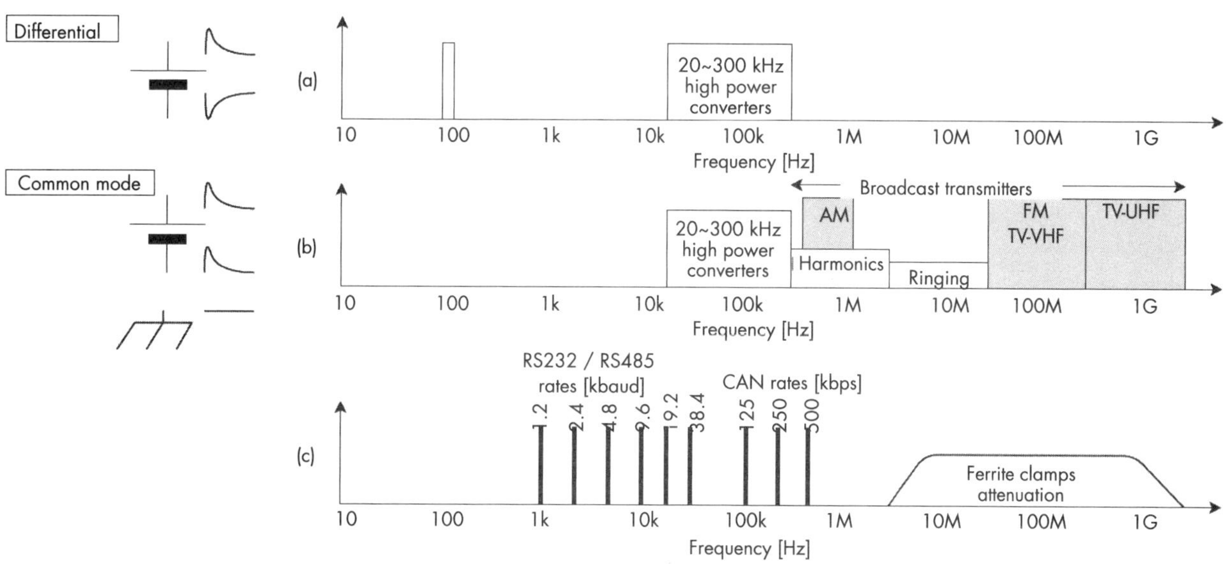

FIGURE 4.64 (a) Differential mode interference sources and frequency ranges, (b) common-mode sources and ranges, and (c) communications data rates.

51. Differential mode noise changes the voltage of one cell terminal relative to the other terminal. The noise voltage on one terminal is different from the noise voltage on the other terminal.
52. Common-mode noise changes the voltage of both battery terminals equally relative to earth ground. The noise voltage on one terminal is the same as the noise voltage on the other terminal.

- When a high-power transmitter is nearby, the BMS is immersed in a strong electromagnetic field (0.5~500 MHz)

4.12.3.2 *Line Frequency*

The charging current from a single-phase AC-powered charger or inverger probably includes a strong 100- or 120-Hz component.[53] This ripple in the current causes a 100- or 120-Hz ripple on the cell voltage.

Since the BMS samples the cell voltages and the battery current at fixed intervals, it may be confused by this ripple, as it may sample at the "wrong" time, resulting in an error in the measurements. A current meter shows steady current, but a BMS may show the current vary at a beat frequency between the line frequency and the sampling frequency.

The BMS should be able to time-average the current and the cell voltages, in hardware, before they are sampled in software. For example, the current sensor should include a low-pass filter with a time constant of about 20 ms, so that whenever the BMS samples the current, it sees the average current.

Without such a filter, the readings fluctuate at the beat frequency between the line frequency and the BMS's sample rate. In particular, the beat frequency between the precise sampling frequency and the slightly variable line frequency results in wobbles. In this example (Figure 4.65), the beat frequency varies between 4 and 8 mHz.[54]

A filtering time constant of 20 ms is not too slow, assuming that the fastest sampling is every 100 ms (which is fast enough even for a buffer battery that is discharged over a few minutes).

However, some BMSs can be configured to sample as fast every 10 ms; a hardware filter would interfere with the BMS's ability to sample at such a high rate.

4.12.3.3 *Internal Communication Data Rates*

A BMS that consists of multiple assemblies that communicate digitally may use standard data rates (e.g., 1.2 to 9.6 kbps for RS-232; 125, 250, or 500 kbps for a CAN bus). The sources listed above may interfere with these digital links if both operate over the same frequency ranges (Figure 4.64(c)).

A BMS that communicates at baud rates (1.2 to 9.6 kbps) that are below the range of switching converters (20~300 kHz) would gain some noise immunity. However, this rate may be too slow to carry the required amount of information. Of course, you have no control over what data rates a BMS uses internally; I only mention this as it could be an issue in a BMS.

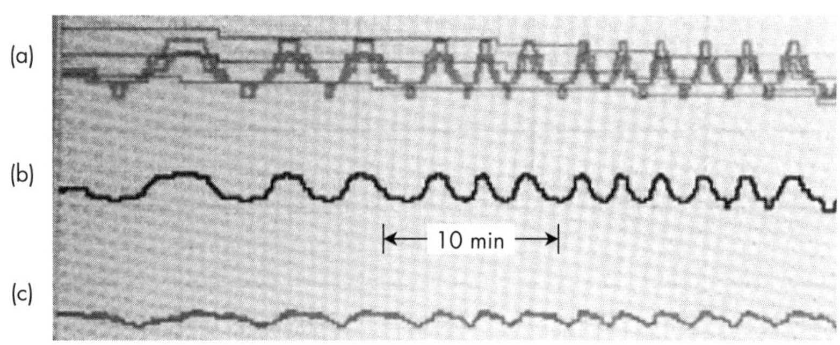

FIGURE 4.65
Wobble in the beat between the line frequency and BMS sampling: (a) cell voltages, (b) battery voltage, and (c) current.

53. Twice the line frequency, due to the rectification process.
54. That's millihertz, not megahertz.

4.12.3.4 Switching Frequency

A switch-mode power device uses high frequency (20~300 kHz) to convert voltages (see Volume 2, Section B.6). In so doing, it generates a lot of EMI.

It's unclear whether a wired BMS or a distributed BMS is more immune to switching noise:

- A wired BMS can more easily include a filter to remove the noise at the switching frequency. However, it senses the voltage at a cell plus the high-power connection between cells, which has some inductance and develops a voltage across it proportional to the change in current.
- The cell board in a distributed BMS senses the cell voltage directly, so the voltage reading is not affected by the noise voltage across the power connection between cells. However, this noise voltage does affect the communication between cell boards (unless each cell board is isolated) and to the BMU.

4.12.3.5 Radio Frequencies

It's relatively easy for a BMS designer to make it immune to interference at radio frequencies (1 MHz~ 3 GHz) since they are so far above the measurement rates and communication data rates, and because ferrite clamps[55] work well in this range (Figure 4.64(c)). BMSs with a poor RF immunity experience sudden misbehavior, an offset in the readings while a transmitter is powered[56] or even a breakdown in communications.

4.13 BMS SOURCING

Selecting an off-the-shelf BMS is not quite as critical as selecting cells, though it requires no less care.

4.13.1 BMS Selection

Consider these criteria when selecting a BMS, in order of importance:

- Number of cells in series;
- Minimum expected cell voltage;
- The physical layout of your battery;
- Cell format;
- BMS technology;
- Special functions your application may require;
- A BMS designed specifically for your application;
- Any regulatory certifications required in your industry.

4.13.1.1 Number of Cells in Series

Check that the BMS can handle the number of cells in series for your battery.

If your battery uses a series-first arrangement with P strings in parallel, it requires a BMS with P times as many sense inputs as it would for a parallel-first arrangement. It also requires a BMS that can manage parallel strings.

55. The bulbous cylinder you see at times on a computer cable is a ferrite clamp.

56. The RF is rectified by any semiconductor junction exposed to it, and filtered by the parasitic capacitance to ground, resulting in a DC level proportional to the RF intensity, in effect creating an AM receiver.

4.13.1.2 Minimum Cell Voltage

Make sure the input voltage range of the BMS includes the minimum cell voltage in your application. For example, the voltage of LTO cells and cells in race car applications drops below the range of a distributed BMS.

4.13.1.3 Battery Physical Layout

Select a BMS topology that matches your battery (see Section 4.2.2). For example, do not use a wired BMS with a modular battery. Do not use a distributed BMS to manage a block of pouch cells with a balance connector.

If your battery consists of separate modules, make sure the BMS can support several banks. The number of banks should match the number of modules; each bank should support the number of cells in series in each module. Note that a 12-cell bank may support a module with only eight cells, but not one with only two cells. Check the minimum number of cells supported by a bank.

4.13.1.4 Cell Format

Check that the BMS supports the format of your cells. The cell boards in a distributed BMS fit a particular cell format. Some BMSs only support large prismatic cells.

4.13.1.5 Technology

A simple and cheap analog protector may be sufficient for a small battery; other batteries use a digital BMS.

If the battery current is less than about 50 A, a protector BMS may be appropriate; otherwise, a BMU that controls a separate protector switch is required.

4.13.1.6 Special Functions

Check if a BMS supports any special function you may require, such as SoH evaluation, mid-balancing, isolation detection, ModBus (see Section 5.10.3.8), or sealed enclosure.

An off-the-shelf BMS that meets all your requirements may not exist. If so, consider a BMS that meets the critical requirements (listed above) and consider ways to work around other, less critical requirements. For example, if the BMS is not sealed, enclose it in a sealed case. If the BMS does not evaluate SoH, you can use its data to calculate SoH in a separate computer.

4.13.1.7 Application

A BMS explicitly designed for your application by a company that specializes in your industry is more likely to meet your needs than a one-size-fits-all BMS. Some BMS companies specialize in passenger automotive applications, some in marine applications, some in golf carts.

Each "Application" chapter lists BMSs advertised as appropriate for a given application (see the "Applications" chapters in Volume 2).

Specifically, if the BMS must communicate with products in the application (a controller, a charger, an inverger), make sure that it can, or be prepared to design a gateway between them.

4.13.1.8 Hardware Certifications

Your application may require that the BMS hardware be certified by a regulatory agency, such as ISO26262 for automotive (hardware and software) and DO-178B for aviation (software only). If you require a specific certification, be prepared to pay for this testing because the BMS manufacturer is unlikely to pay for testing that only benefits your application.

Generally, standards include

- All electronic products must meet the *unintentional emitter* regulations (e.g., FCC part 15);
- They also should meet immunity standards such as ESD, RF immunity, fast transient;
- You may also require a standard specific to BMSs (e.g., CE, UL, CSA);
- You may also require a standard for your particular application (e.g., aviation, automotive).

Note that the *complete product* must meet CE certifications, not a subcomponent's such as a BMS. It's OK if the BMS is not CE-certified because a BMS is not a complete product; the product may still be CE-certified. A CE-certified BMS does help the CE certification process of the complete product. Conversely, using a CE-certified BMS doesn't automatically certify your product.

There is much more to be said about these certifications, which I would do in a book about BMS design. The above information should be sufficient to you as a BMS consumer.

4.13.1.9 Software Certifications

There are also standards for software:

- You may require that the software development process occurred in a strictly regulated environment (regardless of the resulting quality of the code); for example, the day's software is committed every evening, regardless of its state;
- Or you may require that the software meets precise rules to avoid risky constructs (regardless of the environment in which it was developed); for example, do not write an infinite loop without a time-out.

Neither type of certification ensures that the code is bug-free (see Section 4.12.2).

4.13.2 BMS Manufacturer Selection

Selecting a BMS manufacturer (see Volume 2, Section A.5.6) is as critical as selecting a BMS. Here are some criteria to consider in that process.

4.13.2.1 BMS Manufacturer Longevity

Off-the-shelf BMUs are made in small batches by small companies that provide personalized support and may customize their products for your application. Consequently, their prices are high.

There is a valid fear that such a small company may decide to stop selling BMSs or may go out of business. I have been tracking manufacturers of off-the-shelf BMSs for 10 years [8]. Of 28 companies I listed:

- Eight have offered off-the-shelf BMSs for at least 5 years—123 Electric, Elektromotus, Elithion, Ewert, JTT, John Elis, Lithium Balance, REC;
- One has only offered off-the-shelf products for a short time but has a long history of BMS design—Nuvation;
- Ten no longer sell BMSs to the general public;

- Nine are out of business.

Yet, in my company's experience, the opposite is true: some large and well-funded companies expressed concerns about the expected longevity with my small company. Ironically, each of these companies has since failed:

- B. Automotive —EVs, $17 M, Indiana, gone February 2012
- A. Systems[57]— LFP cell manufacturer, Massachusetts, $500M, gone October 2012
- E. — EV batteries, Israel, $10 M, gone December 2012
- B. Place — EV charging infrastructure, Israel and California, $800M, gone May 2013
- C. Automotive — EVs, California, $100M, gone May 2013

4.13.2.2 BMS Manufacturer Location

A consideration is the location of the BMS manufacturer:

- Deservedly or not, European and North American companies have a good reputation for proper BMS design, manufacturing quality, and product availability;
- Other origins suffer from a poor reputation, often undeservedly;
- Australian manufacturers suffer from the high cost of product shipment to the rest of the developed world;
- The inadequate documentation from some Chinese manufacturers may give a (possibly undeserved) impression of low quality.

You should give higher consideration to companies on your continent, due to lower shipping costs, faster turnaround on repairs, and better tech support hours.

Brazilian and Indian customers suffer regardless of where the BMS comes from because of business-unfriendly and bureaucratic customs departments in their country. To overcome this, an Indian company is now offering a BMS to Indian customers.

4.13.2.3 BMS Manufacturer Tech Support

Do consider the level of tech support provided by a BMS supplier. But do *not* judge a company based on whether it can support you with matters unrelated to the BMS:

- Can you recommend cells?
- How do I design my battery?
- Can you help me troubleshoot communications to my charging station?
- What does this error code mean in my inverter?

You may find that a BMS company is more responsive than a large manufacturer of inverters; this doesn't mean that the BMS company can answers questions that only the inverter manufacturer knows how to answer.

57. Today's A. Systems Inc. is a different company, which bought the original company's assets, indirectly, through a third company.

4.13.3 BMS Vendors

At times, it's more convenient to buy a BMS from a vendor rather than directly from the manufacturer.

4.13.3.1 PCMs

PCMs are manufactured by the thousands by Chinese companies and are distributed by hobby companies, ephemeral Chinese companies, AliExpress, and eBay.

Reliable supply is a concern because tomorrow you may not be able to buy the same unit that you bought yesterday. For medium production levels, try to talk directly to the manufacturer (see Volume 2, Section 1.6.1, "Cell Selection").

4.13.3.2 BMUs

The most likely source for a BMS for the new battery designer is the one-stop Li-ion shop that sells cells, chargers, and contactors. Specifically, hobby stores for small batteries and EV-component stores for large batteries.

Just know that these stores offer little technical support, and you'll end up talking directly to the manufacturer.

4.13.4 Switching to a Different BMS

At times people call my company looking for a replacement for a competitor's "crap BMS." However, I understand that our competitor receives similar calls from people wanting to switch from our "crap BMS" to theirs, which tells you something.

Maybe the problem is not the BMS; maybe the problem is a mismatch between the BMS and the application, or maybe it's you (sorry). If the problem is that you didn't install or configure the first BMS correctly, you won't be happy with it. Switching a different BMS won't help as you won't be happy with it either.[58] When someone comes to our company complaining about someone else's BMS, we try to look into it. We may discover that the real problem is because it's not being used correctly.

Switching to a BMS that is a better match for your application *does* make sense. For example, if you are working with a ready-made battery block with a balance connector, you are correct to switch from a distributed BMS to a wired BMS.

References

[1] Zdravkova, Liliana, "Fiber Optic Sensor for In-Situ State-of-Charge Monitoring for Lithium-Ion Batteries," UWSpace. http://hdl.handle.net/10012/9059.

[2] U.S. Patent US20130335094.

[3] Andrea, D., Battery Management Systems for Large Lithium-Ion Battery Packs, Norwood, MA: Artech House, 2010, Section 3.2.4.3.

[4] Anderson, R., R. Zane, G. Plett, et al., Life Balancing–A Better Way to Balance Large Batteries, SAE Technical Paper 2017-01-1210, 2017, doi:10.4271/2017-01-1210.

[5] Plett, G., Battery Management Systems, Volume I: Battery Modeling, Norwood, MA: Artech House, 2015.

[6] Fundaro, P., "Impedance Track ™ Based Fuel Gauging," Texas Instruments, Tech. Rep., 2007.

58. And, frankly, we'd rather they would be our competitor's headache, not ours.

[7] "Worn-Out Flash Memory Is Suddenly Bricking Tesla Cars," https://www.vice.com/en_us/article/qvgxqp/worn-out-flash-memory-is-suddenly-bricking-tesla-cars.

[8] "Li-Ion BMS Options," http://liionbms.com/php/bms_options.php#Digit._balancers.

CHAPTER 5

BATTERY DESIGN

5.1 INTRODUCTION

Now that we discussed all the pieces (cells, cell arrangement, BMS) it's time to put them together into a battery, which is covered in multiple chapters:

- This chapter covers battery design in a general way, independent of the application;
- Chapter 7 discusses assembling a battery, installing it, and configuring it, also in a general way;
- Chapter 8 discusses common pitfalls, plus troubleshooting and repairing any issues that may arise;
- The "Applications" chapters in Volume 2 will go into some of the details that apply to batteries for specific applications.

5.1.1 Tidbits

Some interesting items in this chapter include:

- A Li-ion battery whose BMS cannot interrupt the battery current is like a trapeze artist performing over lines painted on the concrete floor below to look like a safety net (5.12.1);
- A single cell is not a battery (5.1.3);
- Avoid pouch cells if you're not an expert in their use (5.3.2.1);
- If it sparks every time you connect it, you need precharge (5.13.1);
- If it sparks when you disconnect, you need to add kickback protection (5.11.3);
- For safety, a battery should be isolated from ground during assembly (5.14.1.1);
- Thermal insulation lets you control when to exchange heat with the ambient (5.17.4);
- Procrastination is good—store the heat so you don't have to deal with it yet (5.17.7);
- Plan for thermal runaway to mitigate the damage (5.18.3).

5.1.2 Orientation

This chapter starts by defining a battery, classifying its types, and giving an outline of the design process. It discusses components selection. It describes how to design a battery for each cell format. It talks about wiring the BMS, from sensing the cell voltages to checking for loss of battery isolation. It talks a bit about chargers and the effects of electrical noise. It gives an overview of thermal management, mechanical design, standards, and regulations.

5.1.3 Battery Definition

A battery is a collection of two or more cells in any arrangement and possibly other components (Figure 5.1(a)); a single-cell plus other components—a protector BMS, a thermistor, a case or a connector—is also considered a battery (Figure 5.1(c)). A cell by itself is not a battery (Figure 5.1(b)) (see Sections 2.1.3 and 1.2.2.2).

A battery has

- One current;
- One state of charge;
- One protector switch (or a set of switches, for charging and discharging separately);[1]
- One set of power terminals (or two sets if charging and discharging separately).

If a "battery" has more than one current (and therefore more than one SoC) it's not a battery, it's a battery array. Specifically, a split battery (with a center tap), is not one battery; it's two batteries that happen to share a terminal (Figure 5.1(d)) (see Section 6.5).

5.1.4 Battery Use Classification

Although we think that a battery is discharged from 100% SoC down to 0%, this is rarely the case. In practice, a battery is often operated in a narrower range of SoC levels.

Roughly speaking, batteries can be classified based on how they are used—how quickly and within what SoC range (Figure 5.2):

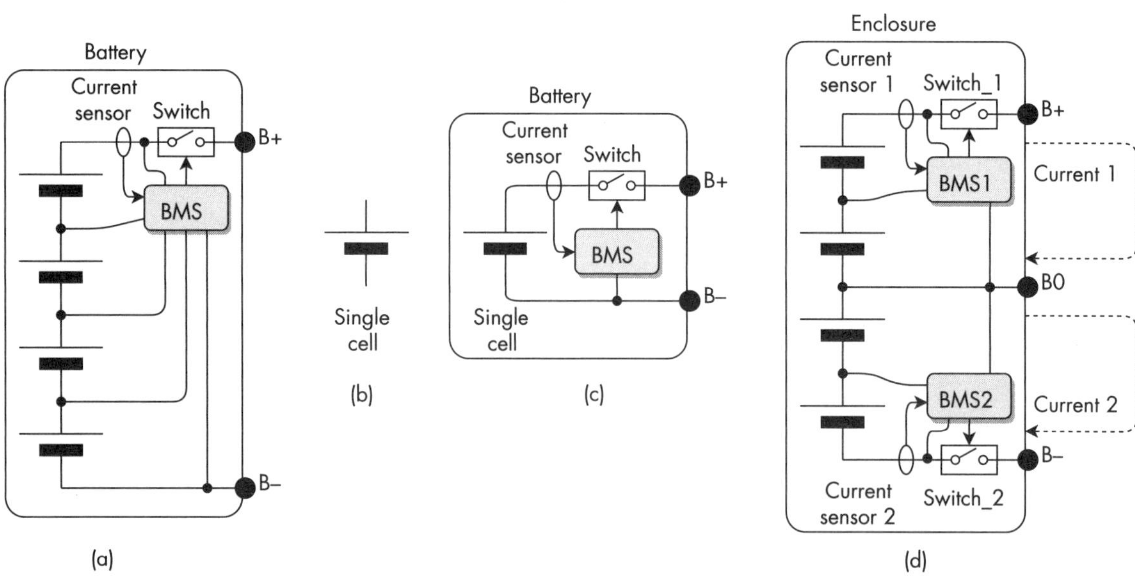

FIGURE 5.1 Definitions: (a) multicell battery,(b) single-cell, (c) single-cell battery, and (d) split battery: two batteries.

1. A lead-acid battery doesn't have a protector switch. A Li-ion battery without a protector switch is unsafe.

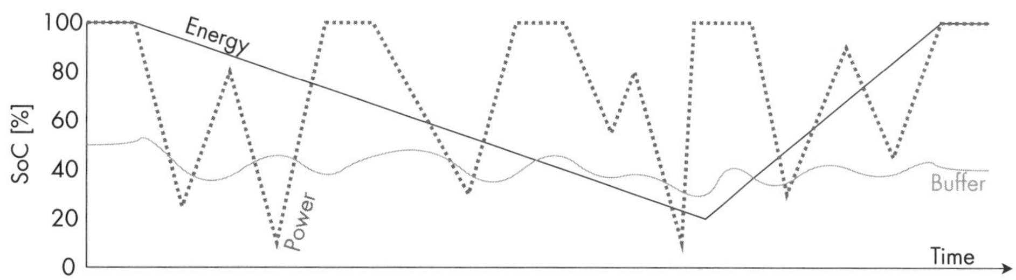

FIGURE 5.2
Plot of SoC for the three classes of batteries.

- Energy: Slow and deep;[2]
- Power: Fast and deep;
- Buffer: Fast and shallow;
- Long-term buffer: Slow and shallow.[3]

The point of this classification is to help us when discussing solutions, as when saying: cells with an MPT of more than 100s should only be used in energy batteries.

Table 5.1 describes each battery type and its characteristics. Batteries in the range between 0.5 and 2C are ill-defined; they could be considered either energy or power batteries.

Table 5.2 presents the same information, arranged as a matrix based on current and SoC range.

I said "roughly" because there is no clear delineation in this classification:

- Is a PHEV (see Volume 2, Section 3.1.3.3) battery still an energy battery when used in charge sustain (see Volume 2, Section 3.2.3)? Yes.
- Is a traction battery still an energy battery, even though, when still new, it's used only between 80% and 30% SoC? Yes.
- Is a battery used for peak shaving (see Volume 2, Section 4.3) still a buffer battery, even if occasionally it's used for backup power? I'd say yes.
- Take a battery that starts from 100% SoC, and is used at 1C; is it an energy battery or a power battery? I don't know.

TABLE 5.1
Comparison of Battery Types

Type	SoC Range	Current	Balancing	Cells	Typical Applications
Energy	100% down toward 0%	< 0.5 C	Top	Energy	Battery electric vehicles Backup power
Power	100% down toward 0%	> 2 C	Top	Power	Power tools UAVs (drones)
Buffer	~50% ± ~20%	> 2 C	Mid	Power	Hybrid electric vehicle Grid stabilization
Long-term buffer	~50% ± ~50%	< 0.5 C	Mid	Energy	Capacity firming for renewable resources

2. You have a dirty mind.
3. Long-term buffer batteries may be used for capacity firming of renewable resources. As the same objectives can be achieved effectively with a combination of energy batteries and long-term management of resources, this book does not discuss them in any depth. (See Volume 2, Section 4.3.2).

Table 5.2
Comparison of Battery Types as a Matrix

			Current		
			< 0.5 C	**0.5 ~ 2 C**	**>2 C**
			Uses Energy Cells	**Use Medium-Duty Cells**	**Uses Power Cells**
SoC Range	**100% Down Toward 0%**	**Top-balanced**	Energy battery: Battery electric vehicles Backup power	(Could be defined as energy or power battery)	Power battery: Power tools UAVs (drones)
	Around ~50%	**Mid-balanced**	Long-term buffer battery: Capacity firming for renewable resources		Buffer battery: Hybrid electric vehicle Grid stabilization/ frequency regulation

Note that it's the application that keeps a battery SoC within a desired operating range, not the BMS. The BMS only ensures that the cell voltages remain in a safe range.

5.1.4.1 Energy Batteries

Most rechargeable batteries are energy batteries, meaning that they are normally charged fully and they are discharged relatively slowly (Figure 5.3(a)). They use energy cells (see Section 2.2.3) and are top-balanced (see Section 3.2.6.1).

5.1.4.2 Power Batteries

Power batteries (Figure 5.3(b)) are the same as energy batteries, except that they are charged or discharged rapidly and use power cells for improved efficiency and reduced cooling requirements.

5.1.4.3 Buffer Batteries

A buffer battery is operated around a reference SoC that is somewhere in the middle (Figure 5.3(c)). This SoC level is selected based on three criteria (individually or in combination):

- Widest symmetrical range of SoC = maximum headroom in both directions (Figure 5.4(a));
- Equal power capability for charging and discharging = lowest resistance (Figure 5.4(b));
- Minimum degradation = longest lifetime (Figure 5.4(c)).

For the widest symmetrical range of SoC, the battery is operated at a reference SoC of 50% (or maybe 55%, to avoid operating below 10% SoC) (Figure 5.4(a)).

Yet a buffer battery can perform better if it operates around an SoC level where the cells can both accept and generate the most power. This occurs when their resistance is lowest (Figure 5.4(b)). One must consider both charging power and discharging power, and may select an SoC level where those two have the same value, such as at 43% SoC (see Section 2.7.2).

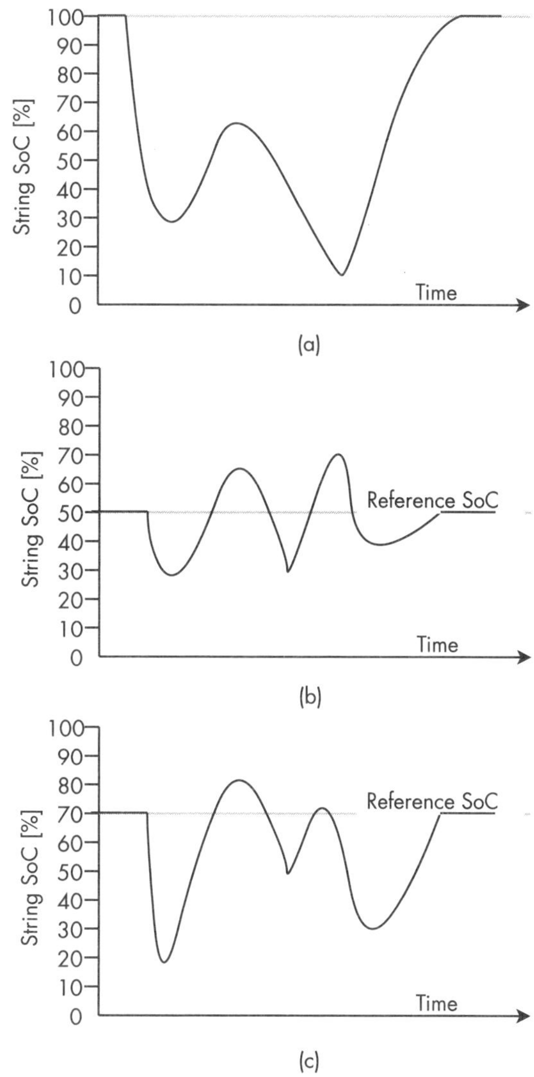

FIGURE 5.3
Operating range: (a) energy or power battery, (b) buffer battery, symmetrical range, and (c) buffer battery, asymmetrical range.

If battery life is more important than performance, the battery may be operated in the range where it has the lowest degradation, away from the two ends, such as 30% SoC (Figure 5.4(c)) (see Section 2.5.3.1).

Choosing a reference SoC level other than 50%, such as at 30%, or 70%, forces the battery to operate over a narrower, symmetrical range (e.g., 70% ± 30%). This usually is not a limitation for an application that uses a buffer battery.

Designing extra capacity into a buffer battery is better than designing a battery with exactly the desired capacity, even if it means that the top and bottom portions of the battery are never used. This is because the specific current is reduced (e.g., 3 C instead of 30 C) and the degrading areas at the two ends of the SoC range are avoided (see Section 2.5.2).

For best performance, a buffer battery should be mid-balanced at the reference SoC.

So far, we assumed that a battery is operated symmetrically around a reference SoC level. This is not necessarily the case because an application may use its battery asymmetrically. If so, the reference SoC should be biased in one direction to account for the asymmetry in the application. For example, a battery may normally sit at

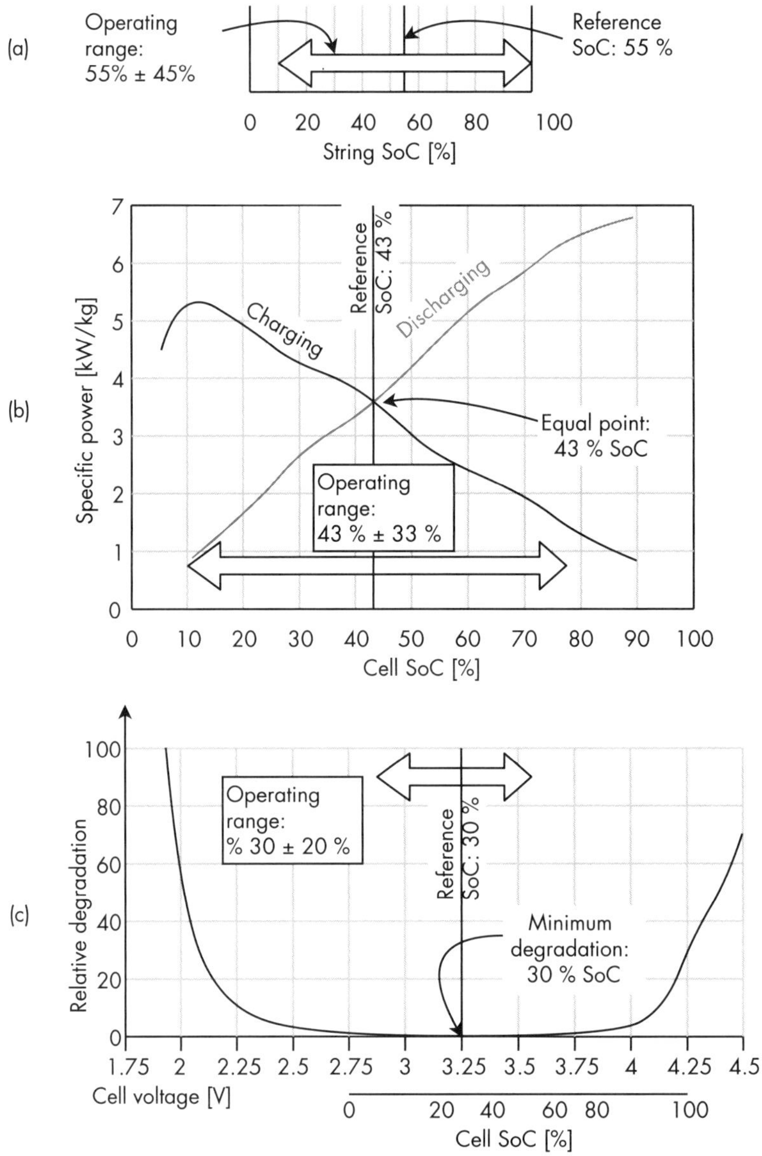

FIGURE 5.4
Selecting the operating SoC for a buffer battery: (a) maximum range, (b) equal charging and discharging power, and (c) minimum degradation.

a reference SoC, be rarely charged to 10% above the reference SoC, and often be discharged to 50% below it (Figure 5.5).

Be aware that SoC evaluation of a buffer battery is difficult. To calibrate the SoC, most BMSs require that the battery be charged completely and regularly (see Section 4.8.1).

5.2 PLANNING

This section discusses the initial planning before making a battery.

5.2.1 Should You Make a Battery?

You may be reading this book because you want to design and make a battery. Before you embark on such a project, ask yourself: Should you?

If you can buy a ready-made battery, I recommend that you go ahead and buy it. If you think that you can make a battery cheaper than you can buy one, you are not

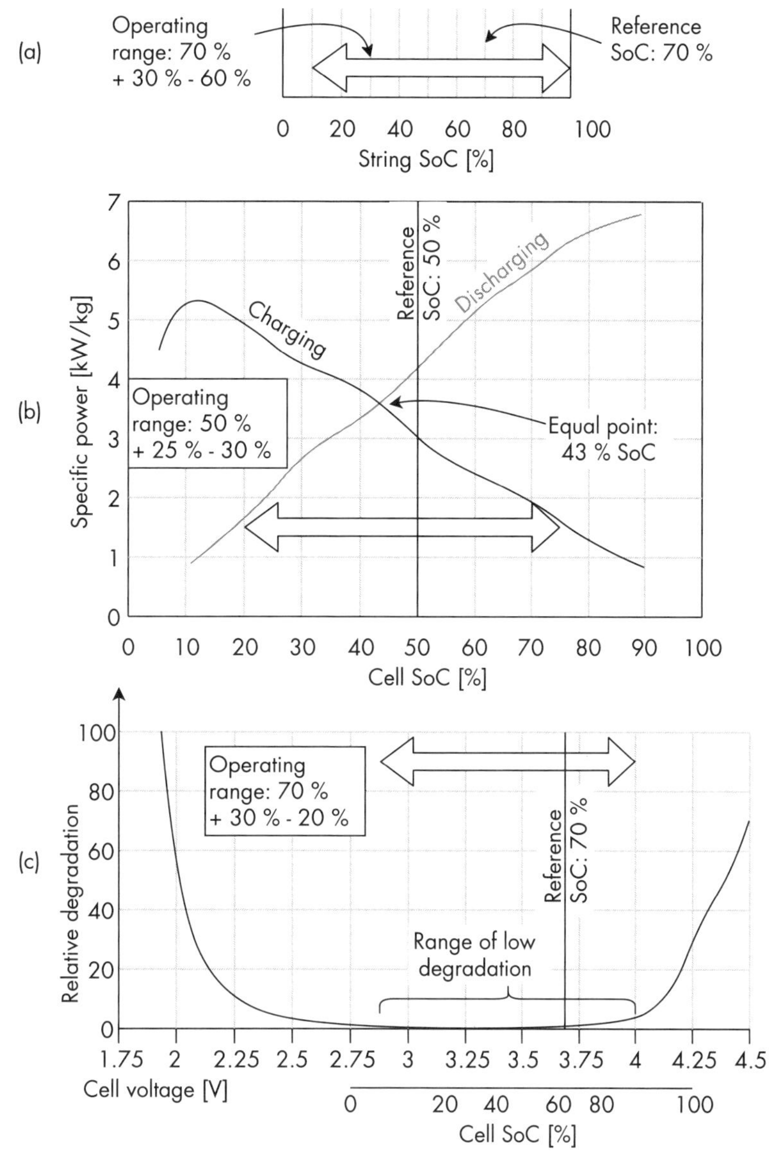

FIGURE 5.5
Asymmetrical operating range: (a) maximum range, (b) equal charging and discharging power, and (c) minimum degradation.

considering the value of your time. Even if your time is free, you're not considering safety and reliability. For example, if you want to make a power bank with cells that you harvested from a laptop, my best advice to you is to buy one instead. Compared to your creation, a manufactured power bank is cheaper, safer, more reliable, guaranteed to work, properly enclosed, and designed to meet regulatory standards.

Maybe a battery is not the right solution to your problem.

5.2.2 The XY Problem

Watch out for the dreaded XY problem. An XY problem occurs when someone asks about their proposed solution (*X*) rather than stating the problem (*Y*) [1].

For example, a post titled "Soldering 18650s" is submitted to a forum.

How do I soldier 18650 batteries?????

You don't solder 18650 cells; you weld them with a specially designed spot welder.

I ain't got no welder I got a soldiering gun.

What are you trying to do?

I want to make a battery with some 18650s I got from laptops.

I mean, what is the *ultimate* problem you wish to solve?

I wanna make a battery.

Yes, I understand. I am asking what the *ultimate* problem is; what is the battery for?

To power stuff.

What stuff?

Me and my brother gonna throw allot of raves in a field.

You want to power a concert with a battery made of recycled 18650 cells?

Yeah how do I soldier to 18650 batteries????

Never mind that: you can't power a rave from laptop batteries. Rent a generator.

Well if you don wanna help screw you ill ask somebody else.

Figuratively, we tend to get stuck in the weeds, keeping us from raising our point of view to see the entire forest. Start by trying to identify your ultimate problem (powering a rave in the middle of a field) rather than your proximate problem (making a battery) or your suggested solution (soldering cells). When asking for help:

- State the ultimate problem;
- Then, present your suggested solution;
- List any solutions that you ruled out and state why;
- Be available to respond to requests for additional information.

If your suggested solution makes sense, others can help you with it. If it doesn't, they can direct you to a better solution.

Let's now look at it from the point of view of the person offering help. Often, it is harder to get someone to identify the problem that needs to be solved than it is to solve the problem. You don't help a person by limiting yourself to their proposed solution (X). You help them by finding out what the ultimate problem is (Y) and then addressing *that* problem.

5.2.3 Battery Design Checklist

You may use this guide as you prepare to design a battery.

5.2.3.1 Design Steps

Generally, a battery design goes through these steps:

- Establishing application requirements (see below);
- Produce battery specifications (see below);
- Propose two or three solutions, and select one based on cost and cell availability;
- Main component selection:
  - Cell selection;
  - BMS selection.
- Cell mounting design;

- High-voltage side design:
  - Protection switch topology, components, and precharge;
  - High-power circuits—fuses, circuit breakers, safety disconnect;
  - Sensing—cell voltage taps, current, battery voltage;
  - Connections—output connector, charging connector;
  - Isolation.
- Low-voltage side design:
  - Sensing—temperature, interlock, supply voltage;
  - BMS power supply;
  - Communications.
- Thermal and mechanical design.

5.2.3.2 Establishing Application Requirements

The first step is to prepare a list of the requirements for the battery:

- The application;
- Desired life (years);
- Number of cycles during its life;
- Nominal battery voltage, maximum and minimum;
- A profile of the current during a complete, typical cycle, and for an extreme cycle;
- Required energy density;
- Cost for the first article, cost in quantity.

These need to be actual values, not wishful thinking. "As much as possible" is not a helpful answer; you need to evaluate your dreams in light of the actual costs of flying to the moon.

The nominal battery voltage may be fixed or may be flexible depending on the application:

- For small batteries, the voltage is standard (e.g., 18V for a power tool, 3.6V for a phone);
- For large, low-voltage batteries, it is set by the equipment (e.g., 48V for a residential inverger);
- For traction batteries, the voltage is based on a load resistance of 2 Ω (see Volume 2, Section 3.2.1.1);
- For high-voltage stationary batteries, it is generally set by the AC line voltage (see Volume 2, Section 4.2.1).

5.2.3.3 Create Battery Specifications

From the above, you (or a battery expert who is helping you) can estimate

- Range of battery voltage;
- Nominal capacity required at the start of life;
- Actual capacity at the end of life;

- Range of SoC over which to operate the cells initially;
- Range of SoC over which the cells will operate at the end of life;
- A cell chemistry;
- An appropriate value for the maximum power time for the cells (see Section 1.5.2).

5.2.3.4 Select Cells and Propose Solutions

Select two or more cells, considering their specifications, cost, and availability. Propose two or three different solutions that meet the battery specifications, using the selected cells. Perform a value analysis of each. Finally, select one of the proposed solutions.

5.2.4 Avoiding Pitfalls

Learning how to design a battery is good, but knowing what *not* to do (preferably from other people's mistakes) is just as important. When designing a battery, be aware of the common pitfalls that have resulted in damaged cells, BMSs, and batteries in the past so you may avoid them (see Chapter 8).

5.2.5 Batteries with Capacitors

Having noticed that some batteries are excellent at energy storage and that capacitors are excellent at power storage, many have considered devices that combine the two technologies to reap the benefits offered by each. The effectiveness of doing so depends on the topology (see Volume 2, Section B.2). TL;DR: it's not worth it.

5.2.6 Second-Use Batteries

For all the talk about reusing batteries in a less demanding application after they have reached their end-of-life in the original application, we don't yet see many examples of it.

Of course, one reason is that Li-ion batteries are still relatively new and are still performing well in their original application. A bigger reason is the significant cost of the labor to disassemble an old battery and to build a new one, plus the uncertainty of the quality of used cells. Building a new battery with new cells offers a better value.

Second-use is economical only if the battery is reused as it, with no modifications. Cooperation of the original manufacturer is required to learn the battery's interface. Otherwise, significant reverse engineering is required. For example, now that production EVs have been widely available for a few years, their traction batteries are starting to reach their end of life. Much has been said of reusing them for stationary, residential applications, at a low discharge rate, so that the increase of internal resistance is not an issue. However, most residential applications today use 48V batteries, while these traction batteries have a voltage of 350V. It's too expensive to convert these batteries to 48V.

High-voltage power converters for residential use are now becoming available, which may allow reuse of traction batteries without modifications.

Regardless, dragging a 400-kg traction battery down the stairs to the basement is impractical. Even if installed in a garage, a traction battery that is built like the fuel tank from a WWII tank is rather inelegant (Figure 5.6).

Second-use is not quite there yet, but we should all work toward it.

FIGURE 5.6
Nissan traction battery.

5.3 COMPONENT SELECTION

A critical step in battery design is the selection of appropriate components.

5.3.1 Talking to Suppliers

During this process, discuss your needs with vendors and manufacturers, who are familiar with all sorts of applications, probably yours as well.

Before you talk to a supplier (see Section 7.10.1), understand your needs clearly and from a high-altitude point of view so you may have a productive discussion with a vendor:

- *Effective:* "We're a solar installation company in Hawaii and we use large 48V lead-acid batteries with a Schneider inverter; we want to switch to Li-ion. Do you offer 1,000-Ah LiFePO4 cells?"
- *Ineffective:* "How is your day going today? How is the weather in Chicago? I am calling because I want to know about your company. I used to have an old car that I was going to convert to electric, maybe one day, the wife says to get rid of it. I heard that Lipo's are bad. We use lead-acid, and you know what the problem with lead-acid is? Let me tell you. Anyway, to my question, what do you think? Because, you know, your competitor doesn't know what the heck they're talking about. Oh, one more thing, we'll be connecting 100 mA batteries in parallel for charging and in serial for discharging."

As a supplier, the first prospect feels like someone we can work with. The second one makes me fear that they will waste weeks of our time, buy at most one system, not follow instructions, screw up their battery, and sue my company for lost revenue.

Having a clear idea of your needs before a call and communicating effectively during the call make a fundamental difference in the quality of the support you receive from a vendor. When inquiring with a supplier, please stick to the pertinent facts. In order:

- Battery voltage
- Type of cell—chemistry, format, size;
- Battery capacity or energy, and current or power;
- How many units, when you need it, and where.

Once suppliers know the above, they can either ask for additional information or quickly turn down the request:

- "Sorry, we don't work with such small batteries."
- "Sorry, we don't ship to Easter Island."
- "Sorry, we can't supply such a large quantity in three weeks."

That should be followed by a "... but I can recommend a company that does" helpful advice.

Knowing enough about your plans, the supplier can make recommendations:

- "Yes, we can support five strings in parallel, but let me suggest that you'll save a bundle of money and have a better performance if you connect five cells in parallel first into a block, and then connect such blocks in series."(see Section 3.5.1).
- "Yes, the large prismatic cells you chose are fine (at least for slow discharge applications); however, for your race car, you really must use power cells, and those only come in pouch and small cylindrical formats." (see Section 2.2.3).
- "Yes, the BMS you chose is high quality, but it is not designed to mate to the cells you chose; you'll be happier with a wired BMS."(see Section 4.2.2).

You may feel the need to ask the supplier to sign an NDA[4] before discussing your application; yet, consider that

- Your invention may not be unique because it's likely that someone has already thought about it and your supplier has already heard it. Either it already exists, or the only reason you haven't seen it before is that it is impractical if not impossible.
- Rather than asking for an NDA, limit the information you disclose. The supplier doesn't need to know that the battery is for a secret project.
- Suppliers know that the likelihood of an order is inversely proportional to the hoops they are asked to jump through. The moment you ask for an NDA, the supplier loses some interest.
- In any case, don't worry, because suppliers know how to be discreet, with or without an NDA, as a condition of being successful business people.

5.3.2 Cells and BMS

Often people ask me: "What's the best Li-ion cell?" Of course, the answer is that the best cell is the most appropriate one for *your* application, at the right price, and that

4. Nondisclosure agreement. A mutual, symmetrical, and concise agreement is preferable.

you can actually buy. The same goes for a BMS: the best one is the most appropriate one for *your* application.

Cell sourcing (see Section 2.8) and BMS sourcing (see Section 4.13) are not straightforward and require particular attention. Table 5.3 offers a general guide for the cell format and BMS type, for a battery built from individual cells. In some cases, the suggested components differ for low-volume and high-volume applications.

The "Application" chapters in Volume 2 suggest cell format and list appropriate off-the-shelf BMSs for each application.

Application Class	**Battery Application, Volume**	**Cell***	**BMS Type**	**BMS Topology**	**BMS Format**
Small	Single-cell, low volume	SmCyl	Analog protector	Mounted PCM	Open
	Single-cell, high volume	Pouch	Digital protector	Mounted PCM	Open
	Consumer, multicell, low volume	SmCyl	Analog protector	Mounted PCM	Open
	Consumer, multicell, high volume	Pouch	Analog protector	Mounted PCM	Open
	Power tool	SmCyl	Analog protector	Mounted PCM	Open
	Power bank	SmCyl	Analog protector	Wired PCM	Open
	Block of cells with connector	Pouch	Analog protector	Wired PCM	Open
Low voltage	Auxiliary power unit	Pouch	Digital BMU	Centralized	Open
	Large stationary	Prism	Digital BMU	Distributed	Enclosed
	Massive stationary	Prism	Digital BMU	Distributed master/ slave	Enclosed
	UPS	Any	Analog protector	Wired PCM	Open
	Modular battery	Any	Digital protector	Wired PCM	Open
	Engine starter	SmCyl	Analog BMU	Wired PCM	Open
	Marine house power	Prism	Digital BMU	Distributed master/ slave	Sealed
Traction	UAV (drone)	Pouch	Digital BMU	Centralized	Open
	Small personal, low volume	Prism	Digital protector	Wired PCM	Open
	Small personal, high volume	SmCyl	Analog protector	Mounted PCM	Open
	Small EV	Prism	Digital protector	Wired PCM	Enclosed
	Small industrial	Prism	Digital protector	Wired PCM	Enclosed
	Racing	Pouch	Digital BMU	Centralized	Enclosed
	Marine propulsion	Prism	Digital BMU	Distributed master/ slave	Sealed
	Passenger, EV conversion	Prism	Digital BMU	Centralized or distributed	Enclosed
	Passenger, high volume	Pouch	Digital BMU	Centralized	Open
	Large industrial Public transit	Prism	Digital BMU	Master/slave or distributed master/ slave	Enclosed
High Voltage	Grid tie, high voltage	Prism	Digital BMU	Banked master/slave	Enclosed

* Cell format: SmCyl = small cylindrical; Prism = large prismatic.

TABLE 5.3 Cell Format and BMS Type Suggestions

The folliwng sections provide some additional guidelines for selecting the cell format.

5.3.2.1 *Small Run*

For a single battery or a small production run, use cells that are easier to mount and connect:

- Large cylindrical cells are the easiest to connect;
- Avoid individual pouch cells; hobbyists are likely to use them, but the life of these cells is short given the abuse experienced during assembly and use. Do consider blocks of pouch cells already mounted in a retaining frame.[5]
- If using small cylindrical cells, get them with tabs already welded on them, or work with an experienced company that can weld them for you. Also consider blocks of cells already connected in parallel in a module.[6]

Cell manufacturers are unlikely to talk to you, so you are pretty much on your own, with just the limited information that is published online. Look for guidance from vendors and online forums, not all of which is to be trusted.

5.3.2.2 *Volume Production*

For small batteries, pouch cells and small cylindrical cells are the best choices for both low- and high-power applications.

For large batteries, consider pouch cells to get high performance tailored to your application. Doing so requires a high upfront investment in engineering, to give the cells the proper mechanical, thermal, and electrical support.

Large prismatic cells remain a viable option for stationary energy batteries, low or high voltage.

Small cylindrical cells are generally not used in large batteries, with the notable exception of Tesla's traction batteries and high-voltage residential batteries.

Small prismatic cells are becoming more available. Some of them may be appropriate for power applications.

5.3.3 Connectors

The lowly connector gets no respect, yet it is an essential part of a reliable design of any electrical product.[7] Many battery failures are due to unreliable or inappropriate connectors. In many applications, batteries are exposed to vibrations and corrosive gases that affect connectors more than any other component. Electronic assemblies can be sealed or at least conformally coated. However, connectors cannot be coated and are rarely sealed. Vibrations degrade mating contact areas (*fretting*) and stress wires crimped to terminals.

5.3.3.1 *Large Batteries, Signals*

For large batteries, use automotive-grade connectors:

- They latch together so they won't fall off with vibration;

5. Enerdel Moxie modules.
6. From K2 Energy.
7. I started writing a book about connectors, but I set it aside until I finish this one. One day.

- The terminals use plating that withstands fretting;
- The terminals have two sets of crimp tangs, one set to clamp on the insulation and one set to clamp the strands;
- A terminal position assurance (TPA) confirms that all terminals are seated correctly;
- Can carry a current of up to 2 or 3A.

Automotive connectors are available in various mounts:

- In-line—free floating;
- Panel mount;
- PCB mount.

Sealed automotive connectors are more expensive but are required in marine and some industrial applications.

For connections to the BMS, you're stuck with whatever connectors the BMS uses. Therefore, select a BMS that uses good connectors.

Things to watch out for:

- Fretting—don't use gold-plated terminals for your connectors (the gold rubs away);
- Bad crimps [2];
- Connectors without latching;
- Connectors not rated for the battery voltage (too short a creepage distance).

To increase the voltage handling of a connector, skipping one position increases the allowed voltage between the two positions adjacent to the skipped one. If a contact is present in the skipped position, the voltage handling is doubled. If the skipped position has no contact, the voltage handling is more than doubled.

To increase the current handling of a connector, connect two positions in parallel. The current handling won't be quite doubled, though, because the current won't divide equally. The contact with the lowest resistance carries more than half the current. Also, the connector may not be able to handle the heat generated by both positions.

5.3.3.2 Large Batteries, Power

The connector industry developed connectors specifically for the power output of batteries. These include

- Anderson hermaphroditic connectors—typical of forklifts and golf carts;[8]
- Electric vehicle traction battery connectors—orange, high quality;[9]
- Single-circuit large plugs—generally used for temporary high-power AC connections;[10]
- Ring terminals and chassis mounted insulated studs or junction blocks.[11]

The individual "Application" chapters discuss these connectors.

---

8. Anderson PowerPole SB series.
9. Yazaki.
10. Leviton 16R22 plugs and receptacles.
11. Eaton C1938.

Watch out for the following when using large gauge wires (cables) and ring terminals screwed onto studs:

- Not properly torqued fasteners;
- Battery current flowing through the fastener;
- The weight of the cable places a torque on a fastener in the direction that will eventually unscrew it.

5.3.3.3 Small Batteries

A small battery is exposed to fewer challenging conditions than a large battery. The standard connectors for small batteries (made by JST) are acceptable even though they provide none of the reliability of automotive-grade connectors (see Volume 2, Section 1.3.1.3).

5.3.4 Other Components

The other components used in a battery vary considerably with the size of the battery and the application. The "Application" chapters in Volume 2 discuss components for each of these applications.

5.4 SMALL CYLINDRICAL CELLS

Each cell format has specific requirements for connection and mounting (see Section 2.2.2). Small cylindrical cells give the battery designer flexibility as they can be physically arranged to fit the available space (Figure 5.7).

5.4.1 Physical Arrangement

Arrange the cells to minimize the length of interconnections. Ideally, you want to use only short bus bars, and no jumper cables.

For a single series string, there are principally five physical arrangements:

- *Row:* Cells are placed side by side in an alternating orientation of polarity. Every pair is welded with nickel strips, the assembly is flipped upside down, and the other pairs are welded. Two terminals are welded on the first and last cell (Figure 5.8(a)).
- *Column:* First welded in a row, as above, but then unfolded into a column, which bends the tabs 180° and butts the positive end of one cell against the negative end of the next cell (Figure 5.8(b)).
- *Multicolumn:* The same as above, but in more than one column (Figure 5.8(c)).

FIGURE 5.7
4P3S small cylindrical cluster in a square pattern.

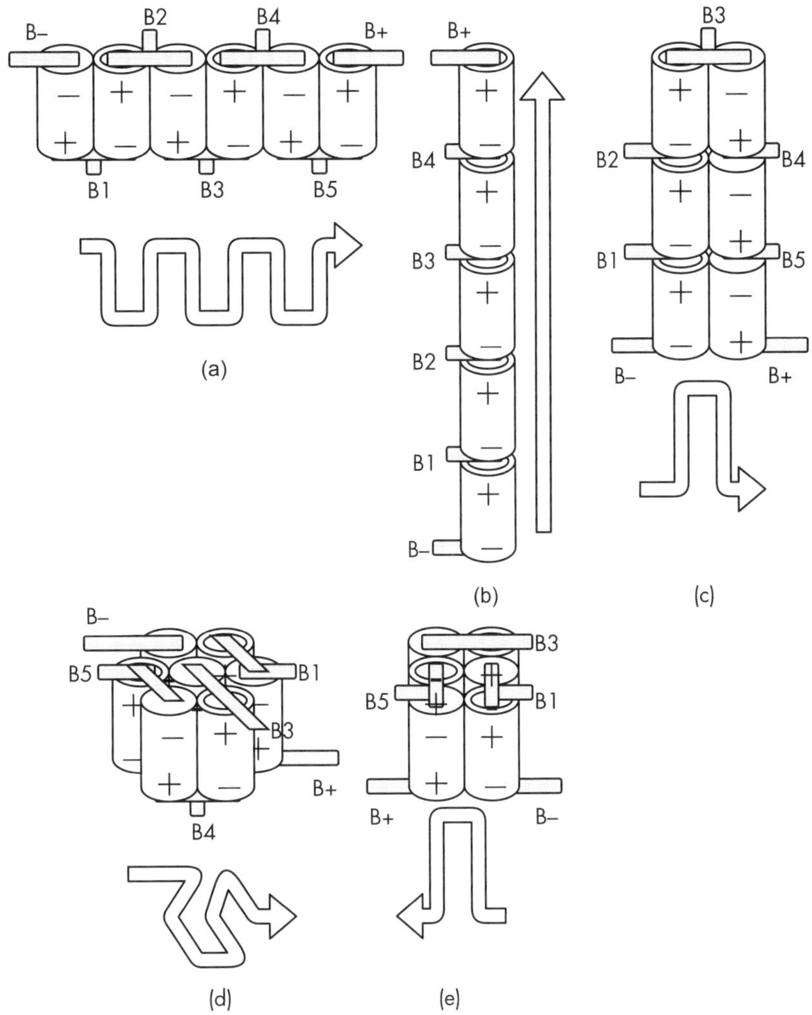

FIGURE 5.8
Small cylindrical, single-series string arrangement: (a) 6S row, (b) 5S column, (c) 6S multicolumn, (d) 7S hexagonal cluster, and (e) 6S square cluster.

- *Hexagonal cluster:* Cells are placed in a hexagonal pattern, carefully oriented. Every pair is welded at the bottom, the assembly is flipped upside down, and the other pairs are welded. Two terminals are welded on the first and last cell (Figures 5.7, 5.8(d)); note that the routing minimizes the voltage drops between adjacent cells for safety.[12]
- *Square cluster:* The same as above, but in a rectangular pattern (Figure 5.8(e)).

Note that each intermediate connection between two cells also includes a stub available as a voltage tap for the BMS.

There are as many physical arrangements for the parallel-first arrangement, including

- *Block:* Cells are placed side by side in a rectangular grid, alternating orientation for every row, and every pair of rows is welded with nickel plates. The assembly is flipped upside down and the other pairs of rows are welded. Two terminals are welded on the first and last row (Figure 5.9(a)).

12. In case of accident and penetration or deformation, an accidental connection would create a short across a lower voltage.

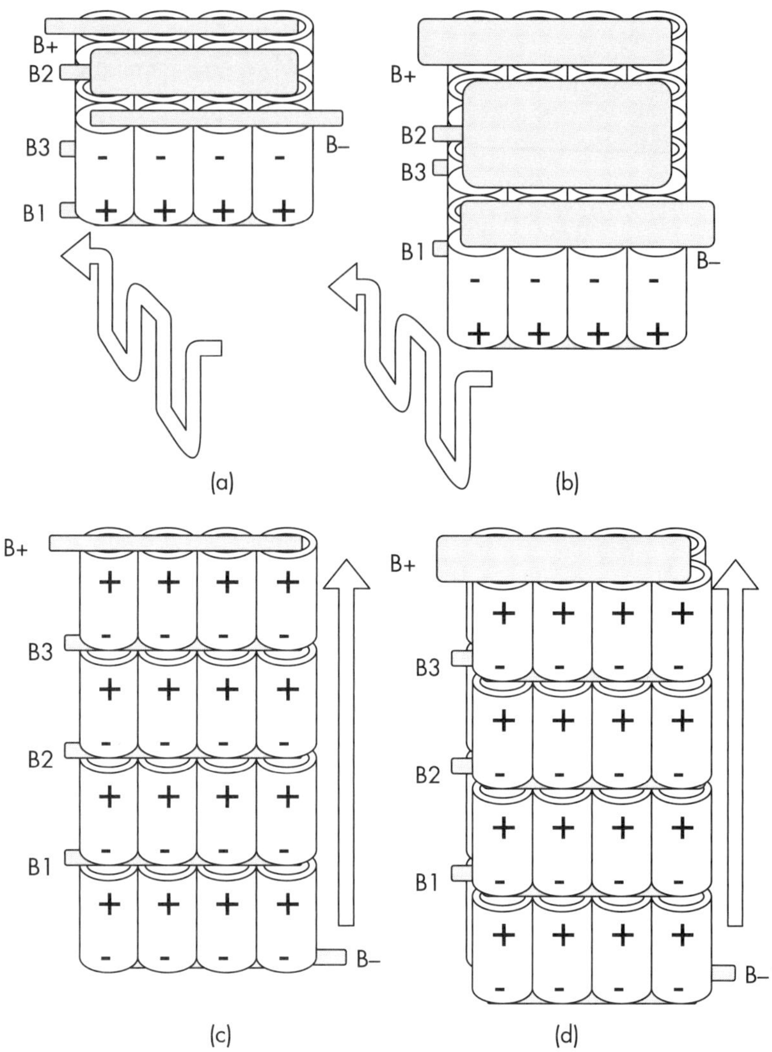

FIGURE 5.9
Small cylindrical, parallel-first arrangement: (a) 4P4S block, (b) 8P4S double-wide block, (c) 4P4S sheet, (d) 8P4S double-wide sheet, (e) 4P4S folded, (f) 4P4S hexagonal cluster, (g) 20P4S hexagonal stack, and (h) 2P4S irregular.

- *Double-wide block:* The same as above, but each parallel block consists of two rows (Figure 5.9(b)).
- *Sheet:* First welded in a block as above, but then unfolded into a sheet, which bends the plates 180° and butts the positive ends of one row of cells against the negative ends of the next row of cells (Figure 5.9(c)).
- *Double-wide sheet:* The same as above, but achieved from a double-wide block (Figure 5.9(d)).
- *Folded:* The same as a sheet, but in two layers (Figure 5.9(e)).
- *Hexagonal cluster:* Cells are placed in a hexagonal pattern, alternating orientation for every row. Every pair of rows is welded and the assembly is flipped upside down, and the other pairs of rows are welded. Two terminals are welded on the first and last row (Figure 5.9(f)).
- *Hexagonal stack*: Large hexagonal clusters are placed on top of each other and connected in series (Figure 5.9(g)).
- *Irregular:* The small size of the cells allows filling the available space with unusual shapes (Figure 5.9(h)).

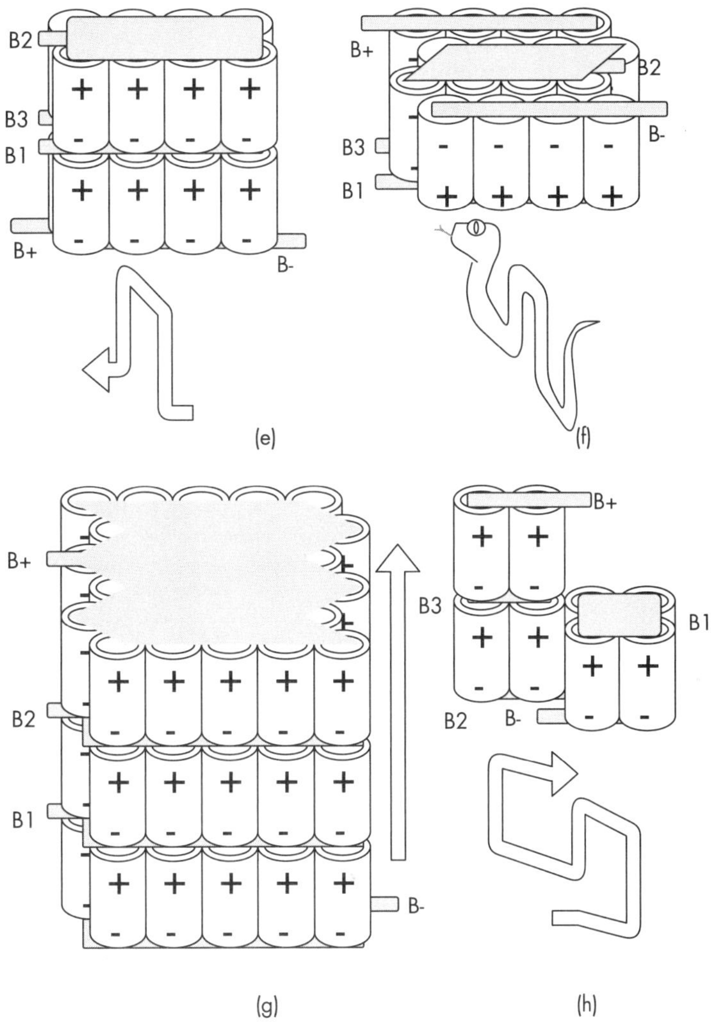

FIGURE 5.9 (continued)

5.4.2 Small Cylindrical Mounting

Small cylindrical cells may be placed in a cell holder, mounted permanently in brackets, or wrapped in heat-shrink tubing.

5.4.2.1 Cell Holder

A single-cell for a small consumer product may be placed in a cell holder.

Cell holders are not appropriate for more than one cell because the user is likely to install two cells at different SoC levels:

- In parallel: Results in a damaging inrush current between the cells;
- In series: Results in a severe imbalance.

Regardless, cell holders are not appropriate for traction batteries because vibrations prevent a consistently reliable contact.

5.4.2.2 Brackets

Commercially available plastic brackets hold many cells parallel to each other in a hexagonal or square pattern (Figure 5.10(a)), after which the cells may be welded

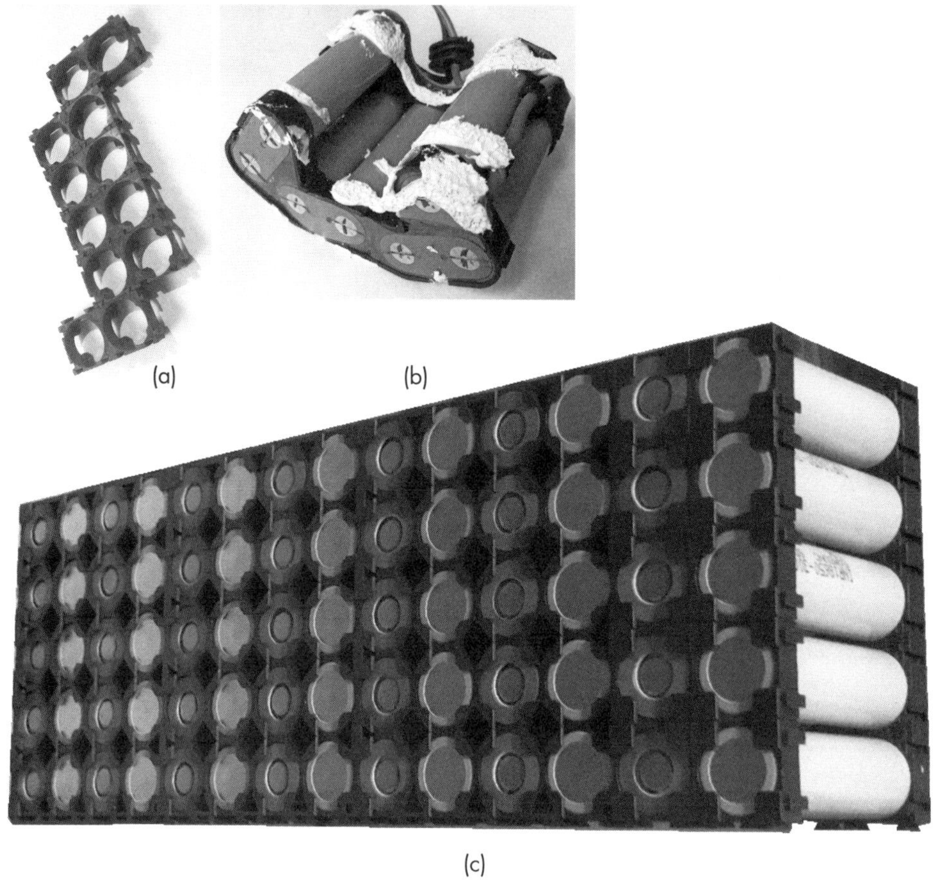

FIGURE 5.10
Brackets for 18650 cells: (a) off-the-shelf, modular, (b) custom, and (c) off-the-shelf modular frame.

(Figure 5.10(c)). Custom brackets (Figure 5.10(b)) may be injection-molded or 3-D printed for small-volume production.

5.4.2.3 *Heat-Shrink Tubing*

A small block of cells may be used without any bracket and shrink-wrapped to maintain its shape (see Section 5.4.6).

5.4.3 Small Cylindrical Interconnection

Cells that come with welded tabs allow a hobbyist to solder to the tab, with some care.

Soldering directly to the cell is not allowed because the heat degrades the cell and may damage it. Plenty of hobbyists solder cells, oblivious to the damage to the cell because they are not equipped to measure the increase in internal resistance and the reduction in battery life.

Just. Don't. Do. It.

Welding a nickel tab or plate is the proper way to connect to the cells. Spot welding is standard, while laser welding is appropriate for high-quality, volume production. Cells are spot welded in four spots for increased reliability and reduced resistance. If welding copper plates for high-power batteries, the spot welder must use copper electrodes rather than the standard tungsten electrodes; laser welding works fine on copper plates.

Take care to avoid blocking the safety vent. The cell spec sheet should indicate an area where welding is not allowed.

5.4.4 Small Cylindrical Sensing

For small consumer batteries, the PCM is placed right against the cell. The interconnecting nickel strips include small tabs that are folded onto the BMS PCB and soldered. Short wires are used for harder to reach tap points.

For large batteries using a wired BMS, wires are routed from the BMS to the tap points along the blocks of cells. This can get rather messy (Figure 5.11(a)). For temperature sensing, a thermistor could be placed against one of the cells in the middle of its long side.

For large batteries using a distributed BMS, cell boards are mounted directly to a cell in each parallel block (Figure 5.11(b)). Cell boards in a distributed BMS sense the cell temperature directly when placed against the cells. Wires or cables connect adjacent cell boards. Communications with the BMU occur through a cable connected to the first and the last cell board in a bank. Cell boards must be easily accessible for troubleshooting. The resulting layout is often cleaner compared to a wired BMS, not to mention safer.

5.4.5 Small Cylindrical Cooling

Small cylindrical cells can be cooled by airflow through the open space between them. To remove heat for the short term, they can be installed inside phase-change material (see Section 5.17.7.2). There are no clean solutions for water cooling.

5.4.6 Small Cylindrical Enclosing

For small batteries, the cells in a block may be secured to each other in various ways to maintain the shape of the block—tape, shrink-wrapping (Figure 5.12(a)), adhesive, heat-seal, mechanical fasteners. A thermistor is usually placed against a cell and included in the wrap.

The ultimate packaging solution for a small battery is to have a custom plastic enclosure that keeps the block of cells in place and protects them from mechanical abuse (Figure 5.12(b)). The enclosure is ultrasonically welded to seal the battery and discourage the user from opening it.

Large blocks of cells may be placed into nonconductive enclosures[13] or isolated with fiberglass sheets and placed into metal enclosures (Figure 5.12(c)).

5.5 LARGE PRISMATIC CELLS

Large prismatic cells are the easiest to install and connect.

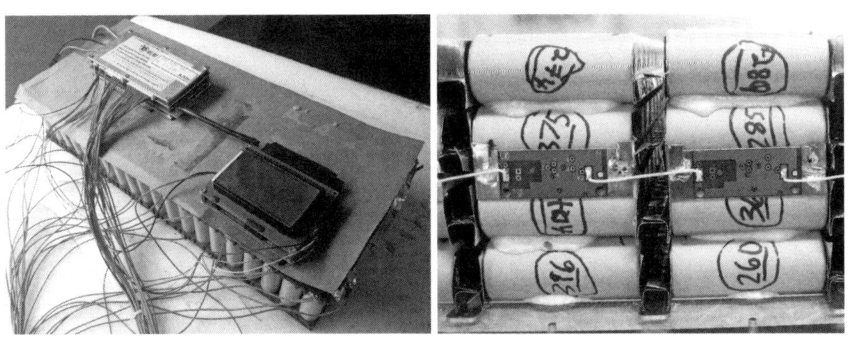

(a) (b)

FIGURE 5.11
Small cylindrical sensing: (a) wired, and (b) cell boards (distributed).

13. Note that carbon-fiber is conductive.

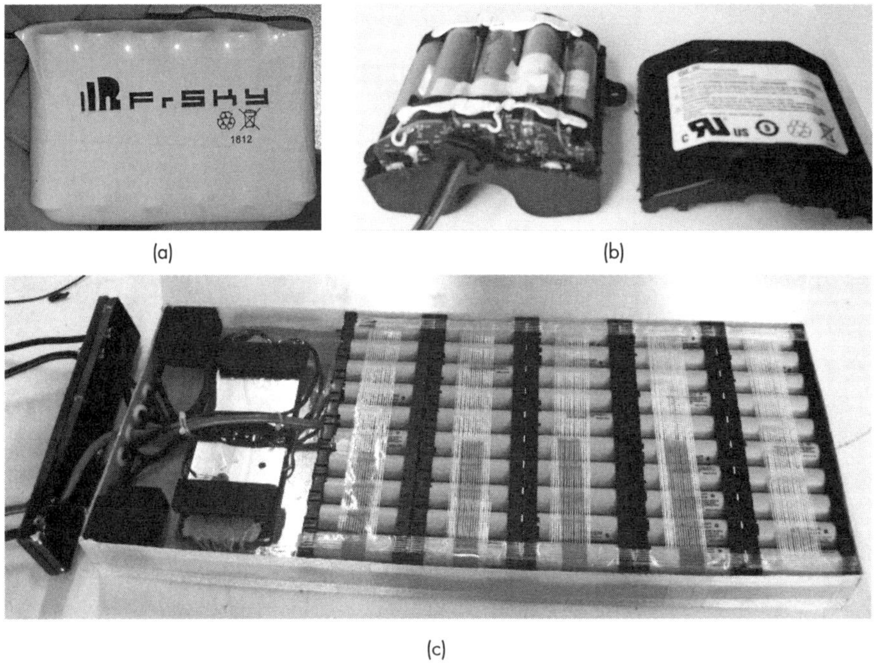

(a) (b)

(c)

FIGURE 5.12
Custom small battery enclosure: (a) heat shrink, (b) custom plastic case, and (c) fiberglass tray and metal case.

5.5.1 Large Prismatic Physical Arrangement

Arrange the cells to minimize the length of interconnections. Ideally, you want to use only short bus bars and no jumper cables.

Using an even or odd number of rows has a significant effect. So does using an even or odd number of cells in each row. For a rectangular box:

- For a single row of cells, the terminals are at opposite ends of the battery (Figure 5.13(a));
- To place the terminals next to each other, use two rows, each with an even number of cells (Figure 5.13(b)):
- To place the terminals on the same face, as far as possible away from each other, use an even number of rows, each with an odd number of cells (Figure 5.13(c));
- To place the terminals at opposite ends of the battery, use an odd number of rows, each with an odd number of cells (Figure 5.13(d));
- For parallel-first arrangement, orient the cells so that they may be connected in parallel first into blocks; then blocks may be connected in series (Figure 5.13(e)).

Of course, many other arrangements are possible (Figure 5.14).

When fitting an existing space with cells, it's more convenient to play with mock cells made of Styrofoam (Figure 5.15(a)).

5.5.2 Large Prismatic Mounting

Large prismatic cells must be mounted in a way that constrains them against expansion.

5.5.2.1 Stand-Alone Block

For a self-contained block, large prismatic cells are placed upright, flat against each other to form a block. The block is sandwiched between two hard metal plates (placed at the two ends of the block), and the entire assembly is strapped with metal or nylon

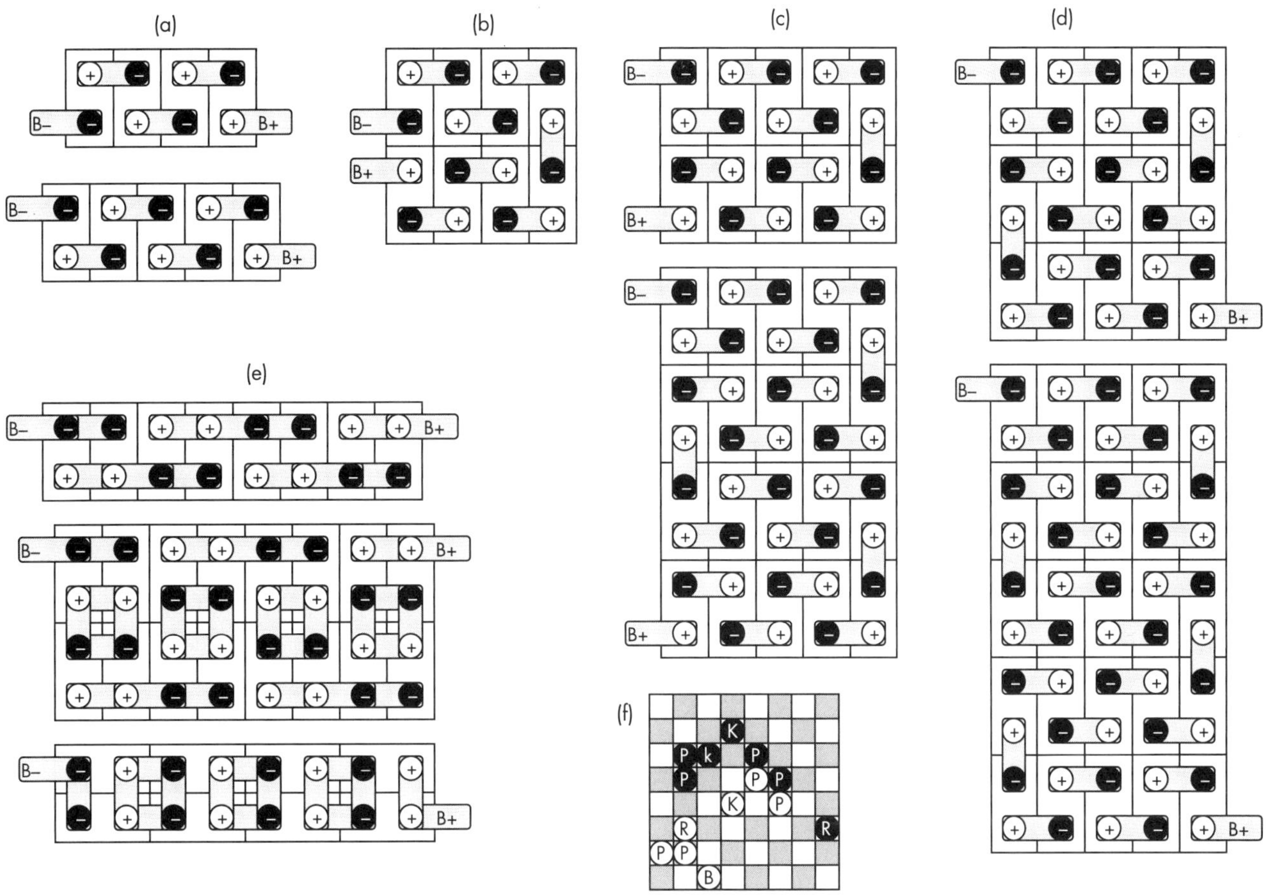

FIGURE 5.13 Large prismatic cell arrangement: (a) four- and five-cell row, (b) even cells per row, even rows, (c) odd cells per row, even rows, (d) odd cells per row, odd rows, (e) parallel-first, and (f) checkmate!

bands to contain swelling (Figure 5.15(b)). Finally, the complete assembly is placed in a battery container that does not need to provide retention.

5.5.2.2 Restraining Container

Individual cells are dropped into a metal container that is shaped precisely for the number of cells for retention against cell swelling. Fiberglass sheets are used to take up any remaining space and to provide extra isolation between the metal enclosure and the cells. An adjustable plate may be used to compress the cells after they are installed (Figure 5.15(c)).

5.5.3 Large Prismatic Interconnections

Before use, clean the oxidation on the terminals with a scouring pad,[14] then coat them with antioxidant paste.[15]

Use heavy-duty interconnects between adjacent cells:

14. Steel wool works, though it leaves metallic particles; Scotch-Brite scouring pads do not.
15. Noalox

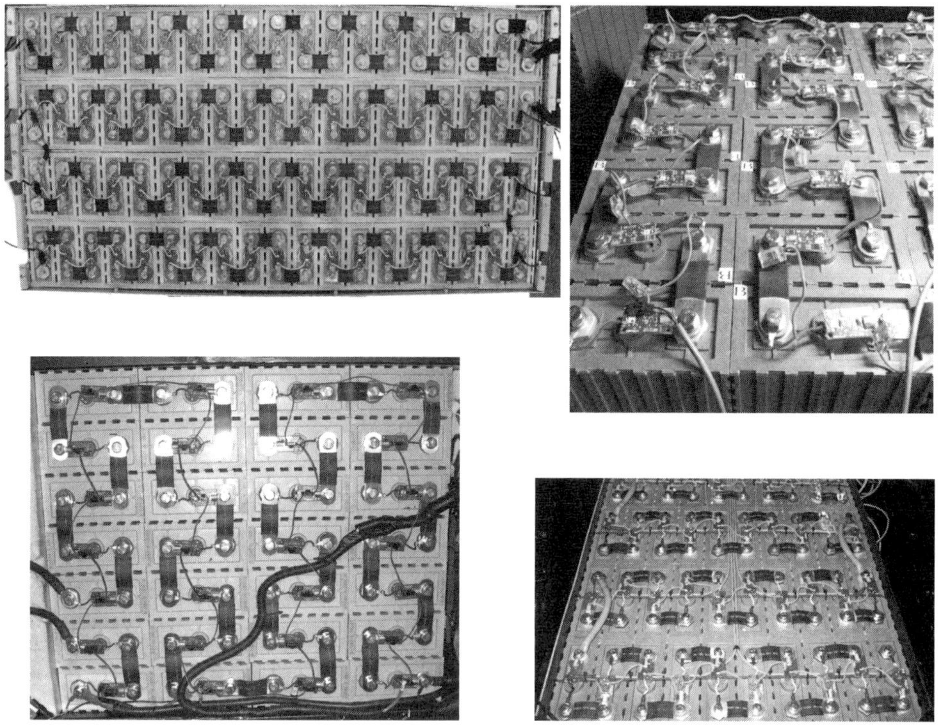

FIGURE 5.14
Examples of large prismatic cell arrangements; note the cell boards for a distributed BMS.

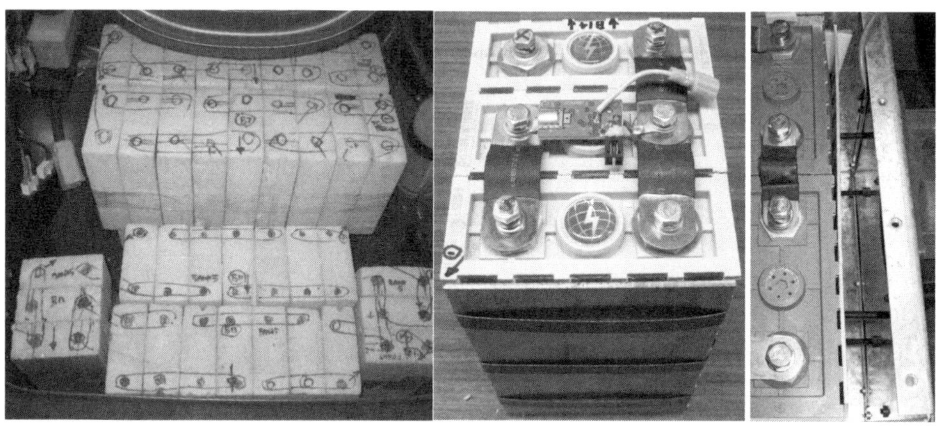

(a) (b) (c)

FIGURE 5.15
Large prismatic: (a) mock-up, (b) retention with end panels and banding, and (c) retention with compression plate.

- Thick bus bar (Figure 5.16(a)) for cells that do not move relative to each other;
- Multiple-hole bus bar (Figure 5.16(b)) for GBS-branded cells;
- Resilient, omega-shaped bus bars composed of multiple thin layers of copper (Figure 5.16(c));
- Large braid terminated with two ring terminals (Figure 5.16(d)).
- Solid bars are not recommended because they stress the terminals in case of vibration, rocking, or thermal expansion.

The proper stack of assembly, top to bottom, is (Figure 5.16(e))

- Bolt;
- (Split washer);
- Flat washer;

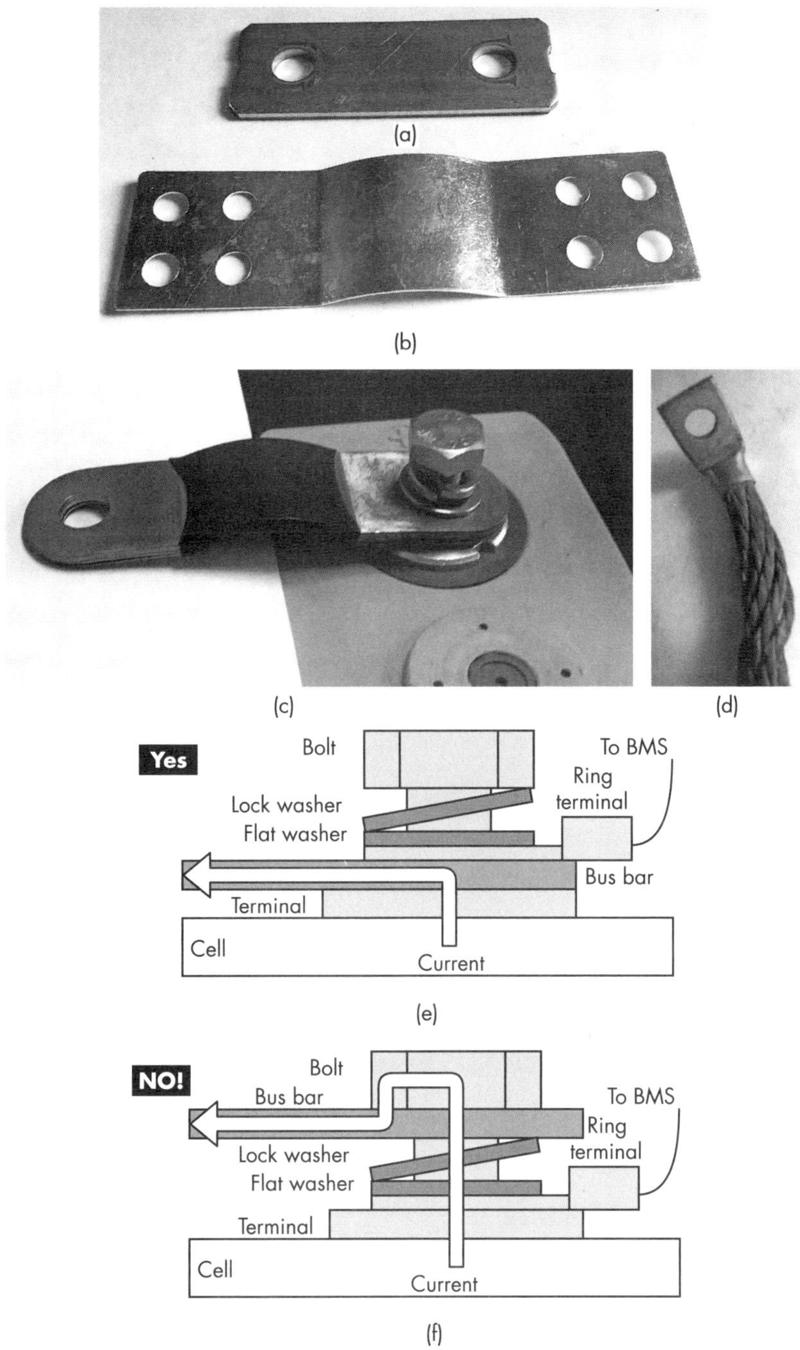

FIGURE 5.16 Prismatic cell interconnection: (a) flat bus bar, (b) bus bar for GBS cells, (c) multilayer bus bar, (d) bus bar mounted on cell, (e) proper order of fastening hardware, and (f) current must not flow through the bolt.

- Ring terminal for voltage tap to BMS;
- Bus bar;
- Terminal.

Cell manufacturers include split washers with the hardware; however, analysis in the automotive industry has determined that split washers do not improve the reliability of the fastening as much as proper torquing does. If resilient fastening is required, a Belleville washer is more reliable and provides uniform pressure throughout its perimeter.

Make sure that the current flows directly from the large diameter body of the terminal (in the body of the cell) into the bus bar or ring terminal (Figure 5.16(e)). That is, make sure that the current does not flow up through the bolt and back down to the bus bar and ring terminal (Figure 5.16(f)). In particular, the ring terminal for the tap wire to the BMS must not be placed between the cell terminal and the bus bar.

Verify that the bolt engages at least three turns when securing the stack-up of the washers, bus bar, and BMS ring terminal. Conversely, verify that the bolt does not bottom out inside the threaded hole in the cell terminal.

Make sure that the nut securing the terminal to the cell is lower than the core of the terminal so that the bus bar contacts the core, not the nut (Figure 5.17(a)). If the nut is too tall, the current tries to flow through the thread-locking compound between the terminal and the nut, offering high resistance (Figure 5.17(b)).

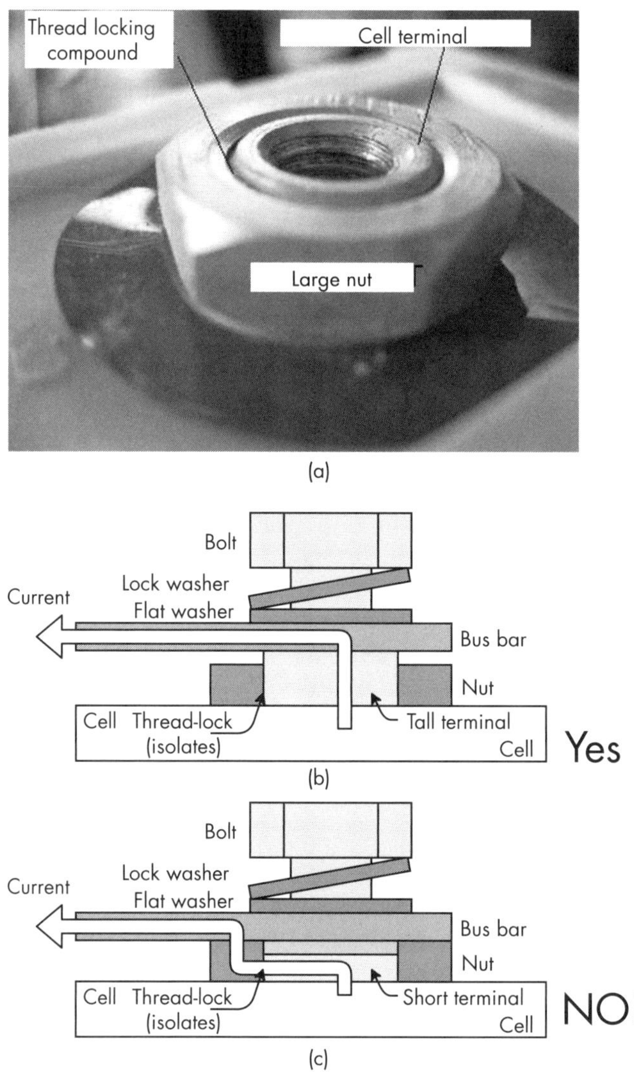

FIGURE 5.17 Prismatic cell terminal: (a) thread locking compound isolates the nut from the threaded terminal, (b) with a tall terminal, current flows directly from the terminal to the bus bar, and (c) with a short terminal, current would try to flow through the nut, which is isolated by the thread-locking compound.

If a ring terminal uses heat-shrink tubing, make sure that none of it extends onto the ring; any heat-shrink tubing between the ring and the cell's terminal will prevent proper contact.

5.5.4 Large Prismatic Sensing

If using a wired BMS, the voltage tap may be connected directly to a cell terminal. The wire is crimped to a large ring terminal, with an internal diameter suitable for the bolt. The ring terminal is placed above the bus bar. Alternatively, the voltage tap could be connected to the bus bar between cells using a small ring terminal.

For temperature sensing, some people place a thermistor against a cell, in the middle of its flat face. The typical cell has large grooves on its face, leaving space for a thermistor. Others place it is against a terminal, which is in direct thermal contact with an electrode inside the cell. The thermistor and its wires must be well insulated to prevent an accidental connection to a cell terminal.

For a distributed BMS, the cell board sits between the two terminals, high enough not to interfere with the vent (Figure 5.15(b)). The cell board senses the temperature of the terminal through its ring terminal, which is mounted directly to the cell terminal. A cell board could have a short-leaded thermistor that reaches to the body of the cell, though I have not seen an example of this solution.

5.5.5 Large Prismatic Cooling

Cooling large prismatic cells is not straightforward. In practice, that is not a problem, because these cells are used only in energy batteries; power batteries would use power cells, which come in a different format.

Still, some do cool large prismatic cells, by running air through the horizontal channels in their cases; liquid cooling is not practical.

5.5.6 Large Prismatic Enclosing

Many large batteries that use large prismatic cells use five-sided metal boxes with barely a cover to protect the top (e.g., EV conversions, large stationary installations) (Figure 5.18(a)).

Products manufactured in high volume (e.g., marine batteries) are more likely to use a rugged plastic box and a sealed top (Figure 5.18(b)).

5.6 POUCH CELLS

Pouch cells require the highest battery design effort, in return for which they can provide optimized performance for a given application. Pouch cells don't last long in a poorly designed battery; allowing pouch cells to expand degrades them (Figure 5.19(b)).

5.6.1 Pouch Physical Arrangement

By and large, multiple pouch cells are stacked face-to-face (Figure 5.19(a)). For a series string, cells alternate polarity (Figure 5.20(a)). For a parallel-first arrangement, several cells are oriented the same way and connected in parallel to form a block. Then, blocks are placed in alternating polarity and connected in series (Figure 5.20(b)).

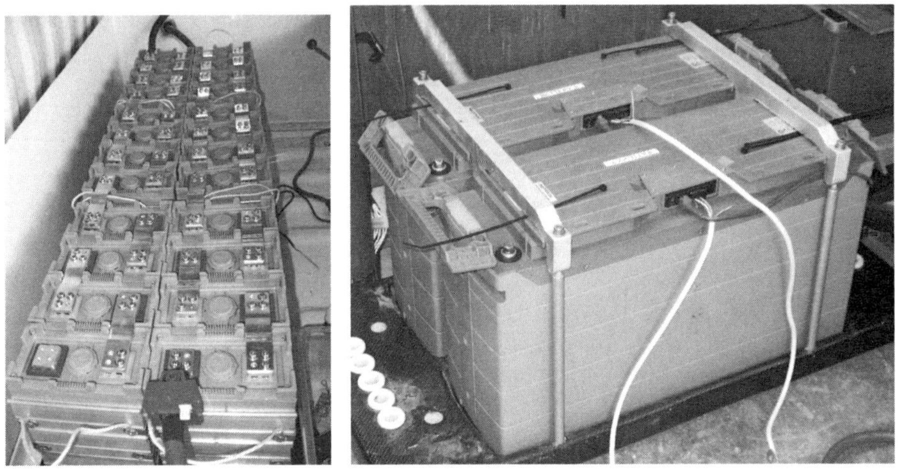

(a) (b)

FIGURE 5.18
Large prismatic cell enclosures: (a) five-sided box, and (b) two marine 24V batteries, each enclosing eight 100 Ah large prismatic cells.

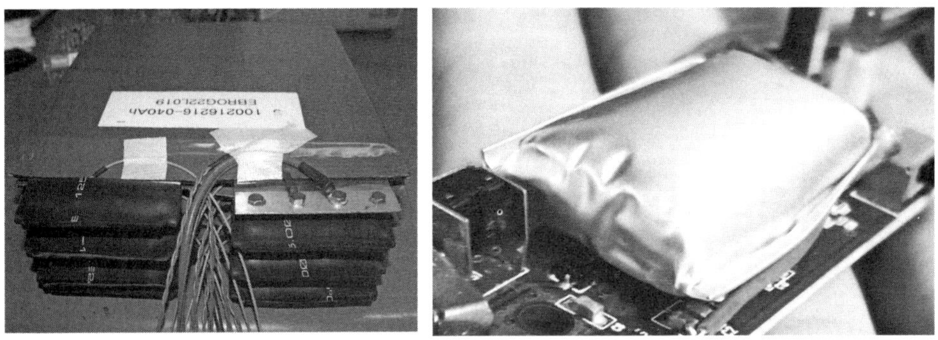

(a) (b)

FIGURE 5.19
Pouch cells: (a) stack, and (b) cell in a cheap consumer product, dangerously expanded.

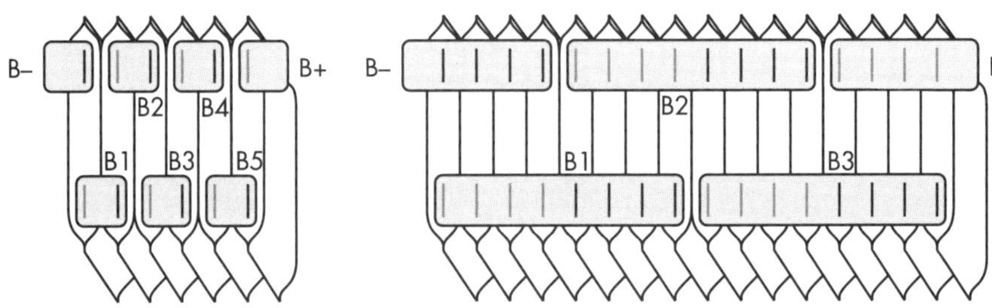

(a) (b)

FIGURE 5.20
Pouch cell physical arrangement: (a) series, and (b) parallel-first.

5.6.2 Pouch Mounting

For a large battery, cells should be placed in frames that ease assembly of blocks, give mechanical support, remove stress on the tabs, and protect the cells from accidental shorts (Figure 5.21(a)).

Any pressure should be evenly applied to the entire flat surface of the cells, without any sharp points or edges. Pressure must not be applied to the edges of a pouch cell. The tabs must not experience any mechanical stress. Several such frames can then be stacked to form a module (Figure 5.21(b)).

The flat sides of a stack of cells must be retained with thin resilient sheets (rubber foam) to allow a small degree of expansion yet prevent excessive expansion.

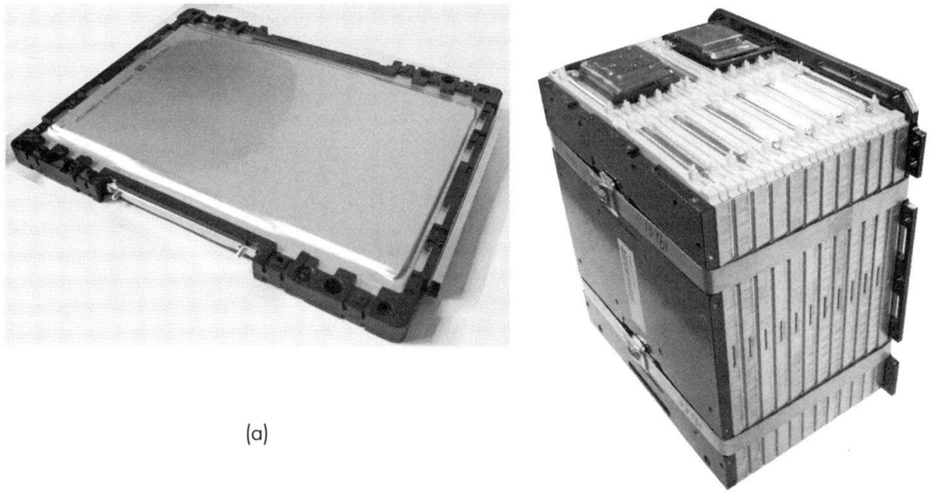

(a) (b)

FIGURE 5.21
Pouch cells: (a) single frame, holding two pouch cells, and (b) block of 2P12S cells, composed of 12 such frames.

Reputable cell manufacturers specify the characteristics of this resilience sheet, but only to qualified customers.

Pouch cells are infamous for being housed in bags that are not completely isolated and being slowly discharged once placed in contact with metal cases or metal cooling plates. Therefore, do not place the body of pouch cells in direct contact with conductors (especially the edges); place dielectric tape along the seams and insulating sheets of the two faces.

5.6.3 Pouch Interconnection

Whenever handling the tabs, wear gloves to prevent contamination of the metal surface. Otherwise, your fingerprints will mark the tabs permanently.

If the tabs are too long, do not use metal scissors, which could cause an accidental short circuit; instead use ceramic scissors that are strong enough to cut both aluminum and copper.

5.6.3.1 Welding

For high-volume production, the terminal tabs should be welded with a laser welder. However, this prevents the replacement of an individual cell in a battery.

5.6.3.2 Clamping

For prototypes and small-volume batteries, tabs of adjacent cells are sandwiched between two plates and clamped together. The tabs may be folded first into a spiral to maximize the contact area. Clamping does allow the replacement of an individual cell.

5.6.3.3 Soldering

With care, it is possible to solder to the copper of the negative terminal tab. Soldering to the aluminum or the positive terminal tab is nearly impossible—it has to be done under oil to prevent oxidation.

Some manufacturers of small cells plate the negative tab to make it solderable. Pouch cells are soldered in small consumer batteries and hobby projects. For a single-cell battery, the tabs may be soldered directly to the PCM.

Soldering time should not exceed 8s for small tabs and 12s for large ones, preferably using lead solder, at 400°C/750°F.

Excessive heat from soldering damages or at least degrades the cell; use a temporary clamp between the body of the cell and the area being soldered to divert some of the heat away from the cell.

5.6.4 Pouch Sensing

The voltage and temperature of pouch cells are sensed with the following methods.

5.6.4.1 Voltage Sensing

If the tabs are clamped, voltage taps may use wires with small ring terminals screwed into the clamping plate. If the tabs are soldered, the BMS tap wire may be soldered there as well. If the tabs are welded, the tap wire may be terminated with a spade terminal that is welded to a tab.

5.6.4.2 Temperature Sensing

For temperature sensing, you may be tempted to place a thermistor in the middle of the pouch cell's flat face. However, doing so puts uneven pressure on the cell, damaging it. A safe placement is against a tab, which is in direct thermal contact with an electrode inside the cell.

A cell board in a distributed BMS senses the tab temperature through the electrical connection to the tab, which must have a low thermal resistance.

5.6.5 Pouch Cooling

Pouch cells may be air-cooled or liquid-cooled.

5.6.5.1 Air Cooling

In a stack of pouch cells, thin plates are placed between adjacent cells to cool them. The plates protrude past the stack and are folded into a channel. The plates draw heat from the full face of the cells and carry it to the channel (Figure 5.21(b), vertical rails on the front side). The plates do not touch each other, to avoid slowly discharging the cells due to the poor isolation of the bag of a pouch cell.

5.6.5.2 Liquid Cooling

Airflow through the channels cools the cells. For liquid cooling, an isolating sheet is placed against all the channels and a cold plate is placed against the sheet (Figure 5.22).

5.6.6 Pouch Enclosing

For small consumer products (e.g., cell phones), a single-cell is mated to a BMS and a connector and is enclosed in a thin yet remarkably sturdy plastic case.

For cheap consumer products (e.g., Bluetooth headphones), a single cell floats in the product's enclosure, battered as the product is treated roughly. The cell doesn't last long.

For UAVs and personal mobility, a stack of cells is wrapped in heat shrink, and that's pretty much it for enclosure. Again, the cells don't last long.

For EV traction batteries, modules (Figure 5.21(b)) are placed in a metal enclosure that includes thermal management (Figure 5.23).

FIGURE 5.22
Liquid cooling of pouch cells.

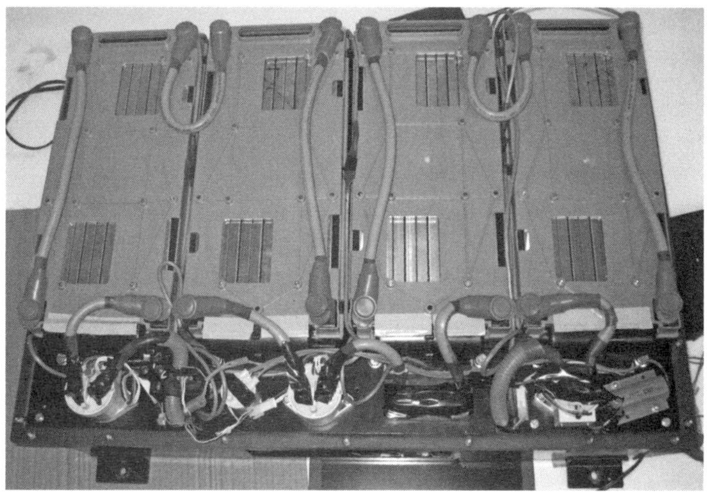

FIGURE 5.23
Pouch cell modules in a traction battery.

5.7 OTHER CELL FORMATS

Large cylindrical and small prismatic cells are mounted and connected similarly to large prismatic cells, with small differences.

5.7.1 Small Prismatic

The guidelines for large prismatic cells apply also to *small prismatic cells*, except that the latter cells may be connected in a slightly different way because instead of threaded holes, they have either have threaded studs, so they require nuts rather than screws (Figure 5.24) or must be welded, like small cylindrical cells.

5.7.2 Large Cylindrical

Large cylindrical cells are mounted and connected like large prismatic cells with a few differences:

- The cells are mounted between two parallel plates (Figure 5.25(a)) or in modular plastic frames (Figure 5.25(b));
- Each cell board in a distributed BMS needs a wire to reach across to the other plate to sense the other end of the cell.

FIGURE 5.24
Small prismatic cells mounting.

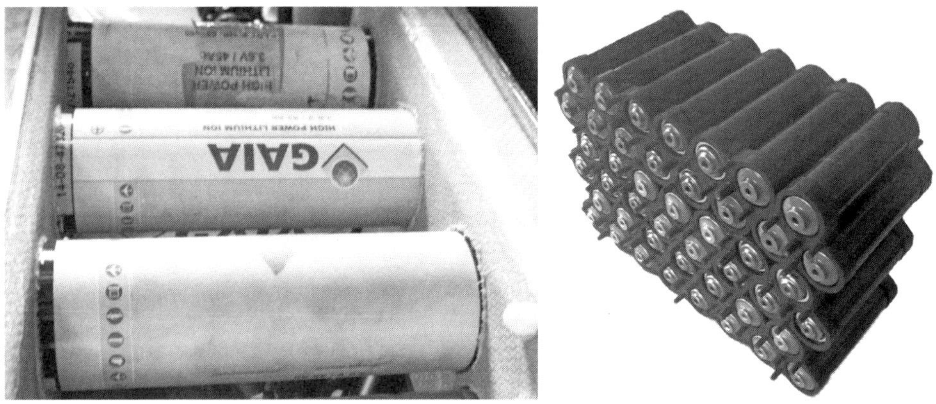

(a) (b)

FIGURE 5.25
Large cylindrical cells: (a) mounted between two fiberglass plates, and (b) on modular plastic frames.

5.8 BMS INSTALLATION

Of course, mounted electronics (cell boards, bank boards, mounted PCMs) must be mounted directly on the cells. However, you do have a choice on where to mount the other assemblies.

For example:

- If a module contains some cells that are managed by a slave, the slave could be mounted inside the module, outside of it, or even far away;
- A master could be mounted inside a traction battery, just outside it, or in the cabin next to other electronic control units, in a place that is convenient for setup, monitoring, and troubleshooting.

A modular BMS topology gives the battery designer more flexibility.

5.8.1 BMU Power Supply Source

The BMS requires power to operate. A PCM is powered by its battery; a BMU for low-voltage batteries may be as well (see Section 4.10.1). Otherwise, the BMU has one or more power supply inputs.

What should power it?

- The string of cells (before the protector switch)? (Figure 5.26(a);)

- The battery's output terminals (after the protector switch)? (Figure 5.26(b));
- Something else? (Figure 5.26(c)).

None of these options is ideal, because

- If connected before the protector switch, the BMS does not have a way to stop current going into its power supply when the battery is left empty for a long time. The battery will eventually be overdischarged and will be damaged.
- If connected after the protector switch, once the BMS turns off the battery, it also removes power from its own power supply, shutting itself down. An external supply must be connected to the output of the battery to revive it.
- Powering the BMS externally shifts the responsibility elsewhere.

Various applications have different solutions. The "Application" chapters in Volume 2 discuss how the BMS may be powered in each application.

The device powering the BMS may be

- None, if the BMS can be powered directly by the available voltage (e.g., a 12V battery or a BMS with a wide range of input voltages);
- A DC-DC converter or a voltage regulator, to reduce the voltage, from the available voltage (e.g., 48 Vdc, 350 Vdc) to the voltage required by the BMS (typically 12Vdc);
- A power supply, powered by an AC power outlet.

5.9 SENSING

This section discusses the devices that allow the BMS sense cell voltages and temperatures and battery current.

5.9.1 Cell Voltage Sensing, Temperature Sensing

The method to sense the cell voltages and temperatures must be designed to minimize the chance of miswiring and short circuits. It must also minimize the susceptibility to electrical noise from nearby power electronic devices.

5.9.1.1 Wired Centralized BMS

Centralized BMSs, wired PCMs, and wired slaves sense the cell voltages through a harness of tap wires.

The harness should be designed as follows:

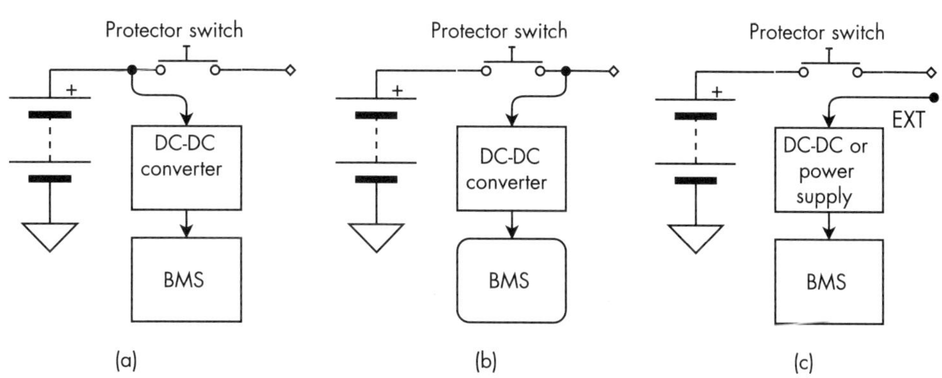

FIGURE 5.26 Possible ways to power a BMS: (a) before the power switch (DON'T!), (b) after the power switch, and (c) externally.

- A block on N cells in series requires N+1 wires;
- To reduce the chances of miswiring, each wire should be the precise length to reach its cell;
- At the BMS end, terminals are crimped onto the wires and then inserted in an N+1 position connector;
- At the cell end, each wire is terminated with a terminal suitable for connection to the cells:
  - Large prismatic cells—a large ring terminal with an internal diameter for the bolt size;
  - Small cylindrical cells—a plain stripped wire, to be soldered to a tab in the nickel strip between cells;
  - Pouch cells, clamped tabs—a small ring terminal to be installed on one of the screws used to clamp the pressure plates that sandwich two tabs together;
  - Pouch cells, welded tabs—a thin, flat terminal, to be welded to the tabs;
  - Pouch cells, soldered tabs—a stripped wire to be soldered to the tab.

A ready-made harness provided by a BMS manufacturer is terminated at one end with a connector for the BMS and at the other end with long and unterminated wires. To adapt it to the battery, the wires are cut to the appropriate length and are terminated as described above (Figure 5.27(a)).

The tap wires must use insulation rated for the maximum battery voltage. Typical voltage ratings for hook-up wire are 150, 300, and 600V. For higher voltages, slip the wire into an insulating tube or other insulation. High-voltage wire is available in 750, 1,000, and 1,500V.

Use small gauge wire, just large enough to carry the balance current. A small gauge allows the wire to act as a fusible link in case of a short circuit. For example, 30 AWG wire can carry 0.86A (more than the balance current of most BMSs) yet fuses at 10A.

Race rules for student races require a fuse for each tap wire, close to the cell; for this reason alone, a distributed BMS, which does not use tap wires, may be better.

A second harness senses the temperature using one or more thermistors (Figure 5.27(b)). Since there are so few thermistors, choose their placement carefully to report the temperature of the entire battery.

The thermistor isolation is critical, especially if the thermistors are referenced to the low-voltage control circuits because a loose thermistor may cause a short circuit. Use double insulation to minimize the chance of such an event.

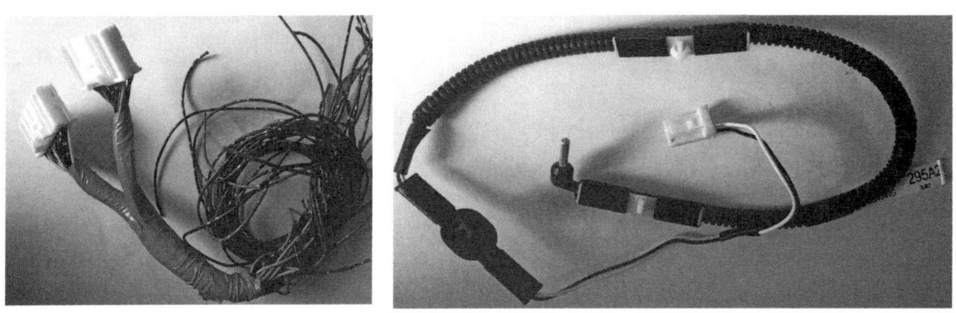

(a) (b)

FIGURE 5.27 Sensing for a wired BMS from a traction battery: (a) voltage sense wire harness, and (b) thermistor assembly.

5.9.1.2 Wired Bank Board

A wired bank board uses individual voltage tap wires without a connector, rather than a harness.

5.9.1.3 Mated Bank Boards and PCMs

Bank boards and mated PCMs are mated directly to the cells to sense their voltages (Figure 5.28(a)). The shape of the board inherently prevents miswiring. A few wires may still be required to sense the voltages at the far end of the bundle of cells.

A thermistor on a long tail is typically included. It must be placed against a cell and secured.

5.9.1.4 Distributed Cell Boards

A cell board is mounted on each cell. Adjacent cell boards are interconnected with a single wire or a multiple-wire cable (Figure 5.28(b)). If a single wire, it should be routed along the corresponding power connection and away from grounded metal (the battery enclosure) to minimize common-mode noise see Section 4.12.3).

5.9.1.5 Banking

A BMS for large batteries divides the cell boards into banks (see Section 4.5.1.8).

Banks are electrically isolated from each other and can handle any voltage between them. Therefore, any of these items should be placed between banks:

- Any item that might open the battery circuit—a fuse, a connector, a safety disconnect, a contactor;
- Any indirect connection—a long cable between modules.

If the battery is divided into modules, assign a bank to each module. If the module has more cells in series than the bank can handle, use two or more banks to handle it. Try to distribute the cells equally among banks. For example, in a 16-cell module with a BMS that can handle up to 12 cells per bank, assign eight cells to each bank.

5.9.1.6 Numbering

If at all possible, number the cells and the banks in the same way as the BMS (see Section 4.5.1.9). This will save many headaches later during troubleshooting. Number

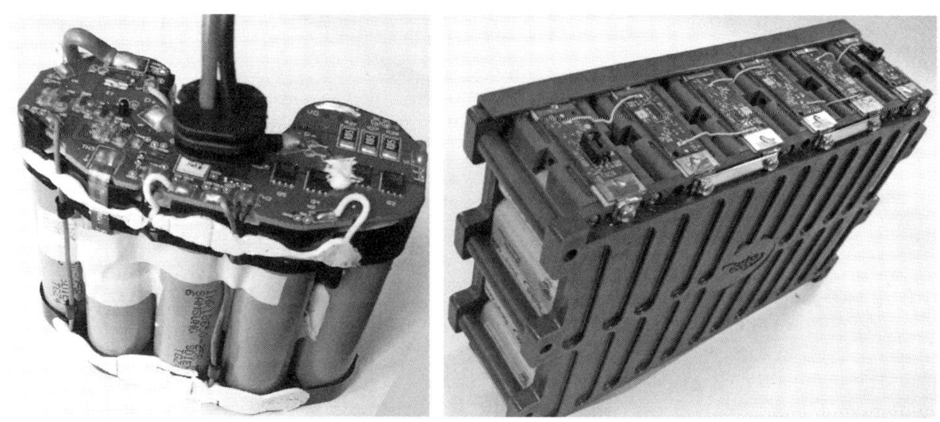

(a) (b)

FIGURE 5.28
Cell voltage and temperature sensing: (a) mated PCM on small cylindrical cells (note the thermistor on second cell from the left), and (b) battery module with Boston Power cells and distributed BMS cell boards.

banks in the same order as their electrical order, regardless of their physical position. Regardless, for dog's sake, be consistent!

5.9.2 Current Sensing

Most BMSs measure the battery current (unless the BMS gets this value from the external system) to protect the battery and possibly to use the evaluation of its state.

Two fundamental technologies can measure battery current: resistive (Figure 5.29(a, b)) and magnetic (Figure 5.29(c)).

5.9.2.1 Resistive Current Sensing

The battery current flows through a shunt resistor, developing a small voltage across it proportional to the current. A differential amplifier increases this small voltage to a level that the BMS can use to measure the current.

The sense voltage is referenced to the battery voltage, which could prevent the BMS from maintaining isolation between the cells and the low-voltage control circuit. This is not an issue in a PCM for a small battery since it's not isolated. A BMS for a larger battery uses an isolated amplifier to keep the battery floating relative to the low-voltage control circuits (Figure 5.29(b)).

The shunt resistor can take various forms, depending on the level of the current:

- PCMs for currents up to 1A use a small, onboard resistor (Figure 5.30(a));
- PCMs for currents up to 100A use one or more wire shunts (Figure 5.30(b)):
- BMUs for large batteries run the current through a large, chassis-mounted shunt (Figure 5.30(c)) capable of handling between 10A and 10 kA.

A chassis-mounted shunt uses a *Kelvin connection*, which is a four-point connection that minimizes errors due to how the battery current is routed. The two outer bolts

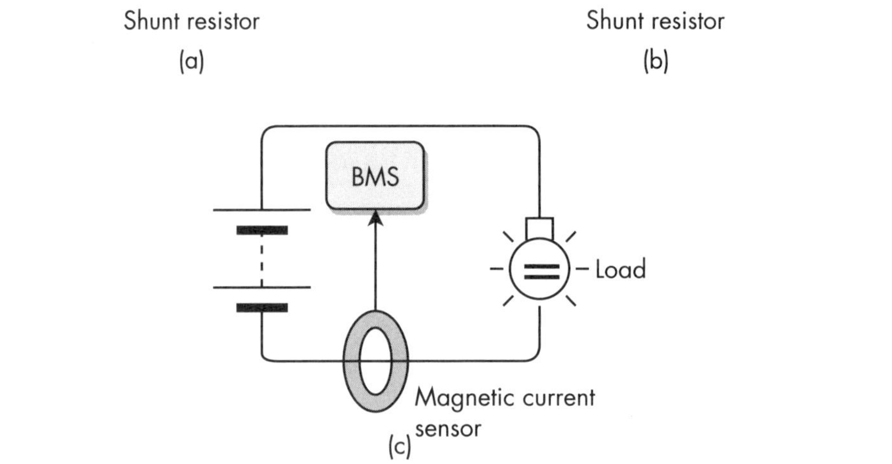

FIGURE 5.29
Current sensing:
(a) resistive, nonisolated,
(b) resistive, isolated, and
(c) magnetic, isolated.

are for the battery current and the two inner screws are to sense the voltage. The resistance of the shunt is on the order of a few $\mu\Omega$. At the maximum current, a voltage of either 10 mV, 20 mV, or 50 mV is developed across the shunt, which requires amplification to a level that the BMS can measure.

The resistive technology is quite accurate, with few limitations:

- The shunt itself has no offset, but the amplifier does;
- The resistance shunt itself can be highly accurate (the tolerance may be as low as 0.1%), but the amplifier may not be as accurate;
- The amplifier is linear, but the shunt resistance is affected by temperature, which increases at high current; a large shunt resistor minimizes the temperature rise.

If the amplifier is mounted directly onto the shunt (Figure 5.30(d)), the output may be analog (0 to 5V) or a digital bus (CAN or LIN). If the amplifier is in the BMS, a shielded, twisted-pair cable carries the sensed voltage from the shunt resistor to the BMS.

5.9.2.2 Magnetic Current Sensing

Magnetic current sensors produce a voltage that is proportional to the current flowing through them.

An advantage of a magnetic current sensor is that its signal is inherently isolated from the battery. Disadvantages include a high offset (which varies with temperature), nonlinearity, and magnetic hysteresis.

Various physical formats are available, depending on how the current is routed:

FIGURE 5.30
Resistive current sensing: (a) resistor, (b) PCB shunt, (c) chassis mount shunt, and (d) complete current shunt sensor.

- Internal: PCB mounted, the current flows inside the sensor's package (Figure 5.31(a));
- Toroid: A conductor carrying the current is routed through an opening in the sensor:
  - Bus bar sensor (Figure 5.31(b))
  - Cable sensor (Figure 5.31(c))

A trick allows a magnetic sensor to measure more current than its rating—the current is divided into multiple identical wires, only one of which is routed through the sensor. For example, splitting the current into four identical wires and running only one through the sensor lets the sensor see only one-fourth of the current (Figure 5.31(d)).

The output of a magnetic current sensor is either a current, a voltage, or a 4~20 mA loop; your BMS needs a voltage output.

Sensors may require one power supply (e.g., 5V) or two (e.g., +15 and -15V). They may be able to measure current in only one direction or both:

- Single supply, unidirectional (Figure 5.32(a));
- Single supply, bidirectional (Figure 5.32(b));
- Dual supply, bidirectional (Figure 5.32(c)).

Some sensors have two outputs for two different ranges (e.g., ±20 A and ±400 A), which is useful in products, such as traction batteries, that sometimes operate at low current, sometimes at high current (Figure 5.32(d)).[16]

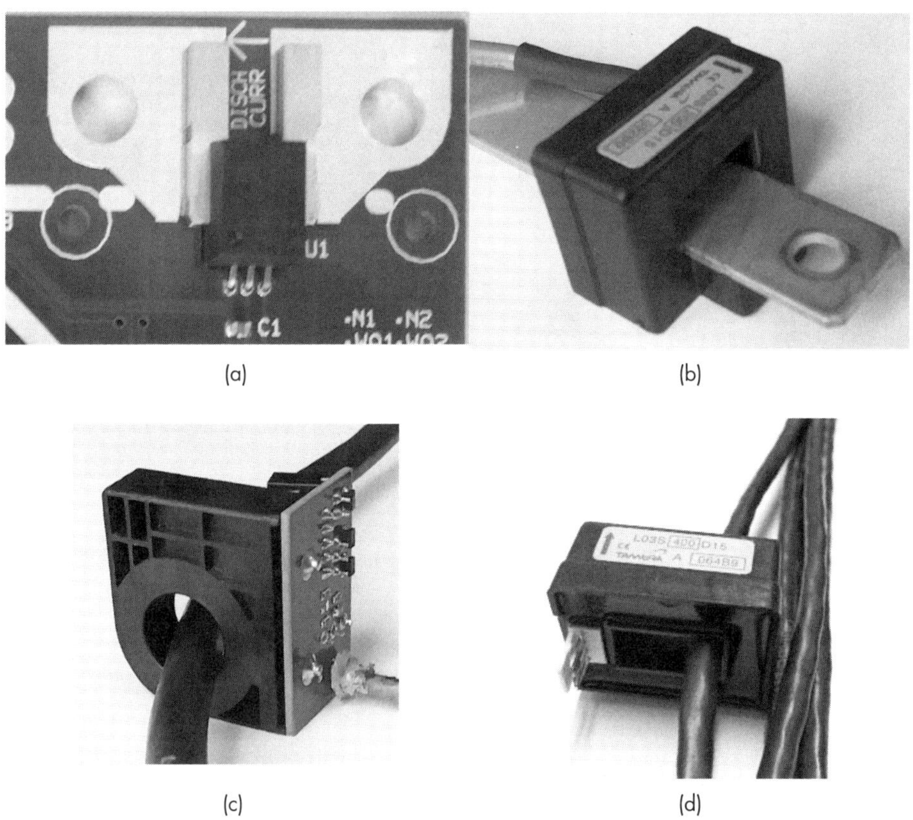

FIGURE 5.31
Magnetic current sensors: (a) internal, (b) bus bar, (c) cable-mount, and (d) split current.

16. LEM DHAB series.

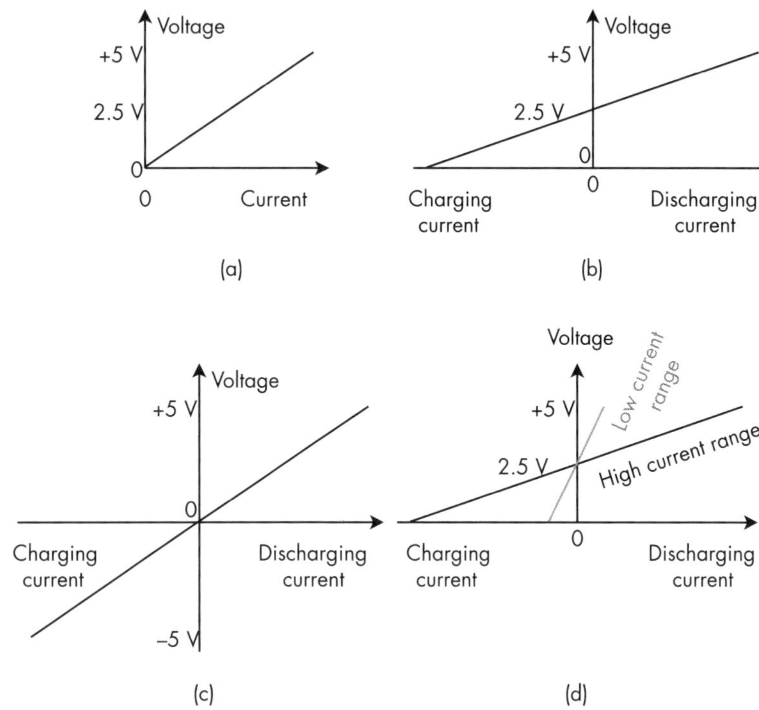

FIGURE 5.32
Magnetic current sensor supply and output: (a) unidirectional, single supply, (b) bidirectional, single supply, (c) bidirectional, dual supply, and (d) dual range, single supply.

These basic technologies are available.[17]

- Hall effect:
  - Open-loop;
  - Closed-loop.
- Fluxgate.

Standard open-loop Hall effect current sensors[18] have mediocre accuracy (a total error of about 1% of full-scale reading). The full-scale input ranges from 25 to 2,500A, unidirectional or bidirectional. They are powered by 5V, 12~15V, or ±12~15V. Some sensors are rated for a much smaller range than they can measure[19] because, over this reduced range, the linearity (and therefore the accuracy) is better.

Closed-loop Hall effect current sensors[20] have improved accuracy (on the order of 0.1%). They generate an equal and opposite magnetic field to cancel the field from the current so that the Hall effect element operates at zero magnetic field regardless of the current being measured. These sensors are more expensive and use more power. The full-scale input ranges from 20 to 2,000A, unidirectional or bidirectional. They require 12V power or higher (not 5V).

Despite the different technology, fluxgate current sensors appear to perform the same as closed-loop Hall effect sensors. They do have a much lower drift in the offset and much better linearity, but they cost a lot more.

17. There are also Rogowski coils and current transformer, but they are only good for AC and therefore can't be used to measure battery current. There are also magnetoresistive sensors, which do work at DC, but they are not that accurate, cannot differentiate charging current from discharging current, and can't handle high current.
18. Available from Honeywell, LEM, Tamura.
19. The LEM HASS 300-S, rated for ±300 A but can measure ±900 A.
20. Available from CR Magnetics, Harting.

Laboratory-grade current sensors using a closed-loop and a zero-flux detector achieve accuracy, linearity, offset, and drift in the parts-per-million range, but they cost more than $1,000!

Here are some usage notes for magnetic current sensors:

- Do not expose to stray magnetic fields, which would affect the output voltage;
- Keep their temperature constant, to minimize drift in the offset;
- If the system knows when there is no battery current (i.e., the protector switch is open), use this chance to calibrate the offset of the sensor.

5.9.2.3 Two-Current Sensors

Applications that have separate charging and load ports may measure the battery current at a single point or two separate ones:

- Single sensor: Placed on a path that is common to both charging and discharging (Figure 5.33(a)).
- Two sensors: One placed on a *charger* port and one on a *load* port (Figure 5.33(b)); this is advantageous if the level of the charging current is different from the load current.

If the battery uses two sensors, you might place a current sensor in a path that sees both the charger and load current (Figure 5.33(c)). This is not a problem if the system has two operating modes (*charging* and *discharging*) and the BMS looks at one sensor or the other one based on the operating mode (see Section 4.4.4). Otherwise, in some cases, the current is seen by both sensors and the BMS reports twice the current because it adds up the readings from both sensors.

Many loads only discharge the battery, so a unidirectional current sensor on the *load* port is sufficient. Yet some loads can also recharge the battery (e.g., regenerative braking and invergers), in which case the *load* port requires a bidirectional current

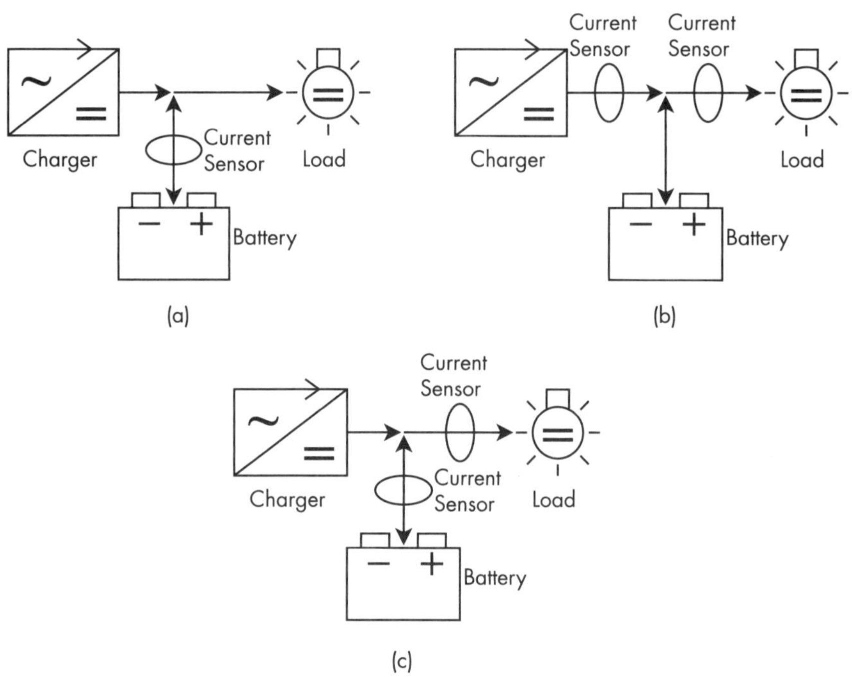

FIGURE 5.33
Current sensing in two-port batteries: (a) single sensor, (b) dual sensor, and (c) problematic sensor placement.

sensor. Similarly, while the typical charger can only charge a battery, some chargers are bidirectional (e.g., vehicle-to-grid, (V2G), to sell electricity back to the power company; see Volume 2, Section 3.9.1.8). If so, the *charger* port requires a bidirectional current sensor.

5.10 COMMUNICATION LINKS

Communications may occur

- Within a BMS: Between various components (e.g., a master/slave BMS);
- Within a battery array: A battery in an array may communicate with an array master to allow it to manage all the batteries as a single unit (see Section 6.1);
- Externally: Between the BMS and the external system (e.g., a charger, a system controller, an inverger).

For normal operation, a BMS doesn't require much data. However, when a user attempts to use a BMS as if it were a scientific data acquisition instrument, buses get clogged with data (e.g., individual cell readings every 100 ms) and reliability plummets.

A BMS may use several links of different types, as discussed below.

5.10.1 Internal Communications

This section discusses the communication links within a BMS.

5.10.1.1 Slave Communications

In a master/slave BMS, data are exchanged between the master and the slaves; power may flow from the master to the slaves. This link may use a bus with multiple drops (Figure 5.34(a)) or daisy chain cables between each pair of adjacent units (Figure 5.34(b)).

Here are some suggestions to increase noise immunity in the slave bus (Figure 5.34(c)):

- Use twisted pairs for a balanced bus;
- Connect the ground wires to one and only one place, the ground connection in the master:
  - Do not connect the ground wire to anything else; in particular, do not connect it to a metal case;
  - If using intermediate connectors through a metal panel, use connectors that are isolated from this panel.
- If shielding is required, ensure that all the shields are connected to each other, grounded only at one point (in the master) and isolated from any other ground, especially metal enclosures.

5.10.1.2 Bank Harness

For a distributed BMS, a bank harness connects the BMU to the most positive and most negative cell boards in a bank.

Here are some suggestions for the bank harness (Figure 5.34(d)):

- Use shielded cables; if possible, use cables with twisted wires;
- The shield must be connected only at the BMU or slave and must be continuous;

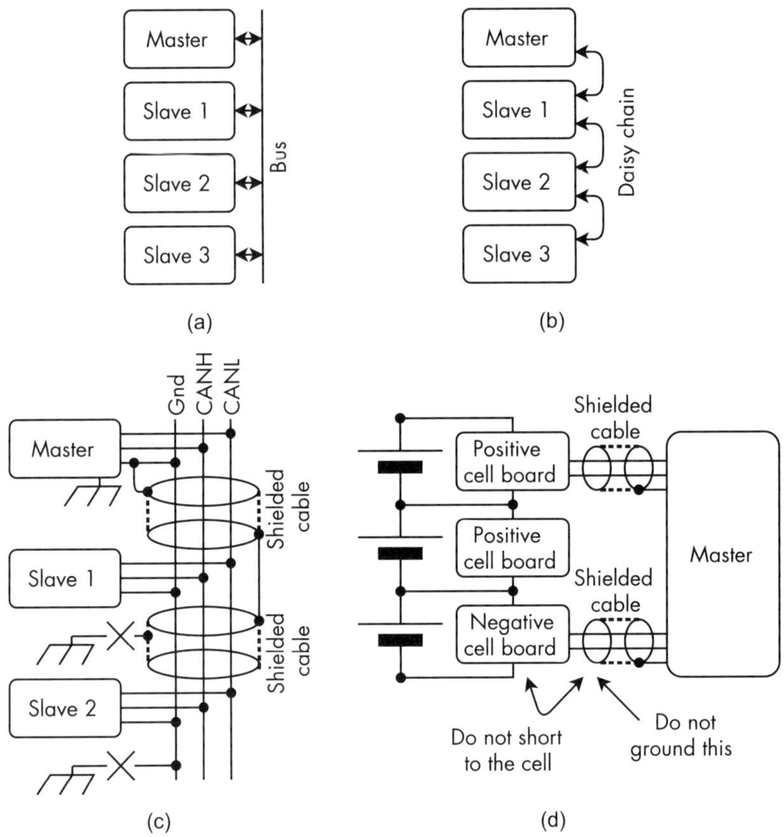

FIGURE 5.34
Slave communications: (a) bus, (b) daisy chain, (c) single-point grounding in bus, and (d) bank harness.

- If going through an intermediate connector, use the same precautions as for a slave bus (above);
- Pay particular attention that the shield is not exposed at the end of the cable that terminates at the cell board; make sure that the shield doesn't touch a cell terminal, which would defeat the battery isolation.

5.10.1.3 Peripheral Bus

Other devices internal to the battery may include a status display (to show the SoC of the battery and other essential data), a current sensor with digital output, a charger, and a peripheral to measure the battery voltage (high-voltage batteries).

Communications to these peripherals may use the same bus used for other internal systems (e.g., for the slaves), or could use a separate link, even one of a different type.

For example, many alphanumeric displays use TTL or RS-232 communications.

5.10.2 External Communications

The BMU may communicate with the external system to

- Control the system, to limit or stop the battery current;
- Possibly control thermal management;
- Report the state of the battery;
- Receive information about the system;
- To a limited extent, to be controlled by the system;

An array master may communicate with its batteries

- To request each battery's state;
- To enable or disable a battery's ability to connect itself to a common power bus.

External communications are not real time in the sense that their timing has limited consequences. Conversely, communications within a BMS are usually time-sensitive because multiple measurements should be taken simultaneously.

5.10.3 Wired Links

Normally, communication occurs over a physical cable, one with copper wires, implementing a serial link such as TTL, RS-485, RS-232, CAN, LIN, LVDS, or variations thereof.

Links may be point-to-point between two devices (Figure 5.35(a)) or multipoint among multiple devices on the same bus (Figure 5.35(b)).

The lines may be

- *Unbalanced* (Figure 5.35(c)): An unbalanced line uses only one wire for each direction. It is referenced to ground; any differential noise between the grounds of the two devices is interpreted as a part of the signal, and may corrupt communications (Figure 5.35(d)).

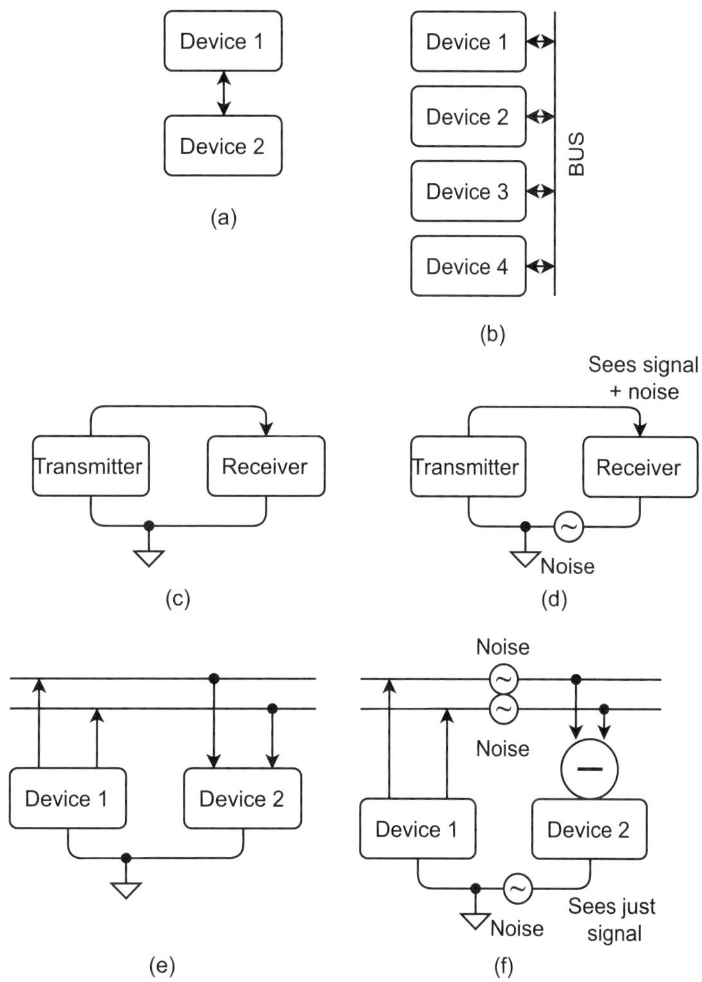

FIGURE 5.35
Communication links: (a) point-to-point, (b) multipoint bus, (c) unbalanced, (d) poor noise immunity with unbalanced, (e) balanced, and (f) good noise immunity with balanced.

- *Balanced* (Figure 5.35(e)): With balanced lines, the transmitter drives the two lines in opposite directions. The receiver looks at the difference in voltage between the two lines to cancel out any common-mode noise (Figure 5.35(f)).

Table 5.4 compares various communication standards. The physical and data layer specify how data are transferred, while the application layer specifies how the data are structured.

5.10.3.1 SMB

System Management Bus is the standard link in small batteries (see Volume 2, Section 1.2.2).

5.10.3.2 TTL

TTL[21] may be used to drive a simple display mounted on the battery or close to it. The components that communicate via TTL are delicate, making this link inappropriate for use outside the battery

5.10.3.3 RS-232

An RS-232 link is unbalanced, and therefore sensitive to noise, making it inappropriate for anything more than short-term communications during battery deployment or during troubleshooting. RS-232 is unreliable in the presence of noisy, high-power conversion products. If an RS-232 link must be used in these conditions, consider adding an RS-232 isolator.

Standard	**Topology**	**Maximum Nodes**	**Wires**	**Line (1)**	**Data Rate**	**Distance [m] (2)**	**Connector**	**Signal Voltage [V]**
SMB	Multipoint, master/ slave	> 2	3	Unbal	10~100 kbps	0.1	(3)	0, 3.3 or 5
TTL	Point-to-point	2	3	Unbal	9.6~230 kbps	30	(3)	0, 5
RS-232	Point-to-point	2	3	Unbal	9.6~230 kbps	30	DE9	± 15
USB	Point-to-point	2	4	Diff	12, 480 Mbps	5	USB	~2 ± 1.25
LIN	Multipoint, master/ slave	16	1	Unbal	2.4~20 kbps	40	(3)	0, 7~12
RS-485	Multipoint, master/ slave	32	2 (4)	Diff	0.1~100 Mbps	1000	(3)	2.5 ± 1.25
Ethernet	Point-to-point (5)	2	8	Diff	10, 100, 1,000 Mbps	100	RJ45	0 ± 1
CAN	Multipoint, peers	64	2 (4)	Diff	0.125~1 Mbps	300	(3)	2.5 ± 1
LVDS	Point-to-point	2	2 (4)	Diff	655 Mbps	30	(3)	1.2 ± 0.17

Notes:
1. Unbal = unbalanced (referenced to ground); Diff = differential (2-wire balanced).
2. Longest distance at slowest rate.
3. Not specified.
4. Typically an optional ground wire is included as well.
5. Extended to a network using a hub.

TABLE 5.4 Comparison of Bus Physical Layers

21. Transistor Transistor Logic.

5.10.3.4 USB

USB[22] is a standard communication link, commonly used to connect a BMU to a standard computer running a GUI[23] application, to monitor the state of the battery, and possibly to configure the BMU. A USB cable connects a computer directly to the BMU during battery deployment or troubleshooting.

A USB link is balanced, improving noise immunity; yet, USB ports tend to freeze when exposed to noise beyond a certain level. USB is not intended for long runs or long-term connections. A USB isolator may be required in noisy environments.

5.10.3.5 LIN Bus

LIN[24] is a data link developed by the automotive industry. It is meant to supplement (though not replace) the CAN bus using cheaper and simpler hardware. It is used to communicate with noncritical devices—mirrors, doors. Since the BMS is a critical device, it cannot use LIN to communicate with the system. However, the BMS could use LIN to talk to noncritical devices such as a display.

5.10.3.6 RS-485

RS-485 is a standard in industrial applications. It is similar to the CAN bus, except that the lines are called A and B, and only the physical layer is defined (the data layer is not defined). To avoid spurious data when the bus is idle, add bias at only one location on the bus: two 680Ω resistors, one between A and +5V, and one between B and ground (Figure 5.36(a)).

5.10.3.7 Ethernet

Ethernet is a standard computer link used in business environments. The version in use today uses cables with four twisted pairs and RJ45 modular connectors. Speed ranges from 10 Mbps (10BASE-T) to 1 Gbps. Data transfer uses the Transmission Control Protocol (TCP).

5.10.3.8 ModBus

ModBus can use either RS-485 (in which case it's called *ModBus RTU*) or Ethernet (*ModBus TCP*). If you have a choice, Table 5.5 may help you decide.

5.10.3.9 CAN bus

The CAN bus[25] is a standard bus for automotive and industrial applications. It is a multipoint, peer-to-peer network for up to about 60 devices, using a two-wire balanced link (high noise immunity). It is suitable for long-distance runs.

Messages are not addressed to a particular recipient. Instead, any node that can make use of any message it sees. If a message requests data, a node that is responsible for those data replies with a response message.

CAN messages can be

22. Universal Serial Bus.
23. Graphic User Interface.
24. Local interconnect network.
25. Control Area Network, developed by Bosh, released in 1986.

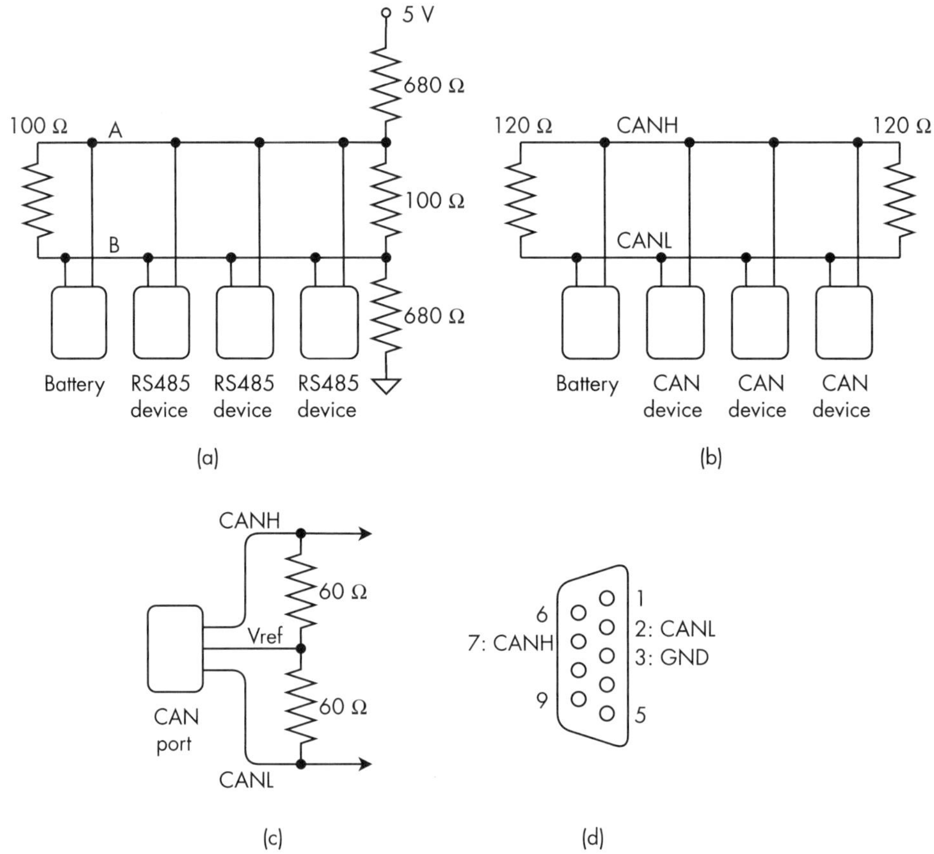

FIGURE 5.36
Balanced buses:
(a) RS-485 bus, (b) CAN bus, (c) CAN bus active termination, and (d) typical CAN DE-9 pin-out.

TABLE 5.5
ModBus Protocol Comparison

ModBus RTU (RS-485)	**ModBus TCP (Ethernet)**
Daisy-chain up to 32 devices on one bus (simple).	For multiple nodes, requires a hub (adds cost).
Plenty fast for such applications.	Much faster, but so what? You don't need it.
Physical bus requires care (terminations, polarity).	Plug-and-play connections.
Simple, low-power processors.	Power hungry, fast processor.
Setup: rate, parity, bits.	Setup: IP address.

https://www.rtaautomation.com/blog/modbus-tcp-vs-modbus-rtu/.

- *Broadcast:* A device reports its state by placing a message on the bus. If any other device on the bus wishes, it can use this information (e.g., the BMS broadcasts battery SoC, CCL, DCL, and flags).
- *Request/response:* A device asks for specific datum, and another device, which is responsible for this datum, replies with its value (e.g., a handheld tool plugged to the OBD II port in a car asks for the oxygen pressure through a PID message,[26] and the VCU[27] responds with another PID message with the pressure).

26. Parameter ID. See Volume 2, Section 3.15.3.5, "OBD II, PIDs."
27. Vehicle control unit.

- *Handshake:* A device requests an action of another device, which replies with a confirmation message (e.g., the VCU ask the traction battery to turn itself on, and the battery responds by reporting the status of the contactors, as they turn on).

A single CAN message (8 bytes) can broadcast all the essential data (Figure 5.37): current limits, SoC, and flags. Conveniently, this leaves enough space for the current (signed) and voltage (offset, such that 0 is equal to the minimum battery voltage). A flag signals a warning, in which case the system can ask the BMS what this fault is, through request/response messages.

Handshakes can get excessive—as seen in Figure 5.38(a), the product specification document for a traction battery lists 14 back-and-forth messages between a VCU and traction battery just to turn on the battery!

I believe that such an elaborate handshake adds problems without safety improvements compared to a KISS[28] protocol (Figure 5.38(b)):

1. The BMS broadcasts a wake-up message,[29] which the VCU receives as confirmation;
2. The VCU requests that the contactor be turned on or off.

That's it!

The VCU knows if there's a problem with the battery because the BMS reports its status. If there are no messages from the BMS, the VCU knows there's a problem with the battery. In either case, it requests that the contactors be turned off.

The BMS knows if the VCU wants the contactors to be off. If there are no messages from the VCU, the EV knows there's a problem with the vehicle and won't turn on the contactors.

CAN messages include an ID, up to 8 bytes of data, and an error correction code. The ID is not an address.

The CAN bus is collision-tolerant—should two nodes start transmitting at exactly the same time, the message with the lower ID prevails as a natural consequence of the hardware design. The other node notices the collision, aborts sending the message, and tries it again later. Therefore, messages with a lower ID have a higher priority.

CAN FD (*Flexible Data-Rate*) is a recently introduced extension to the CAN protocol. It supports a higher data throughput and is *mostly* compatible with standard CAN.

The CAN standard is defined from the Physical layer (layer 1) up to the Transport layer (layer 4): ID, packet size, and error correction).[30] A few communication protocols

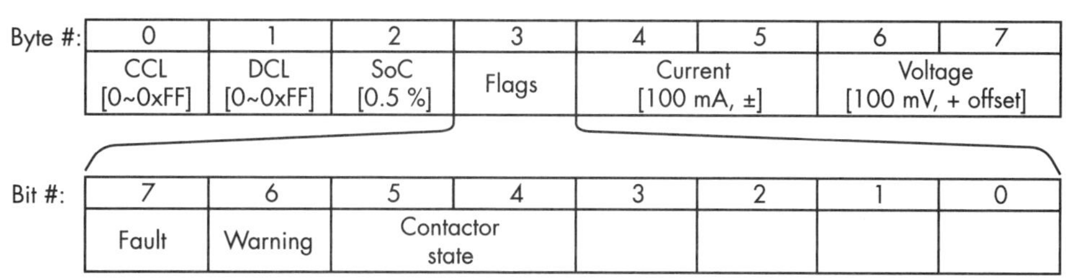

FIGURE 5.37 Most essential broadcast CAN message from BMS.

28. Keep it simple, stupid.
29. Yawn.
30. According to the Open System Interconnection (OSI) model.

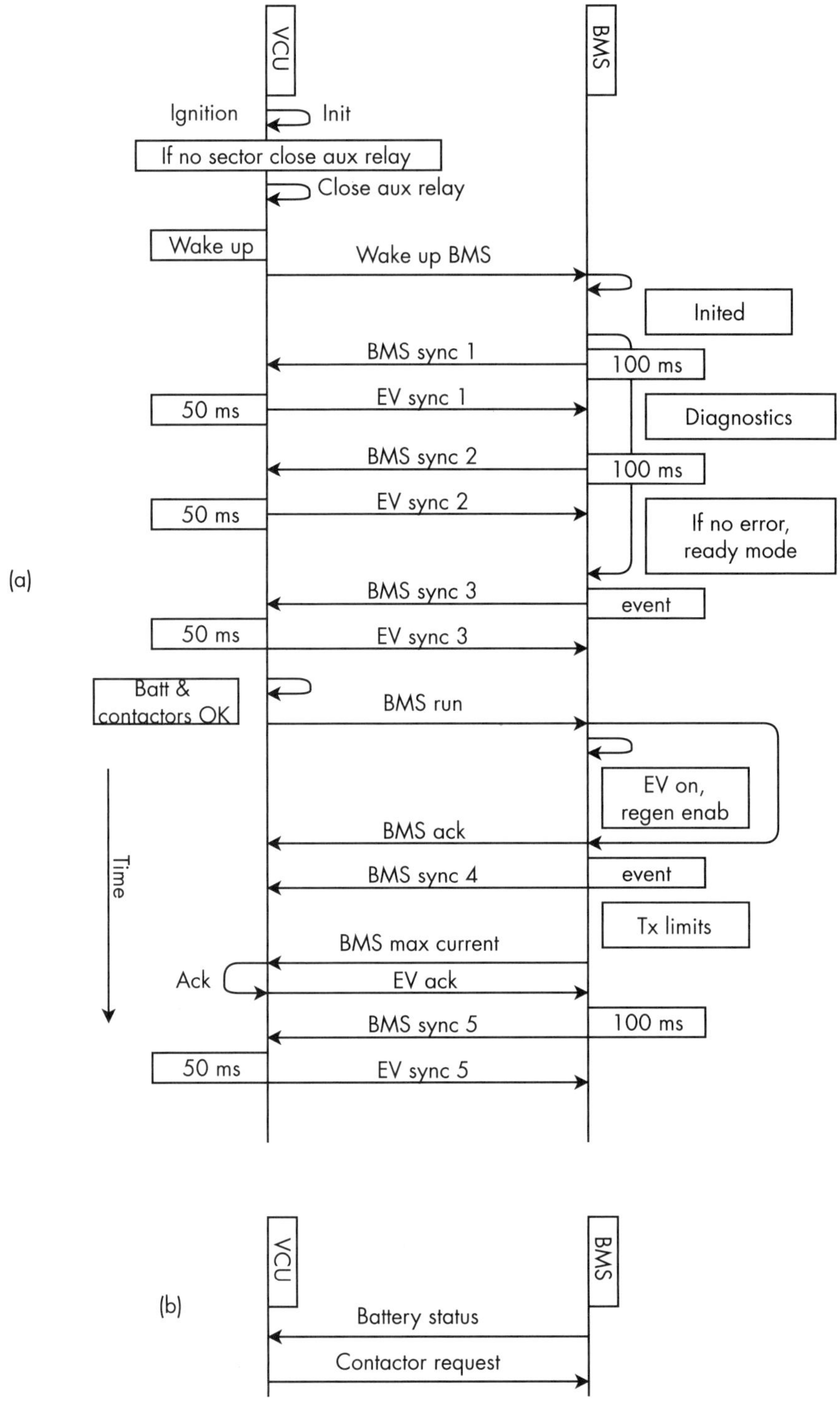

FIGURE 5.38 Traction battery communications. (a) Example of overly complex procedure. (b) Keep it simple.

expand on the CAN bus standard, including CANOpen (commonly used in the automation industry) and DeviceNet. DeviceNet is an open standard while CANOpen isn't; DeviceNet communicates using a single CAN message, while CANOpen may need to use multiple ones to do the same job.

The BMS is unlikely to support these standards, though gateways are readily available to do the conversion. Some programming is required.

SAE[31] developed the J1939 standard for the automotive industry with the intent to standardize messages in a vehicle. It also left the door open for each manufacturer to define custom messages on top of the standard ones. The result is that most J1939 messages in a vehicle are not standard.

J1939 messages are rather verbose:

- They use 29-bit IDs;
- Due to the high overhead, each message contains only 1 or 2 bytes of data, even though CAN supports 8 data bytes.

Therefore, J1939 tends to clog a CAN bus, compared to the same data throughput using simple CAN messages with an 11-bit ID and fully utilized 8 bytes of data.

Certain manufacturers expand these protocols into their proprietary protocols, as described in each "Applications" chapter.

Many BMUs have two CAN buses. A second bus is useful in plug-in EVs, which operate in two modes:

- When the ignition is on, the first CAN bus talks to the rest of the vehicle systems;
- When the vehicle is plugged in for charging, the second CAN bus talks to the charger.

That way, the charger is not burdened by messages unrelated to charging and may operate at a different data rate than the vehicle.

Here are some tips for using a CAN bus:

- Use a twisted pair for CANH and CANL;
- There is no convention for the colors; I like to use white for CANH and gray for CANL;
- Shielding should not be required; if used, ground it only at one location.
- Try to route the bus as a single line, minimizing the length of any stubs extending from it;
- Route the bus away from noisy, high-power lines.;
- Place two and only two 120Ω termination resistors on the bus, one at each end (Figure 5.36(b));
- If a 2.5V *Vref* line is available, you may use *active termination*: two 60Ω resistors to this reference, rather than a single 120Ω resistor (Figure 5.36(c));
- While, theoretically, the CAN bus is a two-line bus, in practice ground is carried as well between devices, either implicitly because all the devices use the same power supply, or by explicitly running a ground wire between devices (do one or the other, not both);
- There is no official standard for a CAN connector, though an unofficial standard uses a DE-9 connector (Figure 5.36(d));[32]
- Devices with an isolated CAN port are more immune from common-mode noise; there is no need to connect their ground line;

31. Society of Automotive Engineers.
32. The technically correct term is DE-9, not DB-9 because E is the shell size of 9-pin D-sub connectors and B in 25-pin D-sub connectors.

- Ensure that all devices are set for the same data rate (e.g., 125, 250, 500 or 1,000 kbps);
- 11-bit (*CAN 2.0A*) and 29-bit address (*CAN 2.0B*) messages can coexist on the same bus.

5.10.3.10 LVDS

LVDS is a recently defined bus used in computing devices. Compared to the CAN bus, it uses much less power (90 mW), it has a much higher data rate (655 Mbps), and it is limited to much shorter distances (~30 m). Off-the-shelf BMSs do not yet use it.

5.10.4 Optic Fiber

Instead of copper cables, links may use fiber-optic cables.

In high-voltage batteries, optical fibers inherently provide a high degree of isolation. However, the higher cost and technical expertise to handle fibers can be a concern. Plastic optical fibers (POF) are cheaper and are appropriate for the short distances within a typical battery.

5.10.5 Wireless

More and more, BMSs are starting to offer wireless monitoring. For safety reasons, only monitoring is allowed, lest someone hacks into the battery and changes the settings.

5.10.5.1 Bluetooth

Chinese digital BMSs for small batteries now offer Bluetooth support, with a small BLE[33] transmitter on the BMS, and an application for Android smartphones (see Volume 2, Section 1.6.2.1). Manufacturers of BMSs for larger batteries are starting to catch up.

5.10.5.2 WiFi

WiFi connectivity is possible in one of two ways:

- The BMS presents itself as a WiFi host, and any wireless-capable device may connect to its wireless network. The BMS serves a web page so that the wireless device uses a standard browser to display the status of the battery.
- The BMS is connected to an existing wireless network. This has the disadvantages that the BMS must be configured to log onto the network and that an application must be installed on the wireless device; it has the advantage that the wireless device can simultaneously remain connected to the internet.

5.11 CONTROL

The previous chapter discussed the BMS's inputs and outputs (see Section 4.10). In this chapter, we discuss how to use them.

33. Bluetooth Low Energy.

5.11.1 Control Inputs

Various devices may drive a BMS control input with a circuit that depends on the device and whether the input includes a resistor (Figure 5.39).

The sources include

- *A dry contact:*[34] A switch or a relay contact;
- *An open-drain or open-collector:* A transistor that is either open or grounded;
- *A logic output:* A TTL level of 0V or 5V.

The BMS input may include

- *A pull-down:* A resistor between the input and ground, so that the line is low by default;

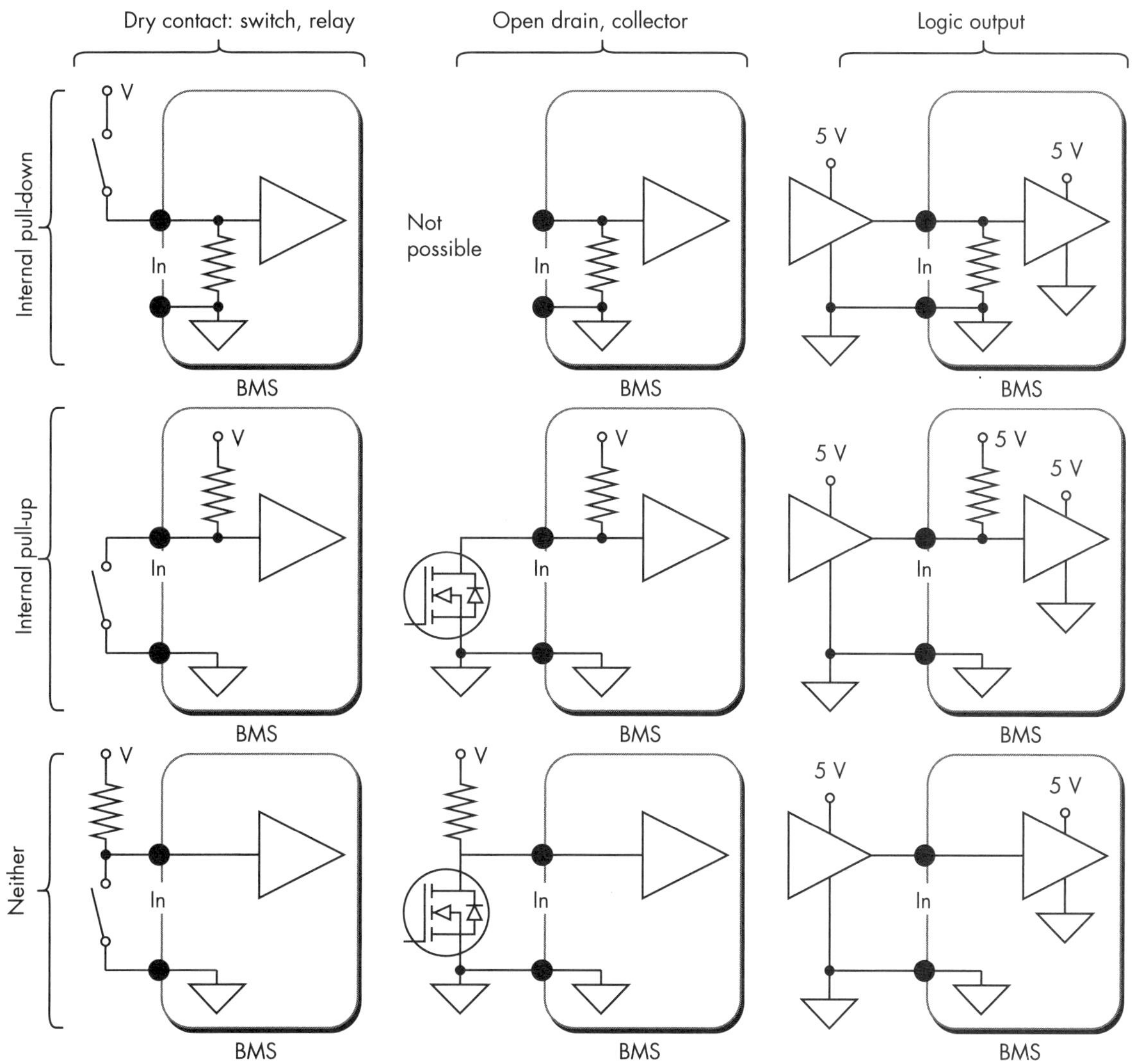

FIGURE 5.39 Various signal sources connected to various BMS input circuits.

34. "Dry" means that it's floating and has no voltage on it.

- *A pull-up:* A resistor between the input and a positive supply, so that the line is high by default;
- *Neither:* If driven by a switch or open-drain an external resistor is required.

For logic lines, the levels are nominally 0 and 5V. For other applications, the pull-up voltage may be 5V, 12V, or some other voltage.

The contacts in switches are categorized roughly as

- *Dry circuit:* Low current;
- *Power:* High current;
- *Both:* Either low or high current.

Using a dry circuit switch to control a power load will ruin the contacts quickly because they are damaged by the heat and by the arc each time they connect or disconnect. Perhaps surprisingly, having a power switch control a low-power load is also a problem because a power switch relies on the small arc that occurs when opening and closing to clean the contacts. A low-power circuit does not create that arc. Therefore, the contacts oxidize more and more over time.

As BMS inputs are low-power loads, they must be driven by switches rated for low power (Figure 5.40(a)). If a high-power switch must be used, two remedies are possible:

- A high-power resistor may be added as a load (Figure 5.40(b)); however, this is wasteful.

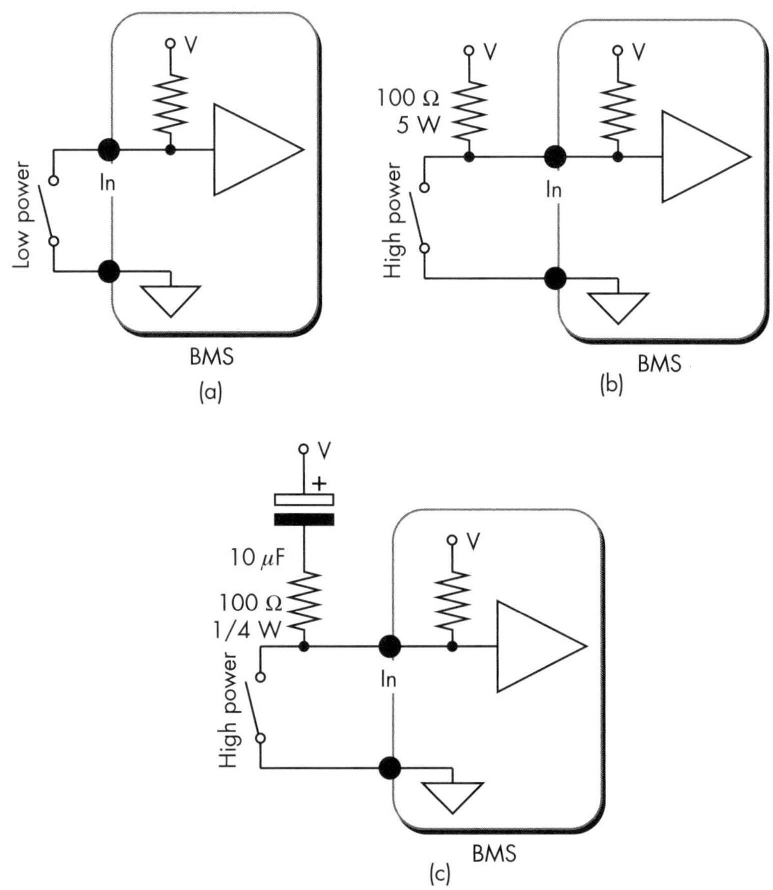

FIGURE 5.40
Switch driving logic input: (a) low-power switch, (b) high-power switch with high-power pull-up resistor, and (c) high-power switch with high power pull-up RC network.

- A capacitor and small resistor in series may be added as a load (Figure 5.40(c)). The capacitor's high charging current cleans the contacts yet lasts only a short time and the resistor limits the charging current. When the switch is open, the capacitor is discharged through the pull-up resistor inside the BMS.

5.11.2 Control Outputs

The BMS may include control outputs. The outputs may be

- Uncommitted relay contacts, normally open or normally closed (Figure 5.41(a));
- A logic output (0 or 5V) (Figure 5.41(b, c)).

The output may drive, for example:

- A logic gate (Figure 5.41(a, b));
- An LED (Figure 5.41(c)).

Note that plain LEDs require a current limiting resistor in series; LED lamps (rated for 5V or 12V) already include current limiting and do not require this resistor.

5.11.3 Open-Drain Outputs

The BMS may include high-power, open-drain outputs. They may be used to drive a contactor coil or other medium-power load requiring a current of up to 1 to 3 A.

The BMS may include a diode to absorb the kickback from an inductive load (e.g., the coil of a relay or a contactor). These outputs may be

- *High side:* A switch between the power supply and the output that is either open or connected to the supply (Figure 5.42(a)).
- *Low side:* A switch between the output and ground that is either open or grounded:

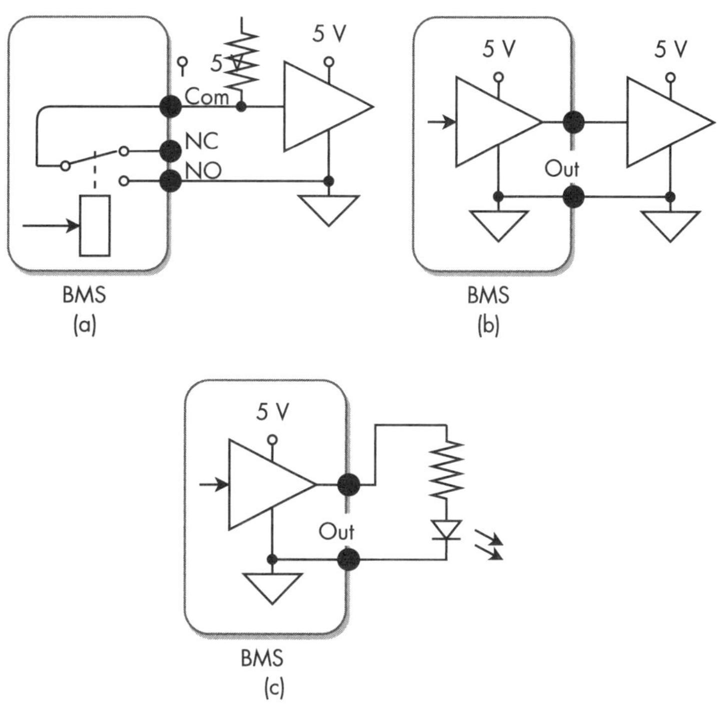

FIGURE 5.41
Control outputs from the BMS: (a) relay driving a logic gate, (b) logic gate driving a gate, and (c) logic gate driving an LED.

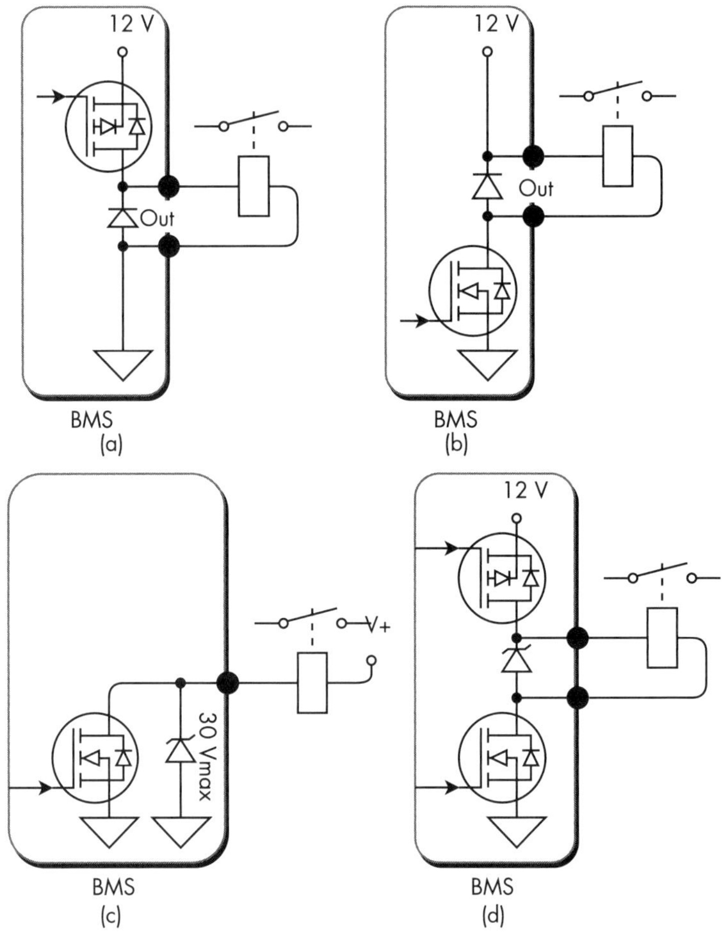

FIGURE 5.42
Open-drain outputs driving a relay: (a) high side, (b) low side with flyback diode, (c) low side with flyback TVS, and (d) redundant drive: high side plus low side.

- Using a diode to the supply rail to absorb the kickback (Figure 5.42(b)). The load must use the same supply so that the diode is directly across the load.
- Using a TVS[35] (Figure 5.42(c)). The load may be powered by any positive supply voltage, as long as it's less than the maximum rating of the diode, as specified by the BMS manufacturer.

- *Both:* A contactor coil is driven by the BMS from both sides (Figure 5.42(d)). This improves reliability through redundancy, by giving the BMS two ways to turn off the contactor should one of the transistors fail as a short circuit or should a line be accidentally grounded

5.12 PROTECTION

The BMS in a Li-ion battery must have a direct or indirect way to interrupt the battery current to protect the cells:

- A protector BMS (PCM) does so directly, with an integral protector switch;
- A nonprotector BMS (BMU) does so by:
  - Telling the external system to stop charging or discharging;
  - Controlling power switches directly.

35. Transient voltage suppressor diode.

From a safety standpoint, it is more trustworthy to use a protector BMS or to have the BMS control contactors directly, rather than having to trust an external system to obey the BMS.

This section discusses the switch topology, switch components, and circuits that a BMS can use.

5.12.1 Protection Is Required

Any BMS that is not wired in a way that it can interrupt the entirety of the battery current (directly or indirectly) is pointless (Figure 5.43).

If you only remember only one thing from this book, let it be this:

TO PROTECT THE CELLS IN A LI-ION BATTERY THE BMS MUST BE ABLE TO SHUT OFF THE BATTERY CURRENT!

Sorry for screaming, but this point is crucial because too many people lost their batteries by choosing to ignore this commandment.

The BMS must be in complete and independent control of battery safety. The responsibility for battery safety may not be shared with other components; yes, additional devices may shut down the battery, but nothing may prevent the BMS from shutting down the battery.

5.12.1.1 Shut-Down Based on Total Battery Voltage Is Ineffective

CCCV chargers and loads with a low-voltage cutoff provide a false sense of security because they cannot protect individual cells. This is because they do not know the individual cell voltages, yet the battery voltage may not be divided equally among the cells (see Section 3.2.2.1).

While it's true that CCCV charging (see Section 1.8.2) limits the top battery voltage, this in no way limits the voltage of each cell. Therefore, CCCV offers no protection; you MUST use a properly installed BMS. Nevertheless, a CCCV charger that has been adjusted to the correct top voltage *does* provide secondary protection, in addition to the protection provided by the BMS, if the battery is top balanced.

FIGURE 5.43
A disobeyed BMS kills the battery.

A load with a low-voltage cutout may help lead-acid batteries, but it is of little use to Li-ion batteries because it does not limit the voltage of each cell. Therefore, you MUST use a properly installed BMS. Nevertheless, a load with a high cutout voltage increases the life of the battery by reducing the operating SoC range.

5.12.1.2 Strategies to Avoid Shutdown

The ability of a Li-ion battery to shut down is good for its cells but not for the application:

- For telecom sites, because they require 100% uptime;
- For vehicles, because it is dangerous to lose traction while in traffic.

To keep a Li-ion battery from shutting down:

- Use well-proven, reliable technology in a correctly engineered battery.
- Use the battery in a way that keeps its cells well within their safe operating area, to avoid:
  - Over- or under-temperature;
  - Overcurrent;
  - Cell under- or over-voltage.

The goal depends on the class of applications:

- *Large, low-voltage batteries:* Avoid shutting down when the battery is full (see Volume 2, Section 2.11);
- *Traction:* Avoid shutting down when the battery is empty (see Volume 2, Section 3.2.4.2);
- *High-voltage stationary:* Avoid shutting down in either case (see Volume 2, Section 4.2.6).

5.12.2 Protector Switch Topologies

The protector switch may use one of the following topologies:

- Internal to the battery:
  - The battery includes a single power switch to interrupt the current regardless of direction (Figure 5.44(a));
  - The battery includes two power switches, one for charging, one for discharging, using a single-port (Figure 5.44(b)).[36]
  - The battery has two ports, one for charging, one for discharging. Each port includes a protector switch (Figure 5.44(c)).
- External:
  - The BMS controls external power switches, one for each set of charging sources and one for each set of loads. The battery still includes a single *fault* protector switch that the BMU can open as a last resort if the external system does not obey its commands (Figure 5.44(d)).

36. Note that the two switches are in series because of the technology used: each switch controls current in one direction but passes current unimpeded in the opposite direction.

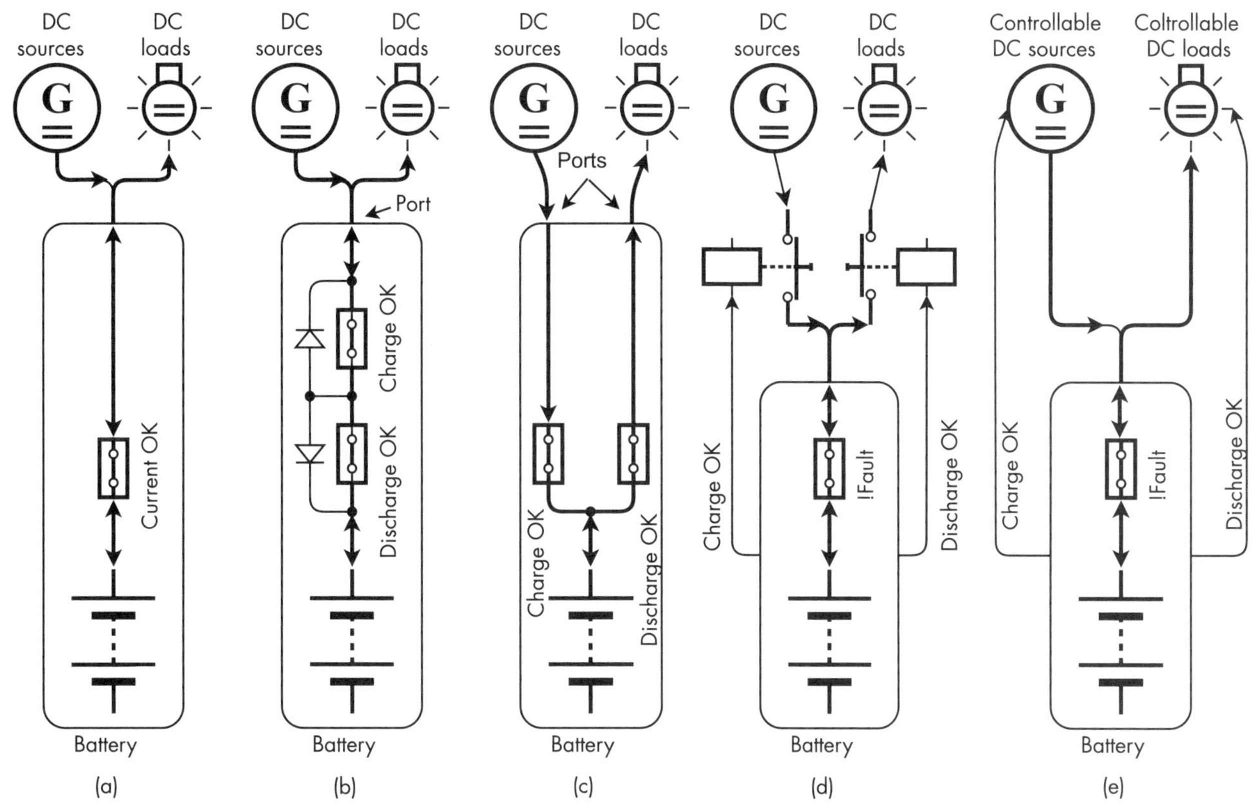

FIGURE 5.44 Protector switch topologies: (a) single-switch, (b) dual-switches, single-port, (c) dual-port, (d) external switches, and (e) external control.

- The BMS controls each external device using signals or communication links to tell it to stop charging or discharging; again, the battery includes a single *fault* protector switch (Figure 5.44(e)).

The BMS typically generates these signals that can be used to control these protection switches:

- *Current OK:* Enables both charging and discharging;
- *Charge OK:* Enables just charging;
- *Discharge OK:* Enables just discharging;
- *Not fault:* Disables both charging and discharging as a last resort.

Let's look at each of these topologies.

5.12.2.1 Single-Switch Topology

Using a single-switch (Figure 5.45(a)) is quite simple, though it is not ideal because if the switch is opened to prevent overcharging, it also prevents the discharging of a full battery (Figure 5.45(b)). Similarly, if the switch is opened to prevent overdischarge, it also prevents charging of an empty battery.

This problem can be solved by using a smart BMS that can tell when it's OK to turn the switch back on to let an empty battery be charged or to let a full battery be discharged (Figure 5.45(d)). Even so, the loads may be unpowered for an instant

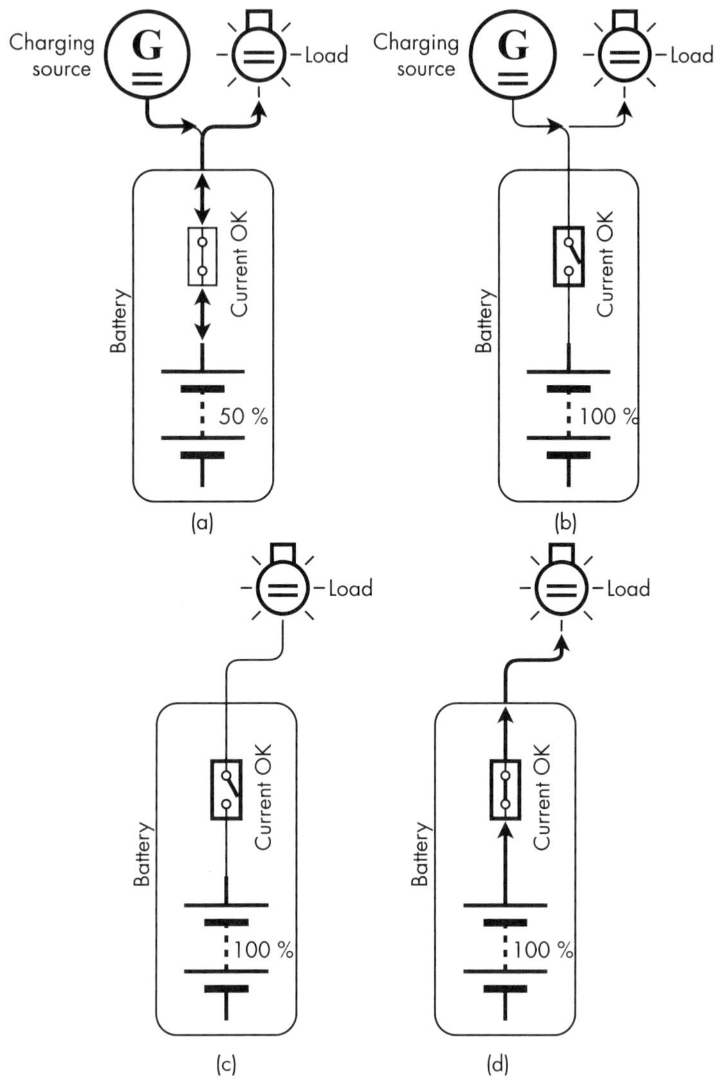

FIGURE 5.45
Single-switch battery operation: (a) normal, (b) battery full, balancing, (c) charging source goes away, and (d) BMS realizes it, turns on the switch.

(Figure 5.45(c)) because it takes time to detect the loss of a charging source and to turn on the switch. A better BMS can minimize this gap by using specialized hardware.

5.12.2.2 Dual-Switch, Single-Port Topology

A dual-switch battery overcomes some of the limitations of a single-switch battery by controlling charging and discharging independently. This allows charging even when discharging is disabled and vice versa:

1. Normally, the charging source is present and powers the load while charging the battery. Both switches are closed; current flows through them, in or out of the battery as required (Figure 5.46(a)).
2. If the battery is full, it may need to open the charging switch, but the load continues to be powered because the charging source is still present (Figure 5.46(b)).
3. If the charging source goes away, the battery powers the load because the discharging switch is still closed (Figure 5.46(c)).

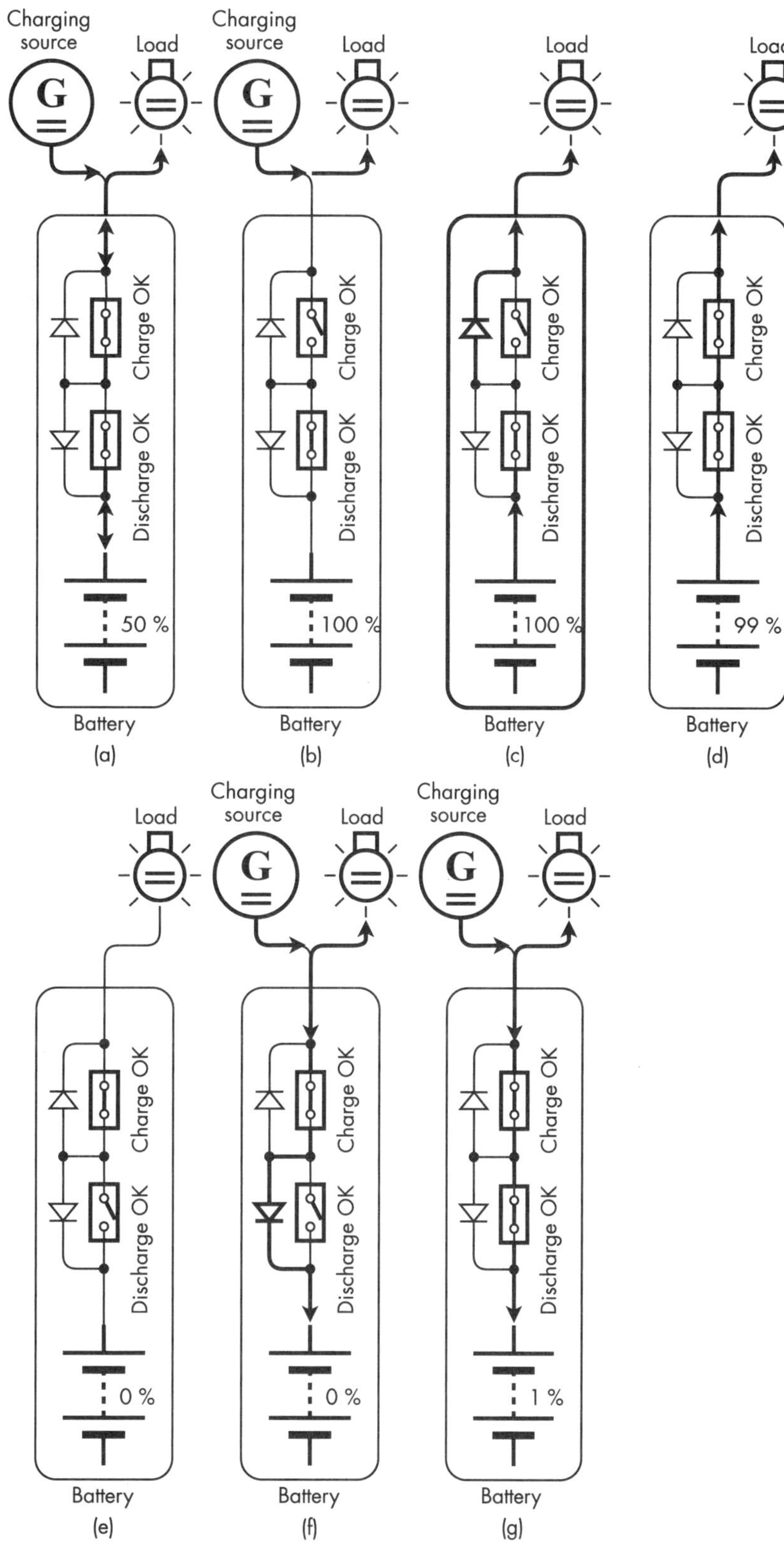

FIGURE 5.46
Dual-switch, single-port operation: (a) normal, (b) battery full, balancing, (c) charging source goes away, (d) slightly discharged, (e) battery low, discharging disabled, (f) return of charging source, and (g) slightly charged.

4. A bit later, the BMS closes the charging switch because the battery is no longer full (Figure 5.46(d)).

5. When the battery is nearly empty, the BMS opens the discharging switch; the load is no longer powered, but the BMS will remain powered for a while longer. The BMS keeps the charging switch closed (Figure 5.46(e)).
6. When the charging source comes back, it recharges the battery through the charging switch that is still closed (Figure 5.46(f)).
7. A bit later, the BMS closes the discharging switch because the battery is no longer empty (Figure 5.46(g)).
8. When the battery is fully recharged, the cycle is repeated (Figure 5.46(a)).

Should the charging source not come back, the battery will be discharged completely and the BMS will shut off its power supply to avoid overdischarging the battery. Applying a charging source cannot charge the battery because the charge switch is open; manual intervention may be required to revive the battery. The various "Application" chapters (in Volume 2) present solutions to this problem.

Using two switches has several disadvantages compared to using a single switch:

- Cost and complexity—not only for the extra switch but also for the components to route the current to one switch or the other;
- Increased voltage drop—the voltage drop is twice the drop for a single-switch solution.[37]

Most protector BMSs with MOSFETs use this topology. They can only handle a relatively low current. For high current, the battery uses a BMU and separate power switches (see Sections 5.12.3 and 5.12.4).

5.12.2.3 Dual-Port Topology

A dual-port BMS can also control charging and discharging independently, though it does so through two separate ports—one switch protects the charging port and another one protects the discharging port.

This topology has the advantage of simplicity (no need for diodes to route the current) and low voltage drop (the current only needs to flow through one switch). The operation is straightforward:

- Normally both switches are closed; current can flow directly from the charging source to the load (Figure 5.47(a));
- If the battery is full, the BMS may need to disable charging by opening the charging switch; the load continues to be powered through the discharging switch, which is still closed (Figure 5.47(b));
- If the charging source goes away, the load continues to be powered through the discharging switch (Figure 5.47(c));
- Once the battery is empty, the BMS opens the discharging switch and the load goes off (Figure 5.47(d)).

There are several disadvantages to a dual-port battery:

- Current cannot flow directly from the DC source to the DC load; it must flow through the battery. Therefore, the current must flow through two switches, resulting in a higher voltage drop. If both switches are open (for example, due

37. With a MOSFET protection switch, when only one of the switches is closed, an even larger voltage drop occurs because current flows through a reverse diode in the MOSFET, which wastes power.

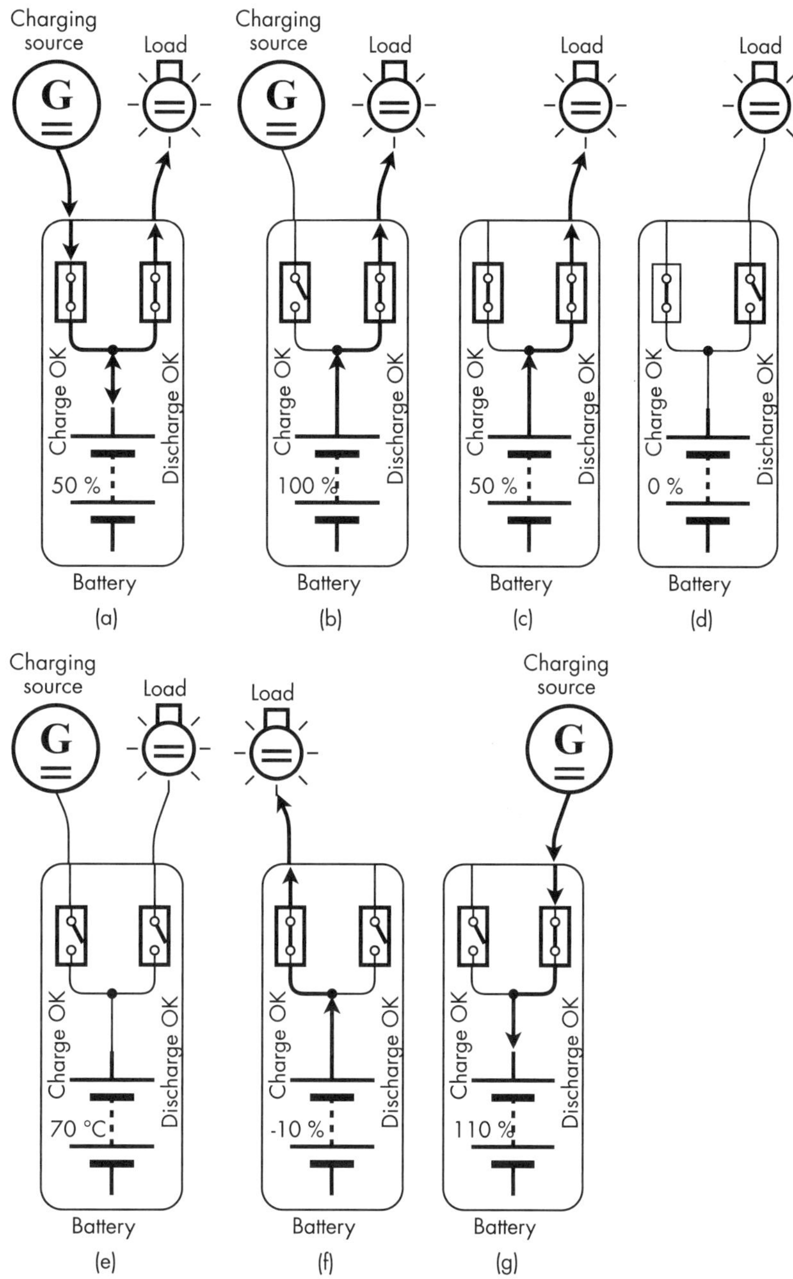

FIGURE 5.47
Dual-port battery: (a) normal, (b) battery full, balancing, (c) charging source gone, (d) battery empty, (e) overheated battery, (f) load on charging port, and (g) charger on discharge port.

to overtemperature), then the source cannot power the load (e.g., even though there is sun, the loads are off) (Figure 5.47(e)).

- Nothing prevents the battery from being discharged through the charging port[38] (Figure 5.47(f)); if the battery is empty, it opens the discharging switch, not the charging switch. A load on the charging port will overdischarge the battery.[39]

38. In particular, if a MOSFET is used as a switch, it is physically unable to block current from flowing in the reverse direction. Ideally, each port should have back-to-back switches to block current in either direction; however, this is never done in practice due to higher cost and reduced efficiency.

39. This disadvantage for the battery is actually an advantage for a solar charge controller, which requires power from the battery to restart charging the battery when the sun returns.

- Similarly, nothing prevents the battery from being charged through the discharging port (Figure 5.47(g)); if the battery is full, it opens the charging switch, not the discharging switch. A charger on the discharging port will overcharge the battery.

To avoid overdischarging through the charge port or overcharging through the discharge port, the battery could include an additional *fault* switch that the BMS can open if any cell voltage goes quite low or quite high. Adding such a switch, however, defeats any advantage of using a dual-port battery.

Important: do not attempt to covert a dual-port Li-ion battery to a single-port one by connecting the two ports. This results in a battery that cannot be controlled.

5.12.2.4 External Switch Topology

If every device in the system is unidirectional (it either charges or it discharges), then, instead of having switches inside the battery, it may be possible for the BMS to disconnect the external devices (Figure 5.44(d)) this way:

- The *charge OK* signal from the BMS in the battery controls power from all the charging sources;
- The *discharge OK* signal from the BMS in the battery controls power to all the loads.

The advantage of this topology is that it may be cheaper and more efficient to control each device with small relays than to cut the total battery current with large contactors. For example:

- An AC relay on the AC input of a charger is smaller and cheaper than a large DC contactor on its output;
- A small relay cutting power to a thermostat is smaller and cheaper than a large contactor on the DC input of a heating and cooling system.

Besides, this approach adds flexibility by allowing critical devices to be turned on or off at a different time from noncritical devices.

This topology has the same disadvantages as the dual-port topology. It also has the disadvantage that the battery has to rely on the external system to obey its commands. Since this cannot be guaranteed, the battery must still contain a *fault* contactor that the BMS can turn off as a last resort.

5.12.2.5 External Control Topology

If every device in the system provides a method for external control (either gradual or on/off), the BMS in the battery may control each device through dedicated wires or communication links. This allows the BMS to control the current and even stop it (Figure 5.44(e)).

This topology has the same advantages and disadvantages as the previous topology, with the additional advantage of being compatible with bidirectional devices (such as invergers); relays and contactors are not required.

5.12.3 Protector Switch Components

Various electrical and electronic components (Figure 5.48) may be used to implement a power switch. They are usually designed for chassis mounting (Figure 5.49).

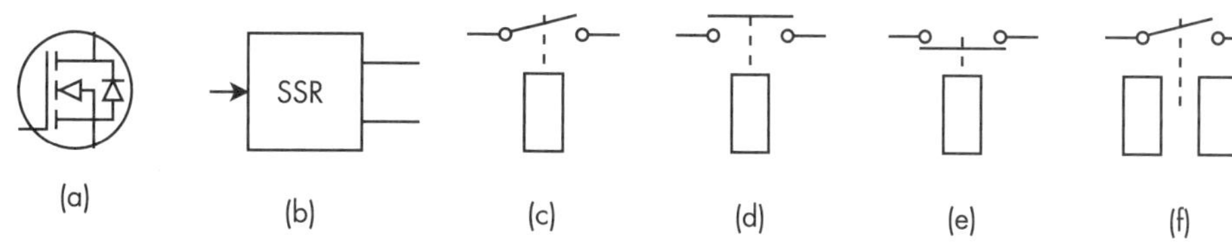

FIGURE 5.48
Symbols of various power switch component:
(a) MOSFET, (b) solid-state relay, (c) relay,
(d) contactor, (e) normally closed contactor, and
(f) dual-coil latching relay.

FIGURE 5.49 Various power switch component: (a) MOSFET, (b) solid-state relay, (c) power relay, (d) contactor, (e) contactor with economizer, (f) cluster of three contactors, and (g) dual-coil latching relay.

Table 5.6 lists those components and compares their advantages and disadvantages. TL;DR: use MOSFETs for up to 50A, dual-coil contactors for above 50A, and do not use latching relays.

5.12.4 Solid-State Protector Switch Circuits

A basic understanding of the circuits commonly used to implement the power switch for a low-voltage battery is useful to the battery designer, whether or not the designer needs to implement one.

5.12.4.1 MOSFETs

MOSFETs are solid-state devices that can be used as a switch, but only in one direction. In the other direction, they are always "on" because they include an intrinsic diode inside their package.

	Device	Pro	Con
Solid-state	**High-power MOSFETs** **Transistors (solid-state)** **(Figure 5.48(a))**	Inherently allow independent control of charging and discharging (two MOSFETs are needed, one to control charging and one for discharging). The drive requires effectively zero power—quite efficient in low power applications. Their nonzero resistance limits the current when a battery is connected to a capacitive load without first precharging.	Nonzero resistance makes MOSFETs less efficient than contactors in high-power applications. Not for currents above 100 A; their heat and cost make contactors more effective. Not as forgiving of abuse compared to contactors. Careful: whenever current is routed through a MOSFET's diode, turn it on, or else the MOSFET overheats.
	dc solid state relays (Figure 5.48(b))	A DC SSR combines a photovoltaic isolator and a MOSFET; convenient when the BMS does not include an isolated gate driver.	Expensive, compared to equivalent MOSFETs.
Mechanical	**Standard DC contactors or relays** **(Figure 5.48(c))**	Straightforward to use. Low-contact resistance: high efficiency at high power.	Coil uses a lot of power: low efficiency in low power applications. Not directional: two contactors and two rectifier diodes are required to control charging and discharging independently.
	Economizer contactors (Figure 5.48(a))	Same as above, but more efficient: they include an electronic circuit to reduce the coil drive some time after activation.	Same as above. Less reliable; generate electrical noise.
	AC contactors or relays **(Figure 5.48(a))**	Cheaper than DC contactors. As they operate at higher voltage, the current is lower for a given power, making them smaller and cheaper still. Control charging and discharging separately in AC applications where the AC power flows in only one direction.	Cannot be used with DC circuits. Not helpful in bidirectional AC power applications (e.g., back-feed to the grid).
	Dual-coil contactors **(Figure 5.48(a))**	Same as above, but switch to a low-power coil after contact is made. More reliable and no electrical noise.	Use a bit of power when on.
	Normally closed contactors **(Figure 5.48(e))**	Use no power during normal operation.	Dangerous: Do not use on the discharge side. When the BMS disables discharge, the contactor coil discharges the battery further until the voltage is too low and the contactor releases, reconnecting the load to the battery and killing the cells.
	Latching relay (Figure 5.48(f))	Require power only when turning on or off. May be considered for batteries that need only a single fault power switch, and for the charge protector switch.	Not fail-safe. Absolutely never to be used to control discharge: if the battery voltage becomes too low, the BMS is unable to turn off the relay. May be used to control charging, but still not advisable, in case the BMS shuts down: charging current is enabled.
	Remote battery switches	Latching (require power only when turning on or off); available as auto-releasing (turn off if power is removed).	They include a manual override knob, which allows the user to override the BMS: Bad!

TABLE 5.6 Comparison of Power Switch Components

Whenever the switch is turned on, there may be a large inrush current to charge the load capacitance (see Section 5.13). The nonzero resistance of the MOSFETs helps to limit this current.

5.12.4.2 Two MOSFETs, Single-Port Topology

Single-port protector BMSs (Figure 5.44(b)) use two MOSFETs connected back-to-back (Figure 5.50).[40] One MOSFET controls charging but lets the discharging current through unimpeded through the intrinsic diode inside its body. The other MOSFET controls discharging but lets the charging current through unimpeded through its diode.

The protector switch operates as follows:

- When neither charging nor discharging are allowed, both MOSFETs are off; current cannot flow through their internal diodes because they face in opposite directions. At any given time, either one diode or the other one is reverse-biased and therefore off (Figure 5.50(a)).
- When both charging and discharging are allowed, both MOSFETs are on; current flows through both MOSFETs, not through their diodes (Figure 5.50(b)).
- When only charging is allowed (discharging is disabled):
  - While connected to a load, the external voltage is lower than the battery voltage; the discharging MOSFET is open, preventing discharge. Its internal diode is reverse-biased, so discharging is not possible (Figure 5.50(c)).
  - While connected to a charger, the external voltage is higher than the battery voltage; current flows through the diode inside the discharging MOSFET, which drops the voltage and becomes really hot (Figure 5.50(d). Therefore, the BMS turns on both MOSFETs, so the discharging MOSFET bypasses its diode and runs cool (Figure 5.50(e)).
- When only discharging is allowed, the same occurs, but the two MOSFETs exchange their functions:
  - While connected to a charger, the external voltage is higher than the battery voltage; the charging MOSFET is open, preventing charge, and its internal diode is reverse-biased, so charging is not possible (Figure 5.50(f)).
  - While connected to a load, the external voltage is lower than the battery voltage; current flows through the diode inside the charging MOSFET, which drops the voltage and becomes quite hot (Figure 5.50(g)). Therefore, the BMS turns on both MOSFETs, so the charging MOSFET bypasses its diode and runs cool (Figure 5.50(h)).

This topology is implemented in single-port protector BMS. It can also be added to a nonprotector BMU, though it's not straightforward as it requires a good understanding of electronics. It requires

- Isolated gate drivers or two solid-state relays with MOSFET output;
- A careful thermal design, lest the MOSFETs melt in thermal runaway;
- Control circuitry to turn on both MOSFETS whenever the battery current is high (even if only charging or only discharging is allowed), to avoid passing current through the intrinsic diode of a MOSFET.

40. In anti-series.

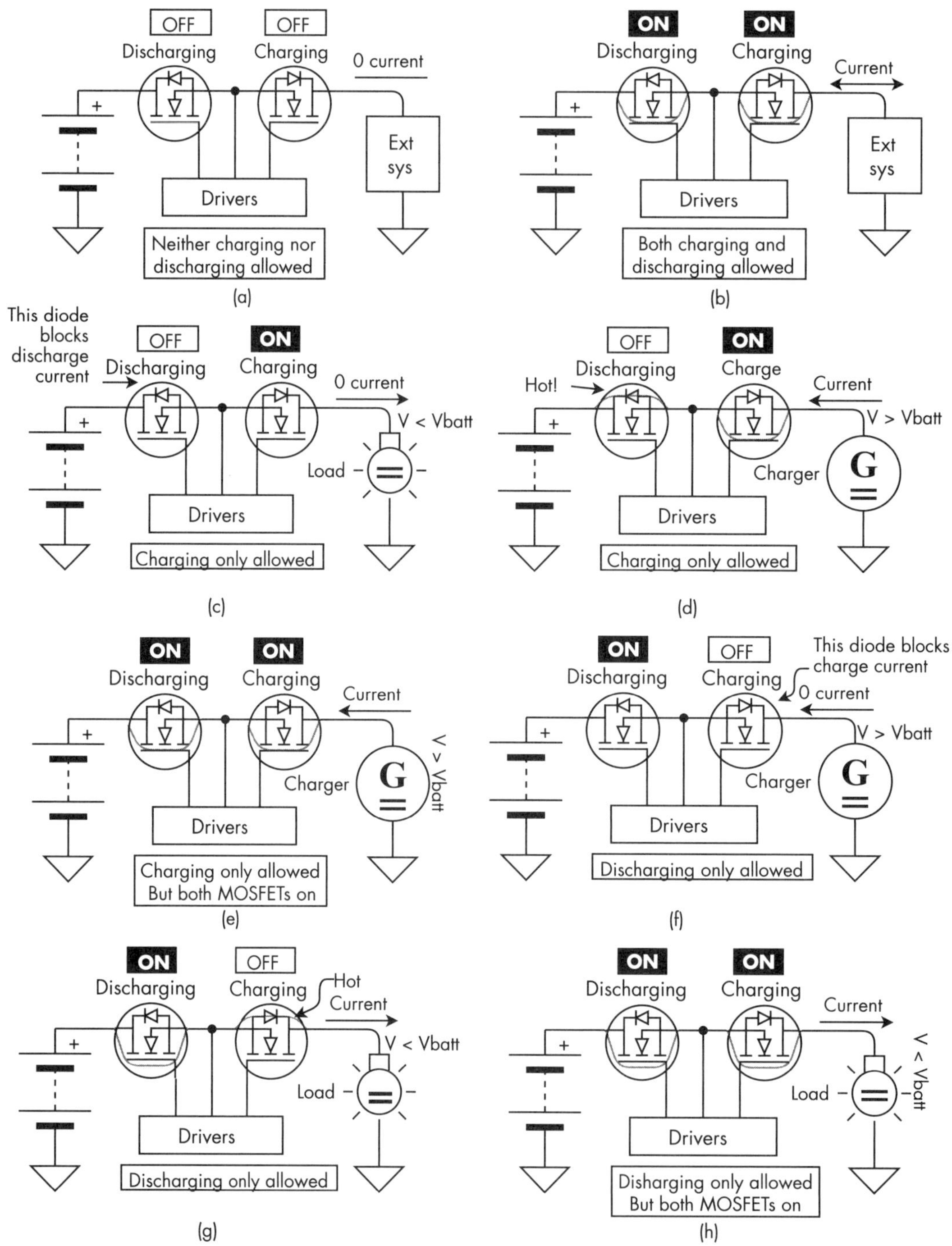

FIGURE 5.50 Single-port solid-state switch: (a) off, (b) fully on, (c) discharging disabled, (d) only charging allowed, through one MOSFET and one diode, (e) only charging allowed, through two MOSFETs, (f) charging disabled, (g) only discharging allowed, through one MOSFET and one diode, and (h) only discharging allowed, through two MOSFETs.

Ideally, the BMS should support external MOSFETs directly by including all the above provisions right out of the box. I know of only one BMS that does.[41]

41. Elithion's Vinci LV.

5.12.4.3 *Two MOSFETs, Dual-Port*

Dual-port protector BMSs (Figure 5.44(c)) also use two MOSFETs, but one is connected to the charging port and the other one to the discharging port (Figure 5.51(a)).

This topology requires that a charger can only source current and that a load can only sink current. In other words, it requires that the charging MOSFET truly does control only the charging current and that the discharging MOSFET controls only the discharging current.

The operation is straightforward:

- When both charging and discharging are enabled, both MOSFETs are on (Figure 5.51(b));
- If only charging is enabled, the charging MOSFET is on, and the discharging one is off (Figure 5.51(c));

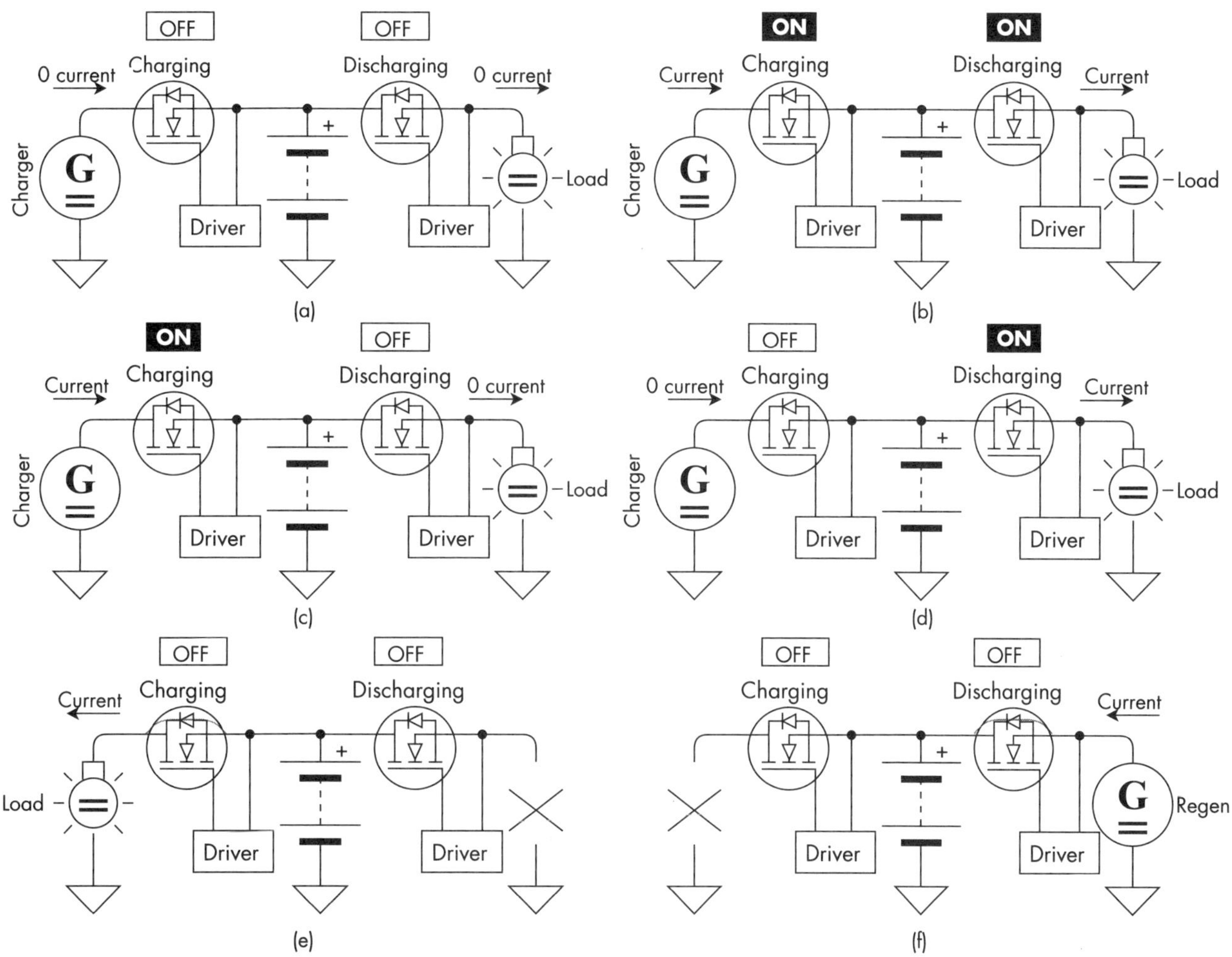

FIGURE 5.51 Dual-port using two MOSFETs: (a) off, (b) both charging and discharging enabled, (c) charging only enabled, (d) discharging only enabled, (e) cannot prevent overdischarging through charging port, and (f) cannot prevent overcharging through discharging port.

- Similarly, if only discharging is enabled, the discharging MOSFET is on, and the charging one is off (Figure 5.51(d)).

This topology can be implemented with any BMS with separate charging and discharging control outputs. Some caveats:

- If the charger is turned off when it is connected to the battery, a large inrush current from the battery charges the capacitor on the output of the charger. To avoid this, turn on the charger before connecting it or use a precharge resistor.
- Avoid this topology in applications where the charger may be connected and disconnected in the field because the battery voltage appears, unprotected, on the charging connector; the battery would power a load connected to the charging connector and the BMS cannot control it because the current flows through the intrinsic diode in the charging MOSFET (Figure 5.51(e)).
- Similarly, do not use this circuit in applications where the load can charge the battery (such as regenerative braking) because the BMS has no way of stopping recharge current from the load, which flows uncontrolled through the intrinsic diode in the discharge MOSFET (Figure 5.51(f)).
- Do not connect two dual-port batteries in parallel because they cannot control the inrush current into their discharge port or out of their charge port (see Volume 2, Section 2.6.3.20).

5.12.5 Contactor Protector Switch Circuits

There are a few ways that contactors[42] may be used in a protector switch.

5.12.5.1 Contactors

High-power applications use contactors (instead of solid-state MOSFETs).

The contacts must be rated for the battery voltage, and for the average battery current (the peak current may exceed the contactor's current rating as long as the current is within the ratings when switching). If the contactor opens under load, the ensuing arc may damage its contacts. The arc continues to carry current (if powering an inductive load) even after it is open. A contactor rated for DC operation avoids this problem.[43]

The contactor must be able to break a fault current that is many times higher than its nominal current rating. For example, in case of a shorted load, the contactors must break the short-circuit current. The breaking current rating of EV contactors is on the order of 1 kA, which is not that high. The short current is set by the total internal resistance of the cells, the wiring, and power components. For a 350V pack with a total resistance of 0.1Ω, the short-circuit current is 3500A, which exceeds the contactor's breaking current rating. Even so, the contactor is rated to be able to break this current only two or three times, after which its performance is reduced.

Because contactors have practically zero resistance, they do not limit the current from the battery to charge a capacitive load when initially closed. This inrush in current may weld the contacts. To avoid this, add precharge (see Section 5.13).

42. Roughly speaking, high-power relays are called contactors. The demarcation line between the two is fuzzy; you could say that above 20 A it's a contactor, but really, we can distinguish a contactor from a relay by its bulky size.
43. The contactor includes a magnet that bends the arc until its path is too long, and the arc is extinguished.

The contactor coil is rated for either the battery voltage or for the low-voltage supply voltage:

- For example, for golf carts or low-voltage stationary batteries, use 24, 36 or 48V coils, the same as the battery voltage;
- Otherwise, use 12V coils, the same as the low-voltage supply.

There are a few approaches to minimizing the power required to drive the contactor coil. The idea is to apply full power to the coil at turn on, and then reducing the voltage, since it takes less voltage to hold the contact closed once it's closed. These approaches include:

- Contactors may include an integral *economizer*, a small DC-DC converter that reduces the voltage to the internal coil some time after turn-on;
- Dual-coil contactors include an internal switch that disables the high-power coil after the contactor is engaged, which is more reliable than an economizer and does not generate any electrical noise);[44]
- A large capacitor with a resistor in parallel are placed in series with a lower-voltage coil.[45]

The contactor may include an auxiliary switch, which closes when the contactor is fully engaged. The BMS may monitor this switch to confirm that the contactor did indeed close.

Contactor manufacturers include

- Curtis/Albright: Open contactor, long-time standard in DIY EV conversions;
- Kilovac: Sealed; standard in low-volume EVs, optional economizer.[46]
- Gigavac: High voltage, high current, sealed for high reliability; dual-coil option;
- Panasonic: Used in high-volume production passenger vehicles;
- Omron: Also used in high-volume production passenger vehicles.

Some BMSs can drive the contactors directly. Otherwise, a small *helper relay* can be used between the BMS and the contactor coil.

5.12.5.2 Two Contactors, Single-Port

For a single-port battery with independent control of charging and discharging (Figure 5.44(b)), two contactors are used (Figure 5.52).

A rectifier diode steers charging current to the charging contactor. Another one steers the discharging current to the discharging contactor (diodes are required because, unlike MOSFETs, contactors are not directional). This topology operates as follows:

- When neither charging nor discharging are allowed, both contactors are off; current cannot flow through the diodes because they face in opposite directions. At any given time, either one diode or the other one is reverse-biased and therefore off (Figure 5.52(a)).
- When only charging is allowed:

44. From Gigavac.
45. Typically, a 100 μF electrolytic capacitor rated for the supply voltage, in parallel with a 100Ω resistor.
46. In my experience, the contactors are fine, but the economizers are unreliable.

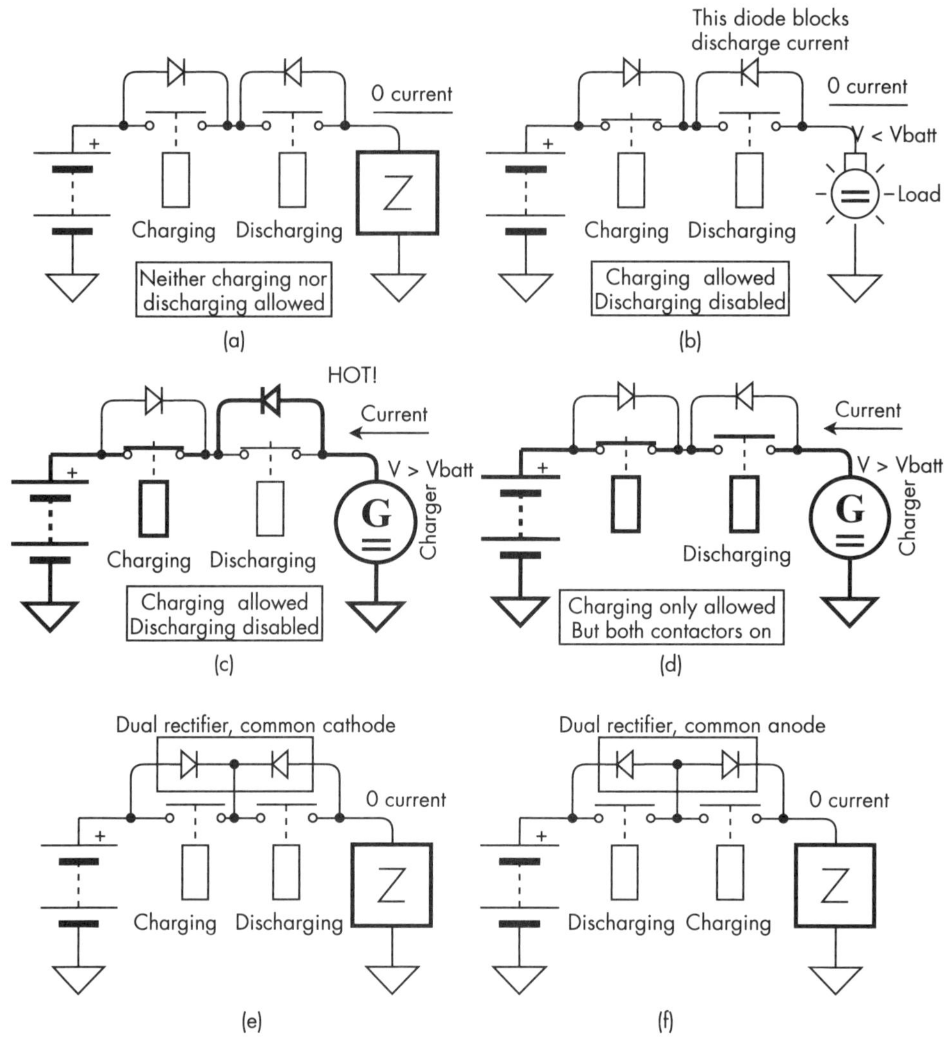

FIGURE 5.52
Power switch with two contactors: (a) off, (b) charging disabled, (c) charging through diode, (d) charging through contactors, (e) common cathode dual diode, and (f) common anode dual diode.

- When connected to a load, discharge is not possible because the discharging contactor is open (Figure 5.52(b));
- While connected to a charger, current flows through the diode across the discharging contactor, which drops the voltage and becomes hot (Figure 5.52(c)); instead, the BMS closes both contactors, so that the discharging contactor bypasses its diode (Figure 5.52(d)).

- Similarly, when only discharging is allowed, the same occurs, but the two contactors and diodes exchange their functions.
- When both charging and discharging are allowed, both contactors are on (Figure 5.52(d)).

It's best if the BMS supports this function directly (to turn on both contactors whenever the battery current is high). Check if the BMSs does that right out of the box. Otherwise, you need to add the logic to provide this function.

For the rectifier diodes, select a part that can handle the peak current and do a careful analysis of the power dissipation to add appropriate heat sinking. Consider dual diodes in a single package, since only one of them can be active at a given time, so a single heat sink works well for both charging or discharging. You can use a *common cathode* dual diode (Figure 5.52(e)). If you need to use a *common anode* dual diode instead, reverse the order of the contactors (Figure 5.52(f)).

5.12.5.3 Two Contactors, Dual-Port

For a dual-port battery (Figure 5.44(c)), rectifier diodes are not necessary because the assumption is that a charger can only source current, and a load can only sink current. Therefore, the charging contactor does control only the charging current, and the discharging contactor does only control the discharging current (Figure 5.53(a)).

Any BMS with separate charging and discharging control outputs supports this topology.

Unlike the dual power circuit using MOSFETs, this circuit with contactors could allow the BMS to control unintended current out of the charging port or in from the discharge port. However, the BMS must include the intelligence to detect that current is flowing out of the wrong port and do something about it—turn off the contactors.

5.12.5.4 Fault Contactor

If the BMS can control the system with external contactors (Figure 5.44(d)) or low-power signals (Figure 5.44(e)), then the battery should include just a single contactor for last-resort fault protection (Figure 5.53(b)).

5.12.5.5 Positive and Negative Contactors

If the battery must be galvanically isolated from the load (in high-voltage batteries), two contactors are used, one on each bus (Figure 5.53(c)) (see Section 5.13.4).

5.12.6 Fusing

A fuse protects the battery in case of excessive current, particularly in case of a short circuit.

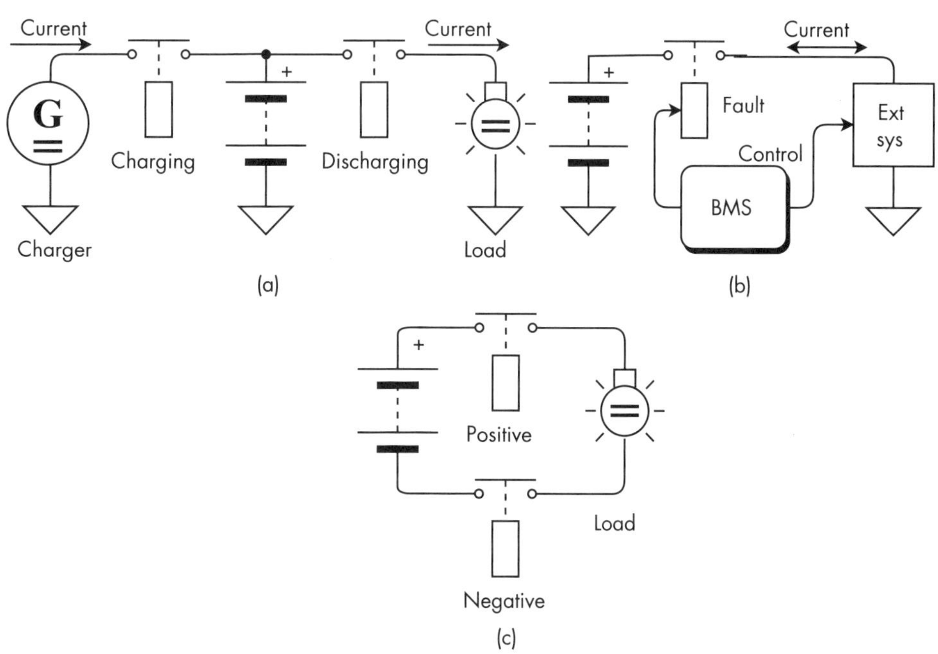

FIGURE 5.53
(a) Dual-port power switch with contactors, (b) fault contactor, and (c) positive and negative contactors.

Some people incorrectly assume that a fuse protects against all sorts of conditions. In reality, a fuse can't protect against conditions that do not result in an excessive current:

- *A runaway motor:* The current is high, but no higher than a hard acceleration;
- *Cell overcharge:* The current does not increase when cells are overcharged;
- *Thermal runaway* (see Section 8.2.1.5): The current is high within the cells, not through the main fuse.

A battery should have at least a main fuse to protect the entirety of the battery current. It may also have smaller fuses to protect just specific, lower-current circuits, such as a charger (Figure 5.54(a)).

A fuse has two current ratings (Figure 5.54(b)):

- *Fusing:* Guaranteed not to blow below this current;
- *Breaking:* Guaranteed not to fail below this current.

The time required for a fuse to blow depends on the current through it (Figure 5.54(e))—the higher the current, the faster it blows. If the current is equal to the rated current, the fuse is guaranteed not to blow; therefore, to blow a fuse requires a current that is about 25% higher than its rating.

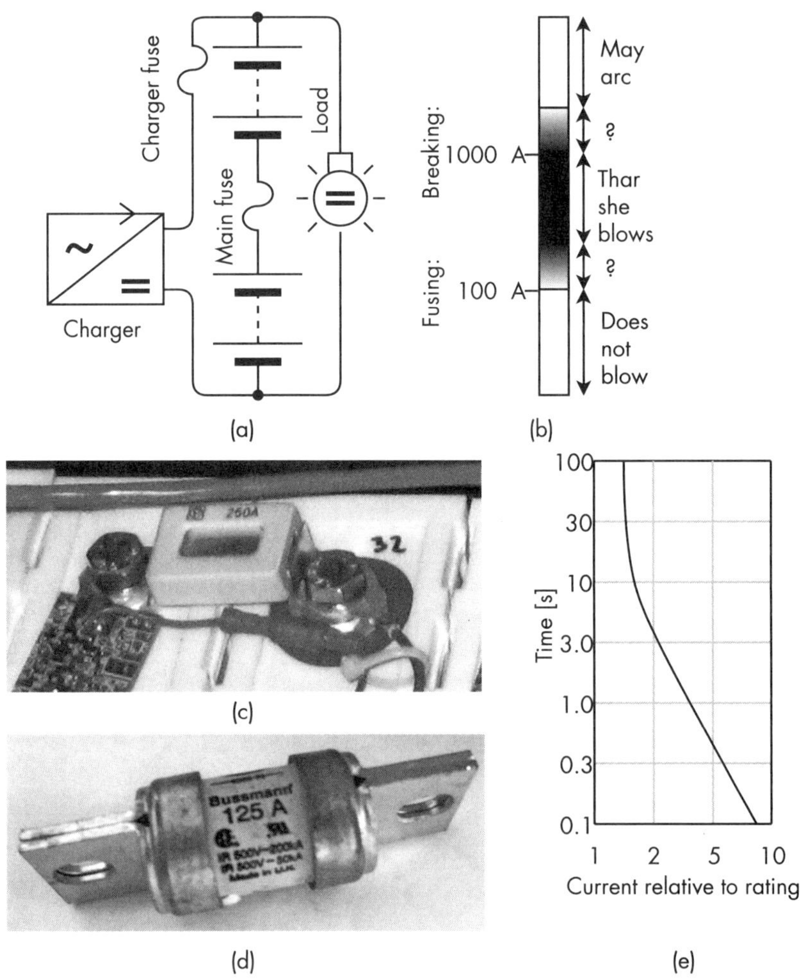

FIGURE 5.54
Main fuse: (a) placement, (b) current ratings of a 100 A fuse, (c) bad mounting, (d) large fuse, and (e) fuse timing curve.

5.12.6.1 Fusing Current Rating

The main fuse (Figure 5.54(d)) must be rated for the lowest current of any of the following devices:

- The maximum current that the battery can deliver, considering both peak and continuous current;
- The average current rating of the current-carrying components in the battery such as wires, connectors, current sensor, contactors; do not consider the current-carrying capacity of any wire past a small fuse (e.g., a 1-A fuse in line with a wire going to a small DC-DC converter);
- The maximum current a load can draw or a charger can provide.

5.12.6.2 Breaking Capacity Rating

In case of a short circuit, the current may be extremely high, higher than the fuse can handle. The fuse may arc and continue to conduct the short-circuit current. A fuse's breaking capacity (a.k.a. *interrupting rating*) is the current below which it is guaranteed to open fully.

This problem is avoided by analyzing the application and estimating the maximum battery current in case of a short and selecting a fuse that is guaranteed to open at this maximum short-circuit current.

The worst case is a short circuit directly across the cell (in a single-cell battery) or string of cells. To estimate the maximum short-circuit current, divide the battery voltage by its total series resistance. If you know the MPT of the cells used in the battery, you can estimate the specific current more easily:[47]

$$\text{short\_circuit\_specific\_current}\,[1/\text{h}] = 3600[\text{s}/\text{h}]/\text{MTP}[\text{s}]/2 \tag{5.1}$$

As a rule of thumb, the specific current through a short circuit is

- Standard cells: ~20C,[48]
- Power cells: ~200C.[49]

For example, for a 100 Ah battery using large prismatic cells, the short-circuit current may be on the order of 2 kA. For a 100 Ah power battery, it may be in the order of 20 kA!

If a fuse with sufficient breaking capacity is not available or too expensive, it may be possible to limit the short-circuit current through the insertion of additional resistance in series, such as by reducing the cross section of the current-carrying conductors.

5.12.6.3 Voltage Rating

The main fuse must be rated for operation at DC[50] and for the total battery voltage—do not use an automotive fuse (rated for 32V) on a 48V battery. Do not use fuses that are only AC rated: they arc instead of opening.

---

47. The factor of 2 is because the cell is discharged into a short twice as fast as the maximum power time.
48. Cells with an MPT of 400 s discharge into a short in 200 seconds, which means that the current is 3600s/200s = 18C.
49. Cells with an MPT of 30 s discharge into a short in 15 seconds, which means that the current is 3600s/15s = 240C.
50. The DC voltage rating of fuses is normally lower than their AC voltage rating. The typical 5 × 20 mm cartridge fuse is rated for 250 Vac but only 125 Vdc.

5.12.6.4 Placement

The most effective placement of the main fuse is mid-battery, where it is statistically most likely to be within the path of a short circuit that occurs within the battery (see Volume 2, Section 3.14.3.1). However, do not place it in the middle of a BMS bank. If the fuse were to blow, the full battery voltage would appear immediately across it, damaging the BMS (see Section 8.3.2.4).

If a battery consists of multiple modules in series, each module must include a fuse to protect the modules in case of an accidental short circuit during manufacture, handling, and installation. These fuses must have a slightly higher rating than the main fuse, so the main fuse will blow first.

5.12.6.5 Mounting

Be aware that the main fuse must not be mounted in a way that applies mechanical stress on it. Do not place in line with a cable and do not mount it between adjacent cells (Figure 5.54(c)). Doing so may damage the fuse. A large battery wire connected to a fuse in mid-air is dangerous, because if the fuse were to blow violently and come apart, the battery wire would become loose and may connect to who-knows-what inside the battery. Instead, mount each fuse in a fuseholder that is designed for that particular fuse and that is, in turn, mounted to a chassis or panel. The fuse holder removes stress from the fuse, retains the wires should the fuse explode, and cools the fuse.[51]

5.12.7 Controlled Fuses

Besides working like a regular fuse, a controlled fuse includes a heater; applying power to this heater blows the fuse. Once the fuse blows, the heater is isolated and thus is shown as two fuses in series and a heater connected to the midpoint (Figure 5.55(a)).

At times, the BMS may need to interrupt the battery current as a last resort

- In a small battery, if the MOSFETs failed as a short (Figure 5.55(b));
- In a large battery that relies on controlling the external system to stop the battery current, if the external system is disobeying orders (Figure 5.55(d)).

A controlled fuse gives the BMS an alternative way of turning off the battery as a last resort, albeit permanently:

- In a small battery, the PCM can blow the fuse[52] by applying a relatively low-power signal (Figure 5.55(c));
- In a large battery, a controlled fuse replaces the main fuse and the fault contactor (Figure 5.55(e)).

Of course, once the BMS blows the controlled fuse, the battery must be serviced to restore operation.

Three types of controlled fuses are available[53]:

- For small batteries, SMD components, rated 5~45A and 36~80Vdc;

51. The fusing current of a fuse is affected by its temperature.
52. A.k.a. chemical fuses.
53. Available from: *Self Control Protector (SCP)* by Dexterial, https://www.dexerials.jp/en/products/compare/c3_list.html. *Thermoprotector with resistors, Thermal links, or Fusing resistors* by Uchihashi Estec Co.,Ltd., https://www.uchihashi.co.jp/en/fuse/Fuse-link by Xiamen SET Electronics Co.,Ltd., http://www.setfuse.com/category/Controlled%20Fuse%20link.html.

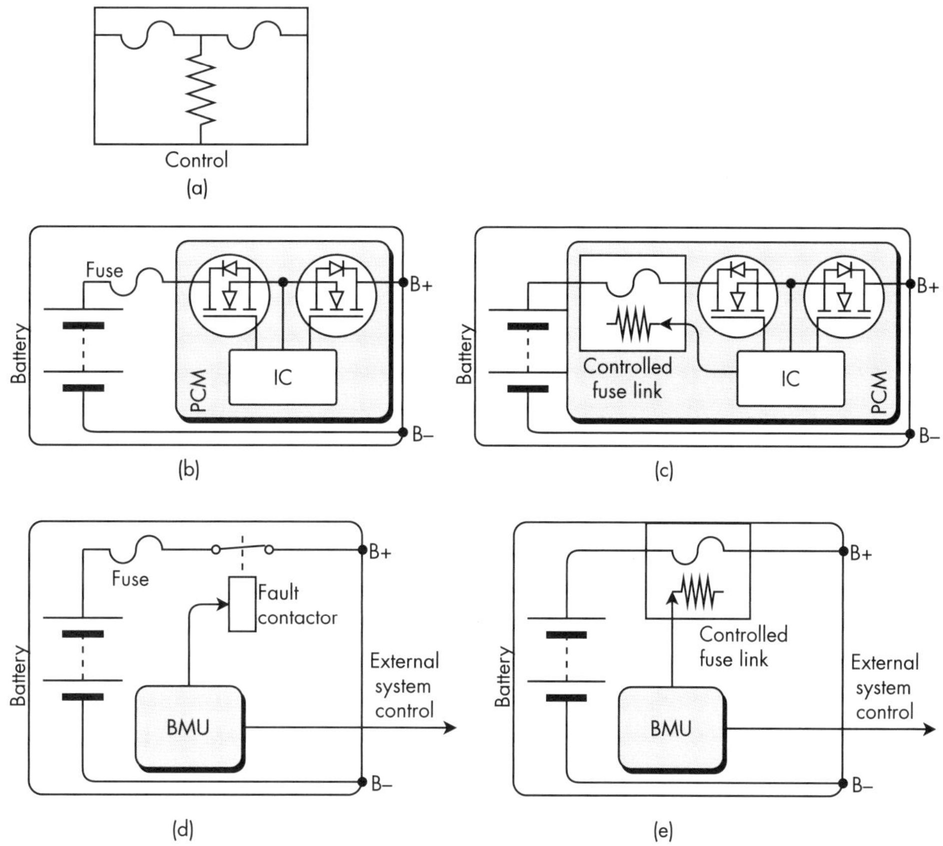

FIGURE 5.55
Controlled fuse link: (a) standard solution with a fuse and a contactor, and (b) solution with a thermal link.

- For medium voltage batteries, small PCB mount leaded components, rated 80~100A and 100Vdc;
- For large stationary batteries, chassis mount components, rated for 75~120A and 50~400V.

5.13 PRECHARGE

The moment a battery is connected directly to a capacitive load, a large inrush current results as the battery charges the load capacitance (Figure 5.56(a)). This current exceeds the ratings of the cells and may cause damage—blown fuse, welded contacts, melted metal in connectors, BMS damage (see Section 8.3.4), and increased internal resistance in cells.

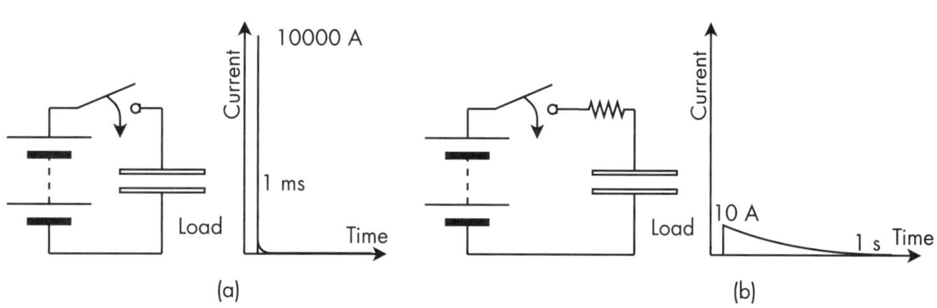

FIGURE 5.56
Connection to devices: (a) without precharge, and (b) with precharge.

Precharge[54] avoids this by raising the load voltage slowly through a current limited path, before making the final, direct connection (Figure 5.56(b)).

5.13.1 Inrush Current without Precharge

Let's start by estimating the size of the inrush current without precharge.

If the final connection is through a contactor, and if there is no fuse, the current is limited only by the internal resistance of the battery and the ESR[55] of the load capacitance. The peak current depends on the chemistry and MPT of the cells, the ratio of the battery capacity over the load capacitance, and the battery voltage.

The following curves are for a typical large, low-voltage stationary battery (Figure 5.57(a)) and for a typical traction battery for a passenger car (Figure 5.57(b)).

For example, for a 1,000 Ah, 48V battery using CALB *CA* cells (with an MPT of 280s), connected to an inverter with an input capacitance of 100 mF,[56] the capacity/capacitance ratio is 1,000 Ah/0.1 F = 10 k [Ah/F]. The graph indicates that the inrush starts at 20C (circle in Figure 5.57(a)); that is, 20 kA!

For a traction battery for a passenger vehicle, the specific current starts even higher: 60C for a standard EV and 300C (!) for a hybrid (two circles in Figure 5.57(b)).

Note that the inrush current for power batteries is exceptionally high. Small values of load capacitance reduce the current due to higher ESR. The current is also lower in large batteries because, for a given actual current, the specific current goes down as the capacity goes up. For large 48V batteries and small capacitors (a ratio of 1 M or more), the current drops to 1~5C, regardless of the MPT of the battery (range A in Figure 5.57(a)). Conversely, traction batteries for passenger vehicles (BEV or HEV) work in a region where the MPT of the battery does matter, with little regard to the load capacitance (range B in Figure 5.57(b)).

The duration of the inrush pulse[57] depends on the MPT and chemistry of cells, the load capacitance, and the battery voltage. These curves assume an MPT of 100s and are for a typical large, low-voltage stationary battery (Figure 5.58(a)) and for a typical traction battery for a passenger car (Figure 5.58(b)).

As the battery capacity increases, the pulse becomes shorter because larger batteries have a lower internal resistance. As the load capacitance increases, the pulse becomes longer because it takes more time to charge the load's capacitors.

5.13.2 Consequences of Skipping Precharge

There are three main consequences of skipping precharge:

- Disturbance from EMP;
- Damage from high current, such as blown capacitors in the load;
- Damage from reversed cell voltages.

---

54. Note that some use the term "precharge" to mean "preconditioning charge": the initial, low-current charging of a deeply discharged cell or battery. This is not what is meant here, and, in any case, a deeply discharged cell or battery should not be recharged. (See Section 5.15.3.1.)
55. Effective series resistance. Assumed to be 10 mΩ for a 1,000 μF low ESR capacitor of 50V or above.
56. mF = millifarad. 100 mF = 0.1 F = 100000 μF.
57. By convention, we use five times the time constant, by which time the pulse is less than 1% away from its final value.

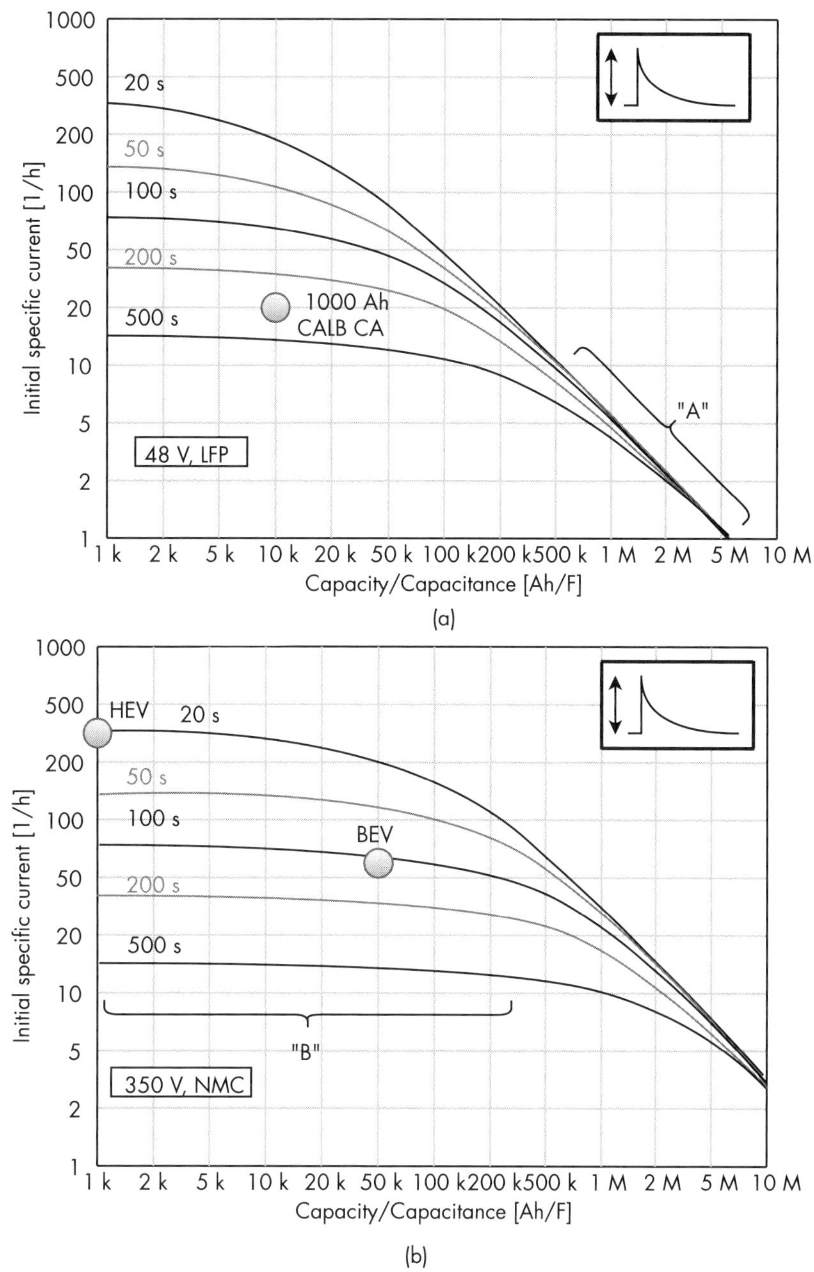

FIGURE 5.57
Initial specific current versus ratio of capacity over capacitance for various levels of cell MPT: (a) 48V battery using LFP cells, and (b) 350V battery using NMC cells.

5.13.2.1 EMP

The leading edge of the inrush pulse is of such short duration[58] and high value that it generates an electromagnetic pulse. This EMP disrupts communications:

- Frozen USB ports in computers; crashed OS;[59]
- Corrupted data in communication buses between the battery and external devices;
- Disruptions in internal BMS communications between slaves or cell boards.

58. The pulse can be as narrow as 60 μs. This is in the audio range, but the sudden beginning of the pulse would be less than 1 μs, except that the inductance in the circuit slows down the leading edge somewhat.
59. Operating System ('blue screen of death' in Windows computers).

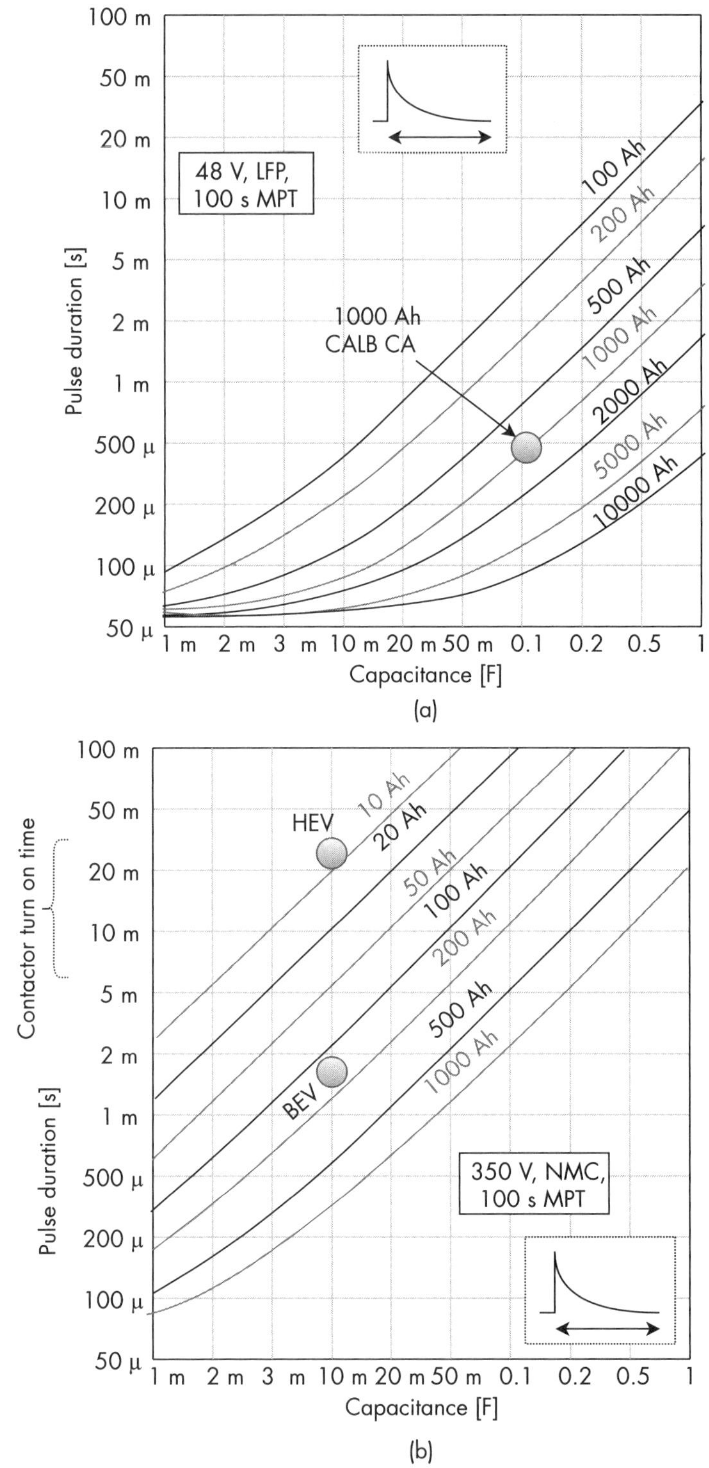

FIGURE 5.58
Pulse duration versus capacitance for various battery capacities: (a) 48V battery using LFP cells, and (b) 350V battery using NMC cells.

It may even cause permanent damage to unprotected electronics, though I haven't seen this myself.

5.13.2.2 Damage

The inrush pulse can be three orders of magnitude faster than the closing time for a contactor. Therefore, its high current flows through the contacts before they are fully mated. The contact area dissipates much power, which may weld the contacts (see Section 8.4.3).

In a series string, the moment the battery is connected directly to a capacitive load, the cells with the highest resistance experience a voltage reversal that damages the BMS (see Section 3.2.12).

5.13.3 Precharge Circuit

The precharge[60]circuit consists of a relay and a resistor (or some other current-limiting device) across the main contactor (Figure 5.59(a)). For solid-state protection switches, transistors are used in place of the relay and the contactor (Figure 5.59(b)).

If the battery must be galvanically isolated from the load, then the protector switch uses two contactors—one for the positive side, one for the negative side (Figure 5.59(c)). The BMU can implement a more complex turn-on sequence to test the integrity of the contactors. My convention is to call the contactors K1 (precharge), K2 (positive), and K3 (negative); I invite you to do the same.

5.13.4 Precharge Operation

Precharge occurs through a sequence of stages (Figure 5.60(a)):

1. Initially, the battery is off; both the relay and the contactor are off (Figure 5.60(b)), and the load voltage is zero (Figure 5.60(a-1)).
2. When there's a request to turn on the battery, the BMU may conduct some tests (such as ground fault detection) (Figure 5.60(a-2)).
3. The BMU turns on the precharge relay first (Figure 5.60(c)), charging the load gradually through the precharge resistor (Figure 5.60(a-3)). The initial current spike is on the order of 5 A[61]; the current decays with a time constant equal to the precharge resistance times the load capacitance (on the order of

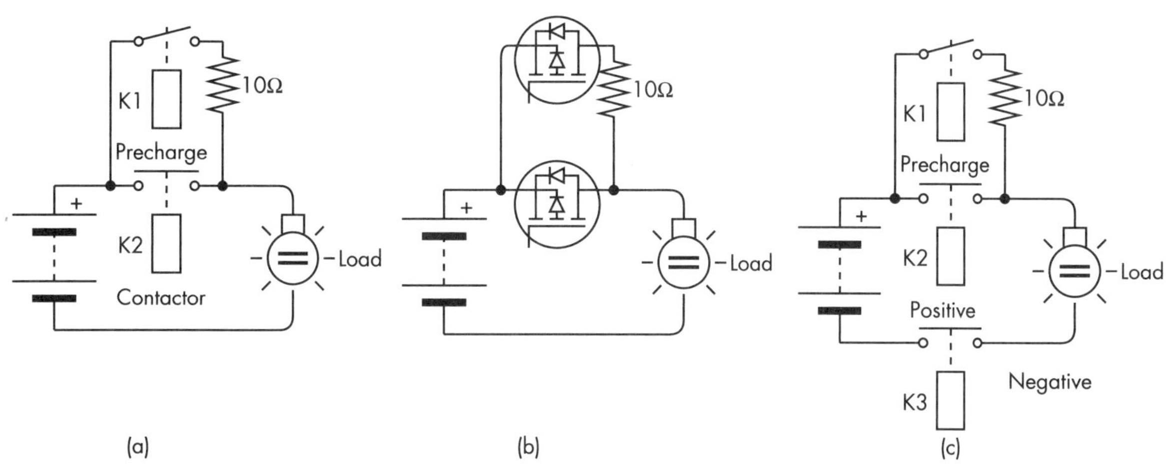

FIGURE 5.59 Precharge circuit: (a) electromechanical, (b) solid-state, and (c) galvanically isolated.

60. A.k.a. soft-start.

61. Assuming 48 V battery and a 10 Ω precharge resistor.

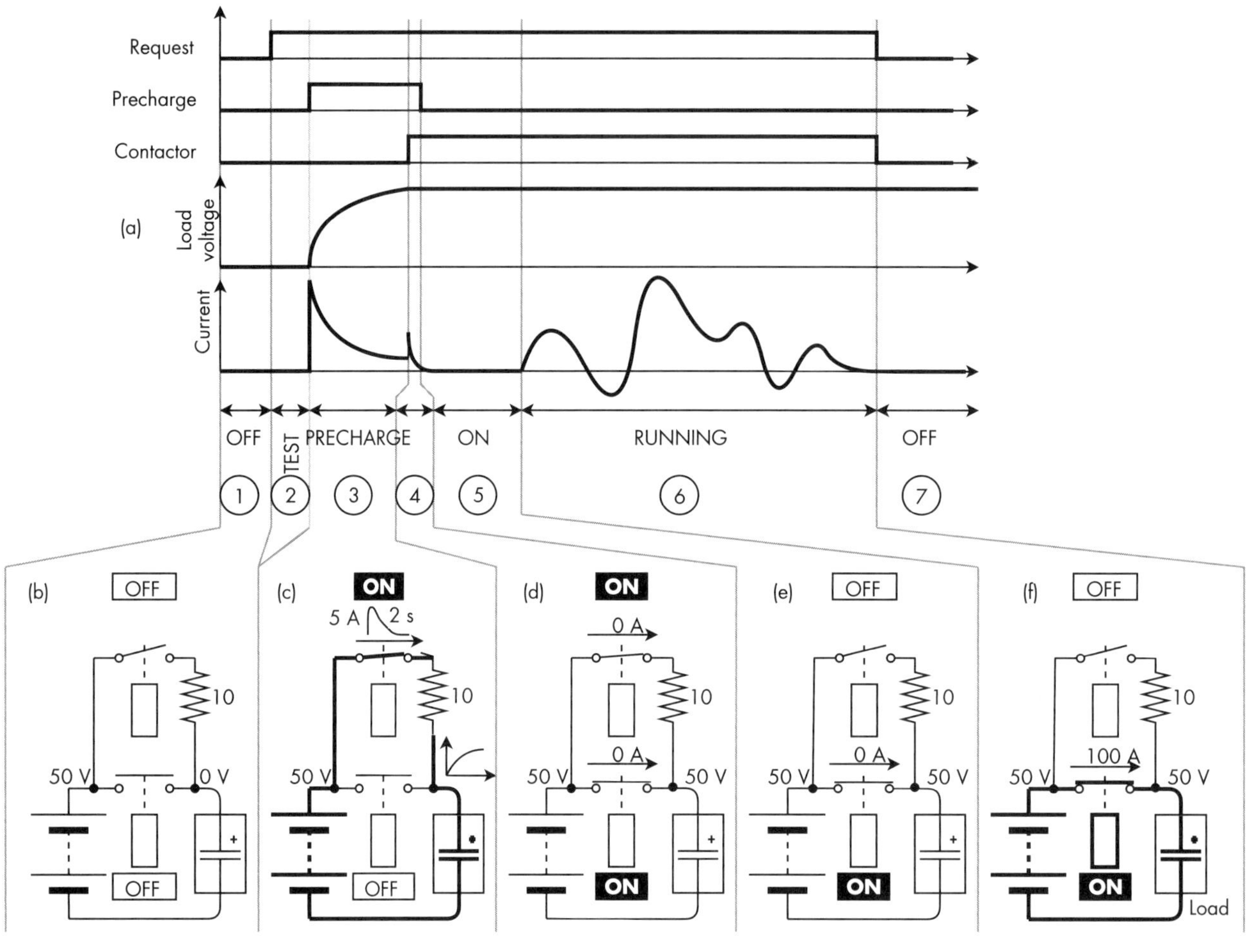

FIGURE 5.60 Precharge operation and circuit: (a) waveforms, (b) off, (c) precharge, (d) direct connection, (e) open precharge relay, and (f) running.

seconds). After some time, the load voltage approaches the battery voltage, the current approaches 0 A, and the voltage across the discharge contactor is nearly 0V.

4. At this point,[62] the BMU turns on the discharge contactor (Figure 5.60(d)); there is a second, smaller and sharper pulse of current because the load capacitance wasn't yet fully charged (Figure 5.60(a-4)), then decays to 0.
5. As the precharge relay is no longer needed, the BMU turns it off (Figure 5.60(e)); the current is 0 (Figure 5.60(a-5)).
6. The load starts operating (Figure 5.60(a-6)), drawing normal current through the closed contactor (Figure 5.60(f)).
7. If the request goes away, or there is a fault, the BMU turns off the contactor (Figure 5.60(b)) and there is no battery current (Figure 5.60(a-7)).

62. By convention, we use 5 times the time constant: $T = 5 \times R(\text{precharge}) \times C(\text{load})$ because after this time the load voltage has reached within 1 % of its final value.

5.13.4.1 *Precharge Testing and Abort*

At each step, the precharge controller may monitor this process and test for a variety of conditions (Table 5.7). The following assumes the galvanically isolated topology

TABLE 5.7
Precharge Tests

Step	Test	Fault
Before precharge	Tested by the ground fault detector (see Section 4.8.7)	Ground fault
	The string voltage appears across the load	The negative contactor is welded as is either the positive contactor or the precharge relay The load capacitance is still charged from the last time the battery was on
	There is precharge current	The negative contactor and the precharge relay are welded and the load is shorted
Turn on the precharge relay	There is precharge current	The negative contactor is welded
Turn on the negative contactor	There is no precharge current and the string voltage appears across the contacts of the positive contactor	Open contacts in the precharge relay, open precharge resistor, or open precharge fuse (if any)
	There is no precharge current and the string voltage appears across the contacts of the negative contactor	Open contacts in the negative contactor
	There is no precharge current and the string voltage appears across the load	The load is disconnected or the main fuse is blown
	There is excessive current	Welded contacts in the positive contactor
Wait 100 ms	The precharge current does not decay	The load is shorted
	The precharge current is reduced but remains constant	The load has been activated too soon
	The precharge current decays but not fast enough	There is an unexpected, additional load
	The precharge current has fully decayed	The full load is not connected; possibly, only a small DC-DC converter is connected
Wait for the end of precharge	Precharge lasts too long	There is an unexpected, additional load or the load has been activated too soon
Turn on the positive contactor		
Turn off the precharge relay	The load voltage decays, especially when the load is activated	Open positive contactor

with two contactors (Figure 5.59(c)) and that the precharge controller can measure the voltage of the B+ and B- load terminals relative to the string voltage.

If the precharge controller detects a problem, it aborts the process, turns everything off, and issues a fault indicating the nature of the problem.

5.13.5 Precharge Components

A precharge circuit includes a current limiter (e.g., a resistor) and a switch (e.g., a relay).

5.13.5.1 Precharge Resistor

We normally let the precharge last until the current drops to less than 1% of its peak. This occurs after five times the time constant formed by the precharge resistor and the load capacitance:

$$\text{precharge\_time [s]} = 5 \times \text{precharge\_resistance } [\Omega] \times \text{load\_capacitance [F]} \tag{5.2}$$

In practice, we wish to determine the value of the precharge resistor that results in the desired precharge time:

$$\text{precharge\_resistance } [\Omega] = \text{desired\_time [s]/load\_capacitance [F]/5} \tag{5.3}$$

For example, if the desired precharge time is 500 ms, and the load capacity is 10 mF, then we want the precharge resistor to have a resistance of 500 ms/10 mF/ 5 = 10 Ω.

The current peaks at the start of precharge, and then decays asymptotically towards zero. The peak current is

$$\text{initial\_current [A]} = \text{battery\_voltage [V]/precharge\_resistance } [\Omega] \tag{5.4}$$

For example, with a 100V battery and a 10 Ω precharge resistor, the peak current is 10 A.

The precharge resistor needs to dissipate as much energy as the energy stored in the load capacitors:

$$\begin{aligned}&\text{capacitor\_energy}[J] = \text{dissipated\_energy}[\text{J}] = \\ &\text{load\_capacitance}[\text{F}] \times (\text{battery\_voltage}[\text{V}])^2/2\end{aligned} \tag{5.5}$$

Note that resistance does not appear in the equation above—surprisingly, the energy dissipated by the resistor is constant and is independent of its resistance. This energy is also equal to the energy stored in the load capacitors.

For example, with a 100V battery and 10-mF capacitors, the energy is: E = 10 mF × $(100\text{V})^2/2$ = 50 Joule. This is the energy in the load capacitance after it is fully charged. It is also the energy dissipated by the precharge resistor during precharge.

The average power dissipated by the precharge resistor during precharge is that energy over the precharge time:

$$\text{average\_precharge\_power [W]} = \text{precharge\_energy [J]/precharge\_time [s]} \tag{5.6}$$

In this example, the precharge power is 50 J/500 ms = 100W.

At the beginning of the precharge, the instantaneous power is quite high:

$$\text{initial\_precharge\_power}[W] = (\text{battery\_voltage}[V])^2/\text{precharge\_resistance}[\Omega] \quad (5.7)$$

In this example, the initial precharge power is $(100V)^2/10\Omega = 1$ kW!

Using a resistor rated for that high a power is impractical. Instead, we use a relatively low-power resistor (on the order of 10 to 100W).

The high, sudden power during precharge stresses the precharge resistor, yet the total energy over the entire precharge is not that high. Therefore, the power rating of the precharge resistor doesn't need to be too high, but the resistor needs to be quite rugged.

Some resistor manufacturers specify the peak power dissipation rating (e.g., "Overload: 5 times rated wattage for 5 seconds."). If so, a 50W resistor can handle 250W (well above the 100W of the example above). You should ask the resistor manufacturer if a particular resistor would work in your application; regardless, you must test the resistor in the actual application.

Often, the precharge resistor is a wire-wound type, encased in ceramic, cement, or extruded aluminum. Inductive resistors are acceptable. For example:

- Tubular wire-wound: Ohmite 270 series, Vishay/Dale NL series;
- Cement wire-wound: Xicon PW-RC series;
- Aluminum extrusion wire-wound: Ohmite 89 series, Stackpole KAL series.

A cooling period between precharge cycles prevents overheating the precharge resistor.[63] A short across the output of the battery, or not leaving enough time between precharges, overheats the resistor and blows it up (see Section 8.4.3).

A fuse may be placed in series with the precharge resistor to protect it in case of a shorted load.[64]

5.13.5.2 Alternatives to Resistor

In the old days, some EV enthusiasts used an incandescent light bulb in place of a precharge resistor; a light bulb's resistance increases as it gets hot, therefore making it able to drive a shorted load without a problem.

Today, for low-voltage batteries, we may use a current source[65] instead of a resistor. This current source should be floating (isolated from ground).

For medium-voltage batteries, one may use specialized ICLs[66] optimized for precharge. An ICL is self-protected because it can withstand a short-circuited load indefinitely (there is no need for a fuse in series).

Specifically, consider the EPCOS B59201J0140B010 (20 Ω) and B59204J0130B010 (100Ω), rated for 10,000 precharge cycles. Table 5.8 specifies the maximum load capacitance for a single ICL.

63. A driver who is used to a standard car may have the habit to crank the engine multiple times. The same behavior in an EV results in multiple, closely spaced precharge cycles, which could blow up the precharge resistor.
64. A shorted load places the entire battery voltage across the precharge resistor, generating far more power than the resistor is rated for.
65. A current source can be as simple as a depletion MOSFET and a resistor between its gate and drain.
66. Inrush current limiter.

TABLE 5.8
Maximum Load Capacitance of a Single ICL in Various Applications

Application	Voltage	Max Temperature	Max Load Capacitance
Stationary battery	48V	45°C	0.10 F
UPS battery	192V	35°C	10000 μF
Passenger car traction battery	350V	85°C	1600 μF
Industrial traction battery	700V	85°C	400 μF

Several ICLs can be placed in parallel to handle more current—two ICLs in parallel handle twice the load capacitance.

5.13.5.3 Precharge Relay

The precharge relay needs to be rated for the full battery voltage because this voltage appears across the contacts before the precharge starts.

The relay needs to be able to handle the peak of the inrush current, but because the average current is low and the current when the relay opens is nearly zero, the current rating of the relay contacts is not critical and DC-rated contacts are not required.

However, in case of a shorted load, the precharge relay may open while carrying the full peak of the precharge current (typically on the order of 3~30A). This event may damage a relay rated for low current or not rated for DC. Therefore, a relay with contacts rated for DC and high current would be more reliable.

A standard power relay can handle these requirements. There is no need to use a contactor.

Adding an RC snubber circuit[67] across the contacts may improve the life of the relay by reducing the wear as the contacts bounce at the start of precharge.

5.13.6 Precharge Responsibility

Precharge is the responsibility of the device that makes the last connection because it is the one device that knows that a connection is about to be made

- If the battery makes the last connection to a bus that is already connected to the load, then the battery is responsible (Figure 5.61(a));
- If a load makes the last connection to a bus that is already connected to a battery, then this load is responsible (Figure 5.61(b));
- If there are multiple loads, then each load is responsible for its own precharge (Figure 5.61(c));
- Finally, if any device may connect at any random time, then each device is responsible for precharge (Figure 5.61(d)).

Unfortunately, few loads and even fewer batteries implement precharge.

5.13.7 Post-discharge

For safety, high-voltage devices should be discharged after use to avoid the risk of shocking a technician. This is called post-discharge. When the battery is turned off, it may post-discharge the load with a resistor (Figure 5.61(e)).

67. A capacitor and a resistor in series, with a time constant on the order of 10 ms.

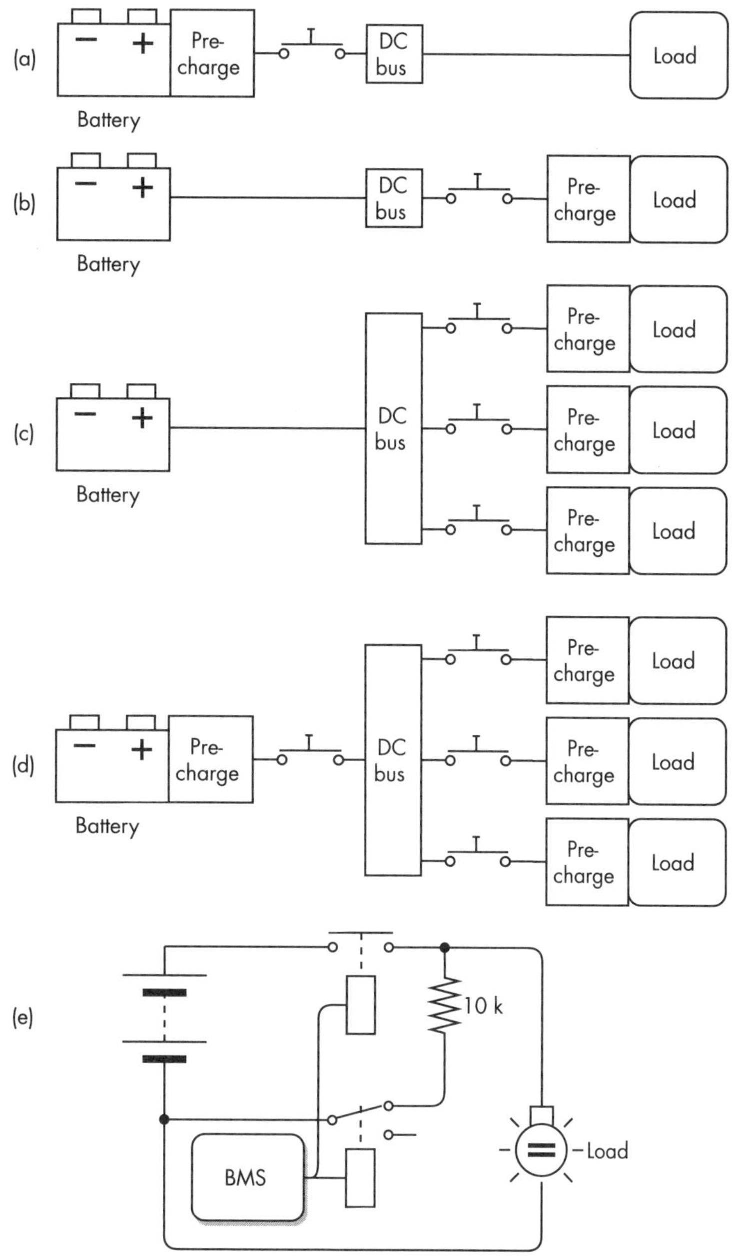

FIGURE 5.61
Precharge responsibility: (a) battery, (b) load, (c) individual loads, (d) everyone, (f) post-discharge circuit.

Post-discharge uses a normally closed relay whose coil is driven by a voltage that is present only when the battery is turned on. BMSs don't normally drive a post-discharge relay.

A simpler solution is to place a resistor across the load permanently. That resistor dissipates power all the time, but that may be acceptable.

5.14 BATTERY ISOLATION AND GROUND FAULTS

Batteries are often isolated from ground or the chassis.

5.14.1 Battery Isolation

For safety and reduction of electrical noise, a large battery should float. That is, it should be isolated from ground—earth ground, chassis, and low-voltage control circuits.

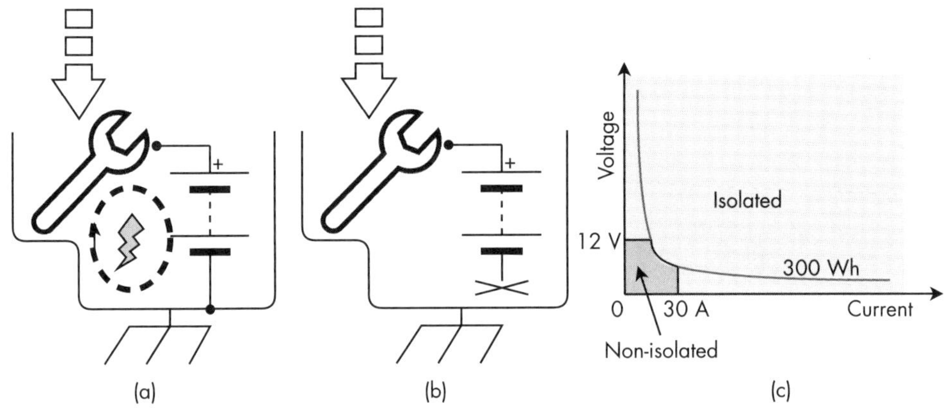

FIGURE 5.62
Short circuit hazard with dropped tool:
(a) floating battery,
(b) grounded battery, and
(c) recommended range of battery isolation.

5.14.1.1 The Case for Battery Isolation

A short circuit requires a complete path. Without it, there is no path for the current to circulate.

With a grounded battery, a tool accidentally dropped into a battery may connect a cell to a grounded metal case, completing a path that results in a short-circuit current (Figure 5.62(a)).

Floating the battery removes one connection; a dropped tool cannot complete the path—no short circuit can occur (Figure 5.62(b)).

Note that isolating the battery does not prevent a short circuit across the battery terminals; a fuse is required to handle this case.

Similarly, if the battery is grounded (Figure 5.63(a)), Mr. Stick Figure is shocked when he touches any live terminal inside the battery. Current flows from the battery, through the body, earth ground, and back to the battery. Floating the battery removes this shock danger by opening this path. Should Mr. Stick Figure touch a conductor inside the battery, there is no complete path for current to flow (Figure 5.63(b)).

An isolated battery with a significant capacitance between the cells and earth ground may also be a shock hazard, though of short duration (Figure 5.63(c)). Current flows through the body for a while until the Y-capacitance is charged. This capacitance could be an actual capacitor[68] or parasitic capacitance formed by the physical proximity of cells to grounded metal (e.g., heat-sink plates placed flat against pouch cells).

5.14.1.2 When to Isolate a Battery

Small batteries are usually not isolated because the voltage is too low. In any case, doing so would be too expensive.

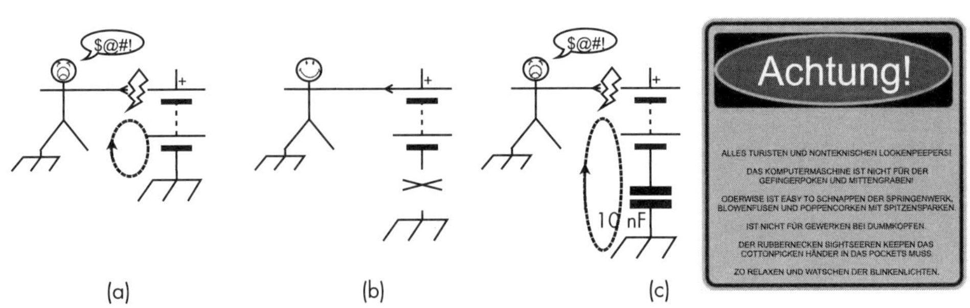

FIGURE 5.63
Shock hazard:
(a) grounded battery,
(b) floating battery,
and (c) battery with large Y-capacitance.

68. Which may be placed there for reduction of electrical noise.

Some safety regulations have picked 40V as the threshold above which extra precautions are required against an electrical shock. There is no voltage threshold below which a short-circuit current does not cause damage, though, clearly, a battery capable of higher current is more dangerous.

My recommendation is that batteries above 12V or capable of delivering more than 30 A or 300 Wh should be isolated (Figure 5.62(c)).

5.14.1.3 Isolated Battery in Grounded Application

Unfortunately, some applications defeat the battery isolation for either regulatory or technical reasons:

- Some regulators have decided that specific battery-powered circuits should be grounded. For example:
  - *Marine:* B- must be connected to the engine block (see Volume 2, Section 2.6.3.1);
  - *Telecom:* B+ must be connected to earth ground (see Volume 2, Section 2.4.2.6,);
  - *Automotive:* The negative of starter batteries must be connected to the chassis (see Volume 2, Section 2.9).
- In some applications, the battery is connected to a load that is not isolated. For example:
  - The neutral line in an inverger connected to the power grid is connected to earth ground;
  - The output of an inverter in a high-power UPS application is grounded.

Even though the application may be grounded, the battery itself should be floating. Doing so

- Reduces the chance of damage during manufacture;
- Reduces the chance of electrical shock (until grounded by the application);
- Reduces issues with electrical noise in communication links;
- Avoids ground loops;
- Allows the application to ground the battery at either end, as desired.

Before the battery is connected to the load, it is still floating and may be tested for loss of isolation (Figure 5.64(a)). Afterward, the battery is grounded and loss of isolation testing would fail (naturally) (Figure 5.64(b)) (see Volume 2, Section 4.9.1).

5.14.1.4 Isolating a Battery

The first requirement to isolate a battery is to ensure that there is no connection between any battery terminal and ground (earth ground, chassis, or low-voltage control circuits). The BMS must have two sections isolated from each other (Figure 5.65) (for how a BMS is isolated internally, see Volume 2, Section A.5.3):

- High voltage: Connected to the cells;
- Low voltage: Connected to the low voltage control circuit.

The battery design must avoid devices that defeat this isolation:

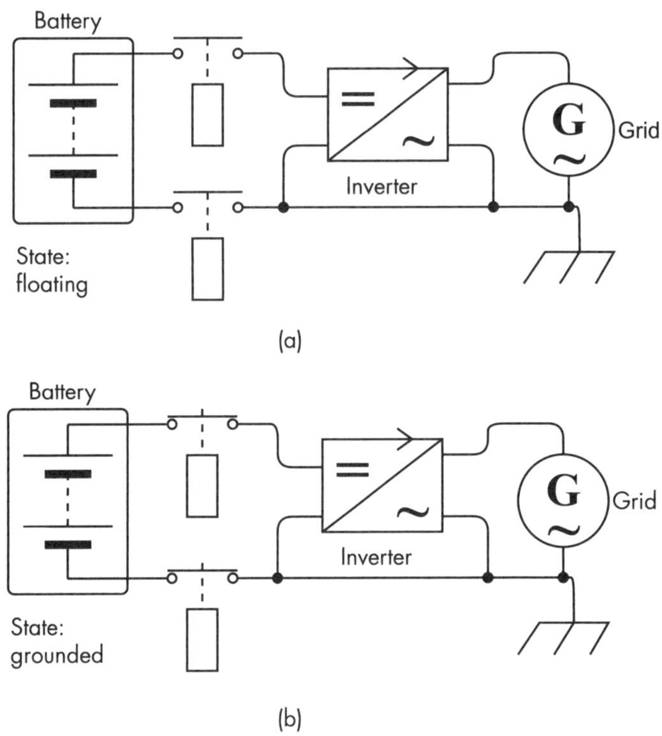

FIGURE 5.64
Isolated battery in grounded application: (a) before connection to grounded load, and (b) while connected to grounded load.

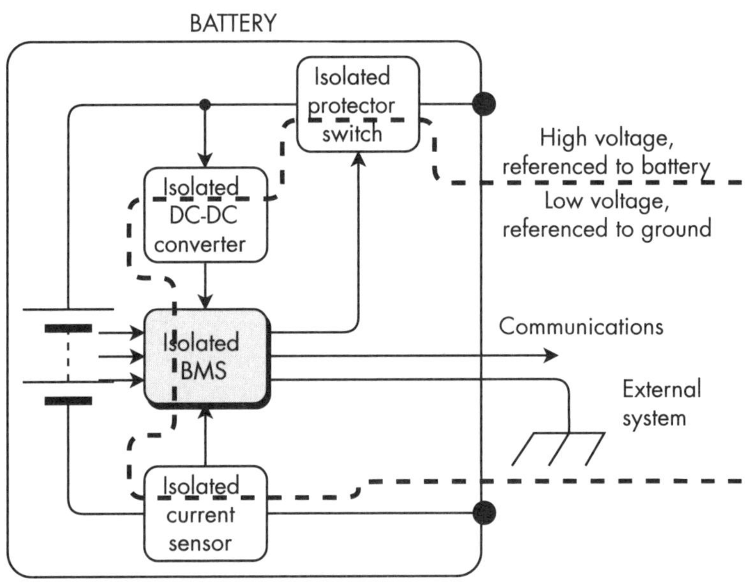

FIGURE 5.65
Battery isolation.

- If the BMS is powered from the battery voltage, it must be through an isolated DC-DC converter;[69]
- If using a current shunt, its amplifier must be isolated;
- The protector switch must be isolated:
  - MOSFETs are not isolated (unless driven by isolated gate drivers);
  - Contactors and relays are isolated.

69. A three-wire DC-DC converter (buck or boost converters) is not isolated.

An isolated battery should be tested periodically during assembly to ensure that its insulation is not compromised by an assembly error or design oversight (see Section 7.7.1).

5.14.2 Ground Faults

This section discusses two types of ground fault, consequences, and detection methods.

5.14.2.1 Types

A hard ground fault is a direct (0Ω) connection between any connection within the battery and any point referenced to ground. It occurs through a good conductor, such as a wire (see Section 1.2.2.9).

A soft fault is similar, though through a high-resistance path, such as carbon fiber, dirt, water, or a squirrel.

5.14.2.2 Causes

At the time of assembly, hard ground faults result from design errors or wiring mistakes.

In the field, hard ground faults result from mechanical damage, metal tools, or any other conductive material dropped in the battery. Soft ground faults result from the collection of dirt, flooding or even dampness, or a person's uncontrollable urge to touch what shouldn't be touched.

5.14.2.3 Consequences

For a floating battery, a ground fault results in loss of isolation. This by itself doesn't result in immediate harm or damage (though it could result in increased noise in communication lines). However, this removes the benefits of isolation—now there is a real danger of an electrical shock or a short circuit from a dropped tool.

For a grounded battery, there were never any benefits from isolation in the first place, so a ground fault doesn't change this. The danger of electrical shock is always there (Figure 5.66(a)).

A hard ground fault (e.g., a metal tool dropped into the battery) causes a short circuit. If the path is through a fuse, it blows (Figure 5.66(b)). Otherwise, conductors melt or the battery may go ballistic (Figure 5.66(c)).

A soft ground fault (e.g., a hot dog dropped into the battery) results in heating of the material that caused the fault. If not addressed quickly, this in turn can cause a fire (Figure 5.66(d)).

5.14.3 Automatic Ground Fault Detection

When and how to look for a ground fault depends on the type of battery and its application.

5.14.3.1 Types

There are two types of automatic ground fault detection, depending on whether the battery is floating or grounded at the time of the test (Table 5.9).

Various methods are used to detect isolation loss and ground current faults (see Volume 2, Sections A.6.1 and A.6.2).

Note that a ground fault detector may report a false positive if another ground fault detector is taking measurements at the same time.

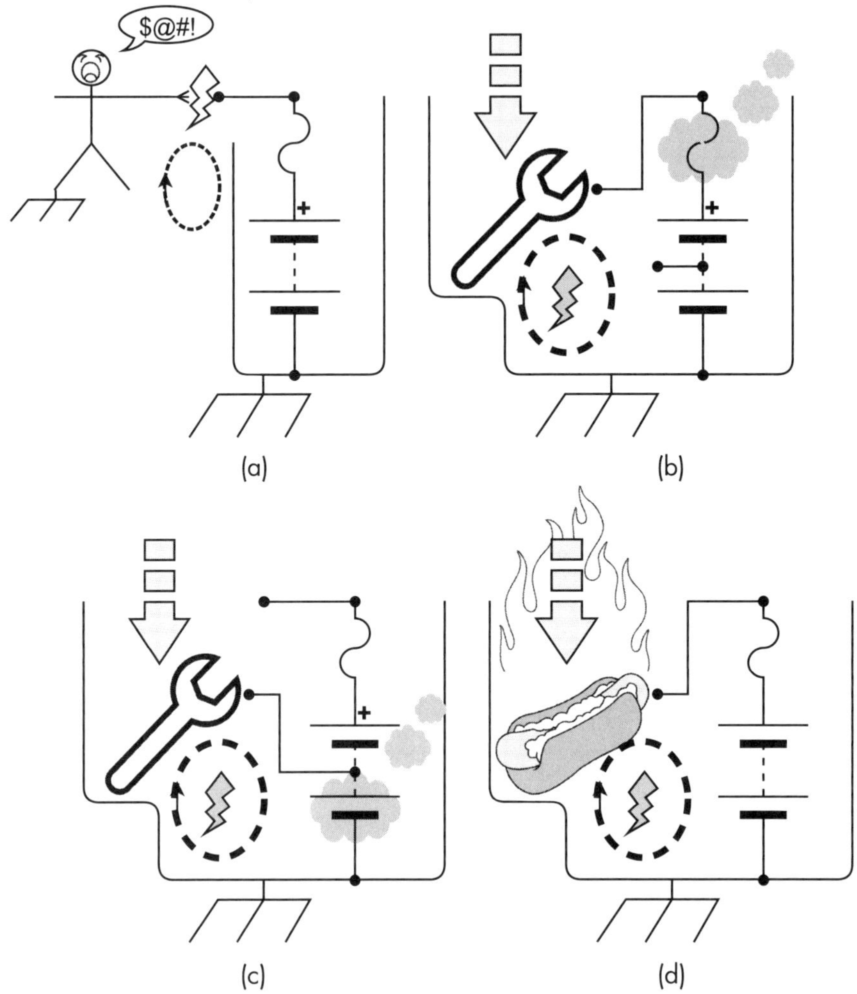

FIGURE 5.66
Grounded battery risks: (a) shock risk, (b) hard ground fault, through fuse, (c) hard ground fault, no fuse, and (d) soft ground fault.

TABLE 5.9
Ground Fault Detection Types

Type	Battery	Detection Time	Detects	Limit
Isolation loss	Floating	Before ground fault accident	Incorrect voltage to ground	0.5 mA
Ground fault current	Grounded	During a ground fault accident	Difference in current in two terminals	~10 W

5.14.3.2 Detection Thresholds

The detection thresholds for loss of isolation are different for a floating battery than for a grounded battery.

For a floating battery, the threshold is high enough to avoid false alarms in the presence of the little ground current drawn by measurement equipment. This limit is exceeded by any current that is sufficient to cause perceptible shock or damage. For example, an electric vehicle traction battery is considered isolated if the current is less than 0.5 mA.[70]

70. In the United States, the Federal Motor Safety Administration's safety standard FMSS305 states 100 Ω/V, which is 10 mA.

There is no standard for the detection threshold for a grounded battery. An analysis of a fire due to a ground fault in a 270V UPS battery revealed that the ground fault was dissipating about 500W, or about 2A of ground-fault current [3].

If the ground fault is through a small, flammable object, it doesn't take much power for it to reach self-ignition temperatures. This is why I suggest using a threshold of 10 W. This corresponds to

- 200 mA for a 48V large stationary battery;
- 10 mA for a 1-kV grid-tied battery.

5.14.3.3 Ground Fault Detection Requirement

The application determines whether ground fault detection is required or recommended and which type of test should be used (Table 5.10).

5.14.3.4 Ground Fault Detectors

Some BMSs include an isolation loss detector; more advanced BMSs not only report that there is a ground fault but also where it is, at what voltage, and the resistance of the ground fault.

No off-the-shelf BMS includes a differential current detector. A dedicated *insulation monitoring device* or *ground fault detector* is available from companies that specialize in this kind of equipment.[71]

5.15 AC-POWERED CHARGERS

This section discusses how to power and control an AC-powered charger, whether it is inside or outside the battery.

5.15.1 Charger Control

The BMS may control a charger to optimize charging the cells in the battery. The method depends on the type of charger:

TABLE 5.10
Ground fault Detection Requirements and Recommendations

Application	Isolation Loss Detector	Ground Fault Current Detector
Small battery	None	None
Large, low voltage, floating battery	Recommended	Not required
Large, low voltage, grounded battery, contactors open	Recommended	Not used
Large, low voltage, grounded battery, contactors closed	Gives false positive	Optional
Small traction battery	Not required	None
Large traction battery	Required	Uncommon
High-voltage stationary, contactors open	Required	Not used
High-voltage stationary, contactors closed	Gives false positive	Recommended

71. The Isometer from Bender Inc.

- *Dumb charger:* The BMS controls it by turning off its input AC power with a relay or solid-state relay (Figure 5.67(a)); or the BMS controls it by disconnecting its DC output with a relay or contactor, precharge is required, due to the large capacitors on the output of the charger (Figure 5.67(b)).
- *Charger with control lines:* The BMS turns the current on or off through a logic line into the charger (Figure 5.67(c).
- *Charger with CAN bus:* The BMS tells the charger the maximum charging current, and reads the actual current from the charger (Figure 5.67(d)).

Disconnecting the DC output is the most complicated solution, but if the user may disconnect the charger from the battery (e.g., the DC charging port in an EV), this is the required solution.

5.15.2 CCCV Charging

The way a battery is charged can be critical, as it has a strong effect on the amount of charge stored and on the cell lifetime.

In an energy or power application (see Section 5.1.4), the cells are charged fully before use (unlike in a buffer application, in which cells are never fully charged).

A CCCV charger is used to charge a single-cell battery (see Section 1.8.2). At any given moment, it operates either in the CC mode or the CV mode. Accordingly, charging occurs in two stages: CC first, then CV (Figure 5.68).

In the CC stage

- The charger limits the current into the cell to a constant value (the CC setting of the charger);
- The cell sets the voltage (which is lower than the charger's CV setting)
  - The OCV of the cell depends on its SoC;
  - The terminal voltage is higher than the OCV due to the IR drop across its internal resistance; when the terminal voltage (not the OCV) reaches the CV setting, the charger switches to the CV mode and the CV stage starts.

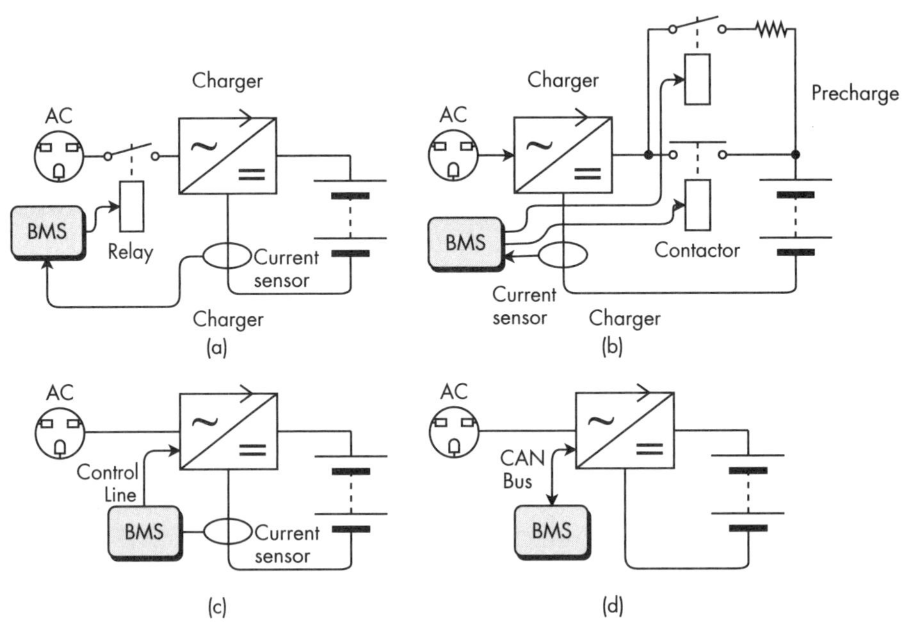

FIGURE 5.67
Charger control: (a) AC relay, (b) DC contactor with precharge, (c) control line, and (d) CAN bus.

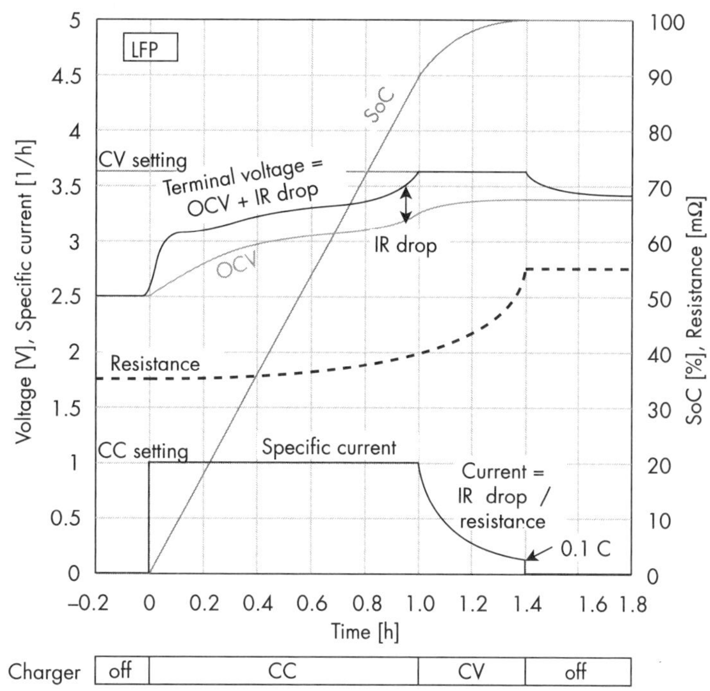

FIGURE 5.68 CCCV charging of an LFP cell.

In the CV stage

- The charger applies a constant voltage across the cell (the CV setting of the charger):
  - The terminal voltage is simply the CV setting of the charger;
  - The OCV of the cell is still lower than the terminal voltage; as the cell is charged, its SoC increases, and so does the OCV, though not linearly;
  - Toward the end of charge, the OCV increases rapidly per the OCV versus SoC curve.
- The cell sets the current (starts at the CC setting and then drops towards zero):
  - The current is proportional to the difference between the terminal voltage (which is the CV setting of the charger) and the OCV (which increases as the SoC increases); it is also inversely proportional to the cell internal series resistance (which increases as the SoC approaches 100%):

current [A] = (CV_setting [V] – cell_OCV [V])/internal_resistance [Ω]

  - Therefore, as time passes, the current drops due to the increase of both the OCV and the resistance.
- Once the current drops below a certain level (typically 0.1 C), the charger turns off and the cell is considered full:
  - The cell terminal voltage drops (because there is no longer a voltage across the cell's internal resistance); it drops further over time, as the terminal voltage relaxes;[72]

72. From an electrical standpoint, this is due to the capacitors in the cell impedance. From a physical standpoint, this is due to the settling of the lithium ions as they slowly diffuse.

- Since there is no more charging, the SoC remains constant, and therefore so does the OCV.

Note that the CV setting (and therefore the terminal voltage) must be higher than the OCV to force current into the cell. The voltage difference between the terminal voltage and the OCV is what drives current into the cell. For example, for an LFP cell that should rest at 3.45V when fully charged, the CV setting is 3.6V, so that the voltage difference (0.15V) appearing across the internal series resistance results in some current into the cell. If the CV setting were 3.45V, when the cell is nearly full the voltage difference would be too low, making the charging current also too low (its OCV is nearly 3.45V). Sure, this would eventually charge the cells, but it would take forever to reach this point.

The above also applies to charging a string of cells in series, as long as it is balanced (see Section 3.2.13).

5.15.2.1 Constant Power Charging

Above, we studied how a cell can be fully charged with a charger, especially a CCCV one; yet, some applications charge the battery at constant power (not constant current). For example:

- Regenerative braking in an electric vehicle, especially hybrids (HEVs) (see Volume 2, Section 3.2.2),
- Buffer batteries in general (not just in HEVs) (see Section 5.1.4).

Without a CV phase, the battery is not charged fully.

5.15.3 Li-ion Profiles

A charger may be advertised to have a "Li-ion profile" designed specifically for Li-ion batteries. Such a profile is helpful when charging a single cell or a balanced string (see Section 3.2.5). However, a Li-ion profile is not helpful for an unbalanced string because the charger does not know the voltage of individual cells. Instead, the BMS is the only device that can decide the appropriate CC and CV settings at a given moment because it monitors individual cell voltages. Ideally, the BMS controls the charger and varies those settings so that the cells are charged in the most effective way and with minimal degradation.

"Li-ion profile" means different things to different chargers. For example:

- *Preconditioning:* Charge a deeply discharged cell at reduced current;
- *Three-stage charging:* Reduced float voltage after the cell is full;
- *Stepped profile:* Reduced CC setting when the cell SoC is above a threshold;
- *Target voltage:* Increased CV setting as the cell is charged.

None of these profiles are recommended, as explained below.

5.15.3.1 Preconditioning

If the cell voltage is too low, chargers with preconditioning start charging at low current and only switch to the full CC setting after the voltage has reached the cell's minimum voltage specification (Figure 5.69(a)).

Slowly charging a deeply discharged cell is acceptable for lead-acid or NiMH batteries, but not for Li-ion—if the cell voltage is still above a specific limit, the cell

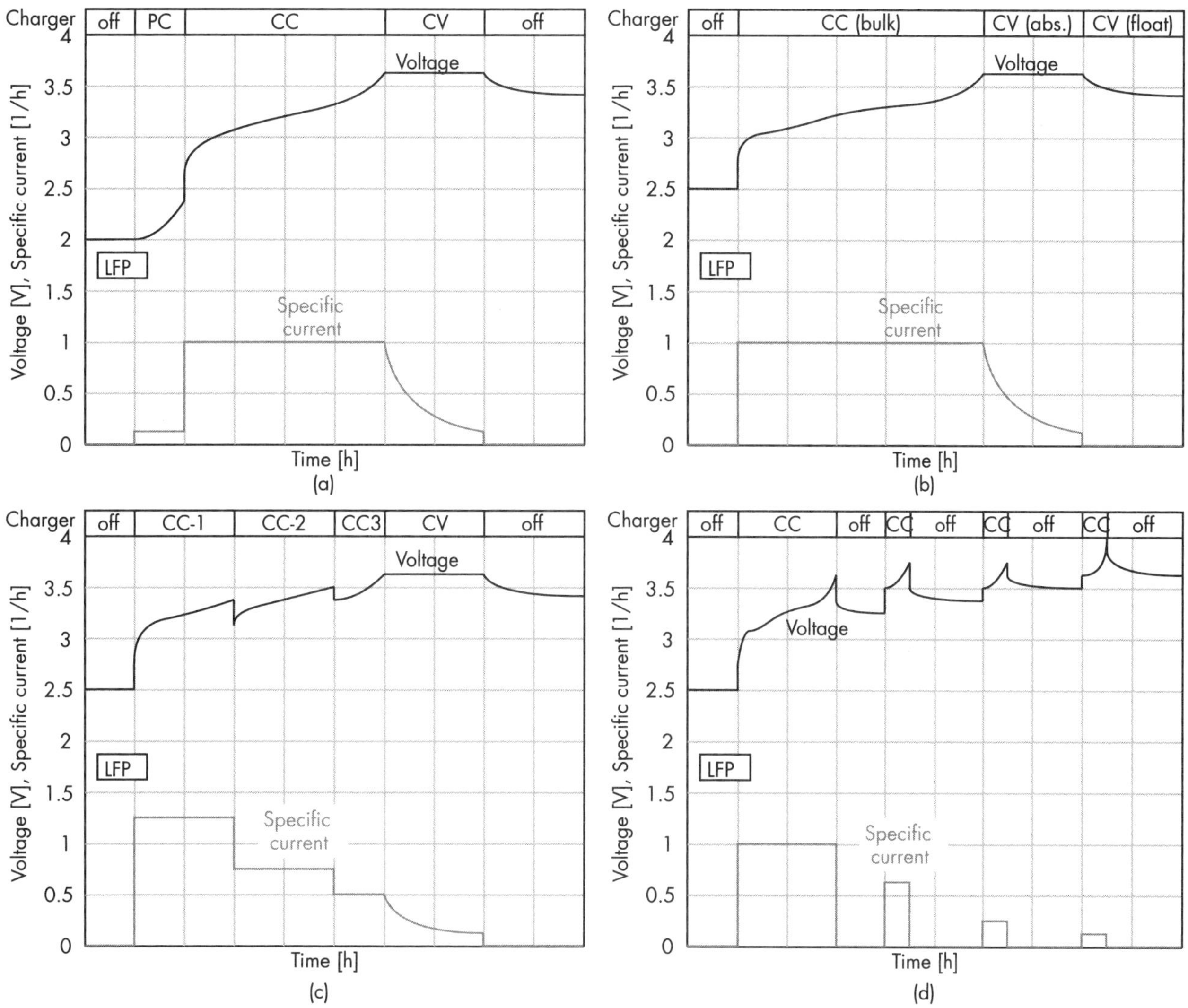

FIGURE 5.69 Other charging profiles: (a) preconditioning, (b) three-stage, (c) step, and (d) target.

can be charged at full current (no need for reduced current); on the other hand, if the cell voltage is below this limit, the cell is damaged and should be discarded.[73]

Unless you understand how a particular cell chemistry is damaged by low voltage, I strongly recommend against this—some NMC cells have self-ignited when recharged after being overdischarged. An NMC cell whose OCV is below its minimum voltage specification should be discarded, not recharged at low current.

5.15.3.2 Three-Stage Charging

A three-stage charger charges the battery in three stages: *bulk*, *absorption*, and *float* (Figure 5.69(b)).

Three-stage charging is appropriate for lead-acid batteries, but the life of Li-ion cells is reduced if kept at a constant voltage in the float stage. Once a Li-ion cell is

73. Yet, Texas Instruments and Linear Technology offer integrated circuits that support preconditioning, which invites users to attempt to charge cells that should be discarded.

charged, the charger must be turned off. If this is not possible, a significantly lower float voltage must be used (see Section 2.4.1.1).

5.15.3.3 Step Charging

With step charging, the cell is charged at full current from 0% to 20% SoC (for example), at half current from 20% to 50% SoC, and at 1/10 current from 50% to 100% SoC (Figure 5.69(c)).

Step charging is presented as a way to prolong cell life, yet tests have disproved this—the battery life is practically the same as when low-power charging is used [4]. In any case, in most applications, the cell is rarely discharged below 20% SoC. Therefore, it is rarely charged at full current. In practice, this means that step charging is effectively the same as low-power charging.

5.15.3.4 Target Voltage Charging

Target voltage charging [5] is a strange beast; I have not been able to discern any advantage to it. It uses a CC charger. There is no CV phase, and the CV threshold is only used to shut off the charger (Figure 5.69(d)).

Initially, the CV is set for the maximum cell voltage (for a series string, the cell with the highest voltage is used). When the actual voltage reaches this CV limit (that is, the instant the charger switches from CC to CV mode), the charger is turned off, and the voltage is allowed to relax. The process is repeated a few times, each time with a higher CV setting and a lower CC setting. In the end, the cell is slightly overcharged (in the sense that its terminal voltage exceeds its ratings), yet it is not quite fully charged (in the sense that its OCV is below rating) because there is never a CV phase.

This charging method does not appear to be advantageous.

5.15.3.5 Adaptive Charging

The only profile that is better for cell life than CCCV is one whose settings are dynamically adjusted based on the conditions of a given cell at a given time. Such algorithms are proprietary and are closely guarded secrets [6].

5.15.4 Multiple Chargers

Chargers may be connected in parallel or in series to achieve the desired power or voltage.

5.15.4.1 Parallel Chargers

Power from a standard AC outlet is limited: 1.5 kW for a 120 Vac outlet, and 3 kW from a 240 Vac one. As a consequence, many chargers are limited to these two power levels. There are many more 3 kW chargers than 6 kW ones. It may be cheaper to use several lower-power chargers than a single high-power charger.

Two or more chargers may be used with their DC outputs connected in parallel to increase the charging power (Figure 5.70(a)).

Multiple chargers may be used to balance the load on an AC line:

- Two chargers may be powered by the two lines of a split-phase AC outlet (Figure 5.70(b));
- Three chargers may be powered from the three lines of a three-phase AC outlet (Figure 5.70(c)).

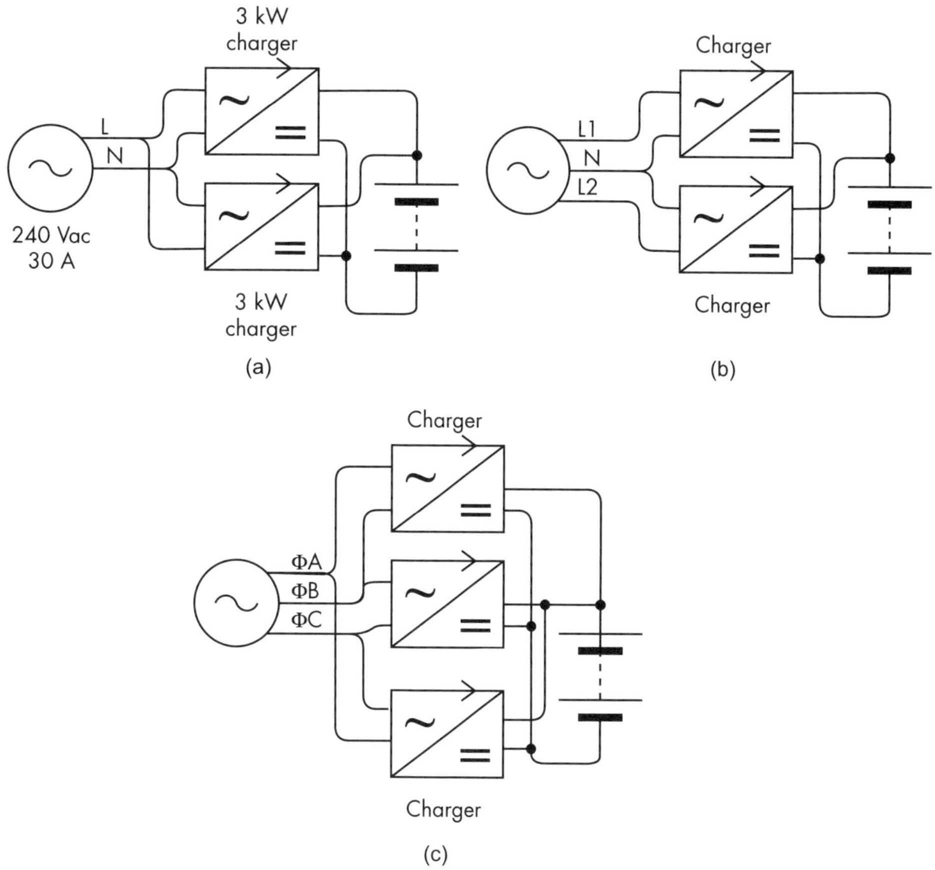

FIGURE 5.70
Multiple chargers: (a) two chargers, single-phase, (b) two chargers, split phase AC, and (c) three chargers, three-phase.

It is safe to connect the outputs of chargers in parallel, as long as they are all configured identically (Figure 5.71). Small variations between the chargers won't usually cause any problems.[74] This is the operation of two chargers in parallel, with Charger 1 having a slightly higher current and a slightly higher voltage:

- *CC mode:* Each charger produces its maximum current, and the load is shared pretty much equally among the chargers.
- *Overlap:* When the battery voltage reaches the CV of Charger 2, this charger switches to CV mode and its current starts dropping. Charger 1 continues in the CC mode as the battery voltage continues to increase a bit.

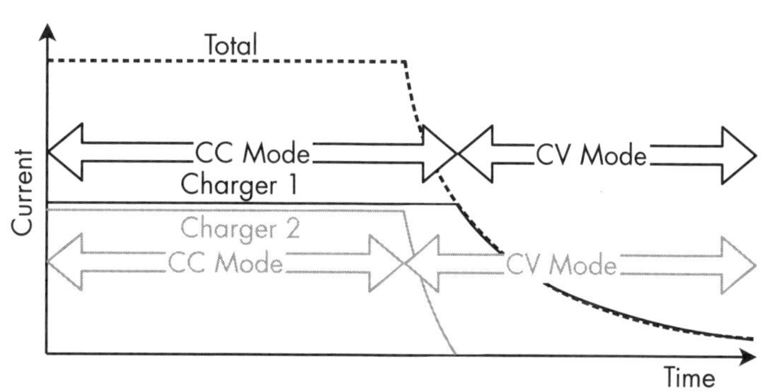

FIGURE 5.71
Load sharing between two chargers in parallel.

74. Note that this may not be true with power supplies which operate only in the CV mode: as they are not current-limited, the load draws excessive current from the power supply whose voltage is slightly higher than the others.

- *CV:* When the current from Charger 1 drops to 0, the battery voltage reaches the CV of Charger 2 and this charger switches to the CV mode as well. Its current starts dropping; Charger 1 is idle.

Chargers that are specifically designed to be used in parallel may implement some form of load sharing, meaning that they all operate at the same current at all times.

5.15.4.2 Series Chargers

Two or more chargers may have their DC outputs connected in series to achieve a higher voltage (Figure 5.72(a)). For example, this may be done to charge a 700V traction battery from two 350V chargers, which are far more common than 700V chargers (see Volume 2, Section 3.3.2,).

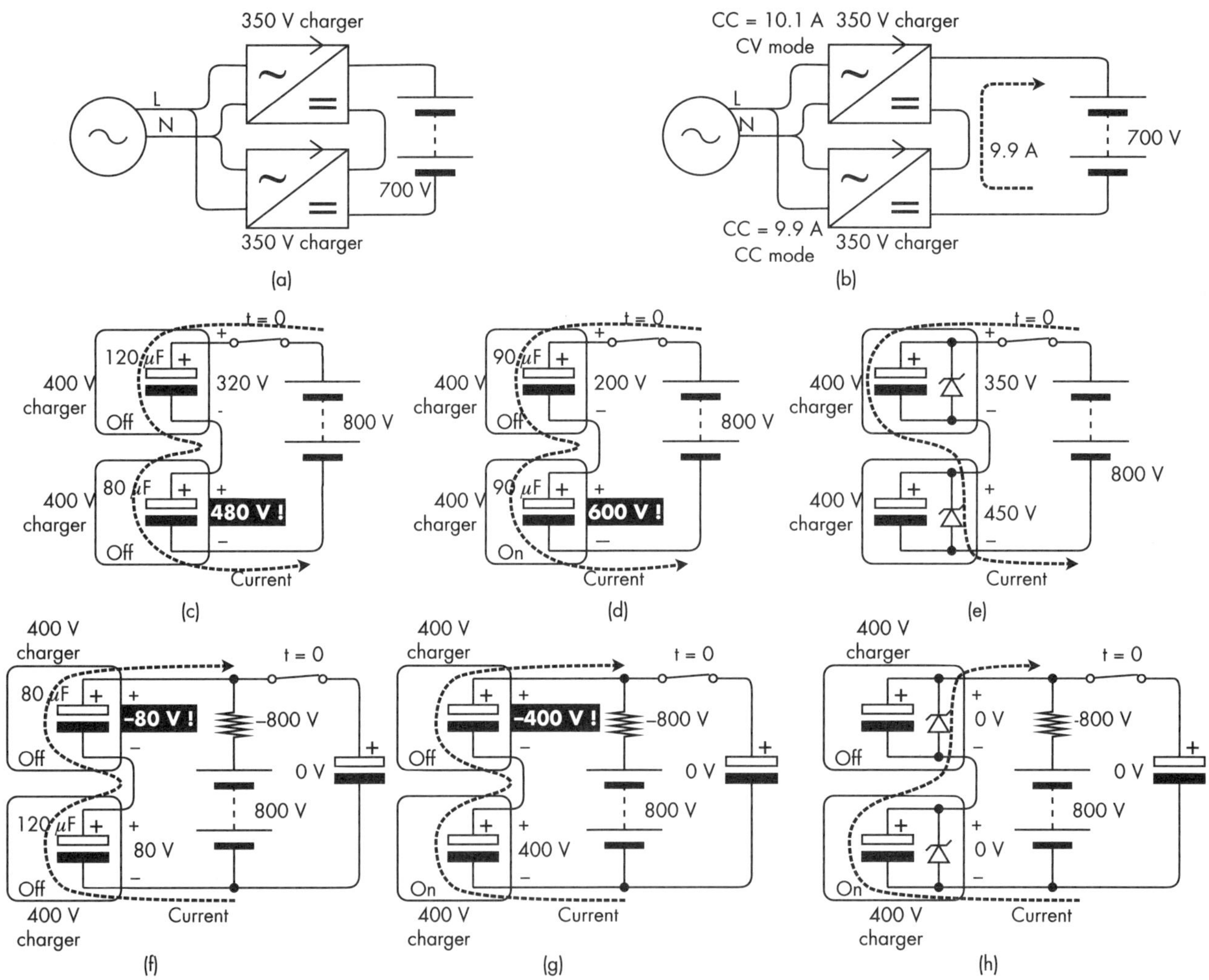

FIGURE 5.72 Chargers in series: (a) connection, (b) uneven load sharing, (c) excessive voltage upon connection to battery when chargers are off, (d) the same, when a charger is on, (e) excessive voltage protection, (f) reverse voltage upon direct connection to capacitive load when chargers are off, (g) the same, when one charger is on, and (h) reverse voltage protection.

The manufacturer may specify whether the chargers may be connected in series. In any case, do not exceed the isolation rating between the AC input and the DC output. That is, do not use twenty 100V chargers in series to get to 2,000V when the breakdown between input and output is only 1,500V.

Connecting chargers in series is problematic under various scenarios. If two chargers are slightly different, the one with the higher current goes to the maximum voltage in the CV mode (and do more than its fair share of work), while the other one remains in the CC mode (Figure 5.72(b)).

Excessive voltage may appear on the output of a charger upon connection to the battery:

- If the chargers are off, current flows from the battery to charge the chargers' output capacitors (Figure 5.72(c)). If those capacitors have different capacitances, the battery voltage is divided unequally among the chargers—the charger with the lowest capacitance sees the highest voltage.
- If one charger is turned off (top) and another charger is turned on (bottom), the current that charges the capacitor in the top charger also flows into the bottom charger (Figure 5.72(d)); that current overcharges the capacitor in the bottom charger, because it was already fully charged before the switch to the battery was closed.

In either case, the excessive voltage on the output could damage the charger. To protect the chargers in either event, there should be a high-voltage, unidirectional TVS diode across the chargers' output (Figure 5.72(e)); this diode clamps the output voltage and routes current through itself rather than through the capacitor.

A reverse voltage may appear on the output of a charger upon connection of a battery to a large capacitive load without first doing precharge. Initially, the load capacitance forces to zero the voltage across the battery and across the series chargers. If the voltage across any charger is above zero, then the voltage across another charger must be negative, to result in a total voltage of zero volt. This occurs in two scenarios:

- If the chargers are off, and if those capacitors have different capacitance, the battery voltage is divided unequally among the chargers—the charger with the lowest capacitance sees the reverse voltage (Figure 5.72(f));
- If one charger is turned off (top) and another charger is turned on (bottom), the bottom charger maintains its output at full voltage, and therefore the voltage across the top charger is reversed to make up for it (Figure 5.72(g)).

In either case, the reverse voltage on the output could damage the charger. That same diode across the chargers' output protects the chargers in either event (Figure 5.72(h)); this diode routes current through itself rather than through the capacitor.

Each charger should include a high-power, high-voltage, unidirectional TVS diode across its output to protect it when it is off:

- As a reverse diode—in case another charger in the series is turned on; used to conduct current from that charger and prevent applying a reverse voltage to this charger;
- As a voltage clamp—in case other chargers in series are also off and a battery is connected to the series of chargers; used to limit the voltage in case the voltage is not shared equally among the chargers (Figure 5.72(c)).

Ask the manufacturer if such diode is already present.

5.15.4.3 Control of Multiple Chargers

Ideally, a BMS would be able to control multiple chargers and receive and compile data from all of them (Figure 5.73(a)).

In reality, most BMSs can control only a single charger. If so:

- Either one of the chargers must act as a master and manage the other ones, combining their data and talking to the BMS as a single unit (Figure 5.73(b));
- Or a gateway is required to manage the individual chargers, compile their data, and talk to the BMS as a single unit (Figure 5.73(c)); this requires programming the gateway.

Few chargers can act as masters. The standard solution is to add a gateway to act as a master.

5.15.5 Charger Selection

Criteria to consider when selecting a charger include

- *CCCV output:* A product without a CC mode is a power supply, not a charger;
- *Profile:* Avoid chargers that use a "Li-ion" profile; instead, have the BMS configure the charger's limits;
- *Output voltage:* Must be equal to the maximum battery voltage or adjustable to it;
- *Output current:* Must be equal to the maximum voltage balanced battery or adjustable to it;
- *Power:* High enough to charge the battery in a reasonable time, but still low enough that it doesn't overload the AC supply;
- *AC input:* Must be able to handle the available AC voltage; select single-phase or three-phase; may need to include a way to reduce the input power to prevent overloading the AC power circuit;
- *Control:* One of the following:
  - Not adjustable: Must be ordered for the correct voltage for the battery ($N \times$ maximum cell voltage, where N is the maximum number of cells);
  - Adjustable with screwdriver adjustments;
  - Remotely configurable (e.g., through the CAN bus): Both the charger and the BMS must be selected to be compatible with each other; otherwise, a gateway must be designed to translate between the two.
- *Multiple parallel chargers:* Prefer chargers that implement some method of load sharing;

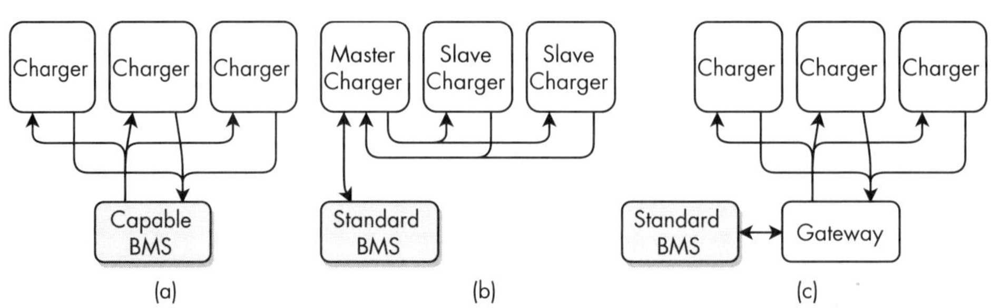

FIGURE 5.73 Multiple charger control: (a) capable BMS, (b) master/slave, and (c) gateway.

- *Multiple series chargers:* Check with the manufacturer if this is possible.

I maintain a list of chargers suitable for large batteries.[75]

5.16 RADIO NOISE, EMI

Batteries are likely to operate in electrically noisy environments and must be able to withstand this noise.

5.16.1 Noise Sources

Batteries are often connected to nearby, high-power switching electronics (e.g., chargers, motor controllers, DC-DC converters, inverters) or may be exposed to radiofrequency emissions.

5.16.1.1 High-Power Switching Devices

High-power switching electronics devices operate typically in the 20~200 kHz range (below the AM broadcast band). The harmonics of the switching waveform and the ringing at each transition are in the MHz region.

If designed to meet regulatory emission standards, these devices should include filters to reduce emissions in the radio bands; many motors drivers and many invergers do not include any filtering.

Few of these devices include sufficient filtering at the switching frequencies, relying instead on the battery itself as the filter. As a result, the current in the battery, far from being pure DC, jumps from 0 to a high level over the span of 10 μs or so. Consequently, the cell voltages jump at the switching frequency (Figure 5.74(a)).

If you're curious why this is, consider, for example, a DC motor controller (Figure 5.74(b)). This controller is simply a step-down DC-DC converter. Internally, it is in effect a single-pole, dual-throw switch, quickly selecting either B+ or B- and sending it to the inductive load of the motor.

The voltage across the inductor alternates between the battery voltage for 25 μs and 0V for 25 μs. The current through the inductance of the motor is steady. Conversely, the voltage across the battery is relatively constant as its current alternates between a high value when the switch selects B+ and 0A when the switch selects B-.

Note how the average output voltage is less than the input voltage, while the output current exceeds the average input current. That is because of conservation of power: the output power is nearly the same as the input power.

A large capacitor across the input of a high-quality DC-DC converter supplies those pulses of current so that the battery only has to supply the average current (Figure 5.74(c)). But large capacitors are big and expensive. Also, a large capacitor on the battery input requires precharge (see Section 5.13). Therefore, manufacturers of power electronics may use small capacitors instead, relying on the battery to act as the capacitor. Consequently, the cell voltage bounces slightly up and down in time with the switching frequency.

A power switching device applies a noisy switched voltage to the output (away from the battery), which emits radio noise (EMI) (Figure 5.75(a)). To reduce radio emission, the product designers may add a filtering capacitor on the output to earth ground. Doing so quiets down this output, yes, but it also transfers the problem to

75. http://liionbms.com/php/charger_options.php.

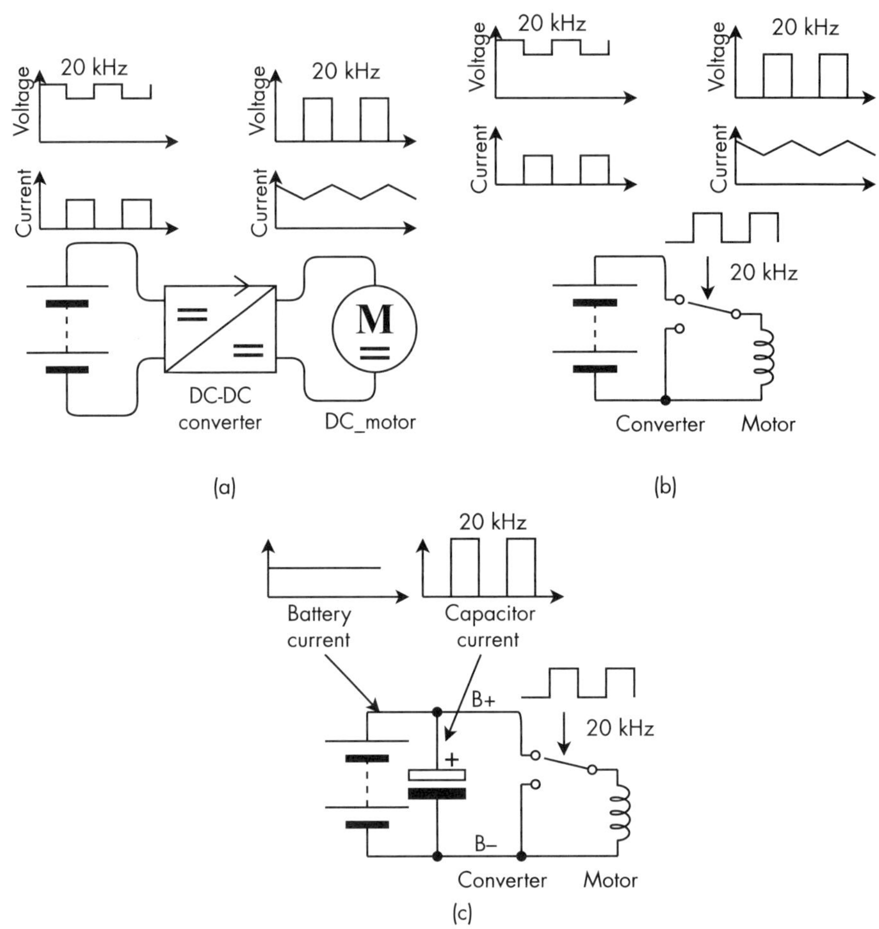

FIGURE 5.74
Switching noise:
(a) effect on cell voltage,
(b) equivalent circuit of a DC-DC converter,
and (c) capacitor across input of converter.

the battery—instead of the output bouncing up and down, the input does (Figure 5.75(b)). Now the battery is a source of EMI.

The simultaneous operation of multiple switching power electronic products results in various beat frequencies. The battery voltage relative to earth ground can be quite ugly (Figure 5.75(c)).

5.16.1.2 Transmitters

A telecom site may share the tower with broadcast radio or TV stations. The telecom equipment (including the battery and its BMS) is exposed to powerful radio frequencies. Even though the broadcast antenna is directional, RF power on the order of 1 kW may still reach the battery.

5.16.2 Noise Immunity

The BMS must be able to operate reliably within such noisy environments. Part of the solution is to choose a BMS with high noise immunity (see Section 4.12.3); part of the solution is a proper installation of the BMS (according to the instructions) and proper shielding in the battery.

5.16.2.1 Switching Converters

Various measures are partially effective when dealing with interference from switching converters:

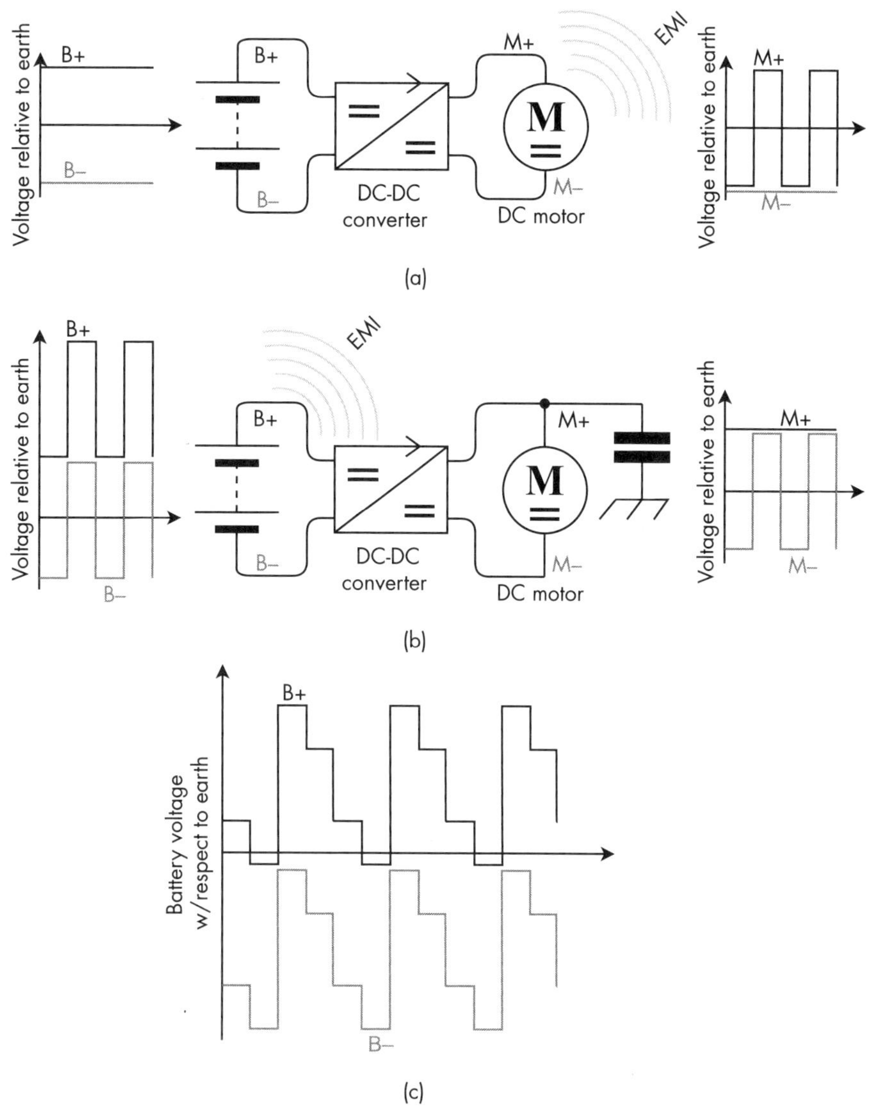

FIGURE 5.75
Bouncing battery voltage: (a) no output capacitor, (b) with output capacitor to earth, and (c) typical battery voltage relative to earth.

- Route CAN bus and other communication wires away from high power conductors (see Section 5.10);
- Route the cell board communications wires next to the corresponding bus bar (see Section 7.5.2.8);
- Add isolators on CAN buses, USB cables, RS-485, or RS-232 links;
- Ground the metal case of a BMS to the chassis;
- Float the power supply voltage to the BMS through an isolated DC-DC converter, or, on the contrary, make sure it's grounded to the chassis (strangely, sometimes one is better than the other).

5.16.2.2 *Radio Transmitters*

In applications near high-power transmitters, ferrite clamps are useful. Clamped on cables, they reduce their effectiveness as receiving antennas, therefore reducing the effect on the BMS electronics.

5.16.2.3 EMI Immunity Testing

An EMC[76] testing lab has the equipment to test a battery's susceptibility to high-level electrical pulses and radio fields.[77] Be aware that an EMC lab may pass a battery that doesn't display gross misbehavior when bombarded by RF. However, the lab is not looking for small errors in measurements that are large enough to be problematic in the application. Check the readings during a test to spot any anomalies.

A quick and easy way to test a small battery, before you go through the expense of an official test, is to use a power supply for a laptop back-light (fluorescent, not LED). This produces a current-limited, high-voltage square wave at a frequency that is typical of switching converters. Connect the hot output of this power supply to a battery terminal to shake it at the switching frequency, and see how it affects the BMS. This method doesn't work as well on a large battery due to its large capacitance.

5.16.3 Emission Reduction

Even if batteries themselves do not include any noisy power electronics, they may act as antennas emitting the noise generated by connected devices. Measures may be taken to reduce these emissions from the battery. In particular, emission reduction measures are taken in traction batteries for passenger vehicles to avoid interfering with the car's radio (see Volume 2, Section 3.15.2).

A large battery may be installed in a *Faraday cage* to contain emissions:

- Enclose the battery in a metal case, paying particular attention to the seams, to make sure that metal-to-metal contact is direct and free of corrosion;
- Use RFI[78] shielded connectors for the external communication cables;
- Use coaxial shielded cables and a shielded power connector for the power routes from the battery to the load and charger,.

When two wires carry a switched current (e.g., from a charger), route them together, close to each other. The wires form a loop antenna that transmits the magnetic component of this EMI. The efficiency of this loop antenna is proportional to the area enclosed by the two wires. If the wires touch each other, the loop area is zero, and therefore the magnetic emissions are minimized. Ferrite clamps are only useful around 100 MHz.

High-power switching converters (and a battery connected to them) emit at the switching frequency (20 to 300 kHz) well below the range where ferrite clamps make a difference. Yes, the harmonics and ringing are at a higher frequency, but they are not as strong. Therefore, ferrite clamps are not that useful to reduce emissions from switching converters.

5.17 THERMAL MANAGEMENT

Thermal management keeps the cell temperature within the desired range by

- Shielding the battery from the ambient temperature;
- Transferring heat between the battery and the ambient, especially heat generated within the battery.

76. Electromagnetic compliance.
77. Radiated Immunity (IEC/EN61000-4-3), Electrical Fast Transient (EFT) (IEC/EN61000-4-4).
78. Radio frequency interference.

5.17.1 Introduction

This section offers just an introduction to thermal management. For an effective design of thermal management for a battery, refer to a thermodynamics textbook, use a thermal analysis computer application, or hire a thermal engineer.

Ideally, a battery is designed so that no thermal management is required:

- It is operated in an environment free of temperature extremes;
- It is efficient, and therefore it generates little heat during operation.

Most batteries do not include any explicit thermal management. Implicitly, such batteries use passive heat transfer.

5.17.2 Internal Heat Generation

As it operates, a battery generates at least some heat.

On the one hand, an energy battery should be designed to be efficient. If operated at room temperature, it should require no explicit thermal management. On the other hand, a power battery may not be designed to be efficient. It may be cheaper to remove the generated heat than to make it more efficient. The most obvious example of a power battery that requires removing generated heat is a traction battery for a hybrid vehicle.

If any other application requires thermal management to remove generated heat, the first question that should be asked is whether the effort wouldn't be better spent on making the battery more efficient.

At design time, the amount of heat generated inside a battery is estimated through thermal analysis. After a battery is built, it may be measured [7].

5.17.2.1 Estimation

The temperature of a cell at a given moment is affected by

- Its previous temperature;
- The rate of heat generated by the cell (affected by current and internal resistance);
- The rate at which heat is extracted from the cell (or externally added to it), which in turn is affected by its temperature.

We consider the heat generated by the cells from both a power and an energy standpoint:

- For continuous operation, we consider the power that the cells release as a rate of heat flow at any given moment;
- For cyclical operation, we consider the energy that the cells release as heat during a cycle.

In the simplest case, there is no airflow, and the following are known: the constant power wasted as heat, the thermal time constant of the system, the thermal resistance between a single cell and the ambient, and the ambient's constant temperature. In this case (Figure 5.76):

- While discharging, the temperature of the cell increases asymptotically toward an equilibrium temperature;
- After the discharge ends, the temperature of the cell decreases asymptotically toward the ambient temperature.

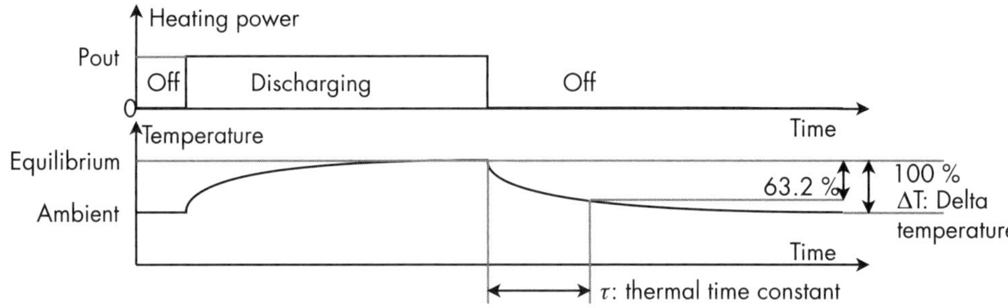

FIGURE 5.76
Plot of cell temperature in the simplest case.

The final, equilibrium temperature is

$$\text{final\_temperature}[°C] = \text{ambient\_temperature}[°C] + \text{delta\_temperature}[°C] \quad (5.8)$$

and

$$\text{delta\_temperature}[°C] = \text{heating\_power}[W] \times \text{thermal\_resistance}[K/W]$$

While operating, the temperature increases asymptotically with the given thermal time constant. Afterward, the temperature decreases asymptotically with the same thermal time constant. This time constant is such that, during this time, the temperature has decreased by 63.2%[79] of the total change in temperature.

In reality, the change in temperature is not as well defined:

- The cell temperature continues to increase after power is removed, as heat from inside the cell reaches its enclosure;
- The heat power is not constant (due to changes in internal resistance) even if the load power is constant (which it rarely is);
- The thermal resistance is not quite constant.

This is an oversimplification of a complex subject for which, again, I would like to direct you to a thermodynamic book.

5.17.2.2 *Measurement*

Once the battery is built, thermocouples may be placed throughout the battery and connected to a data acquisition system to log the temperatures at various locations and under various conditions. A thermal camera is used to look for hot spots anywhere in the battery, not just on cells.

Calorimetry may be used to evaluate the power and energy in the heat generated by the cells so that it may be compared to the estimated level. Significant variations from the predicted levels trigger further study.

79. 63.2% is equal to 1-1/e, where e is Euler's number, 2.71828...

5.17.3 Thermal Management Mechanisms and Techniques

Thermal management uses these mechanisms and techniques:

- Passive heat transfer (heat is naturally transferred from a hotter body to a colder one, toward equalizing their temperatures):[80]
  - Conduction: Direct heat transfer between solids (e.g., to a heat sink) (Figure 5.77(a));
  - Radiation: Heat transfer through electromagnetic emissions (e.g., infrared) (Figure 5.77(b));
  - Convection: Heat transfer through the movement (usually vertical) of a fluid (e.g., air) due to density gradients created by thermal gradients (Figure 5.77(c)).
- Active heat transfer:

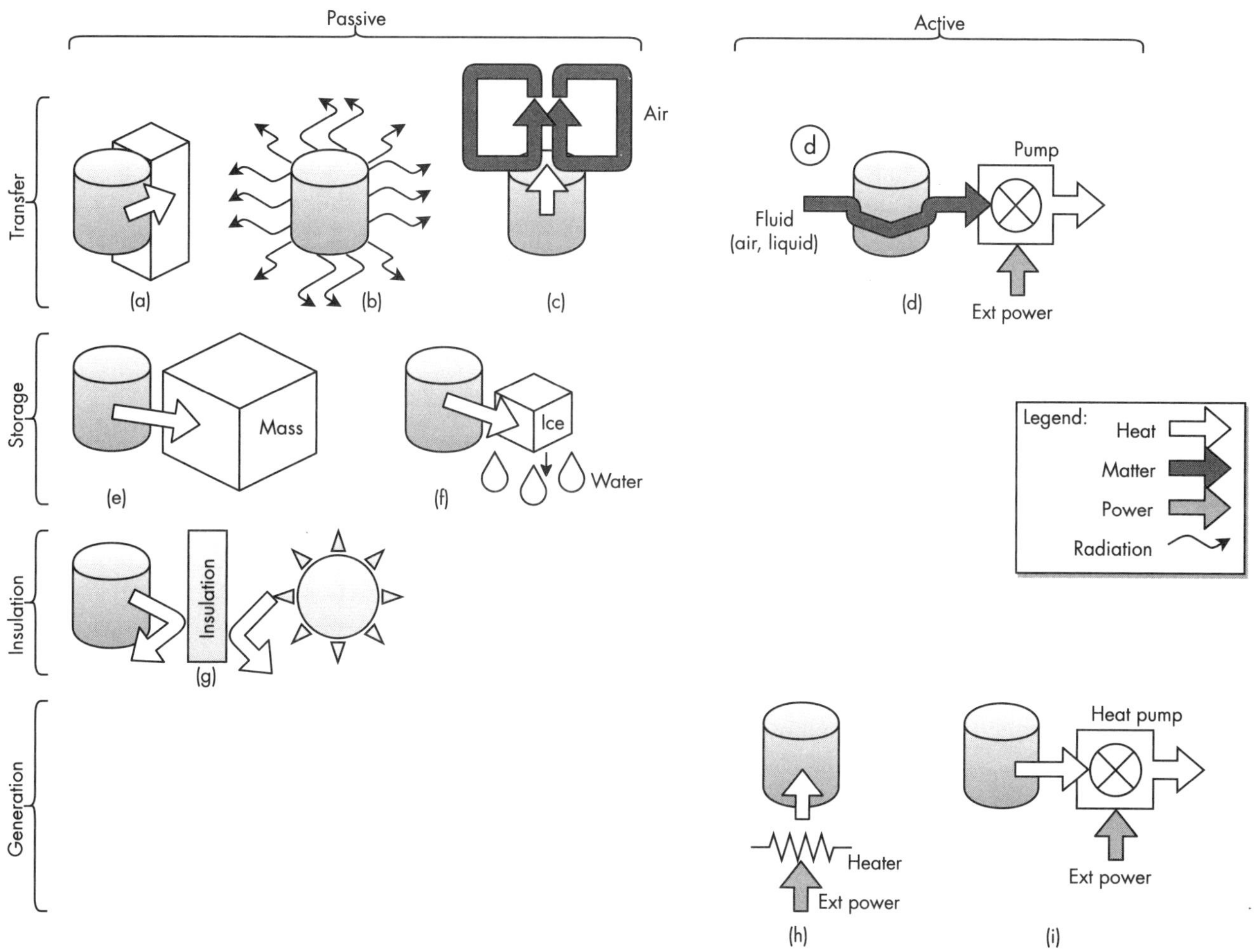

FIGURE 5.77 Heat mechanisms and management techniques; except for heating (h), the cylinder is assumed to be hotter: (a) conduction, (b) radiation, (c) convection, (d) advection, (e) thermal capacity storage, (f) phase change storage, (g) insulation, (h) heating, an (i) heat pumping (including cooling).

80. Some sources also list evaporation, but evaporation itself is a form of phase change, which is a heat storage mechanism, not a heat transfer mechanism. Sure, later, vapor takes heat away, but that's convection, which is already listed as a heat transfer mechanism.

- Advection: Similar to convection, but the movement (in any direction) is due to something other than gradients (e.g., pumped liquid coolant) (Figure 5.77(d)).[81]
- Heat storage:
  - Thermal capacity: For a given heat flow, the higher the thermal mass, the slower its temperature changes (Figure 5.77(e));
  - Phase change (including evaporation): The temperature of the material remains nearly constant during a phase change (e.g., melting ice stays at 0°C) (Figure 5.77(f)).
- Insulation (Figure 5.77(g)):
- Heat generation and pumping:
  - Heating (Figure 5.77(h));
  - Heat pumping (including cooling) (Figure 5.77(i)).

Advection and generation are active (they take external power to operate), while all others are passive (heat is naturally transferred from a hotter body to a colder one).

Typically, multiple mechanisms operate simultaneously, to different degrees.

Note that all of these techniques are also used to control the temperature in your home—a refrigerator is a heat pump, there's insulation in the walls, the walls radiate heat at night.

5.17.4 Thermal Insulation

An uninsulated battery is at the mercy of the ambient temperature. The point of battery insulation is to minimize the heat exchange between the battery and the environment (Figure 5.78). If the battery is at the desired temperature and the environment is not, then insulation can be used to try to maintain the battery temperature. Insulation is passive. Therefore, it is effective even when the battery is off.

There are two mechanisms of thermal insulation:

- Reduction of thermal conductivity (i.e., by surrounding the battery with insulating materials);
- Radiation reflection (e.g., by covering the battery with materials that reflect infrared).

Battery insulation helps minimize temperature gradients throughout a battery that is exposed unevenly to the ambient temperature. For example, a source of heat such as an engine or a charger tends to heat only one end of a battery. Insulation reduces the heat transferred into the battery, while at the same time giving the heat inside the battery a chance to diffuse throughout the battery and equalize the temperature.

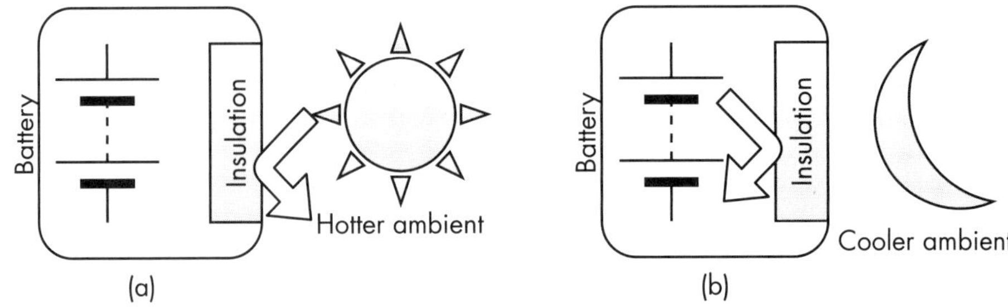

FIGURE 5.78 Insulation: (a) from heat, and (b) from cool.

81. Do not confuse advection with forced convection: the latter includes advection and diffusion due to convection.

Insulation increases the effectiveness of some of the techniques discussed in the following sections: heat storage, heat pumping, and heat generation.

5.17.5 Passive Heat Transfer

A hot battery may be cooled by passively transferring heat to a colder ambient (Figure 5.79(a)); a cold battery may be heated by passively transferring heat from a hotter ambient (Figure 5.79(b)).

Passive heat transfer is helpful if the battery would be in a better state if it were brought closer to the ambient temperature. Specifically, a battery that generates heat benefits from the exchange of heat between it and a cooler ambient.

Passive heat transfer is counterproductive if the ambient temperature is less desirable than the battery temperature (e.g., during a hot summer afternoon in the desert, or a cold night in the Arctic). In such cases, we wish to delay the passive heat transfer between the battery and the ambient, attempting to maintain the previous battery temperature, when the ambient temperature was not so extreme. Insulation and thermal storage (see Section 5.17.7.1) can be used to achieve this.

Passive heat transfer relies on conduction, convection, and radiation, and may be aided by heat sinks, heat pipes, and heat exchangers. As it requires no power source to operate, it is effective even when the battery is not in use. Passive heat transfer uses no moving parts other than fluid motion: natural airflow and the liquid that circulates in a passive heat pipe.

5.17.6 Active Heat Transfer, Advection

Advection is a form of active heat transfer. It is similar to convection, except that an external power source powers it.

A pump (an impeller, fan, or blower) circulates a fluid, such as a gas (e.g., air) or a liquid (e.g., glycol water). For cooling, heat is transferred from the battery to the fluid through advection. The pump brings in cold fluid from outside and carries the hot fluid away from the battery.

5.17.6.1 *Forced Air Ventilation*

Forced air ventilation is an excellent way to help manage the temperature of a battery (Figure 5.79(c)). A fan or blower generates airflow. Air at ambient temperature enters the battery through an air intake. As it flows against the battery, the air exchanges heat with it. Finally, air exits the battery through an air exhaust, taking heat with it. Note that the battery needs both an intake and an exhaust. Without them, the fan generates pressure instead of airflow, and therefore no heat is transferred.

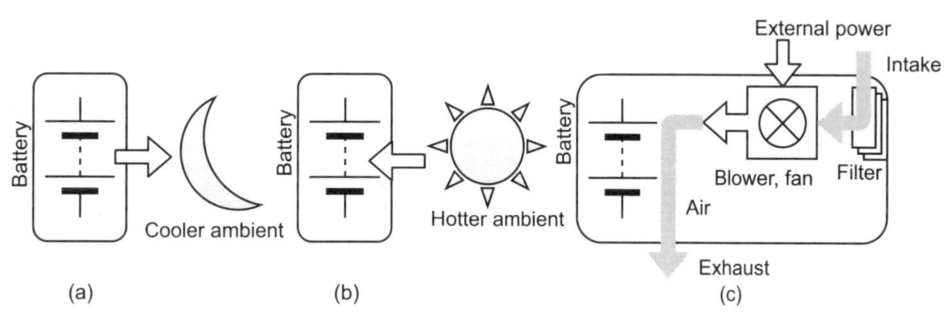

FIGURE 5.79
Heat transfer: (a) passive heat transfer too ambient, (b) from ambient, and (c) forced air ventilation.

5.17.6.2 *External Air Path*

An effective forced air system requires a good design for the external air path.

Maximize the distance between the air inlet and outlet so that you do not get the exhaust air right back into the inlet (Figure 5.80(a)). For the same reason, maximize the volume of air into which the inlet and the outlet are opened. Ideally, at least one of them vents outdoors.

For a rack-mounted battery, draw cold air from the front of the rack, and exhaust hot air to the back of the rack (Figure 5.80(b)). Specifically, server rooms have alternating hot aisles from which the room chiller draws air and cold aisles to which the chiller sends cold air. The racks face the cold aisle, with their backs to the hot aisles. The equipment in the rack draws cold air from the cold aisle and exhausts hot air to the back, into the hot aisle.

A passenger car cools its traction battery by drawing cool, dry, and relatively dust-free air from the passenger compartment. The exhaust is released outside the car (Figure 5.80(c)).

Take advantage of any pressure differential in the external air by placing the inlet in the high-pressure side and the outlet in the low-pressure side. This increases the airflow, relying less on the fan. For example, in a small airplane, take the air from the cowling (the front tip of the plane) and exhaust it by the tail (Figure 5.80(d)).

5.17.6.3 *Airflow Direction*

Suck or blow? One may think that it makes little difference whether the fan is placed on the air inlet or exhaust because, in either case, air flows through the battery. There are some slight advantages to each:

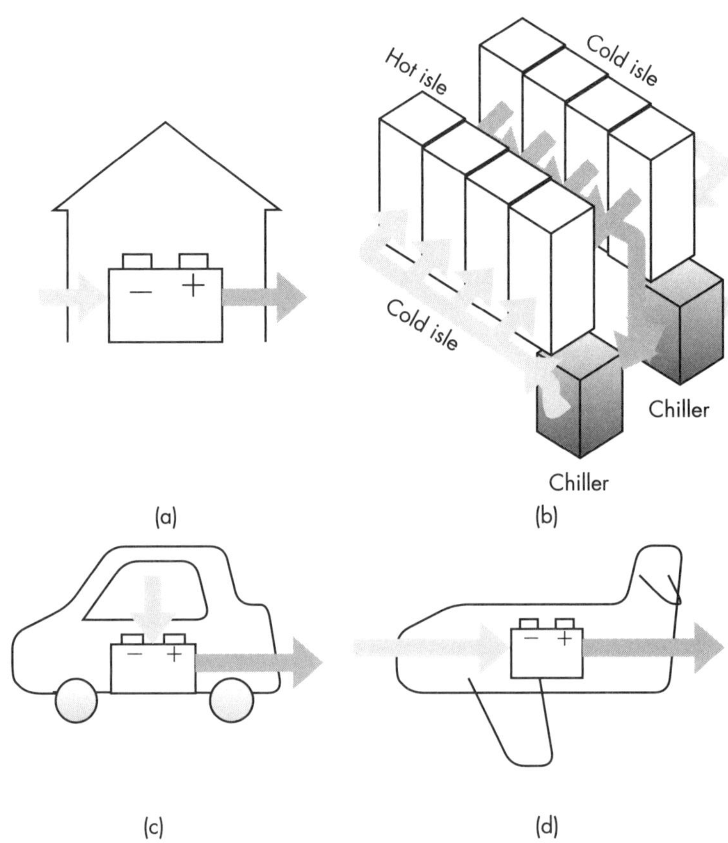

FIGURE 5.80
External air path:
(a) house, (b) server racks,
(c) car, and (d) airplane.

- *On the inlet:* Air at higher pressure has more air molecules so it transfers slightly more heat; positive pressure inside the battery tends to blow dust and moisture out of leaky seams;
- *On the outlet:* Airflow near the fan is more even and therefore so is cooling.

Between the two, eveness of cooling is more important, so place the fan on the outlet.

Try to draw air from a dry and dust-free area to protect the battery from damage and dust. Place an air filter on the inlet so that it may collect dust before it enters the battery.

If possible, draw air from a volume with a relatively constant temperature. The temperature of where you let out the exhaust is not as important and can vary.

If the airflow is vertical, it is best if it flows upward, so that it works in conjunction with convection, not against it.

5.17.6.4 Airflow Speed

An optimal airflow performs sufficient cooling with a relatively low cooling power.

The airflow through the battery depends on the pressure differential generated by the fan and the air path's resistance to airflow. It is not a linear relationship. As a rule of thumb, at sea level, air cooling is nearly maximum at airspeeds of 5 m/s.[82] It does improve somewhat at faster speeds, but the improvement is not as impressive.

A proper thermal analysis estimates how much heat should be removed and selects the proper fan and airspeed to achieve this amount of cooling. Or you can determine this empirically, by working on an actual battery.

At slow airflow speeds, heat transfer is limited by the heat capacity of air—each air molecule can carry off only so much heat.

At mid-airflow speeds, the convective heat transfer coefficient[83] between the cell and the air becomes the limiting factor.

At high airflow speeds, increasing the airflow becomes counterproductive because the power required to generate that flow is higher than the heat is removed.[84]

Overlay the fan's pressure versus flow rate curve on the a parabolic flow resistance curve to determine the airflow (Figure 5.81). Be aware that, depending on the airflow resistance in the battery, a low-speed/high-pressure fan may produce a higher airspeed than a high-speed/low-pressure fan.

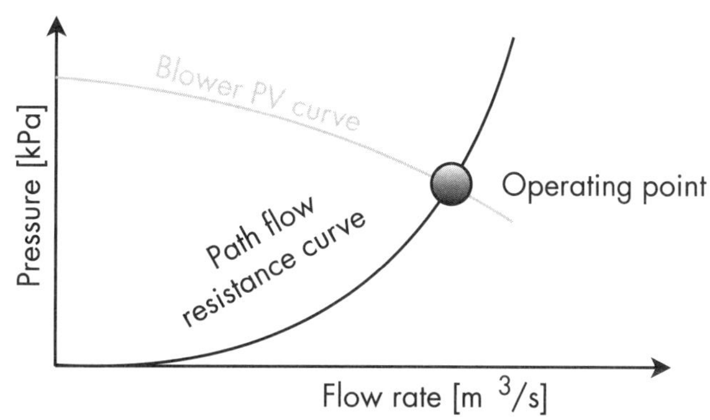

FIGURE 5.81 Overlay of fan's PV curve and flow resistance curve to determine operating point.

82. About 1,200 ft/min.
83. The convective heat transfer coefficient is the rate of heat transfer between a solid and the air per surface area and temperature difference. This measure is akin to the thermal resistance between two solids.
84. Even at high speeds, the heat generated by the friction from the airflow itself is minor.

5.17.6.5 *Temperature Gradients*

A temperature gradient is formed between the air inlet and the exhaust, which can result in uneven heating or cooling that can be harmful to the hottest or coldest cells. The classic solution to this problem is to run the air through the cells in series (one after the other) at high speed to minimize this gradient (Figure 5.82(a)).

The smart approach is to use symmetry to distribute cooling equally to all the cells by running the air in parallel at low speed.

The cooling is the same in both cases, but it is evenly distributed in the second case. The parallel approach uses lower stack pressure but higher total airspeed.

Achieve symmetrical cooling by doing all of the following:

- Use multiple airflow paths in parallel.
- Make all the paths between cells identical. Note that there are no openings between the paths so air cannot jump from the path between two cells to an adjacent path; this prevents any air from following a straight path directly from the intake to the exhaust.

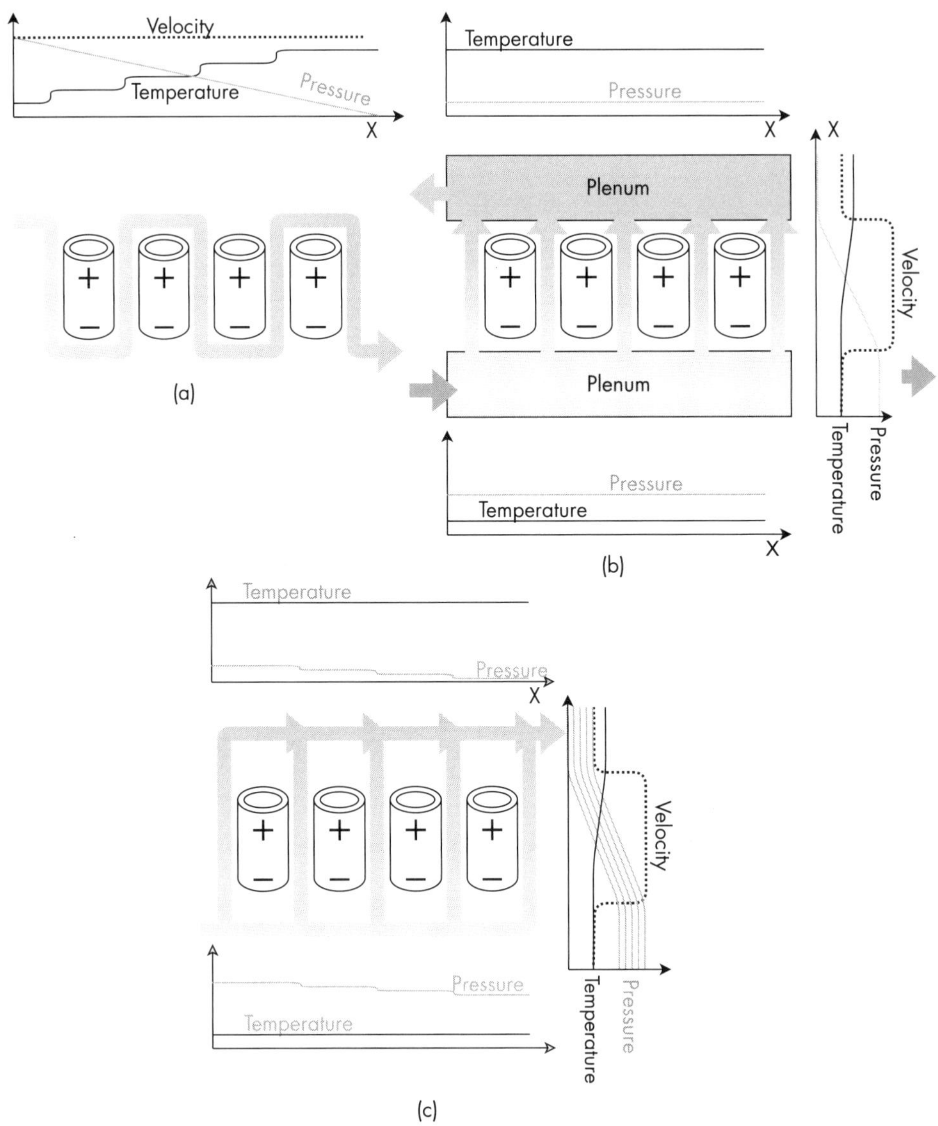

FIGURE 5.82
Temperature gradients:
(a) classic approach,
(b) symmetrical approach with identical paths, and
(c) symmetrical approach with equivalent paths.

- Equalize airflow and temperature through each path through the cells:
  - Use a large air plenum at the intake end to equalize the pressure at each entry point, and another large plenum at the exhaust end to equalize the pressure at each exit point (Figure 5.82(b)).
  - Or use identical total path lengths (some with extra length at the intake end, some with extra length at the exhaust end) by having the intake at the bottom left corner and the exhaust diagonally opposite, at the top right corner (Figure 5.82(c)). Each air molecule travels the same distance, through the same total resistance, regardless of which path it takes.

Yes, there is still a temperature gradient, but it is across each single cell, not among different cells.

In Figure 5.82(c), the pressure is not equal in all the paths: the pressure in the leftmost path is higher than the pressure in the rightmost path. However, the pressure drop across each path is identical, and given that all the air path resistances are the same, the airflow is identical in each path.

Three secondary effects cause a slightly uneven cooling:

- The left-most and right-most paths only cool one cell; therefore, those two cells are cooled more than others; this can be corrected by halving the width of the two end paths.
- High-pressure air (with more air molecules and therefore better heat transfer from the cell) cools better than low-pressure air; therefore, in Figure 5.82(c), the leftmost cells are cooled more than the other ones. This can be corrected by placing more air restriction on the paths on the left.
- If the airflow is horizontal, convection bends the flow vertically, resulting in more airflow at the top of the cell, toward the exhaust, and therefore insufficient cooling in a lower corner of the cell (Figure 5.83(a)). Adding horizontal channels keeps the flow even across the surface of the cell (Figure 5.83(b)).

These secondary effects may be dwarfed by the temperature variations within each cell as it generates heat unevenly..

5.17.6.6 *Liquid Cooling*

Heat may be transferred with a liquid rather than air. Liquid cooling is far more effective in removing heat from a small volume than forced air is, but it is also far more complex and expensive and may be less reliable.

Liquid cooling uses glycol water in a closed loop (Figure 5.84(a)):

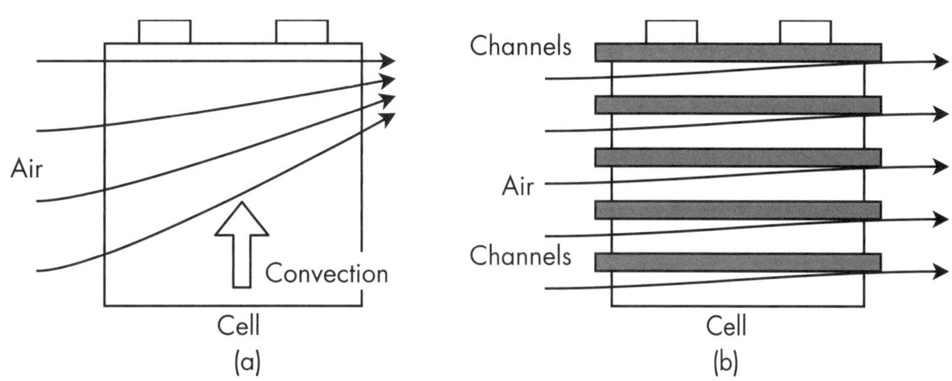

FIGURE 5.83 Horizontal airflow: (a) rising flow due to convection, and (b) solution using horizontal channels.

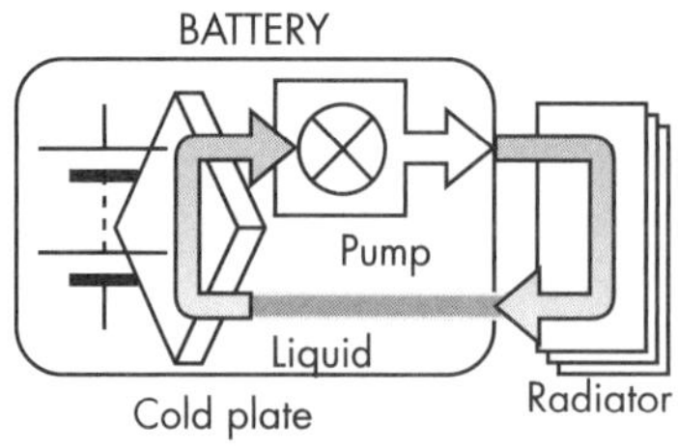

(a)

(b)

FIGURE 5.84
Active heat transfer: (a) liquid cooling, and (b) cold plates on Li-ion battery modules.

- The liquid flows through a *cold plate*, a heat exchanger that transfers heat from the cells to the liquid, mounted against the cells (Figure 5.84(b));
- It also flows through a radiator, a heat exchanger that transfers heat from the liquid to the outside air, mounted externally;
- A pump circulates the liquid;
- A fan may force air through the radiator.

Effectively, liquid cooling moves the problem somewhere else: instead of using air to cool the cells inside the battery, we use air to cool a radiator outside the battery.

5.17.7 Temporary Heat Storage

Instead of transferring heat to and from the battery, it may be possible to temporarily store some heat locally. This is only a short-term solution, which may be all that is needed during times of peak heat generation.

Heat storage may be considered only in cyclic applications, in which heat generation is of short duration. Afterward, there must be sufficient time for the stored heat to be released through cooling so that the battery may be ready for the next cycle. Heat storage is not advantageous in continuous duty applications (e.g., a hybrid EV). In such applications, heat must be removed at the same time as it is generated.

Thermal storage is the thermal equivalent to procrastinating: it doesn't prevent the unavoidable work that must be done eventually. Thermal storage doesn't prevent the need to remove heat, just like procrastinating doesn't avoid an upcoming exam.

There are two techniques for temporary heat storage:

- *Thermal capacity:* For a given heat flow, the more mass a battery has, the slower its temperature changes;
- *Phase change:* The temperature of the material remains nearly constant during a phase change (e.g., melting ice).

5.17.7.1 Thermal Capacity Storage

A cell stores some of the heat it generates in its thermal capacity (Figure 5.85(a)).

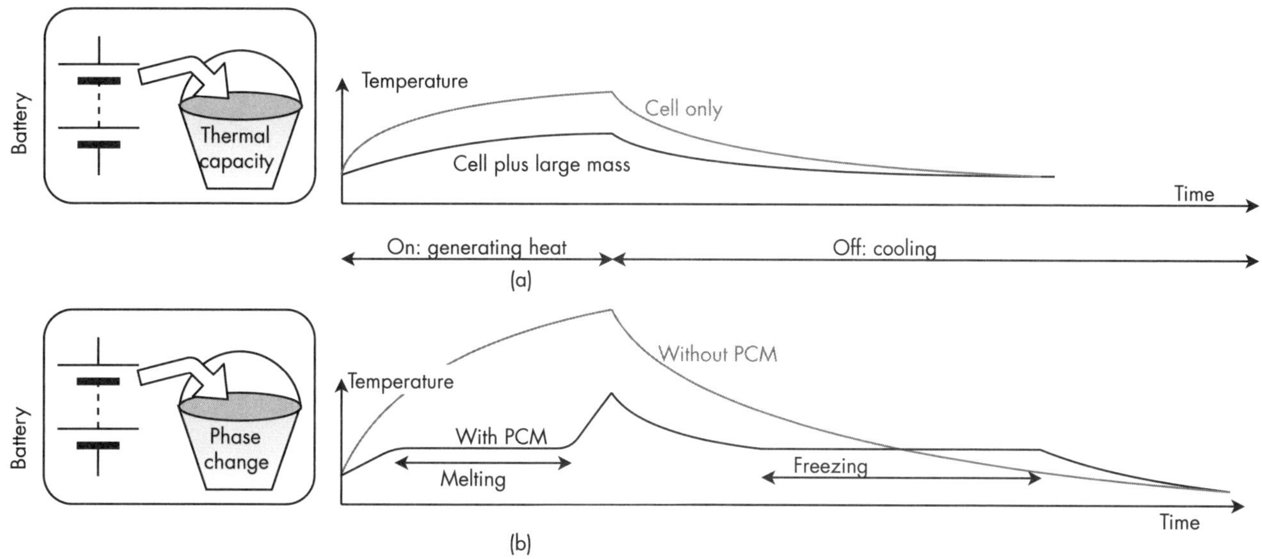

FIGURE 5.85 Temporary heat storage: (a) thermal capacity, and (b) phase change.

Some additional mass may be added against the cells to store heat in its thermal capacity as well. While this slows down the temperature rise, it does decrease the energy density, countering one of the main advantages of Li-ion batteries.

As heat is added, the temperature increases, though more slowly if the mass has a large heat capacity. At higher temperatures, heat is lost through passive thermal transfer, so the temperature doesn't rise quite as fast and may plateau.

If the ambient temperature changes drastically, a combination of battery insulation and thermal capacity storage may be sufficient to maintain the battery's temperature within the desired range:

- During the heat of the day, insulation reduces the heat that reaches the cells, and the little heat that does reach the cells doesn't raise their temperature much because the cells have a large thermal mass;
- By the time the cell temperature starts reaching the maximum, it's a cool evening, and the little heat that makes it through the insulation starts flowing out of the cells;
- During the cold of the night, insulation reduces the heat that escapes the cells, and the little heat that does escape doesn't lower their temperature much because the cells have a large thermal mass;
- By the time the cell temperature starts reaching the minimum, it's a cool morning, and the little heat that makes it through the insulation starts flowing back into the cells.

5.17.7.2 Phase Change Material Storage

Within limits, cells may be placed within phase change material (PCM[85]) to store their heat temporarily (Figure 5.85(b)):

- Initially, the temperature rises slowly due to the mass of the PCM.

85. Not to be confused with PCM for protector circuit module.

- Once the temperature reaches the PCM's phase change temperature (e.g., 60°C), heat goes into melting it rather than raising its temperature, so the temperature remains stable. Of course, this temperature is rather high and will degrade the cells in the long run.
- If the cells continue to operate after all the PCM is fully melted, then the temperature starts rising again. In contrast, if the cell stops operating before this point, the cell temperature never exceeds the PCM's melting temperature.
- When the cells are no longer generating heat, the PCM loses heat through passive heat transfer. As it does, it slowly freezes again, until it's ready for a new cycle.

Dihydrogen monoxide[86] is a familiar PCM. During the ice-to-water transition, and again during the water-to-steam transition, dihydrogen monoxide absorbs heat and converts it to phase change instead of a temperature increase. This is why a glass of ice water remains at 0°C until all the ice is melted. Conversely, during the steam-to-water transition, and again during the water-to-ice transition, dihydrogen monoxide releases heat and converts it to phase change instead of a temperature decrease.

Both transition temperatures for dihydrogen monoxide (0°C and 100°C) are beyond a Li-ion battery's ideal temperature range. Therefore, other materials (waxes) have been developed whose melting temperatures are closer to a Li-ion cell maximum temperature.[87] Blocks of these materials are machined to accommodate small cylindrical cells (Figure 5.86). The material attempts to clamp the cell temperatures to ~60°C.

As a side benefit, in case of thermal runaway, the material reduces the chance of propagation from a cell onto nearby cells. The PCM includes graphite to improve thermal transfer, which, unfortunately, is also a decent electrical conductor. Isolation must be added to keep the PCM from contacting the cell terminals and discharging the cells.

5.17.8 Heating

In applications that are consistently colder than the battery's desired temperature range, the battery should be heated. In particular, since Li-ion cells should not be charged when below 0°C, heating may be used to bring cells above freezing before charging is started (Figure 5.87(a)).

FIGURE 5.86
Phase change material, bored for 18650 cells. (Courtesy AllCell Technologies.)

86. Chemical notation: H_2O.
87. AllCell PCC composite.

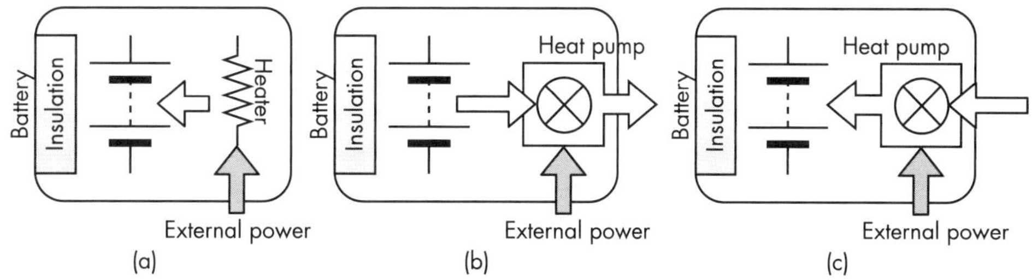

FIGURE 5.87
Generation: (a) heating, (b) heat pump cooling, and (c) heat pump heating.

Typically, a battery is heated only when an external power source is available because using the battery itself as a power source for the heater may not be ideal; at minimum, heating is disabled when the battery is not in use. Insulating the battery makes heating more effective by reducing heat loss to the environment. A heating pad should be placed so that it heats all cells evenly.

Extra precautions must be included to avoid uncontrolled heating, which would damage the cells or even lead to thermal runaway. At minimum, there should be two ways of shutting off heating, one of which could be a thermal fuse or thermal cutoff.

5.17.9 Heat Pumping, Cooling

Again, heat transfer is used when the ambient temperature is preferable to the present battery temperature. It brings the battery temperature closer to the ambient temperature, always transferring heat from the hotter body to the colder one.

Unlike heat transfer, heat pumping transfers heat from a colder body to a hotter one. Specifically, it may be used to cool the battery even colder than the ambient temperature (Figure 5.87(b)) or to heat the battery even hotter than the ambient temperature (Figure 5.87(c).

Heat pumping requires energy from an external source to operate. You are probably familiar with these heat pumps

- Refrigerators and wine chillers;
- Chillers for computer server rooms;
- Peltier coolers;
- Air conditioners;
- Residential heat pumps for both heating and cooling.

While cooling a battery, the side of a heat pump that is against the cells is somewhat colder than the ambient. The other side is much hotter because it is heated by both the heat extracted from the battery and by the energy used to operate the heat pump.

Peltier coolers are solid-state heat pumps. When a current flows through them, one side becomes colder while the other side becomes hotter. While Peltier coolers have significant uses in the industry and even in consumer products (e.g., picnic coolers, computers), they are not practical for battery cooling as they are too expensive and inefficient.

Just as in heating, heat pumping benefits greatly from insulation by keeping the pumped heat from escaping back.

Letting the battery power a heat pump may be undesirable. Therefore, typically, heat pumping is available only when the battery is in use or when an external power source is available.

5.17.10 Internal Equalization

Temperature variations among cells affect specific parameters more than their absolute temperature (see Section 3.2.8).

Internal equalization attempts to equalize the cell temperatures by distributing the heat throughout the battery. Equalization is particularly important if the battery is divided into multiple compartments, especially if they are widely separated. For example, an EV could have part of the traction battery inside the passenger compartment (whose temperature is controlled for the sake of the passengers), part in the trunk (whose temperature varies with the ambient), and part under the hood (where power electronics generate heat). Such a battery would benefit from internal equalization to keep the cells at temperatures close to each other:

- *Parked:* Passive internal equalization to minimize the delta in their self-discharge currents and therefore minimize the unbalancing process;
- *Moving:* Active internal equalization to minimize the delta in their internal resistances so that all cells experience similar voltage sag under load.

5.17.11 Noise Reduction

The noise from a forced-air system may be objectionable in many applications. Indeed, noise levels above 40 dB are objectionable to people in the same location (e.g., in a vehicle's passenger compartment or a server room), requiring some form of noise abatement. Noise may be acceptable in an industrial environment, outdoors, or in a vehicle's engine compartment.

Forced-air systems produce noise through two mechanisms: airflow turbulence and mechanical vibration from the fan's motor and rattling components. Noise abatement is done by reducing the noise at the source and by attenuating the remaining noise.

5.17.11.1 Noise Reduction

Airflow turbulence occurs at the air intake, through finger-guards and filters, along the blades of a fan or blower, at bends in vents, through narrow passages. To reduce turbulence:

- Minimize airspeed by designing large-diameter vents and chambers;
- Provide openings on the fan's mounting surface that match the fan's opening; fan manufacturers may include mechanical drawings for recommended openings;
- Use high-quality fans whose blades are designed to minimize turbulence;
- Remove sharp turns to avoid sudden changes in air direction;
- Avoid running air through small holes;
- Provide smooth transition sections between areas of different cross section.

The motor's mechanical vibrations are transferred to its mounting surface, which acts as a loudspeaker membrane and amplifies it. To minimize this effect

- Use resilient mounts for the fan that absorb the vibration;
- Use high-quality, balanced fans;
- Stiffen the fan's mounting surface.

Multiple fans running at nearly the same speed generate an annoying beat frequency. There are no simple solutions. Things that people have tried include

- Running fans at significantly different speeds from each other;
- Modulating the speed of one of the fans;
- Placing the fans further apart;
- Synchronizing the speeds of the fans with a microphone and a servo loop (!).

5.17.11.2 *Noise Attenuation*

Noise may be attenuated by placing foam along the air path (to reduce the generation of noise) and elsewhere in the battery (to reduce reflections and transfer).

5.18 MECHANICAL DESIGN

The mechanical design of a battery depends on the application (see the individual "Applications" chapters in Volume 2 for a deeper discussion on mechanical design). In this chapter, I mention only a few general points.

5.18.1 Enclosure

The battery enclosure must

- Protect from wandering fingers;
- Offer environmental protection appropriate for the use;
- Be electrically insulated from the battery;
- Support the mass of the battery;
- Be sufficiently sturdy to prevent wear of its contents over time, such as electrical connections and cells;
- Be sufficiently rugged against expected abuse without adding excessive mass.
- Be nonflammable.

5.18.2 Design for Service

When designing a battery, consider ease of access to the cells and the BMS components. A 5-minute service can stretch to 4 hours if the component of interest is buried deep inside the product.

Speaking from experience:

- Getting the traction battery out of a vehicle should take less than 15 minutes; replacing a module should take at most another 15 minutes;
- In a large stationary battery, power cables should not have to be disconnected to measure cell voltages and to repair the BMS;
- An individual section of a grid-tied, high-voltage energy storage system should be accessible by an adequately protected, trained technician without having to take the ESS out of service.

Reliability and low manufacturing costs are more critical than serviceability in a small consumer battery. Therefore, it uses an ultrasonically sealed plastic case.

5.18.3 Thermal Runaway Propagation Avoidance

The battery should be designed to minimize propagation if a cell (*the trigger cell*) should go into thermal runaway (see Section 8.2.1.5).

I derived most of the material for this section from a study for NASA [8]. Thermal runaway occurs in two closely spaced stages:

1. First, the cell's vent opens and fumes escape;
2. Then, the cell ignites, violently ejecting solid fragments and hot gases.

Internal propagation from the trigger cell to its neighboring cells is due to three mechanisms:

- *Heat transfer:* Principally conducted transfer, but also radiated, though to a lesser extent;
- *Current in a parallel block:* The short-circuit current through the hard short heats cells in the same parallel block as the trigger cell;
- *Ejecta:* These are hot gases, solid fragments of the trigger cell and the cell's deformed case.

External propagation is due to external ejecta, possibly through a perforated battery enclosure, which may damage and ignite materials next to the battery (see Volume 2, Section 5.2.1.1).

The probability of thermal runaway has decreased over the years as cell manufacturers have improved their designs and reduced manufacturing defects. At the same time, the probability of propagation has also increased as energy density has increased, partially through the use of lighter materials. This higher density results in a greater energy release in case of thermal runaway.

The only hope to prevent the ignition of the entire battery is to limit the event to just the first trigger cell. If a second cell goes into thermal runaway, the entire battery is lost because the battery temperature increases as each subsequent cell goes into thermal runaway.

Sparks or flames must remain contained within the battery to prevent the catastrophic failure of the product.

Techniques to minimize propagation manage the following:

- Heat transfer;
- Short circuits;
- Internal ejecta;
- External ejecta.

5.18.3.1 Heat Transfer Management

A proven technique to reduce the heat transfer from the ignited cell to the adjacent cells is to separate the cells by 2 mm or more. While the intervening space may be left open, filling it with insulating material further reduces heat transfer.

Use of a heat spreader inside the battery (Figure 5.88(a)) can have mixed results:

- On the one hand, a heat spreader absorbs heat and conducts heat from the ignited cell to the enclosure. This may reduce its temperature below the self-igniting temperature of the adjacent cells.

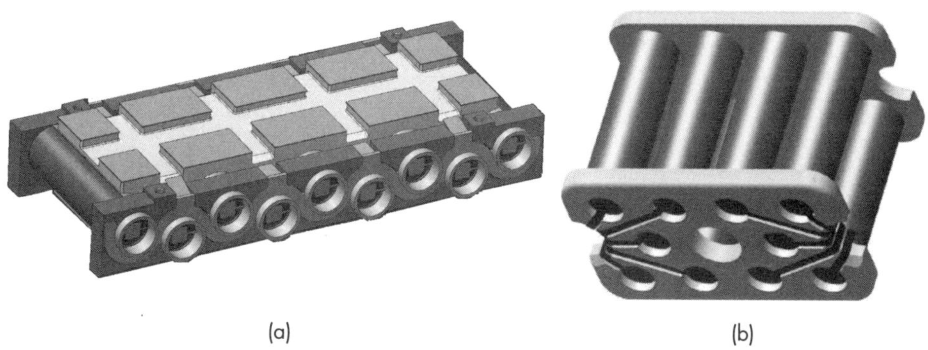

FIGURE 5.88 Management of thermal runaway: (a) heat spreader, and (b) ejecta channeling. (Courtesy of Eric Darcy, NASA.)

- On the other hand, a heat spreader reduces the thermal conductivity between adjacent cells, transferring more heat from the ignited cell to the adjacent ones.

A heat spreader is only effective if it conducts heat more readily from each cell to the enclosure than between adjacent cells. A laminar material may be able to achieve this feat by having a high thermal conductivity in the first direction and high insulation in the second direction.

5.18.3.2 Short-Circuit Management

A thermal runaway event may cause a short circuit

- *Internally:* The trigger cell becomes a hard short;
- *Externally:* Ejecta and deformation form short circuits between adjacent cells.

Cells adjacent to the trigger cells are subject to two sources of heat:

- From the resulting short circuit current;
- Directly from the trigger cell.

The heat from the short-circuit current makes it more likely that the thermal runaway will propagate to the adjacent cells. Therefore, preventing or interrupting the short-circuit current reduces the likelihood of propagation:

- *Prevention:* Through puncture-proof isolating layers that stop the ejecta;
- *Interruption:* Through fuses in series with each cell in parallel (see Section 3.3.12).

For small cylindrical cells in series, a concern is that the enclosure of the trigger cell contacts the adjacent cells, resulting in a short circuit. The standard isolating sleeve on the cells is no match against the violence of thermal runaway. A stronger material, such as Kapton, Nomex paper, Kevlar, or mica paper may be able to withstand penetration from the ejecta and the deformed case of the trigger cell. The same concern applies to pouch cells, which can be isolated with intervening puncture-proof sheets.

In a parallel block of cells, adding a fuse in series with each cell may reduce propagation. Current from adjacent cells into the shorted, ignited cell blows its fuse. Interrupting this short-circuit current reduces the heating in the adjacent cells, keeping them further away from the self-igniting temperature.

For small cylindrical cells, a fusible link can be integrated into the connecting plate. However, experiments have shown that fragments from the blown cell tend to fill the gap in a fusible link, reestablishing the connection, therefore making the fusible link ineffective.

5.18.3.3 *Internal Ejecta Management*

Directing the ejecta from the trigger cell away from the adjacent cells may avoid propagation.

Ejecta exit small cylindrical cells primarily through the cap (typically in the positive electrode). Secondarily, the case may be perforated, and ejecta may exit from the sides. This is increasingly a problem, as manufacturers are using thinner cases to improve the energy density of cells.

Channels may be added to the frame that holds the cells to direct the ejecta externally or at least to a safer area. Individual channels start from the cap at the end of a cell and end in exhaust ports outside the battery enclosure. The channels prevent the ejecta from reaching the adjacent cell (Figure 5.88(b)).

Separators between the cells are designed to block ejecta that exit the sides of the cell's case. The separator must be puncture-resistant; materials that have been used successfully include

- *PCM material:* Also provides thermal storage.[88]
- *Mica paper:*[89] Also helps block ejecta and isolates adjacent cases.
- *Epoxy syntactic foam:*[90] Cells fit snugly in boreholes; also contains the case of the trigger cell, may prevent rupture through the case, even for cells with thin walls.

5.18.3.4 *External Ejecta Management*

The battery case must contain the ejecta to prevent propagation outside the battery. Techniques include

- *Sealing:* The battery is sealed to contain the hot gases that may otherwise ignite materials surrounding the battery.
- *Channeling:* If sealing is not advisable, the hot gases and ejecta may be directed through a tortuous path in a muffler that slows them down and cools them down.
- *Arresting:* The walls of the battery are made of a material that arrests the flames and the ejecta; a carbon fiber composite (CFC)[91] has been used successfully for the purpose, and layers of metal foil have also been used, though they are not as effective.

5.18.4 Wiring

Here are a few tips for safe wiring.

5.18.4.1 *Wire Insulation*

Breach of wire insulation can result in a ground fault or a short circuit; measures must be taken to prevent it:

- Wires must not be squeezed in a way that may breach the insulation;
- Wires must not be placed near sharp surfaces, which may cut through the insulation;
- Wires must be routed away from conductive components of opposite polarity in case the wire insulation should be breached.

88. PCC™ Thermal Management from AllCell Technologies.
89. Silicate minerals in a silicone resin binder.
90. Hollow microspheres in an epoxy binder.
91. Carbon fibers affixed onto aluminum foil; a.k.a. carbon fibercore.

Rugged insulating layers may be placed between the wire and nearby components to prevent this; Kapton tape and fish-paper[92] are commonly used in batteries for this purpose.

5.18.4.2 *Fastening*

An insufficiently torqued fastener in a battery is dangerous:

- A current-carrying connection that is not tightened correctly has high resistance and heats up dangerously under load (see Volume 2, Section 5.2.1.1,);
- If a wire connection becomes loose, the wire may float away and connect to some other point, which may create a ground fault or a short circuit;
- Similarly, if a metal part becomes loose, it may contact a cell terminal or other live conductor, creating a ground fault or a short circuit.

Therefore, all fasteners (whether electrical or not) must be torqued correctly to prevent loosening. The torquing level depends on the fastener's material, its diameter, and whether a lubricant is used.

Some designs include lock washers. Others use only flat washers;[93] regardless, during production, quality control must ensure that the specified hardware is used and the assembly instructions are followed precisely.

5.19 REGULATORY TESTING STANDARDS

Well-intentioned regulatory agencies have devised battery testing standards that could test the patience of a saint with some requirements that can be a bit out of touch with reality. For example, they may require that a battery be tested by crushing the individual cells, even though the cells are contained in a sturdy battery enclosure. Or they may require that the battery be overcharged to see that it doesn't catch fire, even though the BMS shuts down the battery as soon as such a test is attempted.

In one case, a cell must be disassembled, purposely damaged, reassembled, and then reused!

Designing a battery to meet these standards can easily take as much additional effort as the effort to design the battery in the first place. The costs of performing these tests can dwarf the engineering costs of designing the battery. Meeting these requirements can only be justified for high-price batteries or inexpensive batteries manufactured in substantial volumes (100,000 units). It is impractical to design and test any other battery to meet these standards. Battery testing companies provide these services (see Volume 2, Section C.5.2).

There are dozens of standards depending on the type of battery[94] and the country of use, written by at least a dozen regulatory agencies around the world.[95] New standards appear faster than old standards can be harmonized with each other [9].

Table 5.11 provides some of the standards that apply to Li-ion battery design and manufacture [10]. In most cases, you also have to pay for the regulatory agency to come and inspect your factory regularly.

92. A strong, flexible, fibrous, durable electrical insulator made of vulcanized fiber.
93. The automotive industry uses only flat washers, relying only on proper torquing.
94. Such as specialized standards for cell phone batteries, passenger EVs, e-bikes, and lap-tops.
95. UL, SAE, CSA, IEEE, USABC, JIS, IEC, BATSO, UN, CEI, CQC, KC...

Table 5.11
Some Battery Testing Standards

Test	Method	Notes
US	IEC 60950-1 with UL 2054	
EU	EN 60950, EN 62133, IEC 62133	
Canada	CSA 60950-1 with UL 2054	
China	GB 31241-2014	
Japan	DENAN Ordinance, Article 1, Appendix 9	Mandatory.

In general, testing covers these areas:

- Overcharge, overdischarge, external short circuit;
- A short circuit inside the battery or even inside the cell;
- Mechanical abuse, shock and vibration, drop;
- Overtemperature, temperature cycling;
- Low pressure (for airplane transportation).

5.19.1 Transportation

For transportation, the battery must meet the UN 38.3 standard, which in many ways duplicates the requirements of the other tests. If you design and test to only one standard, this would be the one.

UN3480 covers shipping Li-ion batteries by themselves and UN3481 covers shipping products that contain Li-ion batteries.

Various online resources discuss transportation requirements [11].

References

[1] The XY problem, http://xyproblem.info/.

[2] See the excellent guide from Molex: Good crimps, https://www.molex.com/tnotes/crimp.html.

[3] "DC Ground Fault Detection Provided for Uninterruptible Power Supplies," Edward P. Rafter, Tier IV Consulting Group.

[4] Maluf, N., Qnovo, "Why-step-charging-is-misleading"; https://qnovo.com/why-step-charging-is-misleading/.

[5] Weicker, P., A Systems Approach to Lithium-ion Battery Management, Norwood, MA: Artech House, Section 8.1.2.

[6] Maluf, N., Qnovo, "Technology"; https://qnovo.com/technology/.

[7] Santhanagopalan, K. Smith, J. Neubauer, G.- H. Kim, M. Keyser, and A. Pesaran, Design and Analysis of Large Lithium-Ion Battery Systems, Norwood, MA: Artech House, 2014.

[8] Darcy, E., "Driving Factors for Mitigating Cell Thermal Runaway Propagation and Arresting Flames in High Performing Li-ion Battery Designs," NASA-JSC, Electric Aircraft Symposium, Santa Rosa, CA, 2015.

[9] Intertek, "Navigating the Regulatory Maze of Lithium Battery Safety," http://batterypoweronline.com/wp-content/uploads/2013/11/Intertek_Regulatory-Maze-WP.pdf.

[10] Arora, A., S. A. Lele, N. Medora, and S. Souri, Lithium-ion Battery Failures: A Systems Perspective, Norwood, MA: Artech House, 2019, Chapter 7.

[11] PRBA, the Rechargeable Battery Association, "Lithium Battery Transport Information," https://www.prba.org/lithium-battery-transport-information/.

CHAPTER 6

MODULES AND ARRAYS

6.1 INTRODUCTION

At times there is a desire to subdivide a battery for convenience, manageability, scaling, system requirements, flexibility, redundancy, or due to physical constraints. Yet, in the end, the technical complexities and costs of doing so may quench that plan.

This chapter discusses such subdivisions and analyzes their advantages and disadvantages to help you decide if and how to subdivide a battery.

6.1.1 Tidbits

Some interesting items in this chapter include:

- Sorry to tell you, but your great idea maybe ain't so great (6.1.3);
- Just because you could do it with lead-acid batteries doesn't mean you can do it with Li-ion (6.1.3.1);
- If a single battery is good, two batteries are worse (6.1.3);
- A battery with a third wire is actually two different batteries (6.5);
- Yes, you can keep on using your old lead-acid batteries alongside Li-ion (6.6.2.4).

6.1.2 Orientation

This chapter starts by defining an array and a modular battery and listing six ways that a battery could be subdivided. For each subdivision, it talks about its real or perceived advantages and disadvantages, and it suggests implementations. Finally, it discusses using lead-acid and Li-ion batteries together.

6.1.3 "Hey, I Have an Idea!"

People come to me all the time with new ideas; many involve subdividing a Li-ion battery.

"Instead of designing a different battery for every application, let's design and stock a single module. Then we can use it as a building block for any battery of any voltage and capacity, and ship a custom battery in a week instead of six months!"

"Instead of selling different battery sizes, let's sell a single module. Then the user can use it as a building block for any battery, of any voltage and capacity!"

"Instead of a product with a large battery, let's do one with a small battery and then sell expansion modules that the user can add to the product to increase the capacity!"

"Instead of making a large battery, let's make two small ones. If one is empty or bad, the other one would still be OK!"

"Instead of designing batteries of different sizes, let's design a single string. Then we can install the number of strings in parallel required to achieve the desired capacity!"

"Let's make a modular battery so that the customer can replace a bad module!"

"Let's modularize this large traction battery so that the users can recharge a single module at a time in their apartment!"

In my experience, all of these inventors have abandoned their ideas after looking at their associated costs, technical challenges, and dangers. Those who did not give up altogether have reverted to a standard battery design—a single battery of the desired size.

Truth be told, some Li-ion modular solutions have been developed, but they are either unsafe or are sold in a market that is not price-sensitive.

If you have the required engineering skill and your market can bear the higher price, then I encourage you to proceed using the technical help in this chapter.

6.1.3.1 The Lead-Acid Legacy

A modular approach can be implemented with lead-acid batteries due to their higher series resistance and high-voltage hysteresis (which results in a smaller inrush current when first connected in parallel; see Section 3.3.6), ability to operate without a BMS, and lower propensity to catch fire. This approach does not translate into Li-ion, due to its low resistance, need for a BMS, and volatility when abused.

6.1.4 Battery Subdivision

Batteries may be subdivided as

- A single battery versus several interconnected batteries;
- In a single enclosure versus in multiple ones.

6.1.4.1 Single Battery versus Multiple Batteries

Do not confuse a single battery with an array of batteries; recall that (see Section 5.1.3):

- *Single battery:* It has one current, one protector switch, one BMS, one SoC level;
- *Multiple batteries:* Each battery has one protector switch, one current, one BMS, one SoC level.

If a "battery" has more than one protector switch,[1] it's not a battery, it's a set of batteries.

In particular, I would like to emphasize the distinction between a battery array and a modular battery, which at first glance may appear to be the same:

- In a battery array, each battery has a BMS and a protection switch; at a given time a given battery may or may not be connected (Figure 6.1(a));
- In a modular battery, all modules are connected permanently, the modules are not protected, there's only one switch for the entire battery (controlled by a

1. A set of contactors, positive and negative, counts as a single protector switch.

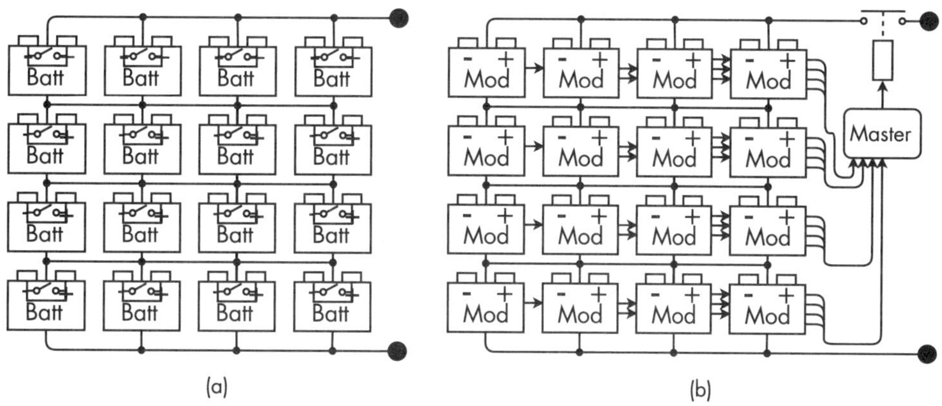

FIGURE 6.1 Subdivision: (a) battery array, and (b) modular battery.

master), and there is a communication cable between each module and a master (Figure 6.1(b)).

6.1.4.2 *Single Enclosure versus Multiple Enclosures*

Deciding between a single enclosure or multiple enclosures is not just a mechanical issue. It also has significant implications from the electrical standpoint: if a battery or array is in multiple, physically separate sections, it uses connectors. Someone or something is likely to make a connection at unexpected times.

In a single battery in a single enclosure, all parallel connections are made only once, when the battery is assembled; presumably, those parallel connections are performed by skilled technicians who know how to check that there is no voltage difference before making the connection. From then on, the strings will remain connected permanently. Therefore, at no other time is there a danger that strings of different voltages are connected in parallel.

On the other hand, in a solution that uses multiple enclosures, the user may reconnect a module at any time. At that time, there may be a significant voltage difference across the open connection, resulting in a potentially damaging inrush current when the connection is closed (see Section 3.5.1).

6.1.4.3 *Four Permutations*

In Table 6.1, I classify these subdivisions into four groups based on the four permutations of single/multiple batteries and single/multiple enclosures.

6.1.4.4 *Subdivision Examples*

There are many ways that a battery can be subdivided, both electrically and physically. These include

- Single, modular battery:

TABLE 6.1 Classification of Battery Subdivisions

	Single Enclosure	**Multiple Enclosures**
Single Battery	Flexible battery with strings in parallel Flexible battery with strings in any arrangement	Modular battery Expandable battery
Multiple Batteries	Split (center-tap) battery Dual battery	Battery array Ganged battery

- Single modular battery with parallel strings: A single battery contains permanently connected strings in parallel; the factory decides the number of strings (Figure 6.2(a)) (or any other arrangement);
- Single modular battery, any arrangement: A single battery contains permanently connected strings in series of parallel (Figure 6.2(b));
- Distributed modular battery: Physically separate modules, permanently connected in series or parallel at the time of assembly (at the factory or in the field) (Figure 6.2(c));
- Expandable modular battery: A battery that can work on its own, or the users can expand it by adding one or more modules, as they wish (Figure 6.2(d)).

- Multiple batteries:
  - Split battery (center-tap battery): Two batteries in series, with three power terminals, to provide a positive and a negative voltage relative to a common terminal (Figure 6.2(e));
  - Dual battery: Two batteries in parallel in a single enclosure (Figure 6.2(f));
  - Battery array: Any number of batteries connected in any arrangement, each capable of connecting to a bus only when it is safe to do so (Figure 6.2(g)):
  - Ganged batteries: Several batteries whose protector switches operate simultaneously, based on the combined input of each of their BMSs (Figure 6.2(h)).

This list is not exhaustive—there are more ways to subdivide a battery. The rest of this chapter discusses each of these subdivisions.

6.2 MODULAR BATTERIES

A flexible battery design consists of modules that may be included in various numbers and arrangements to achieve the desired capacity and voltage. For example:

- Single modular battery with parallel strings: Fixed voltage, variable capacity;
- Single modular battery, any arrangement: Variable voltage and capacity;
- Distributed modular battery: Multiple enclosures;
- Expandable modular battery: Optional expansions may be added to the main unit.

6.2.1 Single Modular Battery with Parallel Strings

A single-enclosure battery may be designed to have a selectable capacity by using two or more series strings permanently connected in parallel at the factory. One string is used for the minimum capacity (Figure 6.3(a)), two strings for more capacity (Figure 6.3(b)), or many, for even higher capacity (Figure 6.3(c)).

This battery is safe because the strings are connected permanently in parallel: there is no switch in series with a string that could be closed at random times, and the user has no access to connectors to disconnect and reconnect. When the battery is first built, a trained factory worker knows how to safely measure the voltage difference before making a permanent connection between strings in parallel. A trained technician knows how to add or replace a string later safely.

The typical BMS doesn't know how to handle strings in parallel: it assumes that the strings are in series, and it reports a total voltage that is N times higher than the

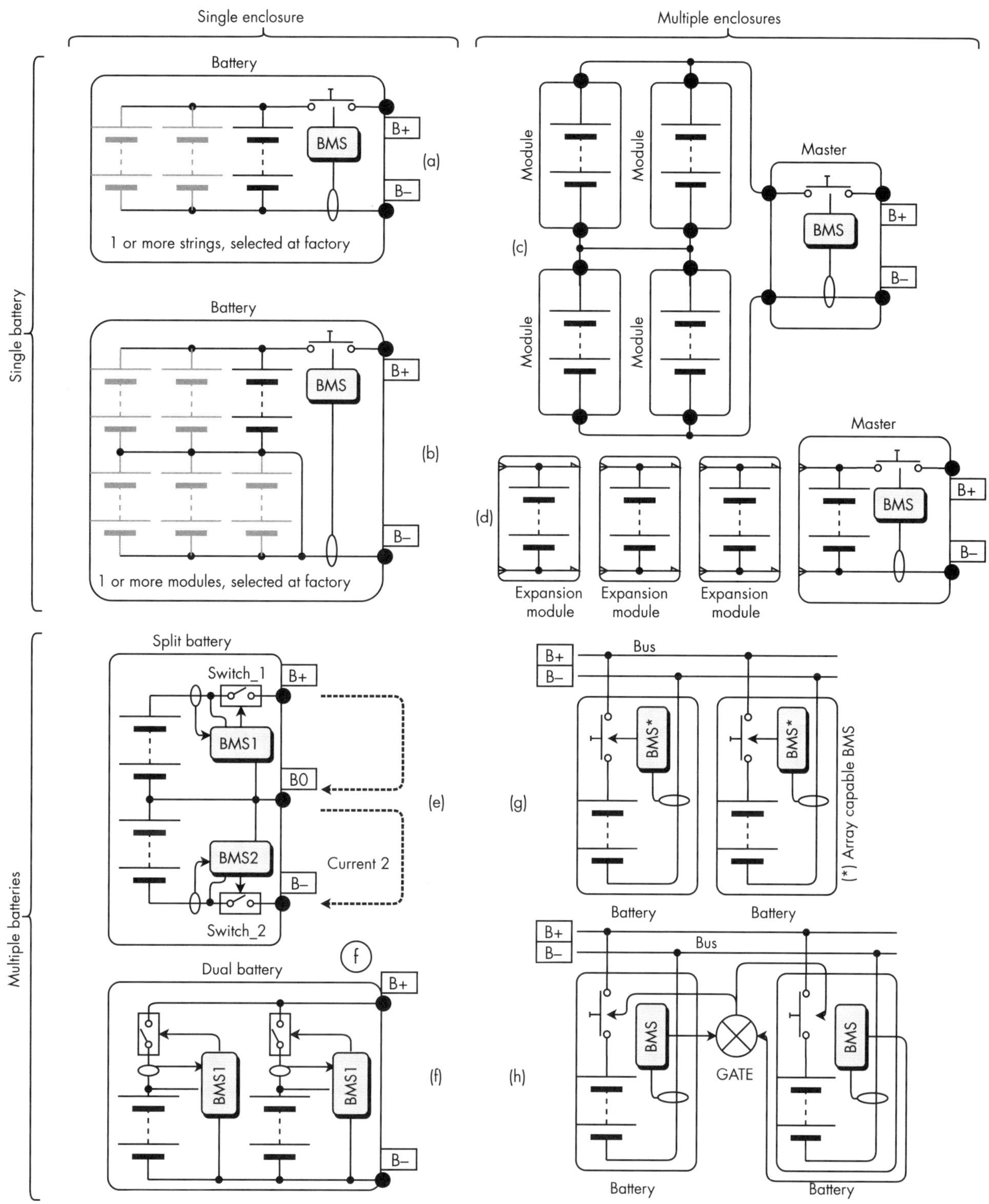

FIGURE 6.2 Subdivision: (a) single modular battery with parallel strings, (b) single modular battery, any arrangement, (c) distributed modular battery, (d) expandable modular battery, (e) split battery, (f) dual battery, (g) battery array, and (h) ganged batteries.

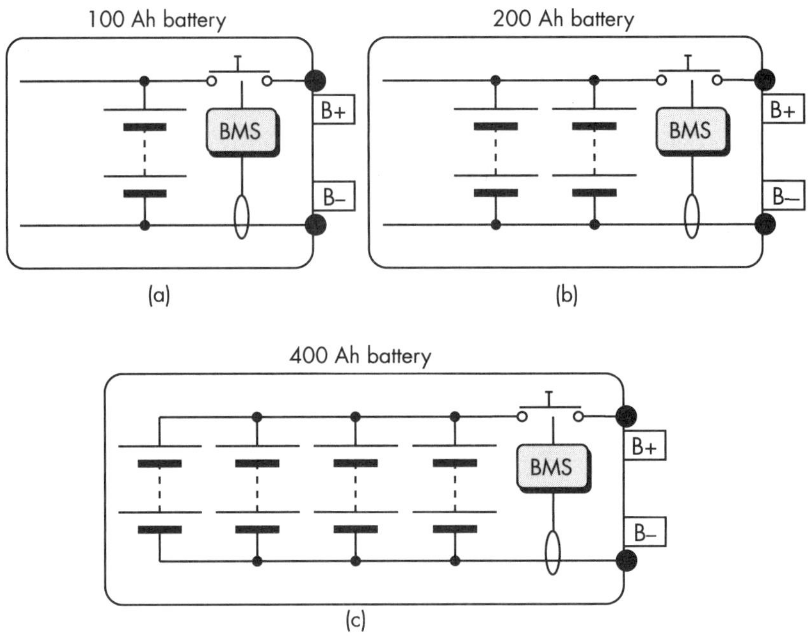

FIGURE 6.3
Single modular battery with strings in parallel: (a) one string, (b) two strings, and (c) four strings.

actual battery voltage (where N is the number of strings in parallel).[2] Regardless, the BMS does protect the battery.

6.2.2 Single Modular Battery, Any Arrangement

This battery is the same as above, except that modules may be connected both in parallel and in series (Figure 6.4). For example, this solution could be used to achieve 12V, 24V, 36V, or 48V.

Having modules in both series and parallel could be even more of a challenge for a BMS. It is also a challenge for the power supply that feeds the BMS because of the varying voltage.

6.2.3 Distributed Modular Battery

A modular battery is the same as above, though in separate enclosures. It consists of two or more battery modules permanently connected in series or parallel at the time of assembly, plus a separate master and protector switch (Figure 6.5).

Each module includes just the sensing and balancing portion of the BMS electronics; it does not include a protector switch. Communications cables go between each battery module and a master.

A modular battery does have some advantages:

- Versatile; can be connected in series and parallel for any voltage[3] and capacity;
- Having just a single protector switch for the entire battery reduces cost.

It also has many disadvantages:

- A range of different protector switches must be made available (voltages, currents);

2. The Lithiumate Pro BMS can handle up to 16 strings in parallel, but each string can only have few cells because the maximum total number of cells is 256.
3. There is a maximum voltage, due to the breakdown of the communication links, but it is likely to be higher than 1 kV.

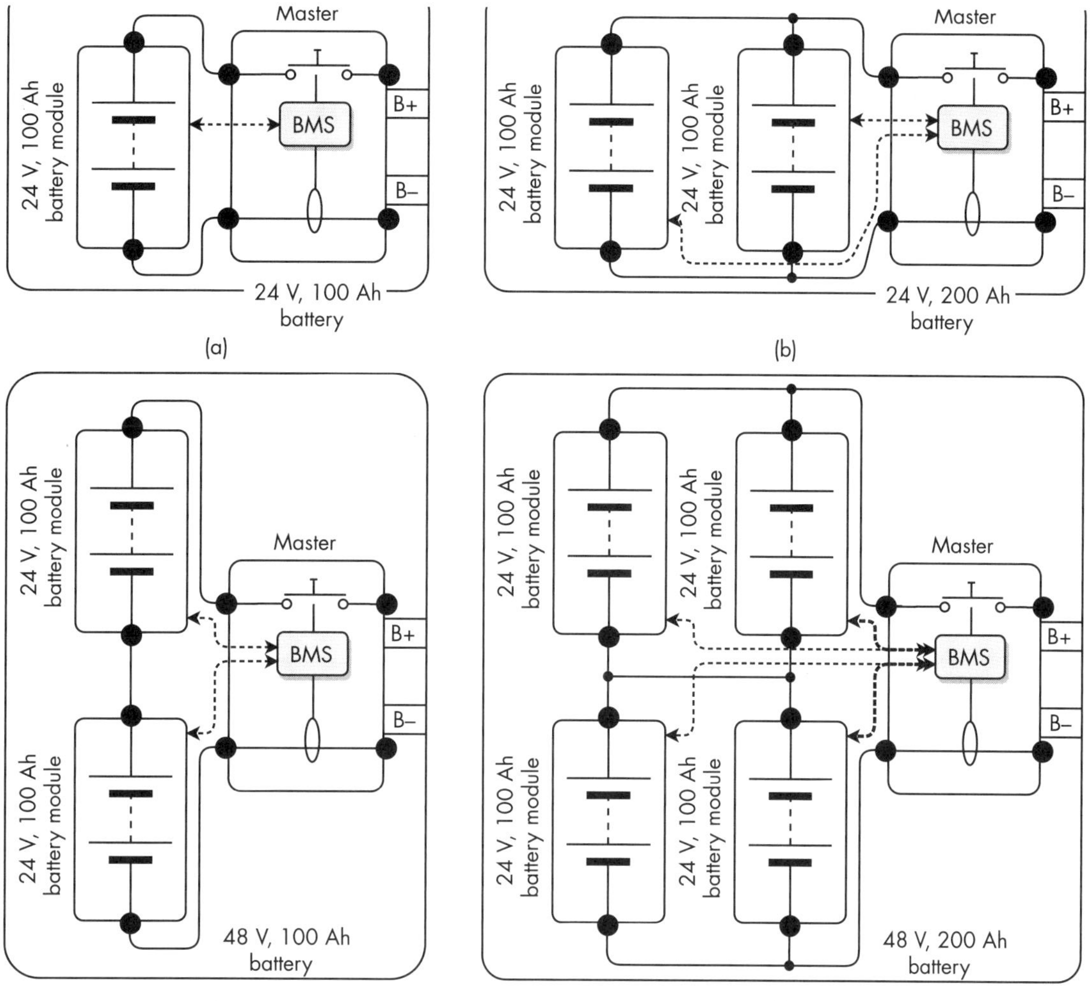

FIGURE 6.4 Single modular battery: (a) single module, (b) modules in parallel, (c) modules in series, and (d) modules in parallel-first.

- A user who is only familiar with lead-acid batteries:
  - May not install a protector switch;
  - May not install a master;
  - May not connect the communications cables;
  - May bypass the protector switch and connect directly to the modules;
  - May connect two series strings in parallel with different voltages.
  - May charge modules individually.
- Replacing an individual module is risky in case of mismatched voltage;
- It is more expensive, because for a given voltage and capacity, two modules in parallel require twice the BMS sensing electronics compared to two modules in series, each with half the voltage.

6.2.3.1 Case Studies

Valence Technology, a manufacturer of LFP cells, developed its U-charge line of 12, 34, and 36V battery modules. It offered an external BMS to manage the modules,

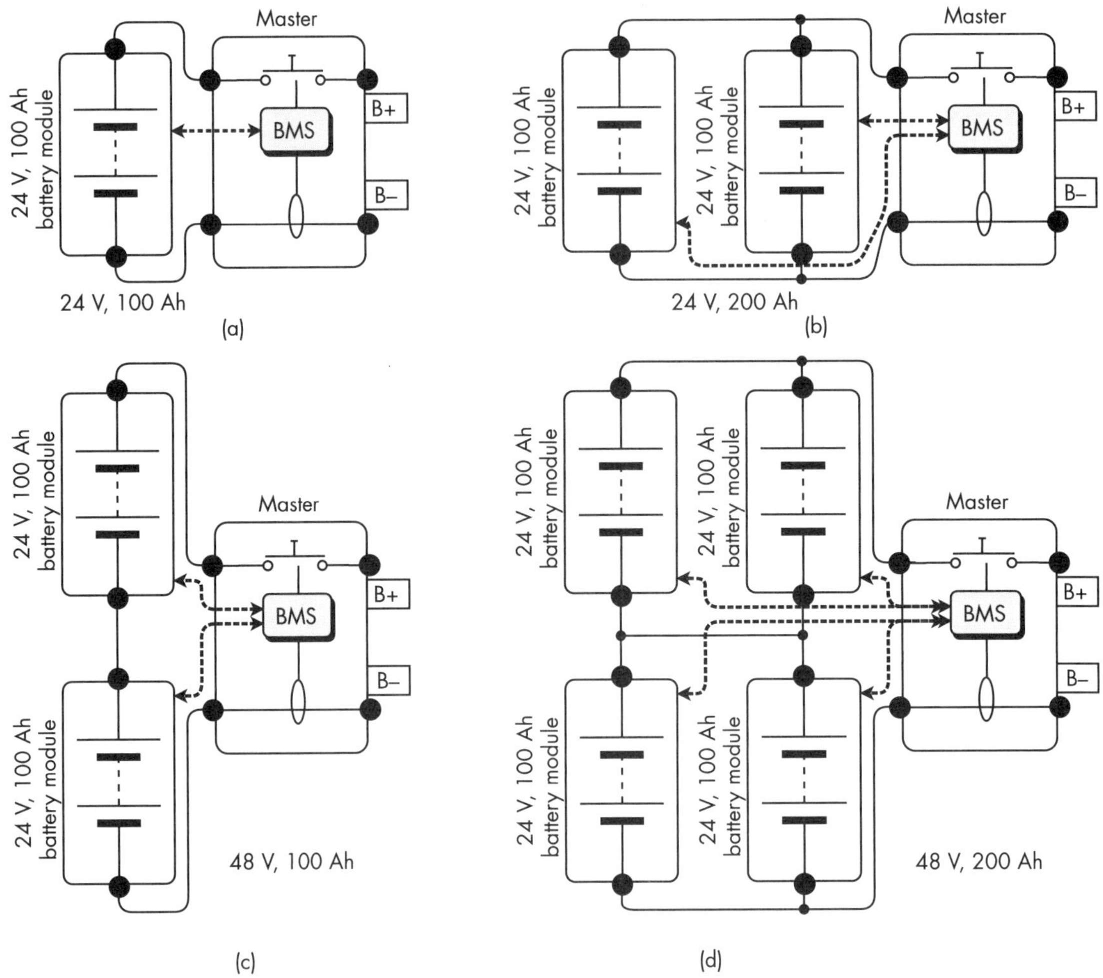

FIGURE 6.5 Distributed modular battery: (a) single module, (b) modules in parallel, (c) modules in series, and (d) modules in parallel-first.

but the implementation of a protector switch was left to the user; in some cases, the protector switch was not implemented, with serious consequences[4] (see Volume 2, Section 5.2.2.2) (Figure 6.6).

6.2.4 Expandable Modular Battery

A base model of a product includes a battery that is as small as possible to keep down the cost and size. It does have provisions for attaching one or more expansion battery modules in parallel with the internal battery.

Customers buy a base product (Figure 6.7(a)); if they wish a longer discharge time, they may add one or more expansion modules, depending on how long they expect to use the product on a given day (Figure 6.7(b)).

The charger can be connected to the main product (Figure 6.7(c)) or one of the expansion modules (Figure 6.7(d)).

From a marketing standpoint, this is a great idea. From a technical one, as described above, not so much:

4. Six months later, in 2012, Valence filed for bankruptcy protection. In 2018 Lithium Werks bought its assets and is still selling these modules.

FIGURE 6.6
Bank of Valence battery modules in a vessel without a protector switch.

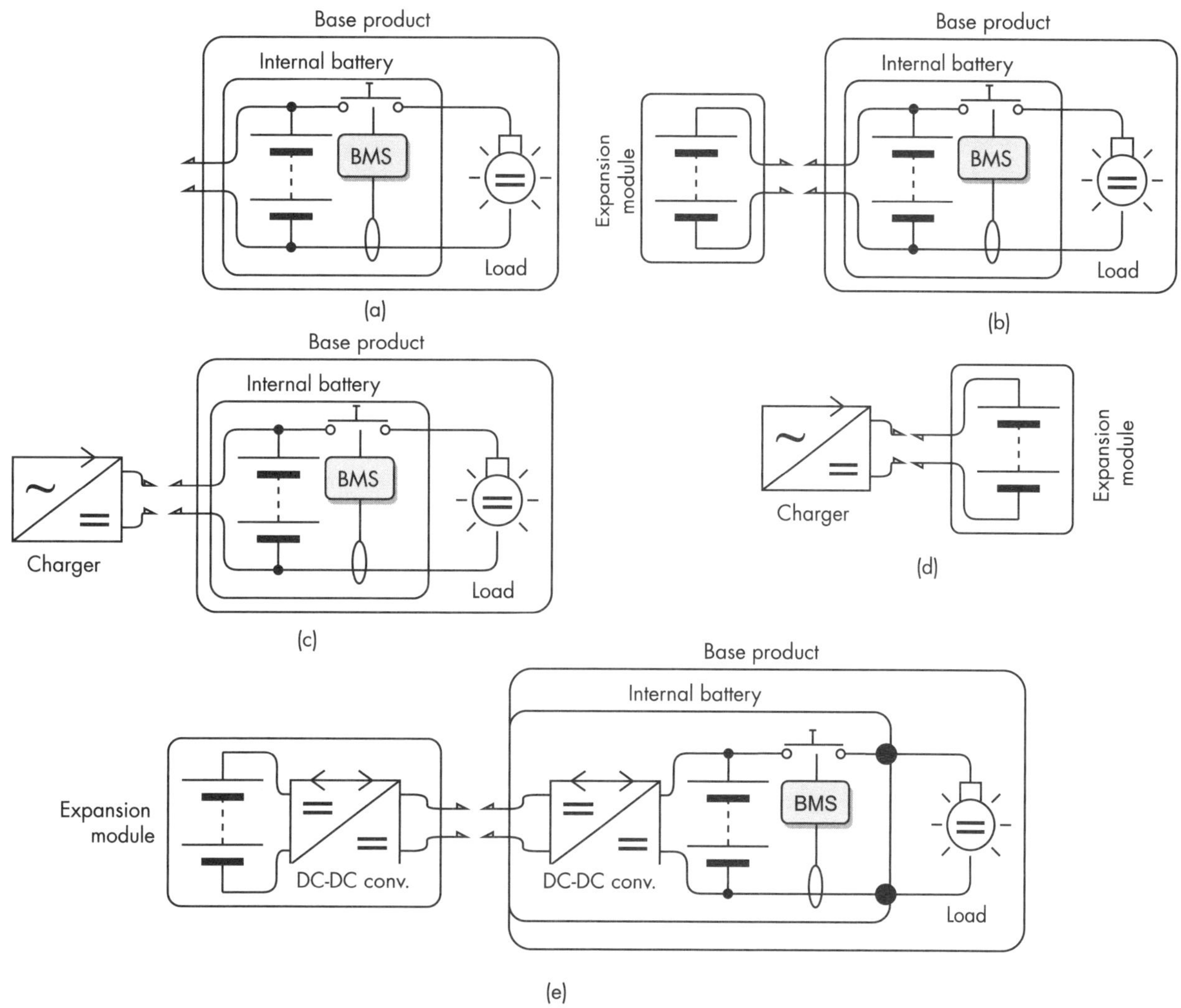

FIGURE 6.7 Expandable battery (DON'T DO THIS!): (a) base product, (b) with expansion modules, (c) charging the base product, (d) charging an expansion module, and (e) a safe solution.

- The expansion modules are not protected, so the user is exposed to the voltage;
- Nothing prevents connection in parallel of modules at different voltages;

- In high-power applications, a single module must be powerful enough to power the application on its own.

A solution would be to use a high-power, bidirectional DC-DC converter between each string of cells and each power connector (Figure 6.7(e)). The converter is current-limited, making able to adapt the terminal voltage safely to whatever happens to be connected to the battery or module. Also, for safety, the converter is turned on only after the expansion module is fully mated.

However, now we have two strings with different currents, and therefore two different SoC levels. How can a single BMS deal with this? Plus, this doesn't change the need for a signal cable between the expansion module and the main battery to let the BMS monitor the cells in the expansion module. This solution is rapidly getting uglier!

The correct way of implementing a product with an expandable battery is by having each expansion module be a complete battery with its own BMS and protection switch. In effect, rather than a modular battery, this product would use a battery array (see Section 6.3).

6.2.4.1 Case Studies

As I write this, a major manufacturer of power tools is exploring the idea of distributing such unprotected battery modules to the general public through a major U.S. chain of hardware stores. The idea is that a user may connect any number of modules to its product line to increase the run time as desired. I strongly, though unsuccessfully, advised that company to reconsider, to look at protected batteries instead, while making it aware of the significant cost of such batteries compared to a nonmodular approach.[5]

6.3 BATTERY ARRAYS

A battery array consists of two or more complete batteries (each with a complete BMS and a protector switch). The batteries in an array may be arranged in

- Series (Figure 6.8(a);
- Parallel (Figure 6.8(b)), most common);
- Parallel-first (Figure 6.8(c)).

The number of batteries in series is limited by the voltage rating of the protector switch in each battery because the entire array voltage may appear across this switch when it opens under load.

If any batteries are placed in parallel, the array must use at least one of these two approaches to protect the batteries from the inrush current that occurs when two batteries at different voltage are first connected in parallel (see Section 3.3.6).

- Array-capable BMS: The BMS in each battery can decide whether it's safe to connect to the bus;
- Array master: A single master knows the state of the bus and each battery in the array and coordinates the connection of each battery to the bus.

A battery array has many advantages:

5. In the next edition of this book, I will update you on whether that company went through with this project.

FIGURE 6.8
Battery array:
(a) series, (b) parallel,
and (c) parallel first.

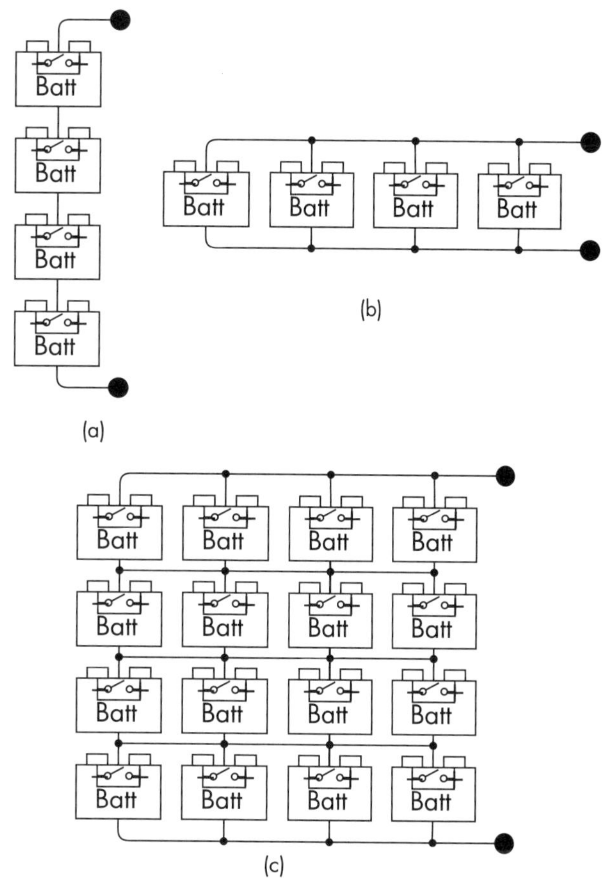

- Versatility: Batteries can be connected in series for any voltage up to the rating of the protector switch in each battery;
- Scalability: Batteries may be connected in parallel for any capacity by adding or removing batteries as needed;
- Flexibility: Different battery technologies may be used together;
- Serviceability: A single battery in a parallel set can be disconnected from the array for testing, service, or repair, and a replacement battery may be installed in its place;
- Safety: Low-voltage batteries, which are intrinsically safe, may be connected in series to achieve a higher voltage;
- Reliability: There is no risk of damage when an array-capable battery reconnects.

However, a battery array is more expensive than a single battery of the same specifications:

- Instead of a single protector switch for the entire system, each battery has a protector switch;
- An *array-capable BMS* commands a premium price.

A typical application for a battery array is in large, stationary installations.

- Low voltage (see the "Large, low voltage batteries" chapter in Volume 2): Batteries are connected to a shared parallel bus, usually 48V.

- High voltage (see the "High voltage batteries" chapter in Volume 2): Typically grid-tied; usually 900 to 1,500V; high-voltage batteries may be connected in parallel, or lower voltage batteries (~500V) are connected in a series-first arrangement.

6.3.1 Array-Capable BMS

An array-capable BMS determines whether it's safe to turn on the protector switch to connect its battery to the external DC bus.

At the most basic level, this BMS measures the voltage of the bus and of its string of cells. If they are close enough to each other, the BMS decides that it's OK to connect its battery to the bus (Figure 6.9(a)). Otherwise, the BMS decides that it has to wait (Figure 6.9(b)).

However, this simple logic prevents any battery in the array from being the first to connect to the bus in two cases:

- There is no charger, so the bus is at 0V; this is lower than the batteries' voltages (Figure 6.10(a));
- There is just a charger, and it's turned on; the bus voltage is at the charger's constant voltage, which is higher than the batteries' voltages (Figure 6.10(b)).

Therefore, the BMS in an array battery must be able to analyze the bus and determine whether it is already connected to another battery (Figure 6.11). If so, it

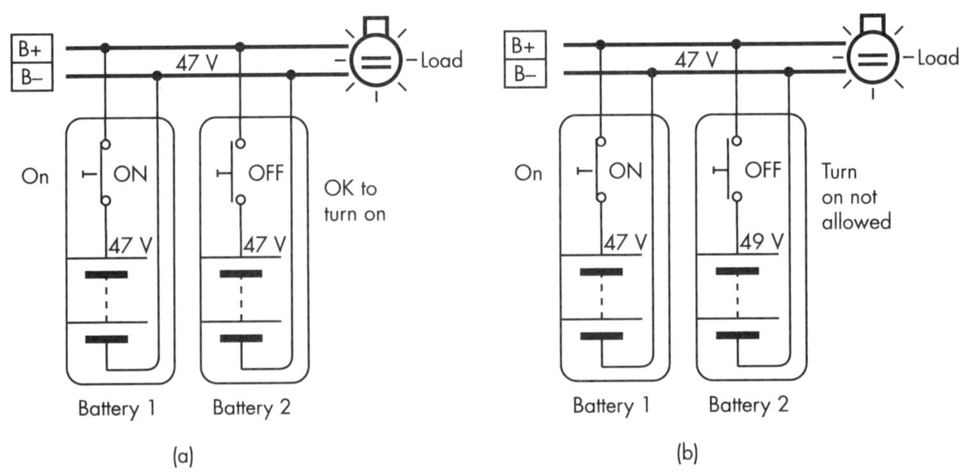

Figure 6.9
Connecting battery to a bus: (a) bus voltage is the same as the battery, and (b) different bus and battery voltages.

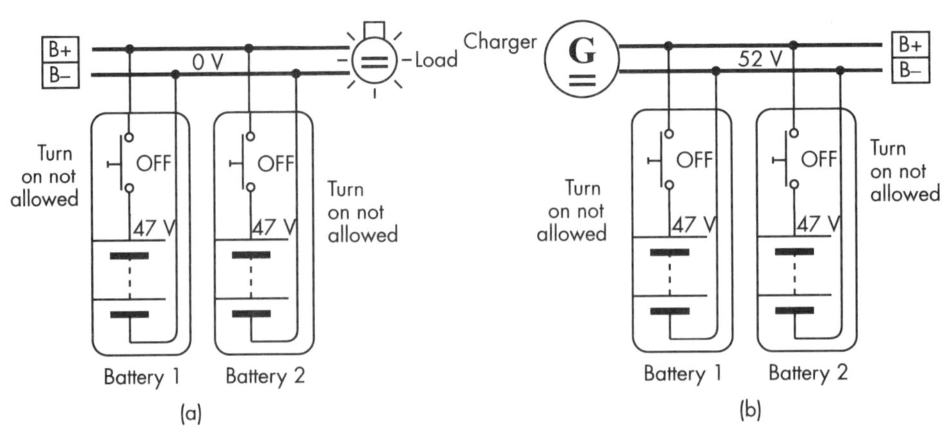

Figure 6.10
BMS won't connect the battery to the bus: (a) bus at 0V, and (b) bus at charger voltage.

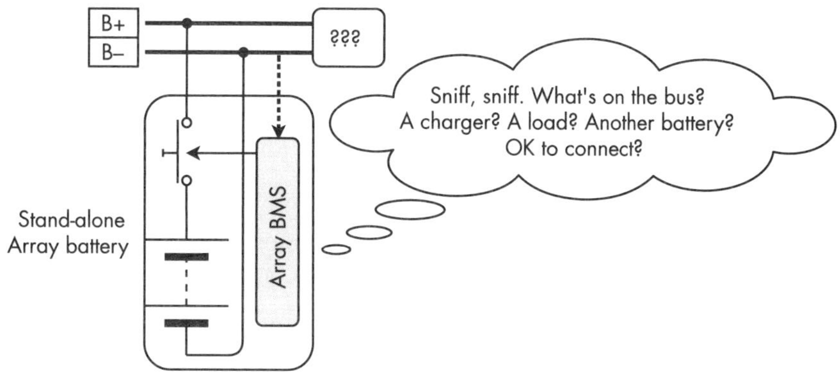

FIGURE 6.11
Thought process in an array-compatible BMS.

must wait for the voltages to be equal. Otherwise, it may be able to turn on its battery immediately. It is rather tricky to distinguish a battery connected to the bus from a power supply that is powering a heavy load because, in both cases, the bus voltage can be the same. I assure you that it's complicated, quite so.

To my knowledge, today only three off-the-shelf BMSs can manage an independent array battery to determine whether it is safe to connect to the parallel bus without the aid of an array master (see Volume 2, Sections 2.12.4 and 4.8.2).

6.3.2 Array Master

An elegant way to manage an array is through an array master (Figure 6.12). Communications cables run between each battery and the array master.

An array master can manage the batteries in an array because it knows the complete status of the array: the internal voltage and state of each battery and what's connected to the bus so that it can allow the individual batteries that may connect to the bus safely to do so.

Besides managing the state of each battery, the array master compiles the data from each battery into a single set of data for the entire array. For example, for a parallel array:

- The capacity of the array is the total capacity of all the batteries that are connected to the bus at the time;
- The SoC of the array is the SoC of any battery connected to the bus;

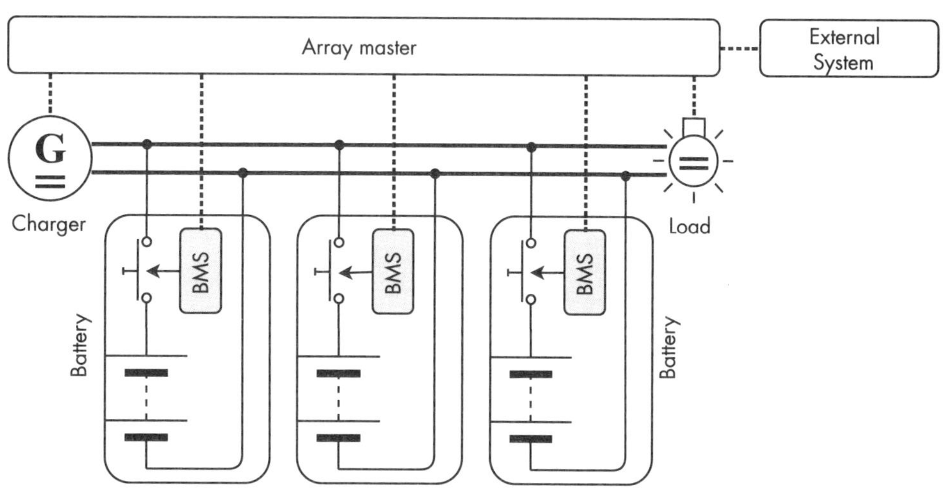

FIGURE 6.12
Array managed by an array master.

- The bus current is the sum of the current of all the batteries;
- The bus voltage is the voltage of any battery connected to the bus;
- The maximum temperature is the maximum temperature of all batteries, whether or not connected to the bus.

Using an array master has these advantages:

- The batteries can use a standard BMS, one that does not need to be able to determine on its own whether it's safe to connect to the bus;
- The external system can treat the array as a single unit, by communicating just with the array master.

I know of only two companies that offer off-the-shelf array masters (see Volume 2, Sections 2.12.4 and 4.8.2). Typically, developers of battery arrays design their own array master to work with the BMS of their choice.

6.3.3 Battery Voltage Equalization

A *straggler battery* is one that is disconnected from the bus because its internal voltage is different. Its charge is unavailable while it waits for the bus voltage to equal its internal voltage. Unfortunately, a straggler battery's charge is unavailable until it can jump on the bus, so to speak.

In applications in which the bus voltage is not likely to change, a straggler battery rarely has the opportunity to jump on the bus.

For example, in a backup system, the bus sits at the maximum voltage all the time (Figure 6.13). A straggler battery will only jump on the bus during a power outage, when the bus voltage finally drops down to its internal voltage. The system's back up time is reduced, compared to if the battery had had a chance to jump on the bus earlier and be fully charged by the time the power outage occurred.

There are some methods to let a straggler battery jump on the bus sooner. If the voltage of the straggler battery is too low:

- Turn off the power source and let the batteries discharge as they power the loads until the bus voltage drops sufficiently and the straggler jumps on the bus, then, turn the power source back on. This is risky because if a power outage happens during this process the backup time is significantly reduced.
- Include a charger in each battery and use it to charge a straggler battery up to the bus voltage (Figure 6.14(a)).

If the voltage of the straggler battery is too high (for example, in a buffer battery, see Section 5.1.4).

- Charge the other batteries up to the voltage of the straggler battery, at which point the straggler battery jumps on the bus;
- Include a dummy load in each battery to discharge a straggler battery down to the bus voltage (Figure 6.14(b)).

My preferred solution, which works in all cases, is to include a bidirectional DC-DC converter in each battery, between the string of cells and the battery terminals, to either charge or discharge a straggler battery towards the bus voltage (Figure 6.14(c)).

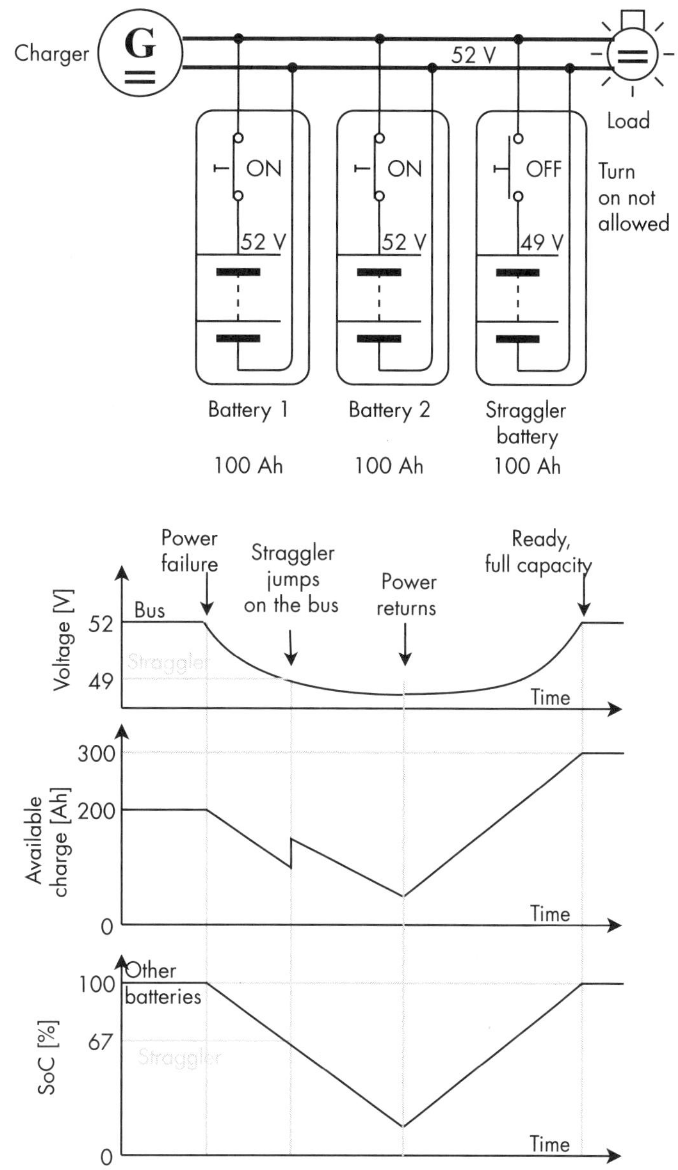

FIGURE 6.13
Straggler battery jumping back on the bus in a backup system.

6.3.4 Array of Small Consumer Batteries

Often hobbyists wish to reuse small consumer batteries (e.g., laptop batteries or power tool batteries) and combine them into a large battery.

This is a problem because each battery has a protector switch. When any of those switches opens or closes, damage can occur. It may be possible to add diodes to prevent this damage. The diodes must be rated for the full load current and the total voltage of the array. Diodes do not make sense for large batteries because the current is too high for standard rectifier diodes or for low-voltage batteries because the voltage drop in the diodes constitutes a large portion of the battery voltage. The following solutions assume a single-port BMS inside the consumer battery.

6.3.4.1 Parallel Array

Diodes may be used to prevent inrush current between batteries in a parallel array when the BMS in one of them turns on its protector switch (see Section 3.3.6).

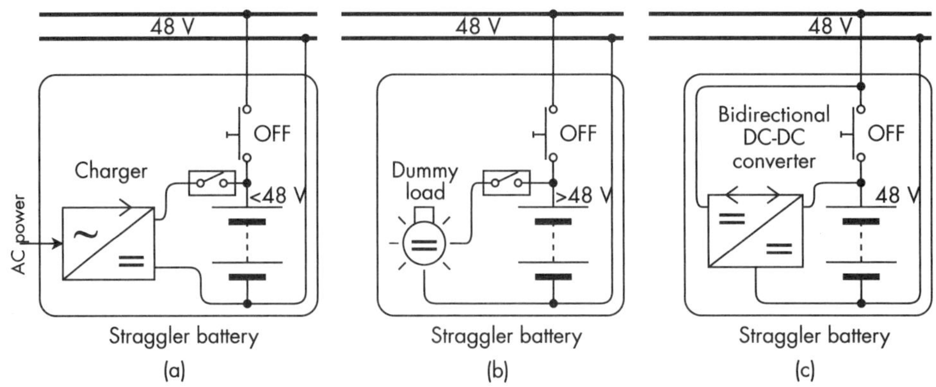

FIGURE 6.14
Methods to bring straggler battery to bus voltage: (a) charger, (b) dummy load, and (c) DC-DC converter.

Two diodes are added to each battery, one to route the charging current, one to route the load current (Figure 6.15(a)). This solution only works for parallel batteries that use a separate charger and load. Assuming that the three batteries start at different voltages, then, when charging (Figure 6.15(b)):

- The battery with the lowest voltage (#1) is charged first. The current flows only into this battery because its diode is the only that is forward-biased (Figure 6.15(c)); the other diodes are off because the voltage across them is reversed.
- Once the #1 battery is charged up to the same voltage as the next lowest battery (#3), the diode for battery #3 turns on, and the current starts flowing into both batteries[6] (Figure 6.15(d)).
- When all three batteries are up to the same voltage, all three diodes are turned on, and the current flows into all three batteries (Figure 6.15(e)).

Similarly, when discharging, the current flows from the battery with the highest voltage. Once the first battery is discharged down to the same voltage as the next highest battery, the current starts flowing from both batteries. When all the batteries are down to the same voltage, the current flows from all the batteries.

Current cannot flow between batteries because there are two diodes between them facing in the opposite direction. At any given time, at least one of them is reverse-biased (off).

6.3.4.2 Series Array

With batteries in a series array that power a load, when one battery switch opens, the entire string voltage appears across this switch, in the reverse polarity, damaging this switch (see Section 8.3.2.3) (Figure 6.16(a)).

A diode across each battery protects the protector switch (Figure 6.16(b)); during regular operation, current flows through each battery, and the diodes are turned off (Figure 6.16(c). If the protector switch in a battery opens, the diode carries the load current and prevents a reverse voltage (Figure 6.16(d)).

Of course, the diodes lower the load voltage, possibly too much.

6.3.4.3 Series-First Batteries

Both sets of diodes are used in an array of batteries connected in a series-first arrangement (Figure 6.17):

6. Half as much current in each battery, because the charger is sourcing a constant current.

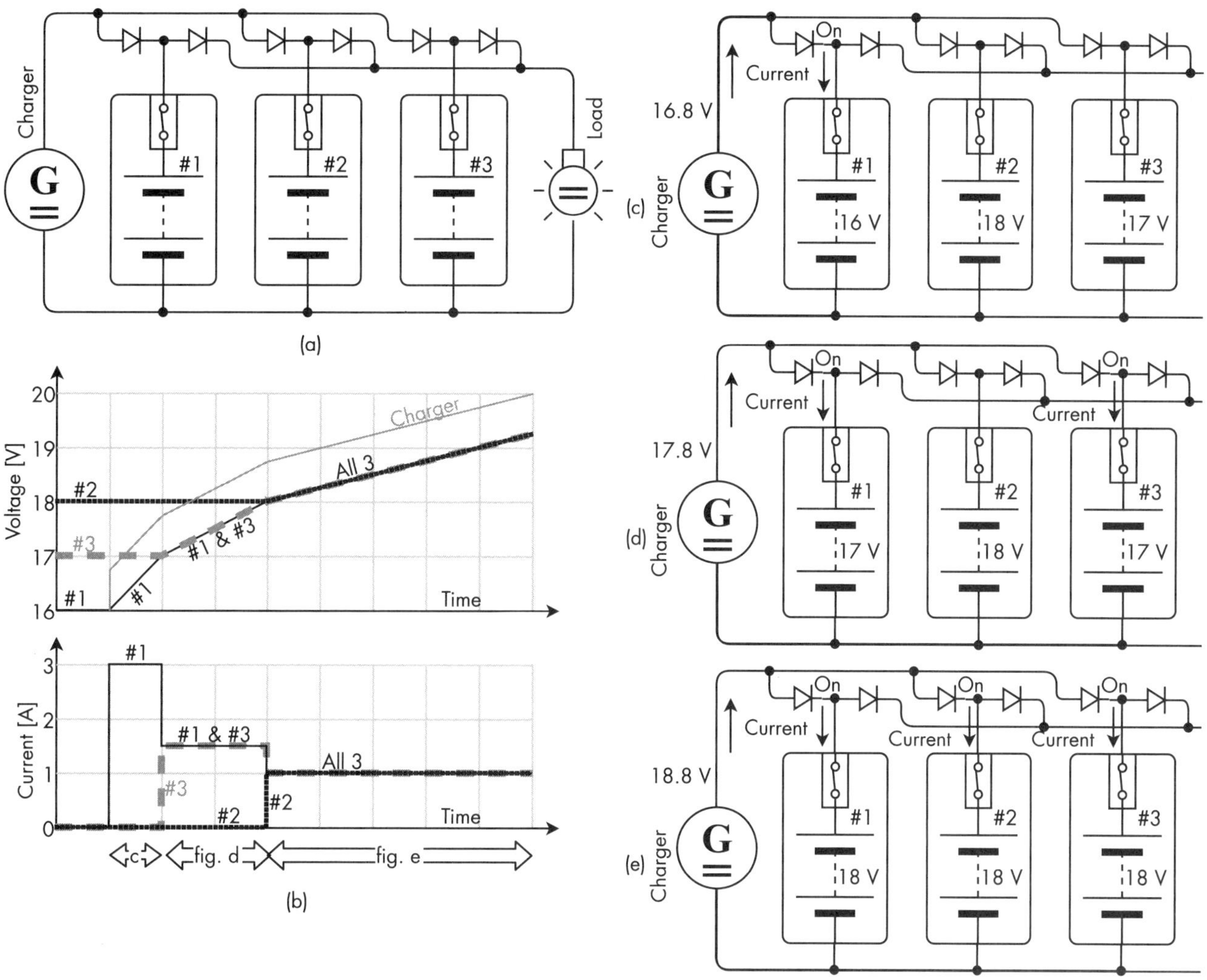

FIGURE 6.15 Batteries connect in parallel through diodes: (a) circuit, (b) plot during charging, (c) charging lowest voltage battery, (d) charging two lowest voltage batteries, and (e) charging all batteries.

- The diodes at the top route the charging and load current while preventing current flow between strings;
- The diode in parallel with a battery protects its protector switch if it's the last one to open.

6.4 GANGED BATTERIES

Ganged[7] batteries are coordinated batteries whose protector switches operate simultaneously. The goal is to keep all of their respective SoC levels the same to avoid the problems associated with using multiple batteries at different voltages SoC:

- *In parallel:* To avoid inrush current when the batteries are connected;

7. Gang in the sense of a band, an organized group. For example, three circuit breakers may be ganged by joining their handles so that they turn on a 3-phase circuit simultaneously.

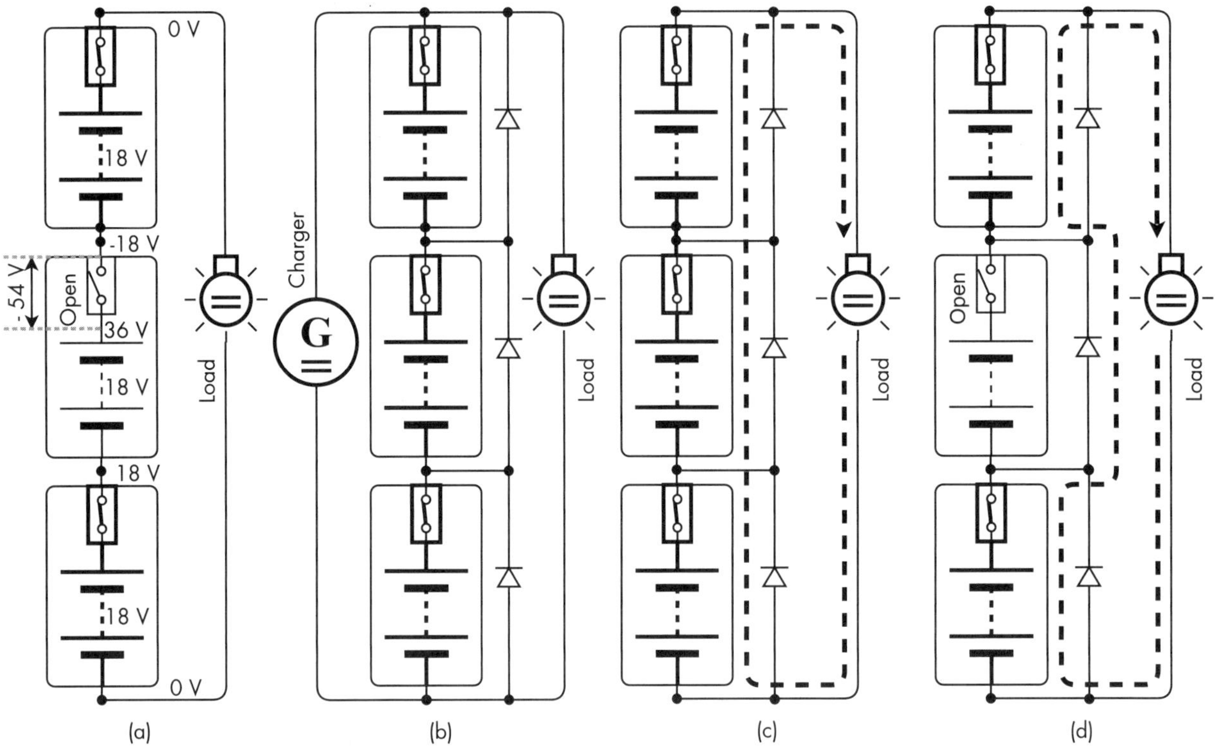

FIGURE 6.16 Series array with protecting diodes: (a) if a battery opens, its switch blows, (b) protector diodes, (c) all batteries are on, and (d) one battery is off.

FIGURE 6.17 Series-first array with routing and protecting diodes.

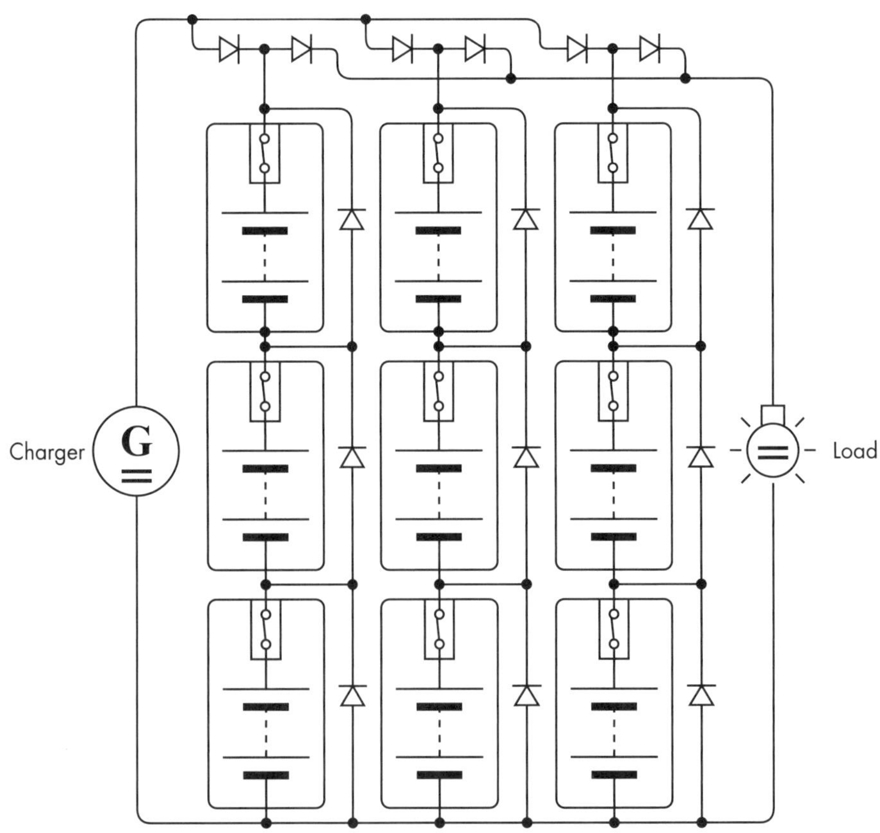

- *In series:* So that the positive and negative voltages of a split battery (see Section 6.5) remain the same.

The state of the protector switches is based on the combined input of each of their BMSs: if any BMS decides that charging should be disabled, then charging is disabled for all the batteries. The same is true for discharging.

An advantage of ganged batteries (compared to a standard array of batteries) is that a standard BMS can be used (there is no need for an array-compatible BMS or an array master). Unlike a standard array, ganged batteries do not provide much redundancy because if one battery needs to shut down, they all shut down.

If at some point and for some reason, a battery must be disabled, it may only be reenabled through a specific procedure performed by a trained technician; this is to prevent inrush current between batteries at different voltages.

6.4.1.1 Applications

Typical applications may use

- *Single-bus:* For example, for an inverger (combo charger/inverter; see Section 1.8.4.2) (Figure 6.18, left column);
- *Dual-bus:* One bus for charging (e.g., for one or more chargers), the other bus for the load (e.g., to an inverter or a motor driver) (Figure 6.18, right column).

They may also be used with

- *In parallel:* To increase capacity (Figure 6.18, top row);

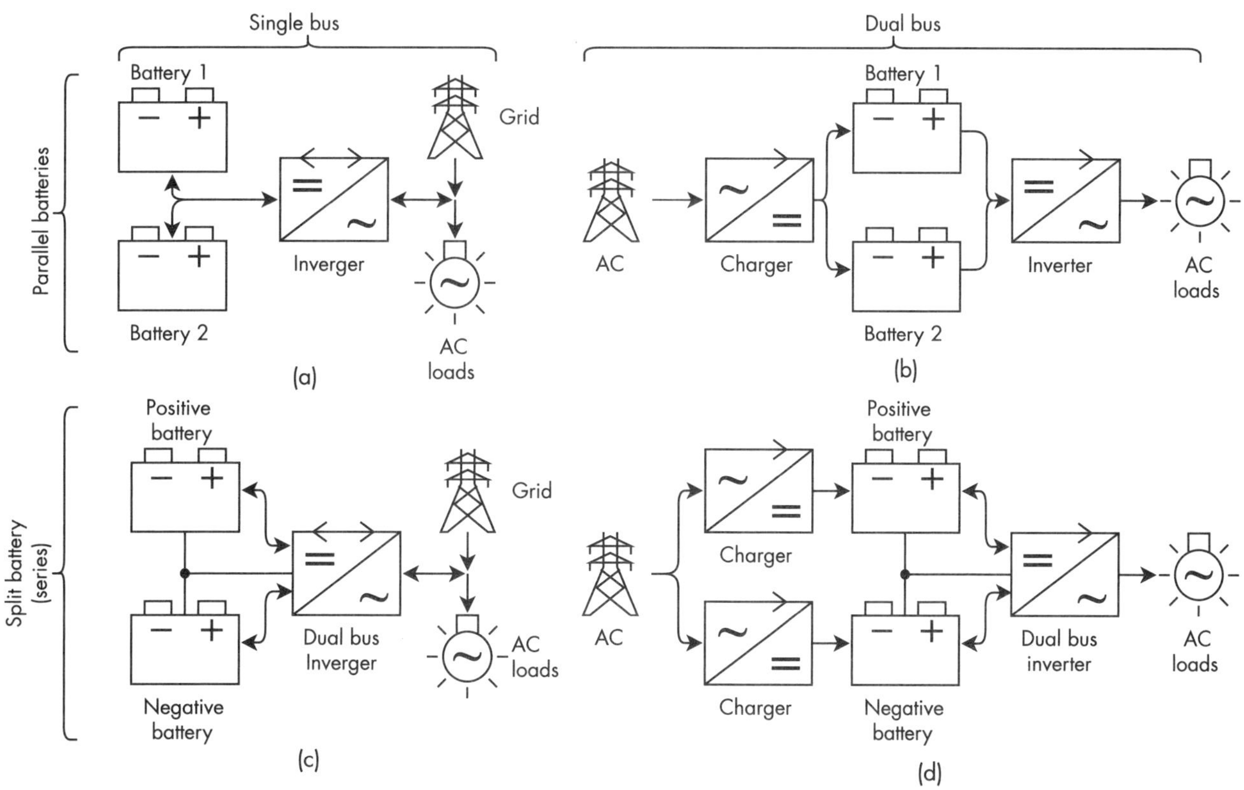

FIGURE 6.18 Applications of split batteries: (a) single-bus, parallel batteries, (b) dual-bus, parallel batteries, (c) single-bus, split battery, and (d) dual-bus, split battery.

- *In series:* A split battery may be used with a motor driver, an inverter, or inverger that requires a bipolar supply (Figure 6.18, bottom row).

6.4.2 Parallel Ganged Batteries, Single-Bus

If a single bus is used for charging and discharging, which is typical when used with an inverger, the ganged batteries try to control charging and discharging by communication to the external system. However, if the external system disobeys, all the batteries disconnect themselves from the bus as a last resort.

6.4.2.1 System Description

Each system includes the following (Figure 6.19):

- 12V supply and ground lines, isolated from the battery voltage and used to power the common devices in the system;
- A *No-fault* line that is normally open, with a pull-up resistor to 12V, grounded by any battery if it detects a fault;[8]
- A *Charge OK* line that is normally open, with a pull-up resistor to 12V, grounded by any battery if it decides that charging is not allowed;
- Similarly, a *Discharge OK* line;
- A *Master on/off* switch, accessible to the user that grounds both the *Charge OK* and *Discharge OK* lines;
- External chargers and loads, which include a *Charge OK* control input and a *Discharge OK* one.

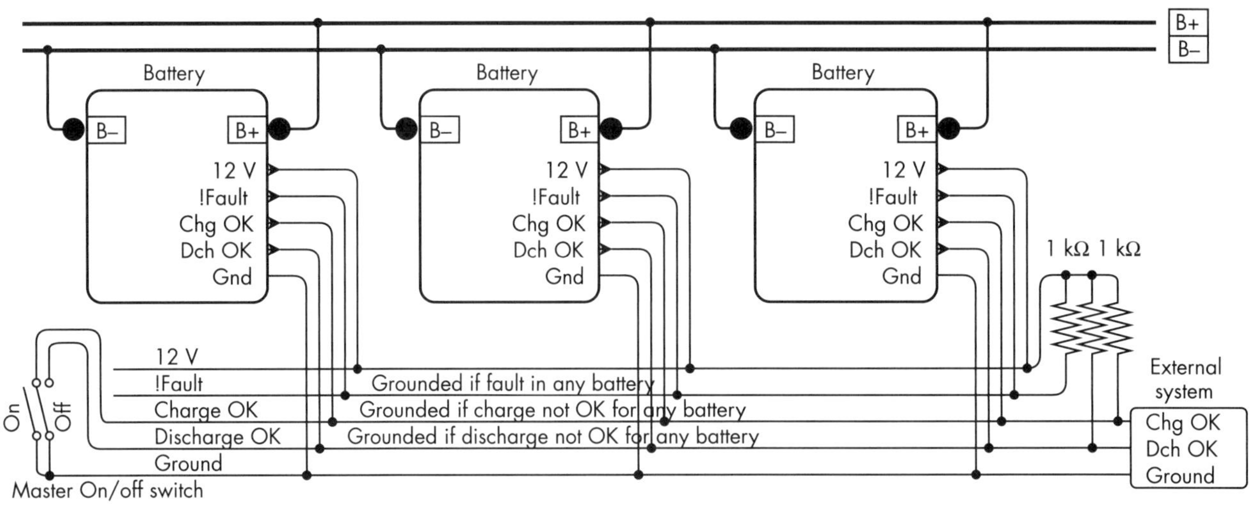

FIGURE 6.19 System circuit for ganged batteries, single-bus.

8. This circuit is called a wired-OR, because it is asserted if the first device asserts it, or the second device does, and so on. A disadvantage is that it is not fail-safe in case of an open wire. A more fail-safe approach would be to add a relay for each battery and then string their normally-closed contacts in series; however, then the circuit would not be fail-safe in case of a short; in any case, the relays could malfunction. In conclusion, making this a truly fail-safe circuit would require a degree of complexity that may in itself introduce more potential failures.

6.4.2.2 System Operation

Normally, the battery operates as follows:

- All the batteries supply the 12V line;
- The *No-fault* line is open because no battery sees a fault, and the *Master on/off* switch is open; this tells all the batteries to connect to the bus;
- If any battery decides that charging is not OK, it grounds the *Charge OK* line to tell the external system to stop charging; otherwise, the pull-up resistor pulls it up to 12V;
- Similarly, if any battery decides that discharging is not OK, it grounds the *Discharge OK* line to tell the external system to stop discharging;
- If the user turns off the battery with the master switch, that grounds both the *Charge OK* and *Discharge OK* lines and the external system stops charging and discharging.

If a battery detects a fault, it grounds the *no-fault* output, and, therefore, the *No-fault* line, which disables all the batteries. For example, if the external system ignores the *charge OK* line and continues to discharge the batteries, as soon as a cell voltage is quite low, its battery issues a fault, and all the batteries open their contactors as a last resort.

An emergency shut-off button (not shown) grounds the *No-fault* line to tell all the batteries to turn off their contactors.

6.4.2.3 Battery Description

This circuit is for a particular BMU,[9] though it can be adapted to many other BMUs.

Each battery includes (Figure 6.20)

- Three switches:
- *Disconnect*:
  - Normally closed push-button;
  - Accessible to the user;
  - Pressed to remove the battery from the bus.
- *Restore*:
  - Normally open push-button;
  - Internal, accessible to a qualified technician (not accessible to the user);
  - Pressed to reenable the battery after making sure it is safe to do so.
- *Disable*:
  - Toggle switch;
  - Internal, accessible to a qualified technician (not accessible to the user);
  - Used to disable the contactor during testing.
- A relay used as a latch to remember that the user removed the battery from the bus;[10]
- An isolated DC-DC converter to power the BMS;

9. Lithiumate BMS.
10. This is a standard relay, not a latching relay.

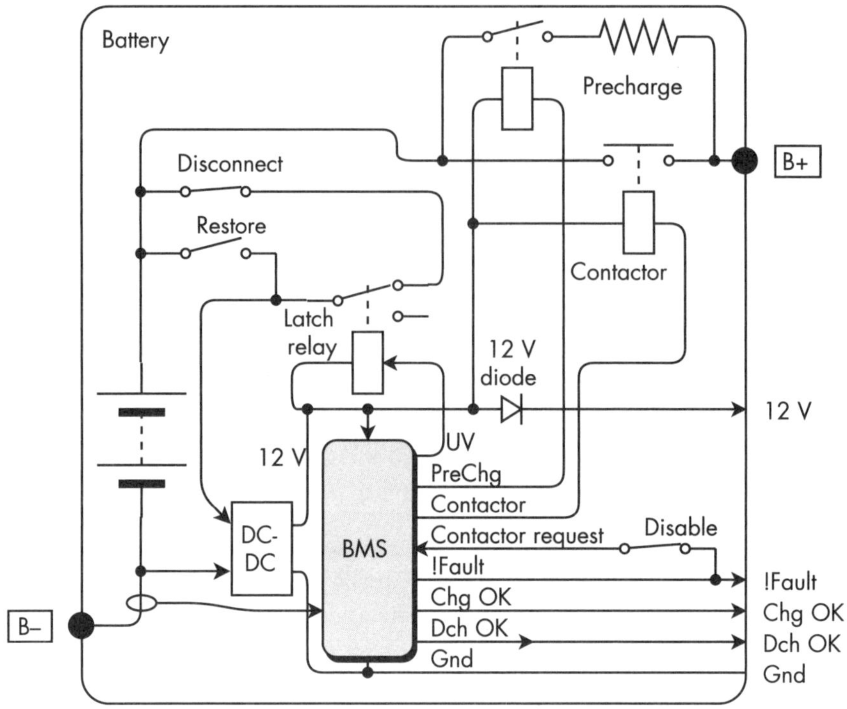

FIGURE 6.20
Battery circuit for ganged batteries, single-bus.

- The BMU, which has:
  - A *No-fault* output that is nor mally open and grounded if the BMS detects a fault;
  - A *Charge OK* output that is normally open and grounded if the BMS decides that charging is not OK;
  - Similarly, a *Discharge OK* output.
  - An *Undervoltage (UV)* output that is normally grounded, but is open if any cell voltage is quite low, lower than the threshold that affects the *Discharge OK* output;
  - A *Contactor* output to drive the main contactor coil by grounding one end of it;
  - A *Precharge* output to drive the precharge relay coil by grounding one end of it;
  - A *Contactor request* input; when high, the BMU first drives the *Precharge* output, then the *contactor* output, and when low, the BMU turns them off.
- A protector switch:
  - A precharge relay and resistor to charge a capacitive load on the bus;
  - A main contactor to connect the battery to the bus.
- A diode to prevent current from flowing from the external 12V bus back into the battery.

6.4.2.4 *Operation*

Normally, this battery operates as follows:

- The latch relay is on because its coil is powered by the DC-DC converter through the UV line that the BMS keeps grounded;
- The cells power the DC-DC converter through the *Disconnect* switch and the contacts of the latch relay.;
- The DC-DC converter powers the BMS;

- The BMS leaves the *No-fault* output open because there is no fault; no other battery grounds this line either, so the line is pulled up to 12V by the external pull-up resistor;
- This line is connected also to the *Contactor request* input through the *Disable* toggle switch (which is on);
- The BMU drives the contactor, and the battery is connected to the bus.

In case of a fault in this or any other battery, the *No-fault* line is grounded. This removes the contactor request, and the BMU turns off the contactor.

When all faults are cleared, the *No-fault* line is pulled back up to 12V. This applies a contactor request, and the BMU starts the precharge sequence, ending with the contactor turned on.

If any cell voltage becomes too low, the BMS opens the *UV* output, removing power to the latch relay, which opens. In turn, this removes power to the DC-DC converter, and therefore to the BMS. The battery is completely turned off and won't be able to turn back on by itself. This sequence also occurs if the user presses the *Disconnect* button.

To restore a battery, a qualified technician goes through this procedure:

- Open the *Disable* switch, to ensure that the battery won't connect to the bus;
- Check if that battery is OK, and do any necessary repairs;
- Press the *Restore* push-button to re-enable the battery;
- Charge or discharge the battery to reach the same voltage as the other batteries;
- Close the *Disable* switch, to allow the battery to connect to the bus.

6.4.3 Parallel Ganged Batteries, Dual-Bus

If the system includes a charger and a separate load, neither of which provides a way to control the current, dual-port batteries may be ganged. The system uses two buses, one for charging and one for discharging.

6.4.3.1 System Description

Each system includes the following (Figure 6.21):

- A *Charge OK* line that is normally open; any battery grounds it if it decides that charging is not allowed;
- Similarly, a *Discharge OK* line;
- A *Master* switch, accessible to the user, which grounds both the *Charge OK* and *Discharge OK* line;
- Charging sources on the *Charge* bus;
- Loads on the *Load* bus.

6.4.3.2 System Operation

Normally, the *Charge OK* line is open because all batteries can charge and the *Master* switch is open. This tells all the batteries to connect to the *Charge* bus. Similarly, the *Discharge OK* line is open because all batteries can discharge and the *Master* switch is open. This tells all the batteries to connect to the *Discharge* bus.

If any battery decides that charging is not allowed, it grounds the *Charge OK* line; all the batteries disconnect from the *Charge* bus yet remain connected to the

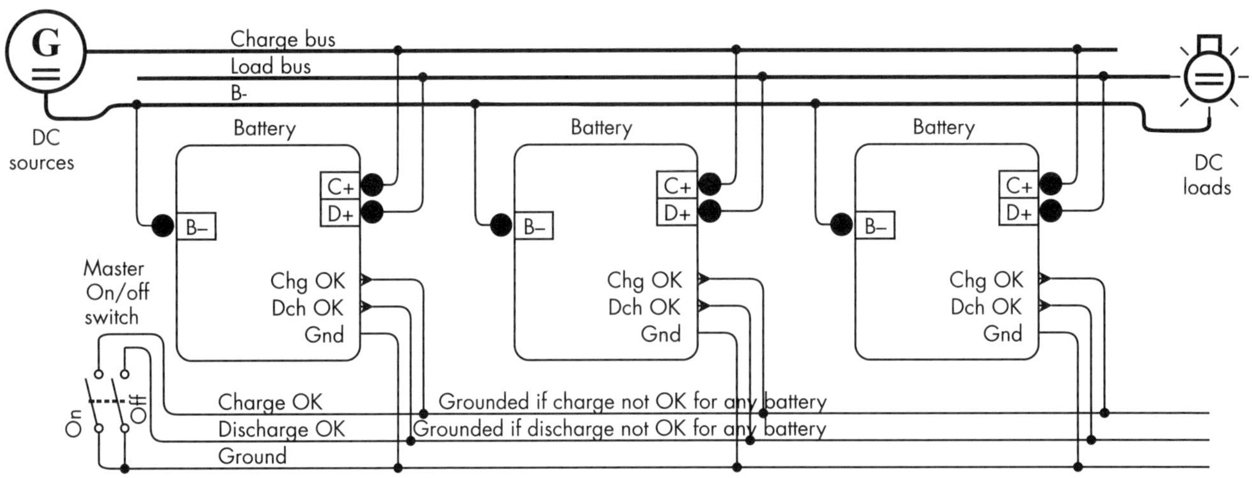

FIGURE 6.21 System circuit for ganged batteries, dual-bus.

Discharge bus. Similarly, if any battery decides that discharging is not allowed, the battery grounds the *Discharge OK* line; all the batteries disconnect from the *Load* bus yet remain connected to the *Charge* bus.

If a battery detects a fault, it grounds both the *Charge OK* and the *Discharge OK* lines; all the batteries disconnect from both buses.

6.4.3.3 Battery Description

This battery circuit works for most BMUs. There is no precharge in this implementation.

Each battery includes (Figure 6.22)

- The same circuit as before, with a *Disconnect* switch, a *Disable* switch, a *Restore* switch, a relay used as a latch, an isolated DC-DC converter to power the BMS;

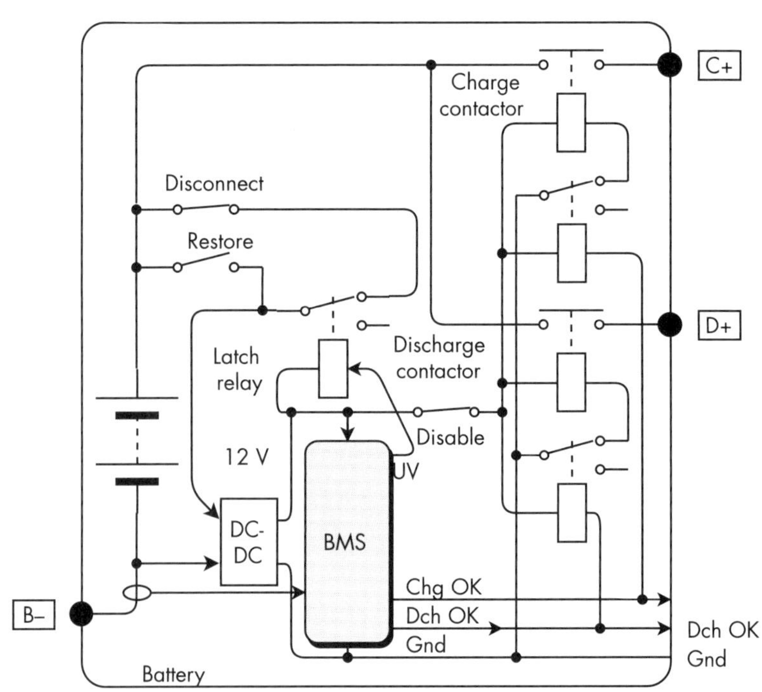

FIGURE 6.22
Battery circuit for ganged batteries, dual-bus.

- The BMU, which needs fewer lines—only the *Charge OK* and *Discharge OK* lines;
- Two normally closed relays to invert the logic of the *Charge OK* and *Discharge OK* lines;
- Two contactors, one to control charging, the other one for discharging.

6.4.3.4 Operation

This circuit works the same as the previous battery circuit, with the following exceptions:

- If this battery or any other battery wants to disable charging, it grounds the *Charge OK* line; this turns on the charging relay, which turns off the charge contactor;[11]
- Similarly, if this battery or any other battery wants to disable discharging, it grounds the *Discharge OK* line; this turns on the discharging relay, which turns off the discharge contactor;
- The BMU is configured to disable both charging and discharging in case of a fault, which disconnects this and all other batteries from both buses.

The process of restoring a battery is the same as for the previous solution.

6.4.4 Series Ganged Batteries, Single-Bus

If batteries in series must operate identically, they may be ganged. For example, a split battery (see Section 6.5) should be ganged so that its load doesn't experience being powered just on one side, and so that the two batteries have mostly the same SoC. If either BMS needs to shut down its battery, both batteries are shut down.

For a single-bus solution (Figure 6.23(a)), all the *Not charge OK* and *Not discharge OK* outputs are connected in parallel. Either BMS may ground this line if it decides that either charging or discharging is not allowed. A single relay for the entire system inverts the polarity and drives the contactors in both batteries. The contactors are turned on if both BMSs decide that both charging and discharging are allowed.

6.4.5 Series Ganged Batteries, Dual-Bus

For a dual-bus solution (Figure 6.23(b)), on the left side, the *Not charge OK* outputs are connected in parallel. Either BMS may ground this line if it decides that either charging or discharging is not allowed. A relay inverts the polarity and drives the *charge* contactors in both batteries. Therefore, the contactors are turned on only if both BMS decide that charging is allowed. Similarly, the same happens for discharging on the right side.

6.5 SPLIT BATTERIES

Some battery-connected products require a dual supply voltage. For example:

11. We could have used a normally-closed contactor, but that would not be fail-safe. This circuit is fail-safe because, if there's no power, the contactors are off (open).

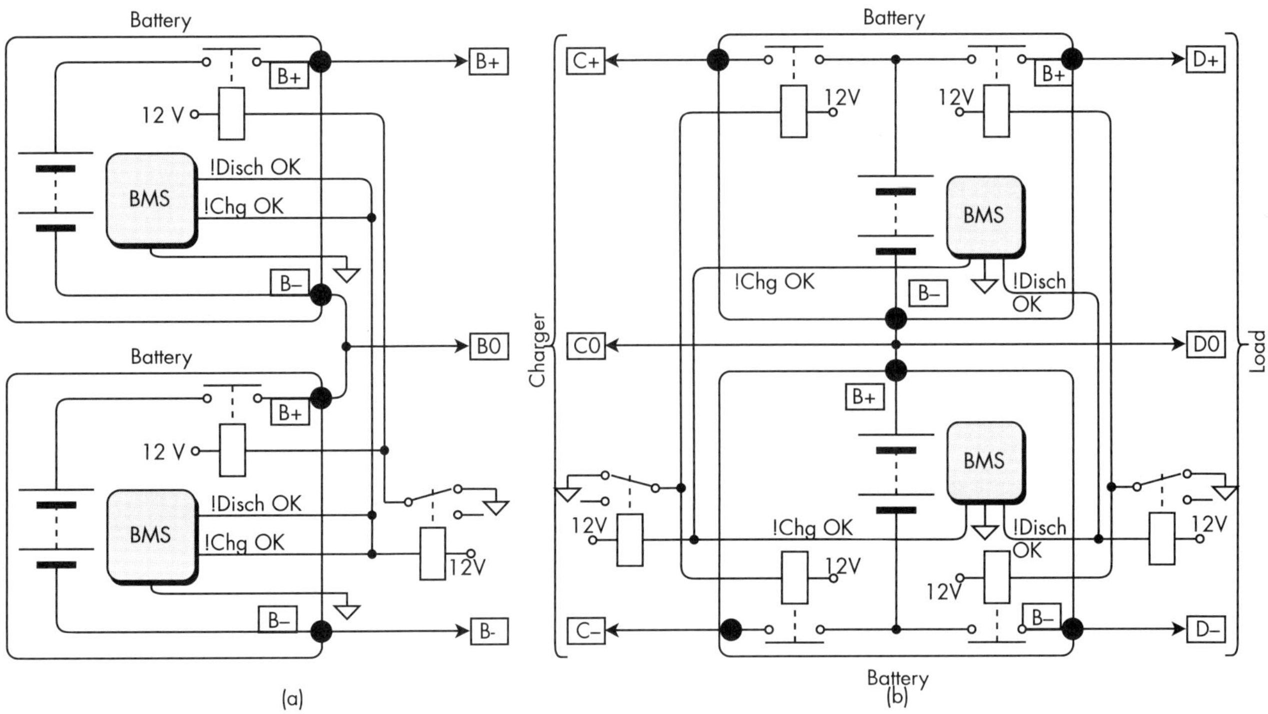

FIGURE 6.23 Ganged batteries in series: (a) single-port, and (b) dual-port.

- High power, high voltage, grid-connected invergers that generate 240 Vac directly; they require +450V and -450V DC (Figure 6.24(a)) (see Volume 2, Section 4.2.1);
- Two 350V chargers in series to charge a 700V traction batteries for industrial vehicles[12] (Figure 6.24(b)).

Adding a tap wire in the middle of a battery results in two batteries, each with half the voltage. This *split battery* (or *center-tap battery*), is not one, but two batteries, which happen to share a common terminal. Each battery has its own current, protector switch and SoC; each one requires its own BMS.

This is also a ganged battery (see Section 6.4.4). The protector switches of the two halves are synchronized: if either BMS decides to shut down the current in its half, the contactors on both halves are turned off. The batteries are turned on only if both BMSs agree that it's OK to do so.

Any imbalance in the charging or discharging currents of the two halves results in a slow drift in the SoC of the two batteries; monitoring just one half tells you little about the other half. Therefore, each half requires its own current sensor and BMS.

6.5.1 Dual-Supply Inverger

While the battery is in use, the current in one half is slightly different from the current in the other half due to small differences in the power devices connected to the battery (see Volume 2, Section 4.2.1). After some time, the SoC of one battery half will be significantly different from the SoC in the other half (Figure 6.25(a)).

12. 350V chargers are more available than 700V chargers. Some chargers do not like to be connected in series without a battery connection between them. (See Volume 2, Section 3.3.2)

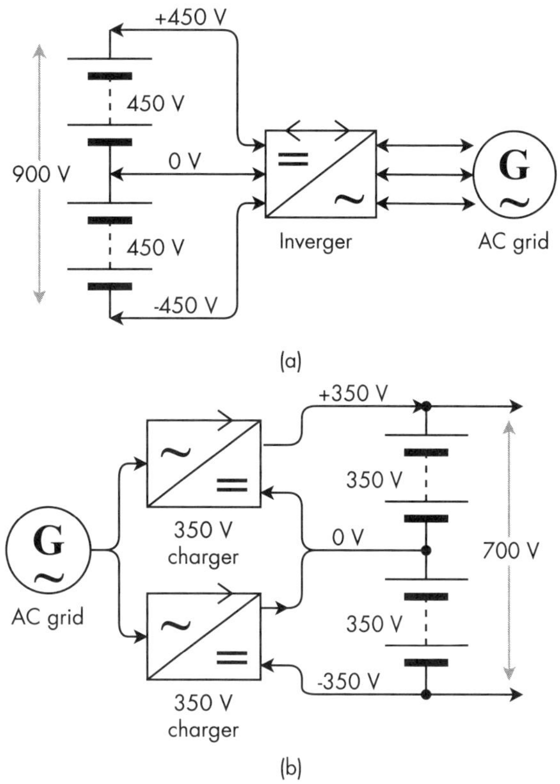

FIGURE 6.24
Split battery applications: (a) dual-supply inverger, and (b) chargers in series.

The inverger is unable to bring both halves to the same SoC because it doesn't have a CV stage (as chargers do), and it has no way of sending more current out of one supply line than the other one. Each battery requires its own charger to bring both halves to the same SoC (Figure 6.25(b)).

6.5.2 Case Studies

The Ford Escape hybrid uses a 350V traction battery. In 2007, I designed a PHEV conversion for it. I designed a Vehicle to Grid (V2G) (see Volume 2, Section 3.9.1.8) inverger powered by a bipolar supply to generate 120Vac and back-feed the grid (see Volume 2, Section 4.2.5). I split the pack in two to get +175V and -175V to power the inverger, but I kept a single BMS for both halves. That was before I knew better.

A customer split a stationary battery pack to work with a grid-connected Princeton Power System inverger. Initially, the battery used a single BMS, until the customer wondered why only one half was always balancing. It was because one half of the inverger was less efficient than the other half, so the currents in the two battery halves were different, and the SoC in one half-battery was higher than the SoC of the other half.

6.5.3 Parallel Charging, Series Discharging

It is possible to charge a high-voltage battery with a single charger whose voltage is half the battery voltage: the battery is split into two halves, which are connected in parallel during charging and in series during discharging. For example, a 700V pack can be split into two 350V half-packs. It is first charged with a 350V charger, and then it is used with a 700V motor driver. We'll discuss this solution when talking about industrial vehicles (see Volume 2, Section 3.3.2).

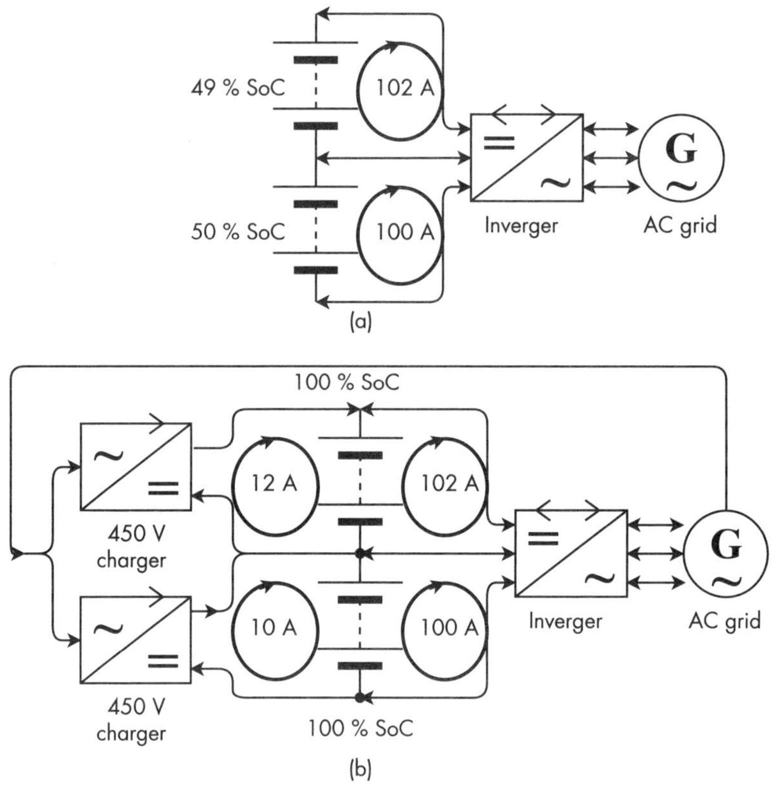

FIGURE 6.25
Imbalance between two halves of a split battery: (a) source of imbalance, and (b) rebalancing with chargers.

6.5.4 Distributed Charging, Balance Charger

Typically, a series string is charged with a bulk charger (Figure 6.26(a)): a high-voltage charger outputs a single current, which flows into the most positive cell, out of that cell into the next one, and so forth, finally coming out of the most negative cell and back to the charger. Each cell sees the same current. When one cell is full, the charger has no way to charge the other cells. Instead, the BMS must remove some charge from the most charged cell to clear some space so that the charger can continue charging (see Section 3.2.9).

An alternative is to use N chargers, for N cells (Figure 6.26(b)). That way, each cell can be charged fully, regardless of the SoC of the other cells in the string. In the end, the string will be top-balanced.

This is an example of parallel charging and series discharging brought to the extreme: instead of splitting a battery into two halves, we're splitting it into many small batteries, each one with a single cell.

Each charger could be in its box, or all the chargers could be in one box. Note that the current in the tap wires between two cells could flow in either direction depending on whether the current is higher in the upper cell or the lower cell. Ideally, the current should be zero because the two cells see exactly the same current, which cancels out in the tap wire. In reality, there is some current in this wire:

- In the CC stage, depending on which charger has a higher CC setting;
- In the CV stage, depending on which cell is more charged.

A balance charger is more common (Figure 6.26(c)). This product includes a high-power bulk charger plus either N small chargers or N bypass balancers, one for each cell. The charger has two connectors:

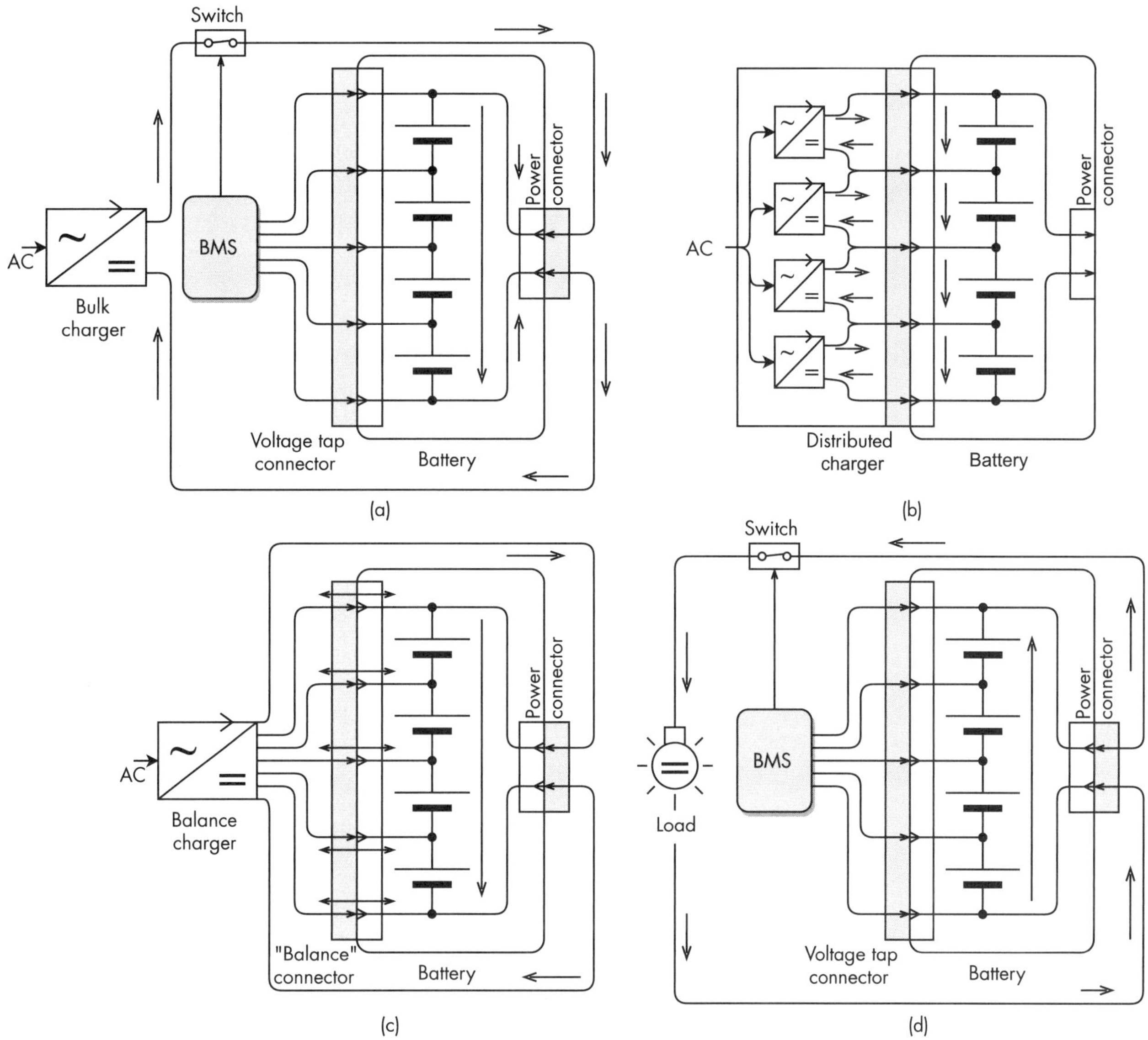

FIGURE 6.26 Spli charging: (a) bulk charger, (b) distributed charger, (c) balance charger, and (d) discharging.

- A high-current connector that is connected to the battery's power connector;
- A balance connector, with $N + 1$ lines that is connected to the battery's balance connector.[13]

The balance connector carries a small additional current to or from the individual cells to ensure that they are all equally charged. At the end of charge, the string will be top-balanced.

In all these cases, the BMS is confused because it doesn't know the charging current. Therefore, its SoC evaluation is scrambled. As a matter of fact, in some cases, the BMS is not even connected to the string while charging! The BMS will be connected later, when the battery is in use (Figure 6.26(d)). Most likely, this BMS

13. Which is another name for the connector at the end of the cell voltage sense harness.

doesn't include a balancing function, since the charger already top-balanced the battery.

6.6 LI-ION AND LEAD-ACID

It is becoming increasingly effective to replace lead-acid batteries with Li-ion ones, particularly in low-voltage, stationary applications. The transition is not seamless due to the requirements of Li-ion. Some may choose to do a complete replacement, while others may prefer to integrate Li-ion with the existing lead-acid batteries slowly. Let's look at both approaches.

6.6.1 Lead-Acid Replacement

Li-ion batteries are not a direct drop-in replacement for lead-acid batteries, as they are somewhat different (see Table 6.2).

As the high-power electronic devices are designed to use lead-acid batteries, they are somewhat incompatible when used with Li-ion; problems with these devices include

- They offer no way to control and stop charging or discharging;
- They may require the presence of a battery to operate;
- They use the same battery port for charging and discharging;
- The charger operates autonomously following a profile designed for lead-acid;
- They do not provide a low-voltage power supply for the BMS.

Let's examine each of these.

6.6.1.1 No Way to Control and Stop Charging or Discharging

Solar charge controllers and invergers may not include any provisions to stop charging or discharging because they are designed to work with lead-acid batteries (see Volume 2, Section 2.11).

Table 6.2
Comparison of Lead-Acid and Li-Ion

Lead-Acid Batteries	**Li-Ion Battery**
Managed by monitoring the battery pack voltage	Managed by monitoring each individual cell voltage
Survives being discharged down to 0V	Damaged if discharged below 2~2.5V/cell
May be trickle charged	Trickle charging overcharges and may start a fire
Benign if abused	Risk of fire if abused
Does not use a protection switch	Requires a protection switch
May be charged below freezing at 0.3°C or less	In general, cannot be charged below 0°C
May be float charged indefinitely	Damaged by float charging; once full, the charger must be disconnected; or, must be float charged at a lower voltage
Cell voltage has -3 mV/°C coefficient	Negligible temperature coefficient

6.6.1.2 Requires the Presence of a Battery to Operate

The control electronics in solar charge controllers and invergers receive power from the battery; the sun and the grid cannot power them. If a Li-ion battery opens its protector switch, these devices cannot recover (see Volume 2, Section 2.2.3)..

6.6.1.3 Same Port for Charging and Discharging

A lead-acid battery is a single-port device: the same terminals are used for charging as for discharging.

A Li-ion battery may have two separate ports, one for charging, one for discharging (see Section 5.12.2.3). Only a single-port Li-ion battery may replace a lead-acid battery directly, preferably, one with two switches.

6.6.1.4 The Charger Operates Autonomously Following a Profile Designed for Lead-Acid

Advanced lead-acid chargers go through a series of charging phases on their own, optimized for lead-acid batteries. But the BMS doesn't appreciate the charger deciding on its own what voltage and current limits to use. The BMS's point of view is: "If anyone's gonna control charging, that's me, doggone it!"

The solution is to use a charger that allows remote control and allows the BMS to control the charger's voltage and current limits in real time.

6.6.1.5 Low-Voltage Power Supply for the BMS Electronics

A lead-acid battery doesn't have a BMS, so it doesn't need 12V power. Therefore, none of the power devices in a typical low-voltage stationary battery provide such a supply. Yet, a Li-ion battery does include a BMS and does require a power supply. A BMS for a low-voltage battery may include a power supply. Others won't. Some partial solutions are possible (see Volume 2, Section 2.12.5.1).

I have seen two types of applications that use both lead-acid and Li-ion:

1. Parallel: Both batteries are connected to the same DC bus;
2. Sequential: The two batteries are used in turn.

Let's look at each type individually.

6.6.2 Parallel Hybrid LA/Li-ion Systems

Faced with replacing aging lead-acid batteries, the manager of a large stationary battery may choose to transition slowly to Li-ion. This entails moving all the good lead-acid batteries to one side of the room and installing Li-ion batteries on the other side of the room.

Connecting lead-acid and Li-ion batteries to the same DC bus has mixed considerations:

- *Advantage:* When all the Li-ion batteries disconnect from the bus for protection, the lead-acid batteries are still there to keep powering the bus;
- *Disadvantage:* It is tricky to reconnect a Li-ion battery to a bus that already has a battery connected to it.

Matching the voltage ranges of the two chemistries can be tricky, especially at lower voltages.

The total capacity is lower than the sum of the capacities of both batteries; that is because one chemistry limits charging to a lower voltage than the other chemistry could handle. Similarly, one chemistry limits discharging to a higher voltage than the other chemistry could handle.

Tables 6.3 and 6.4 offer some recommendations for batteries used in float applications,[14] such as engine starter batteries (see Volume 2, Section 2.9).

6.6.2.1 LFP Cells

LFP cells (Table 6.3) are ideal for lower bus voltages because an integer number of cells in series best matches the voltage of a lead-acid battery. The Li-ion cell voltage ranges from 2.75 to 3.45V.

The lead-acid batteries set the top limit. The Li-ion cells top off at 3.45V instead of 3.6V, yet that results in noticeably longer cell life with only a slight loss of available capacity.

The Li-ion cells set the bottom limit. Typically, the low-voltage cut-out for12V lead-acid batteries is set at about 10V, which is appropriate for a deep discharge battery. Selecting a higher voltage (11V for a 12V battery) reduces the available charge, yet increases the life of the lead-acid batteries. The life of the Li-ion cells increases slightly as well.

6.6.2.2 NMC Cells

Table 6.4 is for NMC cells. It does not list 12V and 24V because there is no integer number of cells in series that can match the voltage of a lead-acid battery.

In one case, the lead-acid batteries set the top limit; in the other, the Li-ion cells do. However, only narrowly—not much capacity is wasted. In both cases, the lead-acid batteries set the low limit, leaving the Li-ion cells still with some remaining charge.

Table 6.3
LiFePO4 and Lead-Acid Batteries in Parallel

Nominal Voltage	LFP Cells in Series	LA Batteries in Series	Voltage Range [V]
12V	4	1	11.0~13.8
24V	8	2	22.0~27.6
36V	12	3	33.0 to 41.4
48 V	16	4	44.0 to 55.2

Table 6.4
NMC and Lead-Acid Batteries in Parallel

Nominal Voltage	Li-Ion Cells in Series	Lead-Acid Batteries in Series	Minimum Voltage [V]	Limited by	Maximum Voltage [V]	Limited by
36 V	10	3	30	LA	41.4	LA
48 V	13	4	40	LA	54.6	Li-ion

14. In which 12V lead-acid batteries are charged at 13.5V float voltage rather than the 14.5V fast charge voltage.

6.6.2.3 Dangers

Mixing battery chemistries carelessly in a parallel array is quite risky: fires have been started when the current was allowed to flow unchecked between batteries of different chemistries, exceeding the safe operating area of some of the batteries. For example, a 4-cell LiFePO4 battery and a 12V lead-acid battery are connected in parallel, balanced with each other, and at rest at 13.2V. Later, the temperature increases rapidly. This affects the lead-acid battery, whose voltage has a negative temperature coefficient—as the temperature goes up, its voltage drops—but not the Li-ion battery. As this happens, current starts flowing from the Li-ion battery to the lead-acid battery. If the internal resistance of both batteries is low, the current can be quite high and could be damaging.

Normally, this is not a problem, as the Li-ion battery would shut down to protect its cells. This *is* a problem if a string of Li-ion cells is connected directly in parallel with lead-acid batteries with no way of disconnecting the two.

6.6.2.4 Load Sharing

Generally speaking, Li-ion cells have a lower maximum power time than lead-acid batteries. That is, for a given size, they have a lower internal resistance (Figure 6.27(a)).

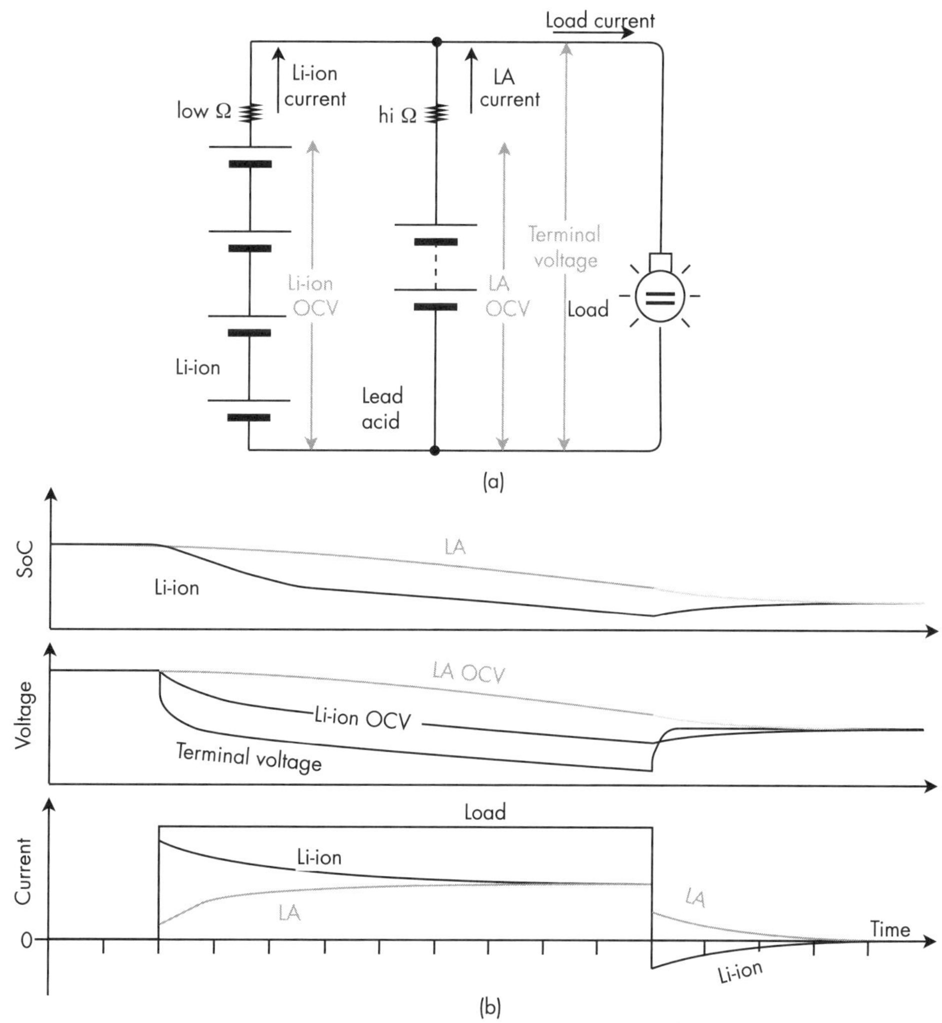

FIGURE 6.27
Load sharing in a parallel hybrid circuit: (a) equivalent circuit, and (b) plot.

Consequently, they supply the majority of the current at the beginning of a discharge or any time the current increases significantly (Figure 6.27(b)).

Under relatively constant current discharge (typical of backup power applications), eventually the lead-acid batteries catch up and start providing more of their share of the current. In the long run, the SoC and the OCV of both batteries decrease at the same rate. As the lead-acid battery provided less current initially, its SoC and OCV are higher. Once the load current stops, the batteries can equalize—current flows from the lead-acid battery, whose OCV and SoC are higher, to the Li-ion battery.

Therefore, in applications where the load current is relatively constant, the Li-ion cells work a bit harder than the lead-acid batteries.

In applications where the load current is changing continuously (e.g., buffer batteries), the current is not shared equally, and the Li-ion cells do most of the work. You may find that the lead-acid batteries are coming along just for the ride and do not contribute significantly to the operation of the system. If the lead-acid batteries are mostly dead weight, why have a hybrid system at all?

Even at constant discharge current, the load may not be shared equally due to a mismatch in the OCV versus SoC curves of the two chemistries. This is particularly an issue with LFP cells (Figure 6.28).

During a full discharge, the portion of the energy delivered by each battery is unequal because the voltage of LFP cells is quite flat over the discharge cycle:

- Initially, the voltage remains relatively constant, and most of the energy comes from the LFP cells;
- About halfway through the discharge cycle, when the LFP cells are almost empty and their voltage starts dropping, the lead-acid cells are finally able to discharge at a high rate, becoming the predominant power source.

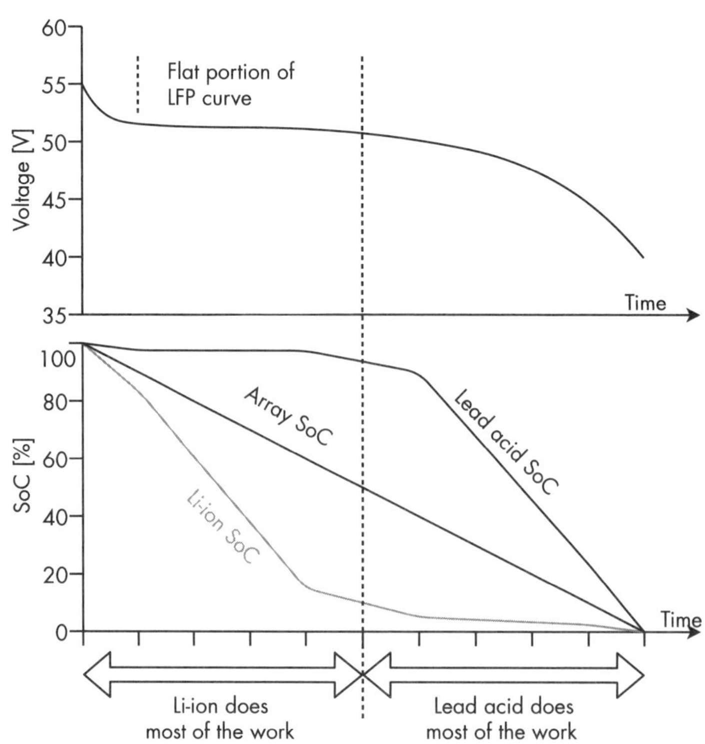

FIGURE 6.28
Proportion of work from each battery during a discharge cycle.

The voltage of LCO cells is not as flat as LFP cells. Instead, it drops with SoC in a way that is comparable to the drop for lead-acid batteries. Therefore, at constant current, LCO cells and lead-acid batteries share the load more equally.

6.6.3 Sequential Hybrid LA/Li-Ion Systems

The designer of a battery to supplement an unreliable power grid may want to leverage the lower cost of lead-acid batteries and the higher endurance of Li-ion batteries. A system may combine a small Li-ion battery used at the beginning of a power outage plus a larger lead-acid battery that is used once the Li-ion battery is depleted (Figure 6.29).

The Li-ion battery handles the majority of outages—the short duration ones—on its own. It can do so without much loss of cycle life. Li-ion batteries are expensive, yes, but the cost of a small battery is not excessive.

Those few times when power is not restored quickly, once the Li-ion battery is empty, the system switches to the much larger lead-acid battery. Lead-acid cannot handle too many discharge cycles, but this battery is not used often because the Li-ion battery handles the more frequent, short outages.

When power returns, the Li-ion battery is recharged first so that it will be ready for a new power outage. The lead-acid battery is recharged afterward.

The lead-acid battery and the Li-ion battery are never connected directly in parallel, so this application does not suffer from the risks and disadvantages of directly paralleled batteries.

There is a degree of complexity in managing the two batteries independently and seamlessly switching between them without interrupting power to the loads. The transition between Li-ion and lead-acid is critical: an overlap would result in a high balancing current flowing from the full lead-acid battery to the empty Li-ion battery; a gap would remove power to the loads. Performing a seamless transition requires careful design of both the electronic topology and the software algorithm.

Here I present two solutions:

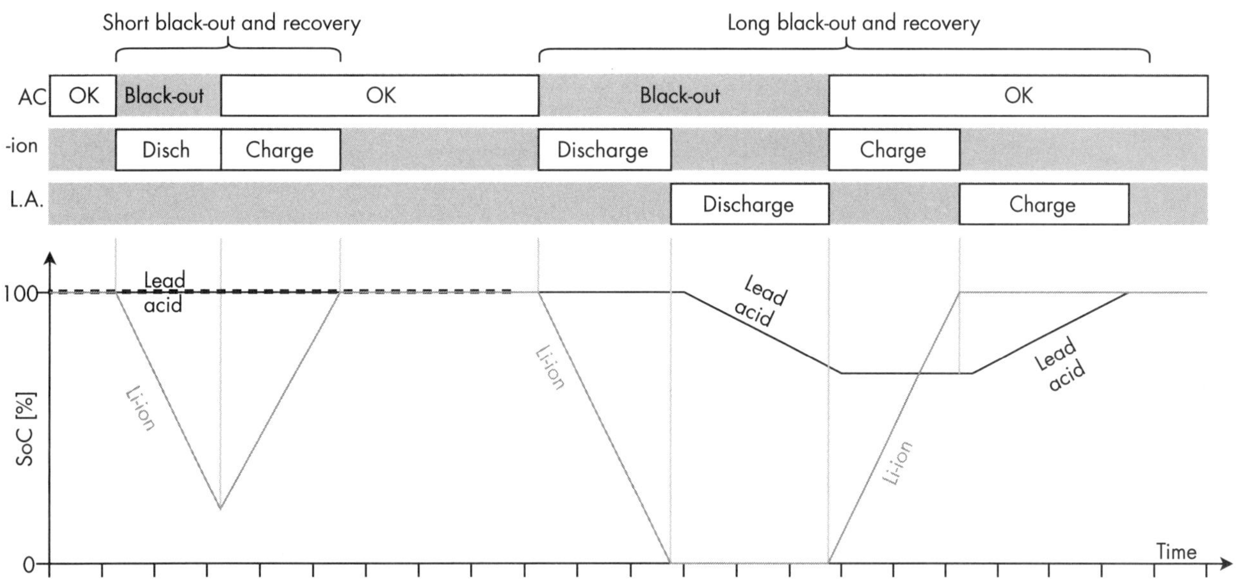

FIGURE 6.29 Operation of a sequential hybrid battery.

- The two batteries use the same DC bus;
- The lead-acid battery is isolated from the DC bus.

6.6.3.1 *Single-Bus Solution*

In the single-bus solution, the two batteries may be connected to the DC bus through contactors (Figure 6.30).

In case of a prolonged blackout, the Li-ion battery is used first, followed by the lead-acid battery. The following numbers match the numbers in the plot:

1. Initially, just the lead-acid battery is connected to the DC bus; the charger powers the load and float-charges the lead-acid battery.
2. At the start of a blackout, the lead-acid battery is disconnected, and the Li-ion battery is connected. As there is an overlap, the load remains powered; during this overlap, not much current flows between the batteries because they are at about the same voltage.
3. The Li-ion battery powers the load.
4. When the Li-ion battery is empty, it is disconnected and the lead-acid battery is connected. There is an overlap, so that the load remains powered; during this overlap, current flows from the lead-acid battery (which is still full) to the Li-ion battery. The current is not excessive because the Li-ion battery is discharged and has a high resistance; in any case, the overlap lasts only about 100 ms.
5. The lead-acid battery powers the load.

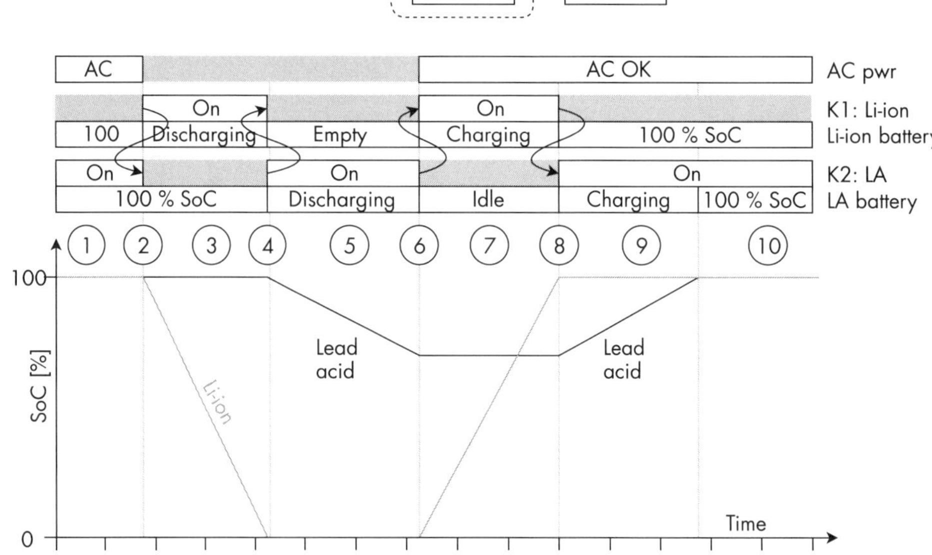

FIGURE 6.30
Sequential Li-ion/lead-acid on a single bus.

6. When the AC power returns, the lead-acid battery is replaced by the Li-ion battery. This time there is no overlap because the load is powered directly by the charger,
7. The charger powers the load and recharges the Li-ion battery (slowly because it is also powering the load).
8. Once the Li-ion battery is full, the Li-ion battery is replaced by the lead-acid battery; again, there is no overlap because the load is powered directly by the charger.
9. The charger charges the lead-acid battery.
10. Once the lead-acid battery is full as well, it remains connected, so the charger can float-charge it.

Once a week, the Li-ion battery is topped off.

6.6.3.2 Isolated Lead-Acid Battery

The *isolated lead-acid battery* topology (Figure 6.31) uses a bidirectional DC-DC converter between the lead-acid battery and the DC bus.

There is no issue of overlap because both batteries can be in use at the same time; no issues arise when power returns. However, the cost of the bidirectional DC-DC converter may be a factor.

Because the Li-ion battery is connected permanently to the charger, the DC bus is at a lower voltage to minimize degradation of the Li-ion cells.

The system operates as follows:

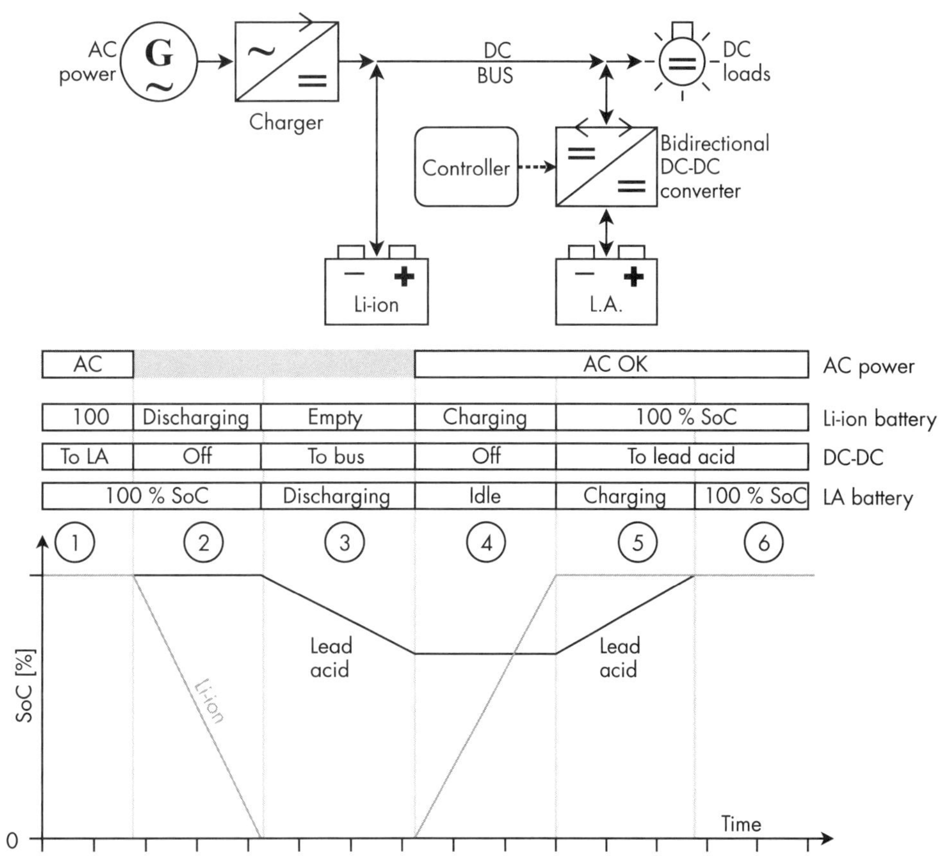

FIGURE 6.31 Sequential Li-ion/ lead-acid with isolated lead-acid battery.

1. When AC power is present, the Li-ion battery powers the load; the DC-DC converter float-charges the lead-acid battery;
2. When the AC power goes away, the Li-ion battery continues to power the DC bus;
3. When the Li-ion battery is fully discharged, the DC-DC converter powers the DC bus from the lead-acid battery, maintaining the previous voltage, so as not to charge the Li-ion battery;
4. When the AC power returns, the charger charges the Li-ion battery, which in turn powers the DC bus; the DC-DC converter is off;
5. When the Li-ion battery is full, the DC-DC converter charges the lead-acid battery from the DC bus;
6. When the lead-acid battery is also full, the DC-DC converter float-charges the lead-acid battery..

CHAPTER 7

PRODUCTION AND DEPLOYMENT

7.1 INTRODUCTION

This chapter covers the steps required to produce and deploy a battery.

7.1.1 Tidbits

Some interesting items in this chapter include:

- Howard Stern can damage your battery (7.2.1);
- Your wedding band can cost you a finger (7.2.2);
- You have 10 seconds to defenestrate[1] a smoking battery (7.2.3);
- Keep one hand in your pocket (7.2.2);
- If you connect cells in parallel one after the other, the last one may not be too happy (7.5.1);
- Cells should be balanced even before a series string is built (7.4.3);
- Test battery isolation as you build it (7.7.1).

7.1.2 Orientation

This chapter starts with a discussion of a safe work environment.

It then goes through the battery manufacturing steps: preparation, battery assembly, balancing, testing, configuration, and functional testing. Finally, it discusses deployment.

7.2 SAFETY

Li-ion batteries can be deadly; a safety-oriented mindset is essential.

7.2.1 Work Environment

Buddhist meditation room: This is what your work environment must be like when working with Li-ion cells: quiet, with no distractions. There must be one thing and one thing only in your mind: the battery you are working on (Figure 7.1(a)). Distractions result in mistakes, and mistakes with Li-ion have dire consequences. Ban chitchat, talk radio, books on tape, annoying bodily sounds. Clear your mind of your worries: stop fretting about your boss, politics, family, health issues, and internet trolls.

1. I am so fond of this word! It means to throw something out the window.

Assemble batteries in an access-controlled room, where only authorized people are allowed. This minimizes the chances of a curious onlooker's fingers poking where they shouldn't. Do not assemble in the middle of a busy factory floor or warehouse.

Become one with the battery. Ω

7.2.2 Tools and Conduct

Wear safety glasses.

Any tool that may come into contact with a Li-ion cell or with live wires must be nonconductive. Nonconductive socket wrenches made of composite materials, calipers, and even nonconductive scissors with ceramic blades are available. Metal tools insulated with electrical tape are not quite as safe but are better than nothing. Inspect tools regularly for insulation damage; for example, multimeter probe wires can become abraded. On that note, for safety, replace multimeter fuses with the correct type.[2]

Wear safety glasses.[3]

When not actively using a tool, put it down; place it at a lower level than the battery terminals. If you place it higher, gravity will invariably make it drop onto the battery and short two terminals together (Figure 7.1(b)).

Wear safety glasses.

Don't wear anything metallic: no rings or wedding bands,[4] no watches with metal wristbands, no metal bracelets, necklaces, or cufflinks, no nipple barbells.

Wear safety glasses.

Depending on the voltage of the section you're working on, you may need to wear insulated high-voltage electrician's gloves. They do make it harder to work, which in itself is a safety concern—you are more likely to drop a washer in the battery if you can't hold it properly—so only wear them if required.[5] Proper fit minimizes chafing and fatigue.

If working on high voltage without gloves, our ancestor's advice still holds: work with just one hand and keep your other hand in your pants' pocket. In case of contact with high voltage, this prevents current from flowing through your heart. Of course,

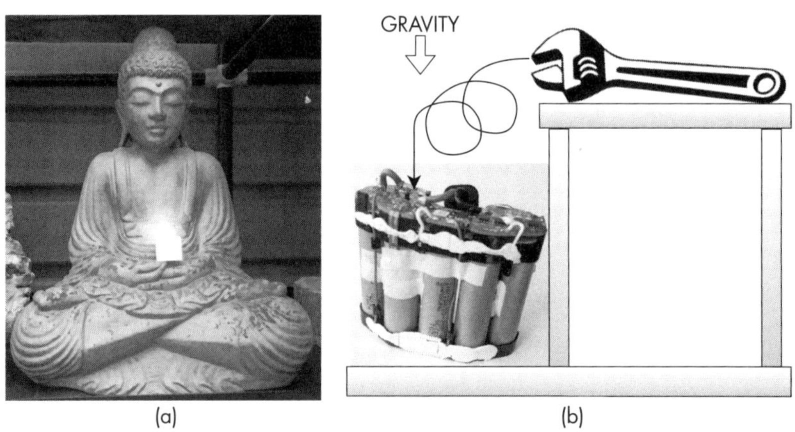

FIGURE 7.1
Li-ion work environment: (a) no distractions, and (b) tool misplacement.

2. I heard of someone who was blinded when a meter fuse exploded in his face when measuring voltage when the meter was set for current.
3. I have had glass shrapnel embedded in my cornea when a fuse blew in my face; that's one reason I insist on wearing safety glasses.
4. I met a man with a permanent groove in his ring finger, where his wedding band used to be. It shorted out across two battery terminals; he could not let go because the ring had become welded to the battery.
5. By one standard, anything above 40V. By IEC standards, 60V.

wear insulating shoes. Even with all those precautions, you may be grounded in unexpected ways: high voltage can arc through clothing; current may jump from a metal table you are leaning against through your trousers and zap your privates.

Wear safety glasses.

Attend a high-voltage safety course (see the "Resources" section in Volume 2).

7.2.3 Emergency Plan

Should a cell start going into thermal runaway, Li-ion cells go through two venting events. If you hear a pop, it is the first event. From this moment, you have just 2~10 seconds to toss the whole assembly into a safe area where it can burn.[6]

Plan for it:

- Designate a the safe area: a location where a battery can burn away safely such as a parking lot or a yard; if this is not possible, build a flame-proof vault next to the assembly area; use an explosion proof blower and externally vented ventilation system to extract smoke and fumes from the building;
- Make sure that a direct exit to the safe area can be opened in less than 1 second;
- Place the battery assembly area next to this exit;
- Place your battery assembly on a cart that can be rolled immediately to the safe area.

Another approach is to have a barrel of water nearby. The battery assembly can be pushed off the bench into the barrel, letting the water absorb the heat from the cell on fire, hoping to avoid propagation to other cells. Make sure the opening of the barrel is wide enough for the battery and that the edge of the barrel is lower than the working counter, so that gravity alone can be used to push the battery into the barrel (no need to lift the battery).

Keep an insulated rod nearby, to push the assembly into the barrel. Have a water hose ready to replenish the water as it steams off.

Be prepared and rehearse.

7.2.4 Safety Training

Production personnel should attend Li-ion safety training.

At a minimum, conduct an in-house session based on the advice listed above. To underscore the seriousness of the matter, show a video of a Li-ion cell undergoing thermal runaway.

Consider attending a Li-ion safety seminar offered in your area (see the "Resources" section in the appendix in Volume 2).

To avoid disappointment, check the agenda to ensure that the seminar truly does discuss battery production safety; be aware that most Li-ion safety seminars focus on topics unrelated to Li-ion production.[7]

6. I have a recurring nightmare of working with Li-ion cells when I hear a pop, and I see white smoke rising; my first thought is: Am I wearing safety goggles? In my latest nightmare, a 18650 cell shot through a thick concrete wall leaving a clean exit wound; I worried about how to hide that hole from the landlord.

7. The morning session of the latest "Battery Safety" seminar from International Batteries consisted of these two topics: "How Does the Electrolyte Change during the Lifetime of a Li-Ion Cell?" and "Uber Elevate–Powering an Electric UberAIR Future"; nothing about battery safety.

7.2.5 Insurance

In developed countries, you need insurance to cover your operations and to protect you in case one of your batteries is present at the scene of an accident (see Volume 2, Section B.10). Coverage for a battery manufacturer is rather expensive due to the safety reputation of Li-ion batteries.

7.2.6 Renting

Landlords are likely to have heard about the dangers of Li-ion. When looking for a shop, a warehouse, or a factory, be forthcoming about the nature of your work. Also, convey how safe and prepared your operation is. Your landlord probably requires coverage by your insurance policy.

7.3 INCOMING QUALITY CONTROL

The quality of incoming components may be checked to varying degrees depending on the type of production. A facility that produces thousands of batteries should be able to set up incoming quality control (QC) for every component. A prototype run relies on the outgoing QC of the vendor.

7.3.1 Li-Ion cells QC

Li-ion cells directly from the manufacturer may be spot-checked. Cells from vendors that do not specialize in Li-ion[8] may be suspect and should be checked carefully.

Typically, testing involves measuring a cell's DC resistance and capacity (see Section 2.9).

Design cell testing judiciously. Use histograms and Shewhart charts (statistical process control charts (SPCs)) for incoming QC data. Look for distinct populations and try to establish the root cause for them. At my first company, incoming QC rejected 1% of the cells because their voltage was 50 mV lower than the rest. This was done under the assumption that those cells were somehow bad. Later, we found out that the cell manufacturer was doing additional testing on 1% of the cells, after which the voltage ended up slightly lower: we were rejecting the cells that had undergone a second test at the factory!

Be suspicious of lots of cells that are characterized by a widely scattered values; cells whose values all fall within a narrow range reflect tighter manufacturing standards. Even cells whose values fit into a few different but narrow windows may be trustworthy, as long as the cause can be established.

Sort cells to build packs from like groupings.

7.3.2 Hardware QC

Check that the bolts shipped with large prismatic cells are the correct length. We have seen different length bolts in the same shipment from a cell manufacturer.

7.4 PREPRODUCTION

Preparing components and assemblies beforehand speeds up production.

8. Alibaba, eBay, Amazon.

7.4.1 Harnesses

Premade wire harnesses can increase assembly speed and quality. These include a cell sensing taps harness for a wired BMS (see Section 5.9.1) and power harnesses. A cable assembly house can assemble these harnesses for you.

7.4.2 Cell Preparation

Cells may be prepared before installation to speed up production and to improve the quality of the battery.

7.4.2.1 Terminal Preparation

Cell terminals that are designed to be fastened—small and large prismatic, large cylindrical—may oxidize, increasing contact resistance. Prepare the terminals by scrubbing with a nonmetallic scouring pad,[9] and immediately coating it with an antioxidant paste.[10] Do the same for the mating connections—bus bars, ring terminals.

7.4.2.2 SoC Preparation

Manufacturers ship cells at about 30% SoC to prolong their life. Whether cells may be used as received or should be prepared depends on the type of battery:

- Single-cell batteries: May be used as received;
- Parallel block: May be connected in parallel with care (see Section 7.5.1);
- Series strings: Should be precharged to the balance point for the string.

7.4.3 Precharging Cells for Series Strings

A battery that uses a series string should be balanced before it is deployed. Otherwise, the BMS will take a long time to balance the string.

This can be done in two ways:

- Precharge the cells before building the string;
- Use specialized equipment to balance the battery after it is built (see Section 7.6.1.2).

This section discusses precharging cells before they are placed in a series string. There are two approaches to doing so:

- Precharge cells individually;
- Connect cells in parallel in a block and precharge the block.

The precharge SoC level depends on the type of battery (see Section 5.1.4):

- Energy or power battery: 100% SoC;
- Buffer battery: The desired balance set point for the battery (e.g., 40% SoC) (see Section 3.2.6).

9. Scotch-Brite™.
10. Noalox.

7.4.3.1 *Precharging Cells Individually to 100% SoC*

Precharge a cell individually by connecting it to a single-cell charger set for the cell's maximum voltage (e.g., 3.6V for LFP cells, 4.2V for NMC and LCO cells). When the current drops below a set point, such as 0.2 C, disconnect the cell.

Using a single-cell charger capable of 0.5 C current, you can charge about six cells in an 8-hour shift. Of course, using more chargers lets you do more cells per day.

7.4.3.2 *Precharging Cells in Parallel to 100% SoC*

To precharge many cells at the same time:

- Connect them carefully in parallel into a block (Figure 7.2(a)) (see Section 7.5.1);
- Connect the block to a high-power charger;

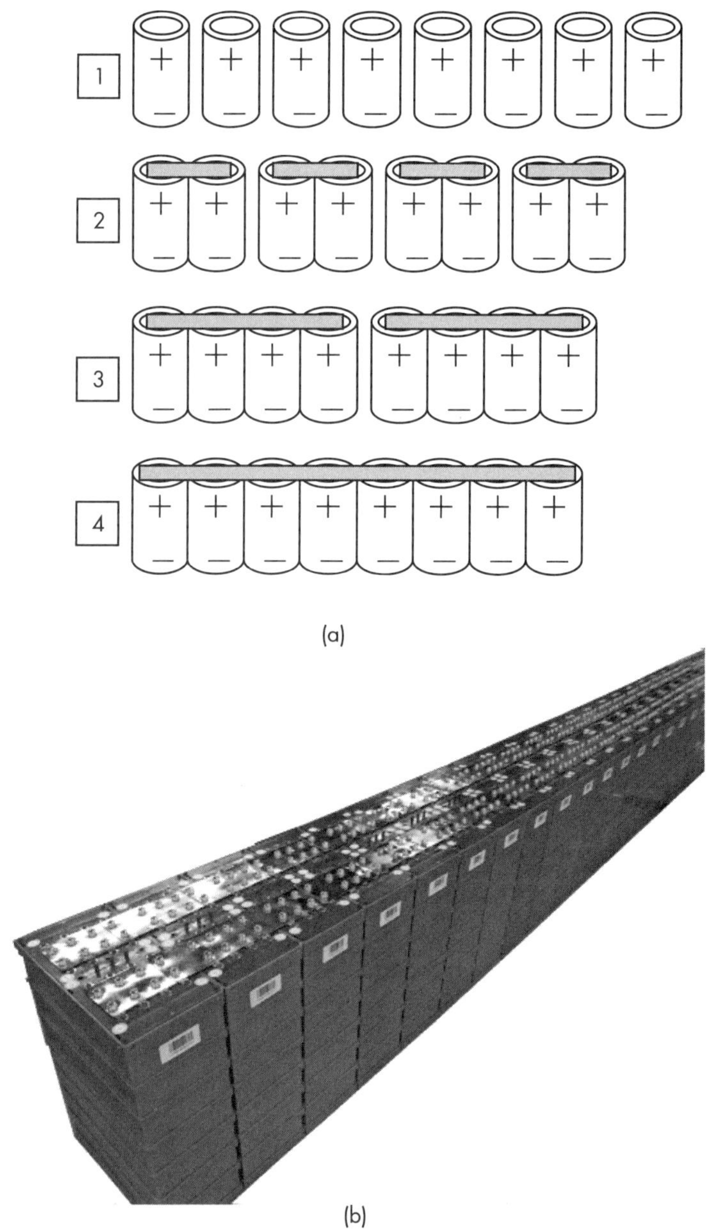

FIGURE 7.2
Prebalancing cells in parallel: (a) connect in binary fashion, and (b) prismatic cells prebalancing.

- Let charge until the current drops to a set point;
- Disconnect the cells and build the string.

Charging time is approximately

$$\text{time}[\text{h}] \cong 0.7 \times \text{cell\_capacity}[\text{Ah}] \times \text{number\_of\_cells}/\text{charger\_current}[\text{A}] \quad (7.1)$$

A single-cell charger of such high power is hard to find. A 5V power supply may not be suited for the job because it is not designed to operate at 3 to 4V—at best, it reduces the output current significantly due to fold-back current limiting; at worst, it shuts down, or it even fails (see Section 1.8.2).

7.4.3.3 *Precharging Cells for a Buffer Battery*

Ideally, the cells in a buffer battery that uses as a series string should be precharged to the desired balance set point for the battery (see Section 3.2.6). In practice, a perfect balance is not crucial for a buffer battery. Therefore, precharging the cells is optional; you may let the BMS do the balancing during use.

If required, cells may be precharged individually or in parallel.

To precharge in parallel, start by connecting all the cells in parallel (see Section 7.5.1) and let the parallel block sit for a day or so, while current flows between them to balance them all to the same SoC. Because the cells are received at about 50% SoC, the block will end-up at about 50% SoC.

This SoC won't be the same as the desired balance set point for the battery. If this is a concern, the block may be charged with a single-cell charger or discharged with a power resistor to the balance set point.

7.5 BATTERY ASSEMBLY

A battery may be installed after it is completely built and tested (e.g., consumer products, traction battery for a factory-built passenger vehicle). Or a battery may be assembled at its final location (e.g., marine house power, large stationary, traction battery for an industrial vehicle, or an EV conversion).

7.5.1 Connecting Cells in Parallel

Care must be taken when connecting cells in parallel, whether to precharge them before reconnecting them in series (Figure 7.2(b)) or when assembling a parallel block.

Connecting cells directly in parallel has to be done with great care to avoid the damage that could result from excessive inrush current between paralleled cells at different voltages (see Section 3.3.6).

First, make sure that voltages are not too different before connecting cells or sets of cells—I suggest 100 mV for energy cells and 50 mV for power cell.

Do not build a parallel block by adding a cell at a time. Doing so degrades the last cell added because it is exposed to the high inrush current from the much lower resistance of all the other cells already in parallel. Instead, connect the cells in a binary fashion (Figure 7.2(a)):

1. Connect the cells in pairs, and wait a while for them to balance;
2. Then connect two pairs and wait;
3. Then connect two quads;
4. And so on.

This method ensures that a given set of cells is connected to a different set of cells of the same size, and therefore of similar internal resistance, which limits the inrush current in each cell.

7.5.2 Cell Placement

A battery is assembled in these stages:

1. Cell placement;
2. Cell interconnection (multicell battery only);
3. BMS installation.

As every battery is different, all that I can offer in this chapter is some general pointers. The actual steps depend on the type of battery and BMS, as seen in the following examples.

7.5.2.1 Single-Cell Battery, Pouch, Open Assembly

This battery is an open assembly, to be enclosed in an inexpensive consumer product, such as a toy or a Bluetooth speaker. It has a 2-wire tail terminated with a 2-pin JST brand connector. There is no retention against expansion. (Figure 7.3).

The typical procedure is

- Solder the two-conductor output cable to the PCM;
- Slip the pouch cell's tabs into the slots in the PCM, paying careful attention to the polarity;
- Solder each tab to the PCM;
- Use nonconductive scissors to cut off the excess tab material;
- Wrap the PCM and the top of the cell with Kapton tape.

7.5.2.2 Single-Cell Battery, Small Cylindrical

This is a replaceable battery in a sealed plastic enclosure, to be mated to a consumer product. The PCM includes a header for the output connector. Its dimensions are 18 × 65 mm.

The typical procedure is

- Weld two nickel tabs on the two ends of the 18650 cell;
- Fold the tabs against the two ends of the PCM, paying careful attention to the polarity;
- Solder each tab to the PCM;

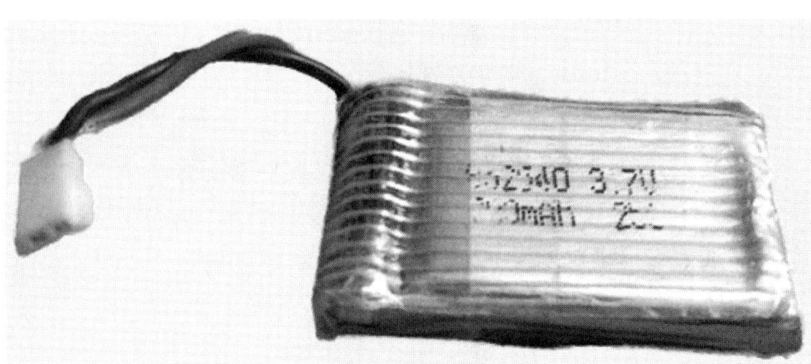

FIGURE 7.3 Single-cell battery, pouch, open assembly.

- Drop in the top shell of the plastic case;
- Test;
- Add the bottom cover and weld ultrasonically.

7.5.2.3 *Small Multicell Battery, Small Cylindrical*

This is a laptop battery or similar replaceable battery for consumer or medical products. The cell arrangement is 2P3S. The BMS includes a header for the output connector (with multiple terminals for power and communications). It supports the SMBus standard.

The typical procedure is

- Place six 18650 cells in a 2 × 3 grid in a rectangular frame, held vertically, oriented with the proper polarity;
- Weld six nickel strips to parallel each pair of cells;
- Weld a nickel strip to connect the first two pairs in series;
- Flip the frame upside down;
- Weld a nickel strip to connect the last two pairs in series;
- Remove the cells from the frame;
- Unfold the cells into a flat sheet;
- Solder the tap wires from the BMS to the nickel strips;
- Place the thermistor against the middle cells and secure with tape;
- Drop in the top shell of the plastic case;
- Test;
- Add the bottom cover and weld ultrasonically.

7.5.2.4 *Self-Balancing Scooter Battery, Small Cylindrical Cells*

This is a battery for a hoverboard with ten 18650 cells in series. The cells are held between two custom plastic frames, each with 10 round openings. The PCM is mounted directly to the cells. The battery has only two terminals for power (Figure 7.4).

The typical procedure is

- Place the top plastic frame (the one that sits against the PCM) on the bench
- Place ten 18650 cells in the top plastic frame (they are vertical), each oriented with the specified polarity (cells are held in the frame by interference fit)
- Place the bottom plastic frame onto the cells;
- Weld five nickel strips to connect five pairs of cells in series;
- Flip the assembly upside down; now the top frame is on top;
- Weld four T-shaped nickel terminals to complete connecting all the cells in series;
- Weld two nickel tabs on the two cells at the two ends;
- Bend the tails of the T-terminals so they point up;
- Place a fish-paper insulator on top of the top frame, slipping the vertical tabs into its slots;
- Drop the PCM onto the assembly, slipping the vertical tabs into its slots;

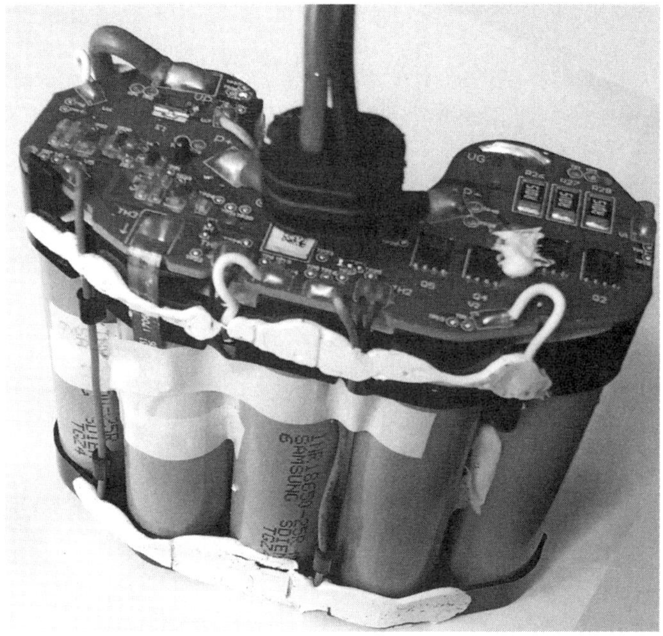

FIGURE 7.4
Self-balancing scooter battery, small cylindrical cells.

- Solder the tabs to the PCM;
- Place the thermistor against the side of a cell and secure with tape;
- Solder five wires to the PCM (tap wires for the other end of the cells);
- Flip the assembly upside down;
- Solder the five tap wires to the five tabs;
- Place the assembly on the bottom cover;
- Slip the output connector into the cover;
- Test;
- Add the top shell of the plastic case and weld ultrasonically.

7.5.2.5 *Medium-Sized Battery, Small Cylindrical Cells*

This is a battery for a small personal EV. A total of 120 18650 cells are arranged as 10P12S. A two-port PCM has Bluetooth communications (Figure 7.5).

The typical procedure is

- Preparation:
  - A rectangular frame, 65 mm tall, with 180 × 216 mm opening.
  - A set of nickel plates:
    - Regular: 11 each, 170 × 30 mm;
    - Thin: two each, 170 × 13 mm.
  - A set of harnesses:
    - A cell voltage sense harness with wires of the exact length, to reduce the chance of miswiring;
    - A power harness.
- Place the cells in the frame, in a 12 × 10 square pattern, orienting them so that the first row of 10 cells is up, the next row is down, and so forth, alternating the polarity of each row;

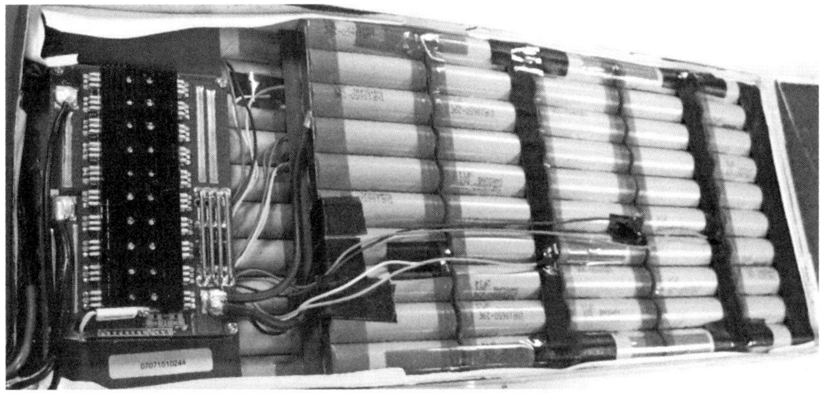

FIGURE 7.5
Medium sized battery, small cylindrical cells; the PCM is on the left.

- Weld the cell on the bottom side:
  - Place a mask to protect cells that are not to be welded yet;
  - Place a regular nickel plate on the first two rows of cells;
  - Weld the plate on the first cell, using four spots;
  - Repeat the process for all the other 19 cells in those two rows;
  - Move the mask over two rows;
  - Weld another regular nickel plate to the next two rows;
  - Repeat the process for all the other pairs of rows.
- Flip the block upside down.
- Weld the cell on the top side:
  - Place a mask to protect cells that are not to be welded yet;
  - Check that there's no voltage between the first two rows of cells, to make sure you're not about to make a short circuit;
  - Place a thin nickel plate on the first row of cells;
  - Weld the plate to the first row of cells;
  - Move the mask over two rows;
  - Weld a regular nickel plate to the next two rows.
- Repeat the process for all the other pairs of rows.
  - On the last row, weld a thin plate.
- Remove the block from the frame.
- Unfold the rows, to turn the block into a sheet, except between the fifth and sixth row, so that you end up with two layers, one of 10P5S, the other one of 10P7S.
- Wrap the block to hold its shape; alternatively, glue cells together;
- Make sure that the voltage sense harness is not connected to the PCM;
- Solder the wires from the voltage sense harness to the corresponding tabs on the nickel plates; mind the other tap wires: keep them from touching something they're not supposed to touch;
- Install the PCM onto the cells with double-adhesive foam tape;
- Secure the thermistors to the cells;
- Check the voltages on the connector (use a test fixture);

- Plug the connector to the PCM;
- Connect the power harness:
  - Negative wire to the PCM's *B–* pad;
  - Positive wire to the B+ terminal of the cell block;
  - A wire between the B– terminal of the cell block and the PCM's *B–* pad.

7.5.2.6 *Small Multicell Battery, Pouch*

This battery is an open assembly, to be enclosed in a toy UAV. The pouch cells are arranged as 7S. The stack has a two-wire tail terminated with a 2-pin high-power connector (Figure 7.6).

It does not include a BMS (it is charged with a balance charger (see Section 6.5.4), and it is abused in many ways during discharge. There is no retention against expansion.

The typical procedure is

- Prepare the harnesses:
  - Output harness: two large gauge wires terminated by a 2-pin power connector;
  - Balance harness: eight small gauge wires terminated by an 8-pin JST connector.
- Slip protecting caps on all the cell tabs;
- Place the cells flat against each other, alternating polarity;
- Remove the caps from two adjacent tabs, fold the tabs against each other, and weld the tabs together;
- Repeat with the remaining tabs (six welds total);
- Solder the two wires of the output harness to the two end tabs;
- Solder the eight wires from the balance harness to the voltage tap tabs;
- Test the voltages;
- Wrap the tab area in Kapton tape;
- Insert the assembly into a large diameter heat shrink tube, shrink.

7.5.2.7 *24V Battery, Large Prismatic Cells*

This is a self-contained, 100 Ah battery, using eight LFP cells in series, with a 50 A PCM (Figure 7.7).

The typical procedure is

- Prepare the harnesses:
  - A voltage sense harness, with ring terminals on one end, and a 9-pin JST connector at the other end;
  - A power harness.

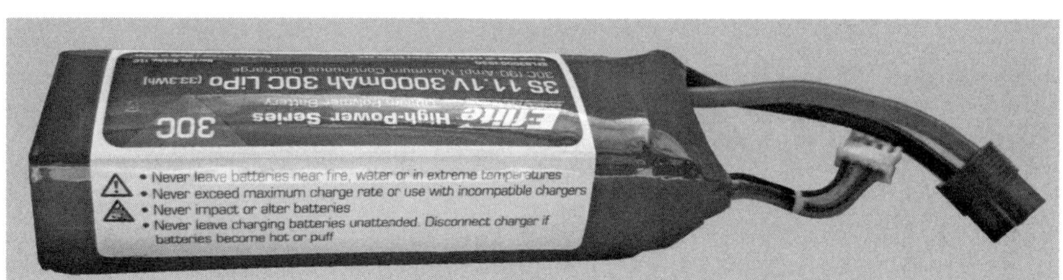

FIGURE 7.6
Small multicell battery, stack of pouch cells.

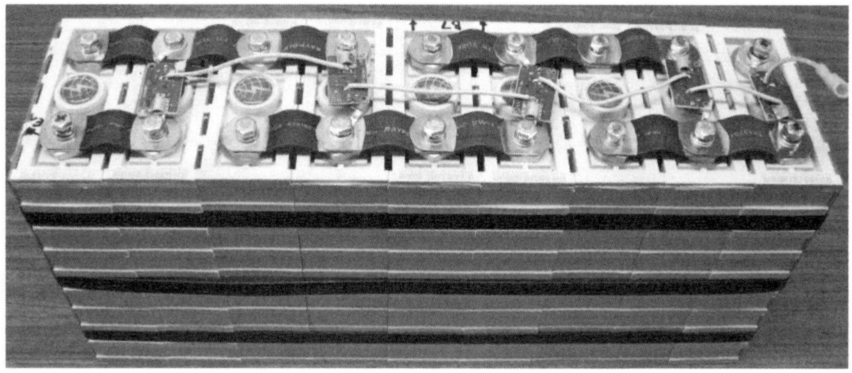

FIGURE 7.7
Large prismatic cell block.

- Align the cells flat against each other, alternating polarity;
- Place two plates at the two flat ends of the block;
- Wrap the assembly with bands, and use a banding tool to retain it securely;
- Drop into the bottom can of the plastic case;
- Make sure the voltage sense harness is not connected to the PCM;
- Clean the aluminum terminals and the bus bars, and spread some antioxidant paste;
- Connect cells in series, with flexible bus bars; include the ring terminals of the voltage sense harness;
- Tighten the bolts with a torque wrench to the specified torque;
- Connect the power harness to the cells and the PCM;
- Check and double-check the voltages on the connector (use a test fixture);
- Plug the connector to the PCM;
- Test it;
- Screw in the top cover.

7.5.2.8 EV Conversion Traction Battery, Large Prismatic Cells

This is a traction battery for an EV conversion, with 50 each, 100 Ah large prismatic cells in series. It uses a distributed BMU and contactors (Figure 7.8).

The typical procedure is

- Drop the cells into their enclosure oriented for the desired polarization;
- Adjust the end of the enclosure to retain the cells against expansion;
- Clean the aluminum terminals and bus bars, and spread some antioxidant paste;
- Make doubly sure that each block of cells is not connected to anything else (no charger, no motor driver, no DC-DC converter);
- Connect cells in series, with flexible bus bars;
- Install a cell board across each cell, making doubly sure that they are not connected backward; follow the proper order (see Section 7.5.3.3);
- If two cells are separated by more than a few centimeters, or if there is a component between cells that may open (a fuse, a contactor, a safety disconnect, a connector), use two different banks, one on each side of the separation;

FIGURE 7.8
EV conversion traction battery, large prismatic cells.

- Use a torque wrench to tighten the terminal bolts to the specified torque;
- Connect the wires between adjacent cell boards;
- Test to make sure the battery is isolated;
- Connect the bank cables from the end cell boards to the BMU;
- Test;
- Repeat with the other banks.

7.5.2.9 *Large Stationary Low Voltage Battery, Large Prismatic Cells*

This is a 48V, 2000 Ah battery for off-grid use. It uses eighty 400 Ah LFP cells in a 5P16S arrangement (Figure 7.9).

The battery is divided into four identical large cases, each 12V, connected in series. It uses a master-slave wired BMS and contactors.

The typical procedure is

- Assemble the first box:

FIGURE 7.9
Large stationary low-voltage battery, large prismatic cells. (Courtesy Apollo Solar.)

  - Drop 60 cells into an enclosure, flat against each other, oriented for the correct polarization for a 5P4S arrangement: five up, five down, five up, five down, and so forth;
  - Adjust the end of the enclosure to retain the cells against expansion.
- Clean the aluminum terminals and bus bars, and spread some antioxidant paste
  - Connect the first five cells in parallel, with bus bars; if they are not prebalanced, be careful with the connection order (to limit the inrush current):
    - First, connect two cells at one end;
    - Then connect three cells at the other end;
    - Finally, connect the two groups of cells.
  - Repeat with the other three groups of five cells each;
  - Connect the blocks in series with flexible bus bars;
  - Make sure the voltage sense harness is not connected to the slave;
  - Slip the five ring-terminals of the voltage sense harness on bolts for each voltage level, making sure the correct tap wire is used at each point;
  - Use a torque wrench to tighten the terminal bolts to the specified torque;
  - Place the slave in the enclosure;
  - Test the voltages on the connector of the voltage sense harness (use a test fixture);
  - Plug the connector to the slave;
  - Install the slave's communication connector onto the panel of the case;
  - Install the power connector onto the panel of the case;
  - Connect the power connector to the end cells through a fuse;
  - Test that the communication cable for the slave is isolated from the cells.
- Repeat with the other three boxes;
- Connect the BMU to the four boxes;
- Test that the BMU sees all four slaves and all 16 cells in series;
- Connect the contactor box to the BMU;
- Connect the four boxes in series and to the contactor box;
- Test the complete battery;
- Disconnect the four boxes, the BMU, and the contactor box;
- Ship as separate boxes;
- Once at the final location, reconnect the boxes and the other items.

7.5.2.10 40V Block, Pouch Cells

This is an Enerdel Moxie module, a block of pouch cells in a plastic frame in a 2P12S arrangement. It uses a wired BMS (Figure 7.10).

The typical procedure is

- Place two cells in a pair of plastic frames to form a plate;
- Fold each cell tab against the other cell tab, to make a parallel connection;
- Prepare 12 such plates;
- Mate the plates flat against each other, alternating the orientation;
- Complete the blocks with end plates to contain expansion;

FIGURE 7.10
2P12S pouch cell block.

- Secure the block with the provided hardware;
- Tighten the fasteners with a torque wrench;
- Make sure the voltage sense harness is not connected to the PCM;
- Connect the first two plates in series by adding a pressure plate;
- Add the correct wire to the voltage sense harness;
- Secure with fasteners;
- Use a torque wrench to tighten the pressure plate to the specified torque;
- Repeat with the other plates;
- Place the end terminals onto the two end cells;
- Secure with fasteners, using a torque wrench;
- Test the voltages on the connector of the voltage sense harness (use a test fixture);
- Plug the connector to the BMS.

7.5.3 BMS Installation

Now that the cells are mounted and interconnected, it is time to install the BMS.

7.5.3.1 *Integrity of Electronic Assemblies*

A BMS is most prone to damage before and during installation.

Most BMSs are designed to withstand electrostatic discharge (ESD), so that is probably not going to be an issue. Still, keep electronic assemblies in static bags until ready for installation.

Not many BMSs survive when their wires touch various connections in a battery. Take the BMS electronic assemblies out of the bag only right before installation, at the assembly location, and take care that wires are not loose and falling onto cell terminals.

7.5.3.2 *Wired BMS Cell Voltage Sensing*

The voltage sense harness in a wired BMS gives you a chance to check voltages before connecting the harness to the BMU.

Procedure (Figure 7.11(a)):

- Make sure that the voltage sense harness is not connected to the BMU.
- Wire the voltage sense harness to the cells:

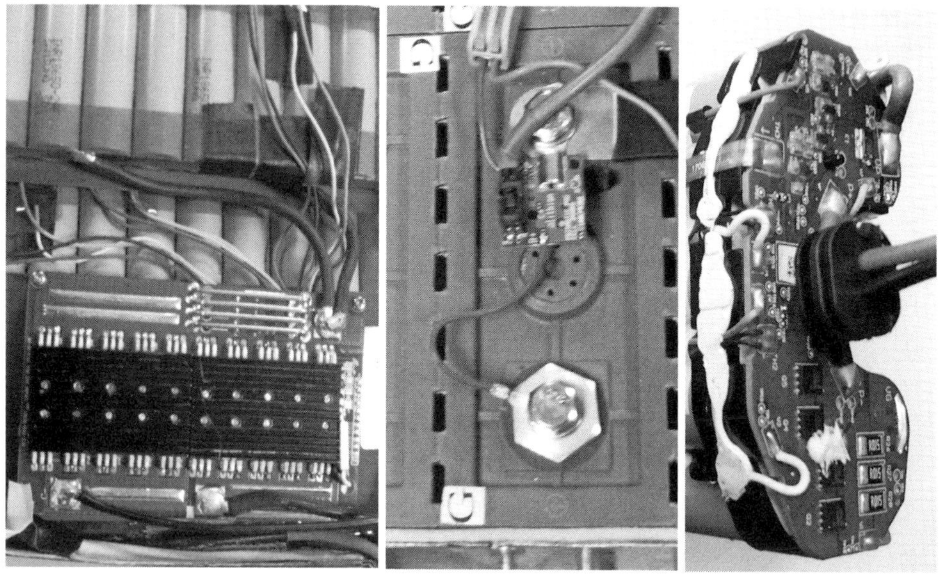

(a) (b) (c)

FIGURE 7.11
BMS installed on cells: (a) wired, (b) distributed, and (c) banked.

- Mind the other tap wires: keep them from touching something they're not supposed to touch
- Check the voltages on the connector using a fixture.[11]
- Only after you're sure they are correct, connect to the BMU.
- Place a thermistor on a cell whose temperature is representative of the set of nearby cells, and secure reliably: a loose thermistor may cause an unprotected short circuit in the battery. If there are multiple thermistors, place in carefully chosen locations.

7.5.3.3 Distributed BMS Cell Boards

With a distributed BMS, you don't get a chance to measure the voltage before connecting a cell board. Therefore, damaging a cell board by connecting it backward is extremely easy to do.

Procedure (Figure 7.11(b)):

- Disconnect the block of cells from everything else (no charger, no load, no DC-DC converter); otherwise, the full battery voltage may appear across a cell board (see Section 8.3.2.3);
- Check the polarity of the cell board and match it to the polarity of the cell;
- Double-check the polarity;
- Mind the communication wires, especially stripped ones; keep them from touching something they're not supposed to touch;
- Connect one end of the cell board to a cell terminal, following the proper order; for a large prismatic cell, from top to bottom (Figure 7.12(a)):
  - Bolt;

11. The Orion BMS offers an optional tool to check the connector voltages before connection to the BMS. (See Volume 2, Section A.8.1.1).

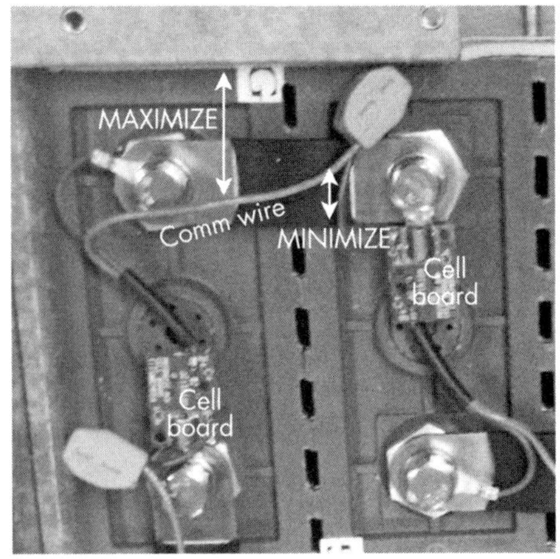

(a) (b)

FIGURE 7.12
Distributed BMS: (a) bus bar and ring terminal mounting, and (b) optimal routing of communication wire between cell boards.

  - Split washer;
  - Flat washer;
  - Ring terminal for BMS sense wire/cell board;
  - Bus bar;
  - Cell terminal.
- Triple check the polarity;
- Connect the other end of the cell board to the other cell terminal;
- When all the cell boards are placed, connect the wires or cables between cell boards:
  - Route the wires close to the corresponding bus bar, which acts as a shield for the signal; you want to maximize the capacitance between the communication wires and the bus bar, and minimize the capacitance between the wires and any grounded metal (e.g., a metal enclosure) (Figure 7.12(b)).

7.5.3.4 Banked BMS Board

The bank board in a banked BMS fits only one way, so there's little danger of misconnecting it and damaging it. (Figure 7.11(c)).

Procedure:

- Disconnect the block of cells from everything else; otherwise, the full battery voltage may appear across the bank board;
- Mount the board directly onto the cells;
- Connect the board;
  - For a mated board, tabs from the cells are soldered to the bank board;
  - For a wired board, individual wires are soldered to the board and connected to the cells; be careful: there is no chance to check the voltages before connection to the bank board;
- Mind the communication cable; keep it from touching something it's not supposed to touch.

7.6 BALANCING

Earlier, we saw how cells should be precharged before placing them in a battery with a series string (see Section 7.4.3). Otherwise, the battery may require gross balancing, which can be done manually or with a gross balancer.

7.6.1 Manual Balancing

First, try to bring the string SoC close to the balance setpoint (i.e., 100% for an energy or power battery, about 50% for a buffer battery).

- To increase a cell's SoC, use a lab power supply[12] as a charger:
  - Set it for the desired voltage (i.e., the voltage of the other cells);
  - Set the current limit to the maximum charging current for that cell, or the maximum for the lab power supply, whichever is smaller;
  - Connect it to the cell whose SOC must be increased;
  - Note that the current starts at the maximum and starts dropping after some time;
  - When the current drops to nearly zero, disconnect the lab power supply.
- To decrease a cell's SoC, use a power resistor:
  - Select a resistor for the desired balance current;
  - Connect a voltmeter directly to the cell terminals;
  - Connect the resistor directly to the cell terminals (not sharing the voltmeter leads);
  - WATCH THE VOLTAGE LIKE A HAWK![13]
  - Disconnect the resistor when the cell voltage drops to the desired voltage.

The time required to balance a string depends on the string capacity, its degree of imbalance, and the balancing current:

$$\begin{aligned}&\text{Balancing-time}[\text{hours}] = \text{capacity}[\text{Ah}]/\text{balancing\_current}[\text{A}]\\&\times \text{State\_of\_Imbalance}[\%]/100[\%]\end{aligned} \tag{7.2}$$

For example, a 1Ω resistor across a cell results in a current of about 3.6A. This removes 1 Ah in about 15 minutes. As the power is about 13W, a 15W or 20W resistor is appropriate.

This graph shows the balance time versus the size of the resistor for various levels of charge to be removed (Figure 7.13).

7.6.2 Top Balance with a Gross Balancer

An effective way to top-balance a battery at the factory is with a high power charger and a *gross balancer* that includes a set of high-power equalizers (Figure 7.14).

Equalizers are voltage clamps that bypass the charging current on a cell if its voltage exceeds a threshold. When used in high-voltage strings of 12V lead-acid batteries (e.g., in an electric vehicle), they are called *regulators* or similar terms. They

12. With adjustable voltage and current settings.
13. Going out to smoke a cigarette is a sure way to smoke the cell too.

FIGURE 7.13
Time required to remove a given charge versus resistor value.

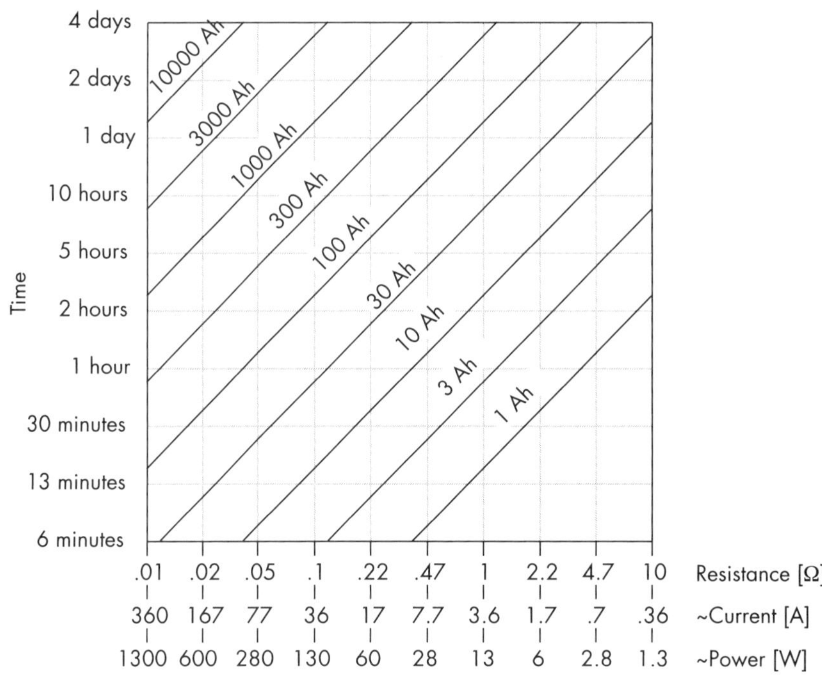

FIGURE 7.14
Sixteen-channel gross balancer: power resistors at the bottom, electronic boards on the rear panel.

are uncommon in Li-ion batteries because, once balanced, they require little current to maintain balance. The BMS can provide this little balance current.

A gross balancer is a useful tool during battery manufacture, as it allows the fastest balancing with little labor. For example, if your factory builds large 48V batteries with 16 LFP cells in series, you can build a gross balancer with 16 equalizers set for 3.6V and capable of bypassing 30A. Cooling is required because the gross balancer needs to dissipate as much as 1,700W.

The gross balancer is used as follows:

- First, build the battery.
- Roll the gross balancer to the battery.
- Connect the cables from the gross balancer to each cell.
- Charge the battery with a bulk charger controlled by the battery's BMS. If the charger current is higher than the equalizer's current, the BMS turns the charger off and on at a duty cycle that results in an average charging current equal to the equalizers' current. For example, if the charger current is 90A, and the equalizer current is 30A, then the BMS turns on the charger 30% of the time (e.g., 3 seconds on, 7 seconds off).

- When all cells reach the maximum voltage, the charging current decays naturally, and all regulators turn off.
- Turn off the charger, disconnect the gross balancer, and move onto the next battery.

7.7 INITIAL TESTING

Now that the battery has been assembled and balanced, we must do some initial tests before we can start using the BMS.

7.7.1 Battery Isolation Test

Unless the battery is nonisolated, test for loss of isolation manually. Not testing for isolation may result in electrical shock, a tool welded inside the battery, a fused wire, a blown BMS, or other nasty surprises. Indeed, it may be appropriate to test multiple times during assembly, between various stages of manufacturing (e.g., test each module after it has been built, and perform this test after all the modules are connected in series and before the BMS is connected. Test after the BMS is connected and before a USB cable is connected to a computer).

Pay special attention that the isolation has not been inadvertently defeated by an unintended connection between the negative of the battery and ground. Even when the battery is assumed to be floating, many times it isn't because

- There is a design error (see Section 8.11);
- Even if the battery by itself is isolated, it is connected to a grounded load;
- There is a ground fault somewhere in the battery.

7.7.1.1 General Test Procedure

In these tests, "ground" means any of the following:

- The common of the low-voltage control circuit;
- The common of a communication cable;
- A metal enclosure (if any);
- The power cord ground (if the battery is meant to be connected to AC power).

Testing for isolation is done in two steps:

1. Measure the voltage between the B+ battery terminal and ground (Figure 7.15 (left));
2. Measure the voltage between the B– battery terminal and ground (Figure 7.15 (right)).

The voltmeter readings reveal whether the battery is isolated from ground:

- Pass:
  - If the battery is floating, both readings are 0V (Figure 7.15(a)).
- Fail:
  - If B– is grounded, the first reading is at the battery voltage (Figure 7.15(b));
  - If B+ is grounded, the second reading is at the battery voltage (Figure 7.15(c));

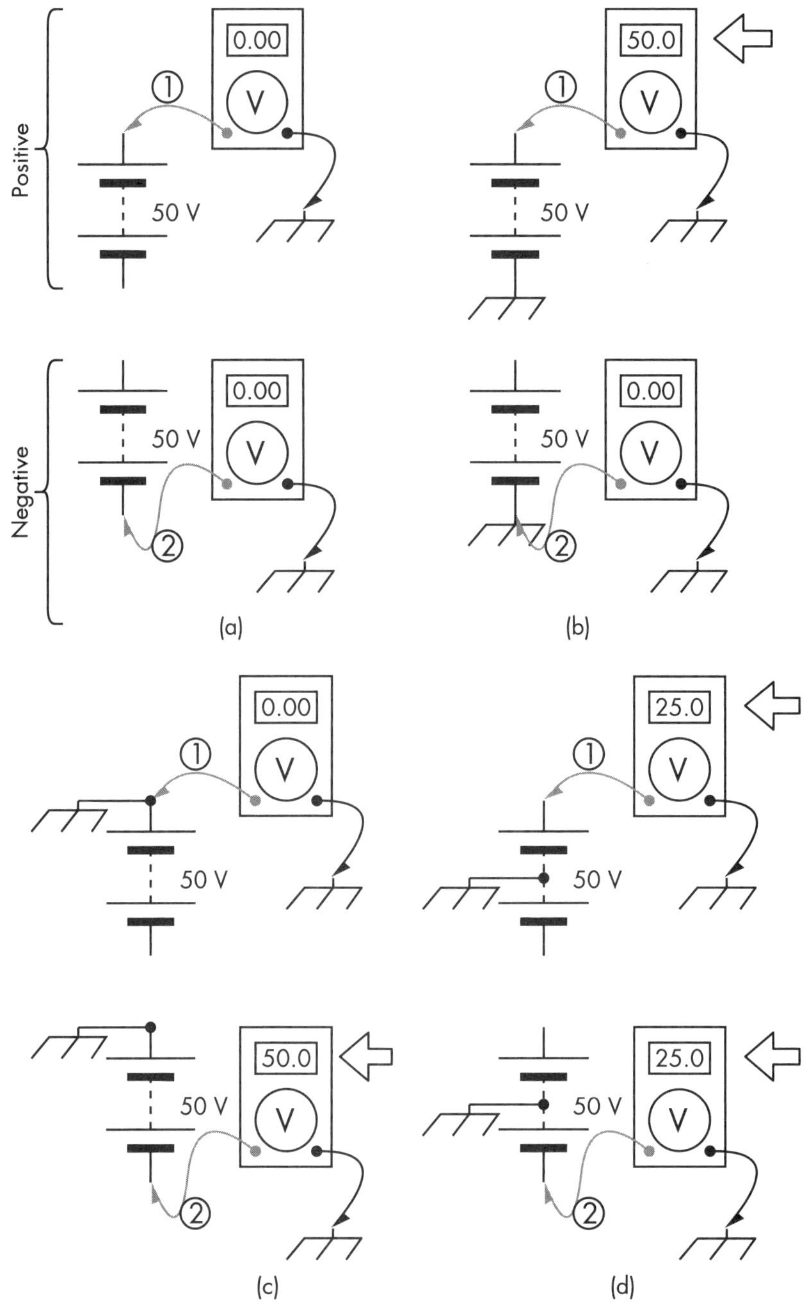

FIGURE 7.15
Manual ground fault testing: (a) floating, (b) B– grounded, (c) B+ grounded, and (d) midpoint grounded.

- If a point in the middle of the battery is grounded, both readings are more than 0V (Figure 7.15(d)).

If the test fails, troubleshoot the battery to identify the cause for the loss of isolation (see Section 8.11).

While testing, be aware of conditions that may result in seeing an unexpected voltage, leading you to believe, incorrectly, that the battery is not isolated.

If the BMS measures the total battery voltage, it may use a circuit that bypasses the isolation with a high resistance. Though there is no danger of shock or short circuit, this testing may produce a false positive. To overcome this, place a 100-kΩ resistor[14] in parallel with the voltmeter leads. If you still see a voltage, there truly is a loss of isolation.

A large capacitance to ground—whether due to an actual capacitor or the physical capacitance between the cells and a metal battery case—results in a decaying voltage reading. Wait for the voltage to decay as the capacitance is charged. A resistor across the voltmeter leads hastens the decay.

The actual test depends on the type of battery. The following test procedures are suggested for some battery types.

7.7.1.2 *Protector BMSs, Powered by the Battery, No Data Port*

A PCM is not isolated from the cells. If the PCM has no data port, there's no risk because there is no path through the BMS to an external ground. Therefore, no test is required.

On the other hand, if there is a data port, the risk is high because there could be an unintentional short between a cell terminal and the external ground. Before connecting a cable to the battery, check that there is no voltage between the two grounds, one from the BMS and one on the cable.

7.7.1.3 *Centralized BMU, Powered by the Battery*

A centralized BMS does have a communication port (e.g., USB), that is powered by its cells, though the data port is internally isolated from them. An unintentional short between a thermistor (referenced to the low-voltage side) and a cell terminal may compromise this isolation, or a design error may unintentionally bypass the BMS's isolation.

Procedure:

- Check that there is no voltage between the B– node (on the cell side, before the protector switch) and any line in the communication port;
- Repeat with the B+ node;
- Check that there is no voltage between the B– node and earth ground and chassis;
- Repeat with the B+ node;
- Check that there is no voltage between the ground of the communication port and earth ground and chassis;
- Check that there is no voltage between the ground of the communication port and the common of a communication cable about to be connected to the battery.

7.7.1.4 *Centralized BMU, Powered Externally*

This BMU is the same as above, but it is powered through a power supply input. Powering the BMS from a nonisolated DC-DC converter may unintentionally bypass the BMS's isolation.

Procedure:

- Do the tests in the previous section;

14. For a high-voltage battery, use a resistor rated for that voltage. You may use several resistors in series, such that each resistor sees a maximum voltage of 200V. Above 150V, use a 1W resistor; above 300V, use a 3W resistor; above 500V, use a 10W resistor; and above 1 kV, use a 25-W resistor.

- Check that there is no voltage between the B– node (on the cell side, before the protector switch) and the negative power supply input;
- Repeat with the B+ node;
- Repeat with the positive power supply input, for both the B– and B+ nodes.

7.7.1.5 *Wired Slave, Powered by the Cells*

Each slave in a master/slave BMS communicates with a digital data link to the BMU. The isolation is inside the slave.

There could be an unintentional short between a thermistor (referenced to the low-voltage side) and a cell terminal.

Procedure:

- Check that there is no voltage between the negative end of the bank of cells and the common in the communication link;
- Repeat with the positive end of the bank of cells.

7.7.1.6 *Wired Slave, Powered Externally*

The slave is the same as above but includes a power supply input.

Procedure:

- Do the tests in the previous section;
- Check that there is no voltage between the negative end of the bank of cells and the negative power supply input;
- Repeat with the positive end of the bank of cells;
- Repeat with the positive power supply input, for both the B– and B+ nodes.

7.7.1.7 *Distributed BMS*

The isolation in a distributed BMS is in the cell boards at the end of a bank. There could be an unintentional short between the bank communications cable (referenced to the low-voltage side) and a cell terminal.

Procedure:

- Check that there is no voltage between the negative end of the first bank of cells and the ground of the BMU;
- Repeat with the positive end of that same bank;
- Repeat with the remaining banks.

7.7.1.8 *Distributed Master/Slave BMS*

This is the same as above, with slaves. Each slave communicates with a digital data link to the BMU, which also brings power for the slave. There could be an unintentional short between the bank communications cable (referenced to the low-voltage side) and a cell terminal.

Perform the tests in the previous section, but relative to the ground in the slave's port that communicates with the master.

7.7.2 Basic Electrical Test

Once you are sure that the battery is isolated, check that the BMS is ready:

- Test for basic operation:
  - If the battery has a user-accessible power switch, turn it on; or, if the battery needs to be woken up, connect it to a charger for a second.
  - Check that the BMS receives power and does so at the correct voltage;
  - Check that the BMS appears to be turned on (if there is any way to tell).
- Check communications (digital BMS only) (Figure 7.16):
  - Check that the battery communicates to the external system;
  - If a GUI is available, check that it sees the BMS.

7.8 CONFIGURATION

Now that there is communication to the BMS, the BMS may be configured (if required). For production, this involves uploading a configuration file to the BMS; that file includes all the settings. Otherwise, the BMS settings must be custom configured (Figure 7.17). These may include

- Measurements:
  - Number of cells—if multiple banks, cells in each bank;
  - Nominal capacity;
  - Current sensors—range, offset, input source if more than one is available;

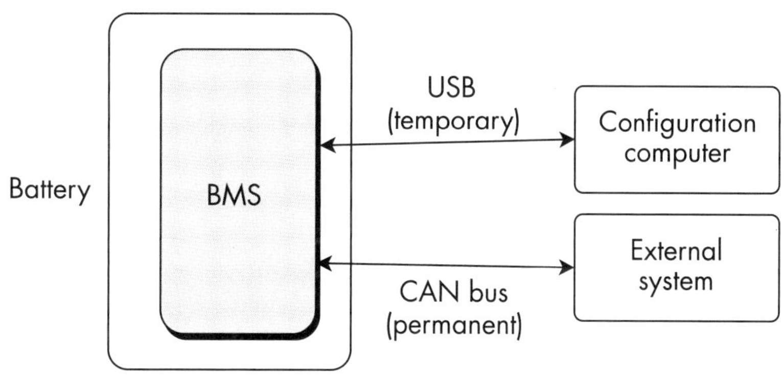

FIGURE 7.16 Communications block diagram.

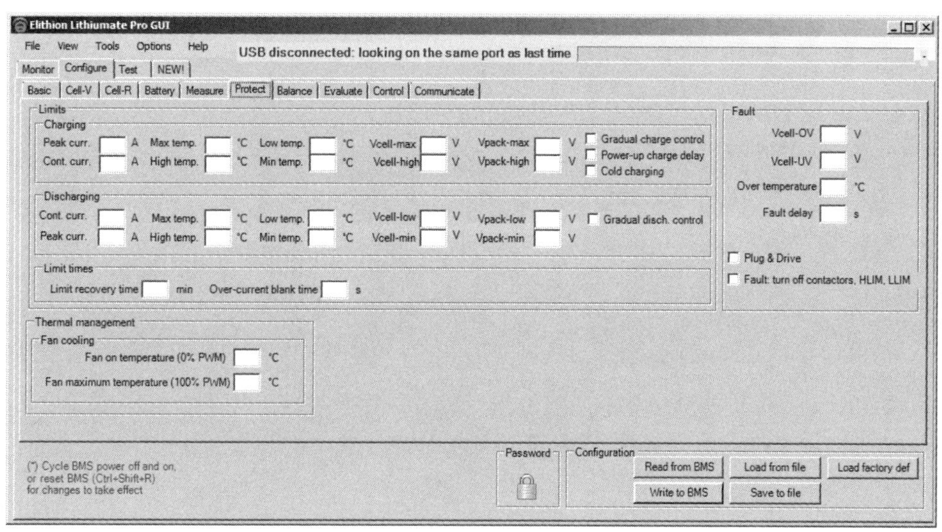

FIGURE 7.17 Configuration screens in a GUI.

- Temperature sensors—type, resistance at room temperature, beta, input source.
- Protection:
  - The minimum and maximum cell voltages;
  - The minimum and maximum charging and discharging temperatures.
- Balance:
  - The minimum cell voltage for balancing.
- Thermal management:
  - The temperatures at which heating and cooling start;
  - The temperatures at which heating and cooling are maximum.
- Evaluation:
  - The OCV versus SoC for the cells;
  - The SoH thresholds;
  - Battery tests—ground fault, contactors, and other components.
- I/O:
  - The function of each input and each output (if configurable);
  - The polarity of digital inputs and outputs;
  - The ranges on analog inputs and outputs.
- Communications:
  - Protocol and rate;
  - External devices;
  - Custom messages.

7.9 FUNCTIONAL TESTING

A functional test determines if the battery functions as designed and safely. The actual test procedure depends on the design of the battery. In general, it should look something like the following.

Check the measurements (digital BMS only):

- Check that the cell voltages are reported correctly;
- Check that the temperatures reported are equal to the room temperature;
- Connect a load to the battery;
- Check that the load is powered;
- Check that the battery reports discharging current and that the value for current is correct, including the sign.

Test the protection:

- Discharge the battery fully, monitoring cell voltages;
- Check that discharge is disabled when any cell voltage reaches the minimum;
- Disconnect the load and connect a charger to the battery;
- Charge the battery fully, monitoring cell voltages;
- Check that charge is disabled when any cell voltage reaches the maximum.

Check the balancing function (if top balanced with bypass balancing): check that the cell with the maximum voltage is being balanced; if none, overcharge a single cell by 50 mV or so, and check that it's being balanced back down.

If any test fails, troubleshoot and repair the battery (see Section 8.5), then rerun the tests.

7.10 DEPLOYMENT

Now that the battery is complete, it is time for its deployment—installation or delivery.

7.10.1 Communicating with the End User

Delivering a battery to the end user, whether a customer or an in-house user, is only part of your obligation; you also need to discuss "care and feeding of your battery" with the end user (see Section 5.3.1) because you are providing a solution, not just a product.

Provide a user's manual, but also go over the main points in person. The actual points depend on the type of battery, of course, but these points may be appropriate:

- Safety:
  - Do not abuse the battery in the following ways (e.g., do not connect batteries in parallel; do not modify and especially do not bypass the BMS' protector switch);
  - Provide safety guidance with applicable dos and don'ts in case of thermal runaway or mechanical damage that ruptures the battery;
  - These are not lead-acid batteries; Li-ion batteries are powerful if treated with respect; do not treat these batteries the way you used to treat lead-acid batteries, or you may cause a fire;
  - Here are emergency numbers.
- Expectations:
  - Over time, the performance of the battery will decrease, and what to expect
  - At times, the battery turns off to protect its cells; this is normal and to be expected and is not a malfunction.
- Support:
  - We're always available to help in you have any questions or something happens; please talk to us first, before you decide to do anything;
  - This is how you can monitor your battery's performance.

7.10.2 Transportation

Because Li-ion batteries can be dangerous, regulations for their transportation are strict and are getting stricter. Here is just a synopsis of options and applicable regulations.

There are a few organizations that created standards [1, 2]:

- ANSI C18.2M, Part 2: Portable Rechargeable Cells and Batteries—Safety Standard;
- BATSO:
  - 01: Light electric vehicles;
  - 02: Stationary batteries.

- CE (Conformité Européenne)—Self-certification
- International Air Transport Association (IATA)—Dangerous Goods Regulations (DGR);
- International Electrotechnical Commission (IEC):
  - IEC 62133 + IEC 61960: For portable applications;
  - IEC 62281:Transportation: IEC version of UN 38.3;
  - IEC 62619, IEC 62620: Industrial.
- IEEE (Institute of Electrical and Electronics Engineers):
  - IEEE 1725: Cell phones;
  - IEEE 1625: Portable computing devices.
- Society of Automotive & Aerospace Engineers International (SAE)—specific to EV traction batteries;
- Telcordia—GR-3150-CORE: Stationary;
- Underwriters Laboratories (UL):
  - UL1642:Test of Li-ion cells;
  - UL1973: Light Electric Rail;
  - UL2054: Household and commercial batteries;
  - UL2271: Light electric vehicles;
  - UL2580: Specific to EV traction batteries.
- United Nations (UN)/Department of Transportation (DOT), United States)—UN38.3:Transportation safety for cells and batteries.

Table 7.1
UN38.3 Tests

		Cell	Battery	Notes
T2	**Altitude**	✓	✓	15,000-meter altitude
T2	**Thermal**	✓	✓	–40°C to +75°C. Exposure during transportation (not power dissipated in use){**AU: degrees Celsius meant?**}
T3	**Vibration**	✓	✓	7 Hz–200 Hz–7Hz in 15 minutes
T4	**Shock**	✓	✓	Half-sine pulse Small cells/batteries: 150 G/6 ms Large cells/batteries: 50 G/11 ms
T5	**External short circuit**	✓	✓	Moot for a battery because its switch is off; still applicable for battery submodules*
T6	**Impact**	✓	—	
T7	**Overcharge**	—	✓	Two times the manufacturer's recommended charge current for 24 hours Moot because the battery switch is off; still applicable for battery submodules
T8	**Forced discharge**	✓	—	

Notes:
Pass means: "Case temperature does not exceed +170° C, no mass loss, leaking, venting, disassembly, rupture or fire, and voltage within 10% of pretest voltage."
"Battery" includes single-cell batteries, but not single cells.
*To meet UN38.3, a low-current fuse is included in the module prior to shipment, which can then be removed and bypassed in actual use.

A Li-ion battery MUST be certified under UN38.3 before transportation, regardless of the shipment method (ground, air, sea). A certification company can perform the tests, for which it requires 16 samples of the battery. The test costs about $5,000~7,000 and takes 4~6 weeks.

UN 38.3 specifies eight tests (see Table 7.1).

This testing is appropriate for a production battery. It is disproportionate for a single prototype battery, in which case there are a few options:

- Ship Ground as Class 9 Hazardous Goods;
- Ship under reinforced conditions as *prototypes*, under Special Provision A88 of the IATA DGR;
- Become a certified company allowed to self-certify its product.

References

[1] EPEC Engineering Technologies, "Battery Pack Certifications. What They Are, What They Cost, How Long Do They Take?" https://www.epectec.com/batteries/battery-pack-certifications.html.

[2] Eurifins, "UN/DOT 38.3: Lithium Battery Transportation," http://www.metlabs.com/battery/un-38-3-transportation-testing-required-for-lithium-battery-safety-during-shipping/.

CHAPTER 8

DYSFUNCTIONS

8.1 INTRODUCTION

This chapter lists some of the common pitfalls that can damage a battery and how to avoid them. Learning how to design a battery is good, but knowing what *not* to do (preferably from other people's mistakes) is just as important. The repair cost and inconvenience of a damaged battery are high, especially if there is an ocean between it and its manufacturer.

8.1.1 Tidbits

Some interesting items in this chapter include:

- RTFM (8.1.4);
- Troubleshooting is not the same as repair (8.1.3);
- When working on a battery, a charger that is turned off can kill your BMS (8.3.2.3);
- Superstition has no place in troubleshooting work (8.1.3);
- Poor placement of a fuse can kill your BMS (8.3.2.4);
- If you see a spark, you may have saved your BMS (8.2.2.3);
- The voltage at one end of a wire is different from the voltage at the other end (8.3.3.1);
- Before you connect your laptop to its AC adapter, make sure it's not at high voltage (8.4.1);
- Don't be surprised that there is no voltage across an open-drain output (8.12.6);
- Don't kill the messenger: if the BMS says something is wrong, don't ignore it (8.14);
- A low-voltage cell is not (necessarily) a bad cell (8.7);
- You're probably looking at the wrong cell (8.6.1).

8.1.2 Orientation

This chapter is divided into three parts:

1. Classic mistakes in the battery design that results in damage to the cells, to the BMS, and the protector switch;
2. Troubleshooting techniques for measurements, mismatched voltages, data evaluation, communications, ground faults, inputs and outputs, power circuits, and fault messages;
3. Typical repair procedures for a battery.

Do read the first part—the mistakes part. I wouldn't expect you to read through the rest, as it is for specific issues: read only the applicable section if you run into problems.

8.1.3 Troubleshooting versus Repair

Do not confuse troubleshooting and repair:

- Troubleshooting comes first—it asks questions and finds answers. When done with troubleshooting, you know the problem, but the symptoms are still there.
- Repair happens last—you can fix issues only after troubleshooting has given you answers. Repair may or may not entail replacing parts.

Troubleshooting is an art, and not everyone gets it. At the same time, troubleshooting is also a science—it must follow a precise sequence so that logical conclusions may be drawn.

Before you start troubleshooting, you must understand what the product does and how it does it. I just read of a car repair shop that spent days trying to fix a car whose engine kept on stopping and restarting. It turns out that it was a microhybrid car. Had the repair people started from this first step—understanding what the product does and how it does it—they would not have wasted all that time.

Typical mistakes include:

- Replacing what "looks wrong" during a visual inspection, For example, hobbyists may replace electrolytic capacitors because there's goop on them; it turns out that this goop is just the glue placed intentionally to hold them in place.
- Believing rumors: replacing all the capacitors, whether electrolytic or not, after hearing about the *capacitor plague*.[1]
- Replacing a fuse without first fixing what made it blow. There's a reason that fuse blew; if a fuse is replaced without first fixing the underlying cause, the new one will blow as well; in the process, the damage worsens.
- Troubleshooting by replacement: replace parts until it works. In practice, inexperienced replacement of good parts is likely to add more faults to the original one.
- Changing more than one variable at a time:
  - Replace a USB cable, use a different AC adapter, and try a different computer;
  - Now it works, therefore "the computer is bad"; not necessarily—it could be the cable or the power supply, or even the AC outlet.
- Being superstitious—if the problem went away when you touched something, then that something must be the problem. Actually, it may be a coincidence: just as you touched that component, the product warmed up and started working,

1. Some batches of electrolytic capacitors manufactured between 1999 and 2007 suffered from early failures.

so it had nothing to do with that component. Repeat the test multiple times to actually prove the correlation before you think you have found the cause.

Specifically, for batteries, typical mistakes include

- Replacing a cell because its voltage is low. Maybe it just needs to be balanced; maybe the BMS is reporting the wrong voltage.
- Replacing a cell board because it reports the wrong value. Maybe it's just a loose connection; maybe you got confused while counting cells and you ended up replacing the wrong cell board.
- Replacing a BMU because you replaced it with another BMU that worked. Maybe the first BMU is configured wrong; maybe there's an intermittent cable that happened to make contact with the second BMU but not with the first one.

Do not replace parts unless you can prove they're bad. Yes, you can try replacing a part temporarily, as a troubleshooting tool. However, afterward restore the original part. Otherwise, you've introduced a variable that you may forget about and will derail your logical train of thought.

Use symmetry as a troubleshooting tool. If there are two identical sections in a product, swap them to see if the problem goes with the sections or the system. For example, if a BMS has two banks, one of which is not reporting, swap the cables to the two banks to see if it's the bank or the BMU. If it's the bank, swap the cables to see if it's the cable or the cells.

8.1.4 Resources

You can save time by working together with people who have experience in various aspects of your battery and are eager to help:

- The battery designer and manufacturer, if someone other than yourself;
- The system designer and manufacturer—the boat-maker, the automotive company, the solar installer;
- The BMS manufacturer, who usually can offer advice on areas other than the BMS itself;
- The cell vendor or even the cell manufacturer;
- Other battery users through a forum; battery information websites (see the "Resources" section in Volume 2).

In many cases, this help is free if not abused. Some of these resources do charge for consulting. Still, a $50 charge for 15 minutes that saves you one day of work is worthwhile.

Of course, RTFM.[2] The troubleshooting section can range from ridiculously simplistic ("turn the power off and on") to hundreds of pages. If the manual is extensive, ask the company's tech support to point you to the right page.

If the manual is online, use Google to find the right page. For example, site:lithiumate.elithion.com/ "baud rate", where "site:" limits you to the manual, and the quotation marks are used to search the exact term.

2. "Read the frigging manual."

8.2 CELL AND BATTERY DAMAGE

This section discusses the most common causes of damage to cells and batteries and how to avoid them.

8.2.1 Cell Damage

As you know, Li-ion cells can be dangerous. They are also expensive, especially when you include the cost of the labor to replace them. To avoid this danger and this cost, know what damages cells and how to avoid this damage.

8.2.1.1 No BMS

All too often, cells are damaged catastrophically when not protected by a BMS that can turn off all the battery current, directly or indirectly. Without a BMS, cells may operate well outside their safe operating area and be badly overcharged, overdischarged, reversed, or overheated.

Even with a BMS, cells degrade more quickly if operated at the edge of their safe operating area. Specifically:

- Cells that are charged to their maximum voltage and held there for an extended period; charging should stop once the maximum voltage is reached. If this is not possible, a lower maximum voltage should be used, such as 3.4V instead of 3.6V for LFP.
- Cells that are charged below freezing temperatures or at excessive current; this may result in dendrite growth, and eventually in a soft short inside the cell (see Section 1.2.2.9).
- Cells that are operated at high current for too long; high current degrades cells more quickly, and indirectly, the resulting high temperature also degrades cells more quickly.
- Cells that are stored at high temperature.

8.2.1.2 Exposure to External Heat Sources

Do not place a battery near high heat sources; be cautious about placing a battery in the engine compartment of a vehicle. Cells exposed to high temperature may reach thermal runaway (see Section8.2.1.5).

Do not place any flammable materials in a battery. Design the battery enclosure to retard external flames.

8.2.1.3 Mechanical Abuse

Cells are damaged irreparably through mechanical abuse:

- Placing localized pressure on a pouch cell, such as by placing a thermistor between the flat surface of a cell and the retaining wall next to it;
- Not containing the expansion of pouch or prismatic cells;
- Stressing the terminals by applying mechanical tension on them, especially in the presence of vibration, rocking, or thermal expansion;
- Damage from improperly designed or secured interconnects.

8.2.1.4 Uncontrolled Expansion

Batteries that do not provide mechanical constraints to the cells (other than cylindrical cells) may result in irreversibly expanded cells (see Section 2.4.6). This is particularly a problem for pouch cells in cheap consumer products (see Volume 2, Section 1.3.2).

8.2.1.5 Thermal Runaway

In general, thermal runaway is a positive feedback process that increases temperature uncontrollably. Once started, it cannot be stopped until destruction due to extreme temperatures. This is what happened to Venus and could happen to Earth if we let climate change continue unmitigated. In particular, Li-ion cells may undergo thermal runaway if abused or damaged.

A Li-ion cell starts a thermal runaway process when exposed to high temperatures from external sources or internal heat due to high current—an internal short or mechanical damage. Self-sustaining exothermic reactions result in an unstoppable, ever-increasing cell temperature until the flammable materials in the cell are exhausted.

The runaway temperature varies considerably with the cell chemistry and, to a lesser degree, with the cell format. This is why chemistries with a high runaway temperature are considered safer. However, the process goes through multiple stages, starting with degradation[3] at a surprisingly low temperature—generally around 120°C, but as low as 80°C—which does not depend so much on the cell chemistry.

Table 8.1 lists these temperatures and rates of temperature increase for some Li-ion chemistries. If it shows a range of values it is because the data are from multiple sources [1, 2].

When exposed to extreme temperatures, cells experience the following, in order:

1. Rapid heating;
2. A first, smaller venting event due to high pressure;
3. A second, larger venting event at the onset of thermal runaway;
4. Ignition, about 10 s after the first venting.

Venting emits white smoke, consisting of the following gases, depending on the cell chemistry [2]:

- H2: ~30%~31%
- CO2: 25%~53%

TABLE 8.1
Thermal Runaway Temperatures

Chemistry	Onset Temperature	Runaway Temperature	Max Temperature*	Rate
LCO	~90°C	120°C ~230°C	720°C	167°C/s
LFP	~90°C	200°C	250°C~420°C	0.02~8°C/s
LMO	~90°C	140°C	590°C	17°C/s
NMC	~90°C	160°C	690°C	42°C/s

*Reportedly, temperatures as high as 1100°C were recorded in some tests, enough to melt aluminum and copper foils.

3. The SEI layer decomposes.

- CO: 5%~28%
- CH4: 4%~9%
- C2H4: 7%~8%
- C2H6: 0%~1%

LFP emits the least amount of gases, LCO the most, and NMC somewhere in between.

The battery should be designed to contain thermal runaway to a single cell by preventing propagation (see Section 5.18.3).

8.2.2 Battery Damage

Bad things happen if cells or BMS taps become disconnected. The most noticeable effect is if the battery no longer works. Otherwise, a less obvious result is that the BMS is damaged. An insidious effect is if the BMS is no longer able to protect the battery, yet the battery remains on.

8.2.2.1 Open Row Connection

If cells become disconnected from their mates on the same row (Figure 8.1(b)), the BMS is only able to monitor the remaining cells in the block. The battery appears to behave the same. However, the disconnected cells may be slightly overcharged or overdischarged, and the BMS won't know.

With only one open connection, there are only two cells (or blocks) in series that are not monitored. Therefore, it is unlikely that those two cells will diverge much from the cells that the BMS is still able to monitor. However, if the self-discharge current in one of those two cells is high, this string of two cells will become significantly unbalanced over time, for which the BMS cannot compensate. At the end of discharge, this cell will be overdischarged.

8.2.2.2 Open Tap

In a wired BMS, if the wire from the voltage sense tap is disconnected, the BMS is no longer able to monitor the voltage of two cells in series (Figure 8.1(c)).

After the sense wire is disconnected, the voltage on the BMS sense input remains the same, and the typical BMS assumes incorrectly that the cell voltage has not changed. An advanced BMS can detect an open sense line (see Section 4.5.1.5).

This is not an issue for a distributed BMS—if a wire to a cell board is disconnected, it is no longer powered, and it does not report, which the BMU detects readily. However, one of the two cell boards (the one that draws the least current) will be exposed to a higher voltage and may be damaged

8.2.2.3 Disconnected Cell and Tap

A cell may be disconnected from its row yet remain connected to the BMS (Figure 8.1(d)). This is dangerous because the BMS does not know the voltage of its former mates on its row.

It is difficult for a BMS to detect this event:

- The BMS may be able to detect that the capacity has dropped. This test is not reliable if there are so many cells in parallel that the change is minimal.

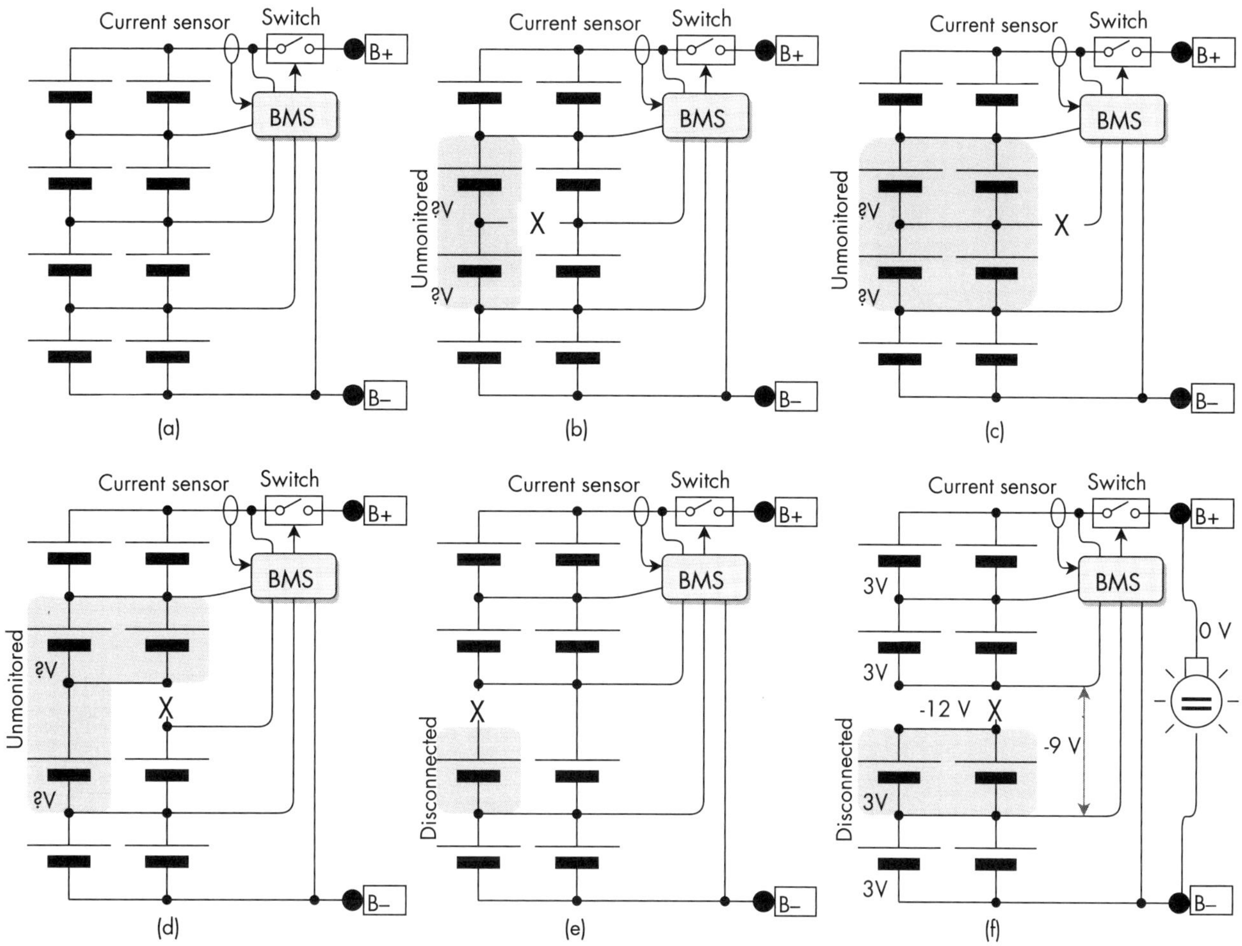

FIGURE 8.1 2P4S battery: (a) functional, (b) open row connection, (c) open tap, (d) disconnected cell and tap, (e) disconnected cell, and (f) disconnected row.

- The BMS may be able to detect that the cell voltage does not change as current flows in and out of the battery, which the BMS evaluates as a cell with 0 resistance.
- The BMS may note that the cell's evaluated SoC never changes, which the BMS sees as a cell with infinite capacity.

In reality, few BMSs are designed to detect this event.

8.2.2.4 Disconnected Cell

If a cell is completely disconnected from its row (Figure 8.1(e)), there is no danger because the cell cannot be overcharged or overdischarged. However, the battery capacity is reduced, which the BMS may report as a sudden reduction in the SoH.

8.2.2.5 Disconnected Row

If an entire row is suddenly disconnected (Figure 8.1(f)), the entire voltage of the battery appears across the two sense wires for that cell, in the reverse polarity. For example, for a 12V battery, -9V appears across the two tap wires.

This event is easily detected because the load is no longer powered; this also damages the BMS (see Section 8.3.2).

8.3 BMS DAMAGE TO CELL VOLTAGE SENSE INPUTS

The BMS is most likely to be damaged on its voltage sensing inputs.

Typically, the cell voltage sensing input of a BMS can handle any voltage from 0 to 5V. One would assume this range is sufficient to monitor a Li-ion cell under any condition because a Li-ion cell's voltage ranges between 1.4 and 4.5V when considering all Li-ion chemistries and any charging or discharging current.

Yet, an excessive voltage may damage the voltage sensing input because sometimes

- The BMS input is not truly connected directly to its cell;
- The voltage at the BMS input is not precisely the same as the voltage at the cell;
- A sudden change in current forces the cell voltage temporarily outside its normal range.

A damaged BMS sense input is baffling because the cell appears to be fine. Initially, you may blame the BMS; after careful analysis, you may discover that, indeed, the BMS input ratings were exceeded. The mechanisms will be covered in the following sections.

To reduce the chance of damage, the BMS input would have to be designed to handle a voltage beyond the standard range of 0 to 5V. While this is possible, it is rarely done because it is expensive, and it reduces the accuracy of the voltage measurement.

If the BMS is installed incorrectly and is not allowed to turn off the entire battery current before overcharge or overdischarge occurred, an excessive voltage may damage a voltage sense input:

- The cell is overcharged, and its voltage exceeds 5V;
- The cell is overdischarged and reversed, so its voltage is negative.

Both the cell and the BMS are damaged, and there is a potential for fire.

8.3.1 BMS immunity to Overvoltage and Reverse Voltage

No BMS is entirely immune to overvoltage or negative voltage on its cell sensing input, though some are partially immune, intrinsically, or by design. In particular:

- Wired BMSs can be designed to handle transitional spikes by including an input filter that slows down the spikes. In practice, this filter has a time constant of no more than 10 ms[4]; the series resistance of a slower filter would be excessive and would affect the accuracy of the voltage reading.
- Distributed BMSs cannot easily include such a filter because the cell board is powered by the cell; the filter would cause a voltage drop due to the current powering the cell board and reduce the voltage reading.
- All multibank BMSs can handle a voltage fault between isolated banks.
- I know of just two off-the-shelf BMSs (one wired, one distributed) that can handle a voltage fault within a bank, but only up to 60V. No BMS can handle

4. Note that 10 ms is a time constant, not a time delay. It does not mean that the BMS is safe for 10 ms. The higher the reverse voltage, the faster it moves through the filter, and reaches the BMS input, even faster than 10 ms.

a voltage fault within a bank above 60V, and no BMS with charge transfer balancing can handle a voltage fault within a bank.

Let's examine some of the conditions that have resulted in these faults to occur in *other people's* batteries, so that you may design and build *your* battery in a way such that those conditions won't occur.

8.3.2 Damage from Disconnection

There are times when a cell voltage sensing input is not connected directly to its cell, whether during installation or in use.

8.3.2.1 During Installation

A BMS is most prone to damage before and during installation:

- The ends of loose wires may land where they should not touch;
- While the BMS is probably ESD safe, take it out of its antistatic bag only when ready to install.

8.3.2.2 Miswired Cell Voltage Sensing

Errors in the cell voltage sense wiring result in excessive voltage (Figure 8.2(a)) or reversed voltage (Figure 8.2(b)) on a BMS input, which results in immediate damage.

Not all BMSs give you a chance to check the wiring before you cause damage.

- Wired BMS—you have a chance to check the voltages on the sense harness before you connect it to the BMU;
- Distributed BMS—you don't get this chance so you must be quite careful; connecting a cell board backward is extremely easy to do.

8.3.2.3 Installing a BMS on a Battery that Is Still Connected

All too often, a BMS is damaged when it's installed on a battery without first disconnecting *EVERYTHING* from the battery—loads, chargers, DC-DC converters, other batteries, and other devices.

In the process of installing the BMS, the connection between two cells may be lifted for just a moment. Typically, this is not a problem because there's no voltage across this gap (Figure 8.3(a)). But if the battery is connected to a load (Figure 8.3(b)),

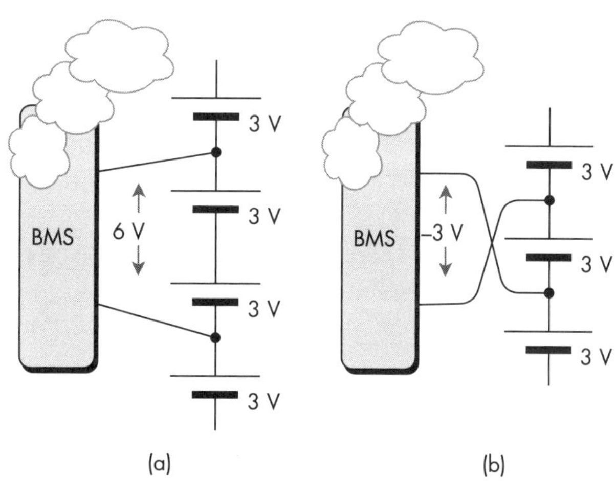

FIGURE 8.2
Miswiring damage: (a) excessive voltage, and (b) reverse voltage.

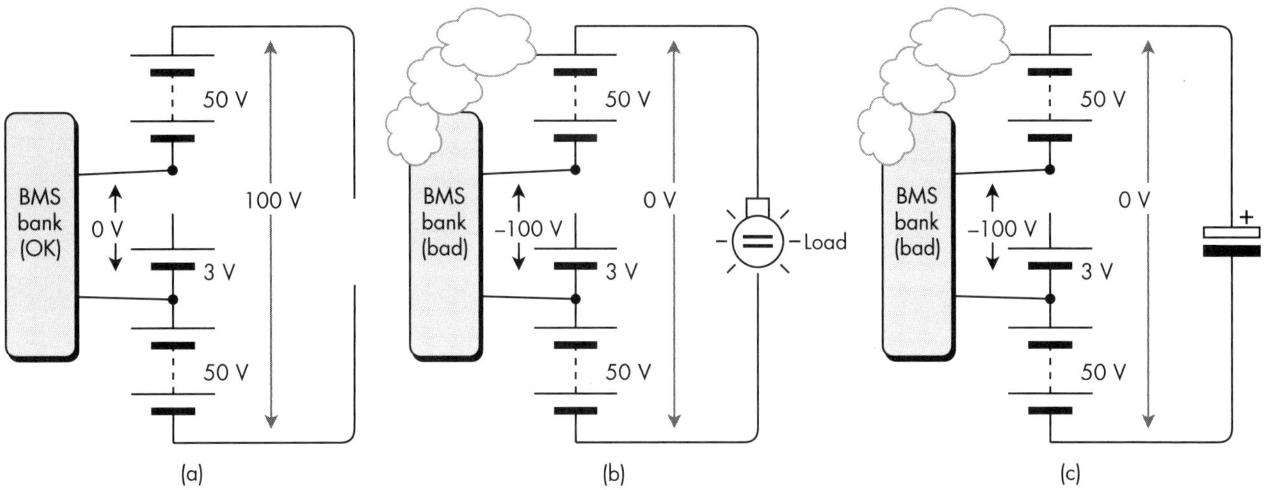

FIGURE 8.3 Voltage across a gap during BMS installation: (a) with no load, (b) with a resistive load, and (c) with a capacitive load.

the entire battery voltage appears across the gap in the reverse polarity. This voltage appears across the gap because there is no current in the system. Therefore, there is no current through the load—the voltage across the load zero. Yet, each of the two sections of the battery, on either side of the gap, has a voltage. This voltage has to appear somewhere: if not across the load, then it has to appear across the gap. The problem occurs even with a capacitive load (e.g., the output of a charger) because initially the capacitor is discharged (Figure 8.3(c)).

With a wired BMS, when installing cell voltage sense wires, if the connection between two cells is open, the entire battery voltage appears across two sense wires, and the BMS is damaged.

With a distributed BMS, when installing a cell board, two scenarios are possible, depending on what is connected first:

- The bus bar: There is a spark as the circuit is completed and current starts flowing. The current is particularly high with a capacitive load; the cell board is fine.
- The cell board: The entire battery voltage appears across the cell board, and it is damaged.

If disconnecting all the loads is not possible, design the battery so that the bus bars have a separate connection point for the BMS, so there is no need to undo a bus bar to connect the BMS. This way, even if the battery is connected to a load, no gap is formed, and the problem of the entire battery voltage appearing across a gap does not occur. The largest prismatic cells include extra terminal that may be used for this purpose.

8.3.2.4 Connection Opens between Cells

The voltage sense inputs of a BMS may be all in one bank or may be divided into multiple banks. A BMS for up to 16 cells in series is likely to have all the inputs in one bank, while a BMS for 48 cells or more is likely to be divided into several banks.

Within a bank, all voltages must increase in the correct order, from tap to tap. As banks are normally isolated from each other,[5] the BMS can handle a high voltage

5. Do check the manual to be sure!

between banks without damage, but can't handle a high or reversed voltage within a bank.

A series string of cells may include a component that could open—a fuse, a safety disconnect, a module connection, a contactor (Figure 8.4). If this device opens under load, the entire battery voltage appears across this open connection and in the reverse polarity, as explained above. If this device is placed within a BMS bank and it opens, the BMS is damaged.

Consider a BMS bank monitoring a group of cells that includes a device that may open (Figure 8.5(a)). If this device does open, the entire battery voltage appears across this gap and blows the BMS bank (Figure 8.5(b)).

On the other hand, if this device is placed between two banks, when it opens, the BMS is not damaged (Figure 8.5(c)).

Plan your battery with the BMS's banking in mind. Match separations between banks with physical separations between battery modules; place a fuse, a safety disconnect, or a midpack contactor only between BMS banks.

8.3.3 Damage from Noise

Noise on voltage sense wires can damage the BMS inputs.

8.3.3.1 Tap Wires Are Antennas

Ordinarily, we assume that the voltage at one end of a wire is the same as the voltage at its other end. Therefore, we assume that a wired BMS sees the correct cell voltage. This is a wrong assumption because long tap wires act as antennas.

Tap wires pick up EMI from nearby switch-mode devises—chargers, motor drivers, invergers—which is then applied to the BMS (Figure 8.6(a)). In particular, the high inrush current that results from connecting a battery directly to a capacitive load without precharge generates an EMP[6] that could be strong enough to damage a BMS input by exceeding its maximum voltage rating.

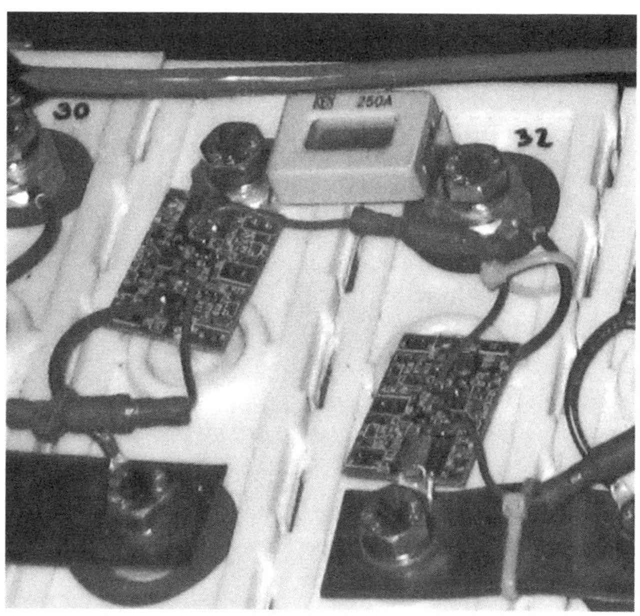

FIGURE 8.4
A fuse installed in the middle of a bank. Don't do this.

6. Electromagnetic pulse.

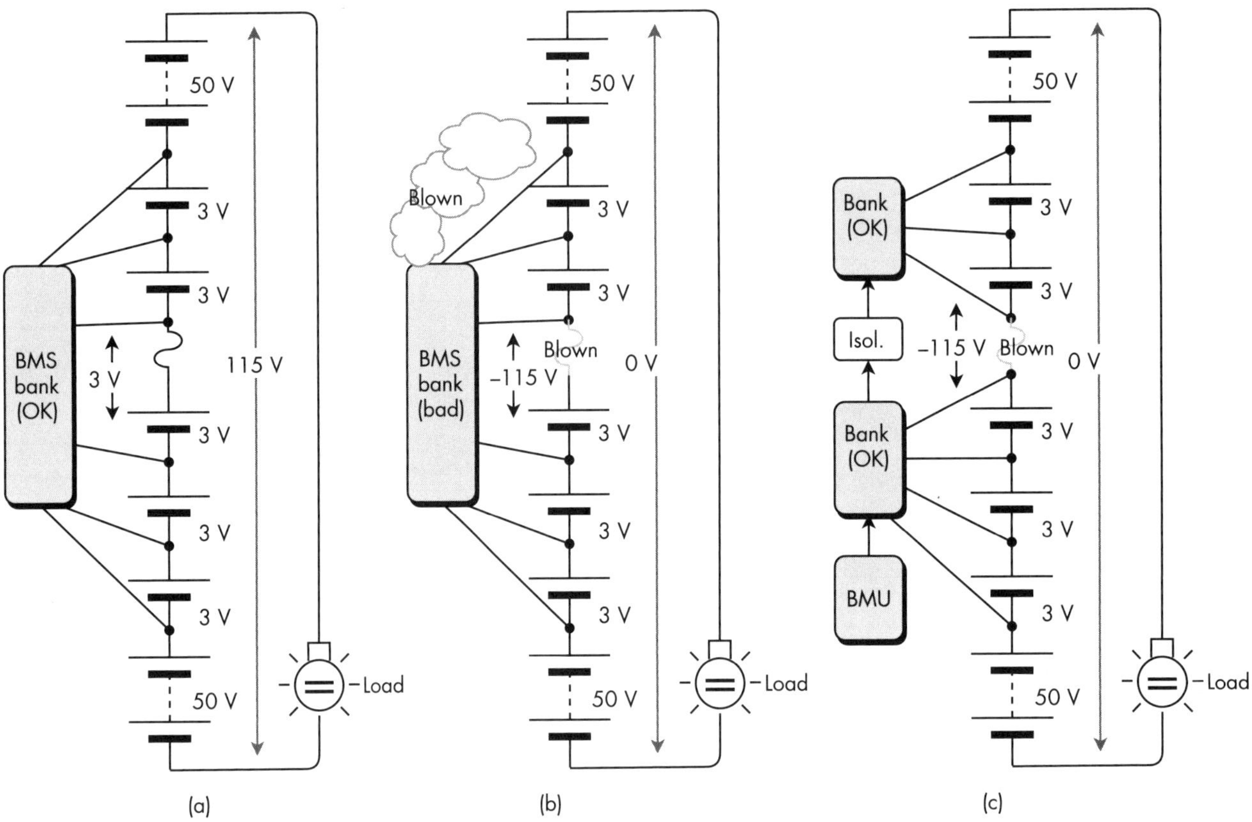

FIGURE 8.5 Placement of component that may open: (a) within a bank, fuse OK, (b) within a bank, fuse blown, and (c) between isolated banks, fuse blown.

FIGURE 8.6
Radiated noise pickup:
(a) long tap wires, and
(b) short, parallel tap wires.

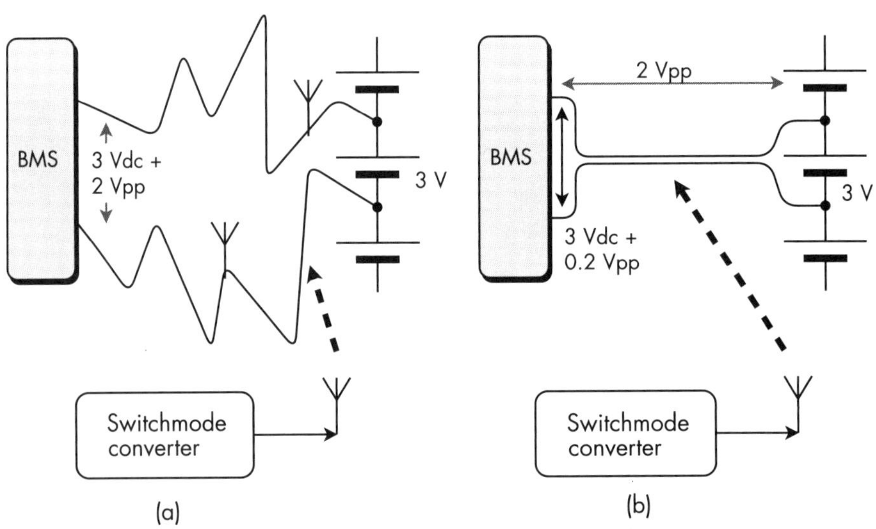

To reduce this noise

- Run tap wires parallel to each other, and even twist a bundle of wires, to achieve a degree of cancellation of common-mode noise (Figure 8.6(b));
- Strive for a battery layout that allows short, direct connections;
- Use a wired BMS with effective noise filtering.

Shielding the tap wires can be problematic, as the shield has to be referenced to the cell voltage, not to ground. This becomes one more conductor that can touch a cell terminal and create a short circuit.

8.3.3.2 Voltage across Bus Bars

We also assume that the voltage at one end of a bus bar is the same as the voltage at its other end. Therefore, we assume that it doesn't matter at which end of a bus bar a tap wire is connected. This is a wrong assumption because, under load, the voltage across power wires and bus bars is not zero; the cell voltage measurement depends on where the BMS is connected.

Often the battery current is not just smooth DC: it also includes a strong AC component due to the switch-mode power devices connected to the battery.[7] The power connection between two cells has nonzero impedance, which converts this noise current to an AC voltage. Long connections (such as between modules) have higher impedance and therefore have higher voltage across them. This noise voltage can have an amplitude of a few volts! This AC voltage is superimposed to the cell voltage and sent to the BMS (Figure 8.7(a)).

To increase immunity to this noise

- Use separate BMS banks on each side of a long connection (Figure 8.7(b)).
- Use a distributed BMS and connect each cell board directly to its own cell's terminals (Figure 8.7(c)). The communication between adjacent banks or cell boards must have high noise immunity using isolation or current sources.

8.3.4 Damage from Cell Voltage Spikes

The cell terminal voltage is not rock-solid because the cell has a nonzero series impedance.

When there is a transition in the current, the cell's terminal voltage jumps (see Section 3.2.12). The spikes may be positive or negative, depending on how the current charged. The duration of these spikes is too short to damage the cells but sufficient to damage most BMSs.

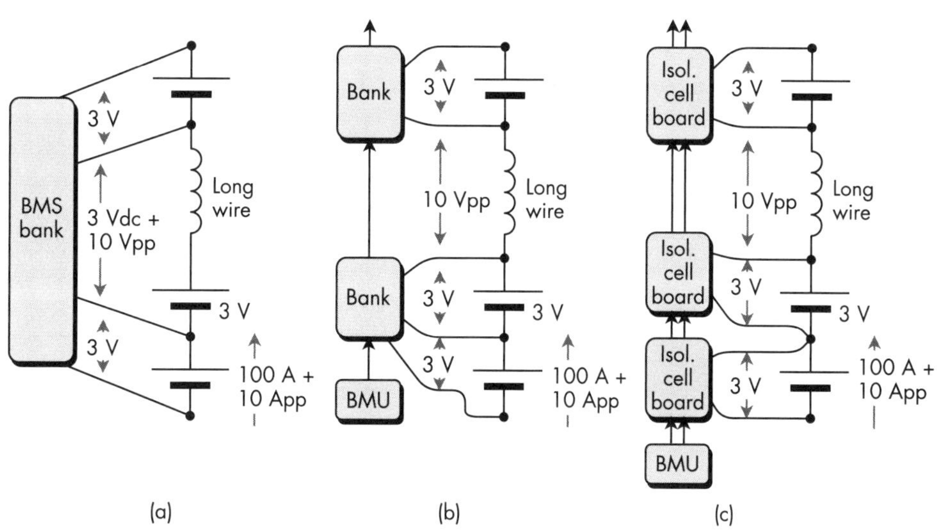

FIGURE 8.7 Power connection with high impedance: (a) same bank, (b) separate banks, and (c) distributed BMS.

7. Motor drivers and inverters operate at about 20 kHz, DC-DC converters at 100~500 kHz. This is only the fundamental frequency: harmonics and ringing are at higher frequencies.

It is technically possible to design a BMS that can survive such spikes, but it is not economical (see Volume 2, section A.9.2).

What's more likely to be a problem, positive spikes or negative ones? On the one hand, negative spikes tend to be taller than positive ones:

- *Negative spike:* From 2.5V for a discharged cell to the bottom BMS limit of –0.5V the headroom is 3V;
- *Positive spike:* From 4.2V for a charged cell to the top BMS limit of 5V the headroom is 0.8V.

On the other hand, there is less headroom for positive spikes than negative ones:

- *3V negative spike:* From 2.5V for a discharged cell to -0.5V for a BMS limit;
- *0.8V positive spike:* From 4.2V for a charged cell to 5V for a BMS limit.

When both are considered, for a given cell resistance, a negative spike is more likely to damage a BMS by a factor of 2.5:1.

8.3.4.1 Problem Reduction

To reduce the effect of spikes in the cell voltage

- Keep the battery balanced and the cells at the same temperature, especially before connection to loads or chargers;
- During manufacture, before making the final connection, use a resistor to precharge a load or a charger;
- For loads connected through a contactor, implement precharge (see Section 5.13) or use a protector BMS with a solid-state protector switch;[8]
- If a charger is included in the product, either design the system so that the charger is connected to the battery directly and permanently (not through a contactor), or implement a precharge circuit, or use a solid-state protector switch;
- If the product uses an external charger, implement a precharge circuit or use a protector BMS with a solid-state protector switch;
- If a capacitive load may be connected at any random time, precharge cannot help; use a BMS that is designed to handle spikes.

8.3.5 Damage from Cell Voltage

Sometimes the BMS is damaged because the cell voltage truly is excessive.

8.3.5.1 Cell Voltage Reversal

With a properly designed and installed BMS, the battery cannot be overdischarged and the cell voltage cannot be reversed because as soon as any cell voltage reaches the low-voltage limit, the BMS disables further discharging. Therefore, the cell voltages remain positive, and no damage occurs to either the cell or the BMS.

Otherwise, overdischarging a battery results in the reversal of the voltage in the cells with the lowest capacity (see Section 3.2.11). This occurs because

- There is no BMS, or the "BMS" is only a monitor;
- The BMS has a design fault and does not disable discharging;

8. A MOSFET acts as a current-limited switch.

- The system designer refuses to let the BMS disable discharging (see Volume 2, Section 3.2.4.1);
- The battery designer left a path that bypasses the protector switch, one that will continue to discharge the battery even after the BMS disables discharging (Figure 8.8(a)).

The result is that the voltage of the cell with the lowest SoC will drop below its minimum safe voltage, then will rapidly discharge down to 0V, and then actually reverse; its positive terminal will become more negative than its negative terminal (see Section 3.2.11).

At this point, both the cell and the BMS are damaged:

- At best, the cell may effectively become a short circuit; at worst, the cell may become an electrochemical mess ready to catch fire later if someone attempts to recharge it;
- Current flows into the BMS input, at best fusing components and at worse overheating them (see Volume 2, Section A.9.1).

To avoid cell reversal

- Select a proven, honest-to-goodness BMS, not a "monitor";
- Do not connect a load directly to the cells (e.g., a small DC-DC converter); do connect all devices to the load side of the discharge protector switch, not to the cell side; and do not connect the cells to anything but the BMS;
- Wire the BMS to allow it to shut off the discharging current to all the loads.

8.3.5.2 Cell Overvoltage

With a properly designed and installed BMS, no cell voltage ever exceeds its maximum rating because, as soon as any cell voltage reaches the high-voltage threshold, the BMS disables further charging. Therefore, the voltage of each cell remains within its safe

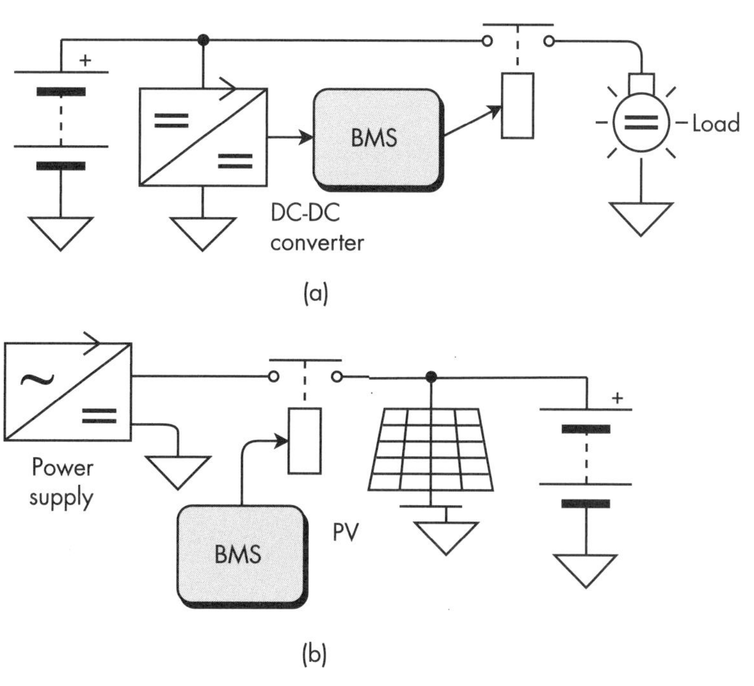

FIGURE 8.8 Uncontrolled devices: (a) uncontrolled load that overdischarges a battery, and (b) uncontrolled source that overcharges a battery.

operating area, and no damage occurs to the cells. As this voltage is less than 5V, no damage occurs to the BMS either.

If the system does not control all charging sources (Figure 8.8(b)), the cell voltages will exceed the safe maximum, then will rapidly reach 5V, damaging the BMS. Current will flow into the BMS input, at best fusing components and at worst overheat them. The cell is degraded as its voltage keeps on increasing. Eventually, the cell will overheat, reach the self-igniting temperature, and go into thermal runaway.

To avoid cell overcharge, in addition to the measures listed in the previous section, pay particular attention to secondary charging sources such as a small solar panel to top off the battery or regenerative braking, especially braking down a long mountain.

8.4 OTHER BMS DAMAGE

While the most likely BMS damage is to the cell voltage sensing inputs, other functions may be damaged as well.

8.4.1 Short Circuits

A BMS may be damaged when short-circuit current flows from the battery through the BMU. This current flows along a complete, closed-loop path, which can only occur if both of the following are true:

- The battery is grounded (whether or not intentionally).
- There is an unintentional short between a cell terminal and a low-voltage wire to the BMS:
  - Distributed BMS: Typically, a poorly build communication cable between the BMU and a cell board has some exposed shield that touches a cell terminal (Figure 8.9(a));
  - Wired BMS: Typically, broken insulation in a thermistor exposing its wire, which then touches a cell terminal (Figure 8.9(b)).

In either case, current flows from a cell terminal, through the unintentional short, through the wire to the BMS, through the PCB in the BMS, to either the 12V supply (if grounded) (Figure 8.9(a)), or to a laptop (if powered by a grounded AC adapter) (Figure 8.9(b)), to earth ground, to the negative of the battery, and back to the cell terminal.

In the first case, we see damage to the BMS inputs from the cells. In the second case, we see damage to the BMS communication port.

If the battery is not grounded, even if there is an unintentional short, no damage occurs initially. It would occur later, when the battery is grounded—accidentally, on purpose, or indirectly through the load. I can relate two examples:

- We have a customer whose BMS blows up when the laptop battery gets low; this is when he plugs in the laptop's AC adapter, completing the path to earth, resulting in a damaging short-circuit current;
- We have another customer whose BMS blows up every time she turns on the inverger, whose high-voltage side is referenced to the AC power line; when it's turned on, it completes the path through the earth ground.

To avoid such damage, one must test the battery as it's being built, and before deployment (see Section 7.7.1).

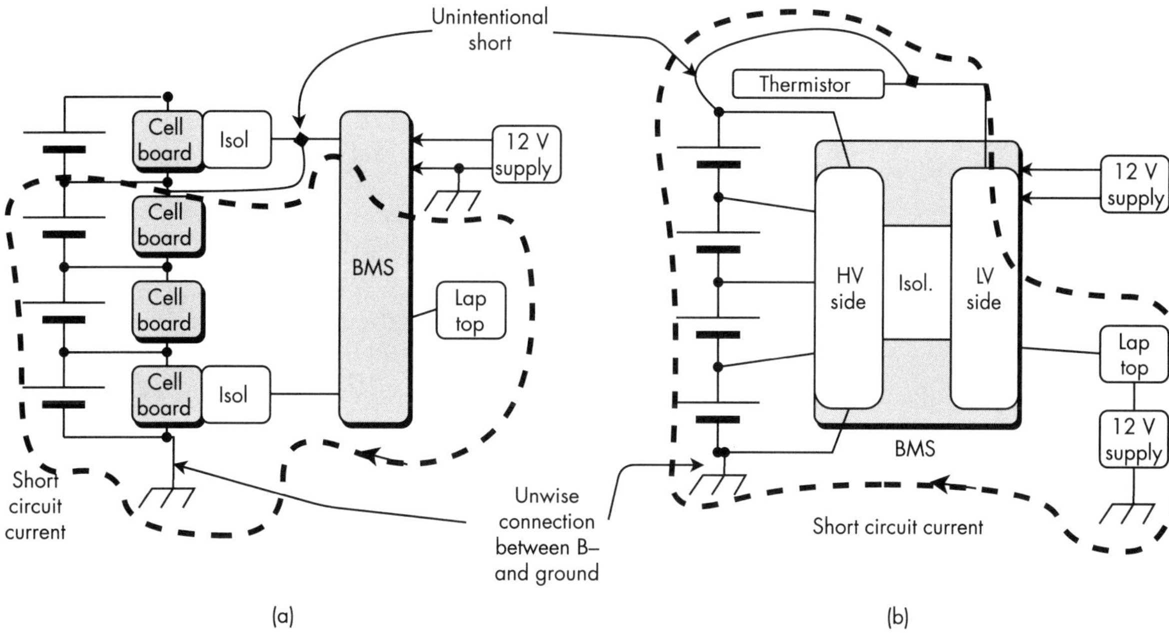

FIGURE 8.9 Short-circuit current path through BMS: (a) distributed BMS, grounded 12V supply, and (b) wired BMS, grounded computer.

8.4.2 Damage to Inputs and Outputs

BMS lines other than the voltage sense inputs may be damaged as well.

8.4.2.1 Power Supply Inputs and Outputs

Typically, a BMS's power supply input accepts a wide range of input voltages. Read the specs and respect the specified limits. If the manual specifies a power supply voltage range of 9 to 16 Vdc, give it 12 Vdc. Don't give it 24 Vdc, 12 Vac, or -12V.

Some BMSs can withstand excessive input voltages and negative input voltages without damage (e.g., BMSs for automotive use). They do shut down, yes, but they recover once the proper voltage is restored. Otherwise, excessive voltages or negative voltages do damage the BMS.

A BMS may include a 5V supply output to power a sensor (e.g., Hall effect current sensor) or peripheral (e.g., a display). This output may be able to handle a short circuit across it. It is damaged by feeding an external voltage into it, such as 12V.

8.4.2.2 Driver Outputs

A BMS may have an output to drive power loads, such as contactor coils. These outputs usually are a *low side switch*[9] and are either open or grounded.

These outputs are damaged by

- Excessive current—using a load that draws more current than the switch is rated for, especially in case of a shorted load;
- Excessive supply voltage—drive a load whose supply whose voltage exceeds the maximum voltage rating of the BMS driver;

9. A transistor to ground (see Section 4.10.7).

- Attempting to use it as a *high side switch*—connecting the output to the positive supply voltage.

Most BMSs are protected against inductive kickback (when an inductive load, such as a relay coil, is turned off). If yours isn't, it could be damaged by this effect.

8.4.2.3 Relay Dry Contacts

The BMS may have normally open and normally closed relay contacts. They are not connected to anything else. Typically there is no protection for the contacts, which can be damaged by excessive current (e.g., short-circuited load) or arcing due to high voltage or turning off an inductive load.

8.4.2.4 Communication Ports

A CAN bus port can be damaged by connecting the CAN lines to a 12V supply or a charger that has excessive leakage from the AC power to the low-voltage side.

8.4.2.5 Signal Inputs

Digital inputs and analog inputs are often damaged by the application of a negative voltage, a positive voltage above 5V, or an AC voltage.

8.4.3 Protector Switch Damage

The protector switch in a battery may be damaged by operation outside its specified limits. A fuse in the battery may help in some of the following cases, but not all.

8.4.3.1 High MOSFET Temperature

A protector BMS shuts off the battery current in case of overcurrent to protect its MOSFETs. However, the overcurrent may be too low to trip the shut-off, yet high enough to overheat the MOSFETs. As a MOSFET becomes hotter, its resistance increases. This positive feedback results in a vicious cycle, which can end-up melting the MOSFET.

8.4.3.2 Multiple Batteries in Series

MOSFETs in a protector BMS are rated for the battery voltage, and not much higher.[10] If batteries with a protector BMS are connected in series, when one battery opens its switch, the total voltage of the batteries in series appears across this switch, and this voltage may be higher than the rating of its MOSFETs. If so, the MOSFETs may be damaged and current may continue to flow.

For example, a 12V battery uses MOSFETs rated for 25V, which is fine. But if four such batteries are connected in series to get to 48V, when one of the batteries opens under load, 48V appears across its discharge MOSFET, blowing it.

8.4.3.3 Shorted Charging Port in a Dual-Port Battery

In a two-port BMS, a short circuit across the charge port damages the switch in this port because the BMS cannot turn it off.

10. This is because high-voltage MOSFETs are not desirable: they are more expensive and have higher resistance.

8.4.3.4 *Contactor*

A contactor may be damaged by

- Direct connection to a capacitive load without precharge;
- A short across the load;
- Connection to a bus that is already connected to another battery;
- The opening an inductive load while operating at full current due to arcing.

The contacts could be damaged and may even weld closed (Figure 8.10(a)).

8.4.3.5 *Precharge Resistor*

Precharge resistors are sized to handle a single precharge cycle. If multiple cycles occur in a row without sufficient time for cooling, or if there is a short circuit across the battery terminals, the precharge resistor overheats and may explode (Figure 8.10(b)). The expelled wires from a wire-wound precharge resistor can do interesting damage to adjacent circuits!

8.4.4 Mechanical Damage

Electronic assemblies can be damaged by stress or intentional modification.

A customer removed the PCB assembly of BMU from its metal case and cut off the four corners of the PCB, in the process cutting off a resistor (Figure 8.11).

8.5 POWER-UP TROUBLESHOOTING

The rest of this chapter discusses troubleshooting techniques, while the last section discusses repairs. You can skip it if you wish unless you have a problem to resolve. If so, use the index to find if your issue is covered and then read the relevant section.

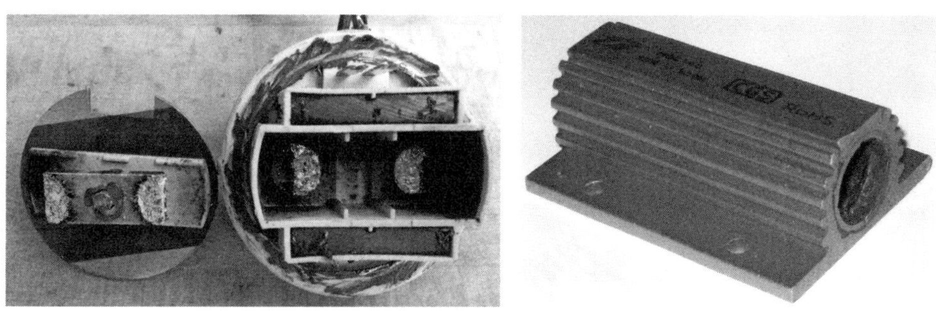

(a) (b)

FIGURE 8.10
Damaged components: (a) contactor with welded contacts, and (b) precharge resistor after violently expelling its guts.

FIGURE 8.11
Damaged assemblies: customer cut a corner out of a PCB, cutting off a resistor.

8.5.1 No BMS Power

The BMS appears to be not powered—if an LED should be lit, it isn't. If it has a power supply output, there's no voltage on it.

Procedure:

- Measure the voltage at the BMS's power supply input relative to the common ground;
  - No voltage:
    - If the BMS is powered through a DC-DC converter, measure its input voltage:
      - No voltage on its input:
        - The power source is off;
        - Blown fuse;
        - Broken wire;
        - Short across this supply voltage.
      - There's voltage on its input:
        - The DC-DC converter is disabled;
        - Open wire;
        - The DC-DC converter is broken;
        - The output of the DC-DC converter is shorted.
    - If the BMS is powered directly:
      - The power source is off;
      - Blown fuse;
      - Broken wire;
      - Short across this supply voltage.
  - Voltage is lower than the minimum supply voltage:
    - Disconnect the BMS:
      - The voltage is still too low:
        - Disconnect other devices connected to this same supply:
          - The voltage is still too low:
            - Check the source of this voltage to see why it's low
          - The voltage goes back to normal:
            - Troubleshoot the other devices.
      - The voltage goes back to normal:
        - Reconnect the BMS power only, keep all other lines disconnected:
          - The voltage remains normal:
            - A load connected to a BMS output is shorted
            - The supply cannot provide enough current to power the BMS loads
          - The voltage goes back to being too low:
            - The BMS is damaged;
            - The supply cannot provide enough current to power the BMS.
  - The supply voltage is in the correct range for the BMS:
    - Bad BMS;
    - Open ground wire.
  - The voltage is higher than the maximum supply voltage (hopefully, the BMS shut itself down to protect itself; otherwise, it would be damaged):
    - A DC-DC converter is required to drop the voltage;

- There's a short between the battery voltage and the BMS power supply.

8.5.2 BMS Power Cycles Constantly

If a load connected to the BMS is shorted, it will cycle off and on continuously, due to this loop:

- The BMS wakes up;
- After a bit, it closes this output;
- But the load on this output is shorted, so the power supply is brought down to 0V;
- That removes power to the BMS;
- The BMS shuts down, opening this output;
- The power supply is no longer shorted and recovers;
- The cycle repeats.

If the power supply cannot supply enough current, it results in the same symptoms. Try powering the BMU with a 12V battery (or whatever voltage it's designed for) and see if this fixes it.

If a power adapter used to power a BMS generates AC instead of DC or produces a poorly filtered DC, the BMS won't be able to remain powered.

8.5.3 Warnings and Fault Troubleshooting

In general, the BMS reports the following faults in one form or another.

- Cell under-voltage:
  - The BMS is not wired to shut down all loads, directly or indirectly;
  - A cell has a high self-discharge;
  - The battery was in storage too long, and cells self-discharged.
- Cell overvoltage:
  - The BMS is not wired to shut down all charging sources, directly or indirectly;
- Voltage tap (wired BMS):
  - A voltage tap wire is open
- Imbalance:
  - Cell voltages differ too much.
- Overtemperature:
  - A cell or other part of a battery is too hot;
  - Open thermistor.
- Undertemperature:
  - Attempted to charge below freezing;
  - Shorted thermistor.
- Thermistor (wired BMS):
  - Open or shorted thermistor.
- Overcurrent:
  - The BMS tried to shut down the battery test but the system disobeyed;

  - The BMS instructed the system to limit the current to a maximum but the system disobeyed.
- Communication error:
  - The CAN bus is stuck;
  - A bank or a cell board is not reporting (distributed BMS);
  - A slave is not reporting (master/slave BMS).
- Ground fault:
  - The battery is not isolated.
  - The test is too fast and does not take into account the battery's capacitance to ground;
  - Another ground fault tester is testing at the same time.
- Contactor or precharge failure:
  - A contactor is welded closed or is not turning on.
  - The precharge did not occur as expected.

All of these are discussed in more detail in the following sections.

In some cases, before issuing a fault, the BMS issues a warning, which gives the system some time to alert the user so that the issue may be corrected. The BMS clears the warning if the problem is rectified. For safety reasons, the BMS doesn't clear a fault. Only human intervention can clear a fault.

8.5.4 Current Limit Troubleshooting

A BMU may communicate a set of limits to the external system, including charging current limit and discharging current limit (see Section 4.6).

If a limit is unexpectedly active (meaning that it's telling the system to reduce the current or stop altogether), the BMS should indicate the reason.

- CCL is < 100%:
  - A cell voltage is high;
  - The total battery voltage is high;
  - One of the temperatures is too close to freezing and the battery is charging;
  - One of the temperatures is too high;
  - The current has been quite high recently.
- DCL is < 100%:
  - A cell voltage is low;
  - The total battery voltage is low;
  - One of the temperatures is too low or too high;
  - The current has been quite high recently;
  - The SoC is low.

8.6 MEASUREMENT TROUBLESHOOTING

This section helps discover the cause of incorrect or unexpected measurements.

8.6.1 Cell Voltage Troubleshooting

- Low cell voltage reading while balancing:
    - Loose connection to the cell; the balancing current generates a voltage drop across the loose connection. Note that some BMSs turn on bypass balancing while measuring, whether or not balancing is required at the time, specifically to detect this problem.
- Low-voltage reading when the battery is discharging at high current and high-voltage reading when the battery is charging at high current:
    - Measure the actual voltage of the cell during this time:
        - The reported voltage matches the measured voltage:
            - Cells have too high a resistance:
                - Energy cells used when power cells should have been used;
                - Cells have degraded over time.
        - The reported voltage does not match the actual voltage:
            - Long (high resistance) interconnection between cells:
                - Use two separate banks, one on each side of this interconnection;
                - Wired BMS: it may be possible to correct for this effect[11];
                - Distributed BMS: connect each cell board directly to its cell.
- Low or high cell voltage reading with no battery current:
    - Measure the actual voltage of the cell during this time:
        - The reported voltage matches the measured voltage:
            - Unbalanced cell (see Section 8.7).
        - The reported voltage does not match the actual voltage:
            - You may be looking at the wrong cell; if the cells or banks are not numbered in the same order as the BMS numbers them, you're going to pull your hair out trying to figure out which cell is which (see Section 4.5.1.9).
            - Poor connection to sense the cell voltage:
                - Some BMSs turn on the balance load while measuring the voltage to detect a poor connection by reporting a low voltage
            - Possibly a bad BMS:
                - For a distributed BMS, try replacing that cell board

8.6.2 Wired BMS Troubleshooting

A wired BMS has its own set of typical failures of the cell voltage sensing.

8.6.2.1 Missing Bank

If an entire bank is not reporting cell voltages:

- The bank's connector is disconnected;
- Inside the BMU, the circuit for this bank is damaged;
- The slave that manages this bank is bad or disconnected from the master.

11. Orion BMS.

8.6.2.2 Slowly Drifting Cell Voltage Reading, Full Scale or 0V

A cell reports a voltage that is slowly drifting. Eventually, it reaches either minimum or maximum reading:

- The cell's tap wire is open;
- A contact in the connector for the tap wire harness is broken;
- A fuse or trace inside the BMS is open.

8.6.3 Distributed BMS Troubleshooting

Each distributed BMS is different, but basically, the BMU communicates with several banks. Within each bank, cell boards communicate with each other. The first cell board receives messages from the BMU, and the last cell board sends messages to the BMU (see Section 4.2.2.7).

8.6.3.1 Troubleshooting through Cell Board LEDs

In some BMSs, the cell boards include an LED, which helps to troubleshoot:

- If no LEDs blink, the message from the BMU doesn't make it even to the first cell board, so look at the communication between the BMU and the first cell board; maybe the BMU or the first cell board is bad.
- If LEDs blink up to a certain point, then either the last blinking board is damaged, or the first nonblinking board is, or the wire between them is open.
- If all LEDs blink, the message doesn't make it back to the BMU. Look at the communication between the last cell board and the BMU; maybe the BMU or the last cell board is bad.
- If only one board doesn't blink, this board is probably damaged or its cell voltage is too low to power it.

8.6.3.2 All Banks Are Missing

If no banks are reporting, follow this procedure:

- Check that the BMU is configured to see the banks;
- Focus on just one bank;
- Check all connectors in this bank;
- Check with a multimeter that the bank harness is wired correctly;
- Check the connections between cell boards;
- Check that the bank is in the correct order: positive cell board on the most positive cell in the bank;
- Check that the cell voltages are OK because cell boards require a minimum voltage.

8.6.3.3 One Bank Is Missing All the Time

If a bank is not reporting, follow this procedure:

- Check that the BMU is configured to see this bank
- Swap two banks:
  - If the BMU reports that the same bank is missing, the problem is the BMU

- Otherwise, the problem is the bank:
  - Swap two bank harnesses.
    - If the BMU reports that the other bank is missing, the problem is the harness:
      - Use a multimeter to check continuity in the harness.
    - Otherwise, the problem is the bank:
      - Check the connections between cell boards;
      - Check that the bank is in the correct order: positive cell board on the most positive cell in the bank;
      - Check that the cell voltages are sufficiently high to power the cell boards.

8.6.3.4 Missing Sequence of Cell Boards

If all the cell boards are missing from a certain point to the end of the bank:

- Check the wiring between the last board that is present and the first one that is absent;
- Try replacing the last board that is present;
- Try replacing the first board that is absent.

8.6.3.5 Missing Cell Board

If only one cell board is missing:

- Check the cell voltage (if too low, it can't power the cell board);
- Try replacing the cell board.

Do not incorrectly diagnose that the last cell board in a bank is damaged. If a bank has 10 cell boards (#1 to #10), and one of them is not reporting, the BMS assigns IDs #1 through #9 to the good boards regardless of their actual physical position. You would be wrong to conclude that cell board #10 is damaged: it's just as likely that #1 is damaged, and the BMS numbered the remaining boards #1 through #9 instead of #2 through #10 because it doesn't know any better.

8.6.3.6 Missing Bank in the Presence of Noise

When a source of electrical noise is turned on, a bank stops reporting.

The noise could be affecting the communications between the BMU and the bank, or between adjacent cell boards. There is no easy way to see which one it is. Try this:

- Check the shielding in the bank harness (visually and with a continuity meter):
  - The shield must be continuous, from the BMU to the cell boards;
  - The shield must be grounded only at the BMU; it must not be grounded to any metal enclosure;
  - Any opening in the shield must be shorter than 2 cm;
  - Inside the shield, wires must be twisted (i.e., twisted pair).
- Check that the battery terminals are isolated from the bank harness; in particular, that the shield of the harness does not contact any cell terminal.
- Check the wires or cables between adjacent cell boards:
  - Route the wires or cables between adjacent cell boards against the corresponding power bus bar and away from any grounded metal surface: you want to maximize

the capacitance to the bus bar (which acts as a shield) and minimize the capacitance to ground (which acts as a noise source) (see Section 7.5.2.8);

- You may need to shield those wires or cables;
- Nonisolated cell boards use a single communication wire and use the power connection between cells as the return wire; make sure that the power connection between two adjacent cell boards is a short bus bar, not a long cable.

8.6.3.7 Missing Cell in the Presence of Noise

When a source of electrical noise is turned on, a cell board stops reporting correctly:

- Make sure that the power connection between two adjacent cell boards is a short bus bar, not a long cable;
- Minimize the wire length between the cell board and its cell;
- Route the wires or cables between it and its adjacent cell boards against the corresponding power bus bar and away from any grounded metal surface.

8.6.3.8 Doesn't Report for a While after the Contactor Closes

There is no precharge. The inrush current generates an EMP[12] that scrambles all communications.

8.6.3.9 Extra Cells

The BMU reports more cells in a bank than expected. Check the following:

- You may be looking at the wrong bank;
- Check the configuration for that bank.

8.6.3.10 One Board Reports Minimum or Maximum Voltage

The cell board does not report the actual cell voltage. Instead, it reports the minimum voltage (e.g., 2V) or the maximum voltage (e.g., 5V). This is probably a bad cell board.

8.6.4 Battery Voltage Troubleshooting

The BMS reports two different values for the total battery voltage, one measured directly, and one by adding the individual cell voltages. This could be due to various causes, depending on the two values:

- The two values do not match:
  - A difference of 100 mV or ± 2% of the full-scale value, whichever is greater, is normal;
  - The BMS is not measuring the cell voltages correctly.
- The value measured directly is half the actual value;
  - There are two strings in parallel, and the BMS does not know how to deal with it.
- The value measured directly is 0V;
  - Disconnected wire;
  - BMS configuration;

12. Electromagnetic pulse.

- The BMS measures the terminal voltage, not the string voltage, and the protector switch is open.

8.6.5 Temperature Troubleshooting

There can be various issues with the reported temperature:

- Reports below freezing temperature: The thermistor is open;
- Reports above boiling temperature: The thermistor is shorted;
- Reported temperature is excessive: Two thermistors are shorted together;
- Reported temperature is off by a few °C: Thermistors are not very accurate, you know?

8.6.6 Current Troubleshooting

Readings for the current may be off in different ways:

- There is battery current:
  - BMS sees no current:
    - The current sensor is disconnected or a wire is broken;
    - The BMS is not configured to look for that current sensor:
      - Configure the BMS for that sensor;
      - If the BMS has multiple operating modes (e.g., *charging* and *ignition*) it may only look at that sensor during one mode, and now it may be in a different mode.
    - The current sensor is bad;
    - Battery current is flowing elsewhere, not through that sensor.
  - BMS reports charging when discharging, or vice-versa:
    - If the BMS allows it, configure the BMS to flip the sign of the current;
    - Otherwise, reinstall the current sensor in the opposite direction.
  - Reported current is twice the actual current (see Section 5.9.2.3):
    - There are two current sensors, and the BMS sees both, so it is adding the two values;
      - Configure the BMS to look at only one sensor at a given time;
      - Move the sensors so that one sees only the charging current, and one sees only the discharging current.
    - The *gain* setting in the BMS configuration is wrong for that current sensor.
  - Normally works fine, but a high current reported much lower:
    - The current sensor is maxed out and doesn't measure a current that high; use a different sensor.
  - The reading of the current is noticeably off:
    - Calibrate the reading by updating the *gain* and *offset* setting in the BMS configuration;
    - Check that the correct shunt resistor or Hall effect current sensor is in use.
  - The reading of the current is slightly off:
    - Hall effect current sensors are not that accurate; still, play with the *gain* and *offset* settings and see if you can improve it;
    - Switch to a closed-loop Hall effect sensor or, if possible, a shunt sensor.

- While a charger is charging the battery, the reading fluctuates even though the current is actually steady:
  - The current from the charger is not continuous (for example, it has a strong component at twice the line frequency, 100 or 120 Hz); you are seeing a beat frequency between that component and the sampling time in the BMS. It's OK: on the average, the BMS does register the correct current.
- There is no battery current:
  - The BMS reports high current:
    - Disconnected current sensor or broken wire from the current sensor;
    - The *offset* setting in the BMS configuration is wildly incorrect.
  - The BMS reports some current:
    - Calibrate the reading by updating the *offset* setting in the BMS configuration;
    - Actually, there *is* current, and you missed a path through which that current is flowing.

8.7 MISMATCHED CELL VOLTAGE TROUBLESHOOTING

This section guides you in understanding why the BMS reports that the cell voltages differ.

8.7.1 Identify the Cause

You would expect all the cell voltages in a string to be the same. If not, you may suspect a problem with balancing, though other causes could have the same effect.

In the following discussion, we assume that the BMS reports whether or not it is balancing the battery.

Mismatched voltages could be due to one of four causes:

1. The BMS decided not to balance;
2. Incorrect BMS measurement;
3. Some cells have low capacity;
4. Actual string imbalance.

Use this procedure to determine which of these four causes is responsible[13]

- The BMS reports that it's not balancing—cause #1;
- The BMS reports that it's balancing:
  - Measure the cell voltage with a multimeter:
    - If the cell voltages are fine and the BMS reports the wrong voltage, it's the BMS—cause #2;
    - If BMS reports the correct cell voltages or if it is not possible to measure the actual cell voltages:
      - Do a full charge or discharge cycle at low current, while logging the cell voltages;
      - Plot the cell voltages versus string SoC:
        - If the curve for one cell is narrower relative to the curve for all the other cells, the problem is that the one cell has a low capacity (Figure 8.12(a))—cause #3;

13. The next section addresses the solution to each cause.

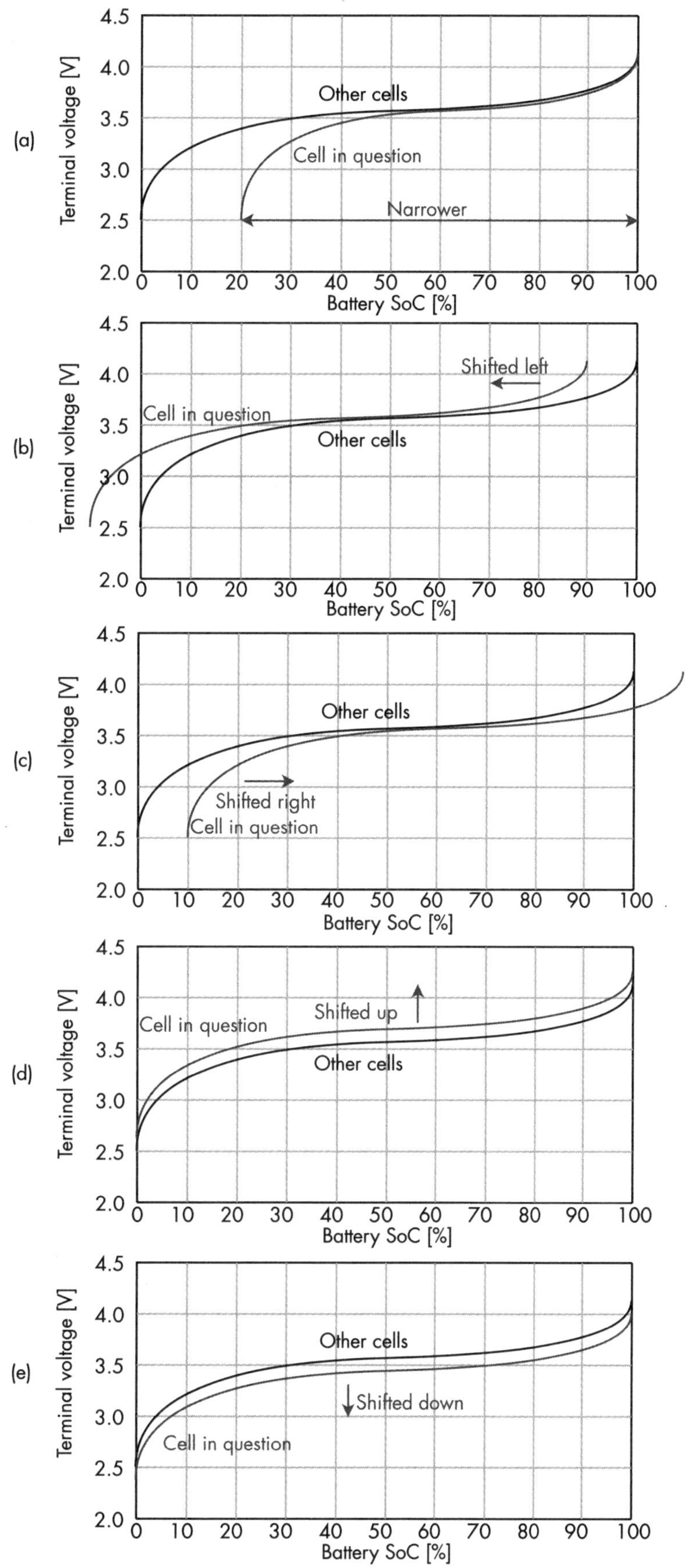

FIGURE 8.12
Mismatched cell voltage identification through plotting voltages versus SoC: (a) one cell has low capacity, (b) one cell has high SoC, (c) one cell has low SoC, (d) BMS reads high, and (e) BMS reads low.

- If the curve for one cell is shifted to the left relative to the curve for all the other cells, the problem is string balance: that one cell's SoC is high (Figure 8.12(b))—cause #4;
- If shifted to the right, the problem is also balance: that one cell's SoC is low (Figure 8.12(c))—cause #4;
- If shifted up, the BMS reads high (Figure 8.12(d))—cause #2;
- If shifted down, the BMS reads low (Figure 8.12(e))—cause #2;
- Note that if this applies to a group of cells rather than for a single cell, the same analysis and causes apply.

8.7.2 Address the Cause

Now that we know which of those four causes is applicable, let's address it.

8.7.2.1 Cause Number 1: Not Balancing

If the BMS reports that it's not balancing, it should tell you why, such as:

- All cell voltages are within a range, so they do not need to be balanced:
  - Accept it;
  - Or reconfigure the BMS to reduce the size of that range.
- Top-balancing and all cell voltages are below a threshold voltage:
  - Charge the battery until the voltages are higher than the threshold;
  - Or reconfigure the BMS to reduce that threshold.
- Mid-balancing and all cell voltages are outside a range:
  - Bring the cell voltages into that range;
  - Or reconfigure the BMS to expand that range.
- The battery is not being charged:
  - Charge the battery;
  - Or reconfigure the BMS to allow balancing even when not charging.
- The BMS has not yet determined which cell needs to be balanced:
  - Wait until it does.

8.7.2.2 Cause Number 2: Incorrect Measurement

If incorrect BMS measurement, repair it:

- Check the wiring to that cell, especially look for an open- or high-resistance connection;
- If a distributed BMS, replace the cell board for that cell;
- If a wired BMS, replace the bank board, slave, or centralized BMU.

8.7.2.3 Cause Number 3: Low Capacity

If low cell capacity is the cause, you have two options:

1. Accept it, and be prepared to see diverging cell voltages when not at the balance point (i.e., charge fully for a top balanced battery);
2. Replace the cell, first making sure to match the new cell's SoC with the string SoC before installing it (see Section 8.15.1).

8.7.2.4 Cause Number 4: String Balance

If string balance is the cause, identify the ultimate cause:[14]

- A cell was just replaced without regard to its SoC compared to the other cells:
  - Wait a long time for the BMS to balance the string;
  - Do a manual gross balancing of the string (see Section 7.6.1.1);
- A cell's SoC is low:
  - Measure the cell voltages.
  - Let the BMS balance the string for 24 hours, with little or no battery current
  - Measure the cell voltages again:
    - If the imbalance is worse:
      - The BMS is broken: on that cell, balancing is stuck on;
        - Repair the BMS or replace that cell board, as applicable.
      - The cell's self-discharge current is higher than the BMS's balance current.
        - Replace the cell—it's a fire hazard!
    - If the imbalance is the same:
      - The BMS balancing for that cell is broken;
      - The BMS balance current is too low;
        - Wait longer, or use a different BMS.
    - If the imbalance improved, the BMS is doing its job:
      - Wait a long time for the BMS to balance the string;
      - Or charge the cell with a power supply to the same voltage as the other cells.
- A cell's SoC is high:
  - Measure the cell voltages;
  - Let the BMS balance the string for 24 hours, with little or no battery current;
  - Measure the cell voltages again:
    - If the imbalance improved, the BMS is doing its job;
      - Wait a long time for the BMS to balance the string.
      - Or reduce the cell voltage to the same voltage as the other cells with a resistor (see Section 7.6.1.1).
    - If the imbalance is the same:
      - The BMS balance current is too low;
        - Wait longer, or use a different BMS;
        - The BMS uses charge transfer and is broken: on that cell, balancing is stuck off;
    - If the imbalance worsened:
      - The BMS uses charge transfer and is broken: on that cell, balancing is stuck on, transferring charge to that cell.
- There's a tap that draws current from only half the battery:
  - Remove that tap and use a DC-DC converter to power that line instead;
  - Or split the battery into two batteries (each with its own BMS) and add a *battery balancer* (see Volume 2, Section 2.6).
- The BMS causes it:

14. This assumes bypass balancing and top balancing.

- It is misconfigured:
  - It tries to do mid-balancing without reliable data:
    - If possible, reconfigure it for top-balancing;
    - Change to a BMS that can do mid-balancing effectively;
- It is poorly designed and doesn't balance correctly:
  - Change the BMS;
- It does not provide enough balance current for the application:
  - Change the BMS;
  - Add Balance Boosters™ to increase the balance current.

8.8 DATA EVALUATION TROUBLESHOOTING

This section addresses issues with the evaluation of the state of the battery.

8.8.1 State of Charge Troubleshooting

If the state of charge evaluated by the BMS is off, use these instructions:

- The SoC does not change, even though the battery is in use:
  - The BMS is basing SoC evaluation on battery current and doesn't see any battery current (see above);
  - The BMS is basing SoC evaluation on cell voltage, and the voltage is steady right now;
  - The BMS algorithm is incorrect.
- The SoC changes in the wrong direction:
  - The current measurement is backward (see above).
- The SoC does change, and in the correct direction, but it's always wrong:
  - The BMS is configured for a different cell chemistry.
- The SoC changes too slowly:
  - The BMS is configured for a much larger battery capacity than the actual one
  - The battery has lost a lot of capacity (degraded cells or highly imbalanced)
  - The BMS is basing SoC evaluation on cell voltage, and the voltage is not changing by much right now.
- The SoC changes too fast:
  - The BMS is configured for a much smaller battery capacity than the actual one;
  - The BMS realized that its SoC evaluation is wrong and it's correcting it quickly.
- The SoC jumped:
  - The BMS realized that its SoC evaluation was wrong and corrected it suddenly.
- The SoC is stuck at 100%:
  - The BMS sees the charging current but not the discharging current.
- The SoC is stuck at 0%:
  - The BMS sees the discharging current but not the charging current.
- The BMS forgets the SoC when powered down:
  - Broken BMS.
  - The BMS requires a constant supply voltage, even when turned off.

8.8.2 Actual Capacity Troubleshooting

If the BMS evaluates the actual capacity too low, it's probably because during a discharge cycle it had to shut down discharge due to a low cell voltage, and defined the total charge as the value of the actual capacity. This could be due to:

- The battery may not be a good match for the application: it is not able to deliver the power that the application demands;
- There may be a weak cell—high resistance or low capacity.

8.8.3 Actual Resistance Troubleshooting

If the BMS reports an unrealistic value for cell or battery resistance, it's probably because resistance calculation is tough. It is nearly impossible at a steady current.

8.8.4 State of Health Troubleshooting

If the BMS reports an unexpectedly low SoH, it could be because

- The BMS evaluation of other parameters (especially actual capacity or actual resistance) is unreliable;
- The configured value for nominal capacity or nominal resistance is incorrect.

8.9 CAN BUS TROUBLESHOOTING

Troubleshooting CAN communications can take some effort and skill. As with all troubleshooting, using logic and following a guide step-by-step helps.

8.9.1 No Communications

The most common complaint is lack of CAN communications. While it helps to have a CAN to USB adapter to monitor the messages on the bus, most problems can be solved with a simple multimeter. An oscilloscope is not required. On the contrary, you can waste a lot of time trying to interpret a scope trace instead of using a multimeter.

In most cases, the problem is configuration, a miswired bus or an open connection. Rarely, a CAN port is damaged.

8.9.1.1 Check the Configuration

First, check the CAN bus system setup:

- Are there at least two devices on the bus? A CAN to USB adapter may be one of the devices.
- Are all devices using the same CAN rate?
- Is the CAN rate too high for the physical length of the CAN bus?

8.9.1.2 Ohmmeter Testing

Next, use an ohmmeter to test the bus with the power off:

- Power off all the devices on the CAN bus;
- Measure the resistance across the two CAN lines; it should be 60 Ω;
- If 0 Ω, the two CAN lines are shorted to each other;

- If 40 Ω, you have three termination resistors (you must have exactly two resistors);
- If 120 Ω, you either only have one terminator resistor or a wire is broken;
- If infinite resistance, either you don't have any termination resistors, or a wire is broken, or you didn't set up the meter correctly;
- Check that there is no short between each line and ground;
- In case of abnormal readings, isolate the cause.

8.9.1.3 Voltmeter Testing

Then, use a voltmeter to test the bus with the power on:

- Power just one device on the CAN bus;[15] make sure it's transmitting
- Measure the voltage of the two CAN bus lines relative to their common ground:
  - If CANH is at 4V or higher and CANL is at 1V or lower, this test passed, go to point A, below;
  - If CANL is at 4V or higher and CANH is at 1V or lower:
    - The CAN lines are interchanged;
    - You have confused CANH and CANL;
    - The probes in your meter are plugged into the wrong jacks.
- If both are 4V or higher:
  - The CANL low line is disconnected;
  - The CAN driver on that device is broken.
- If both are 1V or lower:
  - The CANH line is disconnected;
  - The CAN driver on that device is broken;
  - Another device is pulling the CAN lines low:
    - Disconnect all other devices until you identify the one that is pulling the lines low.
- If one line is at 0V:
  - The line is disconnected.

A—Power down this device, and power-up another device;

- Repeat the tests with this second device; if this test passes, go to the next point;
- Power the first device back up; now two devices are powered;
- Measure the voltage of the two CAN bus lines:
  - If both lines are at about 2.5V, this test passed; go to point B, below;
  - If CANH is at 4V or higher and CANL is at 1V or lower, the two devices cannot communicate:
    - Different baud rate;
    - Too long a line;
    - Two identical devices are sending the same message at the same time:
      - Change the ID of the messages from one device.
    - The receive settings of one device are inappropriate:

15. Be aware that some products include two CAN devices in one enclosure.

- This is somewhat complicated; talk to the manufacturer.

B—Measure the voltage between CANH (red probe) and CANL (black probe) of the two CAN bus lines:

- The voltage is positive, on the order of 100 mV, depending on how much data are on the bus:
  - This test passed; exit.
- The voltage is 0V:
- Check that messages are being sent.
- The voltage is negative:
  - Check that you have the positive probe on CANH.

8.9.1.4 *CAN Adapter Testing*

If you have a CAN to USB adapter and a computer, use this procedure:

- Connect the CAN adapter to the CAN bus and be mindful of the polarity;
- Connect it to a computer, and start the monitor application at the expected CAN rate;
- Reset the CAN adapter, in case it had seen a *bus fault;*
- Note that the CAN adapter does not report any bus problems;
- Power one device on the CAN bus that does not transmit messages;
- Have the CAN adapter transmit a dummy message:
  - If the CAN adapter does not report bus problems, this test passed; go to point A, below;
  - There are bus problems:
    - The CAN bus speeds are different.
    - The CAN bus has physical issues: start from the "Ohmmeter Testing," above.
- Have the device transmit messages;

A—if the CAN adapter sees the messages:

- This test passed; go to point B, below.

B—if the device can report CAN transmit errors, see if it reports any errors:

- If none, go on to point C, below;
- If it does, check the physical bus: start from the "Ohmmeter Testing," above.

C—turn on one more device on the CAN bus:

- If the CAN adapter does not report bus problems, this test passed; go to point D, below;
- If there are bus problems, the new device has the wrong CAN speed or its CAN port is damaged.

D—repeat with all other devices on the bus.

8.9.2 Poor Noise Immunity

The CAN bus is excellent for noisy environments because it is balanced, and therefore has a high rejection of common-mode noise.

However, this common-mode rejection is effective only if the noise level is not excessive. Otherwise, messages are lost (error frames) and, in the worst case, the

CAN bus quits (*bus heavy*). An isolated CAN port can handle much higher levels of common-mode noise.

Such situations occur when inexperienced users install components haphazardly, with little regard to EMI considerations, or manufacturers design switching devices that are strong sources of EMI.

8.9.2.1 Troubleshooting

Use one of the devices or a CAN adapter to monitor the CAN errors while performing the following:

- Turn off switching power devices (chargers, inverters, DC-DC converters, motor drivers, etc.) whether or not they are connected to the CAN bus until you identify the source of interference;
- See if grounding the device affects its noise emissions;
- Turn off a CAN device at a time, to see which devices cannot handle the noise;
- Disconnect the CAN bus from each device, to see where the noise enters the CAN bus;

8.9.2.2 Minimize EMI Sensitivity

Try the following:

- Route the CAN bus wires away from any noisy device and its cables.
- Shield the CAN wires. Ground the shield at only one point; do not ground to the metal chassis.
- Reduce the CAN rate of all devices.
- Add a CAN filter between the port of the device that introduces the noise on the bus and the rest of the bus.
  - Add a CAN isolator[16] on the port of the device that introduces the noise on the bus.
- Use *active termination* (see Section 5.10.3.9).

8.9.2.3 Minimize EMI Emissions

Various publications cover ways to design and build systems with reduced EMI emissions in depth [3]. This discussion is beyond the scope of this book.

8.9.3 Poor Data Throughput

When a user mistakes a BMS for a scientific data acquisition instrument and requests individual cell readings every 100 ms, the CAN bus becomes clogged with messages. This impedes the transmission of essential data, which may have safety implications. If you want to be a scientist, buy a data acquisition instrument and let the BMS do its job. Data acquisition is beyond its pay grade.

8.10 TROUBLESHOOTING OTHER COMMUNICATIONS

This section discusses communication links other than a CAN bus.

16. ICP DAS I-7531, Peak GC-CAN-OPTO-ISO.

8.10.1 Windows GUI Troubleshooting

A BMS may communicate with a computer running a GUI application. Communication is through a USB cable. These instructions assume a Windows computer:

1. Open the *Device Manager;*
2. Plug the USB cable into the BMS and the computer;
3. Note in Device Manager that a new item appears or that the computer is looking for or installing a driver; if so, wait for the USB device to be ready, then go to point 4; otherwise:
   - The computer port may be hung up; shut down the computer completely, then restart it;
   - The cable is bad; try a different USB cable;
   - The BMS is bad.
4. Make sure the BMS is powered.
5. Run the application that communicates with the BMS; if it sees the BMS, the test passed; otherwise:
   - The GUI application is looking at the wrong COM port;
   - The BMS or the GUI is misconfigured:
     - Check the baud rate and other communications settings;
   - The BMS is bad.

It helps to have duplicate devices: an extra BMS master, an extra computer, and extra USB cable: then, by substitution, it may be possible to isolate the problem to the BMS, the cable, or the computer.

8.10.2 RS-232

The procedure is the same as above, but the BMS has an RS-232 port, and an RS-232/USB adapter is used.

- It's a *Prolific* adapter:
  - Use an authentic FTDI adapter instead.
- The adapter is bad;
- Requires a null modem to exchange the TX and RX lines;
- Turn off the GUI, run a terminal emulator application (e.g., PuTTY), try connecting pins 2 and 3 of the DE-9 connector on the computer side of the RS-232 cable; then type something in the terminal:
  - If you see what you type, the computer and the cable are fine:
    - The problem is with the BMS;
    - Mismatched settings between the BMS and the computer.
  - If you don't:
    - The problem is with the computer.
    - Bad cable.
- Check the voltages on the DE-9 connectors, both on the computer side and the BMS side: ~ 0V on the RX pin, and -5~-15V on the TX pin; depending on which side, pins 2 and 3 can be either TX or RX, and pin 5 is ground.

8.10.3 RS-485

Electrically, this bus is similar to the CAN bus, so troubleshooting the bus uses a lot of the same steps and criteria. In addition to those pointers, check the following:

- Ensure that all devices use a 2-wire bus or a 4-wire bus; most implementations are 2-wire.
- Add bias resistors to set the bus level when no data are present.
- If using RTS control, the RTS line should be set high during the message and restored low afterward.
- If using Send-Data control, make sure to wait for one character length before sending a new message. If less, a character may be missed; if more, a device may switch mode and miss characters.
- If the devices include LEDs, use them to pinpoint the problem; the Tx LED in the master flashes sending a command and the Rx one when receiving a response.

Some insist that a common ground line is required; others propose that a ground line should be avoided as it may introduce a ground loop.

Troubleshooting the messages is more complicated, especially if the bus uses a protocol that requires many messages:

- Check that there is one and only one master on the bus;
- Check that each slave address is unique;
- Set an appropriate time-out in the master for receiving a response from a slave; 0.5 s is typical and is independent of bus speed;
- Set an appropriate delay in the master for transmitting a new message after receiving a response from a slave; 0.05 s is typical;
- Check that the address range supported by all devices.

8.10.4 Command-Line Terminal

The BMS may use a generic terminal emulator application (e.g., PuTTY), and a command-line user interface, through a USB port or an RS-232 port and an adapter.

- Garbled data:
  - Wrong serial port settings (baud rate, number of bits, handshake).
- No data:
  - Disconnected cable;
  - The BMS may be waiting for input:
    - Try pressing some keys: ESC, 0, 1, Home, Return, Ctrl-C...
  - Wrong COM port (in Windows).
- If through an RS-232 port, check the items for RS-232 (just above).

8.10.5 Slave Communications

The BMU doesn't see a slave:

- Check the configuration to make sure that the BMU is set-up to see the slave;
- Check the continuity of the cable to the slave;

- Check that there are no shorts on the bank cable;
- Try swapping the slave with a known good one;
- If it uses a CAN bus or an RS-485 bus, follow the test procedures for that bus, above.

8.11 GROUND FAULT TROUBLESHOOTING

Note that an automatic ground fault detector may produce a false positive:

- If another ground fault test is occurring at the same time;
- If it doesn't test long enough to account for high spurious capacitance to ground;
- Some products that measure the total battery voltage introduce a high resistance to ground.

Troubleshooting procedure:

- Disconnect the battery from the load, which may be grounding the battery
- Test the battery isolation manually (see Section 7.7.1);
  - The isolation is OK:
    - The problem is with the load.
  - If there is an isolation loss:
    - The voltage measured during the test tells you where the ground fault is: on B-, B+, or somewhere in between; use this information to focus your search in that area;
    - The problem could be unintentional:
      - Loose wires;
      - Exposed cable shielding that is in contact with a cell terminal;
      - Water, dirt;
      - When a cover is closed, it touches or pushes a cell terminal or other conductor;
      - A bus bar touching the metal case of the battery;
      - The battery case uses carbon fiber, which is conductive;
      - A voltage sense tap wire touches a low-voltage wire;
      - A thermistor's isolation is broken.
  - The problem could be in the battery design:
    - Nonisolated BMS;
    - Nonisolated DC-DC converter powering the BMS;
    - The power supply for the BMS is connected to the battery;
    - The BMS drives a load that is powered by the battery, and the BMS output is referenced to the low voltage control ground;
  - Disconnect one item at a time, methodically, until can narrow down the fault area; eventually, you should be able to identify the cause.

8.12 TROUBLESHOOTING INPUTS AND OUTPUTS

This section discusses issues with the inputs and outputs of the BMS.

8.12.1 Digital Input Troubleshooting

The BMS doesn't see a digital input change. Do the following:

- Measure the input voltage and see if it changes:
  - The voltage stays a 0V:
    - The input line is grounded;
    - Broken wire to the BMS;
    - The input is driven by a switch to ground, open-collector, or open-drain:
      - The BMS does not provide a pull-up on that line, and there is no pull-up resistor external to the BMS;
      - The switch is broken, permanently closed.
    - The input is driven by a switch to 5V or 12V:
      - The switch is broken, permanently open.
    - The input is driven by a logic output (TTL):
      - Troubleshoot the device that generates that output.
  - The voltage stays at high voltage (e.g., 5V or 12V):
    - The input line is shorted to the supply voltage;
    - Broken wire to the BMS;
    - The input is driven by a switch to ground, open-collector, or open-drain:
      - The switch is broken, permanently open.
    - The input is driven by a switch to 5V or 12V:
      - The switch is broken, permanently closed;
      - The BMS does not provide a pull-down on that line, and there is no pull-down resistor external to the BMS.
    - The input is driven by a logic output (TTL):
      - Troubleshoot the device that generates that output.
  - The voltage changes, and ranges from less than 0.8V to at least 2.8V:
    - The BMS is not configured to look at that input.

8.12.2 Analog Input Troubleshooting

The BMS doesn't read an analog input correctly:

- The BMS doesn't see an analog input change:
  - Measure that input and see if it changes:
    - The voltage stays a 0V:
      - Broken wire;
      - Grounded line;
      - Troubleshoot the device that generates that output.
    - The voltage stays at 3 or 5V:
      - Troubleshoot the device that generates that output.
    - The voltage changes:
      - The BMS is not configured to look at that input.
- The BMS sees an incorrect change:
  - Check the gain and offset settings;
  - Check the polarity setting;
  - The analog input may exceed the input voltage range of the BMS.

8.12.3 Logic Output Troubleshooting

A BMS logic output is misbehaving:

- The output stays low:
  - The BMS is not configured to use that output;
  - The parameter that controls that output remains off;
  - That line is grounded externally;
  - Broken BMS.
- The output stays high:
  - The parameter that controls that output remains on;
  - Line shorted to the 5V supply;
  - Broken BMS.
- The output changes in the opposite direction:
  - Configure the BMS to reverse the polarity;
  - Use a small helper relay to invert the polarity.

8.12.4 Relay Output Troubleshooting

There is no voltage on the relay outputs because relays are just switches. They do not generate a voltage; you can only see a voltage if the relay output is used in a circuit (connected in series with a power supply and a load) (see Section 1.3.3).

8.12.5 Analog Output Troubleshooting

An analog output is misbehaving:

- The output is 0V:
  - The BMS is not configured to use that output;
  - The parameter that controls that output is at zero;
  - Open wire;
  - Grounded line;
  - Broken BMS.
- The voltage is non-0, but does not change:
  - The parameter that controls that output is steady;
  - Broken BMS.
- The voltage changes in the wrong direction:
  - Configure the BMS to reverse the direction;
  - With a floating load, instead of referencing it to ground, reference it to 5V.
- The voltage changes, but it is too low:
  - The line is loaded by a low-resistance load, too low for that output, and it can't drive the load.

8.12.6 Open-Drain Driver Troubleshooting

If the open-drain output is not yet wired to a load:

- There is no voltage (see Section 1.3.3):

- This is normal; it's because open-drain outputs are just switches to ground. They do not generate a voltage; you only see a voltage after the output is connected to a load, and the other end of the load is connected to a power supply voltage:
  - When measured without a load, with an ohmmeter between the output and ground (Figure 8.13(a)):
    - The meter shows 0 Ω when the output is on;
    - Is shows infinite Ω (actually, "OL" for *overload*) when the output is off.
  - When measured without a load, with a voltmeter between the output and ground (Figure 8.13(b)):
    - The meter shows 0V regardless because there is nothing to generate a voltage.
  - When measured with a load connected to a 12V supply, with a voltmeter between the output and ground (Figure 8.13(c)):
    - The meter shows 0V when the output is on;
    - It shows 12V when the output is off.
- To convert an open-drain output to drive a logic input, add a pull-up resistor (for example, 10 kΩ) between the output and a 5V supply (Figure 8.13(d)); note that the polarity is inverted: on gives 0V and off gives 5V.

If the open-drain output is already wired to a load, powered by a power supply:

- The output doesn't change:
  - The BMS is not configured to use that output;
  - The parameter that controls that output is not changing.
- The output is always on (grounded):
  - Grounded line;
  - Broken BMS.
- The output is always off (open):
  - Broken wire;
  - Broken BMS.

8.13 TROUBLESHOOTING POWER CIRCUITS

This section discusses problems with the high-current carrying components in the battery.

8.13.1 Contactor Troubleshooting

A contactor is not operating as expected:

- The contactor does not come on:
  - The coil is not driven:
    - There is no supply voltage to the coil;
    - Open wire.
    - The BMS is not driving the coil:
      - Not configured correctly;
      - The parameter that controls that coil output remains off;
      - The open-drain driver in the BMS is broken (open).
  - Open coil
  - For a contactor with an economizer, the economizer is not working;

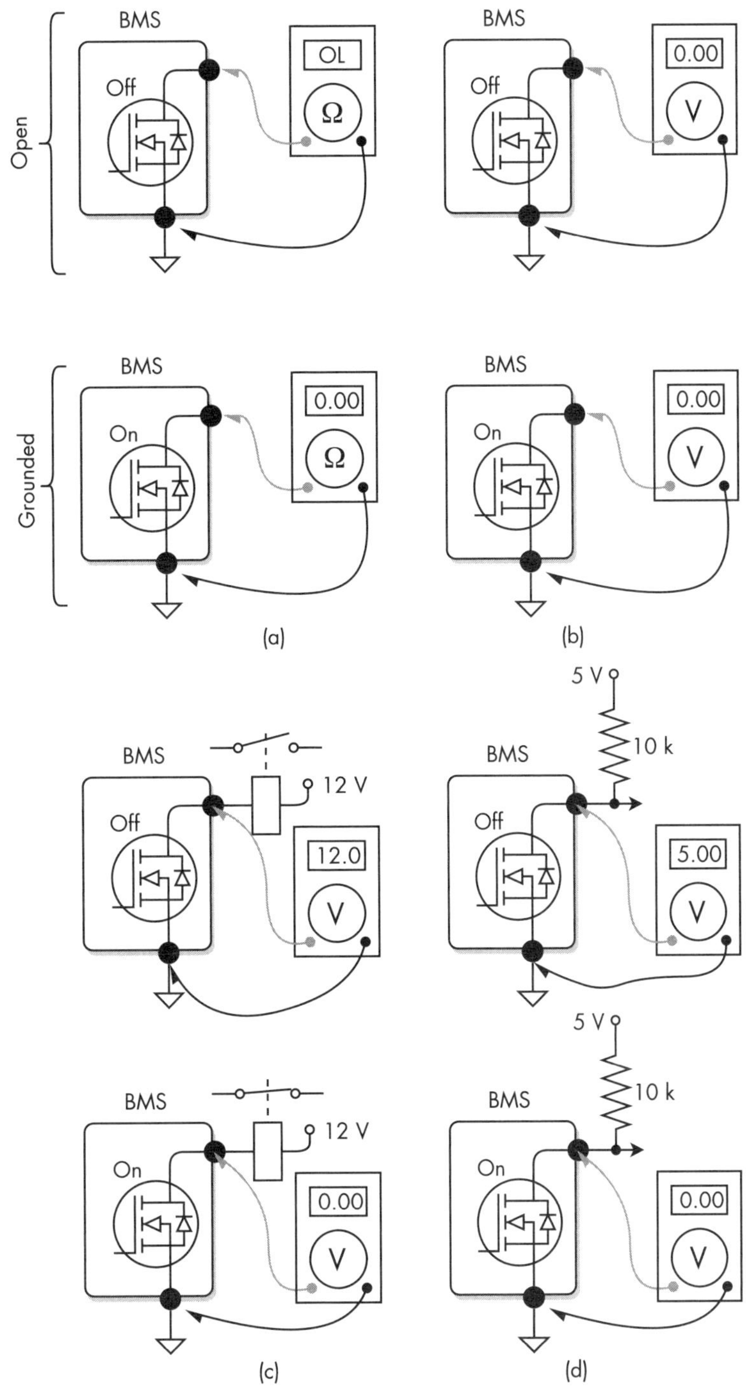

FIGURE 8.13
Open-drain output measurement:
(a) ohmmeter, no load,
(b) voltmeter, no load,
(c) voltmeter, with load,
and (d) resistor load to drive a logic input.

- Damaged contacts or plunger stuck inside the contactor;
- The contactor does come on, but its auxiliary switch is not reporting it, or the BMS doesn't see the state of the auxiliary switch.

- The contactor is always on:
  - Contactor contacts are welded;
  - The line driving the coil is grounded;

- The BMS drives the coil continuously:
  - The BMS decided to do so;
  - The parameter that controls that coil output remains on;
  - The open-drain driver in the BMS is broken (shorted).
- The contactor is hot even with no battery current:
  - For a standard contactor, that's normal;
  - For a contactor with an economizer, the economizer is not working.
- The contactor is hot with high battery current:
  - Contactor contacts are in bad shape.
- The contactor chatters (turn off and on, off and on, rapidly):
  - The power supply cannot provide enough current for the contactor coil;
  - The power supply voltage is too low for the contactor coil;
  - The BMS cannot carry enough current for the contactor coil;
  - The electronic economizer in the contactor is broken;
  - Too many contactors are turned on at the same time;
  - Strong vibration may cause the contactor to open momentarily.

8.13.2 Precharge Troubleshooting

Precharge is not happening as expected:

- The precharge finishes too early:
  - The current pulse is too short;
    - The main fuse is blown;
    - The precharge resistor is open;
    - The fuse in series with the precharge resistor is blown.
  - The voltage reaches the final value too fast:
    - The load is not connected.
- The precharge current does not decrease, and the load voltage does not rise:
  - Shorted load;
- The precharge lasts too long;
  - The load starts drawing current before precharge is done.

8.14 TROUBLESHOOTING FAULT MESSAGES

If the BMS is reporting a fault, don't ignore and don't disable it. Doing so would be killing the messenger. There is a reason the BMS is reporting a fault, and ignoring it is circumventing the point of having a BMS in the first place. Instead, find why the BMS is reporting a fault and fix the problem.

8.14.1 Battery Shutdown, High Cell Voltage

The BMS issued a "high cell voltage" fault and shut down the battery:

- In a vehicle, hard braking, down a long hill:
  - The battery is undersized for the vehicle, cannot accept that much power;
  - There is a weak (high-resistance) cell;

  - The battery is nearly full;
  - A high current generates high levels of electrical noise, affecting the BMS;
  - The configured limits could be relaxed: increase the maximum cell voltage, increase the fault delay.
- A stationary battery is full:
  - The battery is imbalanced and a cell cannot accept any more charge;
  - The top bus voltage too high.

8.14.2 Battery Shutdown, Low Cell Voltage

The BMS issued a "low cell voltage" fault and shut down the battery:

- In a vehicle, high acceleration, up a long hill:
  - The battery is undersized for the vehicle, cannot provide that much power;
  - There is a weak (high-resistance) cell;
  - The battery is nearly empty;
  - High current generates high levels of electrical noise, affecting the BMS;
  - The configured limits could be relaxed: lower the minimum cell voltage, increase the fault delay.
- A stationary battery is empty:
  - A cell cannot deliver any more charge;
  - The loads do not have a low voltage cutout, or the cutout voltage is too low.

8.14.3 Battery Shutdown, Charge Overcurrent

The BMS issued a "charge overcurrent" fault and shut down the battery:

- At the time the BMS indicated that charging was allowed:
  - The charging current really was too high.
- At the time the BMS indicated that charging was not allowed:
  - The BMS was not wired to turn off the charging source, or the charging source is ignoring the BMS;
  - The BMS sees current, though there is no current (miswired, misconfigured);
  - The BMS disabled the charging source, the current is decaying to zero, but the BMS didn't give it enough time to do so.

8.14.4 Battery Shutdown, Discharge Overcurrent

The BMS issued a "discharge over-current" fault and shut down the battery:

- At the time the BMS indicated that discharging was allowed:
  - The discharging current really was too high.
- At the time the BMS indicated that discharging was not allowed:
  - The BMS was not wired to turn off the load, or the load is ignoring the BMS;
  - The BMS sees current, though there is no current (miswired, misconfigured);

- The BMS disabled the load, the current is decaying to zero, but the BMS didn't give it enough time to do so.

8.14.5 Isolation Fault

The BMS issued an "isolation loss" fault and shut down the battery:

- False positive:
  - The battery is not isolated from ground;
  - The battery has excessive capacitance to ground, and the test is too fast;
  - The electrical noise is excessive;
  - Another device is also testing isolation at the same time.
- False negative:
  - The isolation tester is not connected to earth ground.

8.15 REPAIR

Before working on a battery, take these safety steps:

- Follow the same safety and working environment guidelines used when assembling a battery (see Section 7.2.2): wear goggles, remove jewelry, use isolated tools, wear safety gloves with high voltage, and so on.
- Wear safety glasses.
- Disconnect the battery—for a large battery, remove the safety disconnect or unplug the power connector.

8.15.1 Cell Replacement

To replace a single cell:

- Bring the new cell and the string to the same SoC:
  - Charge or discharge the new cell to the same SoC as the string.
  - Or charge or discharge the string until it has the same SoC as the new cell (see Section 7.6.1.1).
- Protect the BMS:
  - For a wired BMS, disconnect from the BMS the connector for the voltage sense harness that includes the cell to be replaced;
  - For a mated BMS (distributed BMS, banked BMS), disconnect the bank of cells from EVERYTHING: chargers, load, and DC-DC converter.
- Remove the interconnections and the BMS for the old cell:
  - Use nonconducting or at least isolated tools;
  - For a wired BMS, disconnect the cell voltage tap wire and cover it in isolating tape so it doesn't flop around and touch something else;
  - For a distributed BMS, disconnect the cell board's communication wires or cables from adjacent cell boards (this should not be needed, but one cannot be too careful), disconnect the cell board from the cell terminals, and remove the cell board from the battery and store it in a safe place;

    - For a banked BMS, disconnect the bank board from all of its cells; remove the bank board from the battery and store it in a safe place;
    - Remove the power interconnections between the old cell and the adjacent cells.
- Replace the cell.
- Connect the new cell:
    - Reinstall the power interconnections between the new cell and the adjacent cells;
    - Reconnect the BMS:
        - For a wired BMS, reconnect the tap wire, and plug the connector back into the BMS;
        - For a distributed BMS, check, double-check, and triple check the polarity of the cell board and the cell before installing the cell board, then reconnect the communication wires;
        - For a banked BMS, reconnect the bank board to all of its cells.
- Test the battery.

8.15.2 Gross Balancing in the Field

Should the string become badly imbalanced in the field, for example because a cell was replaced with no regard to its SoC, it may be necessary to balance it manually by charging or discharging a particular cell see Section 7.6.1.1.

8.15.3 BMS Repair

These are some of the possible repairs to a BMS.

8.15.3.1 Cell Board Replacement

For a distributed BMS, replace a cell board as follows:

- Disconnect the bank of cells that contains that cell board from EVERYTHING—chargers, load, DC-DC converter;
- Use nonconducting or at least insulated tools;
- Disconnect the cell board's communication wires or cables from adjacent cell boards;
- Disconnect the cell board from the cell terminals;
- Check, double-check, and triple check the polarity of the new cell board and the cell before installing it to the cell terminals;
- Connect the cell board's communication wires or cables to its adjacent cell boards;
- Reconnect the bank to whatever you disconnected before starting the procedure.

8.15.3.2 Bank Board Replacement

For a banked BMS, replace a bank board by following the same instructions as for a cell board (above) except that you must disconnect and reconnect all the cell connections.

8.15.3.3 Component-Level Repair

Normally you would not do component-level repairs to BMS electronics. Even if a BMS is out of warranty, it's cheaper to have the BMS manufacturer repair their product. When a customer attempts component-level repair, there is a significant risk of making an electronic assembly FUBAR.[17]

Yet, there are exceptional reasons driven by desperation:

- The BMS manufacturer is in China, and you're in Brazil;[18]
- The student race is tomorrow.

Component-level repairs may include

- Replacing a blown MOSFET in a protector switch or open-drain driver;
- Replacing a blown current sense resistor in a PCM or an in-line Hall effect current sensor;
- Bypassing broken traces in a PCB that suffered mechanical damage;
- Replacing a surface mount fuse or a ferrite bead blown by a short circuit;
- Replacing a cracked thermistor or balance resistor on a cell board;
- Replace a damaged connector.

A kind soul at the BMS company may be willing to tell you the component value, manufacturer, and part number, so that you may try to find a replacement component locally. You may even steal an unused component from elsewhere on the PCB assembly.

If you don't have a temperature-controlled soldering iron or you have no experience with surface mount components, do not attempt a repair: you'll only make things worse. Look for a shop in town to do it for you, such as a computer repair shop.

References

[1] Lei, B., et al., "Experimental Analysis of Thermal Runaway in 18650 Cylindrical Li-Ion Cells Using an Accelerating Rate Calorimeter," Batteries, Vol. 3, No. 2, 2017.

[2] Andrey W. Golubkov et al., "Thermal-runaway experiments on consumer Li-ion batteries with metal-oxide and olivin-type cathodes", Royal Society of Chemistry.

[3] Chad Lobato et al, "Thermal Runaway Testing of 18650 Lithium Ion Cells", EnerSys Advanced Systems / ABSL Golubkov, A. W., et al., "Thermal-Runaway Experiments on Consumer Li-Ion Batteries with Metal-Oxide and Olivin-Type Cathodes," RSC Advances, Vol. 4, No. 7, 2014.

[4] Williams, T., EMC for Product Designers, self-published.

17. Fouled up beyond all recognition.
18. We weren't able to return a repaired product to Brazil, thanks to their G-damned customs agents. The box was shipped three times, back and forth between the United States and Brazil. In the end, we were able to ship it to a teammate who was vacationing in the United States, who hand-carried it back to Brazil.

INDEX

Q

R

X

Recent Artech House Titles in Power Engineering

Andres Carvallo, Series Editor

Advanced Technology for Smart Buildings, James Sinopoli

The Advanced Smart Grid: Edge Power Driving Sustainability, Second Edition, Andres Carvallo and John Cooper

Battery Management Systems, Volume I: Battery Modelings, Gregory L. Plett

Battery Management Systems for Large Lithium Ion Battery Packs, Davide Andrea

Battery Power Management for Portable Devices, Yevgen Barsukov and Jinrong Qian

Big Data Analytics for Connected Vehicles and Smart Cities, Bob McQueen

IEC 61850 Demystified, Herbert Falk

Design and Analysis of Large Lithium-Ion Battery Systems, Shriram Santhanagopalan, Kandler Smith, Jeremy Neubauer, Gi-Heon Kim, Matthew Keyser, and Ahmad Pesaran

Designing Control Loops for Linear and Switching Power Supplies: A Tutorial Guide, Christophe Basso

Electric Power System Fundamentals, Salvador Acha Daza

Electric Systems Operations: Evolving to the Modern Grid, Second Edition Mani Vadari

Energy Harvesting for Autonomous Systems, Stephen Beeby and Neil White

GIS for Enhanced Electric Utility Performance, Bill Meehan

Introduction to Power Electronics, Paul H. Chappell

Introduction to Power Utility Communications, Dr. Harvey Lehpamer

IoT Technical Challenges and Solutions, Arpan Pal and Balamuralidhar Purushothaman

Lithium-Ion Batteries and Applications: A Practical and Comprehensive Guide to Lithium-Ion Batteries and Arrays, from Toys to Towns, Volume 1, Batteries, Davide Andrea

Lithium-Ion Battery Failures in Consumer Electronics, Ashish Arora, Sneha Arun Lele, Noshirwan Medora, and Shukri Souri

Microgrid Design and Operation: Toward Smart Energy in Cities, Federico Delfino, Renato Procopio, Mansueto Rossi, Stefano Bracco, Massimo Brignone, and Michela Robba

Plug-in Electric Vehicle Grid Integration, Islam Safak Bayram and Ali Tajer

Power Grid Resiliency for Adverse Conditions, Nicholas Abi-Samra

Power Line Communications in Practice, Xavier Carcelle

Power System State Estimation, Mukhtar Ahmad

Renewable Energy Technologies and Resources, Nader Anani

A Systems Approach to Lithium-Ion Battery Management, Phil Weicker

Signal Processing for RF Circuit Impairment Mitigation in Wireless Communications, Xinping Huang, Zhiwen Zhu, and Henry Leung

The Smart Grid as An Application Development Platform, George Koutitas and Stan McClellan

Smart Grid Redefined: Transformation of the Electric Utility, Mani Vadari

Synergies for Sustainable Energy, Elvin Yüzügüllü

Telecommunication Networks for the Smart Grid, Alberto Sendin, Miguel A. Sanchez-Fornie, Iñigo Berganza, Javier Simon, and Iker Urrutia

For further information on these and other Artech House titles, including previously considered out-of-print books now available through our In-Print-Forever® (IPF®) program, contact:

Artech House
685 Canton Street
Norwood, MA 02062
Phone: 781-769-9750
Fax: 781-769-6334
e-mail: artech@artechhouse.com

Artech House
16 Sussex Street
London SW1V 4RW UK
Phone: +44 (0)20 7596-8750
Fax: +44 (0)20 7630-0166
e-mail: artech-uk@artechhouse.com

Find us on the World Wide Web at: www.artechhouse.com